Preface

The Nature and Growth of Modern Mathematics traces the development of the most important mathematical concepts from their inception to their present formulation. Although chief emphasis is placed on the explanation of mathematical ideas, nevertheless mathematical content, history, lore, and biography are integrated in order to offer an overall, unified picture of the mother science. The work presents a discussion of major notions and the general settings in which they were conceived, with particular attention to the lives and thoughts of some of the most creative mathematical innovators. It provides a guide to what is still important in classical mathematics, as well as an introduction to many significant recent developments.

Answers to questions like the following are simple and will be found in this book:

Why should Pythagoras and his followers be credited with (or blamed for) some of the methodology of the "new" mathematics?

How do modern algebras (plural) generalize the "common garden variety"?

What single modern concept makes it possible to conceive in a nutshell of *all* geometries plain and fancy—Euclidean, non-Euclidean, affine, projective, inversive, etc.?

What is the nature of the universal language initiated by a thirteenth-century Catalan mystic, actually formulated by Leibniz, and improved by Boole and De Morgan?

How did Omar Khayyám solve certain cubic equations?

What are the common features of any boy's "engagement problem," the geishas' pantomime of baseball, and modern engineering decisions?

Who are the "Leonardos" of modern mathematics?

How did Queen Dido set a precedent for mathematicians and physicists?

Why should isomorphism, homomorphism, and homeomorphism be an intrinsic part of the vocabulary of every mathophile?

How did Maxwell's "demon" make the irreversible reversible?

Why did the mere matter of counting socks lead a millionaire mathematical genius to renounce mathematics in favor of finance?

What are the beautiful "ideals" formulated by Richard Dedekind and advanced by Emmy Noether?

Proceeding from illustrative instances to general purposes, let us state that the author's objectives are:

1. To survey the entire field of mathematics, with emphasis on twentieth-century ideas.

2. To furnish the type of exposition that should make it possible for a layman who is educated but not a specialist in mathematics to gain insight into the manifold aspects of modern mathematics, including its essential relationship to all areas of scientific thought. This objective was formulated because the "new" mathematics which has become the vogue in our schools is not really new, and those who seek popular treatment of contemporary mathematics can find only occasional superficial articles in periodicals. Books offering fuller exposition, including the author's *Main Stream of Mathematics*, have generally terminated with material from the early decades of the present century. In the potentially democratic world which men of good will envision, the man in the street must be entitled to more mathematical stimulation than the puzzle column in a Sunday newspaper, an occasional profile of a Nobel prize winner, an enigmatic summary of some recent discovery in applied mathematics—whether that man is an engineer in a remote village in India who is seeking to fill loopholes in his mathematical knowledge, a retired physician anxious to convince himself he is a mathematician *manqué*, a high school senior in quest of a research topic for a science talent contest, a nun whose objective is to inspire her students with an account of the accomplishment of women mathematicians, or a stockbroker eager to indulge in some pure mathematics. (Such are some of the individuals who have corresponded with the author *in re* mathematical information.)

3. To stress "human interest" and thereby to reveal mathematics as a living, growing endeavor, holding a strong place in man's culture. This aim was conceived because there is danger to the humanities in the present educational crash programs designed to produce a large number of mathematicians, physical scientists, engineers, and technical workers. Our times make such programs a necessity, and the leaders who suggest the concentrated curricula or write the texts are rendering an inestimable service. Although these men understand the value of the pure mathematical content of the courses of study they prescribe, can the same be said of the students or of the teachers in elementary and secondary schools, or of the general public? They must have the opportunity to realize that there is more purpose to the "new" mathematics than recounting the tale of Little Red Riding Hood in terms of set theory or computer language. Thus, part of the third objective of the present work is to supply material which can serve as a cultural background or supplement for all those who are receiving rapid, concentrated exposure to recent advanced mathematical concepts, without any opportunity to examine the origins or gradual historical development of such ideas. Hence, although designed for the layman, this book would be helpful in courses in the history, philosophy, or fundamental concepts of mathematics.

At this point I should like to express my indebtedness to a number of persons. First and foremost, there is my husband, Dr. Benedict T. Lassar, through whose advice and assistance the manuscript has received a "fondest father's fondest care."

THE NATURE
AND GROWTH
OF MODERN
MATHEMATICS

THE NATURE AND GROWTH OF MODERN MATHEMATICS

by

EDNA E. KRAMER

PRINCETON UNIVERSITY PRESS
Princeton, New Jersey

Published by Princeton University Press, 41 William Street,
Princeton, New Jersey 08540
In the United Kingdom: Princeton University Press,
Guildford, Surrey

Text designed by Vincent Torre

First edition, Hawthorn Books, 1970
First paperback edition, Fawcett Books, 1974
First Princeton Paperback printing, with corrections, 1982

LCC 82-47630
ISBN 0–691–08305–3
ISBN 0–691–02372–7 pbk.

Printed in the United States of America by
Princeton University Press, Princeton, New Jersey

Thanks are due to the Oxford University Press of New York for permission to reprint,
in revised and updated form, the following pages from the 1951 edition of *The Main
Stream of Mathematics* by Edna E. Kramer, on which Oxford University Press holds
the copyright: 16, 17(1.1–21), 99(1.13–26), 215–239, 240(1.1–5), 259, 260, 264–284,
293(1.17–40), 297–301, 302(1.1–15).

To

My husband,

Benedict Taxier Lassar

I appreciate the many valuable suggestions made by Linda Allegri, who served as mathematics editor, as well as those of C. B. Boyer. The author is especially grateful to the mathematicians who gave their kind permission to quote from important articles they have written. Exact references to those mathematicians and the particular papers are provided at appropriate points in the body of the work. Thanks for the special permission must also be extended to the organizations sponsoring the publications in which the articles appeared, namely, the Royal Society of London, the Université de Paris, the American Mathematical Society, the Mathematics Association of America, and to the editors of *Biometrika*, *Econometrica*, the *Annals of Mathematical Statistics*, the *Journal of the American Oriental Society*, *Scripta Mathematica*. Oxford University Press graciously gave permission to quote from the author's previous book, *The Main Stream of Mathematics*. A most valuable form of assistance was rendered by those who encouraged the author to initiate or to continue (through long years) the writing of this book. They are Hanna Neumann, Bernhard H. Neumann, Peter M. Neumann, Jean Leray, Christie J. Eliezer, Sigekatu Kuroda, John M. Danskin, Stanislaw Ulam, Wilhelm Magnus, Fred Kerner, the late E. B. Wilson. Finally, thanks are due to William Taylor for the careful preparation of the diagrams, and to Beth Goldberg for the translation of barely decipherable scribbling into neat, accurate typescript, for noting errors, and for making suggestions.

New York   E. E. K.
January, 1970

Contents

21

East Meets West in the Higher Arithmetic   497

22

The Reformation of Analysis   528

23

Royal Roads to Functional Analysis   550

*group theory, differential equations, eigenvalues and eigenfunctions,
relativity, differential geometry, quantum mechanics, logic and
foundations, philosophy of science | "Those were the happiest months
of my life" | Zurich and Göttingen | Protest against Nazi removal
of his colleagues | Change to United States mathematical environment |
Details of Weyl's discoveries | His "intuitionism"*

27

Twentieth-Century Vistas—Analysis   639

*Calculus and the measurement of (abstract) lengths, areas, volumes,
masses, pressures, etc. | Borel, Lebesgue, and the problem of measure
for general point sets | Association with theory of integration |
Mengoli-Cauchy integral once more | Riemann integral | Lebesgue-
Young integral | Average temperature as a Riemann integral | As a
Lebesgue integral | Where Riemann fails but Lebesgue succeeds |
Lebesgue measure | Measure as a function | All Borel sets as domain |
Lebesgue's enlargement of the domain | He illustrates nonmeasurable
sets | Placid existence of a revolutionary | Denjoy, Perron, and
Stieltjes integrals | Colorful biography of a co-inventor*

28

Twentieth-Century Vistas—Algebra   656

*Emmy Noether, a major influence on recent developments | Early years
at Erlangen | Later work at Göttingen | The "Noether boys" | Her
last years in the United States | Her ability at abstract formulation |
A disciple of Dedekind | Her personality | Ideals as special subrings |
The ring of integers and its modulo 3 image | Homomorphism of
rings | Kernel of a homomorphism | Principal ideals | Analogy between
inclusion and divisibility | Factorization of ideals | Failure of
uniqueness | Integral domains and fields reexamined | Homomorphisms
of the ring of integers | Noether and hypercomplex systems | Brauer
on Artin | Artin on Wedderburn | Climactic memoir on hypercomplex
numbers | Linear associative algebras | Discovery by Grassmann and
Hamilton | Nilpotency | Division algebras | General structure
theorems | Peirce-Frobenius theorem | The Adams-Bott-Kervaire-Milnor
generalization | Manipulations with quaternions | Applications to
physics | Limited scope of Grassmann algebras | Grassmann, the man |
Wedderburn's structure theorems and his curriculum vitae*

29

Twentieth-Century Vistas—Logic and
Foundations   684

*Axiomatics as a partial cure for paradoxes | Gödel's incompleteness
theorem | Metamathematics | Absolute consistency | Gödel's proof |
Richard paradox | Arithmetization | An undecidable proposition |*

Side effects | Essential incompleteness | Impossibility of internal consistency | The need for "finitistic" methods | Wavre on Brouwer | Taboos on law of excluded middle, existence proofs, use of defining properties | Rejection of Cantor's continuum | Summary of intuitionist dogma

30

Introduction

The era in which we live has variously been described as the age of the common man, the age of anxiety, the nuclear age, the space age, and so on. To this list of appellations one might well add the *age of mathematics*. At no other time in history has the man in the street been informed emphatically and repeatedly that the further progress of mathematics concerns him vitally because it is sure to have effects, direct or indirect, on his material comforts, his mode of thought, and his very survival. The nonspecialist cannot help wondering at this sudden prominence of a subject whose higher branches have always seemed to belong to a world apart, a sphere where physical scientists find advanced mathematics a useful tool, while philosophers and numerical geniuses indulge in it as an art or as an intellectual hobby. The present phenomenon may have been precipitated by the political and economic circumstances of our day, but the major role of the new mathematics is no fortuitous overnight occurrence. It represents the cumulative effect of some four thousand years of scientific thought.

Circles of Thought

Most interest must center on the outcome of this long period of fruition, and therefore emphasis must be placed on mathematical activity as it exists today. However, since many classic aspects of mathematics have stood the test of time, they are actually part of the present and must be considered in relation to it. Some of the older notions pave the path for recent developments, and others clarify the new replacements or generalizations. In order to integrate the traditional with the modern in mathematics, a "spiral" organization is used by the author in tracing the progress of concepts from their origins in antiquity or some period prior to the twentieth century up to their ultimate formulation. In some instances the first presentation of a topic may be compared to a trip around one of the small inner whirls of a spiral. After this initial full circle, the subject may be dropped until later chapters where new convolutions are examined and the original journey continues through one or more additional broader circles of thought.

Something Old, Something New

The use of the spiral sequence necessitates the juxtaposition of "something old, something new" in a number of chapters. For example, since certain Pythagorean problems are similar to issues handled by Dedekind and Cantor in the late nineteenth century, it seems logical to discuss some aspects of the earlier and later viewpoints side by side. If Euclid's discoveries in the higher arithmetic were first generalized by the great Russian mathematician Tchebycheff, there seems no reason to defer all mention of the nineteenth-century analytical number-theorist simply because he lived much later. If Euler, the most prolific of mathematicians, was "analysis incarnate" but nevertheless employed intuitive methods with grave logical defects, why not reveal at once just how these faults were remedied a couple of centuries later, after they had produced a veritable crisis in mathematics? In the same way, the games of chance played by primitive man and his more civilized successors are linked with the more sophisticated games of military, industrial, or economic strategy, whose theory has been studied by Borel, Wald, and above all, by von Neumann, heroic figures in recent mathematical history.

Thus the mode of organization makes it impossible to adhere to strict chronological order. Continuity of subject matter seems a more important criterion than sequence in time. When, for example, Newtonian mathematical physics is specified as a set of "deterministic" laws, an obvious way to secure emphasis and clarification is to proceed to the "indeterminacy" in a vast major modern area. Therefore a treatment of probabilistic or statistical thought follows the discussion of the deterministic "differential equation type" of mechanics produced by Newton, Euler, Lagrange, Laplace, and Hamilton. Thus, relatively early in the book there appear the names and discoveries of some recent probabilists, statisticians, and researchers in statistical mechanics, whereas mathematical ideas which developed earlier are deferred for later presentation.

Treatment of Biography and Mathematical Lore

As for the biographical material, it is not necessary, in a layman's cyclopedia, to present details of the lives of all the mathematicians in history, and a few of the more usual favorites are given only brief mention in order to give more space to the newer names. It is inevitable, moreover, for an author to have individual enthusiasms and personal preferences. Again, history has left more detailed information about the lives of some mathematicians, or else they were particularly fortunate in having had friends or contemporary biographers to record facts before they were lost forever. In a few instances where the biographical material on record may be apocryphal or, on the other hand, has been repeated again and again as part of the anecdotal stock-in-trade of mathematical writers, the present author has employed an imaginative narrative instead of a biography. For example, there is the story of Hypatia, influenced inevitably by Charles Kingsley's Victorian version.

Integration of Exposition with Source Material

Popularization need not, and should not, be vulgarization, and the quotation of original sources is one method that is often used to maintain a scholarly level of exposition. But since recent mathematical research is exceedingly advanced and abstract in nature, the actual symbolic or technical format in which the creators clothe their ideas might baffle a general reader. If first formulations cannot be used on account of their difficulty, it is sometimes possible to examine how the foremost mathematicians of our era present popularization of ideas related to their own. Some of their narrative or expository writings, quoted occasionally in the present work, were not originally directed to the man in the street. Nevertheless they constitute popularizations in the sense that they were prepared for an audience of biologists, psychologists, economists, engineers, philosophers, chemists, physicists or for mathematical readers who were not specialists in the field under discussion. This sort of secondary source serves the dual purpose of indicating the clarity of such exposition and bringing recent mathematical leaders into the story by having them speak for themselves as well as for those whose research they are evaluating or eulogizing.

In making use of such source material, it is not feasible, in most cases, to quote material *in toto*, with assistance merely from prefatory or supplementary notes. Instead, one may need to combine statements by a number of mathematicians or else interrupt the course of a discussion in order to provide clarification or amplification at the exact points where difficulties may arise. What is done resembles the plotting of a detective story. When a Scotland Yard sleuth is congratulated on his solution of a case, he will respond modestly: "Oh, but if not for your questions, I should never have been able to find the answers, for you have a perfect talent for failing to understand at the very instant when clarification is most needed."

Special Features

As in the matter of biographies, so in the question of topics to be discussed, it is necessary to make some choice. The writer of a popular work on mathematics must select subjects at varied levels of difficulty and present these with varying degree of completeness in order to appeal to a wide audience. That is why, in a few instances, the life and times of a creative thinker are given more space than the advanced mathematical theories to which he gave birth. Again, an informal, anecdotal, even discursive introduction may be the best approach to certain topics. As for variety in exposition, a general work can enjoy a freedom which mathematical textbooks have never had. It seems to the author that even in books for students at school much would be gained by banishing a pattern that makes each chapter a "pedagogic" stereotype. *The Nature and Growth of Modern Mathematics* avoids such a pattern, and hence a discussion is sometimes initiated with a biography or a narrative while at other times the story of the discoverers is told last, not first. Or if mathematicians have worked in the same or related fields, it may be interesting to consider their biographies simultaneously. Where ideas are particularly dramatic, one should plunge into them without preliminaries. The flashback technique, used so frequently in nonmathematical literature, is employed occasionally in the present work in order to emphasize the culminating result of decades, even centuries, of preliminary theorizing, or to give the most general concepts priority over the special cases from which they evolved, or to reveal the goal of a particular chapter or portion of the book.

An Overall Picture

It is the author's opinion that one can provide an appreciation of the development of mathematics and its position in our present culture without according full technical treatment to every mathematical specialty in existence at the present time. To attempt more than brief reference to certain areas would be impracticable and might, moreover, have the effect of confusing or overpowering a general reader. Nevertheless, the origins and elementary aspects of the most important subjects are presented in broad outline. Also there is explanation of why the "multivalence" of modern pure or completely abstract theories makes them all-encompassing so that, in actuality, consideration of a single theory implies a vast content in the specific models to which it applies. This is one reason why the mathematical present can be properly termed the age of abstractness. But the author does not limit the material or adjust the style in accordance with the current emphasis on pure mathematics, and the book deals with much that can be described as "applied mathematics." Although the present abstract trend may have potent effects on the future of mathematics, we are where we are just now because "applied" problems, especially in astronomy and mechanics, are part of our cultural heritage and were, in the past, the motivating factors which stimulated the growth of mathematics toward its present magnitude and condition.

In any comprehensive presentation of mathematics as it exists today, it is inevitable that certain phases of "higher" mathematics must be treated. A reader need not be fearful on this account, however, since part of the present picture is the fact that some of the new advanced mathematical ideas are easier to comprehend as well as far more stimulating than many technical fine points of traditional elementary mathematics.

THE NATURE
AND GROWTH
OF MODERN
MATHEMATICS

1

From Babylonian Beginnings to Digital Computers

"Schoolboy, where did you go . . . ?"
"I went to school."
"What did you do in school?"
"I read my tablet, ate my lunch,
  prepared my tablet, wrote it, finished it; then . . .
  upon the school's dismissal, I went home,
  entered the house, [there] was my father sitting.
  I spoke to my father . . .
  read the tablet to him, [and] my father was pleased;
  truly I found favor with my father. . . .
  [I said]: 'I want to go to sleep;
  wake me early in the morning,
  I must not be late, [or] my teacher will cane me.'
  When I awoke early in the morning,
  I faced my mother, and
  Said to her: 'Give me my lunch, I want to go to school. . . .'
  My mother gave me two 'rolls,' I went to school.
  In the tablet house, the monitor said to me: 'Why are you
      late?' I was afraid, my heart beat fast.
  I entered before my teacher, took [my] place.
  My 'school-father' read my tablet
  . . . [and] caned me. . . .
  The teacher, in supervising the school duties,
  looked into house and street in order to pounce upon
      someone,
  . . . [and] caned me. . . .
  [He] who was in charge of drawing [said]: 'Why when I was not
      here did you stand up?' [He then] caned me.
  [He] who was in charge of the gate [said]: 'Why when I was not
      here did you go out?' [He then] caned me."

The above is taken from the opening passages of a short story* recently
deciphered by the noted Assyriologist Samuel Noah Kramer, who gives 2000 B.C.
as its probable date of creation. It must have been a popular work in ancient

---

* S. N. Kramer, "Schooldays, A Sumerian Composition Relating to the Education of a
Scribe," *Journal of the American Oriental Society,* Vol. 69 (1949), pp. 199 ff.

Mesopotamia, for numerous cuneiform copies of the original tablet have been found and its title was listed in an early Sumerian literary catalogue.

Babylonian mathematical tablets of approximately the same era as "School-days"* will be discussed in this chapter. Hence it will be interesting to examine the rest of that tale not only for its intrinsic charm and its indication of the invariance in human characteristics through four millennia, but also for its picture of a cultural setting in which mathematics was able to thrive.

The story of the young scribe continues with his recital of added woes. In all, he is caned three times by the teacher and half a dozen times by other members of the school staff. When the anxious father seeks a solution for the alleged juvenile delinquency, the son suggests that the teacher be invited to dinner and, if possible, engaged for extracurricular tutoring.

The father consents, summons the instructor, uses diplomatic finesse to bring about better teacher-pupil rapport. Having placed the pedagogue in the seat of honor at the dinner table, the Sumerian parent thanks the "school-father" (parent surrogate) for educating the boy in the "recondite details of the tablet-craft, counting and accounting." Subsequently a banquet is arranged where the servants are ordered to

> "Pour out for him [the teacher] good wine, . . .
> Make flow the good oil in his vessel like water . . ."
> They pour out for him good date-wine, . . . [then]
> he [the father] dressed him in a new garment, put a band about his hand.
> The teacher with joyful heart gave speech to him:

> "Young man, because you did not neglect my word, did not forsake it,
> May you reach the pinnacle of the scribal art, achieve it completely. . . .
> May Nidaba, the queen of the guardian deities,
> . . . show favor to your fashioned reed. . . .
> Of your brothers, may you be their leader,
> Of your companions, may you be their chief,
> may you rank the highest of [all]."*

Whether or not this particular scribe ever fulfilled his teacher's hopes, it is certain that Babylonian mathematics did "rank the highest of all" in the world of 2000 B.C. and that Sumerian computers were leaders of all their ancient "brothers" in the field. The mathematical historians Otto Neugebauer and F. Thureau-Dangin (1872–1944) have deciphered and interpreted cuneiform tablets which show that during the two millennia before the Christian era the Babylonians evolved a remark-able body of arithmetical and algebraical procedures and, in the last centuries of that epoch, an incredible mathematical astronomy.

Whenever Babylonian mathematics is mentioned, one must realize that the term "Babylonian" is a generic one. The civilization from which the mathematical cuneiform tablets come was preceded by an earlier culture, the Sumerian. According

---

* S. N. Kramer, *loc. cit.*

to some theories, the Sumerians are supposed to have migrated from the Ural-Altaic region of north-central Asia to the more fertile Tigris-Euphrates valley. Whatever their country of origin, a non-Semitic people did live just north of the Persian gulf for many centuries prior to the days of Babylonian mathematical achievement. These early inhabitants of lower Mesopotamia founded the Chaldean cities of Ur, Nippur, and Babylon; they developed a culture in which weights and measures, bills, receipts, legal contracts, promissory notes, interest (simple and compound) were commonplace. Semites who occupied Akkad (Assyria), the upper portion of the land between the rivers, built Nineveh as a rival to Babylon, as well as other important cities like Assur, Calah, and Arbela. Eventually the Assyrian Semites conquered their neighbors to the south, and the combined kingdom of Sumer and Akkad was formed. When Hammurabi became supreme ruler of that united kingdom, he chose Babylon, a Sumerian metropolis, as his capital, and from the renown of this city the entire country came to be known as Babylonia.

Historians disagree on the date of Hammurabi and give it variously as some time between 2000 B.C. and 1700 B.C. His era was the golden age of "Old Babylonian" culture. In the field of mathematics, quadratic equations were being solved by the equivalent of the formula we use today; all sorts of tabulations assisted the accountant, the statistician, the algebraic theorist; some cuneiform tablets were akin to our logarithmic tables, and interpolation was used to "read between the lines"; a considerable knowledge of practical geometry was available for application to engineering and architecture.

There is a saying that no subject loses more when divorced from its history than mathematics. Certainly mathematics does not flourish in a vacuum, and some would have us believe that its subject matter developed to solve the everyday problems of advancing civilization. But this is only part of the story. Even in ancient Babylonia a considerable body of theoretical mathematics was built up without any thought of immediate practical application. Whether the Mesopotamian "pure" mathematicians held the belief that their discoveries would ultimately be useful, we do not know. But if we realize that Babylonian creativeness extended to other fields—astronomy, religion, government, and law—we can gain some idea of the cultural level which a society must reach for mathematics to be more than a tool, for mathematics to advance along the lines that have made it the logical, philosophical, and universal subject of modern times.

Details of the truly advanced status of mathematics in ancient Babylonia will be presented at many points in this book. We shall subsequently see that great ideas formerly credited to the Greeks were undoubtedly of Mesopotamian origin, that Babylonian algebra was probably an influence in India as late as the twelfth century A.D., and that the western world did not contribute notably to the subject before the time of the Renaissance. In the present chapter we shall first examine those mathematical ideas that form the logical background for the sophisticated mathematical activity of Hammurabi's day (and that of the suffering schoolboy). Then, in order to provide a specific example of the mathematics of that era, we shall explain the Babylonian concept of a "positional" system of numeration, an idea essential for the efficiency of computation (whether by reference to cuneiform tablets or by use of digital computers) and one which has important logical by-products, such as a pattern for the decisions of an "electronic brain."

Since Mesopotamian number mastery was not a sudden development, it would be proper to consider origins, but here one is handicapped by a lack of documentary evidence. Nevertheless, latter-day ethnological studies suggest the sort of gradual evolution of ideas that may have taken place among the Sumerians or their predecessors or the even more remote ancestors of the Babylonians. We may conjecture that those earliest forebears, like some aborigines whose mores modern anthropologists have studied, had a rather limited conception of number. Some tribal languages, in fact, show a complete lack of number words. A time must come, however, when even a tribesman may want to keep track of his wives, children, clubs, canoes, flocks. The idea of number arose from his need to count those particular aggregates that were important to him.

Reference has just been made to various collections of objects which may occur in a primitive environment. We shall have occasion to consider many different *sets* of things related to fundamental mathematical ideas, both in this chapter and throughout our book. The logic of modern mathematics indicates that the notion of *set* must play a basic role because so many other concepts can be defined or explained in terms of it. As we have just seen, the idea is involved even in the earliest use of numbers.

A *set* is a collection of objects, empirical or conceptual, which are called *elements* or *members* of the set. As synonyms for the word "set," there are the terms *aggregate, class, ensemble, collection*. Thus one can refer to aggregates of people, classes of students, collections of paintings, sets of ideas, and so on. We shall even consider sets whose members are sets, for example, a set of sets of china, the set of all classes in a school, a collection of boxes of matches, the set of all twins in a town.

Finite sets can be specified by listing their members. But one cannot provide a full roster of the numbers used for counting. We shall call that collection the *natural number aggregate* and state that it is symbolized as $\{1, 2, 3, 4, \ldots\}$ (where the three dots after 4 are to be read as "and so on") to signify that the sequence of symbols is unending. Although we are unable to list all elements of the set in question, we have defined it by giving a property, "are used for counting," shared by the numbers in the set and *not* characteristic of other numbers. The use of a defining attribute is the only means of specifying the membership of an aggregate whose roster is unending or "infinite." But one can proceed in the same fashion even for finite sets. For example, reference to the first two presidents of the United States defines a unique pair of individuals, and it is not necessary to list their names.

For many mathematical purposes, a set and a property defining it are used interchangeably. But there is an essential difference in the fact that a particular collection is a unique entity, whereas its elements may share many different common characteristics. In the instance of an aggregate defined in the previous paragraph, one might specify the same set as "the United States presidents elected during the eighteenth century." Or if, say, Samuel Jones has written a recent best-selling book about George Washington and John Adams, one might define the aggregate in question as the "subjects of Jones's study."

Let us now return to the issue that launched our discussion of the set concept, namely, primitive man's need to enumerate particular sets in his immediate environment. Anthropologists have found that, among many tribes, if a savage wishes to state that he has felled *two* trees, has *two* children, owns *two* canoes and *two* clubs, he may use a different word for "two" in each of the four connotations. He has not yet arrived at a *pure* or *abstract* concept of "twoness," a property common to many aggregates, for example, a man's eyes, hands, feet, ears, twin children, and so on.

Primitive man progresses toward a more abstract notion of number when he applies one and the same word for "two" to *any* aggregate of objects that matches some standard collection symbolized by {. .} or {‖}. He may think of the things pictured within the braces as pebbles or sticks. In fact, ‖ or = may ultimately become his written symbol for "two."

The procedures for matching and counting described above can be formalized by means of a suitable definition: Two sets are said to be in *one-to-one correspondence* if and only if the elements of those two aggregates can be paired (matched) so that every member of either set has a unique partner in the other set. In that case, the two sets are said to contain the same *cardinal number* of objects. For example, the set of seats in a theater is in one-to-one correspondence with the set of tickets printed for a particular performance and, in a specific case, the cardinal number of either set might be symbolized by the numeral 1250.

In modern mathematical philosophy the cardinal number of a given set is sometimes described as an attribute which that set shares with all sets that can be put into one-to-one correspondence with it. That description would interpret a particular cardinal number as a defining property for a whole class of matching sets. Or, from the point of view of Bertrand Russell, who credits the same idea to the German logician Gottlob Frege (1848–1925), the cardinal number of a given set *is* the aggregate of all sets that can be matched with it. We observe that the individual members of that collection are sets, one of which is the given set (since it can be matched with itself). Russell's conception is an instance of a fact mentioned earlier, the interchangeability of a set and one of its defining properties.

It may seem an anachronism to consider the counting procedures of early man as the basis for the modern, sophisticated Frege-Russell conception of cardinal number. But here we are reminded of Molière's *Bourgeois Gentilhomme*, who discovered belatedly that he had been talking *prose* all his life. Ordinary prose and tribal arithmetic may, of course, lack elegance. Nevertheless, children learn to talk by imitating the former, and we shall do well to use primitive counting in the further analysis of the meaning of number.

To determine the size or cardinality of a given set by means of matching (one-to-one correspondence), our remote forebears made comparisons with ever-ready standard models—the eyes for *two*, the legs of an animal for *four*, the fingers of one hand for *five*, etc. Let us, for the sake of uniformity, imagine a tribe that used pebbles to form the collection of standard sets, {.}, {..}, {...}, and so on. Then if a flock of sheep was to be counted, the tribesman would attempt to match the flock with some set of pebbles in the collection. The matching might fail because there were either more or fewer sheep than pebbles in the standard set selected. Then it would be necessary to try matching with some other standard set, and the procedure would be repeated until a one-to-one correspondence was obtained.

The method of counting just described would be a good one in a primitive society where all aggregates of interest are small, for then matching would call for only a few standard sets and not too many trials. Otherwise, just as a draftsman or mechanic arranges his tools neatly in a kit, modern man must put some law and order into his aggregate of standard sets. Then, too, if one is to be able to make ready reference to some cardinal number, he will want to name it by a word or a symbol.

Instead of trial and error with a few standard sets, primitive or modern man may count the elements in a given set by a process of tallying, that is, by placing pebbles or sticks on the ground, one by one, which is the same as putting ink dots or pen strokes on a piece of paper. In this way one can actually construct standard sets in an orderly fashion by starting with one dot or stroke and adding one more repeatedly until he arrives at the standard set matching the aggregate he wishes to count. If he does this with dots, he will be building up the standard sets we have suggested above. If he does his tallying with strokes, he will simultaneously be recording successive numerals in the match-stick notation used by the Chinese, later adopted by the Japanese and used by them as recently as the eighteenth century. Or, if he does his counting orally, he will say "one, two, three, four," etc. Here "two" signifies "one and one," "three" means "two and one," "four" signifies "three and one," etc. Thus, by the process of starting with the cardinal number we call "one" and adding "one" repeatedly, the standard sets are built up in the orderly fashion of Figure 1.1, where names in the English language, match-stick numerals, and the Hindu-Arabic numerals we use today are recorded for the first few sets.

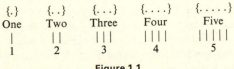

Figure 1.1

If we are to be truly modern, we should precede the standard sets in Figure 1.1 by { }, called the *null set*, or *empty set*. The special symbol $\emptyset$ is usually used to designate this vacuous aggregate, that is, $\emptyset = \{ \ \}$. To the cardinal number of $\emptyset$ we give the name *zero*, represented by the numeral 0. Although we do not usually count "zero, one, two, three," we actually begin with the empty set in the process of tallying, since placing the *first* pebble or stick on the ground might be considered as adding one object to the empty set in order to obtain the next standard set in the sequence of standard sets.

In logic, there is much emphasis on the distinction between an entity and the word or symbol used to name it. Thus the word "Henry" is not identical with the man named. Also, he may be designated in many ways, for example, "Hal" or "Mr. Brown" or by his social security number, and so on. Again, in the sense of Frege and Russell, the cardinal number of {. . .} is the collection of sets matching {. . .}. Surely that aggregate is not the same as the symbol "three" or the numeral 3. In practice, however, constant emphasis on the distinction between a name and the object named makes for pedantry. Hence, once the matter is understood, a mathematician will habitually indulge in a pardonable "abuse of language" by speaking of the "cardinal number 3" instead of the "cardinal number symbolized by the numeral 3." We shall, in fact, use the simpler form of expression in the very next paragraph.

The cardinal numbers, 1, 2, 3, 4, . . . , are called *counting numbers* or *natural numbers*. But from a more sophisticated point of view, *zero* should be included as a counting number and should precede 1 in the sequence above. Thus the natural numbers would be identical with the set of (finite) cardinal numbers. However, since zero has not been traditionally associated with the process of counting, the aggregate, {0, 1, 2, 3, 4, . . .}, is sometimes described as the class of *whole numbers*.

When the standard sets in Figure 1.1 are generated in orderly fashion by adding one "pebble" repeatedly, each set can be matched (placed in one-to-one correspondence) with a part or *proper subset* of its immediate successor which can, in turn, be matched with a proper subset of the next set, and so on. Therefore a standard set can be matched with a proper subset of any set further to the right in the sequence. Or the cardinal number of any standard set is *less than* the cardinal number of any set further on in the sequence. For example, {. . .} can be matched with part of {. . . . . . .}, and hence $3 < 7$ (3 is *less than* 7).

One would say that the relation "is a proper subset of" gives some structure to the collection of standard sets by arranging them in order. The relation $<$ establishes a similar order among the natural numbers. Therefore, as far as order is concerned, the ordered standard sets are said to be *isomorphic* to the natural numbers arranged in increasing order, which means that the two systems have the same *form* or are *abstractly equivalent*. Isomorphism is an exceedingly important concept in modern mathematics, since reasoning in a single system establishes corresponding facts in all systems that are isomorphic to it, and sometimes it is easier to handle a particular system rather than one of its abstract equivalents.

We have described two isomorphic systems that provide two equivalent ways of counting. One method is tallying, that is, building up standard sets, and the other is recitation of number words in sequence—one, two, three, four, etc. Either technique is satisfactory if one has only small aggregates to count. Otherwise, think of the quantity of pebbles, sticks, or pen strokes necessary to tally a million objects, or the need to memorize, in order, a million different number words! As far as we know, the Babylonians were the first people in history to realize that such suffering is unnecessary, and that only a few symbols can suffice to represent all counting numbers however large. Again, it is fairly certain that their ideas were anticipated, in a small way, by primitive predecessors. Economy of words and symbols is practiced by the savage who counts on the fingers of one hand, then ties a knot in a rope and starts counting again, repeating the *same number words* used originally. When he reaches "five" once more, a second knot is tied, and he starts anew. We would say that he has chosen *five* as the *base* or *radix* of his system of numeration. Because fingers and toes are used in counting, *five* or *ten* or *twenty* is the most usual choice of radix.

The ancestors of the Sumerians must have proceeded in just the way we have described except that their choice of base was *ten*. This is indicated in the Sumerian-Babylonian numerals. Symbols for the first nine numerals are akin to match-stick notation, except that the vertical strokes are wedge-shaped or *cuneiform*, ▌ , ▌▌ ,

... , ▐▐▐ . Then (in lieu of knotting a rope), the primitives in question made a mark, ⟨ , and started the count anew. When they arrived at another count of *ten*, they made another mark, so that ⟨ ⟨ represented twenty and, with the orderly

condensation of cuneiform symbols, ⟨⟨ represented fifty. The repetitive scheme

of numeration continued with fifty-one, ⟨⟨ ▌ , and went on up to fifty-

nine, ⟨⟨ ▐▐▐ Only when larger numerals had to be represented

was there evidence that *sixty* was used as the radix of a "positional" system whose nature will be explained below.

Since primitives have only small sets to count, they do not carry a good idea like counting by tens to a general conclusion. On the other hand, if we were faced with the task of counting a large set of cards or coins, we would almost automatically place them in packages of ten until we had ten such sets, when we would fuse the objects into a single set of one hundred. Then we would start counting out sets of ten until we had another hundred, and so on.

Our system of numeration with Hindu-Arabic numerals is a *positional* one with radix *ten*. This signifies that there are *ten*, and *only ten* symbols:

$$0, 1, 2, 3, 4, 5, 6, 7, 8, 9$$

With the use of these symbols in different positions, all finite cardinal numbers can be represented. The same few ciphers have different significance according to *position*. In counting, as we have illustrated, objects are grouped into tens, and tens of tens (hundreds) and tens of tens of tens (thousands), etc. In "place-value" notation, the grouping is indicated by the position of the number symbols. Thus, 8,439 means 8 thousands, 4 hundreds, 3 tens, and 9 units.

If man had had six fingers, *six* might have been the radix of our number system. The written symbols would have conveyed different meanings, but the abstract mathematical concepts and procedures would have been the same. We would have used six symbols and only six—0, 1, 2, 3, 4, 5—to write all numbers. Then *six* would be written as 10 (1 six and 0 units); twelve would be written as 20 (2 sixes and 0 units); eighteen would be written as 30; thirty-five would be 55 (5 sixes and 5 units). Since this is the largest two-digit number we can write with the six symbols available, thirty-six, the next higher number, would have to be written as 100. Continuing in the same way, 555 (5 thirty-sixes + 5 sixes + 5 units) would be the largest three-digit number, and the next number would have to be written as 1000. It would be equivalent to our six × six × six.

On the other hand, if man had had six fingers on each hand, we might have had *twelve*, not six, as base. Although the Aphos of Benue, an African tribe, do not have twelve fingers, they use twelve as their number base and thereby seem to indicate some remarkable arithmetic intuition in the choice of a superior base. For the use of the base twelve, we would need twelve symbols,

$$0, 1, 2, 3, 4, 5, 6, 7, 8, 9, *, \#$$

and we should write twelve as 10 (1 twelve and 0 units); thirteen would be 11, twenty-four would be 20 (two twelves and 0 units); one hundred forty-three would be ## (eleven twelves and eleven units). As this is the largest two-digit number that can be represented with the twelve symbols available, the next number, one hundred forty-four, would be written 100.

The reader may amuse himself by checking the following sample exercises.

System with Base Six

35	Five + two = seven, written as 11 in this system.
+ 52	Put down 1 and carry 1.
131	Three + five + one = nine, written as 13 in this system.

43	Three × three = nine, written as 13 in this system.
× 3	Put down 3 and carry 1.
213	Three × four = twelve. Twelve + one = thirteen, written as 21.

System with Base Twelve

8*	Ten and one are eleven. Eight and four are twelve.
+ 41	
10#	

35	Eleven fives = fifty-five, written as 47 in this system.
× #	Put down 7 and carry 4. Eleven threes = thirty-three.
317	Adding four, the result is thirty-seven, written as 31 in this system.

In our decimal system, positions to the right of the decimal point represent tenths, hundredths, thousandths, etc. Thus 0.462 stands for the *sum* of 4 tenths, 6 hundredths, and 2 thousandths. Using the customary exponential symbolism of algebra, we can summarize the nature of the decimal system by stating that successive places to the left of the decimal point have the values 1 (or 10^0), 10^1, 10^2, 10^3, 10^4, etc., whereas those to the right have the values $1/10$ (or 10^{-1}), $1/100$ (or 10^{-2}), $1/1000$ (or 10^{-3}), etc. Thus 7084.91002 means

$$7\,(10^3) + 8\,(10^1) + 4\,(10^0) + 9\,(10^{-1}) + 1\,(10^{-2}) + 2\,(10^{-5})$$

In general, if the base of a number system is represented by b, successive places to the left of the "decimal" point have the values b^0, b^1, b^2, b^3, b^4, etc., and those to the right have the values $b^{-1}, b^{-2}, b^{-3}, b^{-4}$, etc.

In the Sumerian-Babylonian notation sixty was the radix for natural numbers above fifty-nine and also for the sexagesimal fractions. Now in a modern sexagesimal system of numeration the symbol 54 would signify 5 sixties and 4 units. It would be the symbol for our decimal number three hundred four. Again, in a sexagesimal system, 8.49 would mean the sum of 8 units, 4 sixtieths, and 9 thirty-six hundredths. As a matter of fact, our method of angular measure is derived from the Babylonian sexagesimal system. In order to avoid confusion, however, we would *not* write 8.49 but instead would employ the symbol

$$8° \ 4' \ 9''$$

and read it as 8 degrees, 4 minutes, 9 seconds. This is identical in meaning with the above 8 units, 4 sixtieths, 9 thirty-six hundredths, since a *degree* is the unit of angular measure and a *minute* and a *second* are one sixtieth and one thirty-six hundredth, respectively, of the angular unit.

The cuneiform symbol for sixty was the same as for *one*, namely, ▌. The only way to realize that sixty was meant, and not one, was from the context. For example, in a textbook tablet where the natural numbers were listed in order, one could judge that the single vertical wedge succeeding fifty-nine represented sixty. Again, the group of cuneiform symbols, ▌▌◁◁▌▌▌ , might be interpreted as our $2(60)^1 + 23(60)^0 = 143$ but *other* interpretations are possible, for example, $2(60)^2 + 23(60)^1 = 8580$. What the Babylonians needed was a symbol for *zero*. Ultimately such a sign was developed, for it appears in the Seleucid mathematical tablets and there is some evidence that it may possibly have been created as early as 700 B.C. But even in the final phase of Babylonian writing there were no zero symbols at the ends of numbers. To put this weakness into modern dress, let us imagine our decimal system without a zero. We write

$$1 \quad 2 \quad 3$$

Is this one hundred twenty-three, or ten thousand two hundred three, or one thousand two hundred thirty? In the first case, the Babylonians would have avoided difficulties by writing the symbols close together. The second number would have been symbolized by 1-2-3, that is, by using a zero symbol that is akin to our dash. But in the third instance, the ambiguity was never removed.

The Babylonians' choice of radix sixty is now attributed to the adjustment of their arithmetic to their monetary units. In the earliest days the basic medium of exchange was barley, but pieces of precious metals like silver were used later on. Since coinage had not yet been invented, value was measured according to weight. In the table of weights, sixty *shekels* were equivalent to one *mina* and sixty *minas* to a *talent*. This counting of weights (and ultimately of minted coins) in groups of sixty was the origin of the sexagesimal system. Formerly, the use of sixty as base was erroneously attributed to the Babylonian estimate of the year as 360 days, involving the subdivision of the apparently circular orbit of the sun into 360 equal parts. Today it is known that this phase of Babylonian astronomy occurred in the last centuries B.C., whereas the use of a sexagesimal number system is found in the earliest cuneiform tablets.

How did the Babylonians handle arithmetic computation? They seem to have had little trouble with addition and subtraction. For multiplication and more advanced arithmetic operations, they made extensive use of *tables*. Whereas the modern schoolboy memorizes tables for multiplication by 2, 3, . . . , 9, the Babylonian scribe would have consulted a cuneiform tablet like the one now on view in the Brussels collection. In that table, the products of 7, 10, 12½, 16, 24, by 2, 3, . . . , 9, 10, 20, . . . , 50 are listed.

The Babylonians were very modern in their idea that it is *not* necessary to introduce a division operation providing one is able to carry out multiplication with whole numbers and common fractions. For example, one need not consider $7 \div 2$ because one can instead perform the multiplication, $7 \times \frac{1}{2}$. The thought is that division by a number is equivalent to multiplication by its *reciprocal* or *multiplicative inverse*. Moreover, except for *zero*, every number x of elementary arithmetic has a multiplicative inverse, $1/x$, that is, a number such that $x \cdot 1/x = 1$. Thus the multiplicative inverses or reciprocals of 3, 4, 1/5, 2/9 are respectively 1/3, 1/4, 5, 9/2.

Many Babylonian tables of reciprocals have been unearthed and deciphered. When transcribed into our symbols, the first entries in one of those tables would appear as follows:

Number	Sexagesimal Reciprocal
2	30
3	20
4	15
5	12
6	10
8	7, 30
9	6, 40

Here the multiplicative inverse of each number in the first column is expressed as a sexagesimal fraction in the second column. Interpreting that column in terms of angular measure with degrees, minutes, and seconds may serve as a helpful artifice.

Thus the reciprocal of 2 is 1/2 and $(1/2)° = 30'$. The reciprocal of 3 is 1/3 and $(1/3)° = 20'$, that for 8 is $(1/8)° = 7'30''$, etc. The transcription of the Babylonian procedure for obtaining $7 \div 2 = 3\ 1/2$ and $9 \div 8 = 1\ 1/8$ is

$7 \times (1/2)° = 7 \times 30' = 210' = 3°\ 30'$, which is equivalent to our 3 1/2
$9 \times (1/8)° = 9 \times 7'30'' = 63'\ 270''$
$\qquad\qquad = 63' + (4'30'')$
$\qquad\qquad = 67'\ 30''$
$\qquad\qquad = 1°\ 7'\ 30''$, which is equivalent to our 1 1/8.

We observe that 7 and its reciprocal were omitted from the Babylonian tabulation. We can conjecture why this is so if we try to convert $(1/7)°$ into minutes and seconds, using only whole numbers. This cannot be done, since the result is $8'34\ 2/7''$. Babylonian tables listed only "regular" reciprocals and omitted "irregular" ones like 1/7, 1/11, etc., with the statement that these involve "impossible" divisions. However, in such cases, sexagesimal approximations like 8,34 for 1/7 were used much as we might use the decimal 0.14 as an approximation of 1/7. Some of the reciprocal tablets are very extensive; one from the Seleucid period contains about fifty pairs of numbers. A row near the end of this tablet bears an entry which we transcribe as 2, 59, 21, 40, 48, 54 with reciprocal 20, 4, 16, 22, 28, 44, 14, 57, 40, 4, 56, 17, 46, 40. This means that the number in question should be interpreted by us as

$$2(60)^5 + 59(60)^4 + 21(60)^3 + 40(60)^2 + 48(60)^1 + 54(60)^0$$

and its reciprocal as equal to

$$20(60)^{-6} + 4(60)^{-7} + 16(60)^{-8} + 22(60)^{-9} + 28(60)^{-10}$$
$$+ \ldots + 46(60)^{-18} + 40(60)^{-19}$$

We must admit that this is fearless arithmetic. But then tables of squares and square roots, cubes and cube roots, as well as exponential and compound interest tables had been a commonplace 1500 years before this reciprocal table was prepared!

As far as the pure mathematician is concerned, it is the "place-value" *concept* that is an all-important contribution of the Babylonians to arithmetic; the fact that they based their system on sixty is immaterial. If they had used a radix of four or five or twelve or twenty (or any other number that is not too large), their creation would have been equally valuable as a means of condensing the tasks of enumeration and computation. In algebraic terminology, a numeral in *any* positional system of notation represents a number

$$a_n b^n + \ldots + a_2 b^2 + a_1 b^1 + a_0 b^0 + c_1 b^{-1} + c_2 b^{-2} + \ldots + c_m b^{-m}$$

where b represents the number chosen as radix, and an a or a c is either zero or an integer between 1 and $b - 1$. (The subscripts are a customary algebraic device for indicating position.) The great advantage of using any positional system lies in the fact that only b symbols are required. The formula above tells the entire story in a nutshell.

Is there possibly a best choice of radix? Pure mathematicians have indicated that a *prime* base (one like 7, 11, 13, 17, etc., with no divisors but 1 and itself) would have led to greater uniformity in arithmetic procedures. Practical reformers, on the other hand, have been in general agreement that *twelve* would have been a better base than ten, since it has the divisors 2, 3, 4, and 6, a fact that would have made work with fractions easier than it is with base ten (divisors 2 and 5). The learning of only two additional symbols would be worthwhile, compared to the tremendous saving in other arithmetic effort. Charles XII of Sweden was supposed to have been contemplating, at the time of his death, the abolition of the decimal system in all his dominions, in favor of the duodecimal. A universal change to the base twelve, or to any base, however, is just as impracticable as the general use of Esperanto. If a choice were possible, a duodecimal or a prime system would be the scale of civilization, but quinary, decimal, and vigesimal systems are the scales of nature.

In the United States in colonial days, the Reverend Hugh Jones (1692–1760), professor of mathematics at the College of William and Mary, was an ardent advocate of the *octary* system (radix *eight*), which is, in fact, used today in connection with certain electronic computers. Jones was a reformer who argued that ordinary arithmetic had already become "mysterious to Women and Youths and often troublesome to the best Artists." Without going into detail, it can be asserted that the base eight makes fractional work simpler because octary fractions are just a matter of halving again and again. Thus, one-quarter is just one-half of a half, and one-eighth is just one-half of a half of a half, etc. Moreover, computation is facilitated because the radix eight is a perfect cube, that is, $8 = 2 \cdot 2 \cdot 2$, and four, which is one-half the radix, is a perfect square ($4 = 2 \cdot 2$). Still another advantage of repeated halving is seen in the fact that arithmetic is made consistent with the British system of weights, in which an ounce is one-sixteenth of a pound, and so on. The Reverend Jones showed that, as far as England and its colonies were concerned, many arithmetic difficulties arose from the use of the British system of weights, measures, and coinage.

Whereas the Reverend Jones saw advantages in octary arithmetic, and British mathematicians* today are still arguing about the virtues of decimal versus duodecimal systems, one of the mathematical immortals, Gottfried Wilhelm Leibniz (1646–1716) had advocated the binary or dyadic system, the type of numeration with two as base, which, according to anthropologists, was used by the most primitive of primitive groups. Among such people a tribesman would count "one, two," then "two and one," "two and two," "two and two and one," "two and two and two." At about this point the strain became too great and every higher number was called "heap," a lot. Then what advantages could Leibniz see in the primitive dyadic system? For one thing, it appeared to hold mystic significance for him. But mathematicians today are inclined to view his imaginative statements merely as a

* See B. H. Neumann, "On Decimal Coinage," *Matrix*, Melbourne University Mathematical Society, 1964, pp. 16–18.

poetic formulation of aspects of the algebra of sets (Chapter 6). From still another point of view, Leibniz realized that the utter simplicity of binary arithmetic must hold definite practical advantages, and in this, as well as in his thoughts on symbolic logic, he anticipated present developments in applied and pure mathematics.

As a preliminary to the explanation of the recent applications of numeration with base two, a few decimal numbers and the corresponding binary numerals are listed below.

Decimal Number	Binary Representation
Zero	0
One	1 (one unit)
Two	10 (one "two" and no units)
Three	11 (one "two" and one unit)
Four	100 (one (two)^2)
Five	101
Six	110
Seven	111
Eight	1000
Nine	1001
Ten	1010
Eleven	1011
Twelve	1100
Thirteen	1101
Fourteen	1110
Fifteen	1111
Sixteen	10000

To obtain the decimal equivalent of any binary number one need only apply the basic formula for the meaning of a number in a positional system of numeration (page 12). Here, if we think in terms of our decimal system, $b = 2$ and each a will have either the value 0 or the value 1, that is, the only symbols needed in a binary system are 0 and 1. Successive places proceeding toward the left, starting at the decimal point, have the values 2^0, 2^1, 2^2, 2^3, etc. Thus the binary number 10111 has, according to the fundamental formula, the decimal value

$$1(2)^4 + 0(2)^3 + 1(2)^2 + 1(2)^1 + 1(2)^0 =$$
$$16 \ + \ 0 \ + \ 4 \ + \ 2 \ + \ 1 \ = 23$$

and 11.1001 has the value

$$1(2)^1 + 1(2)^0 + 1(2)^{-1} + 0(2)^{-2} + 0(2)^{-3} + 1(2)^{-4} =$$
$$2 \ + \ 1 \ + \frac{1}{2} \ + \ 0 \ + \ 0 \ + \frac{1}{16} \ = 3\frac{9}{16}$$

The first advantage of a binary system of numeration is that only two symbols are needed, namely 0 and 1. Another is that children studying binary arithmetic would have very few tables to memorize. The addition table reduces to

$$0 + 0 = 0 \qquad 0 + 1 = 1 \qquad 1 + 1 = 10$$

All multiplication tables are embodied in the three statements

$$0 \times 0 = 0 \qquad 0 \times 1 = 0 \qquad 1 \times 1 = 1$$

But, on the other hand, there is the *length* of a numeral! Our thousand would have to be written as

$$1,111,101,000$$

In our modern electronic "children"—our digital computers or mechanical brains, and so on—a lengthy binary numeral would, however, constitute no serious problem. With 1 and 0 represented by "on" and "off" respectively, a long numeral merely means a more extensive bank of telegraph-type relays or radio-type tubes. To see how this occurs, let us suppose, for simplicity, that one wishes to register or signal any number from zero to seven. Then, using the binary representations, and a row of three electric light bulbs with "on" for 1 and "off" for 0, the signals are as follows:

Decimal Number	Binary Representation	Signal
Zero	000	0 0 0 (off, off, off)
One	001	0 0 ● (off, off, on)
Two	010	0 ● 0 (off, on, off)
Three	011	0 ● ●
Four	100	● 0 0
Five	101	● 0 ●
Six	110	● ● 0
Seven	111	● ● ● (on, on, on)

The binary representations are the same as those previously given, except that 0's have been added on the left in the case of the first four numbers so that every number may have three places.

One might think, alternatively, that one of the numbers from zero to seven is to be fed into a computer by means of a punch card with three dots where 1 and 0 are represented by punching or not punching the dot. Then the above "signals" show the appearance of the punch card for the different numbers. Similarly, one could use a punch card with ten dots, or a row of ten bulbs to register all binary numerals up to 1,111,111,111. Some decimal names and binary equivalents would be as follows:

Decimal	Binary	Signal
Zero	0,000,000,000	0 0 0 0 0 0 0 0 0 0
One	0,000,000,001	0 0 0 0 0 0 0 0 0 ●
Two	0,000,000,010	0 0 0 0 0 0 0 0 ● 0
...	....	....
...	....	....
...	....	....
Thousand	1,111,101,000	● ● ● ● ● 0 ● 0 0 0
...	....	....
1023	1,111,111,111	● ● ● ● ● ● ● ● ● ●

To understand the extreme importance of Leibniz' favorite radix *two* in the present age of automation, one must realize that the name "computer" is misleading, and "thinking machine" or "giant brain" are substitute terms that give a better idea of the remarkable services the electronic devices in question can render. It is true that binary numbers like those above may be fed into a machine or stored in its registers or signaled, and so forth. But these binary figures (or their decimal equivalents) are often merely code numbers for names or words or messages or mathematical propositions or general statements of some sort.

A digital computer receives its problems through "input" devices like punched cards, punched paper tape, magnetic tapes, magnetic wire, or photographic film. The reason for the almost universal use of binary representation is that the equipment generally employed in digital computers has *two* states. The punched cards and tapes are either punched or not in each position, and magnetic equipment distinguishes a pulse from a nonpulse; relays are open or shut, and electronic circuits are built with two stable states. Computers are designed to do much more than calculation. They can make selections or *decisions* providing questions are put to them properly. The binary computers must be asked to make comparisons between *pairs*, much as the last rounds in a sequence of basketball games compare teams in order to select the winning team.

All this may seem a long way from Hammurabi's era and Babylonian mathematics. Digital computers, however, are direct descendants of the cuneiform tablets, linked by the dominant idea of the place-value notation, which is basic to all schemes of simplifying the practical calculations required in each period of history. Practicality must be the concern of the applied mathematician, and therefore he sees a positional scheme as a superior one, and points with pride to the speedy, painless performance of electronic computers where astronomic figures are involved. The point of view of pure mathematics, however, provides unity for the subject by pronouncing ancient and modern systems merely different examples of a *single* abstract pattern, in other words, by calling the systems *isomorphic* with respect to the arithmetic operations of addition and multiplication.

That a one-to-one correspondence can be established between two systems of numeration with different radices, or between a set of numerals and a set of electric signals, is indicated by previous tabulations (page 15). Isomorphism or structural identity exists because there is matching not only of number symbols, but of operations (the addition and multiplication operations in the two systems) and the results of operations. Thus the decimal statements

$$2 + 3 = 5 \qquad 2 \times 3 = 6$$

correspond to the binary

$$10 + 11 = 101 \qquad 10 \times 11 = 110$$

and, again, these match, symbol for symbol, operation for operation, with certain computer signals and operations.

What we have been discussing, then, is the matter of varied interpretations of a single pure or abstract system, the one usually described as the additive and multiplicative arithmetic of the whole numbers and the common fractions. This particular structure is not the entire story, however, and this makes it inadequate for important theoretical issues associated with measurement in the physical world. The nature of certain geometric problems arising from questions of measure will now be considered; it will be seen that it was not Babylonian but rather Hellenic science that revealed and succeeded in solving related numerical paradoxes, thereby making possible the logical derivation of formulas for lengths, areas, and volumes.

2

Mathematical Method and Main Streams Are Launched

At the beginning of our story, a Sumerian document was seen to present a picture of what the life of a schoolboy was like in 2000 B.C. The ancient story was a case of fiction imitating fact. But what is the life of today's schoolboy like, almost four thousand years later? Are the blows of the cane being replaced by stinging intellectual impacts? Are parents still involved in the backlash of pedagogical methods?

To face only the mathematical aspect of these questions, one must consider the United States educational phenomenon called "the new mathematics." Perhaps the evidence we shall adduce will be a case where facts appear to imitate fiction, since we now propose to show that basic features of the "new" program must be given a date at least as early as the sixth century B.C. As we have seen, pebble devices for picturing cardinal numbers go back to prehistoric man. But school children who are now using such representations in order to "discover" patterns or mathematical laws are emulating procedures followed by the ancient Pythagorean Society.

That organization, part secret brotherhood, part institute for advanced study, had a membership of wealthy men whom Pythagoras (*ca.* 550 B.C.) had gathered about him in Crotona, a Greek city in southern Italy. The avowed purpose of the Pythagorean school was to carry on scientific research in number theory, geometry, music, and astronomy. It is known that Pythagoras made many trips to the Middle East, and therefore it is not surprising that some of the ideas he presented to his disciples had a strong Babylonian flavor. But in one respect the Babylonians exerted no influence, since the Society showed no interest in techniques of practical computation, and devoted itself instead to the logical and philosophical aspects of the *theory* of numbers.

Our own objective in the present chapter is to show how discoveries of the Pythagorean brotherhood led to the "real number system," the rock on which most of mathematics, old or new, is built. So great were the Pythagorean birth pangs that they could only be alleviated by the panacea of a second major conception, one which crystallized for all time the method of *proof* that is used in mathematics. Although the Babylonians and Egyptians based mathematical conclusions on experiment and observation, the Pythagoreans discovered that certain intuitive ideas about measurement will produce an apparent paradox. Therefore Pythagoras decided, once and for all, that empirical observations cannot furnish pure mathematical *proof*, which must be provided instead by deductive chains of reasoning. Hence he went forward with a task that Thales (*ca.* 600 B.C.), his teacher, had initiated, and which Hellenic thinkers up to the time of Euclid (300 B.C.) were to pursue. The program was the organization of geometry into a logical pattern in which, after a few assumptions and definitions are laid down, all further statements are deduced.

"Number rules the universe," said Pythagoras. To some extent this was an expression of the deep emotion with which mathematicians react to their specialty, but mainly it was an assertion of the major dogma of the Pythagorean brotherhood, the belief that all of mathematics and science could be based on the natural numbers, {1, 2, 3, . . .}. It was in their study of some of the elementary properties of the natural numbers that the Pythagoreans set the example for the "method of discovery" in the "new mathematics." Since *pattern* was the objective, and *geometry* was a fundamental interest, the brotherhood formed *figurate numbers*, designs in the form of *polygons*. Thus, by counting the dots and observing the shape in each of the diagrams of Figure 2.1, we see why 10, 16, 20, 22, and 28, are described as

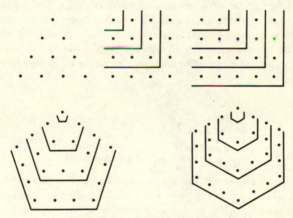

Figure 2.1   Figurate numbers

triangular, square, rectangular, pentagonal, and *hexagonal* numbers. The Society used such diagrams in the derivation of formulas related to certain number series.* But here we shall be more concerned with how today's schoolboy is applying figurate numbers.

Our kindergartners, like the Pythagoreans, may construct figurate numbers by adding *one* pebble at a time. The Society held that *one* is therefore a *source* of numbers but is *not* itself a counting number, since there is no need to count in a set

* See, for example, E. E. Kramer, *The Main Stream of Mathematics*, Oxford University Press, New York, 1951, pp. 196 ff.

containing a unique, solitary pebble. In the course of building numbers, our children may recite appropriate rhymes or ditties. The Pythagoreans chanted prayers that incorporated oriental mystic beliefs about the natural numbers.

The gist of the faith was that *one* represented reason, *two* stood for woman, *three* for man, *five* represented marriage, since it is formed by the union of two and three. *Four* stood for justice, since it is the product of equals. All the *even* numbers were regarded as soluble, therefore ephemeral, *feminine*, pertaining to the earth; *odd* numbers were indissoluble, *masculine*, partaking of celestial nature. *Ten*, inevitably a number *base*, was pictured with *four* rows of pebbles as in Figure 2.1 and hence called the *tetraktys*. It was considered the sacred "fourfoldness," because its four rows represented a totality of the reason and justice of man and woman, and also, in Pythagorean metaphysics, cosmic creation through the four basic elements—fire, water, air, and earth. The ritual of the brotherhood included a special prayer to the "holy ten, the keyholder of all."

Mathematics was mixed with mysticism in the activities of Pythagoras' followers, and later in the present chapter we shall see how the "eternal feminine," that is, the characteristics of the even numbers, led them to the most startling of their discoveries. Opinion to the contrary, we do not bring sex into the kindergarten today and hence do not visualize counting numbers as male and female. Instead, we use the term *even* for 2, 4, 6, . . . and all natural numbers that are multiples of 2, and we refer to the other natural numbers—1, 3, 5, . . .—as *odd*. But we can agree with the Pythagoreans in picturing an even number as a rectangle with two columns (Figure 2.2.*a*) and an odd number as such a rectangle supplemented by a single pebble (Figure 2.2*b*).

$$\begin{matrix} \bullet & \bullet \\ \bullet & \bullet \\ \bullet & \bullet \end{matrix} \qquad\qquad \begin{matrix} \bullet & \bullet & \bullet \\ \bullet & \bullet \\ \bullet & \bullet \end{matrix}$$

Figure 2.2*a*
Even number 6

Figure 2.2*b*
Odd number 7

Any sophisticated use of Pythagorean figurate numbers presupposes that the meaning of *addition* and *multiplication* of natural numbers is understood. Therefore today's school children are being encouraged to "discover" the meaning of $3 + 4$, for example. They know that 3 is the symbol for the (cardinal) number corresponding to the set $\{. . .\}$, and 4 is the number for $\{. . . .\}$. If they unite these sets by placing the pebbles or dots in the second aggregate to the right of those in the first, they are constructing the united set or *union* of the two sets, namely, $\{. . . . . . .\}$. The sum $3 + 4$ can then be defined as the cardinal number of the union, and that number is 7. The "discovery" is that $3 + 4 = 7$.

Children are told that the above united set or union might have been pictured as

$$\left\{ \begin{matrix} \bullet \\ \bullet \end{matrix} \quad \bullet \quad \begin{matrix} \bullet \\ \bullet \end{matrix} \quad \bullet \right\} \quad \text{or as} \quad \left\{ \begin{matrix} \bullet \\ \bullet \\ \bullet \end{matrix} \quad \bullet \quad \begin{matrix} \bullet \\ \bullet \end{matrix} \quad \bullet \right\}$$

or in a variety of different ways. No particular pattern is required in forming the union of two sets. That being the case, the diagrams we have just indicated (or the original picture) might just as well correspond to the sum $4 + 3$. Therefore $4 + 3 = 7$ and $3 + 4 = 4 + 3$.

In passing, let us remark that, in modern mathematics, the symbol $=$ can be read as "equals" but is understood to signify that the words or symbols on either side of the equals sign are names for the same object. Thus $4 + 3$ is not identical with the symbol 7. Nevertheless the same thing is named in either case—specifically, a certain cardinal number.

A child may enjoy repeating the steps of his "discovery" with suitable variations for the sake of interest. If he wishes to "discover" the meaning of $3 + 3$, say, he will wish to unite $\{...\}$ and $\{***\}$. (We recall, from the previous chapter that 3 is the cardinal number of *many* different matching sets.) Here what should be emphasized is that in defining addition of cardinal numbers, the sets that are united must be *disjoint* or nonoverlapping, that is, must have no members in common. That fact enables one to say that $3 + 3$ corresponds to the picture,

$$\left\{ \bullet \quad \bullet \quad \bullet \quad * \quad * \quad * \right\} \quad \text{or} \quad \left\{ \begin{matrix} \bullet \\ * \quad * \quad \bullet \\ \bullet \quad * \end{matrix} \right\}$$

and so on, and therefore $3 + 3 = 6$.

After uniting many varied pairs of disjoint sets in order to find different sums, the child is supposed to make more significant "discoveries," namely, those that are laws or general rules for *all* pairs of natural numbers. Thus the fact that $3 + 4 = 4 + 3$, $1 + 2 = 2 + 1$, $2 + 5 = 5 + 2$, etc. should suggest the "discovery" that the order of terms is immaterial in adding a pair of natural numbers. Here we have the *commutative law* of addition of natural numbers: If a and b are any two natural numbers, than $a + b = b + a$.

Our illustrative examples have indicated that addition can be described as a *binary operation* because basically it combines only *two* numbers at a time. Now if one is to drive a car or operate any machine, it is always advisable to check in advance that it is in good operating order. By crude analogy, if addition is to be considered a proper, well-defined binary operation on the natural numbers, one should know, at the start, that it is possible to perform it with *every* pair of natural numbers. There should be no stalling or disastrous results where a sum is some "monster" unknown to the world of counting numbers. Nor should one arrive at a dilemma where there are two possible natural numbers that might serve as the sum. This is all so obvious that the reader may wonder why it is mentioned at all. The point is that one is not going to stop with addition. One will subsequently define and carry out $8 - 2$, $6 - 3$, etc. But what about $2 - 8$, and $3 - 6$? Will a child in the lower school grades be asked to "discover" how to remove 8 pebbles from a set of 2? Evidently one cannot say that subtraction is a well-defined binary operation on the natural numbers. One way of handling such a situation in the newest "new mathematics" is to say that subtraction of natural numbers cannot be honored with the title of binary operation. An older method starts all studies of a particular operation with a "closure" law.

Returning to *addition* of natural numbers, the latest style would merely summarize the situation by stating: Addition is a *binary operation* on the set of natural numbers. The older fashion would provide the implicit content of this statement by asserting the closure law: The aggregate of natural numbers is *closed* under addition. This means that the sum of any two natural numbers is a unique natural number.

In a *logical* treatment of the arithmetic or algebra of natural numbers, the above closure law will naturally precede the commutative law of addition or any other law relating to addition, for how can one say that $a + b = b + a$ for *all a* and b unless one is sure that $a + b$ and $b + a$ have meaning in the aggregate under consideration? Because a closure law should logically come first, one finds very often that teachers and textbooks expect such a law to be one of the first "discoveries" made by children. Thus young students who carry out a dozen or more pebble additions corresponding to $1 + 1$, $3 + 4$, $4 + 3$, $7 + 5$, etc. are coaxed into "discovering" that one can *always* (sic) carry out the pebble process and obtain a unique result. The fact is that only one very young pupil in a million would realize that there is any need to demonstrate the closure property of addition, and that pupil would realize the "discovery" to be a hasty generalization based on limited evidence. For our own information, the closure law is a postulate or assumption. It could only be demonstrated if we made other assumptions. As far as the average schoolboy is concerned, the logical sequence is not always the pedagogical one. It seems to this author that the time to "discover" the closure law for addition is *after* the failure of closure in the case of subtraction and division of natural numbers has been "discovered." Such counterexamples as $2 - 3$ and $2 \div 3$ can motivate reconsideration of addition of natural numbers.

Granted that addition of natural numbers has closure or is a *binary operation*, how is the child to assign a meaning to $3 + 4 + 2$, the sum of *three* (not two) numbers? It is a simple exercise to draw pebble diagrams that "associate" the first two numbers and then combine their sum with the third. In other words, $(3 + 4) + 2$ is pictured, and after the first binary addition is performed, one has $7 + 2$, a second binary addition, yielding the result 9. Alternatively, children consider the interpretation $3 + (4 + 2) = 3 + 6 = 9$, which is the same as the answer previously obtained. After carrying out repeated additions of three terms with the two different interpretations, pupils "discover" the *associative* property (or law) of addition of natural numbers: If a, b, and c are any three natural numbers, then

$$(a + b) + c = a + (b + c)$$

which says that the result of adding the third number to the sum of the first and second is identical with the outcome of adding the first to the sum of the second and third.

Since so many of the "discoveries" may seem obvious or intuitive, it is always important to provide counterexamples by varying the set or the operation. Thus there is noncommutativity in "adding" to one's feet "shoes + stockings" rather than "stockings + shoes" and nonassociativity in the sums, "(baking powder + milk) + other cake ingredients" and "baking powder + (milk + other cake ingredients)."

As in the case of addition, Pythagorean figurate diagrams are helpful in defining the *multiplication* of natural numbers and discovering its properties. In fact, multiplication can be interpreted as repeated addition, although a more general definition that mirrors a "multiplication" of sets will be given in Chapter 7. Here, we shall say that $2 \times 3 = 3 + 3$ and $3 \times 4 = 4 + 4 + 4$, as pictured in Figures 2.3a and 2.3b. Hence, in general, if a and b are natural numbers, then $a \times b$, usually

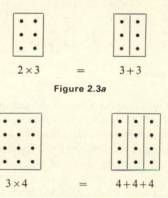

$$2 \times 3 \qquad = \qquad 3 + 3$$

Figure 2.3a

$$3 \times 4 \qquad = \qquad 4 + 4 + 4$$

Figure 2.3b

written as ab, can be pictured as a rectangular figurate number with a columns containing b dots each. It is then a simple matter to "discover" the closure, commutative, and associative laws for the multiplication of natural numbers.

Perhaps the most interesting law is the one that connects the two binary operations on the natural numbers. It is called the *distributive law* and it states:

$$a(b + c) = ab + ac$$

where a, b, and c are any natural numbers. Figure 2.4 pictures a pebble diagram for a special case.

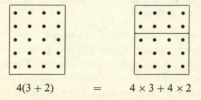

$$4(3 + 2) \qquad = \qquad 4 \times 3 + 4 \times 2$$

Figure 2.4

Products which were of special interest to the Pythagoreans were those that could be pictured as *square numbers*, namely, 1×1, 2×2, 3×3, 4×4, etc., or 1^2, 2^2, 3^2, 4^2, etc. Children can diagram these squares and then, after many transformations like those in Figures 2.5a and 2.5b, they can make the "discovery," that is, arrive at the inductive generalization: The square of an odd number is odd and the square of an even number is even.

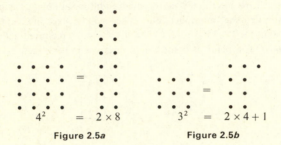

$$4^2 \quad = \quad 2 \times 8 \qquad 3^2 \quad = \quad 2 \times 4 + 1$$

Figure 2.5a   **Figure 2.5b**

Because the "discovery" just described was actually made by the Pythagoreans who verified the validity of the induction by providing a *deductive* proof, let us do likewise. We shall, however, employ modern *algebraic* symbolism and state that $2x$ and $2x + 1$ represent even and odd numbers if x is replaced by any one of the whole numbers $\{0, 1, 2, 3, \ldots\}$. Then the *square* of an even number is symbolized by $(2x)^2$ and the square of an odd number by $(2x + 1)^2$. But $(2x)^2 = 4x^2 = 2 \cdot 2x^2$, an *even* number, and $(2x + 1)^2 = 4x^2 + 4x + 1 = 2(2x^2 + 2x) + 1$, an *odd* number.

If the reader should weary of pebble diagrams and Pythagorean preoccupation with even and odd numbers, he will be interested in the following dialogue from one of the comedies written by Epicharmos (540–450 B.C.), dramatic poet, philosopher, and member of the Pythagorean school:

"When you have an even number or, for all I care, an odd number, and someone adds a pebble or takes one away, is the original unchanged?"
"May the gods forbid it!"
"Now if some one adds to a length or cuts off a piece, will it measure the same as before?"
"Why, of course not."
"But look at people. One may grow taller, another may lose weight. . . . Then by your argument, you, I, and others are not the same people we were yesterday and we shall be still different individuals in the future."

Epicharmos may have been poking fun at his fellow Pythagoreans, but the dialogue does reveal some of their beliefs. The two questions asked indicate that the brethren distinguished between numbers for counting (the even and odd natural numbers) and numbers for *measuring* various magnitudes—lengths, areas, volumes, weights. When we think of modern measurements like 2.43 cm., $5\frac{1}{8}$ sq. ft., etc., fractions make their appearance. But now we must ask whether such results of measurement did not, somehow, conflict with the Pythagorean belief that *natural numbers* must suffice in all situations.

For example, when the brotherhood discovered that, other conditions being equal, sounding musical strings measuring 1/2, 2/3, 1 (octave, fifth, fundamental, or *do, sol, do*) in unison or in succession produces a pleasing or "harmonic" effect, it would seem that common fractions were being used. But, to the Pythagoreans and their Hellenic successors, a fraction was a *ratio* (hence our term *rational number*) of lengths or areas or volumes, that is, of geometric magnitudes of the same kind. The fraction 1/2, for example, might mean that if the length of a particular musical string is taken as the unit of measure, and another string measures 2 units, then the ratio of lengths is 1 unit : 2 units. This is more like our 1 foot : 2 feet than the "pure" or abstract fraction 1/2, a symbol which can have many different concrete interpretations. An abstract logical definition of a ratio or common fraction would describe it as an *ordered pair* of natural numbers, for example, (1, 2) or (5, 7), in the order (antecedent, consequent) or (numerator, denominator), and this conception would in no way conflict with the Pythagorean faith in the natural numbers.

In a modern logical treatment of arithmetic, one would define addition and multiplication of the number pairs representing common fractions and also show that closure, associative, and commutative laws apply to such addition and multiplication. The natural numbers are part of the new system only when they are used for comparison or measurement. Then instead of $\{1, 2, 3, \ldots\}$ one would have ordered pairs like $\{(1,1), (2,1), (3,1), \ldots\}$. There is need to consider the special point that a *number*, whether it is cardinal or fractional, is an abstract entity that can be symbolized or named in many different ways. We have already pointed out that III or 3 or $2 + 1$ or *drei* are merely different names for the *same* cardinal. In the case of fractions, we may say that

$$\frac{1}{2} = \frac{3}{6} = \frac{70}{140} = \frac{5}{10} \text{ or } 0.5, \text{ etc.}$$

that is, $(1,2) = (3,6) = (70,140) = $ etc., because these symbols, although not identical, name the same entity.

Although the Pythagoreans and their Hellenic successors considered some of the special properties of natural numbers and their ratios, the name of Pythagoras is more likely to be associated with geometry than with arithmetic. The abstract conception of fraction may seem novel to some laymen, but they will all recall school days when they, like Gilbert's "model of a modern major-general," were

> teeming with a lot of news
> with facts about the square of the hypotenuse.

Those facts were known to the Babylonians even before Hammurabi's day and possibly to the Egyptians and Chinese of a somewhat later date. But it is traditional to ascribe to Pythagoras the first logical, deductive proof of the theorem which asserts: In any and every right triangle, *the square on the hypotenuse is equal to the sum of the squares on the other two sides.* Like Gilbert, we shall consider the classic proposition in its algebraic form and refer to the square "of" rather than the square "on" the hypotenuse. Then the theorem can be expressed algebraically as

$$c^2 = a^2 + b^2$$

where c stands for the length of the hypotenuse, a and b for lengths of the "legs" or "arms" of any right triangle.

Now we come to that historic day when the Pythagoreans, in attempting to apply "facts about the square of the hypotenuse" to an apparently simple problem of measurement, experienced a trauma for themselves, a crisis that challenged their beliefs. Some say that the fateful day occurred *after* Pythagoras had given his proof and made the theorem a favorite with the brotherhood. But it is more likely that the "facts" were first accepted on Babylonian authority and that when they produced a crisis in elementary measurement, doubt as to their validity provided the motivation for examining the whole issue of what constitutes *proof.* Probably failure produced the moment of truth when Pythagoras, rejecting inductions based on observation, called for deductive chains of reasoning and gave an example of the latter kind of demonstration by proving his famous theorem.

Before or after that decisive moment, Pythagoras proposed to find the length of the diagonal of a square whose side measures 1 unit. Figure 2.6 indicates that the

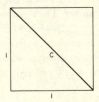

Figure 2.6

diagonal is the hypotenuse of a right triangle whose legs each measure 1 unit and hence

$$c^2 = 1^2 + 1^2$$
$$c^2 = 2$$

In accordance with Pythagorean beliefs it was now merely a matter of expressing the length c as a common fraction. Since $1^2 = 1$, $2^2 = 4$, the fraction should be larger than 1 and smaller than 2. We can try to guess or approximate its value, but the Pythagoreans had an *algorithm* or standardized procedure for doing so. If we should guess $1\frac{1}{2}$ and multiply it by itself, the result would be $2\frac{1}{4}$. Therefore the value of c must be smaller than $1\frac{1}{2}$. If we next try 1.4, then $(1.4)^2 = 1.96$. Hence our approximation is too small. We may then try 1.41, 1.42, etc., and find that $(1.41)^2 = 1.9881$, $(1.42)^2 = 2.0164$. No matter how we try, our estimates will apparently always be a little too small or a bit too large. Is it possible that our decimal scheme of approximation is at fault? Perhaps if we were to use ternary fractions or sexagesimal fractions (Chapter 1), we might be able to express the length of c exactly. Or maybe perfect measurement will only be possible by employing a fraction with some very special, huge denominator, like 10,032,571, for example. In effect, what we are asking is whether, with the side of the square as our "foot rule," we may not, perhaps, be able to hit on a scheme of exceedingly fine subdivision into tiny parts of suitable fractional denomination and thus (in theory) be able to measure the length of c. Our questions are a modern version of what the Pythagoreans asked themselves. For the answer they resorted to *deductive* proof. They were horrified when they were able to demonstrate that *no* common fraction (decimal or otherwise) will give the value of c, that is, measure the diagonal of the unit square. In other words, if the side is the unit of measure, the diagonal is *incommensurable*.

A deductive proof that no common fraction can possibly give the value of c in $c^2 = 2$ is usually attributed to the Pythagoreans. At any rate, it was known to Aristotle (384–322 B.C.) and included in a more general theorem in Euclid's *Elements* (X-9). The argument is as follows: Suppose that c is actually equal to a common fraction. Then that fraction can be reduced to lowest terms so that its numerator and denominator have no common divisor. Then

$$c = \frac{m}{n}$$

where m and n are natural numbers with no common divisor. Hence $c^2 = 2$ becomes

$$\frac{m^2}{n^2} = 2$$

Therefore

$$m^2 = 2n^2$$

From this point on, the proof applies Pythagorean theorems about "feminine" numbers. Thus $2n^2$, being a multiple of 2, is "feminine" or *even*. But the above equation implies that m^2 names the same *even* number.

What about m itself? If it were odd, then m^2, its square, would be an odd number by virtue of the special property proved earlier in this chapter. But we have just established that m^2 is, in fact, even. Therefore m cannot be odd, and hence must be even. Let us then represent it by $2x$ and subsitute this new form in the equation above. The result is

$$4x^2 = 2n^2$$

or

$$2x^2 = n^2$$

Here $2x^2$ is a multiple of 2 and hence even. Thus n^2 is even, and by an argument similar to the previous one, we conlude that n is an even number.

The net result is that m and n are both even, that is, are both exactly divisible by 2. But this *contradicts* the fact with which we started, namely, that m and n have no common divisor whatsoever. We have a *reductio ad absurdum*! Therefore the assumption leading to the contradiction must be false, and it is *not* true that the length of *c* is a common fraction, the ratio of two natural numbers.

"Natural numbers and their ratios rule the universe," Pythagoras believed. Hence what he had proved was a challenge to his faith. The numbers which he worshiped were ineffectual in a simple situation! If the side of a square is the unit of measure, the diagonal of the square is *incommensurable*. In terms of number, the length of the diagonal is "without ratio" or *irrational*. If $c^2 = 2$, then, employing modern algebraic symbolism, $c = \sqrt{2}$. Our proof has demonstrated that this symbol is *not* just another name for some common fraction, the ratio of two natural numbers. The symbol $\sqrt{2}$ represents something new. It measures a definite length, but should it be called a "number"? That question plagued the Pythagoreans, the more so since the irrationality of $\sqrt{2}$ implies the existence of an endless number of other irrationals. It is merely a matter of doubling, trebling, quadrupling, . . . , bisecting, trisecting, . . . , etc. the side of the unit square, to arrive at squares whose diagonals correspond to $2\sqrt{2}, 3\sqrt{2}, 4\sqrt{2}, \ldots, \frac{1}{2}\sqrt{2}, \frac{1}{3}\sqrt{2}, \ldots$. Also, the square-root spiral (Figure 2.7) indicates how repeated application of the Pythagorean

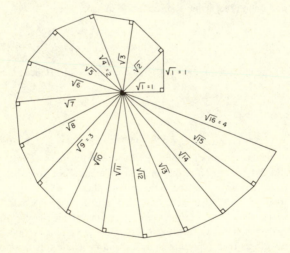

Figure 2.7   Square-root spiral

theorem gives rise to other square-root irrationals like $\sqrt{3}, \sqrt{5}, \sqrt{6}, \ldots$. (The reader can establish the incommensurability or irrationality of those lengths by imitating the proof for the irrationality of $\sqrt{2}$.)

The Greeks used the term *logos* (word, speech) for the ratio of two natural numbers. Hence when incommensurable lengths were described as *alogon*, the term had a double meaning—"not a ratio" and "not to be spoken." The latter meaning, *unutterable*, was applied with dire consequences to a particular member of the Pythagorean Society. Legend has it that when Hippasus uttered the unutterable by revealing the brotherhood's difficulties with the handling of incommensurables, he was assassinated. After that, the brethren were free to worry about the *alogon* in secret. They never resolved their difficulties in that area, but later in this chapter we shall see how Eudoxus (*ca.* 370 B.C.) gave a brilliant answer to that most challenging question: What is number? For the moment, however, let us rest with the assumption that, given any line segment, there is some "number" that will describe its length.

As mathematics developed, it became apparent that not all incommensurable lengths are square roots. There are cube-root, fourth-root, fifth-root, irrationals, etc., and irrationals not expressible by roots of any order. A proof similar to that for $\sqrt{2}$ can establish the incommensurability of $\sqrt[3]{5}$, for example. But there are more complicated root irrationals, such as

$$\sqrt[7]{6 - \sqrt[4]{5} + \sqrt{11}}$$

Furthermore, there are irrationals that cannot be expressed as finite combinations of root irrationals. These may be solutions of algebraic equations of fifth degree or higher, or they may be numbers like the π of geometry, which is *transcendental*, completely beyond algebra, so to speak.

Although the Pythagoreans progressed from the natural numbers to their ratios and then made the most profound of all number discoveries, they failed to extend the number concept in a way that seems almost trivial compared to the giant step that revealed the irrationals. Although Pythagoras made so many trips to the *east*, and then traveled back to the *west*, he never gave direction to the numbers describing his journeys. Today we would find it natural to say that he traveled $+1000$ miles and subsequently proceeded -1000 miles back to his starting point, as indicated by $(+1000) + (-1000) = 0$. But Pythagoras did not recognize zero as a number, and never seemed to experience any theoretical motivation for giving direction to the numbers he used for measurement. One wonders whether, in this respect, he once again rejected the wordly practicality of mathematicians he met in the Middle East. There is some evidence that they may have understood "signed" numbers, since one interpretation of cuneiform tablets (from 1600 B.C.) now at Yale University points to algebraic equations in which negative numbers appear.

Suppose, then, that we have accepted zero (0) as a number, and that signs or directions have been given to other numbers previously considered. Then the natural number aggregate can be extended to the set of *integers*, $\{\ldots -3, -2, -1, 0, +1, +2, +3, \ldots\}$, where each natural number has produced a pair of "opposite" integers like $+1$ and -1, $+2$ and -2, etc. For convenience, one often identifies the natural numbers $\{1, 2, 3, \ldots\}$ with the *positive integers* $\{+1, +2, +3, \ldots\}$. The set of all rationals now includes both $+1/2$ and $-1/2$, and the irrationals also come in directed pairs like $+\sqrt{6}$ and $-\sqrt{6}$, $+\sqrt[3]{5}$ and $-\sqrt[3]{5}$, $+\pi$ and $-\pi$, etc.

All the numbers we have discussed thus far—the integers, the rational fractions (positive or negative), and the directed irrationals—are described as *real numbers*. The reason for the appellation "real" will become apparent in Chapter 4.

To conceive of the *totality* of all real numbers, to put some law and order into that aggregate, and to hold on to our notion of a "number" as the length of a line segment, we shall now picture such lengths as distances on a *number line* (Figure 2.8), or as *coordinates* of points on such a line. For this purpose we can use any straight line, $X'X$ (Figure 2.8), as number line, and set up a "system of coordinates"

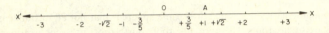

Figure 2.8   Number line

on that line in the fashion first given marked emphasis by René Descartes (1596–1650) and Pierre de Fermat (1601–1665). We begin by selecting O, any point on the number line, as the *origin* of coordinates, or representative of the number *zero*. The origin divides the number line into two *rays* or "half-lines," OX and OX'. On OX we select any point A (distinct from O) and call the length of OA the *unit distance*. We think of A as the representative of the integer $+1$, or alternatively, in accordance with our previous conception, we consider $+1$ as the directed distance from O to A. The ray OX passing through O and A gives the *positive* direction on the number line, and the other ray OX' gives the *negative* direction. In Figure 2.8 a *number scale* is formed by starting at O and marking off unit distances successively to right and left. The points labeled $0, +1, +2, +3, \ldots$ and $-1, -2, -3, \ldots$ can be considered to represent the numbers named, or alternatively, one can picture the numbers as the directed distances from the origin to the representative points. These directed distances or the numbers measuring them are called the *coordinates* of the respective points.

To obtain a point whose coordinate is a rational number that is not an integer, for example, $+3/5$, we can, if we wish, carry out a Euclidean construction. Before Euclid had incorporated the geometry of Thales, Pythagoras, and later Hellenic mathematicians into the *Elements*, Plato had exerted an influence that caused Euclidean constructions to be limited to those that can be carried out by straight-edge (line without a scale) and compasses. It would be possible (in theory) to use such tools to divide OA [Figure 2.8] into five equal parts. Then the directed distance from O to the terminal point of the third part would be $+3/5$, or that terminal point would represent $+3/5$. (One might, instead, divide the segment from O to the point $+3$ [Figure 2.8] into five equal parts and take the terminal point of the first part.) If, starting at O, the length of $+3/5$ is measured off on OX', the point whose coordinate is $-3/5$ is obtained. Any rational number, positive or negative, can be constructed in analogous fashion.

If we consider the points plotted on the number line thus far, we can observe certain important properties. First, there are the *integral* points which convert the number line into a *number scale*, the points corresponding to the integers arranged in order of size, namely, $\{\ldots -3, -2, -1, 0, +1, +2, +3, \ldots\}$. These points are spaced on the number line in such a way that each integer has a unique predecessor as well as an immediate successor. For example, -5 is preceded by -6 and followed by -4. This characteristic is described by saying that the integers or integral points form a *discrete* set. The property of discreteness no longer holds after the constructions described in the previous paragraph are carried out, that is, when all the fractional points are added to the picture. In the enlarged set of *all rational numbers* ordered according to size, we can name neither a specific antecedent nor a consequent to a particular number like 0.5, say. If we examine a fraction slightly larger than 0.5, for example, 0.50001, this is not the next number of the set, since the arithmetic mean of the two numbers, 0.500005, is still closer to 0.5. In turn, this new number is not the immediate successor to 0.5, since 0.5000025 is still closer, and we can continue the process of averaging *ad infinitum*, obtaining at each step a rational number closer to 0.5 than the fraction under consideration, so that the latter is *not* consecutive to 0.5 in the set of ordered rationals. Thus we see that it is hopeless to try to find, among the rational numbers, a next number after 0.5. Similarly, no fraction can be singled out as the immediate predecessor of 0.5.

The characteristic that between any pair of rational numbers there are always others (in fact, an infinite number of others) is usually described by saying that the rational numbers are *dense*. Now in spite of the close packing of integral and fractional points everywhere on the number line, there are *gaps*, an infinity of them, as we shall soon see. Since $\sqrt{2}$ is *irrational*, it is *not* included in the dense rational set. But a Euclidean construction can locate a point whose coordinate is $+\sqrt{2}$. With the Platonic tools one can construct a unit square (where the unit is OA in Figure 2.8), and then, starting at O, the length of the diagonal of this unit square can be measured off on OX. Using that same length, one can readily construct the points whose coordinates are $+2\sqrt{2}, +3\sqrt{2}, +4\sqrt{2}, \ldots, -\sqrt{2}, -2\sqrt{2}, \ldots$. All of these points fill gaps in the dense rational aggregate. The square-root spiral would enable us to fill other gaps by constructing $+\sqrt{3}, +2\sqrt{3}, +3\sqrt{3}, \ldots, -\sqrt{3}, -2\sqrt{3}, \ldots$, and, in fact, an "infinity" of points whose coordinates are multiples of $+\sqrt{5}$ or any other square-root irrational. But the Platonic tools are too limited for the construction of most irrational distances. Still we can imagine that there might nevertheless be a point whose coordinate is $+\pi$, and a point whose coordinate is $-\sqrt[5]{1 + \sqrt[3]{2}}$. It is all a matter of the definitions or rules which we lay down.

Therefore, if we are to give an exact definition of "real number," we must provide a formulation that will make it possible (in theory) to locate a point of the number line (measure off a directed distance from the origin) for every such number. In reverse, we would like to be able to provide a real number for any specified point or distance from the origin, since we want no gaps in the number line. We must avoid the Pythagorean predicament of having no numbers available for the exact measurement of certain line segments. What we seek then is a definition that will put the real numbers into *one-to-one correspondence* (Chapter 1) with the points of the number line so that the numbers will form a "continuum" (page 38).

To provide an exact definition of "real number" we shall follow the Pythagorean example of defining new numbers in terms of old. First there were the natural numbers and then the fractions (positive rationals) became pairs of these. Then a fraction or rational number led to an approximation of an irrational, $\sqrt{2}$. In our own struggle with $\sqrt{2}$ we used approximations that were *decimal fractions*, and we shall now employ such fractions once again because they are familiar to us. We remark that our formulations could be made equally well in terms of binary, ternary, sexagesimal, etc. fractions, that is, in a system of numeration whose base is not ten.

The division algorithm of school arithmetic would enable the reader to express any rational fraction in decimal form. Thus $5/16 = 0.3125$. When the fraction $1/7$ is converted into a decimal, it becomes $1/7 = 0.142857142857142857\ldots$, where the figures 142857 are repeated over and over again *forever*. Every rational fraction must have one of these two types of decimal expression but *cannot* have a nonterminating or infinite decimal expansion that is not repeating. To see why this is so, consider the division for decimalization of $5/16$,

$$16 \overline{)5.0000}^{\,0.3125}$$

and then that for $1/7$,

$$7 \overline{)1.0000000\ldots}^{\,0.1428571\,\ldots}$$

In the former case the process ceases when there is no remainder. In the latter the first six steps yield the quotients, 0, 1, 4, 2, 8, 5 and the remainders 1, 3, 2, 6, 4, 5. The seventh step yields a quotient of 7 and a remainder of 1. This is a repetition of the first remainder obtained in the conversion to a decimal. Therefore, beginning with the eighth step the original steps will be repeated up to the point where a remainder of 1 occurs again when the same original steps will be repeated once more, and so on. In general, in the case of decimalization of *any* rational fraction whatsoever, division by an integer n can yield at most a *finite* number of different remainders, namely $1, 2, 3, \ldots, n-1$ (not necessarily in this order). Some or all of these remainders may occur in the first steps of the decimalization, but eventually, if the process does not terminate, one of the above remainders must *recur* because the variety of possible remainders is not infinite, and when this remainder recurs, there will be a repetition of some or all steps in the decimalization, and so forth.

Conversely, terminating and repeating decimals can always be converted into ordinary fractions. This is obvious for the terminating case. For example, $3.0217 = 3\,217/10,000$ or $30,217/10,000$. To see how the conversion might be carried out for a repeating decimal, consider

$$x = 0.217217217\ldots \text{ forever}$$

Multiplying this equation by 1000 gives

$$1000x = 217.217217\ldots \text{ forever}$$

or
$$1000x = 217 + 0.217217 \ldots \text{ forever}$$
or
$$1000x = 217 + x$$
and
$$999x = 217$$
$$x = \frac{217}{999}$$

Consider once more the case where a rational fraction is expressed as a terminating decimal, for example, $2/5 = 0.4$. We can convert this terminating decimal into a *nonterminating* repeating decimal in two ways. There is the obvious way of writing $0.40000 \ldots$, where the zeros after the 4 are repeated forever. But it is *not* obvious that $0.4 = 0.3999 \ldots$, where the 9's are repeated forever. Let us prove that this is the case, by stating that

$$0.39999 \ldots = 0.3 + 0.09999 \ldots$$
$$= 0.3 + x$$

where

$$x = 0.09999 \ldots$$
$$10x = 0.9999 \ldots$$
$$100x = 9.999 \ldots = 9 + 0.999 \ldots = 9 + 10x$$
$$90x = 9$$
$$x = \frac{1}{10} = 0.1$$

Therefore
$$0.39999 \ldots = 0.3 + 0.1 = 0.4$$

We can proceed similarly in other cases. Therefore every rational fraction can be expressed as a *nonterminating* repeating decimal. If our division algorithm leads to a terminating decimal like 0.4 or 5.2018, we shall substitute the *non-terminating* decimal where zeros are repeated, for example, $0.4000 \ldots$ or $5.2018000 \ldots$. Although we might write $0.3999 \ldots$ or $5.2017999 \ldots$ we shall, for the sake of uniqueness, exclude the latter type of representation where the repeating part consists of 9's.

Since all rational numbers can be expressed as *nonterminating, repeating* decimals, and conversely, such decimals (positive or negative) can be considered as *defining* what is meant by a rational number, the only type of nonterminating decimal excluded by this definition is that which is *nonrepeating*. Therefore it is possible to define an *irrational* number as a *nonterminating, nonrepeating* decimal. The rationals and irrationals constitute the *real numbers*.

We have finally arrived at one goal. At long last, the *real numbers* can be *defined* as the set of all positive or negative nonterminating decimals.

It remains to indicate that this definition makes it possible to establish a one-to-one correspondence between the real numbers and the points of a number line, that is, there must be a point for every nonterminating decimal and such a decimal for every point. Suppose, then, that a real number is defined by some nonterminating decimal which we can carry out to as many places as we please, and that the first few places are given by 2.6314. . . . The decimal gives us a sequence of *rational* approximations to the real number, namely, 2, 2.6, 2.63, 2.631, 2.6314, ... In other words, the first approximation in the sequence places the real number in the interval (2, 3), and then 2.6 gives the approximating interval (2.6, 2.7), etc. Thus we have the sequence of *nested intervals*, (2, 3), (2.6, 2.7), (2.63, 2.64), . . . , illustrated in Figure 2.9. The adjective *nested* describes the fact that each interval lies within

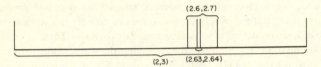

Figure 2.9   Nested intervals

the preceding one. We observe also that the lengths of successive intervals are 1, 0.1, 0.01, 0.001, 0.0001, ... Since we are considering a *nonterminating* decimal, the nest of intervals will ultimately contain an interval of length 0.000 000 001 and then there will be still smaller intervals, so that interval length shrinks toward zero. As the innermost intervals get smaller and smaller, one can imagine their bounding walls approaching collision or, at any rate, getting close enough to "trap" a point of the number line. It is postulated, that is, *assumed*, that there is a unique point contained in all intervals of the nest. If there is such a point, we see that it must be unique, for if there were another distinct point, it would be separated from the first by some distance, 0.000 01, say. But ultimately some interval of the nest will be smaller than that number, and the first point must be contained in that very small interval. Then the second point would be too far away to be inside the interval and hence would not be contained in every interval of the nest. Since every nonterminating decimal will give rise to a sequence of nested intervals like the one described, there will always be a unique point of the number line corresponding to every real number.

Conversely, if we are given any point P on the number line, then we can show that there will be a nonterminating decimal or real number corresponding to this point. If P is at one of the integral positions on the line, like 3 or -4, then the corresponding nonterminating decimal would be 3.0000 . . . or -4.0000. . . . Now suppose that P is not an integral point but lies on the number line somewhere between 1 and 2. Divide the interval (1, 2) into ten parts each measuring 0.1. If, by chance, P should be at the right end-point of the third small part, then our task would be finished, for the corresponding nonterminating decimal would then be 1.3000. . . . But let us suppose instead that P lies within the seventh interval (1.6, 1.7). Divide that interval into ten parts each measuring 0.01. If P lies within the fourth of these intervals, (1.63, 1.64) we can subdivide that interval into ten equal parts and perhaps locate P in the interval (1.638, 1.639), etc. We can keep on subdividing each new interval into ten equal parts. If, at any step, P is at the end-point of some interval, the corresponding real number is rational. Thus P might turn out to be 1.6382, that is, 1.6382000. . . . If P is not the end-point of any interval, the corresponding real number is irrational, and the left end-points of successive intervals give the corresponding decimal expansion. Thus, if ultimately P is found to be in the interval (1.6382704, 1.6382705), the real number corresponding to P is 1.6382704. . . . Hence, in this case, and in every case, to a point of the number line there will correspond a nonterminating decimal (real number). Having argued both ways, we have indicated a one-to-one correspondence between points of the number line and real numbers as represented by nonterminating decimals.

To provide some structure for the real number system, one can define addition and multiplication in this system and show that closure, associative, commutative, distributive, etc. laws are satisfied. We shall merely suggest how this is done. Thus, to conceive of the *sum*, $\sqrt{2} + \sqrt{3}$, one would consider that

$$\sqrt{2} = 1.41421 \ldots$$
$$\sqrt{3} = 1.73205 \ldots$$

leads to the two sequences of nested intervals:

$\sqrt{2}$	$\sqrt{3}$
(1.4, 1.5)	(1.7, 1.8)
(1.41, 1.42)	(1.73, 1.74)
(1.414, 1.415)	(1.732, 1.733)
(1.4142, 1.4143)	(1.7320, 1.7321)
etc.	etc.

Then adding lower boundaries of the corresponding intervals, and doing the same for upper boundaries, the sum $\sqrt{2} + \sqrt{3}$ is defined by the following sequence of nested intervals: (3.1, 3.3), (3.14, 3.16), (3.146, 3.148), (3.1462, 3.1464), etc. In similar fashion, one can define a sequence of nested intervals for $\sqrt{2} \cdot \sqrt{3}$ and in general, for sums, differences, products, quotients of real numbers, whether these are rational or irrational. Only division by zero is excluded.

The formulation of real numbers by using nested intervals or nonterminating decimals is relatively new mathematics—late nineteenth century. The idea is due to Georg Cantor (1845–1918), whose profound conception of *infinite* sets will be discussed in a later chapter. It is apparent that nested intervals and Cantor's theory are a far cry indeed from the Pythagorean despair on first encountering the *alogon*. But while Greek geometers of 500 B.C. regarded irrationals with dread, Eudoxus (*ca.* 370 B.C.) created a brilliant method for handling incommensurables and thus defining the meaning of and operations on the real numbers. The Eudoxian ideology, which is incorporated in Book V of Euclid's *Elements*, is in essence the same as the theory of Richard Dedekind (1831–1916), one of Cantor's contemporaries. The point of view of Eudoxus and Dedekind is mathematically more sophisticated than Cantor's conception of real number.

Dedekind observed that when the rational numbers are represented as points on the number line, any one of these points, for example, 0.6, produces a "cut" in the number line that divides it into two distinct parts, every rational point in one portion lying to the left of all rational points in the other. The point 0.6 can be considered either the last point in the left section or the first in the right. In general, then, *any* rational number effects a *Dedekind cut* in the aggregate of rational numbers by dividing that set into two subsets, A and B, so that every rational in A is less than every rational in B. The rational number producing the cut may be either the largest rational in A or the smallest in B. We observe that it is *impossible* to have *both* a largest rational a in A and a smallest rational b in B. For, as we have seen in establishing the density of the rationals (page 31), $\frac{1}{2}(a+b)$ would lie midway between a and b, and being greater than a and less than b, it would not belong to either A or B, which is impossible.

When Dedekind had observed that every rational number effects a cut in the set of rationals, he considered the converse situation. This involves the question: If, by some criterion or other, the rational aggregate is divided into two subsets A and B so that every number in A is less than every number in B, is there always a greatest rational in A or a smallest rational in B? The answer to that question is in the negative, as we shall now illustrate. But Dedekind, desiring *continuity* of the number line, postulated (assumed) that in *every case* the cut (A, B) *defines* or, from a purely logical viewpoint, actually *is* a real number. When A has no maximum rational and B has no minimum rational, there is a gap in the rational series, a puncture in the number line which must be filled. In that case, the cut (A, B) is said to define (or to be) an *irrational* number.

For illustrative purposes, suppose that the criterion for forming a cut in the rationals is as follows: Positive rationals whose square is greater than 2 are to be assigned to class B, and all other rationals to A. Then -5, -3, 0, $+1$, $+1.4$, $+1.41$, $+1.414$, for example, will be in class A, and $+1.415$, $+1.42$, $+2$, $+3$, etc. will be in B. There is no maximum in A and there is no minimum in B, and the cut (A, B) is irrational. It is, in fact, that classic irrational $\sqrt{2}$.

To emphasize the nature of irrational cuts and to avoid belaboring $\sqrt{2}$, the reader should seek other criteria for bringing about a cut (A, B) where A has no last member and B has no first. For example, all rationals whose cubes are less than 7 can be assigned to A, and all others to B. Then $(A, B) = \sqrt[3]{7}$.

For Dedekind, then, the real number system was the set of all possible cuts (A, B) in the rational aggregate. He was able, in terms of his conception, to define addition and multiplication of real numbers, that is, of cuts (A, B), and to establish that those operations satisfied the fundamental laws of arithemetic—closure, commutativity, associativity, etc. To illustrate Dedekind's formulations, there is his definition of addition: Given any two cuts or real numbers (A_1, B_1), and (A_2, B_2), if C is the set of all rationals of the form $(a_1 + a_2)$ where a_1 is in A_1 and a_2 in A_2, and D is the set of remaining rationals, then the real number defined by the cut (C, D) is said to be the sum of the given real numbers.

We shall not give further details of the Dedekind treatment of the real numbers except to emphasize once more that it is in essence, if not in symbolism or wording, a modern-dress version of Eudoxus' theory. We shall return to Dedekind in later chapters. Before leaving Eudoxus, however, we must indicate that his theory of the real numbers is not the only contribution which places him among the mathematical "greats." The Pythagoreans may have begun the emphasis on deductive demonstration which led ultimately to the modern axiomatic method. But it was Eudoxus who fostered such methodology by making suitable assumptions and furnishing deductive proofs of numerous theorems, all for the purpose of resolving the Pythagorean crisis with $\sqrt{2}$. In still another anticipation of modern mathematics, Eudoxus established a method of finding areas and volumes that, perfected by Archimedes, was tantamount to the basic notion of Newton's seventeenth-century integral calculus. As if all of this were not enough, we have the Eudoxian theory of planetary motion which, revised and elaborated by Hipparchus in 140 B.C. and Claudius Ptolemy about 150 A.D., was to dominate astronomical theory until the time of Copernicus.

We are more certain about the scientific contributions of the mathematical giant of the fourth century B.C. than we are of the details of his personal life. Eudoxus was a native of Cnidus in Asia Minor, and some biographers give his life span as 408-355 B.C. In early youth he moved to Athens, where he became a protégé of Plato. Legend has it that the great philospher and the young mathematician were companions on a journey to Egypt but parted company on their return to Athens. The termination of their friendship is attributed by some to Plato's envy of Eudoxus' brilliance, while others feel that the younger man's belief in pleasure as the *summum bonum* aroused the ire of Plato, just as later on it provoked the criticism of Aristotle.

Eudoxus, influenced by his teacher Archytas and his pupil Menaechmus, made the suggestion that geometry would be more of a pleasure if the number of mechanical tools permitted were increased. It was Plato, we recall, who had set the precedent of limiting all geometric constructions to those possible with straight-edge and compasses. When Archytas, Eudoxus, and Menaechmus proposed the use of other curves, Plato inveighed against them with great indignation and persistence. He accused them of employing "devices that require much vulgar handicraft." At any rate, such is the statement of Plutarch, who recounts further that "in this way mechanical devices were expelled from geometry, and being looked down upon by philosophers, were used to promote military science."

Whatever the reason, Eudoxus was unpopular in Athens and ultimately returned to Asia Minor, this time to Cyzicus, where, emulating Plato, he established an academy of his own. He communicated to the students his discoveries in mathematics, astronomy, and physics; he also found time to enter local politics and practice medicine. Specific details concerning his activities as a physician are lacking, but one might conjecture that he picked up medical knowledge on his trips to Egypt. He may have been more of a medicine man than a physician, in the sense of the latter term today. In the fourth century B.C. and for almost two millennia thereafter, astronomers were usually astrologers as well, and often engaged in medicine as a sideline. They treated patients in accordance with astrological rules rather than from any real knowledge of the human body. This was true even of Copernicus in the sixteenth century, who, during his lifetime, was more renowned as a physician than as an astronomer.

Eudoxus, Dedekind, and Cantor all developed theoretical concepts for the purpose of filling the gaps in the ordered set of rationals so that the final geometric picture is a straight line which is *continuous* or unbroken, in the intuitive sense. For this reason, one speaks of the real number *continuum*. More formally, Dedekind considered a one-dimensional or *linear continuum* to be, like a line segment, a *dense* aggregate with *no gaps*. (Cantor was more stringent and required an additional property.)

It may seem that we have wandered far afield in arriving at the Cantor and Dedekind concepts of the real number continuum when the original point of departure was the Pythagorean procedures which today's "new mathematics" emulates. But it was a Pythagorean discovery that motivated exact definition of the continuum and, in fact, produced a sort of schism in mathematical activity that was to last up to the present day. One school of thought has been devoted to working out the implications of Pythagoras' original belief that the framework of the universe is to be found in the natural numbers: $\{1, 2, 3, 4, \ldots\}$. This sequence and also the set of integers ordered according to size, are described as *discrete*. That adjective is applied to *any* sequence and also to any ordered set in which every term (except the first, if any) has a unique predecessor, and every term (except the last, if any) has an immediate successor. For example, any finite set (arranged in any order) is discrete.

Because of their devotion to the natural numbers, the followers of Pythagoras are sometimes called *discretists*. In contrast to their point of view, there are those who would base science on the real number continuum. Although certain mathematicians have been polemic in their advocacy of the discrete rather than the continuous, or vice versa, mathematical history indicates that it is more a case of specialized interest in, or a talent for, one type of mathematics rather than another. The discretists have had a flair for the classic theory of numbers, algebra, and logic. Geometers, analysts, and physicists, on the other hand, are likely to be devotees of continuity. Although physicists tend to use continuous models, there are important exceptions. Matter has been thought of as discrete or atomistic since the day of Democritus (*ca.* 400 B.C.), and modern quantum theory is the example par excellence of the physics of the discrete or discontinuous.

The present domain of mathematics is so vast and varied that today no one attempts to define the subject precisely. In the past, however, mathematics was sometimes described as the "science of discrete and continuous magnitude," and its evolution was pictured as the progress of two dichotomous mainstreams, with Pythagoras and his associates as the source of both. Since an influence that lasts over 2400 years is bound to give rise to legends, it seems fitting to close the present chapter with some of the lore that has gradually accumulated around the ancient master's name.

Of Pythagoras' early life we know little except that he was born on the Greek island of Samos and traveled extensively in Egypt and surrounding countries, where it is almost certain that he was exposed to both the excellent Babylonian algebra and the crude Egyptian beginnings of mathematics. In administering the famous brotherhood, Pythagoras, according to some stories, was in favor of disseminating mathematical knowledge freely and was willing to lecture to any audience interested in what he had to say. But a different spirit was present among the brethren, who felt that the Society should be aristocratic in nature and that the members should keep their discoveries secret from the general public.

The brotherhood was pledged to vegetarianism, and it has been said that animal food was taboo because it was believed that men and beasts are kin. The Pythagoreans lived an ascetic and disciplined existence to purify their souls and fit them for the hereafter of which Pythagoras told when he preached the doctrine of the immortality and transmigration of the soul.

The Pythagorean interest in music is supposed to have started when Pythagoras, passing a blacksmith's forge, listened to the harmonius tones produced by the impact of the hammer on the anvil. In the field of astronomy as well as in music and mathematics, Pythagoras led his followers toward notable advances. It is generally believed that they were the first to assert that the earth is a sphere suspended in space without support of any kind and that it revolves with all the other planets around a "central fire." Some scholars claim that the Pythagoreans knew the central luminary to be the sun but hesitated to make such an assertion lest they be persecuted. Such fears were probably justified. Anaxagoras, shortly after the day of the Pythagoreans, was banished from Athens because he had publicized his astronomical theories on the sun, which he described as a huge mass of burning metal. He had also written on eclipses, and had proposed a "nebular hypothesis" that anticipated the eighteenth-century cosmogony of Laplace. Such views came into conflict with the prevailing polytheistic religion of the Greeks, and one legend has it that only the intervention of Pericles saved the rationalistic astronomer from a death sentence.

To return to the Pythagorean theory of planetary behavior, Copernicus expressed the opinion that it contained the germ of his own heliocentric hypothesis. But the brotherhood forced the "all-comprising, all-bounding, never-tiring holy ten" into their cosmological scheme. The sun, moon, and five known planets totaled seven, the earth and the heaven of fixed stars two more. To obtain the mystic "holy ten" the Pythagoreans invented a "counter-earth" that, like the central fire, is invisible, because the side of earth on which we live always faces the other way.

To combine their cosmogony with their musical theory and their belief that nature is modeled on the natural numbers, the Pythagoreans assigned to the distances of the various planets from the central fire the same ratios that produced harmonious combinations in the musical scale. The planetary motion produces sublime music, Pythagoras said, inaudible to us because we are like the blacksmith and his assistants who cease to be aware of sounds which they hear constantly and cannot contrast with silence.

How romantic the personality that could see nature in such a light! The myth-makers, however, created the legend of a more conventional, personal romance. They related that Pythagoras fell in love with one of his young students, Theano. After their marriage, the couple might have lived happily ever after if the Society had not become involved in politics. But having regulated its astronomic theory so as not to fall afoul of religious bigotry, the brotherhood encountered another pitfall —the lust for power. The democratic populace of Crotona objected to the Pytha-gorean control of local affairs. The group was disbanded, and Pythagoras, accom-panied by the loyal Theano, went into exile. Tragedians would have us believe that enemies followed him and slew Pythagoras in the very presence of his young wife. For us, such fact or fiction is unimportant; the Pythagorean legacy to mathematics gives the master an eternal role in that drama we call the "glory that was Greece."

3

Mathematical Reasoning from Eudoxus to Lobachevsky

After the method and two main streams of mathematics had been launched by the Pythagoreans, Greek mathematical activity shifted away from cities in Italy. Athens and centers in Asia Minor then became important, but ultimately Hellenic scientific activity was concentrated in the metropolis of Alexandria. Here Euclid (*ca.* 300 B.C.), the most classic figure in mathematics, lived and worked. Archimedes, the greatest of all Greek mathematicians, was in constant touch with Alexandrian scientists.

In an address* before the Mathematical Association of America, Professor Rudolph E. Langer (1894–1968) described the historical and geographical setting of this famous city.

... On the southern shore of the eastern Mediterranean Sea, not far from the westernmost mouth of the great river Nile, a narrow ridge of limestone separates the sea from a lake named Mareotis. Not far offshore and lying like a breakwater parallel with it, is a long narrow island called Pharos. In the year 331 B.C. Alexander the Great saw in this spot unusual potentialities for the site of a city. He conceived upon it a great city which should serve at once two missions which he had set himself, namely, the spread of Hellenic influence over the world, and the return of the ancient land of Egypt to a former greatness and glory.

Upon the death of Alexander, Egypt fell under the governorship of his general Ptolemy. There was much mutual good fortune in this turn of events. For while Egypt, a fabulously fertile country, meant power and almost limitless wealth for Ptolemy, he in turn proved himself worthy of it. The distinctive culture of the country, with traditions extending into the interminable past, had always fascinated the Hellenic mind, and Ptolemy was not unreceptive to this charm. His rule, first as governor and later as king, was guided by intelligence and statesmanship, by a fine natural artistic taste, and by an appreciation of the dignity and substantial worth of intellectual attainment. His dynasty was to rule the land, now for better, then for worse, over a period of two hundred and fifty years.

From early times in the history of the Greeks there had existed so-called philosophical schools, which were in reality communities of scholars. These schools were

---

* Rudolph E. Langer, "Alexandria, Shrine of Mathematics," *American Mathematical Monthly*, Vol. 48, No. 2 (February 1941), pp. 109 ff.

frequently organized as brotherhoods, dedicated, as in the cases of the Pythagoreans, to the cult of the Muses. Their housings, therefore, came quite generally to be known as *museums*. Ptolemy conceived the ambition to establish such a museum at Alexandria. He envisaged the city as a center of Greek culture, not merely as a trading post, and with large revenues at hand he thought to make association with his school attractive by the then novel means of granting salary stipends, as well as board and residence, to prominent scholars of his choice. Not unnaturally this plan was an immediate success, and at about the year 300 B.C. the Alexandrian Museum had become an actuality. It included in its membership intellectuals of all sorts—poets, philosophers, grammarians, mathematicians, astronomers, geographers, physicians, historians, artists, and many others. The mathematician in this initial galaxy was the immortal Euclid.

Ptolemy's constructiveness, which was manifested in his incorporation of the Alexandrian Museum, was matched again both in originality and grandeur by his founding of . . . the Alexandrian Library. . . . The Library was established almost simultaneously with the Museum and adjacent to it. . . .

Close to the harbor and connected by colonnades with the palaces stood the fine white buildings of the Great Library and of the Museum. Such were the environs in which the scholars of Alexandria lived. In a great basilica they ate their meals together, and had their dormitories and lecture halls. In groves of palms and under arcades adorned with classical sculptures they carried on their discussions with each other or with their students and disciples. Contemporary Alexandrian satire liked to depict them as costly birds fed and treasured in a golden cage.

It would be difficult to think of Euclid as a bird in a gilded cage, but one would like to believe that he enjoyed the natural beauty of Alexandria, its artistic wonders, its advanced culture, and all those happy features set forth in Professor Langer's address. None of the personal details of Euclid's life have come down to us, however, and we must limit ourselves to his mathematical activities. Because his name is usually identified with geometry, there is little realization that he contributed to other subjects as well. He wrote on astronomy, optics, and music; his greatest treatise, the *Elements*, contains not only geometry but considerable number theory (Chapter 21) and geometric algebra as well.

Even more important than the comprehensive content of Euclid's classic is its logical format, which has served as the prototype of all modern deductive systems constructed by the *axiomatic* or *postulational* method. Euclid must not be given all the credit for this standard pattern, however, since it seems to have evolved gradually, as we have already related, through the ideas of his Greek mathematical predecessors, among whom were Thales, Pythagoras, and Eudoxus (Chapter 2). The basic idea of the axiomatic method is that the initial content of geometry or any scientific subject should consist of a set of assumed propositions, called *axioms* or *postulates*, and that other propositions, called *theorems*, should be derived from the basic assumptions by applying the rules of deductive logic.

Since mathematics is the domain par excellence where facts are proved, it is often a shock to the uninitiated to learn that the fundamental postulates of a mathematical science are *unproved* propositions. To see why an argument or a debate or a "deductive system" must start with assumptions, let us suppose that in some mathematical science proof is demanded for a particular statement, which we shall call Proposition *A*. Now this proposition may be a logical consequence of propositions *B*, *C*, and *D*. Suppose next that one is asked to demonstrate these propositions and that they are collectively dependent on propositions we shall label *E*, *F*, *G*, *H*, *I*, *J*, *K*, and *L*. If proof is demanded once more, we shall have to demonstrate eight more propositions, and so on. If we continue to demand proof, we shall have an infinite sequence of backward steps in our reasoning, unless we decide to call a halt at some point and accept at least one proposition without proof.

A parent attempting to reason with a four-year-old is familiar with the threat of an infinite regression in logical argument. "Why must I go to sleep?" Johnny asks. "Because you need rest," says Dad. "*Why* do I need rest?" Johnny continues. "Because it will give you the strength to play games with your friends tomorrow." "*Why* should I play games with my friends tomorrow?" "Because a little boy needs exercise." "Why do I need exercise?" At this point, Dad says, "*Because!* And that's that! Go straight to bed and *no more questions*!"

Dad is declaring as *axiomatic* the proposition that a child requires exercise; that is, he is stating that Johnny must accept this proposition *without proof*. It is a necessity to stop at some point in a regressive argument and "lay down the law." Even if one does not attempt to answer an endless chain of "whys," there is a certain logical danger in permitting a *finite* sequence to become too lengthy. Let us imagine that Dad decides to argue a little longer. When Johnny asks "*Why* do I need exercise?", Dad may answer, "Because it provides healthy fatigue." "*Why* do I need healthy fatigue?" "Because it will make you *sleepy and ready to go to bed right now*." Dad is right back where he started. He has just involved himself in a "vicious circle" or he is "begging the question," since if we fuse all Dad's answers into one long chain, we find that Johnny must go to sleep because he must go to sleep.

Just as one gets into an endless chain of argument by trying to prove all propositions, so is there a similar difficulty in trying to define everything. In some of the older texts one may find a point defined as "that which has no dimensions," a line as "something with one dimension," and so on. Then, what, pray, is a dimension? And when you have defined that, we shall have more questions. So make your choice of a starting place. You will just have to leave some things undefined. In this way you will avoid the endless chain and also, as in the case of propositions, the danger of circularity. A dictionary may define frugality as economy and economy as frugality, but a mathematician must not do so. He must initiate a deductive science with *unproved* propositions about *undefined* terms.

If the scientific subject under construction in postulational-deductive fashion is to be practical, realistic, and purposeful, the axioms are usually selected so as to approximate or idealize actual experience. Thus Euclid may have been the founder of postulational thinking, but he did *not* conceive of his postulates as mere assumptions. Instead, he described them as "common notions." Later textbooks called them "self-evident truths" or "matters of empirical fact," and Euclid's axioms about "points," "lines," and so forth, do actually appear to tally with the drawings we make by the use of pencil and straight-edge. Also, histories of mathematics customarily suggest that the origins of geometry are to be found in the work of

Babylonian and Egyptian surveyors, just as we have explained that the Pythagorean "theorem" was an early empirical discovery (Chapter 2). All that we have just stated gives the point of view of *applied mathematics*, which can be described roughly as consisting of those mathematical sciences which deal with situations in the real world, for example, statistics, physics, astronomy, biology, economics, psychology, etc., or deductive systems derived from portions of those subjects.

The attitude of modern *pure* or abstract mathematics is quite different from that just described. In it one has the right to choose the content of axioms somewhat arbitrarily, subject only to certain logical criteria which we shall explain presently. Then, in theory, a pure mathematician may found a deductive system by assuming what he pleases. He is unlikely to take advantage of such latitude, however, nor does he prepare a list of postulates out of a clear sky, with no special background or specific purpose. Such an aimless system could serve only as the basis for a mental exercise or as the set of rules for some esoteric game. Although his undefined terms are meaningless symbols as it were, and he is permitted to state whatever he wishes about them in his postulate system, he may actually be making assertions with which he is already familiar from other mathematical subjects, for example, statements about cardinal numbers (Chapter 1), algebraic x, y, z's, and the like. A common method of forming postulate systems, as we shall illustrate shortly, is to use a standard set of axioms, like those for ordinary algebra, or those for Euclidean geometry, and alter or omit one or more of the axioms.

But, one may say, if the pure mathematician is not required to furnish intuitive or experimental justification for his postulates, will these propositions be "true"? If not, will his theorems be valid? One answer customarily given to such questions is that *if* the postulates are true, and *if* the theorems are derived in accordance with the laws of logic, then the theorems are true. The theorems thus become *relative*, not absolute truths. A second answer is to say that in modern logic, "truth" has a different meaning from the usual one (Chapter 6). Since neither of these explanations may seem entirely satisfactory, we shall now present a third point of view, one which focuses on the role of the undefined terms used in the axioms of a system.

In modern improvements of Euclid's postulate system, "point" and "line" may be listed as undefined terms. Suppose that one includes among the axioms Euclid's postulate: There exists one and only one "line" containing any two "points." Although "point" and "line" are undefined or meaningless, it is difficult to divorce the terms from their usual connotations. It is of no use to declare that the "truth" of the postulate is *not* a pertinent issue, for someone may insist, "But the statement in question is obviously true." He may change this judgment, however, if we replace the two undefined terms by x and y, respectively, in order to emphasize that they are meaningless, empty symbols. The postulate becomes: There exists one and only one y that contains any two x's. Is this statement true or false? We cannot answer if x and y are "unknowns."

If the revised statement of the postulate were presented to someone unacquainted with Euclidean geometry, he would be unprejudiced, and if asked whether it is true or false, would say he could not tell. Or in order to render a verdict, he might seek to interpret x and y, that is, to assign specific meanings. Suppose he recalls that his three children have formed a miniature club with committees as follows: {James, Harold}, {James, Martha}, {Harold, Martha}. With this picture in mind, he decides to interpret x as meaning any one of his children and y as representing any one of the three committees. Then the postulate is converted into the statement: There exists one and only one committee that contains any two children. Thus he is able to pronounce the postulate true for his *particular* interpretation.

On the other hand, his wife, who is preparing to do some shopping, is thinking of a "system" consisting of various coins and the purses in which they can be placed. She interprets x as "coin" and y as "purse." She transforms the postulate into the statement: There exists one and only one purse that contains any two coins. She pronounces the postulate false for her interpretation by indicating that there are several coins inside a change purse which is contained in a larger purse. Thus, for pairs of these coins it is *not* true that there is only one purse containing them. Or she may have placed two coins on a table. For these coins it is not true that there exists a purse containing them.

The illustrations above are meant to indicate that one cannot judge the truth of an abstract postulate containing undefined terms. Such an axiom becomes true or false only for specific, concrete interpretations of these terms. For a pure or abstract science the postulates are like rules in a game. No one judges chess rules to be either "true" or "false." One merely observes these rules in playing the game.

Clarification of the issue of "truth" is not the only advantage in conceiving of the undefined terms of a deductive system as mere symbols akin to the x, y, z's of ordinary algebra. In that subject, letters are usually considered to be *variables*, that is, to represent many different numbers simultaneously, as in statements like $x^2 - y^2 = (x + y)(x - y)$, which is true when x and y are replaced by *any* of the numbers used in common algebra. This suggests that the undefined x's and y's of postulate systems also be conceived as variables with a potential for many valid interpretations. Thus, if the statement of some postulate should assert that

$$x \text{ is a believer in } y$$

then it may be true that "Mr. Smith is a believer in Zen-Buddhism" but this is not necessarily the only valid interpretation. Or if x, y, and z are undefined terms in a postulate,

$$z = x + y$$

then the following and many more specific statements are valid interpretations:

$$5 = 2 + 3$$
$$\frac{5}{6} = \frac{1}{2} + \frac{1}{3}$$
$$(-9) = (-6) + (-3)$$
$$10\sqrt{3} = 8\sqrt{3} + 2\sqrt{3}$$

In fact, if, in the postulate $z = x + y$, the symbols $+$ and $=$ are also undefined, we might offer "and/or" as an interpretation for $+$ and "is equivalent to" as a meaning for $=$. Then, interpreting x, y, and z in terms of requirements for the Ph.D., one might arrive at a true statement: Russian is equivalent to German and/or French.

In summary, then, the use of undefined terms or empty symbols in the postulates of a deductive system* has the advantages of all abstract formulation, namely economy and generality. A single concept or statement or proof, even an entire deductive system, may simultaneously fit numerous different situations. A small example of this fact occurred in connection with positional systems of notation, where arithmetic statements in any one system provided the abstract form for procedures in all other systems, including computer systems. In the same way, some abstract deductive system may have valid interpretations as a plane geometry, a hemispherical geometry, an algebra of number pairs, or a science that is not our idea of either a geometry or an algebra. Although the specific content of these subjects would vary, they would all have the same logical pattern, namely, the abstract postulates about x, y, z's and the theorems deduced from those postulates.

Thus, by using undefined terms or meaningless symbols in the postulates of a pure science, one accomplishes a double purpose by avoiding issues of "truth" and at the same time furnishing the potential for numerous valid interpretations. But if a pure mathematician need not be concerned with "truth," is he limited in any other way in setting up a postulate system? The answer is to be found in the logical criteria to which allusion has already been made. Logic furnishes only one imperative, but there are several other standards which are desirable under certain conditions. The primary characteristic which logic demands of a postulate system is that it be *consistent*. This signifies that it must not be possible to deduce from the postulates a pair of theorems which contradict one another.

How can a mathematician tell whether or not a system of axioms meets the requirement of consistency? If, in the course of deductions, he arrives at two contradictory theorems, he can pronounce the axiom system *inconsistent*. He must then discard or repair it. But suppose that he has derived all the theorems which he requires for some particular purpose and has found no contradictory pair among them. Is he to conclude that the postulate system is consistent? Decidedly not, for if he were to continue to deduce theorems, whether or not he is interested in such new propositions, he might ultimately arrive at a theorem contradictory to one previously established, and therefore *inconsistency* would be the proper verdict. The procedure of deducing more and more theorems is evidently not a practicable one. Can we ever reach a point where we can state with certainty that no more theorems are deducible? We need *all* theorems before we can render a judgment of

---

* Throughout the present chapter the following terms should be considered synonymous: deductive system, deductive science, mathematical system, mathematical science, pure science or theory, abstract theory. Each of these terms signifies an aggregate of the following ingredients: (1) undefined terms, (2) postulates involving the undefined terms, (3) definitions involving the undefined terms, (4) theorems deduced from the postulates.

consistency. For this reason, mathematicians usually accept the following working criterion for consistency: A set of postulates is said to be consistent if there exists an interpretation of the undefined terms which converts all the postulates into true statements.

Certain subtle logical issues form the rationale behind the working criterion for consistency, and we shall not go into them at this point. We shall, however, indicate the logical consequences of accepting the criterion. In order to avoid verbosity in explaining these logical fine points, we shall employ special vocabulary. In the first place, we shall consider *only* those interpretations of the undefined terms which convert the postulates into true statements. In mathematical logic the result of such interpretation, that is, the concrete set of true statements, is called a "model" of the abstract postulate system. Since the postulates are transformed into true statements of the model, all theorems deduced from the postulates must also be true statements of the model. Thus the abstract deductive system, consisting of postulates and theorems, is said to be transformed by interpretation from an *abstract theory* into a *concrete theory*. The word "model" is sometimes used to apply to the entire concrete theory (and not merely to its postulates) and sometimes to the relation of the concrete theory to the abstract theory. Since "model" is a favorite word of mathematicians, they use it with still other shades of meaning.*

To return to the working criterion of consistency, suppose that it is to be applied to some abstract postulate system. This requires that a "model" of the system be found, that is, an interpretation of the undefined terms which converts every postulate into a true statement. If a mathematician has abstracted the postulates from experimental situations, he will have a model at hand. All he has to do is reverse the process of abstraction which led to his assumptions and offer the original empirical propositions as the model. The applied mathematician might accept such a proof of consistency without any qualms. But others would become involved once more in that troublesome issue of "truth." Empirical propositions have no absolute validity, limited as they are by our senses and observational tools. Statistical inductions based on numerous careful repetitions of an experiment may have a high probability of being general truths, but they are still hypotheses, and future evidence may show them to be false.

Suppose, therefore, that empirical models are not to be accepted as proofs of the consistency of a postulate system. What other type of model is permissible? It is customary to find a model in some branch of mathematics that is already well established. When a pure mathematician seeks such a model, he may, in the fashion

---

* For example, one usage appears to reverse things by taking the abstract system as the model of the concrete one (see Chapters 10–14). In that case, a mathematical or abstract model is described as a deductive system in which axioms and undefined terms represent certain aspects of the the observed realities. See Carl B. Allendoerfer, "The Narrow Mathematician," *American Mathematical Monthly,* Vol. 69, No. 6 (June–July 1962), pp. 462 ff.

  To straddle both points of view, and to provide proofs of consistency, another description of a model makes it the interpretation of one mathematical theory with the assistance of another. See Nicolas Bourbaki, *Éléments d'histoire des mathématiques,* Hermann, Paris, 1960, p. 35.

already described for the applied mathematician, return to the *source* of his postulate system. Whereas the applied mathematician uses empirical sources, the pure mathematician selects propositions (either axioms or theorems) from some well-established mathematical theory. Then part (or all) of that theory provides the model which demonstrates the consistency of his postulate system. Later we shall see that parts of Euclidean geometry, a long-established mathematical theory, are used in models establishing the consistency of different *non*-Euclidean geometries. Again, if we wished to demonstrate the consistency of a postulate system for a one-dimensional geometry of "points," we might proceed as in Chapter 2, and interpret the undefined "point" as a *real number*. If our postulates are thereby converted into accepted postulates or valid theorems about real numbers, we have demonstrated the consistency of our geometric system. In fact, if the same idea is extended to two-dimensional geometry, it turns out that Cartesian analytic (algebraic) geometry is the model that demonstrates the consistency of Euclidean geometry. (Although Descartes lived much later than Euclid, the arithmetic and algebraic foundations of the Cartesian subject are more secure, as it were.)

This sort of consistency proof, although accepted, merely displaces the problem of consistency from one mathematical theory to another. It leads to the inquiry: In the theory furnishing the model, is the underlying postulate system consistent? Its consistency is essential, for otherwise it may imply contradictory theorems, and some theorem used in the model (and hence considered "true") may be contradicted by an as-yet-undeduced theorem of the well-established subject. This makes the usual proof of consistency *relative*, not *absolute*. Such a proof shows that *if* the theory furnishing the model is consistent, then the model establishes the consistency of the postulate system which it interprets. All such considerations and still other difficulties into which we have not entered have provoked the comment: There is, in fact, *no* way of demonstrating the consistency of a system (Chapter 29). If such a judgment is too drastic, it is evident, nevertheless, that consistency proofs form a delicate logical issue. It is true, moreover, that some of the greatest minds of the recent era have carried on profound research related to the problem, and have indicated the difficulty of obtaining an absolute species of consistency proof. Kurt Gödel (page 54) has, in fact, shown the *impossibility* of such a proof for a comprehensive, formal (completely abstract) system (Chapter 29).

Consistency is the only essential property of a set of axioms, but a pure mathematician may ask for other qualities in certain postulate systems. With economy always in mind, he inquires: Have I assumed too much? Could I dispense with one or two axioms? Whereas an orator may achieve emphasis by describing a man as honest, upright, reliable, and so on, such redundancy does not appeal to the mathematician. Even concealed repetition seems to the mathematician a blemish which destroys the perfection of a set of axioms. Therefore he may require his postulates to be mutually *independent*. This means that no postulate of the system may be a logical consequence of other postulates. If it were deducible from the other axioms, then it could be stated as a theorem and there would be no need to assume it. If, paraphrasing Lewis Carroll, we assume that

(1) All babies are illogical,

(2) Illogical people cannot manage crocodiles,

then we need not also assume that

(3) Babies cannot manage crocodiles,

since proposition (3) is a logical consequence of (1) and (2), or is *dependent* on these propositions.

Let us emphasize again that independence is not an essential attribute of a set of postulates but is only called for by an ideal of logical perfection. It may be advisable, for psychological or pedagogical reasons, to use postulates that are not mutually independent. Readers may recall, from school geometry, that various propositions related to the congruence of triangles were *postulated* (after an experimental proof). These propositions are actually deducible from Euclid's postulates, in other words, are *theorems* of his geometry. But the proofs are exceedingly difficult for the beginning student, and hence textbooks add the propositions to the axiom set, which, as a result, fails to possess the quality of *independence*. This and other examples up to this point have all been negative in that they have exhibited systems lacking independence. We shall temporarily postpone positive illustrations because they involve an explanation of how independence is customarily established. The independence proof which will eventually be cited involves the most dramatic incident in the history of axiomatics, the turning point from the classic to the modern. It was Nicholas Lobachevsky (1792–1856), a great Russian mathematician, who was responsible for this revolutionary turn in mathematical events.

By contrast with assuming too much in founding a mathematical science and thereby failing to have independent postulates, there is the question of whether one can assume too little. A postulate system may be *incomplete* in that additional assumptions are required in order to prove certain theorems. As an example, the reader may recall, from school days, a "proof" that purported to deduce the proposition: Every triangle is isosceles. The demonstration involves the drawing of various auxiliary lines, and the proof is ultimately termed "fallacious" because the drawing is "careless" and the positions of various points and lines are incorrect. If there are any faults in the proof, however, they must be attributed to Euclid, since there are no postulates in his system which require points and lines to take the allegedly correct positions. Put otherwise, there is nothing in his system to rule out the "faulty" diagram. What this means is that one cannot deduce from his postulates either of the following contradictory statements: (1) Every triangle is isosceles. (2) Not every triangle is isosceles. This illustrates that Euclid's postulate system is *incomplete*. It also suggests a possible definition for completeness: A postulate system is called *complete* if it is possible to deduce from it either a proof of any proposition about elements of the system (undefined and defined entities) or a proof of the negation of that proposition.

Categoricalness is another attribute which may be sought for certain postulate systems, because it limits their models to a sort of uniformity (similar to that of the *isomorphic* arithmetic systems of Chapter 1), and in this way insures that the single abstract theory arising from the postulates will provide the total pattern for *all* applied theories which interpret it. A postulate system is *categorical* if, in every pair of models of the system, elements are in one-to-one correspondence, and all relations among elements are preserved.

In defining the properties of postulate systems and in explaining how these can be established, the role of the model or interpretation has appeared repeatedly. This gives the impression that specific, concrete subject matter is used only as a preliminary to formulating the axioms of an abstract theory, or for purposes of testing the consistency and other attributes of these axioms. But this is not the case. After the theorems of the pure theory have been deduced, they are often reinterpreted to check with the sources that inspired the postulate system, or to make "predictions" in some physical model of the system, or to provide proofs of theorems in concrete mathematical models, etc. The process will be illustrated later on when we discuss relativity. Einstein abstracted his axioms from observations made by physicists, deduced his abstract theory, reinterpreted it in terms of "planetary orbits," "Doppler effects," etc. so that certain theorems became "predictions." The fact that the latter were confirmed by observation shows that the abstract theory of relativity has a "real" model.

And now let us return to Euclid, who, with his illustrious Greek predecessors, founded the axiomatic method. We appear to have presented only negative aspects of his postulate system by pointing out that it was incomplete, and that Euclid himself did not realize that he had constructed a *pure* science, not limited to surveyors' "points" and "lines" or founded on "common notions." More satisfactory postulate systems for Euclidean geometry were devised by a number of mathematical leaders of the recent era. Among these there is a postulate system due to David Hilbert (1862–1943) and another formulated by Oswald Veblen (1880–1960). Both men will appear frequently as *dramatis personae* in our story because their discoveries were not limited to geometry or to the logical aspects of founding mathematical theories.

But how and when did the mathematical world first come to realize that Euclid's notion of postulates was inadequate, and that his set of axioms required repair? The most famous axiom in mathematical history, namely, *Euclid's parallel postulate*, will provide the answer. This axiom can be stated in many equivalent forms. If we define two "lines" that lie in the same "plane" but never meet as parallel, a statement equivalent to Euclid's parallel postulate is: *Through a point P outside a line l there is one and only one line parallel to l.*

Euclid's attitude toward the parallel postulate differed from his opinion of the other axioms. Although he considered the latter to be "common notions," he did not find the parallel postulate "self-evident." Because substantial empirical justification was lacking, he realized, in this one case only, that he was making an assumption. Since he dared not *assume* the parallel postulate, he tried to prove it, that is, to deduce it from the other axioms. In doing this, he established a precedent for centuries to come (twenty-one of them) for, until the day of Lobachevsky, mathematicians were constantly struggling for proofs of the parallel postulate. None of the Russian's predecessors considered the possibility that the postulate is *not* deducible from Euclid's other axioms. To use twentieth-century terminology, the parallel postulate may possibly be *independent* of the other Euclidean axioms.

If one is to show that a given postulate is independent of the other axioms of a system, he must demonstrate that it cannot possibly be a theorem deduced from those axioms by logical methods. The procedure for accomplishing such a demonstration today is strongly reminiscent of the technique for establishing consistency by the use of a model (interpretation) in which postulates and theorems of an abstract deductive system are converted into true statements of the model. Let us then suppose that we have already provided such a model to demonstrate the consistency of some deductive system. But, as we have seen, there can be many different ways of interpreting the undefined terms of a system. With this fact in mind, let us proceed further by removing the postulate whose independence is under scrutiny and then seeking models for the residual set of axioms. In any such model those axioms will be converted into true statements, and so will any theorems deducible from them. Now if we can exhibit a model for the residual axioms such that the deleted postulate is converted into a *false* statement, then that postulate cannot possibly be a theorem deducible from those axioms, for otherwise it would be true in the model. In summary, then, in a consistent postulate system, any given postulate can be proved to be independent of the other axioms of the system by exhibiting an interpretation in which the given postulate is false and the other postulates are true.

Lobachevsky's ideas included the above procedure. Specifically, he removed the parallel postulate from Euclid's list, and substituted a postulate equivalent to the following assumption: *In a plane, through a point outside a line l, there are an infinite number of lines which do not intersect l.* Then, on the basis of the revised postulate system, he constructed his *non*-Euclidean geometry. That Lobachevsky's geometry is (relatively) consistent was demonstrated by the Italian geometer Eugenio Beltrami (1835–1900), who used a model, namely, the "pseudosphere" (page 53). But more important for our present purpose is the fact that geometry on the "pseudosphere" provides an interpretation in which Euclid's parallel postulate is *false*, while his other axioms are *true*. Therefore the parallel postulate *cannot* possibly be a theorem deducible from those axioms, but is independent of them. That is why the twenty-one centuries of search for a proof of the parallel postulate had proved fruitless.

Thus Lobachevsky's geometry solved the problem of the parallel postulate for all time. But much more than that was accomplished by his bold procedure. It showed that postulates ought not to be conceived as "self-evident" truths, since it seems possible to replace a particular axiom by a radically different one without producing logical nonsense. Because one may do this, an axiom is a somewhat arbitrary statement, neither self-evident nor something to be proved. In other words, it is, as constantly emphasized, a mere assumption. Therefore, in the creation of Lobachevskian geometry and its implications, we have the modern point of view in a nutshell. After Lobachevsky's initial step, mathematicians were led to a reconsideration of the nature of postulates and eventually to the point of view we have set forth in this chapter.

Credit should also be given to Janos Bolyai (1802–1860), a Hungarian mathematician, who worked out the notion of non-Euclidean geometry simultaneously with Lobachevsky, but independently. Since Lobachevsky's publication antedated Bolyai's, it is customary to name Lobachevsky as the creator of the concept. Gauss, the great mathematical giant of the nineteenth century, is said by some to have discovered results similar to those of Lobachevsky and Bolyai before either, but to have lacked the courage to publish facts so startling. As a matter of fact, Girolamo Saccheri (1667–1733), an Italian priest, had attempted a proof of Euclid's parallel postulate in 1733, in the course of which he actually discovered Lobachevskian geometry without being aware of it. Thus the way was well paved for Lobachevsky or Bolyai or Gauss.

After Lobachevsky had constructed his own particular brand of logical non-Euclidean geometry, the German mathematician Bernhard Riemann (1826–1866) constructed another. Riemann subsituted for Euclid's parallel postulate the assumption: *Through a point P outside a line l there is no line parallel to it; that is, every pair of lines in a plane must intersect.* When the term *non-Euclidean geometry* is used in mathematical literature, the geometries of Lobachevsky and Riemann are always meant, although the term can well be applied generically to any geometry that denies one or more of Euclid's axioms.

Lest the assumption of Lobachevsky and Bolyai seem too bizarre for the practical man, let us ask him to picture the infinite number of parallels to a line through an outside point as a sheaf so thin that it is not distinguishable from a single line. Riemann's postulate is easier to imagine, for may not two lines that seem to be parallel meet at an infinite distance? If this interpretation of the two non-Euclidean geometries is too much for the reader to swallow, let us offer another. We know that the modern point of view does not require us to interpret "point," "line," etc. in the ordinary way. We must first explain a preliminary concept or two. One of Euclid's axioms is: *A straight line segment is the shortest distance between two points.* That is the assumption for a *plane* surface, but what, then, is the shortest distance between two points on a spherical, cylindrical, or ellipsoidal (egg-shaped) surface? You can find out experimentally by stretching a string taut between two points of the surface. On a spherical surface, the shortest distance is a *great-circle arc*, the path followed by navigators in seeking shortest routes. It is a portion of a circle cut from the sphere by a geometric plane passing through the center of the sphere. Any plane section of a sphere will produce a circle, but only a plane through the center will produce a great circle. The shortest paths on other surfaces will be curves of various types. The mathematical name usually assigned to a shortest route is *geodesic*.

Since the geodesic for a plane is a line, this suggests interpreting the undefined term "line" as geodesic. Let us then consider Riemann's axiom first and interpret it in this way: *Through a given point P outside a geodesic l there is no geodesic which*

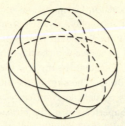

Figure 3.1   Geodesics (great circles) on a sphere

does not meet l. If we picture a *spherical* surface, this axiom will satisfy the demands of common sense. All geodesics (great circles) on a sphere must meet one another, for their planes all go through the center of the sphere and must cut one another (Figure 3.1). In order to satisfy other postulates of Riemann's geometry, it is better, in fact, to use a *hemisphere*. Thus, on a hemisphere, geodesics (great circles) will meet in one point only. Then part of Euclidean geometry, namely, geometry on a hemisphere, is a model for Riemann's non-Euclidean geometry.

Let us next give Lobachevsky's axiom the interpretation: *Through a given point P outside a geodesic l there are an infinite number of geodesics which do not intersect l.* In 1868 Beltrami showed that there is a surface on which this makes sense (Figure 3.2). The diagram pictures a *tractrix*, whose revolution about its axis forms a sort of double-trumpet surface, Beltrami's "pseudosphere," so named by analogy with the sphere of Riemann's geometry. Then part of Euclidean geometry, namely, geometry on the pseudosphere, is a model for Lobachevsky's non-Euclidean geometry.

Having discussed the geometries of Euclid, Lobachevsky, and Riemann, one might ask: Which one of the three is the geometry of everyday life? Which one fits observed data best, involves the least computation and the simplest mathematics? As far as agreement with the result of experiment is concerned, the surprising fact is that all three geometries are equally good in the small finite domain of ordinary existence. In Euclidean geometry there is the oft-quoted theorem, *The sum of the angles of any triangle is* 180°. The corresponding theorem of Lobachevskian geometry demonstrates that this angle sum is less than 180°, whereas Riemann's geometry holds that the sum is greater than 180°. In an attempt to ascertain which is the "true" theorem, Gauss performed an experiment in which a huge triangle with

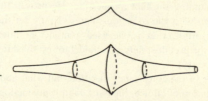

Figure 3.2   Tractrix and pseudosphere

vertices on three mountain peaks in Germany was measured, and after that stellar measurements were brought into play. Even if it made no practical difference which geometry was the true one for small figures, it was felt that it might make some difference if the sides of a triangle were so huge that it would take centuries for light, traveling 186,000 miles a second, to traverse them. All experiments failed to bring about a decisive conclusion. The sum of the angles observed was always so close to 180° that the excess or deficiency in each case could have been readily caused by the unavoidable imperfection in the measuring techniques. Even if the three geometries fit experimental facts equally well, they are not equally simple for computational purposes. For ordinary everyday measurement on earth the Euclidean system has the easiest formulas, and hence we use it, not because it is the absolute and only truth, but because it is more readily applicable.

Presently we shall give considerable space to a biography of Lobachevsky because we consider him such a very important figure in the development of pure mathematics. Not only did his non-Euclidean geometry initiate the tradition of axiomatics, which is the essence of formal mathematical thinking, but it also led to patterns for establishing *consistency* and *independence* which are imitated in the most important *current* issues associated with the foundations of all mathematics.

To illustrate this fact, we must point out that although we have treated the notion of a postulate system, we have not as yet considered the logical apparatus by which theorems are to be deduced from the basic axioms. That question will be discussed in Chapter 6. However, we have indicated almost from our first page that the concept of *set* is basic in mathematics. Chapter 6 will emphasize the algebra of finite sets, but infinite aggregates will be discussed much later (Chapter 24), and credit will be given to Georg Cantor for developing the theory of such sets. Set theory is an essential part of the logical foundation of pure mathematics, and hence it has been provided with a formal postulate system by Cantor's successors.

Now just as the parallel postulate was a thorn in the side of mathematicians prior to Lobachevsky (because they were attempting to deduce it as a *theorem* from the other Euclidean postulates, when in fact it was consistent with, but *independent* of, those postulates), there is a similar situation in abstract set theory. In that subject there is an assumption called the *axiom of choice* which is essential to logical reasoning in many branches of mathematics but which some mathematicians for special reasons (Chapter 24) find hard to accept. The first step in making it palatable to them occurred in 1938 when Kurt Gödel, one of the foremost leaders in the study of the foundations of mathematics, showed that there is nothing inconsistent about assuming the fact in question. He proved that the troublesome postulate is *consistent* with the other postulates of pure set theory. Gödel gave a relative consistency proof, that is, he used a model or interpretation (thereby imitating the use of the pseudosphere as a model for establishing the consistency of Lobachevskian geometry).

Gödel took care of the consistency question, but what about the *independence* of the bothersome axiom of choice, the issue completely analogous to the problem of the parallel postulate? That has been handled only since 1963, in the work of the young American mathematician Paul J. Cohen, a living Lobachevsky, who has presented us with a consistent *non-Cantorian* set theory in which he uses one or another form of the *negation* of the Cantorian axiom. Gödel had previously done this within a certain special class of sets, the "constructible sets," as he called them. But Cohen did not limit himself to such sets. By using models (exceedingly subtle in nature in contrast with such elementary pictures as the sphere and pseudosphere of the non-Euclidean geometries), he established independence from the other axioms for the axiom of choice and, in fact, for another challenging postulate (which Cantor and his immediate successors hoped to deduce as a theorem), called the *continuum hypothesis* (Chapter 24).

At a meeting of the International Congress of Mathematicians, held in Moscow in August 1966, Cohen was awarded a Fields Medal in recognition of all his achievements in attacking major unsolved problems in the foundations of mathematics. He had previously received the Bôcher Prize of the American Mathematical Society in 1963 and part of the special RCA award for scientific research.

To return to Lobachevsky, the man who started the sort of thinking Cohen is now carrying on at a much more advanced level, we have seen that two millennia of mathematical history formed a prologue to his bold replacement of the parallel postulate. But what sequence of events in Lobachevsky's own intellectual life led to that revolutionary step? Although the man in the street may picture Archimedes or Einstein or Cohen arriving at a theory as the result of a sudden brilliant inspiration, the inspired moment, if it exists, is just the culmination of a lengthy sequence of considerations on the part of the scientist, and this is subsequent to all the thought contributed by his predecessors. What was the course of the creative process with Lobachevsky? To answer, we must reveal some events from his personal life.

The year 1956, the centenary of Lobachevsky's death, was the occasion for special lectures to commemorate the fact that his discovery marked a turning point in mathematical history. Under the auspices of the University of Paris, one of the memorial addresses was delivered by Sophie Piccard, Professor of Mathematics at the University of Neuchâtel, Switzerland. Her lecture on Lobachevsky was delivered on December 1, 1956, at the Palais de la Découverte, one of the principal science museums of Paris. Professor Piccard was born in Russia, where she attended a lycée and later the University of Smolensk. Her understanding of Russian as well as her mathematical scholarship enabled her to draw on some hitherto untapped sources and the memorial address was an excellent popular account of Lobachevsky's work. What follows is a condensation of the memorial address* freely translated from the French by the present author.

Little is known about the first years of Lobachevsky's life. Only recently have the date and place of his birth been established with certainty. He was born on November 20, 1792, at Nijny-Novgorod. (This date of birth is the one ascertained only recently. It is

---

* S. Piccard, *Lobatchevsky, grand mathématicien russe. Sa vie, son oeuvre*. Conférences du Palais de la Découverte, Université de Paris, Série D, No. 47.

approximately a year earlier than that which was formerly given by historians.) His father, who was of Polish descent, worked for the government as a surveyor. Shortly after his son was born, he became ill and lost his position, with the result that his wife, Praskovia Alexandrovna, and his children spent the ensuing years in dire poverty. After almost a decade of illness, Lobachevsky's father died and in 1802 his mother requested the admission of her three sons, Alexander (aged eleven), Nicholas (aged nine), and Alexis (aged seven), as scholarship students in the "Premier Gymnase" of Kazan.

Situated on a small tributary of the Volga, Kazan, at the beginning of the nineteenth century, was a city of 25,000 inhabitants, about one-fourth of whom were Tatars. Extremely picturesque in appearance with its Tatar quarter and its numerous churches built on hills separated by deep ravines, it had unpaved clay roads that became massive mud puddles after every small shower and from which rose columns of dust when the weather was dry. Although it was an important commercial center of western Russia, it offered little in the way of cultural opportunity. The *gymnasium** attended by Lobachevsky had been founded in 1798 by members of the nobility for the education of their children as well as those of commoners; it offered a four-year course in preparation for civil service and military careers, and its program made heavy demands on the students.

In 1804 Emperor Alexander I signed a law founding a university at Kazan. The following year this university was opened without buildings, equipment, or professors, and was at first merely an annex of the *gymnasium* whose principal, Iakovkine, became University Director and whose teachers became the faculty of the higher institution. In 1805, in an attempt to recruit students, Iakovkine appealed to parents of the students in the *gymnasium* to have their children continue their education by attending the University of Kazan. Free tuition and board were available for indigent pupils, provided they promised to devote six years after graduation to teaching. Lobachevsky's mother was persuaded to consent to this scholarship offer for her three sons, and thus the boys left the *gymnasium* and entered the University of Kazan. Nicholas was admitted in February 1807, at the age of fourteen, having completed his *gymnasium* studies brilliantly a month earlier.

During his first year at the university he was taught mathematics by two student teachers who knew less than their pupils, but in 1808 he had the good fortune to study with Johann Martin Bartels (1769–1836), who had been Gauss's teacher. Thus young Lobachevsky embarked on the reading of research papers as well as scientific works in Latin, German, and French. At this time he decided to specialize in pure science instead of preparing for a career in medicine, which had been his earlier ambition. His reading suggested problems for which he gave remarkably original solutions.

The atmosphere at Kazan was a very serious one. In spite of this earnest setup, or perhaps because of it, Lobachevsky became involved in some student pranks. Having spent five years as a boarding student at the *gymnasium*, where iron discipline reigned, he indulged in a brief breath of freedom at the university. He had the audacity to go to parties which university boarding students were expressly forbidden to attend. Once he prepared a rocket which another student set off in the courtyard of the university. His major blunder, however, was his sarcastic manner toward Kondyrev, the confidant of the principal, Iakovkine. Kondyrev, unpleasant in personality and grotesque in appearance, was only five years older than Lobachevsky. It was his responsibility to supervise the students and to report their conduct to the principal. Kondyrev avenged himself by submitting very bad reports on Lobachevsky, even to the extent of accusing him of atheism, a charge which was not at all justified but which might have had tragic consequences for

---

* A *gymnasium* (German) or *lycée* (French) is a pre-university school on the continent of Europe. Its course of study corresponds to the education provided by the American junior and senior high schools plus the first two years of college, and the students range from ten to eighteen in age.

Lobachevsky. The principal wanted to have him dismissed from the university, and only the energetic intervention of the professors from Germany, who did not wish to lose their best student, permitted him to complete his studies. On July 10, 1811, Nicholas Lobachevsky was called before the council of the school and was upbraided for his bad conduct. He had to apologize, admit his guilt, and promise to reform. Only after he had put all this in writing was he granted the master's degree.

Thus, in 1811, at the age of nineteen, he ended his student days at Kazan. He remained at the university, however, in order to receive Bartels' guidance, and at once read two of the greatest mathematical works ever written: Gauss's *Disquisitiones Arithmeticae* and Laplace's *Mécanique Céleste*. These treatises, which have become classics, were less than a decade old at the time; they inspired Lobachevsky's *Theory of the Elliptical Movement of Celestial Bodies* (1812) and his *Concerning the Solution to the Algebraic Equation of the Type $x^n - 1 = 0$* (1813). The first of these papers has not come down to us, but Bartels' report stated that his protégé's study indicated a mastery of Laplace's exceedingly difficult *Celestial Mechanics* and enriched the same subject with original ideas that could be produced only by a remarkable mathematical talent.

Lobachevsky studied the great masters—Gauss, Cauchy, etc.—but strangely, Euler, that prolific genius who had lived in St. Petersburg, scarcely influenced the young Russian mathematician. In 1814 Bartels was influential in obtaining an adjunct professorship for Lobachevsky, and although this was equivalent in rank to an assistant professorship, the young man had to teach arithmetic, algebra, and other elementary mathematics courses because students entering the university had no background whatsoever in mathematics.

Lobachevsky did not resort to textbooks in teaching these elementary courses, but conscientiously prepared his own lecture notes, which he willingly lent to students. Virtue was its own reward, for it appears that pedagogic considerations led to his interest in the theory of parallels. Like his predecessors, he tried to prove Euclid's parallel postulate and presented such a proof in his geometry course of 1815, as the notes of one of his students, P. Temnikov, indicate. The demonstration was ingenious but inevitably erroneous, as Lobachevsky himself soon realized.

The year 1816 brought a promotion to associate professor for the young teacher, but unhappy days ensued for him and his associates. About this time Alexander I came under the influence of men who claimed that the universities were centers of atheism and revolutionary activity, and hence should be shut down one and all. In 1819 the dull-witted Magnitzky was assigned to investigate the setup at Kazan. He promptly recommended that the university be closed, but Alexander did not accept this suggestion. Instead he ordered the malevolent Magnitzky to institute reforms. The latter lost no time in setting about the task; he dismissed seven professors including Iakovkine; he established a chair of Greek Orthodox theology; he laid down the law on methods of instruction, insisting that they be religious and completely autocratic; he "purified" the library, and burned all the books he condemned; he instituted monastic discipline for the students. One result of this terror was that all the outstanding foreign scholars felt insecure and departed from the University of Kazan.

Lobachevsky was ill and had missed the first round of Magnitzky's activities. But when Bartels left in 1820 and Lobachevsky was appointed in 1821 to succeed his friend in the chair of pure mathematics, as well as in the direction of the entire physico-mathematical section, Magnitzky summoned him to St. Petersburg and kept him busy there for almost a year with tasks that were either dull or distasteful—for example, preparing statistical data on Kazan, directing the acquisition of library books and laboratory equipment, finally, examining the lecture notes of the natural science professors at St. Petersburg, Raupach and Herrmann, to see if these men of German origin were not subversive. Lobachevsky rendered a mild, noncommittal report but never forgave Magnitzky for burdening him with such an ugly mission.

Starting in 1822, when Lobachevsky was teaching ten courses, more and more duties were assigned to him. He was placed in charge of the university library, and then made chairman of the committee on the construction of university buildings. Next he was asked to help in the establishment of a scholarly scientific periodical for the physico-mathematical section of the university, and then to supervise the founding of an astronomical observatory. Little by little he became the central figure of the university. He was an indefatigable worker, a man of action who enthusiastically undertook all tasks that were entrusted to him, straightforward, just, poised, calm, and a keen diplomat. He knew how to get the whole university council to reach unanimous agreement. He was warmly loved and venerated by the students.

Since he did not flatter Magnitzky, however, the latter considered him subversive and eventually had him watched, along with other suspects. Thus it came about in 1823, when Lobachevsky submitted the manuscript of his *Geometry* for publication at government expense, that Fuss, the permanent secretary of the St. Petersburg Academy, criticized the work severely. He even objected to Lobachevsky's use of the metric system, because it was "created by the French Revolution"! The result was that the *Geometry* was not published during Lobachevsky's lifetime, was mislaid, then found in 1898, and printed for the first time in 1909.

When Nicholas I became czar, Magnitzky was removed from his position, and his duties at Kazan were taken over by General Jeltoukhine on February 21, 1826. Two days later there occurred an event of major importance in the history of modern mathematics: Lobachevsky presented before the physico-mathematical section of the University of Kazan a report written in French and entitled *A Brief Statement of the Principles of Geometry with a Rigorous Demonstration of the Theorem of Parallels.*

He asked the opinion of his colleagues on the merit of this work. They submitted it to a committee of three professors, whose respective subjects were astronomy, chemistry, and applied mathematics. No one of the three was able to understand the report, and the group was unanimous in its disapproval. Nevertheless Lobachevsky included the same material in a larger work, the *Elements of Geometry*, published in 1829. This treatise contained the foundations of his new geometry, but neither this nor any of Lobachevsky's subsequent works was appreciated in Russia. On the contrary, he was severely and unjustly criticized, even by the St. Petersburg Academy. He was not discouraged, however, and from 1825 to 1838 wrote additional papers including his great treatise, *New Elements of Geometry with the Complete Theory of Parallels.* Because he was accused of lack of clarity, he decided to write a short expository work on the more elementary aspects of his new geometry, and this was published as a pamphlet in Berlin in 1840. Although it was extremely lucid, a veritable mathematical masterpiece, which was to rouse the scientific world a quarter of a century later, Lobachevsky became the butt of German criticism of a type similar to that offered in his native land.

Fortunately, his personal life was happy during this period of professional frustration. In 1832 he had married Varvara Moiseieva, a wealthy young woman of property. They had several children, of whom Lobachevsky was most fond. He purchased an estate on the Volga, near Kazan, where he spent his vacations, such as they were. He would rise at 7 A.M., permit himself an hour's walk, and then start a day of work. He was an enthusiastic gardener and built a beautiful park covered with Siberian cedar trees. The Lobachevskys were hospitable, and among their many guests there were always half a dozen needy students on holiday.

From 1827 to 1846 Lobachevsky was rector of the university, but in 1846 was not reelected. After thirty years of service to the university, he was forced to leave and was assigned by the minister of public instruction to the position of "adjunct superintendent" of the school district of Kazan. His chair in pure mathematics was taken over by his pupil Popov, and Simonov, the professor of astronomy, replaced him as rector of the university.

Lobachevsky did not get along well with his superior, the superintendent of the school district, a General Molostvov. The mathematician's heart was not in his new work. His interest was in the university, which to this day maintains both the external appearance and the internal policies which he gave to it.

His last years were unhappy ones not only on account of his removal from academic activities but also because of the death of his son Alexei, the image of his father, who had been his student and excellent in mathematics. The youth died of tuberculosis at the age of nineteen. Added to these sorrows were grave financial difficulties resulting from Lobachevsky's loss of his job at the university. Next came arteriosclerosis, blindness, and the loss of his position as adjunct superintendent. Nothing deterred him, however, in his complete devotion to science. In 1855 he decided to contribute a paper to the collection commemorating the fiftieth anniversary of the founding of the University of Kazan, but his blindness made it necessary to dictate it to his pupils. This memoir, Lobachevsky's last, was called *Pangeometry*.

He died in February 1856, just thirty years after he had presented the concept of his new geometry to the physico-mathematical section of the university. When Professor Boulitch delivered a glowing eulogy, he was denounced as a freethinker and removed from his position. Such was the tragic close to the life of a mathematician whose genius revolutionized the pattern of all scientific thought.

The present chapter has treated postulational reasoning, Hellenic style, along with Lobachevsky's discovery that there is one grave imperfection in the Euclidean point of view, namely, the failure to realize that the fundamental axioms in any mathematical science are assumptions. The matter at issue is the nature of the foundation of a branch of mathematics or any abstract theory. Euclidean geometry (in modern dress) has been cited as an example of such a theory, but arithmetic or algebra can also be constructed on the basis of postulates containing undefined terms. Thus, in contrast with the informal, heuristic presentation of our previous chapter, Giuseppe Peano (1858–1932), one of the first and foremost contributors to modern axiomatics, based his pure, logical theory of the natural numbers on the following five postulates, in which "natural number" and "successor" are undefined terms:

(1) 1 is a natural number.
(2) Every natural number has a successor which is a natural number, and this successor is unique.
(3) No two natural numbers have the same successor.
(4) 1 is not the successor of any natural number.
(5) Principle of finite induction: Any property that belongs to 1, and also to the successor of any natural number that has the property, belongs to all natural numbers.

If the reader will examine the first four of these axioms, he will see that they are satisfied by the "counting numbers"—1, 2, 3, 4, ... of elementary arithmetic, although Peano's use of "natural number" as an undefined term leaves the way open for other interpretations. The fifth of Peano's postulates is a subtler assumption, which has its roots in what algebra textbooks call the method of mathematical induction. At any rate, from his five simple assumptions, Peano deduced, in rigorous fashion, theorems concerning the behavior of the natural numbers. The importance of Peano's formal foundation for the arithmetic of the natural numbers will become apparent only much later in our story. But we have already seen how and why, even for the purposes of Euclidean geometry, it became necessary to extend the natural aggregate so as to include other types of number. Still further number extensions will be presented in our next chapter, along with the postulate systems governing the arithmetic of the new numbers.

4

Algebra from Hypatia to Hamilton

"His Excellency, the Prefect of Alexandria." The announcement was made by a young slave to the group of half a dozen men seated in one of the lecture rooms of the museum.

A minute later, a striking-looking young man made his appearance. He was attired in elegant senatorial robes; elaborate metallic and jeweled ornaments hung from his neck, and a heavy odor of Oriental perfume surrounded him.

"Welcome, honored Orestes. Hypatia will rejoice to see you back in Alexandria and present at her lectures once more." Theon, the oldest of the group, was the first to speak. His pale, drawn face and his simple philosopher's robe contrasted sharply with the splendid attire of the young Roman prefect.

Palladas the poet, Wulff the Goth, Petrus the Christian, Hesychius the Jew, Kalliphonos the astronomer, and Euoptius, the emissary of Synesius of Cyrene, greeted Orestes in turn. All were fellow students in the classes of that leader of Alexandrian thought, Hypatia, daughter of Theon.

The prefect told of his travels, while the others spoke of affairs at home. A sudden silence fell on the group when the door of the adjoining room opened, and Hypatia mounted the tribune. All eyes were focused on the figure of the young woman who was about to address the group. She was clad in a loosely flowing Ionic robe, whose simplicity was relieved by the touches of color in the purple stripes marking Roman citizenship, the gold-embroidered sandals, and the wide gold band of the headdress.

"In honor of Your Excellency's return, and in order that you may know what we have discussed in your absence, my good father, Theon, has consented to read a brief summary of this month's lectures." Hypatia descended from the platform and motioned to Theon to take her place.

He began to read: "The primeval Being, the One, is the source of life, the only real existence, and the supreme Good, All things are brought forth by it. All things are divine, being reflections of the One." Theon read on and on. Orestes, attentive at first, was ultimately distracted by the restlessness of others in the group.

Barely waiting for Theon to complete his summary, the astronomer Kalliphonos voiced his criticism: "In this document Theon has failed to mention my objections to our abject subscription to the doctrine of Plotinus. Even if you yourself cannot agree with any other point of view, Hypatia, you should devote some lectures to the thought of Democritus, Epicurus, Lucretius. My fellow students are entitled to hear at least once the words of those who believe in the virtue of the material world. Give us the chance to be eclectic. We must have the right to choose from all philosophies."

"Then you shall be the one to present such a point of view." Hypatia's countenance had become suffused with color and her eyes were flashing. "I cannot teach what I do not believe. Still I am willing to have my followers exposed to opinions which differ from my own."

The Christian Petrus had seemed impatient during the conversation between Kalliphonos and Hypatia. He gave the speakers no opportunity to prolong the discussion, but voiced his own complaint. "Theon has not recorded my opinion, either —that we ought to discuss the philosophy of the religion which Orestes and I have embraced. Let us see whether our fellow students will choose Lucretius or Plotinus after they hear the teachings of Jesus Christ."

Hypatia trembled visibly as she answered, "Petrus, you also shall have the right to speak. I can see where the beliefs of Lucretius are in contrast with ours. But those of the Christians—never! Do we not, like you, teach that charity is stronger than the sword, that man must lose his life to gain his life, that self-sacrifice is the highest virtue? Do we not believe in the mortification of the flesh for the immortality of the soul? Let those who see greater beauty in the Christian creed embrace the cross. But do not force us to desert the faith which seems to us the equivalent of your Christian philosophy."

A troubled look came over the face of Orestes. As he rose to speak, the dignity of the Roman governor seemed to replace the youthfulness of the dandy. "These riddles of 'Who am I? What am I?'—shall we, here, try to settle them, when they are as old as thought? Never! We shall quibble from now until the end of time, if once we get started. Why not spend these precious hours in which Hypatia shares her wisdom with us, on pure mathematical questions whose abstract nature will prevent quarrels, and whose logical format will be indisputable? I recall that last year when Synesius visited us, we had a series of lectures on the *Arithmetica* of Diophantus. I think that I personally would enjoy a review and a continuation of this mathematical work."

The tension was relieved. Hypatia smiled once more. "Why, just this last week, Synesius sent an epistle through Euoptius, requesting further information on the construction of an astrolabe and also on those very questions of Diophantus to which your excellency alludes. I can read from the response I am sending to our good friend, the Bishop of Cyrene."

For an hour or longer Hypatia spoke, explaining the techniques of Diophantus for the solution of indeterminate problems of various types. She told of the novel Diophantine symbolism which would facilitate expression and resolution of such problems. Her students interrupted every now and then to suggest particular answers. Then she would indicate that general formulas were Diophantus' objective

and that he had obtained them for some of the more elementary types. She promised that she would devote future meetings to the "quadratic" problems which
Diophantus had solved most ingeniously. With this, Hypatia concluded her lecture.
Then there was the inevitable after-class discussion mingled with a bit of social
gossip. When it was all over, Orestes still remained.

With a glance through the window to make sure that all the others were on
their way, he turned to Hypatia. "As usual I feel intellectually stimulated by your
brilliant lecture. But you know very well that I come not to hear these arguments of
materialism versus idealism, nor even to study the work of Diophantus, but rather
to seek the guidance of the best brain in Alexandria concerning affairs of state. I
come once more about the same problem. My absence from Alexandria has not
improved matters."

"And what is the nature of this problem?" Hypatia questioned.

"You know well enough Cyril is ever the same. Although he is Bishop of
Alexandria, his conscience does not deter him from stirring up trouble. What
stand am I to take in these latest riots?"

"That of justice tempered with mercy." Hypatia's voice resounded through the
lecture room.

"A fine abstract doctrine, dear teacher, for your lectures, but an impractical
cliché outside the classroom. You must modify or elaborate it." Orestes paused as if
awaiting some further suggestion from Hypatia. But she remained silent. After a
moment or two, Orestes seemed both anxious and hesitant to make another request.
"And, Hypatia . . . ," he started slowly, "I am tempted to repeat an oft-spoken
plea. May I once again entreat you to embrace Christianity as a practical safeguard?
At this very moment Petrus is probably in Cyril's study, reporting on your lecture
and my presence here. By this evening there will be new whisperings everywhere.
Cyril will see to that. The words 'heathen' and 'sorceress' will circulate freely
on the tongues of the ignorant. General opinion does not favor this free association
of men and women to which you are accustomed. Come then, accept my counsel
in exchange for your own. Then you will be safe from the bishop's persecution, and
free to carry on your researches with peace of mind."

A long silence followed Orestes' speech. Hypatia then turned her head, and
her dark expressive eyes met those of the young prefect. "Charmingly put, my
practical Roman. I wish that I could agree with you, but, good friend, I can only
reiterate what I have said several times before. I cannot compromise with my beliefs,
my ideals, my devotion to a particular philsophy."

The prefect was still undeterred, and renewed his plea. "Hypatia, if I can make
no other appeal to you, let me plead for those very ideals and activities which you
cherish. Loyal follower of Plotinus, your very life is in danger. With Cyril as
bishop, what chance will there be for you to pursue studies or lectures in
safety?"

The rejoinder came quickly. "Shall I, a disciple of Plotinus, delay the joyous
reunion of my soul with the highest Spirit? Christianity should have taught you,
too, Orestes, not to fear death."

"Forgive me, dearest Hypatia, if I have been too emphatic with regard to your welfare. You must not blame me."

And so it was that Orestes gave up this battle as lost.

Orestes' warning was prophetic. Not long after it was delivered, Hypatia was barbarously murdered by the Nitrian monks and a fanatical Christian mob. She was torn from her chariot, dragged to the Caesareum, a Christian church, stripped naked, slashed to death with oyster shells, and finally burned piecemeal. In the eloquent, beautiful, and brilliant high priestess of mathematics and philosophy, the bigoted could see nothing but a pagan influence on the prefect of Alexandria.

After Hypatia's martyrdom, Hellenic mathematics came to an end, and we are fortunate that the mathematical tradition was kept alive in India. The great Alexandrian library had been burned by Roman soldiers before Hypatia's day, and during her lifetime the only remaining library, in the temple of Serapis, was sacked by an Alexandrian mob. This will give some idea of what Alexandria was like in 400 A.D. at the time when Hypatia lectured there on Neoplatonic philosophy and Diophantine algebra.

Tradition labels Hypatia the first woman mathematician in history, but almost nothing is known with certainty about either her original contributions or her expository treatises, since all her scientific writings are lost. She received her mathematical training from her father, Theon, who is usually credited with the revised and improved version of Euclid's *Elements* that is in use to this very day. In addition, he wrote other valuable commentaries, and so did his daughter. Texts on the *Conics* of Apollonius, the *Almagest* of Ptolemy, the *Arithmetica* of Diophantus are all attributed to Hypatia. But the customary belief is that her mathematical activity was mainly in the field of *algebra* and that is why we began this chapter with her story.

We are now going to discuss certain important aspects of the evolution of algebra from the earliest Babylonian era to the middle of the nineteenth century. In this chapter and in the next, we shall see that what was developed during almost four millennia was mainly the methodology of solving equations of a certain kind, that this entailed an extension of the real number system so that "imaginary" and "complex" numbers would be included, and that further extension produced a turning point in algebra akin to and contemporary with Lobachevsky's revolution in geometry. Then, ultimately, algebra became something very different.

In modern times, one of the most potent creators of the new "abstract algebra" was Hypatia's modern counterpart, Emmy Noether (1882–1935), the greatest woman mathematician in history. She, too, was a martyr of sorts. Fleeing to the United States to escape Hitler's persecution, her health failed, and she did not long survive her arrival on our shores.

To return to Hypatia, it was Diophantus, the greatest algebraist of antiquity, who inspired her. He, too, lived and worked in Alexandria, but his nationality and exact dates are unknown. At present some historians claim that he lived during the first century of the Christian era, about the time of Nero, whereas others cling to the previous opinion, which placed him around 250 A.D. Diophantus' work is his monument, and the only clue to his personal life is found in the following problem from the Palatine or *Greek Anthology*, a compilation of numerical epigrams assembled in the sixth century.

God granted him youth for a sixth part of his life, and adding a twelfth part to this, He clothed his cheeks with down; He lit him the light of wedlock after a seventh part, and five years after his marriage He granted him a son. Alas! lateborn wretched child; after attaining the measure of half his father's life, cruel Fate overtook him, thus leaving to Diophantus during the last four years of his life only such consolation as the science of numbers can offer.

A reader who has studied school algebra might use the literal symbol x to represent Diophantus' age, and then put the epigram into symbolic form, thus:

$$\frac{1}{6}x + \frac{1}{12}x + \frac{1}{7}x + 5 + \frac{1}{2}x + 4 = x$$

$$\frac{75}{84}x + 9 = x$$

$$9 = \frac{9}{84}x$$

$$84 = x$$

Diophantus lived to be eighty-four, and had married at thirty-three. His son was born when the father was thirty-eight and lived to the age of forty-two.

Modern literal algebraic symbolism was used in the equation above, and will be employed throughout this chapter. But such symbolism, launched to some extent by Diophantus, was not in vogue before the sixteenth century, when François Viète (1540–1603), better known by his Latin pen name Vieta, was the first to use letters to represent unknowns. Before the day of Diophantus algebra was *rhetorical*, that is, results were obtained by means of verbal argument, without abbreviations or symbols of any kind. One could, in fact, provide formulas, solve equations, and carry out algrebraic procedures even if one were to write *number* or *numero* or *Zahl* or *res* (the Latin for the word "thing"—a customary usage even as late as the sixteenth century) instead of using x or y or z. But it would make reasoning difficult. Therefore one of Diophantus' major contributions was the "syncopation" of algebra.

Syncopated algebra, as it is called, is more a case of shorthand than completely abstract symbolism, but it is a marked step in the right direction. At any rate, Diophantus was the first mathematician in history to provide any sort of substitute for lengthy verbal expression. He was given to the initial-letter type of shorthand. Instead of our proverbial x, he used a symbol that resulted from the fusion of α (alpha) and ρ (rho), the first two letters of *arithmos* ($\alpha\rho\upsilon\theta\mu o\varsigma$), the Greek word for number. Our x^2 and x^3 were $\varDelta^v$ and K^v from *dunamis* ($\delta\upsilon\nu\alpha\mu\iota\varsigma$), meaning power, and *kubos* ($\kappa\upsilon\beta o\varsigma$), the word for cube.

The degree of influence of Babylonian algebra on the pre-Diophantine geometric algebra of the Greeks, on Diophantus himself, and on algebra in India, is now evaluated as considerable. This may have been brought about through trade relations and, in rather negative fashion, by the frequent foreign invasions of India during the early centuries when Hindu algebra was being developed by Aryabhata the Elder (475–550 A.D.), Aryabhata the Younger (dates uncertain), Brahmagupta (*ca.* 628), Mahavira (*ca.* 850), and Bhaskara (1114–1185). These men, like Diophantus, contributed to the theory of indeterminate equations. They even used a syncopated symbolism akin to that of Diophantus, but they were more imaginative and more thorough. Thus Bhaskara wrote:

> In those examples where occur two, three, or more unknown quantities, colors should be used to represent them. As assumed by previous teachers, there are: *yavat-tavat* (so much as), *kalaka* (black), *nilaka* (blue), *pitaka* (yellow), *lohitaka* (red), *haritaka* (green), *svetaka* (white), *cirtaka* (variegated), *kapilaka* (tawny), *pingalaka* (reddish-brown), *dhumraka* (smoke-colored), *patalaka* (pink), *savalaka* (spotted), *syamalaka* (blackish), *mecaka* (dark blue), etc.*

Another Hindu text lists a few more colors, and also adds a set of *flavors*. Bhaskara used still further types of abbreviation in symbolizing unknowns. Ultimately only initial syllables were used—*ya, ka, ni*, etc. For a known number there was *ru* (from *rupa*), and for a product there was *bha* (from *bhavuta*), written after the factors. Addition was indicated by juxtaposition, and subtraction by placing a dot over the subtrahend. Thus

$$ru\ 9\ ya\ ka\ ni\ 7\ bha\ ru\ \dot{5}$$

(transliterated from the Sanskrit) corresponds to our

$$9 + 7xyz - 5$$

Using their syncopated symbolism, the Indian algebraists achieved great heights in indeterminate analysis, but their discoveries never reached the western world and all their profound efforts were duplicated at a later date by the Europeans Pierre de Fermat (1601–1665), John Wallis (1616–1703), Lord William Brouncker (1620–1684), Leonhard Euler (1707–1783), the first and the last being among the greatest names in the whole history of mathematics.

To illustrate an exceedingly elementary indeterminate problem, let us consider the modern issue of making change for coins. Suppose that a quarter is presented with a request for dimes and pennies in exchange (at least one coin of each kind). This can be expressed by the equation

$$10x + y = 25$$

where *x* represents the number of dimes and *y* the number of pennies. The problem is indeterminate because, as the reader will see, there are *two* solutions, namely,

---

* B. Datta and A. N. Singh, *History of Hindu Mathematics*, Motilal Banarsi Das, Lahore, 1935, Vol. 1, p. 18.

$x = 1$, $y = 15$ and $x = 2$, $y = 5$. If one were *not* talking about dimes and pennies, the abstract equation would have an infinite number of other numerical solutions like $x = \frac{1}{2}$, $y = 20$, $x = 3$, $y = -5$, etc. Diophantus and his Hindu kindred spirits considered both determinate and indeterminate equations. In the latter the unknowns were *always* considered to be natural numbers only, as in the case of our question about dimes and pennies.

The restriction that permits "syncopated" or modern symbols to be replaced only by natural numbers, or the greater freedom that might allow a *ya* or a *ka* or an *x* to represent any real number, is an essential feature of what we call algebra. The *x, y, z*'s need not symbolize unknowns in some practical or recreational problem, but may be interpreted as *variables*, with a specified *replacement set* or *domain*. These concepts are useful not only in solving problems where unknowns may be natural or integral or rational or real numbers, etc., but also in making *general* statements about all members of such sets of numbers, for example, the commutative law, $x + y = y + x$, or the "identity," $(x + y)^2 = x^2 + 2xy + y^2$, where in the two illustrations the domain of both variables, x and y, is the real number aggregate. For this reason, traditional algebra is often described as *generalized arithmetic*. In arithmetic, statements refer only to *constants*, the names of specific things, for example, "three" or IV or 7 or $10 - 2$, etc. In algebraic statements, both constants and variables may be involved.

Since, as we have said in previous chapters, many fundamental concepts of mathematics can be explained in terms of the *basic* notion of a *set* of things, *variable* and *constant* may also be defined in that way. A *constant*, then, is a symbol which names a *specific* member of some set, and a *variable* is a symbol which, in a particular context, may be replaced by *any* member of some set, called the *domain* or *replacement set* of the variable.

Variables and constants are analogous to the pronouns and proper nouns of ordinary speech. If I assert, "*He* is my cousin," I might as well say, "*x* is my cousin." Either "he" or *x* is a variable which is open to replacement by the name of any living male. Any such name, for example, Henry Jones, is a *constant*. Although one is free to replace the variable by an enormous number of different constants, not all replacements will lead to *true* statements. Perhaps in all the wide world I have only two male cousins, Henry Jones and Peter Smith. Mathematical custom would name the assertion containing the variable an *open sentence*. The aggregate {Henry Jones, Peter Smith} would be called the *truth set* or *solution set* of the sentence.

The statements "Michael Brown is my cousin" (in the above example), and "$5 = 2 + 3$" (in arithmetic) are said to express *propositions*. Here as well as elsewhere in mathematics and logic (Chapter 6), an idea that can be pronounced either true or false is called a *proposition*. Logicians call a declarative sentence expressing a proposition a *statement*. This makes a statement a sort of constant that names an idea, namely, the corresponding proposition. Although logicians distinguish between statement and proposition, numeral and number, any constant and the *value* it names, we shall *not* do so in this book. We have indicated here and in previous chapters the distinction between a name and the object named. Belaboring the point makes for pedantic expression but, for the record, we indicate once more that a unique object may have many names. A certain cardinal number may be named "three," "drei," III, 3, $1 + 2$, $1 + 1 + 1$, $8 - 5$, etc. The same proposition is asserted by "If two sides of a triangle are equal, the angles opposite are equal" and "The base angles of an isosceles triangle are equal." For those who know mathematical tradition, *Pons Asinorum* (Bridge of Asses) is still another name for the proposition (because the original proof in Euclid was so difficult that many students could not "cross over" to pass beyond it).

To review some of the things we have said, let us suppose that x is a variable whose domain is the set, $\{1, 2, 3, 4, 5\}$. Then "$x < 4$" is a declarative sentence (read as "x is less than 4") but it is *not* a statement because it does not express a proposition, an idea that can be pronounced either true or false. Because the sentence has the *form* of a statement, however, it would sometimes be described as a *statement form* or *propositional form*. But since the sentence presents an "open question" and the variable x is open to replacement by any value in its domain, it is customary to describe "$x < 4$" as an *open sentence*. Replacement of x by the five values in its domain would lead to five propositions, only three of which would be true. Those values of x which lead to true statements are said to constitute the *truth set* or *solution set* of the open sentence, "$x < 4$." It is readily seen that the truth set is $\{1, 2, 3\}$, a *subset of the domain*.

For other illustrations, suppose that the domain of each of the two variables x and y is $\{1, 2, 3, 4, 5\}$. Then the following are open sentences:

$$x > 2$$
$$x + 1 = 3$$
$$x^2 < 5$$
$$x - y = 2$$
$$x + y = 4 \quad \text{and} \quad x - y = 2$$

The reader can verify, by replacing x in the first three sentences by each of the five values in the domain, that the solution sets of those three open sentences are $\{3, 4, 5\}$, $\{2\}$, and $\{1, 2\}$, respectively. In the remaining sentences one must replace both x and y by values in the domain. The truth set for the fourth sentence is $\{(3, 1), (4, 2), (5, 3)\}$. The fifth sentence, which is *compound*, has the solution set $\{(3, 1)\}$. We observe that only the second and fifth sentences are *determinate*.

If the domain of both x and y is still the same aggregate, $\{1, 2, 3, 4, 5\}$, the following open sentences have solution sets that are special in nature:

(1) $x + 3 = 1$
(2) $3x = 2$
(3) $x + y = 1$
(4) $x + 4x = 5x$
(5) $x \leq 5$ (read "x is equal to or less than 5")
(6) $x + y > 1$

It will be seen that the truth set of the first three sentences is $\emptyset = \{\ \}$, the empty or null set (Chapter 1). On the other hand, sentences (4) and (5) have the entire domain as solution set. Sentence (6) has a solution set containing the *twenty-five* ordered pairs, $\{(1, 1), (1, 2), \ldots, (1, 5), (2, 1), \ldots, (2, 5), \ldots, (5, 1), \ldots, (5, 5)\}$.

We have suggested that all sentences containing variables must be considered *open*. But this is not the case since, for example, the last six sentences can be "closed," that is, converted into statements of propositions if each sentence is prefixed by a suitable *quantifier*. Thus sentences (4) and (5) might be prefixed by "For all x," and sentence (6) might be prefixed by "For all x and for all y." In that case the quantified (4), (5), and (6) would be *true* propositions. If (1) and (2) are prefixed by "For some x," and (3) is prefixed by "For some x and some y," the results are statements of *false* propositions. Thus only sentences containing *unquantified* variables are open.

Our six special sentences as well as the equation $10x + y = 25$, discussed earlier, show that the number of solutions and, in fact, the extremes of possibility or impossibility of solution are dependent on the domain of each variable occurring in a sentence. The relation of the solution set to the domain of variables is closely associated with the history of algebra. Thus when Hellenic mathematicians considered $x^2 = 1$ and $x^2 = 2$, they gave $\{1\}$ and $\{\sqrt{2}\}$ as the respective solution sets. In effect, they limited the domain of x to the *positive* real numbers. In the extended domain of *all* real numbers, the solution sets would be $\{+1, -1\}$ and $\{+\sqrt{2}, -\sqrt{2}\}$. We remark, in passing, that not only Diophantus but all European mathematicians prior to the Renaissance would have balked at including negative numbers in the domain of variables (and hence in a solution set).

But there came a time in the evolution of algebra in India when the "law of signs" for the multiplication of directed (positive and negative) real numbers was known and applied. In fact, quoting Neugebauer as our authority once more, we can state that Babylonian astronomical texts of the third century B.C. made explicit use of the rule: The product of numbers with like signs is positive, while the product of those with unlike signs is negative. Perhaps, somehow, Indian awareness of the law can be traced to Babylonian sources. At any rate, Bhaskara intended to make use of the rule when he raised the following question: What (real) number, multiplied by itself (squared) is equal to -1? In modern terminology, Bhaskara sought the solution set for $x^2 = -1$, where the domain of x is the aggregate of real numbers.

Having posed the problem, Bhaskara labeled it "impossible" to solve, and so did all algebraists, oriental and occidental, before the sixteenth century. Bhaskara argued that any real number is either positive or negative, and when squared, that is, multiplied by itself, will yield a *positive* result by virtue of the "law of signs." In modern terminology, the solution set of $x^2 = -1$ in the real domain is $\emptyset = \{\ \}$, the empty set.

But once again appropriate enlargement of the domain of a variable can make the impossible possible, and the Hindu algebraists, although they did not recognize this fact, had themselves engendered the mechanism that would, so to speak, do the trick. They had solved some equations by a procedure which they called *vilomogati*, usually translated as *inversion*. We shall not explain this method but merely relate that Brahmagupta had given a general rule for it in which square root was recognized as the operation *inverse* to squaring (multiplying a number by itself). Therefore *vilomogati* (inverting or working backwards) required that the solution set of $x^2 = a$ (where a is *any number*) must be $\{+\sqrt{a}, -\sqrt{a}\}$. In the case of $x^2 = 1$ and $x^2 = 2$, Brahmagupta's rule would lead straight to the answers already given. For $x^2 = -1$, the rule would give the solution set, $\{+\sqrt{-1}, -\sqrt{-1}\}$. But the solution set is *meaningless* in the real domain, and Bhaskara was unable to resolve the crisis by finding a suitable extension of the domain.

That task awaited the sixteenth century Italian mathematician Rafael Bombelli (born *ca.* 1530). He elevated $\sqrt{-1}$ to the rank of a "number," and defined arithmetic operations on $\sqrt{-1}$, $\sqrt{-2}$, $\sqrt{-3}$, $\sqrt[4]{-1}$, $\sqrt[6]{-2/5}$, $-\sqrt{-1}$, $-\sqrt{-2}$, etc. and all even roots of negative numbers. He stated that such symbols represent *imaginary* numbers, a derogatory term on a par with those other descriptions, irrational and negative.

Today we still carry out arithmetic with imaginary numbers by using Bombelli's rules, but we have simplified his notation somewhat. We employ the symbol i for $\sqrt{-1}$ and consider it to be the imaginary unit. If we choose an inch as the unit of length, all other lengths are real multiples of that unit. Analogously, all imaginary numbers are real multiples of the unit i. Thus $\sqrt{-25}$ is interpreted to mean $\sqrt{25} \cdot \sqrt{-1} = 5i$, and other examples of imaginaries are $\frac{1}{3}i$, $i\sqrt{2}$, $-i\sqrt[3]{5}$, etc.

Bombelli adjoined the imaginaries to the reals and then defined arithmetic operations on the united set. To have a well-defined addition operation, that is, to have *closure* with respect to that operation, the sum of a real number like 3 and an imaginary like $2i$ must be meaningful. In other words, $3 + 2i$ must be a "number." All such sums, for example, $\frac{1}{2} - i$, $-1 + i\sqrt{3}$, $\sqrt[5]{2} - \frac{1}{4}i$, etc., are described as *complex numbers*. In general, a *complex number* is defined by the formula, $a + bi$, where a and b are any real numbers. Then a real number is a special kind of complex number for which $b = 0$, and an imaginary is the special type for which $a = 0$. In other words, the aggregate of all complex numbers contains the set of real numbers and the set of imaginaries as two proper subsets.

At this point, certain questions seem natural: Are there any more kinds of numbers? Can the complex numbers be embedded in some larger system? We shall answer both these questions presently but, for the moment, it will suffice to state that the problems of classical algebra do *not* require any further types of number. Let us explain why.

In traditional algebra the fundamental expression containing variables was the polynomial, which, equated to zero, gave a polynomial equation, whose solution set was sought. Thus $x^2 + 7x - \sqrt{5}$ and $5x^4 - \frac{1}{2}x^3 + (3 - i)x + i\sqrt[3]{2}$ are polynomials in the variable x, with coefficients that are complex numbers. The degree of a polynomial in x is that of the highest power of x in the polynomial. Thus the illustrative polynomials have degrees 2 and 4, respectively.

In our next chapter we shall consider some of the specific methodology for obtaining solutions of polynomial equations of degrees 2, 3, and 4. But in the present chapter we are concerned only with the *domain* of the variable x which appears in the polynomial equation. If we choose Bombelli's set of complex numbers as the domain of x, will there always be a solution set other than $\emptyset$, the empty set, for the polynomial equation.

$$a_0 x^n + a_1 x^{n-1} + \cdots + a_n = 0$$

where n is a natural number and the coefficients are in the complex set? This question was answered in the affirmative when the great Carl Friedrich Gauss (1777–1855) proved the *fundamental theorem of algebra*: A polynomial equation with complex coefficients has at least one solution in the complex domain.

Gauss's proof of the fundamental theorem was very advanced in nature. But once the basic proposition was established, it was easy to deduce from it the corollary: Every nth-degree polynomial equation with complex coefficients has at most n distinct solutions in the complex domain.

Gauss's theorem and corollary spelled the end of one road in classical algebra, namely, the path of number extension. Since the complex numbers will guarantee one or more solutions for polynomial equations with complex coefficients, why seek new species of number? But classical algebra is not all of mathematics, and hence efforts to generalize the number concept continued. That story will now lead us from the old algebra to the new, from the type of question considered by Hypatia to that created by Emmy Noether.

The history of mathematics indicates that when new numbers are created, they are not accepted with alacrity. The Pythagoreans rejected the irrationals even though they had concrete examples in the form of lengths. Ultimately it was a good abstract theory, namely that of Eudoxus, which gained recognition for the incommensurables. With negatives, the situation was reversed. They made their appearance in algebraic equations but were, for the most part, rejected until such time as they were interpreted as business debts, and operations with them were conceived to be bookkeeping procedures. History repeated itself with the imaginaries and the complex numbers. They achieved full status only in the nineteenth century, when, on the one hand, they were accorded superior theoretical treatment and, on the other, were given concrete interpretation as geometric and physical entities.

An elegant abstract formulation of the algebra of complex numbers and one whose generalization had tremendous impact on all of modern mathematics is due to William Rowan Hamilton (1805–1865), the greatest of Ireland's men of science. Hamilton, like many child prodigies, excelled in languages at an early age. He mastered Latin, Greek, and Hebrew by the age of five, French and Italian before he was eight, Arabic, Persian, Sanskrit, and half a dozen other Oriental languages by the age of thirteen. At the tender age of seventeen he was pronounced the first mathematician of the day. In his adult years he made colossal contributions to mathematics and mathematical physics. In this chapter we shall see how he launched the new era in algebra. Later on we shall indicate that he played a similar role in theoretical physics.

For Hamilton a complex number was just an *ordered pair of real numbers*. Thus $2 + 3i$ was $(2, 3)$ and $3 + 2i$ was $(3, 2)$. A real number like 5, that is, $5 + 0i$, was $(5, 0)$. An imaginary like $4i$, that is, $0 + 4i$, was $(0, 4)$. Hamilton defined addition of the real number pairs and showed that this operation had the properties of closure, commutativity, and associativity. At school the reader learned to add $2 + 3i$ and $-5 + 4i$ by "adding real and imaginary parts separately" to give the sum $-3 + 7i$. In terms of Hamilton's definition, this would become

$$(2, 3) + (-5, 4) = (-3, 7)$$

or, more generally,

$$(a, b) + (c, d) = (a + c, b + d)$$

The sum is always meaningful because $a + c$ and $b + d$ are real sums and real addition has closure. Moreover

$$(a, b) + (c, d) = (c, d) + (a, b)$$

because $a + c = c + a$ and $b + d = d + b$. Thus the commutativity of complex addition follows from the same property of real addition, and it is similar with associativity.

To understand Hamilton's definition of multiplication of complex numbers, we first consider the viewpoint of school algebra, where one is told that since $\sqrt{2} \cdot \sqrt{2} = 2$, it seems logical to define $\sqrt{-1} \cdot \sqrt{-1}$ as -1, that is, $i \cdot i = -1$. With $i^2 = -1$, and the assumption of distributivity, one can multiply any two complex numbers. Thus

$$(a + bi)(c + di) = ac + bci + adi + bdi^2 = ac - bd + (ad + bc)i$$

Instead of the elementary rationale, there is Hamilton's definition, which states that, for two ordered pairs of real numbers,

$$(a, b) \times (c, d) = (ac - bd, ad + bc)$$

From this definition one can deduce (instead of assuming) the closure and associativity of complex multiplication, and its distributivity with respect to complex addition. Once again arithmetic laws for complex numbers can be derived from those for real numbers. Here is just another indication that the latter system is mathematical bedrock.

What Hamilton sought to do next was to generalize his idea of ordered pairs of real numbers, and to consider the possibility of *hypercomplex numbers* which would contain the real and complex systems as proper subsets just as the complex aggregate contains the reals and the imaginaries.

Various modes of generalization will suggest themselves to the reader. Should one think of triples, quadruples, quintuples, etc. of real numbers? Or should one retain the idea of ordered pair, but generalize the nature of the numbers paired, for example, by considering the set of ordered pairs of ordinary complex numbers, $(a + bi, c + di)$? In the latter case, $a + bi$ is the same as Hamilton's (a, b) and $(c + di)$ the same as (c, d). Therefore Hamilton employed the *quadruple* of real numbers (a, b, c, d) instead of the complex pair. Either formulation employs *four* real numbers and hence Hamilton called the hypercomplex numbers he created *quaternions.*

A set of pairs or quadruples is structureless unless one or more operations and/or relations on the set are defined. If one desires an "arithmetic" of hypercomplex numbers in which "numbers" and operations are akin to those in ordinary arithmetic, he must define an addition and a multiplication which obey the customary fundamental laws (closure, commutativity, associativity, distributivity). Now here is where Hamilton made an epochal discovery! After *fifteen years* of thought on the subject, he found that he could *not* formulate a quaternion algebra in which the hypercomplex numbers would simultaneously satisfy both the traditional arithmetic laws and the requirements of a physical science of space. Finally, in 1843, he released a revolutionary but logically consistent algebra in which the commutative law of multiplication was abandoned. Multiplication was defined in such a way that $q_1 \times q_2$ is *not* equal to $q_2 \times q_1$, where q_1 and q_2 are any two quaternions, (a_1, b_1, c_1, d_1) and (a_2, b_2, c_2, d_2).

Hamilton, as we have indicated, considered himself primarily a *physicist*, and not a pure mathematician. If we are to understand the motivation for his definition of quaternion multiplication, we must examine a physical interpretation of quaternions. Before doing that, we shall build up a suitable background for generalization by illustrating concrete geometric and physical counterparts for those more elementary entities, Bombelli's *complex numbers.*

In 1797 a geometric interpretation of complex numbers was given by Caspar Wessel (1745–1818), a Norwegian surveyor. Then in 1806 the Swiss-French mathematician Jean Robert Argand (1768–1822) arrived independently at the same interpretation. The Wessel-Argand representation generalizes the use of coordinates on a number line. In a previous chapter, a one-to-one correspondence was established between the aggregate of real numbers and the points of a straight line. In similar fashion, one can match real number *pairs*, that is, *complex numbers*, with the points of a plane. Thus the complex number $3 + 4i$ has the Hamiltonian interpretation (3, 4) which corresponds to the point whose *Cartesian coordinates* (Chapter 7) are 3 and 4, respectively (Figure 4.1). If the plane is considered as a rectangular network of streets and avenues, then (3, 4) corresponds to the corner where "street 3" meets "avenue 4." Some real and imaginary numbers (conceived as real number pairs) are represented by points in the diagram of Figure 4.1. It will be seen that the real numbers correspond to the points of the real number line (Chapter 2), usually called the X-axis, the imaginaries to points of a second number line, the Y-axis. In this picture of the imaginary numbers, the American mathematician Arnold Dresden (1882–1954) saw a way of removing the stigma of their name. He suggested that they be called *normal* numbers because the term "normal"

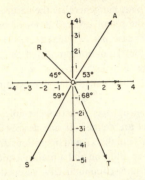

Figure 4.1

is synonomous with *perpendicular*. Thus, imaginaries are *normal* because their picture, the *Y*-axis, is perpendicular or *normal* to the *X*-axis of real numbers.

We have stressed the fact that new number types did not seem to gain general acceptance until the abstract or pure number symbols were given interpretations in the real world. The Wessel graphic representation lends itself to such an interpretation of the complex numbers. By a slight addition to the Wessel diagram (Figure 4.1) one can picture any complex number as a *directed line segment* or plane *vector*. Thus the complex number $3 + 4i$ can be pictured as the arrow OA in Figure 4.1, the *directed line segment* whose initial point is $(0, 0)$ and whose terminal point is $(3, 4)$. The term "vector" is derived from the Latin word *vectus*, which means carried. If one is *carried* from $(0, 0)$ to $(3, 4)$ by car or plane (where a unit is one hundred miles), the journey can be pictured by the vector OA in Figure 4.1.

The vector interpretation of a complex number was designed for application to physics. For the physicist a vector is an entity that has both magnitude and direction, for example, a displacement, a force, a velocity, an acceleration. Specifically he may cite an *upward* force of 3 lb., a velocity of 20 mi. per hr. *southwest*, a *downward* acceleration of 32 ft. per sec. per sec.

For another way of representing a vector, let us find the *length* (size) and *direction* of OA in Figure 4.1. By the Pythagorean theorem,

$$\mathbf{r}^2 = 9 + 16 = 25$$
$$\mathbf{r} = 5$$

Physicists might say that 3 and 4 are the components of the vector in the horizontal and vertical directions, respectively. Or they might say that the *resultant* of two component motions is 5 ft. per sec. in a direction that can be found by protractor to be 53° approximately (or more precisely by trigonometry). Thus, a force or velocity can be described by a vector, an entity having both *size* (5 lb. or 5 ft. per sec.) and *direction* (53°). The description $(3, 4)$ or $3 + 4i$ is called the *rectangular* form of this vector or complex number, and $(5, 53°)$ is called the *polar* form (Chapter 7). Rectangular and polar representations of other vectors in Figure 4.1 can readily be obtained.

The geometric vectors representing complex numbers are called *centered* or *bound* vectors because they all initiate at *0*, the origin. In physics, however, vectors are *free* since they may start at any point. Thus the geometric vector represented by $3 + 4i$ or (3, 4) or (5, 53°) describes simultaneously an infinite number of vectors, for example, all forces of 5 lb. acting at an angle of 53° with the horizontal, but applied at different points in a plane.

Since a plane vector is just a real number pair, plane vectors can be added by Hamilton's definition,

$$(a, b) + (c, d) = (a + c, b + d)$$

To obtain the geometric analogue, two plane vectors (complex numbers) and their sum or resultant are pictured in Figure 4.2. This diagram suggests what the physicist

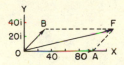

Figure 4.2

calls the *parallelogram law* for the addition of plane vectors. To him a resultant means the combined effect of two forces or velocities or accelerations, etc. The triangle of Figure 4.2 is just half of a parallelogram in which the vector representing the first force or velocity is one side OA, the vector representing the second force or velocity forms the adjacent side OB, and the resultant or sum is the diagonal OF.

Physical forces, motions, etc. are not confined to a plane, but may occur in space. Hence the abstract idealization of such "3-dimensional" physical vectors should be an ordered *triple* of real numbers, like (3, 4, 12), say. To picture a physical analogue for this particular triple, let us imagine that a particle at point O in Figure 4.3 is moved 3 ft. east, then 4 ft. north, then 12 ft. up to position P. Its total displacement from point O, measured "as the crow flies," can be described by the directed line segment or vector OP in Figure 4.3. This is the vector (3, 4, 12), which is the *resultant* of the *component* displacements in the three directions, "east," "north," "up." The vector OP is described by the ordered triple (3, 4, 12) but it could be described by another ordered number triple in which the numbers are its length or magnitude, 13, and two angles giving its direction. Thus, by protractor or trigonometry (as in the case of $3 + 4i$), angle $COA = 53°$ (approximately). This angle gives the direction of OC in Figure 4.3. Then OC and OZ determine the plane in which vector OP lies, and again experimental measurement would give the angle between OP and the Z-axis as 23° (approximately). Hence vector OP can also be described as the number triple (13, 53°, 23°).

If, in Figure 4.3, the particle at point P were subsequently to be moved 5 ft. east, 2 ft. north, 6 ft. up, this displacement would correspond to the *vector* (5, 2, 6).

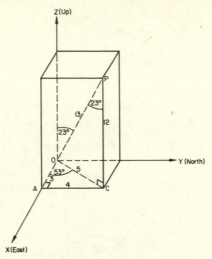

Figure 4.3

Obviously the final displacement from O would be the vector $(3 + 5, 4 + 2, 12 + 6)$ $= (8, 6, 18)$. This suggests that Hamilton's rule for addition of complex numbers or two-dimensional vectors can be extended to vectors in space so that the latter, like the former, are added by summing their respective components.

Today mathematicians generalize the vector concept so that an abstract vector can have any number of components. Thus $(3, -5, 0, -2/7, 6)$ is a vector of order 5, or a "5-dimensional" vector. Later on we shall be able to furnish significant concrete interpretations of such vectors.

If "structure" is to be provided for sets of vectors, it will be necessary to define arithmetic operations on them. Since physical vectors in space (3-dimensional vectors) do not seem essentially different from plane vectors, it is not surprising that one can define addition for them and in fact for all higher-dimensional vectors merely by extending Hamilton's definition for the plane. Hence

$$(a_1 a_2, \ldots, a_n) + (b_1, b_2, \ldots, b_n) = (a_1 + b_1, a_2 + b_2, \ldots, a_n + b_n)$$

The sum of two vectors is a vector whose components are equal to the sums of the components of the addends, respectively.

One can also define an operation on vectors (of all dimensions) which we shall, for the moment, call a change of scale. To illustrate it, consider the displacement vector $(3, 4, 12)$ pictured in Figure 4.3. The same graph might well represent the vector $(6, 8, 24)$, whose components are twice as great, or the vector $(1, 4/3, 4)$, whose components are one-third as great. It would be merely a matter of changing the scale indicated on the drawing. Even if the diagram remains the same, the physical meaning would be different in each case. In Figure 4.3 the *resultant* velocity OP was equal to 13 ft. in the direction indicated (by the angles 53° and 23°). When components are multiplied by 2 or 1/3, the direction would be the same but the magnitude of the resultant would be 26 feet or 13/3 feet, respectively. In other words, if V is the original vector, the two changes in scale lead to the vectors $2V$ and $1/3$ V, respectively. The change in scale is thus tantamount to *multiplying a vector by an ordinary number*, and is accomplished by *multiplying each* component of the vector by this number.

This rule for multiplying a vector by a number has been illustrated for positive rational numbers only. It has more general validity, however, but the description as a change in scale may not seem appropriate if one multiplies the vector by zero, in which case it is converted into a null or zero vector. Again, to multiply the vector by a negative number would require a change of scale that reverses the signs of numbers on the X-, Y-, and Z-axes of a diagram like Figure 4.3. In such cases, an alternative interpretation of the multiplication is possible, where instead of changing the scale one changes the vector. Then multiplying it by a positive number would *stretch* or *contract* it, according to whether this number is greater than or less than one, and multiplication by a negative would not only stretch or contract it, but would also reverse the direction of the vector. For the use of alternative viewpoints, see the discussion of "alias" or "alibi" on page 411.

The reader may find the description of multiplication of a vector by a number confusing if the vector is also considered a sort of "number" albeit a complex or *hypercomplex* number. Therefore physicists apply the term *scalar* to the single "ordinary number," like 2 or 1/3, which effects the change in scale. Then what we have been talking about is the multiplication of a vector by a *scalar*, that is, multiplication of a complex or hypercomplex number by a single "ordinary number."

We have dwelt on the *addition* of vectors of *any* dimension and have given Hamilton's definition for the multiplication of plane vectors, but we have not discussed the geometry or physics of that multiplication. We have nowhere defined the multiplication of vectors of dimension 3 (or of any higher dimension). The reader must not be confused by our discussion of "scalar multiplication," a term which we could not avoid since it is used so frequently in the literature of mathematics and physics. As far as vectors are concerned, scalar multiplication is a *unary* operation because it is carried out on *one vector only*. It is a change of scale in the components of that one vector or, in the alternative interpretation, a stretching or shrinking of that vector.

Now to consider a *binary* operation, how can one define the multiplication of two 3-dimensional vectors, that is,

$$(a_1, a_2, a_3) \times (b_1, b_2, b_3)$$

where the letters represent real numbers? That was the question which Hamilton kept asking himself during the fifteen years from 1828 to 1843. The reason the issue presented great difficulty was that no matter what "common sense" definition Hamilton attempted, he could not get the resulting vector "multiplication" to satisfy certain essential criteria as well as all the fundamental laws of ordinary algebra. In the course of his reasoning he decided to embed the 3-dimensional vectors in the set of quaternions, that is, to consider those vectors as the ordered *quadruples* (0, b, c, d). Then if quaternion multiplication is defined, it will apply to those special quaternions which are 3-dimensional vectors.

But if quaternions are to include the real numbers, now defined as (a, 0, 0, 0), and the complex numbers (plane vectors), now defined as (a, b, 0, 0), quaternion multiplication must be so defined that it will reduce to real and complex multiplication for the special quaternions just illustrated. This factor, plus the needs of space physics, plus the fundamental laws of algebra, gave Hamilton a very exacting set of conditions. Day after day, week after week, year after year, he would sit for hours in his study, forgetting about meals, brooding, struggling and striving to formulate the elusive definition. If he appeared at the breakfast table, his wife and son would immediately inquire, "Have you succeeded in multiplying quaternions yet?"

Finally, and most unwillingly, Hamilton recognized what to do. He could satisfy some of the most important requirements of physics and algebra by defining a *noncommutative* multiplication. His action in doing so was analogous to Lobachevsky's procedure in abandoning Euclid's parallel postulate and substituting a different axiom.

Full details of Hamilton's quaternion algebra will appear in a later chapter. Here we shall merely explain his definition of multiplication. Now, a plane vector has the alternative forms (a, b) and $a + bi$, where the latter form is the sum of real multiples of the "units," 1 and i. If $b = 0$, one obtains the real numbers, a, and if $a = 0$, one has the imaginaries, bi. Similarly a quaternion can be expressed as an ordered quadruple of real numbers or as a hypercomplex number, $a + bi + cj + dk$, where the latter form is the sum of real multiples of the "units," 1, i, j, and k. If $b = c = d = 0$, one obtains the real numbers. If $c = d = 0$, one has the complex numbers. If a = 0, one has the 3-*dimensional vectors*, $bi + cj + dk$.

For agreement with complex multiplication, it is necessary to have $i^2 = -1$. Hamilton considered the units j and k to be akin to i and therefore defined $j^2 = -1$ and $k^2 = -1$. Suppose that one is to carry out the quaternion multiplication

$$(1 + i - j + k)\,(3 - i + j - k)$$

Then application of the distributive law will lead to a product containing terms like ij, ji, jk, kj, etc. If quaternion multiplication is to be *closed*, all such terms must be defined to be of the form $a + bi + cj + dk$. Hamilton accomplished this by providing the following multiplication table.

	1	i	j	k
1	1	i	j	k
i	i	-1	k	$-j$
j	j	$-k$	-1	i
k	k	j	$-i$	-1

He had poured so much of his mind and heart and soul into quaternion multiplication that one day, when he was out walking with his wife, he felt impelled to carve the above tabulation on the stone of a bridge near Dublin.

In the table, the quaternion expression for the product ij can be found in "row i" and "column j." Thus, $ij = k$. Reading the entry at the intersection of "row j" and "column i" reveals that $ji = -k$. Here is our first example of the noncommutativity of quaternion multiplication. The product ji is not equal to ij. In fact, $ji = -ij$. The reader can consult the table to verify that $ik = -ki$, $kj = -jk$. Also, by using the table he can carry out some simple quaternion multiplications like $(2 - k)(i + j) = 3i + j$ and $(i + j)(2 - k) = i + 3j$ to show that reversing the order of quaternion factors can lead to two completely different products.

Again, it is easy to show that if the domain of the variable x is the set of quaternions, then the polynomial equation

$$x^2 + 1 = 0$$

has a solution set containing the quaternions $\pm i, \pm j, \pm k, \frac{1}{2}\sqrt{2}j + \frac{1}{2}\sqrt{2}k$, and an infinite number of other solutions. The corollary to the fundamental theorem of classical algebra permits a maximum of *two complex solutions*, and those are the answers $\pm i$. But infinitely many *hypercomplex* solutions exist.

More of the physics of Hamilton's quaternions will be discussed later (Chapter 28). Pure mathematicians, however, were less interested in applications than in extending his ideas further—to other types of hypercomplex number, and in examining the effects of relinquishing algebraic postulates other than the commutative law of multiplication. For example, the British mathematician Arthur Cayley (1821–1895), generalizing Hamilton's notion of pairing, considered hypercomplex numbers that are *ordered pairs of quaternions*, or ordered "8-tuples" of real numbers. Cayley defined a multiplication of these hypercomplex numbers that was neither commutative nor associative, thus abandoning another fundamental law of common algebra.

It should also be mentioned that in 1844, just a year after Hamilton published the details of his quaternion algebra, the German geometer Hermann Grassmann (1809–1877) formulated a far more general concept of algebras of hypercomplex numbers of all orders (Chapter 28). Grassmann's algebras have a multiplication that is noncommutative. In fact, Hamilton's quaternions became just one special algebra among the many Grassmann species.

Because Hamilton holds chronological priority in the creation of nontraditional algebras, he is often considered the founder of modern "abstract algebra." That term applies not only to hypercomplex number algebras like those of Hamilton, Cayley, and Grassmann, but to all systems where postulates differ from the rules of the game of classical algebra, by being either less or more restrictive. But it is not this feature that makes the new algebra "abstract," but rather the fact that it is conceived from the axiomatic point of view discussed in our previous chapter, and suggested right from the start of this book in all our discussions of the fundamental properties of arithmetic operations. But the idea of treating algebra as a pure postulational-deductive science did not really begin until the period just before Hamilton. Then it was gradually developed by members of the British algebraic school, namely, George Peacock (1791–1858), Duncan Farquharson

Gregory (1813–1844), Augustus De Morgan (1806–1871), and George Boole (1815–1864).

The modern abstract point of view requires a pure science, whether it is an "algebra" or a "geometry" or something else, to be founded on postulates (assumptions) about undefined elements, which are *not* necessarily "numbers" or "points" but *abstractions*, potentially capable of varied interpretations consistent with the basic assumptions. With this in mind, we can gain some idea of the variety of modern algebraic systems or "structures" if we "purify" the postulates for common algebra, and play what we shall call Hamilton's game* of relinquishing some postulates, or replacing them by different ones. In the previous chapter we saw that Lobachevsky played a more restricted "game" in changing only one postulate of Euclidean geometry.

One might begin with systems in which S, a set of undefined elements, is given, as well as two binary operations on these elements, $\oplus$ and $\otimes$. These symbols are used to suggest some kinship with ordinary addition and multiplication, although in different concrete realizations the interpretations might be considerably different from the usual ones. Then our postulate set can include those "laws" or "properties" we have mentioned so frequently; that is, we can assume closure, commutativity, and associativity for both $\oplus$ and $\otimes$, and also distributivity of $\otimes$ with respect to $\oplus$. We have seven postulates thus far. To make sure our fundamental set contains elements akin to *zero* and *one*, the additive and multiplicative "identities," we shall use the symbols Z and I, and make two more assumptions:

(1) S contains an element Z such that $a \oplus Z = a$, if a is *any* element in S.
(2) S contains an element I (different from Z) such that $a \otimes I = a$, if a is *any* element of S.

Then to avoid the necessity for introducing "subtraction" and "division" operations, we introduce additive and multiplicative inverses by postulating:

(3) For each element a in S, there exists an element $\bar{a}$ in S such that $a \oplus \bar{a} = Z$.
(4) If a is any element of S with the exception of Z, there exists an element a^{-1} such that $a \otimes a^{-1} = I$.

Any system satisfying the eleven postulates of common algebra is called a *field*. Because it is associated with common algebra, the field is the most familiar of algebraic structures. Thus the rationals or the reals or the complex numbers each constitute a field if the binary operations are ordinary addition and multiplication. The reader might try his hand at showing that the set of all numbers of the form $a + b\sqrt{2}$, where a and b are rational, is a field under $+$ and $\times$. But the postulates for a field are not categorical (Chapter 3); there are fields where the fundamental set S is *finite* and, in addition, $\oplus$ and $\otimes$ are *not* ordinary addition and multiplication. Such finite fields are called *Galois fields*, in honor of Évariste Galois (1811–1832),

---

* This must not be confused with the "Hamiltonian game" discussed in books on mathematical recreations and described in Chapter 25 of the present work.

who first studied their properties and who, as the next chapter will indicate, accomplished the liberation of algebra in a way essentially different from Hamilton's.

If, playing Hamilton's game, we delete one or more of the postulates for a field, we shall arrive at a system that is less restricted or more general. If, for example, we remove (2) and (4) above, the nine remaining axioms define what algebraists today call a *commutative ring*. If, further, the commutativity of multiplication is relinquished, a more general type of ring results. We shall illustrate such more general rings later and shall also indicate the nature of certain special *subrings* studied by Emmy Noether, to which Dedekind had given the subtle and beautiful name *ideals*.

Returning to familiar territory, let us examine the system consisting of the integers, $\{\ldots, -3, -2, -1, 0, +1, +2, +3, \ldots\}$, and the binary operations of ordinary addition and multiplication which, as we can see, satisfy all the field postulates except (4), the assumption of multiplicative inverses within the integral set. Thus the system is *not* a field, because only $+1$ and -1 have multiplicative inverses that are integers. Therefore one cannot carry out, within the set of integers, the Babylonian scheme of substituting a suitable product for any quotient of number pairs in the set. But we know that there are integral answers for some quotients like $12 \div 3$, $30 \div 5$, etc. Putting this fact into algebraic form, $3a = 12$ or $3 \times a = 3 \times 4$ yields $a = 4$, and if c is any integer except zero, $ca = cb$ yields $a = b$ by a process of *cancellation*. Then the addition and multiplication of integers obey the first ten field postulates and a postulate permitting cancellation.

Therefore, playing Hamilton's game, we define an abstract system consisting of a set S and binary operations $\oplus$ and $\otimes$ which are governed by the first ten field postulates and the

> *Cancellation Law:* If $c \otimes a = c \otimes b$, where a, b, and c are any members of the set S such that $c \neq Z$ (the "zero"), then $a = b$.

Such a structure is called an *integral domain* (because of its kinship to the integers).

One can make more drastic deletions from the field postulates, for example, by removing one of the binary operations and all the postulates which refer to that operation. Suppose that we remove $\oplus$ and limit ourselves to $\otimes$. Also, in true Hamiltonian spirit, let us abandon the commutative postulate for $\otimes$. What is left is a set S, with an operation $\otimes$ satisfying closure, associativity, and the postulates we have labeled (2) and (4), namely, existence within S of a multiplicative identity (*unit element*) and multiplicative inverses. (Because commutativity has been relinquished, the statements for these postulates must be enlarged to include $I \otimes a = a$ and $a^{-1} \otimes a = I$.) Any system with a single binary operation satisfying the four specified postulates is called a *group*. The group concept is of inestimable importance in modern mathematics and physics, so much so that the great Henri Poincaré (1854–1912), whose ideas will be discussed at many points of our book, once brashly said, "The theory of groups is *all* of mathematics." At any rate, we shall be alluding to groups in so many later chapters that we shall not give specific references at this point.

We remark that there are *commutative groups* in which $\otimes$ satisfies the commutative law. Such groups are described as *abelian* in honor of the Norwegian Niels Henrik Abel (1802–1829). The fact that the adjective is not capitalized is testimony of the frequency of its occurrence in mathematical literature. If, instead of adding a fifth postulate to the four group postulates, we were to relinquish (2) and (4), we would be defining a *semigroup*, a system consisting of a set S and a single binary operation which satisfies closure and associativity. If one also abandons associativity, the system becomes a *groupoid*.

If Hamilton had never succeeded in multiplying 3-dimensional vectors, he would still have been able to add them and perform *scalar* multiplication on them, and the same would have been true for vectors of higher dimension. Moreover, the operations would have satisfied the traditional properties. Thus there is a theory of *vector spaces*, which are systems with two fundamental sets, V (the vectors) and F (a field of scalars), with a vector addition and scalar multiplication satisfying the same postulates as those for physical vectors and scalars. If a third operation is defined on a vector space, namely, a *vector multiplication* (which may possibly be neither commutative nor associative), one arrives at an *algebra* in the spirit of Hamilton or Grassmann or their successors.

After our brief excursion into modern abstract algebra, let us now recall that this new area of mathematical thought is just the ultimate outgrowth of the number extensions of ordinary algebra. Even if abstract algebra is all-embracing and is probably the favorite speciality of pure mathematicians today, classical algebra is still of practical importance. Therefore our next chapter will recapitulate slightly, and then examine both the practical procedures and the additional generalizations associated with the solution of polynomial equations.

5

Equations, Human and Inhuman

To make good the promise of the previous chapter, we shall now consider the evolution and nature of techniques for solving polynomial equations. Throughout our discussion it will be assumed, unless a statement to the contrary is made, that wherever constants a, b, c, etc. are mentioned, they specify numbers in the complex domain and also that the replacement set for all variables x, y, z, etc. is the same complex domain. There will be no quaternions, Cayley numbers, or other "hyper-complex" entities in this chapter.

Since decipherment of cuneiform tablets is providing increasing evidence that all mathematics started with the Babylonians, who, in addition, had a particular penchant for algebra, one would expect to hear that they contributed to the methodology of solving polynomial equations. They did, in fact, evolve formulas for the solution of first- and second-degree equations, apply some truly advanced ideas to cubic (third-degree) equations, and solve some systems of equations in two unknowns. They even found solutions in the real domain for *eighth*-degree equations like

$$x^8 - 17x^4 + 16 = 0$$

and we shall subsequently examine their procedure for doing so.

Sometimes one can guess or estimate solutions, and such hunches can always be checked by substitution for the variable. For example, it is easy to verify that $x = +1$ and $x = -1$ are solutions of the above equation. Are there other solutions? The corollary to the fundamental theorem of algebra (Chapter 4) tells us that there may possibly be as many as eight distinct solutions. Perhaps some readers can guess the other roots of the above equation. But, in general, how can we guess or estimate answers, with the vastly infinite real number continuum (Chapter 2) from which to make a possible selection and the much greater *embarras de choix* if one makes guesses that are complex numbers, that is, *pairs* of real numbers? In most equations, to track down solutions without specific clues or standardized procedures would be infinitely worse than searching for the proverbial needle in the haystack!

But to descend from the sublime to the elementary, it would seem logical to begin by solving equations of *first* degree. In the case of such equations, for example,

$2x - 1 = 0$ or $3x + 2 = 0$, it is easy to guess and check the solutions $x = 1/2, x = -2/3$, respectively.

One can readily generalize such results to obtain a formula for solving *all* polynomial equations of the first degree. If b is any constant in the complex domain and a is any such constant except zero, then

$$ax - b = 0$$

has the solution

$$x = \frac{b}{a}$$

This is verified by the substitution

$$a \cdot \frac{b}{a} - b = b - b = 0$$

Moreover, the solution $x = b/a$ is *unique*, that is, the complete solution set for the above equation is $\{b/a\}$. This is true by virtue of the corollary to the fundamental theorem of algebra. That corollary states that a polynomial equation of first degree cannot have more than one root. Finally, as examples of how our formula can be applied, we state that the solutions of $5x + 3 = 0$, $3x - 2\sqrt{7} = 0$, $\frac{1}{2}x + 3i - 4 = 0$ are respectively $-3/5$, $2\sqrt{7}/3$, $8 - 6i$.

If all polynomial equations of first degree can be solved simply by formula, the reader may well wonder why so much time is given to first-degree equations in school algebra. To see what may be involved, consider the following problem, which the Hindu Mahavira (*ca.* 850 A.D.) addressed to his students: Of a basket of mangoes, the king took 1/6, the queen 1/5 of the remainder, and the three oldest princes 1/4, 1/3, and 1/2 of the successive remainders. There were then only three mangoes left for the youngest member of the royal family. Tell me, all you who are clever in such problems, how many mangoes there were in the basket.

We see that in such problems, practical or fanciful, the first step, which may not be easy, is to express the question in algebraic form. If we let x represent the unknown number of mangoes, then $1/6\ x$ will represent the number taken by the king, and $5/6\ x$ the remaining number. But the queen takes 1/5 of this remainder or $1/5 \cdot 5/6\ x = 1/6\ x$, and after her share (and the king's) have been removed, only $4/6\ x$ mangoes remain. The first prince takes $1/4 \cdot 4/6\ x = 1/6\ x$. It is readily seen that the other princes also take $1/6\ x$ each. Totaling the five shares, we find that $5/6\ x$ mangoes have been removed, leaving only $1/6\ x$ for the youngest member of the royal family. Therefore

$$\frac{1}{6}x = 3$$

We can now guess that $x = 18$ mangoes.

But suppose that a verbal problem leads to a first-degree equation like

$$7x - 2 = 4x - 11$$

Then it may not be so easy to guess the solution. If one wishes to solve by formula, it will be necessary to deduce from the given statement an equation in the standard form

$$ax - b = 0$$

Until recently there was a tendency to mechanize the deductive process in algebra by using certain rules concerning "transposition and collecting similar terms." In fact, the word "algebra" derives from just such procedures carried on, of all places, at the court of Harun-al-Rashid, Caliph of the Arabian Nights. This Moslem ruler and his son Al-Mamun encouraged scientific activity. In the royal entourage there was the mathematician Al-Khowarizmi, who wrote a text called *al-jabr w'almuqabalah*, meaning transposition and condensation of terms in an equation. The abbreviated title, *al-jabr*, became our *algebra*.

Instead of using mechanical *al-jabr w'almuqabalah*, let us see how one can deduce the transformation of $7x - 2 = 4x - 11$ into the "canonical form," $ax - b = 0$. The idea is to obtain *zero* in the right member of the equation, and we shall do this by adding 11 to both sides (to eliminate -11 on the right) and then adding $-4x$ (in order to eliminate $4x$). This is permissible because the given equation states that $7x - 2$ and $4x - 11$ are symbols for the same number. Hence if we add $11 - 4x$ to that number, there will be a unique sum (by the closure property of addition of real numbers) even though the two members of the equation lead to different expressions for that sum. Hence, adding 11 first, we have

$$(7x - 2) + 11 = (4x - 11) + 11$$

By virtue of the associative law of addition the above equation can be expressed as

$$7x + (-2 + 11) = 4x + (-11 + 11)$$

or

$$7x + 9 = 4x$$

Adding $-4x$ to each member (or adding each member to $-4x$, which is the same thing in accordance with the commutative law of addition),

$$-4x + (7x + 9) = -4x + 4x$$

Once more the associative law permits a change in the left member to yield

$$(-4x + 7x) + 9 = -4x + 4x$$

or

$$3x + 9 = 0$$

which is a first-degree equation in "canonical form."

The early mathematicians in general, as well as students and laymen throughout the ages, have been as much interested in recreational problems as in the practical uses of algebra. Nevertheless, at an early era, some algebraic equations arose from more serious issues. Second-degree or *quadratic* equations played an

important role in even the simplest questions of buying and selling in ancient Babylonia. This was the result of the Mesopotamian method of quoting prices, which
was the reverse of our own. Our authority in the matter is the eminent Dutch
scholar, Professor E. M. Bruins.

We would quote the price of ground or "pearled" barley as 20 cents per package, while the Babylonians spoke of 5 sacks of barley per shekel. We point to the
package first and then give a certain amount of money, the number of cents depending on the particular item. In Mesopotamia the purchaser indicated the money first
and then received a certain amount of merchandise, depending on the amount
of money offered. As a result, when *exact* answers were required for some
elementary questions of profit and loss, this called for the solution of quadratic
equations.

Thus, suppose that experience had taught a Babylonian merchant that he
could dispose of 120 sacks of barley at market time. Then he might have asked
himself, "What can I afford to pay the farmer for barley if I am to profit 10 shekels
on 120 sacks with a difference of 2 sacks per shekel between cost and selling
price?"

Using modern algebraic symbolism, let

$x =$ the "cost," that is, the number of sacks received by the merchant for
each shekel he gave the farmer

$x - 2 =$ the "selling price," the number of sacks received by a customer in the
market for each shekel he gave the merchant

$\dfrac{120}{x} =$ the number of shekels spent by the merchant

$\dfrac{120}{x - 2} =$ the number of shekels spent by the customers and received by the
merchant

Then the fact that receipts are to exceed expenditures by 10 shekels can be put in
the form of the equation

$$\frac{120}{x-2} - \frac{120}{x} = 10$$

Having simplified this by various algebraic manipulations, we arrive at the quadratic
equation

$$x^2 - 2x - 24 = 0$$

Algebraists, starting with the Babylonians, developed a formula for solving
quadratic equations, and subsequently we shall see how such a formula can be
derived and applied. But in the above equation one can proceed simply by expressing it as

$$(x - 6)(x + 4) = 0$$

This form asserts that the product of two (unknown) numbers, $x - 6$ and $x + 4$, is zero.

If we were asked to illustrate a zero product using *known* numbers, we might cite $0 \cdot \frac{1}{5}$, $-2 \cdot 0$, $0 \cdot \sqrt{-3}$, $0 \cdot 0$, etc. These examples illustrate a law governing the arithmetic of real and complex numbers: The product of two numbers is zero if and only if at least one of these numbers is zero.

Therefore at least one of the above unknown numbers must be zero, that is,

$$x - 6 = 0 \qquad \text{or} \qquad x + 4 = 0$$

and

$$x = 6 \qquad \text{or} \qquad x = -4$$

Both 6 and -4 are solutions because they will check when substituted in the quadratic equation. But only the positive answer $x = 6$ would be meaningful in the Babylonian merchant's problem because x stood for the number of sacks of barley he obtained in exchange for each shekel he gave to the farmer. The solution $x = 6$ says that he did receive 6 sacks per shekel, and since he acquired 120 sacks in all, he must have spent 20 shekels. For every shekel, his customers received $x - 2 = 4$ sacks. Hence the customers' expenditures, or the merchant's receipts, would have been 30 shekels, and the latter's profit would have been 10 shekels.

The method of solution of a quadratic by factoring and equating factors to zero is due to Thomas Harriot (1560–1621), who, as a young man, had been a member of one of Sir Walter Raleigh's expeditions to Virginia. After Harriot had completed an extensive geodetic survey in the new land, he returned to England and gave his scientific efforts to algebra and astronomy.

Both before Harriot's time and after it, algebraists actually used his method and, in fact, applied it to polynomial equations of higher degree. Thus, by Gauss's day, there was the following theoretic result, a corollary to his fundamental theorem (page 71): *A polynomial of the nth degree with complex coefficients is factorable uniquely as follows:*

$$c(x - r_1)(x - r_2) \ldots (x - r_n)$$

where c, r_1, r_2, ... , r_n are complex numbers.

Formerly this corollary was often stated in the form: Any polynomial equation of the nth degree with complex coefficients has n complex solutions. This form of the proposition is sometimes confusing, however, since the n solutions may not all be distinct. Thus a cubic polynomial may be factorable as

$$(x - 2)(x - 2)(x - 2)$$

and the corresponding cubic equation,

$$(x - 2)(x - 2)(x - 2) = 0$$

has as solutions

$$x = 2 \qquad x = 2 \qquad x = 2$$

The cubic polynomial does satisfy the first statement of the corollary, since there are three factors. However, these factors are all the same, and hence the equation has only one solution. Therefore we have, in the previous chapter, stated the corollary in the alternative form: *Every nth degree polynomial equation with complex coefficients has at most n distinct solutions in the complex domain.*

But to return to Harriot's technique, it would not be readily applicable to quadratic equations in general. For example, the reader might find it difficult to factor the left member of $x^2 + 2x - 4 = 0$. To solve such equations, other methods are preferable. The cuneiform textbook tablets gave verbal instructions that are equivalent to the quadratic formula of today's algebra. We do not know how the Babylonians arrived at this formula, but it is generally believed that they derived it by the method of *completing the square*, which was also used by Greek and Arab mathematicians. To illustrate this technique, we consider the equation that would be difficult to solve by Harriot's method.

The quadratic $x^2 + 2x = 4$ signified to the Greeks that the area of a certain figure was 4 square units. That figure was pictured (Figure 5.1) as composed of a

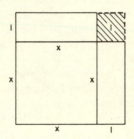

Figure 5.1

square with unknown side x (and area x^2), and two rectangles, each with length x and width 1 unit. The area of each rectangle is x square units. Hence the area of the entire figure is $x^2 + 2x$. Now to complete the figure so as to form a square, the shaded area must be added to it. This shaded area is a square whose side is 1 unit, and whose area is 1 square unit. Then the area of the completed square is $4 + 1 = 5$ square units. Thus, when 1 is added to each member of the original quadratic equation, the result is

$$x^2 + 2x + 1 = 5$$

Referring to Figure 5.1 once more, the side of the completed square is $x + 1$, and therefore its area is $(x + 1)^2$, making

$$(x + 1)^2 = 5$$

If the area of the large square is 5 square units, its side must have the irrational length $\sqrt{5}$ linear units and therefore

$$x + 1 = \sqrt{5}$$
$$x = \sqrt{5} - 1$$

The answer just obtained is the only one that Greek and Arab algebraists would have given. But when negative numbers were understood and, in addition, solution was no longer dependent on geometric pictures, there was the alternative,

$$x + 1 = -\sqrt{5}$$

and

$$x = -\sqrt{5} - 1$$

The second solution is both *negative* and *irrational*.

Further liberation from geometry is called for if one is to solve the quadratic equation $x^2 + 4x = -5$. One can, of course, be helped by making a diagram (Figure 5.2). But the figure is merely an auxiliary scheme, since no physical or geometric

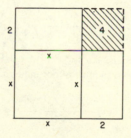

Figure 5.2

square can have a total area measuring -5 square units. But, in the abstract algebraic sense, one can "complete the square" to obtain

$$(x + 2)^2 = -5 + 4$$

or

$$(x + 2)^2 = -1$$

Therefore

$$x + 2 = \sqrt{-1} = i \qquad x + 2 = -\sqrt{-1} = -i$$

and

$$x = -2 + i \qquad x = -2 - i$$

The solutions are *complex numbers*.

Although the later Babylonians had some understanding of negative numbers, they never considered imaginary or complex numbers. But they apparently completed the square in the *general* equation

$$x^2 + Bx + C = 0$$

In other words, they carried out algebraic manipulation equivalent to

$$x^2 + Bx = -C$$
$$x^2 + Bx + \frac{B^2}{4} = \frac{B^2}{4} - C$$
$$\left(x + \frac{B}{2}\right)^2 = \left(\frac{B}{2}\right)^2 - C$$
$$x = -\frac{B}{2} \pm \sqrt{\left(\frac{B}{2}\right)^2 - C}$$

This is a formula for solving all quadratic equations. If the given equation is $2x^2 - 2x - 1 = 0$, one need merely divide both sides by 2 in order to obtain $x^2 - x - 1/2 = 0$, to which the Babylonian formula is applicable. But if one desires a formula for solving

$$ax^2 + bx + c = 0$$

where $a \neq 0$, he can divide by a to obtain

$$x^2 + \frac{b}{a}x + \frac{c}{a} = 0$$

Then substituting b/a for B and c/a for C in the Babylonian solution yields the quadratic formula that appears in most of our modern elementary texts, namely,

$$x = \frac{-b \pm \sqrt{b^2 - 4ac}}{2a}$$

With the Greeks and those continuing their tradition, problems in algebra involved unknown lengths, areas, and volumes—our x, x^2, x^3. There were no geometric pictures for x^4, x^5, x^6, etc. But the Babylonians were not limited by diagrams and were therefore not afraid to consider the eighth-degree equation cited at the beginning of this chapter. What they did was reduce the problem of solving

$$x^8 - 17x^4 + 16 = 0$$

to the more elementary question of solving a quadratic. By the substitution, $x^4 = z$ (and hence $x^8 = z^2$), the above equation is transformed into

$$z^2 - 17z + 16 = 0$$

We use Harriot's method and obtain

$$(z - 1)(z - 16) = 0$$
$$z = 1 \qquad z = 16$$

or

$$x^4 = 1 \qquad x^4 = 16$$

What the Babylonians did was to cascade quadratics, that is, repeat the previous type of transformation, by setting $w = x^2$, whence

$$w^2 = 1 \qquad w^2 = 16$$

Therefore

$$w = +1 \qquad w = -1 \qquad w = +4 \qquad w = -4$$

which means

$$x^2 = +1 \qquad x^2 = -1 \qquad x^2 = +4 \qquad x^2 = -4$$

Finally there are the *eight* roots

$$x = 1, \; -1, \quad i, \; -i, \quad 2, -2, \quad 2i, \; -2i, \qquad \text{where } i = \sqrt{-1}$$

(The Babylonians would not have recognized the imaginary solutions.)

Our illustration was an exceptionally simple instance, produced *ad hoc*, as were all the Babylonian eighth-degree equations. We shall see that, in general, eighth-degree equations *cannot* be solved in the elementary sense. In fact, most polynomial equations of degree higher than four cannot be solved in finite terms. What we mean is that there are no answers expressible in terms of the coefficients of the equation if one is limited to a *finite* number of additions, subtractions, multiplications, divisions, and root extractions performed on the coefficients. But we are getting ahead of our story, for what we have just stated was not given a complete and rigorous proof until the nineteenth century, when the young genius Évariste Galois furnished a profound and elegant demonstration (Chapter 16).

Solving an "easy" eighth-degree equation is just a matter of cascading quadratics, as we have just seen. But what issues arise between second- and eighth-degree equations? What about third-degree or cubic equations, since they are next in rank after quadratics? Great algebraists as they were, the Babylonians were naturally interested in the question. Their tablets show many cubic equations of a specialized type illustrated by

$$x^3 + x^2 = 80$$
$$x^3 + x^2 = 12$$
$$x^3 + x^2 = 1100$$
$$x^3 + x^2 = 36$$

If these problems are verbalized, the statement of all of them is seen to be the same except for the final word: Find a number such that if its cube is added to its square, the sum is—(80 or 12 or 1100, etc., as the case may be).

The Babylonians had tables of squares and cubes and merely combined them to form tables of $x^3 + x^2$, thus:

Number	$x^3 + x^2$
1	2
2	12
3	36
4	80
5	150
6	252
.	.
.	.
.	.
10	1100

Then the answers to the illustrative cubics can be read directly from the table and found to be 4, 2, 10, 3, respectively. But suppose that the problem assigned is

$$x^3 + x^2 = 10$$

Evidently 10 is not in the table of values of $x^3 + x^2$, but since 10 is smaller than 12 (the value of $x^3 + x^2$ for $x = 2$), therefore a number slightly below 2 will solve $x^3 + x^2 = 10$. We might use "linear interpolation" to read between the lines of the table, and say that since 10 is 8/10 of the way from 2 to 12, we estimate crudely that the corresponding value of x will be 8/10 of the way from 1 to 2, or $x = 1.8$. Since the Babylonians carried out interpolative processes freely in their other tables —square root, interest, astronomical data, etc., it is conjectured that they probably did so in the tables of $x^3 + x^2$. Now 1.8 is only an estimate. If we substitute 1.8 for x, then $x^3 + x^2 = 9.1$ approximately. If we had the exact answer, the sum would be 10, not 9.1. Hence 1.8 is too small an estimate. If we try a slightly larger approximation, $x = 1.9$, then $x^3 + x^2 = 10.5$, which is closer to 10 than 9.1. The true value of x lies between 1.8 and 1.9 but if we are approximating to the nearest tenth, $x = 1.9$ might seem to be a better approximation than $x = 1.8$.

To see how a simple practical problem may lead to a cubic equation, consider that an open rectangular box with a square base is to be constructed from 27 sq. in. of cardboard and is to have a volume of 13.5 cu. in. Let

x = the length of the base edge

$\dfrac{13.5}{x^2}$ = the height of the box (Height × length × width = 13.5, and therefore height × x × x = 13.5)

$(x)\left(\dfrac{13.5}{x^2}\right)$ = the area of one lateral face of the box

The fact that the total surface consisting of the square base and the four lateral faces measures 27 sq. in. leads to the equation

$$x^2 + 4\left(\frac{13.5}{x}\right) = 27$$

or

$$x^3 - 27x + 54 = 0$$

It is easy to obtain the root $x = 3$ by trial. Substitution in the equation will verify this answer. Therefore the dimensions of the box should be 3 in., 3 in., 1.5 in. When Harriot's method is applied to the above cubic, it is expressed as

$$(x + 6)(x - 3)(x - 3) = 0$$

and hence there are two distinct answers, namely, $x = -6$ and $x = 3$. The negative root is not a suitable answer, however, for the box problem.

The above illustration was selected to indicate that geometric problems where volume is involved may lead to cubic equations. However, the need to solve cubics is not limited to such issues. There are many situations in modern applied mathematics where such equations must be solved. To give a single example: In physical chemistry, Van der Waals' equation for the specific volume, x, of carbon dioxide, when the numerical value of the pressure is 70 atmospheres and the absolute temperature is $300°$ C, is as follows:

$$7x^3 - 64x^2 + 224x - 260 = 0$$

The reader can check by substitution that $x = 2.3$ is an approximate root of this equation.

Neugebauer believes that the Babylonians went far beyond the elementary phase of cubics, but other scholars in the field do not agree. It is possible, if one knows sufficient algebra, to perform a manipulative stunt that will transform *any* cubic into the "canonical" form

$$x^3 + x^2 = k$$

Then the Babylonian tables, if extended sufficiently, can be used to obtain exactly or approximately any positive roots that such an equation may have. For example, with the cubic

$$4y^3 + 20y^2 + 33y - 22 = 0$$

the substitution $y = 1/2\, x - 3/2$ will transform the equation into

$$4\left(\frac{1}{2}x - \frac{3}{2}\right)^3 + 20\left(\frac{1}{2}x - \frac{3}{2}\right)^2 + 33\left(\frac{1}{2}x - \frac{3}{2}\right) - 22 = 0$$

When this is simplified, it becomes

$$x^3 + x^2 = 80$$

and the tables indicate that $x = 4$ is an answer. Then

$$y = \frac{1}{2}x - \frac{3}{2} = \frac{1}{2}(4) - \frac{3}{2} = \frac{1}{2}$$

should be an answer to the equation in y. A substitution of $y = 1/2$ in that equation will verify that this is so.

But the facts involved in deciding what transformation will bring a general cubic to the above canonical form were first developed, as far as we know, in the sixteenth century A.D. The Babylonian mathematicians may have used sheer intuition and ingenuity, but if they possessed the necessary theoretical knowledge, their mastery of algebra was even more incredible than that revealed in the cuneiform tablets deciphered to date.

If we solve a cubic by modern methods, it is easier to transform it to the standard form

$$x^3 + ax + b = 0$$

where the x^2 term is absent (not the x term as in the preceding normal form). Before we examine how this is accomplished, however, let us discuss some of the early history of cubic equations.

Among Greek mathematicians who solved special cubic equations by the use of geometric algebra, there were Menaechmus (*ca.* 350 B.C.), tutor to Alexander the Great, and Archimedes (287–212 B.C.), whose name is known to all. The Persian poet, mathematician, and astronomer, Omar Khayyám (*ca.* 1100) is the next name in the story of cubics. Although he used geometric algebra and never arrived at a general cubic formula, Omar bin Ibrahim al-Khayyami created techniques of solution for thirteen types of third-degree equation. For this and the applied mathematics by which he devised a calendar superior to our own, Omar is considered to be one of the most original of the Moslem mathematicians. As for the the algebra of cubics, his ideas were among the greatest contributions between the fifth and sixteenth centuries.

The geometric methods of Menaechmus, Archimedes, and Omar are still valuable for special problems, and we shall examine some of them in a subsequent chapter. What mathematics demands, however, is a technique that will solve any and all cubics. The algebraic activity of Italian mathematicians in the sixteenth century provided such a method and produced the formula

$$x = \sqrt[3]{-\frac{1}{2}b + \sqrt{R}} + \sqrt[3]{-\frac{1}{2}b - \sqrt{R}} \text{ where } R = \left(\frac{b}{2}\right)^2 + \left(\frac{a}{3}\right)^3$$

for the solution of

$$x^3 + ax + b = 0$$

where a and b are any *real* numbers. Earlier in this chapter it was shown how a general cubic can be transformed into the Babylonian "canonical form." It is even simpler to change a general cubic to the standard form above, which is the one needed before the general formula can be applied. For example, in the complete cubic

$$y^3 - 3y^2 - 2y - 2 = 0$$

the substitution $y = x + 1$ will transform the equation into

$$(x + 1)^3 - 3(x + 1)^2 - 2(x + 1) - 2 = 0$$

When this is simplified, it becomes

$$x^3 - 5x - 6 = 0$$

Then if -5 is substituted for a and -6 for b in the cubic formula, the root

$$x = \sqrt[3]{3 + \sqrt{\frac{118}{27}}} + \sqrt[3]{3 - \sqrt{\frac{118}{27}}}$$

is obtained. Because $y = x + 1$, a root of the original complete cubic can be found merely by adding 1 to the answer for x. There are two other roots for either cubic and a modification of the above formula will produce these. Since the slight change in formula involves imaginary numbers and may appear difficult to a reader who is not well acquainted with their manipulation, we shall not indicate the procedure here. The appearance of the exact value for x is rather formidable, but the whole point is that this root is *exact*, and not an approximation interpolated in a table or read from a graph. While the pure mathematician desires precise formulation, a practical worker might still prefer a good approximation, and he could obtain this by substituting values obtained from a table of square roots and cube roots in the above formula. If a reader wishes to try this, his answer should be $x = 2.7$ (to the nearest tenth) and hence $y = 3.7$, answers which can be checked by substitution in the equations above.

A strange fact will now appear—namely, when a cubic is specially easy in the sense that its three roots are all *real* and distinct, the general formula leads to an "irreducible" result. This is one of the reasons why a geometric approach may be more revealing and why we shall return to it again in a later chapter. When all is said and done, modern physicists, engineers, statisticians, and applied mathematicians in all fields usually solve all but the simplest cubic equations by *graphic* methods or "algorithms" equivalent to such methods. In American textbooks the two most favored algorithms are due to Newton and William George Horner (1786–1837). Although Horner does not have status as a mathematician, his name has become well known through its association with school mathematics. He was a teacher at Bath, England, when he discovered his algorithm. We now know that a similar method had been used in China and that Paolo Ruffini (1765–1822) had formulated virtually the same technique around 1804. But Horner discovered it independently, presented it in a paper to the Royal Society in 1819, and had full details published (of all places!) in the *Ladies' Diary* for 1838.

The geometric spirit of the Greeks continues in modern (approximate) methods of equation-solving, but the analytic geometry of Fermat and Descartes (Chapter 7) has made it possible to provide geometric pictures not only for cubics, but also for equations of higher degree, and even for "transcendental" (nonalgebraic) equations. Therefore, as far as concepts and practicality are concerned, nothing novel has been added to the approximate geometric solution of cubics since the days of Omar.

　　But the achievement of the general cubic formula, from which a general formula for fourth-degree (quartic or biquadratic) equations soon followed, was more than a milestone in algebra. It spelled *finis* to the classic period in that subject, although mathematicians were not to realize until the late eighteenth and early nineteenth century that it was futile to search for formulas that would solve algebraic equations of degree higher than the fourth. The reason for this will appear in Chapter 16. To return to the cubic formula, if R is negative, then $\sqrt{R}$ will be imaginary. There is no objection to this except that in the theory of equations it is proved that R will be negative when the cubic has *three real roots* that are distinct. In this "easy" case, the general formula furnishes the answer in terms of the cube roots of imaginary numbers, and there is no simple algebraic method for extracting such cube roots. For example, one can readily verify that -6 is a root of

$$x^3 - 63x - 162 = 0$$

Yet the general formula furnishes the answer

$$x = \sqrt[3]{81 + \sqrt{-2700}} + \sqrt[3]{81 - \sqrt{-2700}}$$

If one is a good guesser and is apt at the manipulation of imaginaries, one can actually find the cube roots and thus obtain

$$x = (-3 + \sqrt{-12}) + (-3 - \sqrt{-12}) = -6$$

For this and other reasons the formula is a pure mathematical triumph but, as we have said, other methods are needed in applied situations. Thus, if one were really able to guess -6 (and special propositions of the theory of equations would assist his guess), he could complete the task of solution by using Harriot's method of factoring. He might consider his guess, $x = -6$, as if it were the result of equating the factor $x + 6$ to zero. Then, dividing the above cubic polynomial by $x + 6$ to obtain the quotient $x^2 - 6x - 27 = (x - 9)(x + 3)$, he would say that $x - 9 = 0$ or $x + 3 = 0$, and the remaining roots are $x = 9$, $x = -3$.

　　It is difficult to assign credit for the cubic formula to any particular one of the sixteenth-century Italian algebraists. From their own writings, which are not considered reliable since each author was so biased in his own favor, one gathers the following approximation of the truth: Around 1535 a mathematics contest was proposed by Antonio Mario Fior of Bologna. Each contestant was to deposit a certain stake with a notary, and whoever could solve the problems in a collection of thirty propounded by his opponent was to get the stakes, thirty days being allowed for the solution of the questions proposed. Fior had learned to solve a special type of cubic from his teacher, Scipione del Ferro (1465–1526). It is believed that Del Ferro may have obtained his method from Arab sources. Fior's opponent in the contest was a Venetian mathematics professor, Niccolo Fontana (1500–1557), commonly known as Tartaglia, a nickname meaning "stammerer" but nevertheless adopted by Fontana, whose impediment was due to an injury suffered during the French sack of Brescia, his native town. Tartaglia suspected that the questions would all be cubics, and so developed a formula for solving cubic equations. He answered all the questions put to him, and in return gave Fior questions on cubics of a type the latter could not handle. Thus Tartaglia won the contest and composed some verses to commemorate his victory. Each stanza described a step in the derivation of the general formula for solving cubics. In his *Quesiti et invenzioni diverse* (1546), dedicated to Henry VIII of England, Tartaglia recorded these particular verses, stating, "If this poem is not very good, I don't care. It helps me to remember the rule."

Tartaglia's happiness, however, was to be short-lived. He planned to keep his method secret, but another Italian mathematician, Girolamo Cardano (1501–1576) (Cardan), professor of medicine at Milan and author of the algebraic masterpiece *Ars Magna* (1545), wheedled the mnemonic poem out of his stammering friend. Having secured the facts, which he promised to keep secret, Cardano nevertheless included in the *Ars Magna* Tartaglia's method for solving a cubic. He gave full credit to Tartaglia, but in many modern texts the method is still referred to as "Cardan's Solution of the Cubic."

The cubic equation incident was only one of many that reveal why Cardano is considered the "bad boy of mathematics." That his career was no path of ease or virtue is set forth in his *Book of My Life*, written at the age of seventy-four, a sort of Rousseau's *Confessions*. Though the book lacks the literary and philosophic qualities of the latter, it is equally lusty, in the true spirit of the Italian sixteenth-century scene. To place Cardano in the correct historic epoch, we need only state that he was a contemporary of Benvenuto Cellini. The extravagance of Cardano's autobiography might be attributed to senility except that it is known that his life-long behavior lacked balance. On the one hand, he holds a high rank in the history of medicine and he was truly the leading algebraist of his era. His mathematical creativeness is only now being fully recognized. For example, he was one of the founding fathers of modern probability theory (Chapter 13). On the other hand, he was an astrologer, a gambler, and a man of dubious ethical principles. Perhaps he was slated for neurosis from birth, since he was the illegitimate son of a professor of jurisprudence and medicine in Milan.

According to Cardano's own account, his later life was as unhappy as his early years. He relates that his older son was deaf and hunchbacked but, worst of all, had a psychopathic personality which led him to poison his worthless wife, a crime for which he was put to death. The younger son of the great mathematician was also a ne'er-do-well of violent type. His daughter, Cardano comments, was no trouble except in the matter of dowry. After all his efforts in providing this sum and getting her married, he complains, she disappointed him by failing to provide grandchildren.

Although Cardano derived no joy from his own children, he took great pleasure in the accomplishments of his favorite pupil, Lodovico Ferrari (1522–1565), who discovered a general method for solving any *quartic* equation. Cardano published Ferrari's formula for the quartic, along with Tartaglia's for the cubic, in the famous *Ars Magna*. These formulas can be considered, as we have said, the terminal performance in classic algebra.

In the present chapter we have considered polynomial equations involving a single unknown number. But in the previous chapter we solved some equations in two unknowns. The solution was not always determinate; that is, mere specification of the domains of the variables was not sufficient to fix a unique solution (pair of values satisfying the equation) or even a *finite* number of solutions. If a polynomial equation contains several unknowns, a finite solution set will require further limitations on the variables, for example, additional equations or reduction of the domains to subsets of the original replacement sets. The effect of such conditions and also graphic methods of solving systems of simultaneous equations will be discussed in a later chapter, but we remark that in many cases algebraic techniques make possible the elimination of all but one of the unknowns and provide a polynomial equation for that single variable so that the whole problem is reduced to the classical issue of solving a polynomial equation.

In the previous chapter we mentioned the name of Évariste Galois in connection with finite fields, and later we shall discuss Galois' theorem of 1831. We shall see how it put an end to traditional algebra, Italian style, where one provides a neat formula in terms of coefficients, which is applicable to all polynomial equations of a certain degree. For two hundred years prior to 1826, when Abel dealt the first deathblow to such equation-solving, algebraists tried in vain to solve the general equation of the fifth degree,

$$ax^5 + bx^4 + cx^3 + dx^2 + ex + f = 0$$

In the year named, Abel showed that even though Gauss guaranteed a solution when coefficients are real or complex, it is *impossible* to express this solution using only a *finite* number of rational operations and root extractions on the coefficients a, b, c, d, e. Galois gave a more elegant proof of the same fact, but then established a general theorem indicating the impossibility of finite algebraic formulation of solutions for polynomial equations of *all* degrees greater than 4.

Galois' great theorem states: *A polynomial equation is solvable if and only if its group is solvable.* This theorem, whose date is 1831, mentions the term "group," which is due to Galois himself, and refers to specialized types like the "group of an equation" and a "solvable group." Now all this was mentioned a dozen years in advance of the date when Hamilton's quaternions launched the "game" which, by deletion of one or more of the postulates for a field, leads to the various modern abstract algebraic structures, including the group. The idea (if not standard name) of a group structure had, in fact, started in the eighteenth century and had been developed by a number of mathematicians before Galois. His great discovery began the new era in two ways very different from Hamilton's initiation of new algebraic systems by changing the rules of the traditional game. In the first place, Galois put an end to equation-solving and, secondly, his work provided motivation for a general abstract approach in which the elements of an algebraic system are *not* necessarily numbers. Galois, and Augustin-Louis Cauchy (1789–1837) before him, studied finite groups whose elements are *mappings* (Chapter 16) with a finite domain.

As a final comment, we remark that Galois' theorem did *not* shut the door to *nonalgebraic* solutions requiring an "infinite" number of operations on the coefficients of a polynomial equation. But just what does "infinite" mean, in an exact mathematical sense? From the point of view of common sense, it would seem that one could, at best, approximate a solution requiring an infinite number of steps by taking a great many of these steps so as to get closer and closer to some ideal goal. This requires concepts from *analysis*, the area of mathematics which is associated with the "infinite" real continuum and "infinite" processes in general. Some of those concepts were treated in Chapter 2, and others will be developed later in our story.

6

A Universal Language

"The human race, considered in relation to its own welfare, seems comparable to a battalion that marches in confusion in the darkness, without a leader, without order, without any signal or command to regulate its march, and without any attempt on the part of individuals to take cognizance of one another. Instead of joining hands to guide ourselves and make sure of the road, we humans run hither and yon, and merely interfere with one another."

The simile is not part of some current plea for world federation or increased activity on the part of the United Nations, but a thought recorded almost three hundred years ago by Leibniz, who believed that the panacea for the lack of human cooperation lay in the formulation of a universal language (*characteristica universalis*) and an algebra of reasoning (*calculus ratiocinator*). To people today this suggestion for the solution of man's woes must appear over-optimistic, just as it did to Voltaire, whose *Candide* satirized Leibniz' "Everything is for the best in this best of all possible worlds." Nevertheless, Leibniz did initiate a "universal language," namely, a form of *symbolic logic*.

The illustrious creator of that "Esperanto" was always one to give credit where credit was due, and hence he pointed out that the Catalan mystic Ramón Lully (1235–1315) had anticipated his idea to some extent. Lully's *Ars Magna* contained formulas so utterly mystic as to be completely unintelligible, but it also presented tabulations resembling the multiplication table for quaternions (page 78) except that the entries were not quaternion "units" but instead ideographs representing the primitive concepts which Lully planned to combine in order to express all other ideas and to solve all problems of science, religion, and philosophy.

It is true that Leibniz' motivation was similar to that of Lully. But Leibniz provided a lucid formulation of his universal tool, so that later logicians were able to develop and improve his creation. If twentieth-century philosophers are not as sanguine as Leibniz, they feel nevertheless that part of his dream has been realized, for a universal *scientific* language, mathematical logic, has been devised, and there is every hope that the future will witness its further progress, the solution of its unresolved problems, and its application to all situations calling for scientific thought.

Such a program for logic can be considered the ultimate outgrowth of an essay, *De arte combinatoria*, written by Leibniz in 1666, when he was "barely out of school" (*vix egressus ex Ephebis*), to use his own words. In that year he laid the foundation for "a general technique by which all reasoning can be reduced to mere calculation." "This method," Leibniz wrote, "should serve at the same time as a sort of universal language, whose symbols and special vocabulary can direct reasoning in such a way that errors, except those of fact, will be like mistakes in computation, merely a result of failure to apply the rules correctly." In later writings Leibniz enlarged on these ideas, stating that "the method should be an Ariadne's thread, a medium that will guide the mind in the fashion that geometric lines guide the eye." He explained that the language of logic should be ideographic, each symbol representing a simple concept, but should differ from languages like Chinese by combining symbols in order to compound ideas, instead of having a vast number of different characters corresponding to different things. Thus while it may require a lifetime for a foreigner to master Chinese completely, any one should be able to perfect himself in the *characteristica universalis* in a few weeks. Leibniz himself provided only a sketchy outline of his proposed language, but said that leading scientists ought to study the question, and select the "alphabet of human thought," the catalogue of the simplest ideas and associated symbols, and then rewrite all of science in terms of this standard alphabet.

Leibniz' idea was regarded as nothing more than a dreamer's fantasy until the time of Giuseppe Peano (1858–1932), who actually did devise one species of scientific Esperanto. Then, in 1894, with a number of collaborators, he started on the rest of Leibniz' program. The result was the five-volume *Formulaire de Mathématiques* (1895–1905), in which all of mathematics was rewritten in terms of the Peano language. The Italian logician continued to propagandize the *Formulaire* during the rest of his life. Maria Cinquini-Cibrario, an Italian analyst and one of the few women mathematicians of our day (or any day), started her own career by serving as an assistant to the aging Peano, and for any one who would like to see a sample of the Peano language, there is one of Professor Cinquini-Cibrario's early papers, *Proposizioni universali a particolari e definizione di limite.**

Some of Peano's symbols will appear in the present chapter, but we do not plan to expose the reader to the rigors of his *Formulaire*. Instead, we shall show how Leibniz' great ideas experienced a rebirth in the writings of George Boole and Augustus De Morgan, logicians whose names have been mentioned as contributors to the task of placing traditional algebra on a sound postulational foundation. Boole's *Laws of Thought* (1854) was written in apparent ignorance of Leibniz' ideas on the same subject. In a sense, Boole's unawareness of the work of his predecessor was fortunate, because his own formulation was so much simpler and clearer. He became the "re-founder" of symbolic logic, and all subsequent contributions to the subject go back to his 1854 treatise. Some of the outstanding mathematical logicians after Boole were the American Charles Sanders Peirce (1839–1914) and the Germans Ernst Schroeder (1841–1902) and Gottlob Frege (1848–1925). The ideas of Frege and Peano were carried further in the *Principia Mathematica* (1910–1913) of Bertrand Russell and Alfred North Whitehead (1861–1947), a treatise

---

* See *Atti della Reale Accademia delle Scienze di Torino*, Vol. 44 (1929).

often pronounced the greatest twentieth-century contribution to both a universal scientific language and an algebra of reasoning.

Symbolic logic, in the opinion of Russell, consists (like ancient Gaul) of three parts—the algebra of classes, the propositional calculus, and the calculus of relations. In the present chapter, the first two of these branches will be considered as two species of "Boolean algebra." We shall also touch on the question of logical relations, but shall postpone amplification to our next chapter.

It must be emphasized that we shall be treating only certain selected aspects of symbolic logic, mainly because a complete, rigorous presentation would baffle the general reader, but also because we wish the present chapter to be properly contiguous with the chapters immediately preceding. We have just completed a discussion of how symbols were manipulated mechanically for the purposes of classical algebra. Now Boole's conception will make it possible to proceed in somewhat analogous fashion with literal symbols which are capable of far more general and more important interpretations than the *x, y, z*'s of common algebra. Again, in discussing the formal aspects of a "Boolean algebra," that is, its axiomatic foundation, we shall be continuing the discussion of algebraic structures, a topic initiated in Chapter 4. In one sense, the present introduction to symbolic logic might have formed an immediate sequel to Chapter 3, where a pure mathematical science was seen to be based on a postulate system from which the theorems of the science are deduced. But just how is the process of deduction to be carried out? Evidently, in addition to the fundamental axioms, there must be agreement at the outset on acceptable logical laws and rules of proof. Such principles become additional postulates, as it were, although they may be theorems of another science, namely, some type of logic—the classic Aristotelian variety or some modern system of symbolic logic.

Boole, as we have said, was a member of the British algebraic school that gradually "purified" common algebra so that emphasis was placed not so much on the results of operations (addition, multiplication) as on their *formal properties* such as commutativity, associativity, distributivity. Thus algebra became a pure science where the *x, y, z*'s are abstractions and need not necessarily be interpreted as representing numbers. This point of view gave Boole his opportunity. In his *Laws of Thought* he stated that he would "exhibit logic, in its practical aspect, as a system of processes carried on by the aid of symbols having a definite interpretation, and subject to laws founded upon that interpretation alone. But at the same time these laws are identical in form with the laws of the symbols of algebra, with a single addition, that is, the symbols of logic are further subject to a special law." In other words, Boole was able to assume commutativity, associativity, distributivity for his "logical addition" and "logical multiplication," and thereby reduce certain types of reasoning to the sort of algebraic manipulation one carries out in school algebra.

Having given a capsule description of Boole's contribution, let us now expand our brief outline. We shall gradually develop the concepts and laws of the algebra of classes in informal, intuitive fashion, prior to formulating a postulate system for a pure (abstract) Boolean algebra. Thus our procedure will be somewhat analogous to that of the ancient surveyors who provided the physical observations which were ultimately abstracted in the axioms of pure Euclidean geometry.

To review what has been said in earlier chapters, a *set* can be described as a *collection* of things that are called *elements* or *members* of the set. Some logicians use the term "class" in a somewhat different sense from the term "set," but we shall not do so, and in order to avoid constant repetition, any of the following words will be used as a synonym for set: class, collection, aggregate, ensemble.

In traditional arithmetic and algebra, one carries out various operations (+ and ×) within sets of numbers. In Boolean algebra one operates within sets of classes. We have already illustrated some sets of classes, for example, the aggregate of classes in a particular school, a collection of sets of china, the set of bunches of sticks which primitives used for counting. But it is possible that such sets of classes may not be suitable for a Boolean algebra because Boole's "logical addition" or "logical multiplication" may lack closure on the given collection.

To illustrate the last remark, we may as well reveal that we have in fact already performed "logical addition" without giving the operation that name. The *logical sum* of two sets is merely their *union* (page 20). Now if we unite two sets in one of the collections described in the previous paragraph, we may fail to obtain a set of the original collection. Thus, combining two sets of china will yield an enlarged set but not necessarily a set identical with one in the original collection, and a similar statement applies to the fusion of two classes in a school.

To avoid possible failure of closure, and also for other reasons that will appear presently, a Boolean algebra of classes starts by specifying a fundamental domain, that is, a *universal set* or *universe of discourse*. (This idea is due to De Morgan.) Then the collection of all *subsets* of the selected universe is an aggregate on which a *closed* "addition" and a closed "multiplication" can be defined. For purposes of illustration we proceed informally once again and select a small universe of discourse, namely, one made up of three men. Let I symbolize that universe, where

$$I = \{\text{Ames, Brown, Grant}\}$$

Then, recalling that a set X is said to be a subset of a set Y if and only if every member of X is also a member of Y, we see that $A = \{\text{Ames, Brown}\}$, $B = \{\text{Ames, Grant}\}$, $C = \{\text{Brown, Grant}\}$, $D = \{\text{Grant}\}$, etc. are subsets of I.

If we think of the subsets as committees, then, in addition to one-man and two-man committees, there will be a "committee of the whole," namely, I, which is an *improper* subset of itself, a *proper* subset of I being a subset that is not identical with I. At the other extreme, it is convenient for logical purposes to consider that $\emptyset = \{\ \}$, the empty (null) set, is a subset of any set, and hence is a subclass or "committee" of I.

If the universe of discourse selected should be the set of natural numbers, say, then finite subsets can be specified by roster, but the membership of an infinite subset could never be listed completely. In the latter case, one would have to define the subset by specifying some property common to its members and possessed by no other natural numbers. In that way one can define the subset of *even* natural numbers by stating that they are the natural numbers exactly divisible by 2. The symbol {2, 4, 6, . . .} is suggestive but it is *not* a list. The use of a defining property in place of a roster is also advisable for large finite subsets of any universe.

Since we are discussing *symbolic* logic and indicating similarities between Boolean and common algebra, let us now introduce the symbol $\subseteq$ for "is a subset of," and point out that its properties are similar to those of $\leq$ for numbers. In our illustrative three-member universe, $A \subseteq I$, which can be read either as "A is a subset of I" or as "A is *included in* I." In the same universe, $B \subseteq I$, $I \subseteq I$, etc. But we can also state that $A \subset I$, read either as "A is a *proper* subset of I" or as "A is *properly* included in I."

Just as $3 \leq 3$ (read as "3 is either equal to or less than 3") and $x \leq x$ (where x is any number), $X \subseteq X$ where X is any set; that is, every set is a subset of itself. One describes this characteristic by saying that $\leq$ and $\subseteq$ are *reflexive* relations, that is, relations which a thing can have to itself.

It will seem obvious to the reader that if $x \leq y$ and $y \leq z$, then $x \leq z$, where x, y, and z are numbers. It may seem just as clear that a subset of a subset of a universal set is also a subset of that universe. In general, if X, Y, Z are any three sets such that $X \subseteq Y$ and $Y \subseteq Z$, then it is also true that $X \subseteq Z$. The property just considered is described by saying that $\leq$ and $\subseteq$ are *transitive* relations.

If one states that two sets are equal, that is, $X = Y$, then the use of the "equals" sign is the same as that throughout mathematics. As already explained in Chapter 2, the sign $=$ means that the symbols on the two sides of the sign are both names for the *same* thing. When sets are defined by properties, it may sometimes be difficult to recognize that different properties define the *same* set, and in this connection a special attribute of $\subseteq$ may be helpful. If, in fact, X and Y do name the same set, that is, $X = Y$, then every member of X is a member of Y and vice versa, which signifies that $X \subseteq Y$ and $Y \subseteq X$. It is customary to postulate the converse fact (and apply it to test equality of sets): If $X \subseteq Y$ and $Y \subseteq X$, then $X = Y$.

But let us proceed to other concepts of Boolean algebra. In the universe of natural numbers,

$$I = \{1, 2, 3, 4, 5, 6, . . .\}$$

there is the subset of *even* natural numbers, those exactly divisible by 2,

$$E = \{2, 4, 6, . . .\}$$

and the subset of *odd* natural numbers, those not exactly divisible by 2,

$$O = \{1, 3, 5, . . .\}$$

The subset O is the *complement* of the subset E in the sense that O contains all those members of the universe I which are not in E. We can also say that E is the complement of O.

In the small universe containing three members, $I = \{$Ames, Brown, Grant$\}$, the subset $D = \{$Grant$\}$ is the complement of the subset $A = \{$Ames, Brown$\}$. Modern symbolic logic uses A' or $\bar{A}$ as the symbol for "the complement of A." In the present instance, $D = A' = \bar{A}$. Also, if $B = \{$Ames, Grant$\}$, then $B' = \bar{B} = \{$Brown$\}$. If $C = \{$Brown, Grant$\}$, then $C' = \bar{C} = \{$Ames$\}$. Thus the subsets of any universal set occur in complementary pairs. In any universe of three elements, $I = \{\alpha, \beta, \gamma\}$, it is readily seen that there are eight and only *eight* subsets which can by symbolized as follows:

$$
\begin{aligned}
I &= \{\alpha, \beta, \gamma\} & \emptyset &= \{\ \} \\
A &= \{\alpha, \beta\} & A' &= \{\gamma\} \\
B &= \{\alpha, \gamma\} & B' &= \{\beta\} \\
C &= \{\beta, \gamma\} & C' &= \{\alpha\}
\end{aligned}
$$

We observe that the universal set I and the null set $\emptyset$ are complementary, a fact obviously true in any universe. Also, in any universal set, if subset A' is the complement of subset A, it is also true that A is the complement of A'. Symbolically, $(A')' = A$, or $\bar{\bar{A}} = A$. The complement of the complement of a subset is the original subset.

In our miniature Boolean algebra there are eight subsets. These are the objects which will be "added" and "multiplied" presently. Since we already know that "logical addition" means union, a brief inspection of our list of eight subsets will indicate that uniting any pair of them will produce some subset of I. Hence "logical addition" is closed on the collection of eight subsets and we shall see shortly that the same is true of "logical multiplication." Therefore we shall be studying a *finite* algebra of eight objects, in strong contrast to traditional algebra, where the basic class of objects is usually infinite—the set of rational numbers or the set of real numbers or the complex aggregate. If an algebra of eight elements seems petite, let us remark that if one were to select a universe of two elements, $I = \{\alpha, \beta\}$, then there would be *four* subsets and a Boolean algebra of four objects. If one selects a universe containing a single element, $I = \{\alpha\}$, there would be only *two* subsets, namely I and $\emptyset$. We shall see that the *binary* Boolean algebra of two objects has exceedingly important interpretations. It is the smallest nontrivial Boolean algebra. To obtain a smaller algebra one would have to select $I = \{\ \}$, that is, $I = \emptyset$, and all procedures would lead to this vacuous set.

We have already carried out some arithmetic on the eight subsets of $I = \{\alpha, \beta, \gamma\}$, since forming the complement of a set can be considered a *unary* operation (on *one* set) or else a limited sort of subtraction where the minuend is always I but the subtrahend may be any subset. Since we have already explained that "logical addition" means *union*, let us now consider Boole's interpretation of $\otimes$.

Boole defined the *logical product* of two sets as their *intersection*, that is, as the aggregate of members common to both sets. Thus, referring to the subsets of $I = \{\alpha, \beta, \gamma\}$ as listed on this page, we can state that the logical product or intersection of $A = \{\alpha, \beta\}$ and $B = \{\alpha, \gamma\}$ is $C' = \{\alpha\}$, that is,

$$AB = A \otimes B = C'$$

Today we would use Peano's symbol for intersection, $\cap$, and symbolize the above multiplication as

$$A \cap B = C'$$

The reader can verify that $A \cap C = B'$, $A \cap I = A$, $A \cap A' = \emptyset$, etc. But the best way for him to obtain some practice with logical multiplication would be to check the products in the following multiplication table, where a product like $B \cap C$, for example, appears in row B and column C.

Multiplication Table for Subsets of $I = \{\alpha, \beta, \gamma\}$

$\otimes = \cap$	I	A	B	C	C'	B'	A'	$\emptyset$
I	I	A	B	C	C'	B'	A'	$\emptyset$
A	A	A	C'	B'	C'	B'	$\emptyset$	$\emptyset$
B	B	C'	B	A'	C'	$\emptyset$	A'	$\emptyset$
C	C	B'	A'	C	$\emptyset$	B'	A'	$\emptyset$
C'	C'	C'	C'	$\emptyset$	C'	$\emptyset$	$\emptyset$	$\emptyset$
B'	B'	B'	$\emptyset$	B'	$\emptyset$	B'	$\emptyset$	$\emptyset$
A'	A'	$\emptyset$	A'	A'	$\emptyset$	$\emptyset$	A'	$\emptyset$
$\emptyset$	$\emptyset$	$\emptyset$	$\emptyset$	$\emptyset$	$\emptyset$	$\emptyset$	$\emptyset$	$\emptyset$

Since every entry within the above table is one of the eight subsets of I, logical multiplication is *closed* on the collection of subsets. That logical multiplication will be closed on the subsets of *any* universal set results from the definition of that operation. For if X and Y are any two such subsets, then their intersection Z will contain the elements common to X and Y. Thus every element of Z is also a member of X, that is, $Z \subseteq X$. But $X \subseteq I$. Hence, by the transitivity of $\subseteq$, $Z \subseteq I$. In other words, the intersection (logical product) of two subsets of I must be a subset of I.

That logical multiplication is commutative and associative in the present example can be verified by means of the table. That the properties of $\cap$ in the present instance also hold for subsets of any universal set can once again be demonstrated by applying the definition of logical multiplication.

But the above multiplication table indicates other algebraic properties. For example, when $\emptyset$ and I are "factors," they act like the numbers 0 and 1, respectively, in ordinary algebra. Just as $a \times 0 = 0 \times a = 0$, and $a \times 1 = 1 \times a = a$, the table shows that $A \cap \emptyset = \emptyset \cap A = \emptyset$ and $A \cap I = I \cap A = A$, etc. On the other hand, our miniature Boolean algebra lacks the property that made possible the solution of polynomial equations by Harriot's method (Chapter 5). Whereas in common algebra the product of two numbers is equal to zero only if at least one of these numbers is zero, our multiplication table indicates that $B' \otimes C' = \emptyset$, $A' \otimes B' = \emptyset$, $A \otimes A' = \emptyset$, etc. It is possible for a logical product to be "null" without having either of its factors "null."

Again, there will *not* be perfect analogues in Boolean algebra for the "powers" of ordinary algebra, that is, for $a \times a = a^2$, $a \times a \times a = a^3$, etc. From our Boolean multiplication table,

$$A \otimes A = A$$
$$A \otimes A \otimes A = (A \otimes A) \otimes A = A \otimes A = A, \text{ etc.}$$

and if we "multiplied" A by itself (or B by itself or C by itself) any number of times, the logical product would always yield the original set, as we can verify from the table, or obtain by reasoning that the intersection of any set with itself is that set. Logical multiplication is said to be *idempotent*, signifying that all "powers" of a set are the *same*, and hence there is no need for the exponential symbolism of common algebra. When Boole stated that "the symbols of logic are further subject to a special law" (page 102), he amplified his assertion by indicating that the special law is $AA = A$, that is, $A \cap A = A$, the *idempotent* law we have just illustrated.

The reader can readily develop most of the properties of "logical addition" after recalling that the *logical sum* or *union* of any two classes, X and Y, is the set of elements which belong to X or to Y or to both X and Y. Thus, in our illustrative universe, if the committee $A = \{\text{Ames, Brown}\}$ is united with the committee $B = \{\text{Ames, Grant}\}$, the committee of the whole, $I = \{\text{Ames, Brown, Grant}\}$ results. Both common sense and the definition of "union" permit the names in the union to be listed in any order and also cause Ames's name to be listed only once in the fused set. Symbolically,

$$A \oplus B = A \cup B = I$$

(The symbols $\cup$ and $\cap$ for union and intersection are due to Peano.) Using the definition of A, A', B, B', etc., on page 105, the reader can check the entries in the following addition table. He can use it to verify that closure, commutativity, and associativity are properties of logical addition in our miniature Boolean algebra and can also apply the definition of $\cup$ to show that the properties named will hold in general.

Addition Table for Subsets of $I = \{\alpha, \beta, \gamma\}$

$\oplus = \cup$	I	A	B	C	C'	B'	A'	$\emptyset$
I	I	I	I	I	I	I	I	I
A	I	A	I	I	A	A	I	A
B	I	I	B	I	B	I	B	B
C	I	I	I	C	I	C	C	C
C'	I	A	B	I	C'	A	B	C'
B'	I	A	I	C	A	B'	C	B'
A'	I	I	B	C	B	C	A'	A'
$\emptyset$	I	A	B	C	C'	B'	A'	$\emptyset$

The reader can also verify that just as $a + 0 = 0 + a = a$ in common algebra, $A \oplus \varnothing = \varnothing \oplus A = A, B \oplus \varnothing = \varnothing \oplus B = B$, etc. in the table. He can see that if X is any set, then $X \oplus \varnothing = \varnothing \oplus X = X$. Again, just as there is an idempotent law for logical multiplication, there is a similar law for logical addition. The table shows that

$$A \oplus A = A$$
$$A \oplus A \oplus A = (A \oplus A) \oplus A = A \oplus A = A, \text{ etc.}$$

and if one "adds" A to itself any number of times, the logical sum will always be A. Boolean algebra has no need for the coefficient symbolism of common algebra.

In traditional algebra, multiplication is *distributive* with respect to addition, that is,

$$a(b + c) = ab + ac$$

where a, b, and c are any numbers whatsoever. That the analogous law holds in Boolean algebra, in other words, that

$$A(B \oplus C) = AB \oplus AC$$

or

$$A \cap (B \cup C) = (A \cap B) \cup (A \cap C)$$

is suggested by Figure 6.1, which exhibits one type of *Venn diagram*, named after

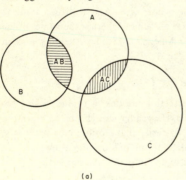 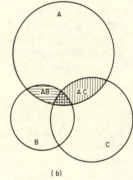

(a)   (b)

B and C have no common elements   **B and C overlap**

Figure 6.1 $A(B \oplus C) = AB \oplus AC$

the logician John Venn (1834–1883). But in Boolean algebra there is a second distributive law. It turns out that logical addition is distributive with respect to multiplication. The second distributive law can be obtained from the first by substituting $\oplus$ for $\otimes$ and $\otimes$ for $\oplus$, that is, $\cup$ for $\cap$ and $\cap$ for $\cup$ in the first law. Then

$$A \oplus BC = (A \oplus B)(A \oplus C)$$

or

$$A \cup (B \cap C) = (A \cup B) \cap (A \cup C)$$

That this law holds is suggested by the Venn diagram of Figure 6.2.

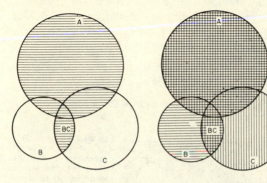

Figure 6.2 $A \oplus BC = (A \oplus B)(A \oplus C)$. The left member of the equation is represented by the shaded region in the first diagram. In the second diagram, horizontal shading represents $A \oplus B$, vertical $A \oplus C$. The doubly shaded region (intersection of the two sets) is the same as that shaded in the first diagram.

To see that the second distributive law does *not* apply to the ordinary addition and multiplication of numbers, let us show that it is *not* true that $a + bc = (a + b)(a + c)$ for all numerical values of a, b, and c. Thus if $a = 1$, $b = 2$, $c = 3$, the left member would have the value $1 + 2 \cdot 3 = 7$, and the right member would be equal to $(1 + 2)(1 + 3) = 12$.

Not only the second distributive law but also all theorems deduced from it fail to have counterparts in common algebra. Let us prove one such Boolean law by starting with the fact that

$$A = A \oplus \emptyset = A \oplus (\emptyset \otimes B)$$

Now, by virtue of the second distributive law, we can say that

$$A \oplus (\emptyset \otimes B) = (A \oplus \emptyset)(A \oplus B) = A(A \oplus B)$$

The result of all this is that the original set A is identical with $A(A \oplus B)$, or

$$A(A \oplus B) = A$$

that is,

$$A \cap (A \cup B) = A$$

which is called a *law of absorption* because (if B is not empty) the law states that the part common to class A and a larger class including A, namely, $A \oplus B$, is the first class. The larger class is "absorbed" into the smaller. In common algebra there is no such law of absorption. That, in general,

$$a(a + b) \neq a$$

when a and b are numbers, is indicated by the counterexample, $a = 2$, $b = 1$, which yields $a(a + b) = 2(2 + 1) = 6$. Now $6 \neq 2$.

The second distributive law was obtained from the first by replacing $\oplus$ by $\otimes$ and $\otimes$ by $\oplus$. That particular interchange of logical addition and logical multiplication illustrated a general *principle of duality* which asserts: In any Boolean law, the result of replacing $\oplus$ by $\otimes$, $\otimes$ by $\oplus$, $\emptyset$ by I, and I by $\emptyset$ is also a law.

Thus, applying the duality principle to the commutative law of logical addition, $A \oplus B = B \oplus A$, yields the corresponding law for logical multiplication, $A \otimes B = B \otimes A$, that is, $AB = BA$. Applying the same principle to the law of absorption, $A(A \oplus B) = A$, produces a *second law of absorption*, namely

$$A \oplus AB = A$$

which can also be proved to be true.

That there is a duality between logical sums and logical products is implicit

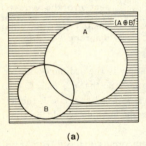
(a)

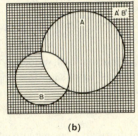
(b)

Figure 6.3 $(A \oplus B)' = A'B'$. If the rectangle represents I, the universal set, the shaded region in (a) is identical with the doubly shaded region in (b), thus illustrating De Morgan's first law. Also, (b) illustrates De Morgan's second law.

in De Morgan's laws (Figure 6.3) which assert that

$$(A \oplus B)' = A'B' \text{ and } (AB)' = A' \oplus B'$$

The complement of a logical sum is equal to the product of the complements of the addends, and the complement of a logical product is equal to the sum of the complements of the factors. Application of the De Morgan laws enables one to convert any "sum" into a "product" and vice versa so that it would actually be possible to perform Boolean algebra by using only one of these operations (and complementation).

In order to summarize all that has been said thus far, the set of fundamental laws for a Boolean algebra of classes is listed below, where dual laws are paired. Some of these laws can be taken as *postulates* and the remainder derived from them as theorems, but it will be simpler for the reader to accept them all, in order to be able to advance to other issues of Boolean algebra. To indicate similarities with common algebra, we have used juxtaposition of letters to indicate logical multiplication, and have used $\oplus$ instead of $\cup$. The letters A, B, C symbolize *any* subsets of *any* universe I.

Closure laws:	$A \oplus B$ is a uniquely defined subset of I	AB is a uniquely defined subset of I
	A' is a uniquely defined subset of I	
Commutative laws:	$A \oplus B = B \oplus A$	$AB = BA$
Associative laws:	$A \oplus (B \oplus C) = (A \oplus B) \oplus C$	$A(BC) = (AB)C$

Distributive laws:	$A(B \oplus C) = AB \oplus AC$	$A \oplus BC =$
		$(A \oplus B)(A \oplus C)$
Idempotent laws:	$A \oplus A = A$	$AA = A$
Complementation laws:	$A \oplus A' = I$	$AA' = \emptyset$
Laws of De Morgan:	$(A \oplus B)' = A'B'$	$(AB)' = A' \oplus B'$
Double complementation:	$(A')' = A$	
Laws involving $\emptyset$ and I:	$A \oplus I = I$	$A\emptyset = \emptyset$
	$A \oplus \emptyset = A$	$AI = A$
	$I' = \emptyset$	$\emptyset' = I$
Laws of absorption:	$A(A \oplus B) = A$	$A \oplus AB = A$

To give an especially simple, and hence somewhat artificial, instance of how some of the above laws can be applied, consider the problem of two librarians who are told to sort a pile of books just returned by borrowers. The first librarian is told to collect all political works by American authors and all books over 500 pages by foreign authors. The second is told to take political works exceeding 500 pages and novels by Americans, provided they are not political in nature. Will there be any books claimed by both librarians? If

I is the class of books in the pile
A is the class of books by Americans
B is the class of books over 500 pages
P is the class of books of a political nature
N is the class of novels
A' is the class of books by foreign authors
B' is the class of books of 500 pages or less
P' is the class of nonpolitical works
N' is the class of books other than novels

then the first librarian will call for $AP \oplus A'B$ and the second will claim $BP \oplus ANP'$.

Both will claim the intersection of these two sets, namely, $(AP \oplus A'B)(BP \oplus ANP') = ABPP \oplus AANPP' \oplus A'BBP \oplus AA'BNP'$.

Since $PP = P$, $BB = B$, and because complementary classes have no members in common, $PP' = \emptyset$ and $AA' = \emptyset$, therefore the set of books claimed reduces to $ABP \oplus A'BP$. This equals $BP(A \oplus A') = BPI = BP$ since the elements common to the universe and any class in it constitute the class itself. Hence both librarians will claim political books of over 500 pages.

We abstracted the fundamental laws for a Boolean algebra from the behavior of the subsets of a universal set. As we have stated, some of those laws can be treated as postulates (assumptions) and the remaining laws can be deduced as theorems. Now if we divorce the Boolean postulates and theorems from their original association with classes, that is, if we consider the symbols I, $\emptyset$, A, B, C, $\ldots$, $\oplus$, $\otimes$, $'$, as *abstractions*, which need *not* be interpreted as sets, union, intersection, complementation, then we have the fundamental laws of a *pure* Boolean algebra. There are, in fact, many *essentially different* concrete interpretations of those laws, as the present chapter will show. That a postulate system for a Boolean algebra is *not* categorical (Chapter 3), that is, does *not* have all its interpretations isomorphic (abstractly identical) is indicated by the fact that there are Boolean algebras of 2 elements, 4 elements, 8 elements, 16 elements, etc., so that matching of elements would be impossible in such interpretations and *a fortiori* they could not be isomorphic. Nevertheless any manipulations we perform or theorems we deduce must apply to all Boolean algebras, in particular to the algebra of classes, and also to two other Boolean algebras which we shall consider, namely, the algebra of propositions, and the algebra of switching circuits.

One way of emphasizing two important points, namely, that Boolean algebra applies to things other than sets and that its structure differs from the *field* (Chapter 4) of common algebra, is to describe a Boolean algebra as a *lattice*, an abstract system created and studied by the American mathematician Garrett Birkhoff. He defines a *lattice* as a set of elements (of any kind), closed under two binary operations which are commutative, associative, idempotent, and which satisfy the two absorption laws. If, in addition, the two distributive laws hold, the system is called a *distributive lattice*. Then a Boolean algebra is a *special* kind of distributive lattice, the specialization being brought about by the laws governing I, $\varnothing$, and the unary operation of complementation. One reason pure mathematicians may prefer the term *Boolean lattice* to the traditional "Boolean algebra" is that an "algebra" in modern usage often refers to a structure like Hamilton's quaternion system where one has a "vector addition," a "vector multiplication," and a "scalar multiplication" (Chapter 4).

Having discussed the nature of a pure Boolean lattice, we shall proceed to further applications. The surprising fact is that the most important interpretations, from the point of view of logic, are related to the smallest nontrivial Boolean lattice, where operations are performed on *two objects* only. In the algebra of classes we obtained such a tiny Boolean lattice by selecting a universe containing a single element, namely, $I = \{\alpha\}$. Then the only subsets are $I = \{\alpha\}$ and $\varnothing = \{\ \}$. The reader will readily verify that the entire arithmetic of this binary Boolean lattice is governed by the following tables.

Addition			Multiplication			Complementation	
$\oplus = \cup$	$\varnothing$	I	$\otimes = \cap$	$\varnothing$	I	Set X	Complement X'
$\varnothing$	$\varnothing$	I	$\varnothing$	$\varnothing$	$\varnothing$	$\varnothing$	I
I	I	I	I	$\varnothing$	I	I	$\varnothing$

But these tables are associated with a binary algebra of classes. To "purify" them, let us avoid the class symbols, $\varnothing$ and I; but, recalling that those classes behave somewhat like the 0 and 1 of common algebra, let us use the latter symbols, *not* to represent numbers, but to stand for abstract elements (of any nature whatsoever) which obey laws abstracted from the above. Then for a pure binary Boolean algebra, we have the following tables.

Addition			Multiplication			Complementation	
$\oplus$	0	1	$\otimes$	0	1	x	x'
0	0	1	0	0	0	0	1
1	1	1	1	0	1	1	0

These tables emphasize that the operations of a binary Boolean lattice (algebra) are almost like those of common arithmetic. The sums and products, *with one exception*, are those that would result from performing ordinary addition and multiplication with numbers. The only exception is

$$1 \oplus 1 = 1$$

which involves the "idempotency" of Boolean addition and also the fact that I (in this case, 1) in a Boolean lattice is a sort of "upper bound." The Boolean law, $A \oplus I = I$, signifies that no matter what is added to I (in this case, 1), no "higher" entity can be obtained. Complementation is like subtraction from the number 1.

Before we relate the propositional calculus to the binary Boolean lattice, we shall backtrack slightly in order to explain, by concrete examples, why one should expect a parallelism between the laws of the algebra of classes and those of the algebra of propositions. We might say that the key to the situation lies in the method of specifying a subset of a universal class by means of a defining property.

If, for example, the universe of discourse is $I = \{0, 1, 2, 3, 4\}$, then the property of "being a perfect square" defines the subset $P = \{0, 1, 4\}$. If we use the technical terminology of the two previous chapters, we can assert that "x is a perfect square" is an *open sentence*, where I is the domain of the variable x. Here P is called the *truth set* of the open sentence. This open sentence can also be described as a *propositional form* because substitution of the elements of the domain yields statements expressing *propositions*, that is, ideas which can be pronounced either *true* or *false*. Here, in the very first example, we have a correspondence between a *set P* and a *propositional form*. We shall symbolize the latter by $p(x)$ to indicate that it is a "function of x," a concept which will be discussed at length in the next chapter. Anticipating that discussion, we shall now point out that to each value of x in the domain I there corresponds a unique proposition, so that the "function" consists of the five pairs:

x	$p(x)$
0	0 is a perfect square
.	.
.	.
3	3 is a perfect square
4	4 is a perfect square

In the same universe of discourse, let us consider the *propositional form* which negates $p = p(x)$, namely, "x is *not* a perfect square." In symbolic logic this form might by symbolized as " $\sim p$ " (read as *not-P*) or in the present instance by " $\sim p(x)$." Its truth set is obviously $\{2, 3\}$. We observe that this truth set is P', the complement of $P = \{0, 1, 4\}$. Thus if P corresponds to p, it appears that P' corresponds to $\sim p$. If this is a general fact for all universes, then the unary operation of complementation (') of sets will correspond to a unary operation on propositions, namely *negation* ($\sim$).

In the same universe, the propositional form,

$$q = q(x) : x < 3$$

has the truth set $Q = \{0, 1, 2\}$, as the reader can readily verify, so that set Q corresponds to form q. For the truth set of " p *and* q," that is, "$p(x)$ *and* $q(x)$," we must solve the *simultaneous* open sentences "x is a perfect square" and "$x < 3$." Direct substitution of the five values in I will lead to the truth set $\{0, 1\}$. But one could obtain this set by reasoning that values of x making *both* sentences true would have to belong to P, the truth set of $p(x)$, and also to Q, the truth set of $q(x)$. In other words, the solution set would have to be the intersection of P and Q, namely $P \cap Q$. We can verify that $P \cap Q = \{0, 1\}$. Thus "p and q," the *conjunction* of two propositional forms, symbolized by "$p \wedge q$," corresponds to the *intersection* of two sets, $P \cap Q$, a fact that holds in more general universes than the one under consideration.

We leave the reader the task of continuing the analogy further by showing that, in the present instance, $P \cup Q = \{0, 1, 2, 4\}$ is the truth set of "p *or* q" = "$p(x)$ *or* $q(x)$." Here the connective "or" has the significance of the legal term "and/or." In other words, to be in the solution set a number must be a perfect square or less than 3, but there is no objection to its having both properties as, for example, in the case of 1. In logic, the form "p *or* q" is symbolized as "$p \vee q$" and, even in more general universes, corresponds to the union of two sets, $P \cup Q$.

In our special universe I, the truth set of "x is an integer" is the universal set I. Hence the given sentence is a *law* for our universe. We can state that law as a *closed sentence* by *quantifying* (page 69) the open sentence as follows: *For all x in I, x is an integer*. The reader can readily find other laws for our universe, for example: *For all x in I, x is not greater than 4*. At the other extreme from the universal truths there are propositional forms like $x < 0$, $x^2 = 2$, $x^3 = 5$, which have $\emptyset$ as truth set. What is the use of such universal falsehoods? Recalling that *negation* of forms corresponds to *complementation* of sets, we see that the negation of any one of the above forms will have I as truth set, that is, will be a universal truth. Thus *for all x in I, x is not less than 0, $x^2 \neq 2$, $x^3 \neq 5$*.

One might go further with our particular example to show how commutativity, associativity, etc. of union and intersection have analogues in the same properties of propositional forms, so that in an algebra of such forms in any universe, the Boolean laws for ', $\cup$, $\cap$ will transfer to identical laws for $\sim$, $\vee$, $\wedge$. But a single illustration does not constitute proof. Moreover, in a small, highly specialized universe, the propositional forms and the possible statements they yield are necessarily limited in nature. If symbolic logic is to be the universal language of which

Lully and Leibniz dreamed, it must be possible to talk about things other than numbers in the set $\{0, 1, 2, 3, 4\}$. Of course, one can enlarge or vary the universe of discourse or consider forms like $p(x, y)$, $p(x, y, z)$, etc. where there may be different universes for x, y, z, etc. We shall do just that in our next chapter. But here we must ask whether increasing the number of variables and extending their domains would actually ensure the possibility of making all sorts of statements—about people, money, books, atoms, stars, love, life, etc. Even if the answer should be affirmative, the notation would be somewhat cumbersome if the number of variables is large, to say nothing of the need for numerous substitutions when universal sets are large.

Perhaps then we should free our p's and q's from their dependence on x, y, z's, etc. That is, in fact, what is done in the propositional calculus, where *logical propositional forms* like $p \wedge q$, $p \vee (q \wedge r)$, $\sim p \wedge (q \vee r)$, etc., contain p, q, and r as "independent" variables, *not* dependent on other variables—x, y, z, etc. The greater freedom makes it possible to replace p, q, r, etc., by *any statement whatsoever*, just so long as these statements express propositions. But then will the algebra of classes still hold? What will the "universal truths" be like and how will they be demonstrated? Let us see.

The universal truths of the propositional calculus are called *tautologies*, which are logical forms producing *true* statements no matter what propositions are substituted for p, q, r, etc., that is, for the variables contained in the forms. The calculus studies the conditions under which compound sentences are true solely as a result of their *pattern*, regardless of the nature of the component propositions. For example, the particular statement "3 is an odd integer *or* 3 is *not* an odd integer," is *true*, but so is every statement of the same *form*. In other words, any statement expressing a proposition can be used in place of "3 is an odd integer." Thus "oxygen is a metal *or* oxygen is *not* a metal," "George Washington is now President of the United States *or* he is *not* now President of the United States." All such statements are incorporated in the logical form, "$p \vee \sim p$," which subsequently will be proved to be a tautology, the "law of the excluded middle," as it is called in traditional logic: Either a proposition or its negation must be true. To say that "$p \vee \sim p$" is a universal truth (for all propositions p) is like asserting $P \cup P' = I$ in a Boolean algebra of classes. Once again, $\vee$, p, $\sim p$, and "universal truth" correspond to $\cup$, P, P', and I, respectively.

To start our discussion of the algebra of propositions more formally, we must remind the reader that a *statement* is a declarative sentence which expresses a *proposition*, an idea that can be pronounced either *true* or *false*. We cannot say that a proposition actually is a statement, because any proposition can be expressed in many different ways, for example, in English, French, Italian, etc. This is the same point made in Chapter 1, where it was emphasized that an idea like that of a "number" may have many different names. Having made the point, we shall not be pedantic but instead shall use "numeral" and "number," "statement" and "proposition," etc., interchangeably.

We have already indicated that symbolic logic makes use of the connectives "not," "and," "or," symbolized by $\sim$, $\wedge$, $\vee$, and that when propositions are compounded with the use of those symbols, the question of the "truth" or "falsity" of compounds is the all-important issue. The *postulates* or assumptions about the truth of various composites are usually incorporated into *truth tables*. Thus the assumptions that (1) if any proposition p is true (has *truth value* 1), then $\sim p$ is false (has *truth value* 0), and (2) if p is false, $\sim p$ is true, are tabulated as follows.

Truth Table for Negation

p	$\sim p$
1	0
0	1

Observe that this table of truth values is identical with the table for *complementation* in a binary Boolean algebra (page 113), a fact which is not surprising since we have already indicated (page 114) a correspondence between negation of propositional forms and complementation of sets. Again, a truth table symbolizes four postulates concerning "p and q," that is, "$p \wedge q$," the *conjunction* of any two propositions. Such a conjunction is assumed to be true if *both* p and q are true. In all other cases the conjunction is postulated to be false, as the following table indicates.

Truth Table for Conjunction

p	q	$p \wedge q$
1	1	1
1	0	0
0	1	0
0	0	0

To see that the truth table for the conjunction of propositions is identical with the multiplication table of a binary Boolean algebra, let us present the truth table in the following alternative form.

Truth Table for Conjunction

		Truth Value of q	
$\wedge$		0	1
Truth	0	0	0
Value of p	1	0	1

The truth table for the *disjunction* of any two propositions, that is, for "p or q," symbolized as "$p \vee q$," is given in two different forms below, where the second form is identical with the addition table for a binary Boolean algebra. Either form of the truth table indicates that the disjunction of two propositions is true unless both propositions are false, in which case it is false. Once again, we remind the reader that the symbol $\vee$ has the meaning of "and/or" as used in statements like the following: The residents of Smithtown and/or people who work in Smithtown must pay the income tax levied by that city.

Two forms of Truth Table for Disjunction

p	q	$p \vee q$		$\vee$		0	1
					Truth value of q		
1	1	1		Truth	0	0	1
1	0	1		Value of p	1	1	1
0	1	1					
0	0	0					

What our comparisons have shown is that if we interpret the elements of the binary Boolean algebra as *truth values* of propositions, and the operations ', $\otimes$, $\oplus$ as $\sim$, $\wedge$, $\vee$, then the truth-value algebra of propositions is a binary Boolean lattice. This signifies that all the laws of Boolean algebra hold for the truth values of propositions. Thus the commutative law $P \cup Q = Q \cup P$ for the Boolean algebra of classes has as its counterpart the law that $p \vee q$ and $q \vee p$ are "equivalent" logical forms, that is, must have the same truth tables. Again, one of De Morgan's laws in the algebra of classes states that $(P \cup Q)' = P' \cap Q'$. Therefore, in the algebra of propositions, $\sim (p \vee q)$ and $(\sim p) \wedge (\sim q)$ must be "equivalent" logical forms, that is, must have the same truth table. That this is the case is verified by the truth tables below. In the truth table for $\sim (p \vee q)$, the first three columns are exactly those of the truth table for $p \vee q$, and the last column is obtained by applying the truth table for negation to the third column. In the table for $(\sim p) \wedge (\sim q)$, application of the truth table for negation to the first two columns yields the third and fourth columns. Application of the truth table for conjunction to the third and fourth columns yields the last column, which is the *same* as in the preceding truth table, thus proving the "equivalence" of the logical forms in question. What *both* tables indicate is that either form yields a true statement if and only if *false* propositions are substituted for both p and q.

Truth Table for $\sim (p \vee q)$

p	q	$p \vee q$	$\sim (p \vee q)$
1	1	1	0
1	0	1	0
0	1	1	0
0	0	0	1

Truth Table for $(\sim p) \wedge (\sim q)$

p	q	$\sim p$	$\sim q$	$(\sim p) \wedge (\sim q)$
1	1	0	0	0
1	0	0	1	0
0	1	1	0	0
0	0	1	1	1

As stated in the very beginning of our discussion of the propositional calculus, a major issue is the search for universal truths or tautologies, that is, logical forms which result in true propositions when any propositions whatsoever are substituted for the variables (p, q, r, etc.) which the forms contain. We have already mentioned that the "law of the excluded middle," $p \vee \sim p$, is a tautology. Our "proof" consisted in saying that $P \cup P' = I$ in the Boolean algebra of classes. Then, since p, $\sim p$, $\vee$, 1 correspond to P, P', $\cup$, I, respectively, the truth value of $p \vee \sim p$ must always be equal to 1 and the tautology is established. Since this type of proof may seem a bit too informal, let us now provide a truth-table demonstration. In the table below, the first two columns are exactly those in the truth table for negation. The third column is then obtained by applying the truth table for disjunction to the first two columns. The truth values for $p \vee \sim p$, as listed in the third column, are obtained by using the first two columns and applying the truth table for disjunction.

Truth Table for $p \vee \sim p$ *[Law of Excluded Middle]*

p	$\sim p$	$p \vee \sim p$
1	0	1
0	1	1

Since all the entries in the third column are 1's, $p \vee \sim p$ is *always* true, regardless of the truth value of p, and therefore $p \vee \sim p$ is a *tautology*.

Let us now use the truth-table method to prove that one of the "laws of thought" which Aristotle and Leibniz considered the most important of all, namely, the *law of contradiction*, is a tautology. Its symbolic form is $\sim (p \wedge \sim p)$, and one way of expressing it verbally is: A proposition and its negation cannot both be true.

Truth Table for $\sim (p \wedge \sim p)$ *[Law of Contradiction]*

p	$\sim p$	$p \wedge \sim p$	$\sim (p \wedge \sim p)$
1	0	0	1
0	1	0	1

In the above truth table, the first two columns are the same as in the previous table. The truth values in the third column are obtained by using the first two columns and applying the truth table for conjunction. The truth values in the fourth column are obtained from those in the third column by applying the truth table for negation. Since the fourth column contains only 1's, $\sim(p \wedge \sim p)$ is *always* true, regardless of the truth value of p, and is therefore a tautology.

Thus far only the connectives "not," "and," "or" have been used in combining logical forms. Before we define further connectives, there are two points we wish to emphasize. The first is a reminder that in logical forms any propositions whatsoever may be substituted for the variables p, q, r, etc. Thus a particular substitution in $p \wedge q$ might lead to the statement: The earth is a planet *and* apple pie is nutritious. In ordinary conversation one would be unlikely to combine such unrelated propositions, but then symbolic logic is designed as a *universal* language, and we know that even stranger compounds than the one illustrated do occur in imaginative poetry or in the modern theater of the absurd. The second point is completely analogous to one in the algebra of classes where, by virtue of De Morgan's laws, it is possible to limit Boolean operation to $\cap$, ' or else to $\cup$, '. Thus instead of the logical sum $P \oplus Q$ one can use the equivalent $(P'Q')'$. The corresponding fact of the propositional calculus is that *disjunction* of propositions can be dropped, if one so desires, for instead of "p and/or q," that is, $p \vee q$, one may employ the equivalent "not-(not-p and not-q)," that is, $\sim[(\sim p) \wedge (\sim q)]$. One can see immediately, however, that this would make for lack of intelligibility in verbal expression. It is more lucid and more straightforward to assert that an insurance policy will pay benefits for illness and/or accident than to state that no benefit will be paid to any one who is not ill and has not met with an accident.

The two issues we have raised will have considerable bearing on the logical connective we shall next introduce. In the first place, we shall ultimately see that logic could manage without the new connective by using circumlocutions involving $\sim$, $\wedge$ or $\sim$, $\vee$. But we have just examined a small sample of the sort of verbal difficulties that may arise from such a procedure. Hence we shall now join two sentences by using "if _____, then," a connective used very frequently in everyday linguistic expression. "If p, then q," usually symbolized by $p \rightarrow q$, is described as a *conditional* form, and p is called the *antecedent*, q the *consequent*.

The following is an example of a conditional statement: If $1 + 1 = 5$, then Paris is a city in France. This sentence involves the second point considered above since, once again, two completely unrelated propositions are connected. But, as we have said, such composites are possible, and the only concern of the propositional calculus is whether to pronounce the results "true" or "false." We shall see (from the truth table below) that in modern logic the illustrative conditional is *true*. What is assumed in the truth table is that $p \rightarrow q$ is false only when the antecedent is true and the consequent false. Otherwise the conditional is postulated to be true.

Truth Table for $p \rightarrow q$

p	q	$p \rightarrow q$
1	1	1
1	0	0
0	1	1
0	0	1

To show that the truth table for $p \rightarrow q$ agrees with "common sense," let us analyze an example in which a father promises his son, "*If* your attendance record at school this term is good, *then* I'll give you a bicycle." At the end of the term, the son will pronounce judgment on the truth of his father's conditional statement according to which of the four following possibilities occurs.

(1) The boy actually does have a good attendance record, and his father does give him a bicycle. The boy will feel his father was telling the truth. Here we have the first row of the truth table, namely, truth values 1, 1, 1 for p, q, $p \rightarrow q$, respectively.

(2) The boy's attendance is good, but his father does *not* present him with a bicycle. "Dad, you were *not* telling the truth," the youngster says. Here we have the second row of the table, namely, truth values 1, 0, 0 for p, q, $p \rightarrow q$, respectively.

(3) The boy's attendance is *not* good, but nevertheless his father does give him a bicycle. Since the father made no threat, that is, did not cover the contingency just described, the boy would not question the truth of the original statement, and hence we have the third row of the table, that is, truth values 0, 1, 1 for p, q, $p \rightarrow q$, respectively. (Observe that these truth values also apply to the conditional "If $1 + 1 = 5$, then Paris is a city in France." Hence that conditional statement is pronounced to be *true*.)

(4) The boy's attendance is not good and his father does not give him a bicycle. The boy will feel that his father's behavior is just, and hence will not accuse his parent of making a false promise. Here we have 0, 0, 1 as truth values for p, q, $p \rightarrow q$, respectively (last row of the truth table).

For practice in the meaning of the conditional, the reader can convince himself that the statement "If $1 + 1 = 2$, then swords are plowshares" is *false* whereas the following assertion of Polonius (in which he refers to Hamlet and Ophelia) is *true*:

> "If he love her not,
> . . . .
> I am no assistant for a state."

While we are on the subject of the conditional, let us prove that symbolic logic could do without that logical form since, as the truth tables below *prove*, the forms of $q \vee \sim p$ and $\sim(p \wedge \sim q)$ are both "equivalent" to $p \rightarrow q$, since the new forms have the same truth table as that for the conditional. But if we are never to use $p \rightarrow q$, we shall have to make use of circumlocutions like the following: The father gave his son a bicycle and/or the boy's attendance record was not good.

Truth Table for $q \lor \sim p$

p	q	$\sim p$	$q \lor \sim p$
1	1	0	1
1	0	0	0
0	1	1	1
0	0	1	1

The reader can obtain the third column in the above table by applying negation to the first column. He can then apply the truth table for disjunction to the second and third columns in order to obtain the fourth column.

Truth Table for $\sim(p \land \sim q)$

p	q	$\sim q$	$p \land \sim q$	$\sim(p \land \sim q)$
1	1	0	0	1
1	0	1	1	0
0	1	0	0	1
0	0	1	0	1

In the above table the third column is obtained by using negation on the second. The fourth column results from applying the truth table for conjunction to the first and third columns. The last column is obtained by negating the fourth column. This table, like the truth tables for $p \to q$ and $q \lor \sim p$, shows that a certain composite of p and q is *true* unless p is true and q false.

Although the conditional form $p \to q$ is *not* a tautology since the last column of its truth table does *not* consist entirely of 1's (and, of course, we have illustrated conditional statements that are false), nevertheless certain *specialized* types of conditional form are tautologies. For examples of such special cases, the reader can verify the truth tables we shall next present.

Truth Table for $(r \land s) \to (r \lor s)$

r	s	$r \land s$	$r \lor s$	$(r \land s) \to (r \lor s)$
1	1	1	1	1
1	0	0	1	1
0	1	0	1	1
0	0	0	0	1

In the above table the third and fourth columns are obtained from the truth tables for conjunction and disjunction. The last column is obtained by applying the truth table for $p \to q$ to the third and fourth columns. Since the last column is made up of 1's, the given logical form is a tautology. The law in question signifies that using the conjunction of two propositions as antecedent and the disjunction of the same propositions as consequent will *always* yield a *true* conditional statement. For example, it is not necessary to analyze the following statement for meanings since it must be true by virtue of its form: If Mars is a comet and the reader is a monkey's uncle, then Mars is a comet or the reader is a monkey's uncle.

The truth table below proves that one of the fundamental laws of logic, namely, the *law of the syllogism*, is a tautology: If p implies q and q implies r, then p implies r. The reader should examine the table, column by column, since there is the added difficulty of considering *eight* rows because there are eight triples of possible truth values for *three* propositions.

Truth Table for $[(p \to q) \land (q \to r)] \to (p \to r)$
[Law of the Syllogism]

p	q	r	$p \to q$	$q \to r$	$(p \to q) \land (q \to r)$	$p \to r$	$[(p \to q) \land (q \to r)]$ $\to (p \to r)$
1	1	1	1	1	1	1	1
1	1	0	1	0	0	0	1
1	0	1	0	1	0	1	1
1	0	0	0	1	0	0	1
0	1	1	1	1	1	1	1
0	1	0	1	0	0	1	1
0	0	1	1	1	1	1	1
0	0	0	1	1	1	1	1

If the reader has actually carried out the task of checking the eight rows of the above truth table, he must have observed that the truth-table method may not always be an easy one for establishing laws of the propositional calculus. It is often easier to carry out the corresponding manipulations in the algebra of classes, since the procedures there are purely mechanical and are so much like those in common algebra. Thus one might prefer to establish the law of the syllogism by reasoning as follows. Since, as we have demonstrated, $p \to q$ is equivalent to $q \lor \sim p$, the Boolean class corresponding to the conditional can be symbolized as $Q \oplus P'$. Therefore the classes corresponding to the three conditionals in the law of the syllogism are $Q \oplus P'$, $R \oplus Q'$, $R \oplus P'$. Then the class corresponding to the larger conditional in which the three conditionals are embedded is

$$R \oplus P' \oplus [(Q \oplus P')(R \oplus Q')]'$$

If manipulation (by Boolean laws) can prove that this class is actually I, the *universal class*, then the corresponding logical form will be a *law*. To carry out the proof it will be easier to demonstrate that the complement of the above class is the null class, $\emptyset$, and hence that the class itself is I. In finding the complement of the above class, let us apply the De Morgan law which states that the complement of a sum is equal to the product of the complements of the addends. Then we must show that

$$R'P(Q \oplus P')(R \oplus Q') = \emptyset$$

If we multiply the first two factors and use the fact that $PP' = \emptyset$, the left member becomes

$$PQR'(R \oplus Q')$$

which is the same as

$$PQRR' \oplus PQQ'R'$$

Since $RR' = \emptyset$ and $QQ' = \emptyset$, each term in the above sum is $\emptyset$, and the total is $\emptyset$, which was the fact to be proved.

Our last considerations show that when the truth or falsity of statements is to be demonstrated logically, *both* the truth-table method and the technique of Boolean algebraic manipulation may become quite arduous. Therefore modern logic contains *rules of proof* which, in many situations, make it possible to avoid more cumbersome methodology. One such principle, called *modus ponens* in classical logic, is the *rule of detachment: If $p \rightarrow q$ is a true conditional, and if p is true, then q is true.* Thus when the premises of an argument contain both p and $p \rightarrow q$, then q can be *detached* and asserted to be true.

The rule of detachment can be derived from the truth table for the conditional. Since the rule assumes that p is *true*, only the first two rows of that truth table apply. But the rule also assumes that the conditional $p \rightarrow q$ is true. Hence the second row of the truth table is eliminated as a possibility. Thus only the first row remains and in that row q is true, which was the fact to be proved.

It is customary to symbolize the rule of detachment as follows:

$$p \rightarrow q$$
$$p$$
$$\therefore \ q$$

In this form it is applied to many arguments within mathematics itself, and in other areas as well. For example, one might reason as follows:

$p \rightarrow q$ If Smith was in Europe at the time a burglary was committed in New York City, then he did not commit that crime. (This is a *true* conditional.)

p (There is incontestable evidence that) Smith was in Europe at the time.

$\therefore \ q$ Smith did not commit the crime.

When the rule of detachment is applied to the law of the syllogism, one obtains another rule of proof, namely, the *rule of the syllogism: If $p \rightarrow q$ and $q \rightarrow r$ are both true conditionals, then $p \rightarrow r$ is a true conditional.* This rule can be proved by carrying out the three steps in the rule of detachment. Thus

(1) $[(p \to q) \land (q \to r)] \to (p \to r)$ is *true*. (Law of the syllogism)
(2) $[(p \to q) \land (q \to r)]$ is *true*. (The given rule of the syllogism *assumes* that each of the two conditionals is true. Hence their conjunction is true.)
(3) $\therefore p \to r$ is *true*. (Rule of detachment)

The *rule of the syllogism* is not to be confused with the law of the syllogism. The latter provides only the first step in the above proof. In many cases where the law holds, steps (2) and (3) above may *not* be true. Thus, because the law of the syllogism is a tautology, we may assert, without further analysis, that the following somewhat bizarre statement is *true*:

If $(2 > 1) \to (3 > 4)$ *and* $(3 > 4) \to (2 = 1)$, *then* $(2 > 1) \to (2 = 1)$

It would *not* be permissible, however, to detach the final conditional and assert its truth. It is, in fact, *false* because it contains a true antecedent and a false consequent.

In deductive proofs one may make repeated use of the *rule* of the syllogism. Thus, in school geometry, one may prove that the opposite sides of a rectangle are equal by asserting: If a figure is a rectangle, then it is a parallelogram (by definition), and if a figure is a parallelogram, then its opposite sides are equal (theorem). Therefore, if a figure is a rectangle, then its opposite sides are equal (*rule of syllogism*).

In order to be able to verbalize complex conditionals like those we have been illustrating, mathematicians may read $p \to q$ in many different ways they consider synonomous with "If p, then q." Thus some logicians permit $p \to q$ to be read as "p *implies* q," but others object to this rendition because it suggests a causal relationship between antecedent and consequent. We have made a special point of indicating that no such logical connection need exist. Hence logicians may say that the truth table for $p \to q$ defines *material implication*, and may read $p \to q$ as "p implies q in *material meaning* or *as a matter of fact*." This convention enables them to give *logical implication* a different meaning (but nevertheless one defined in terms of material implication).

The term "logical implication" carries with it the suggestion of logical inference or deduction, a process for which one would like to have *universal validity*. But then material implication, as expressed by $p \to q$, will not serve, since the conditional form is *not* a tautology. The trouble is that the variables p and q have too much freedom, since any propositions whatsoever may be substituted for them. But in relation to special subjects or problems, one is *not* interested in *all* propositions. The field of investigation places considerable restriction on the permissible substitutions for p and q, that is, it prescribes a special universe of discourse for those variables. Then, in relation to such a universe, $p \to q$ may become a *law* or universal truth (a sort of relative tautology).

Common algebra makes us familiar with the kind of situation we have described. Thus, given the open sentence $x^2 - 3x + 2 = 0$, its truth set in the domain of real numbers is $\{1, 2\}$. The given sentence is by no means a law for the specified domain. But if we *restrict* the domain of x to the set $\{1, 2\}$, then, relative to that small universe, we do have the *law* or universal truth: For all x, $x^2 - 3x + 2 = 0$.

To give an analogous illustration for the case of $p \rightarrow q$, the truth-table demonstration on page 121 indicates that if p is restricted to propositions of the form $r \wedge s$, and q to those of the form $r \vee s$, then $p \rightarrow q$ is a tautology relative to the universe thus prescribed.

Whenever the domains of p and q are restricted to special sets of propositions and *all* substitutions for p and q from the domains thus prescribed make $p \rightarrow q$ a true statement, then (according to the usage of some logicians) p implies q *logically* in the specified context. Logical implication is often symbolized as $p \Rightarrow q$.

When a theorem of a deductive science is expressed in conditional form, the statement is always a *logical implication*. For example, "If two sides of a triangle are equal, then the angles opposite are equal" can be symbolized by $p \Rightarrow q$, with interpretations as follows:

$$p = p(x): x \text{ has two equal sides}$$
$$q = q(x): x \text{ has two equal angles}$$

where the domain of x is the set of all triangles in the Euclidean plane. Here $p \rightarrow q$, that is, $p(x) \rightarrow q(x)$, refers to countless diagrams where x may be a small or a large triangle of any shape. There will be many pictures in which p is true and many in which it is false, and the same can be said about q, but there will *never* be a diagram where p is true and q is false. Hence the very numerous statements $p \rightarrow q$ will *all* be true, and thus $p \Rightarrow q$.

Again, let $I = \{1, 2, 3, 4\}$ be the domain of x and

$$p = p(x): x < 2$$
$$q = q(x): x \text{ is a perfect square}$$
$$p \rightarrow q: \quad \text{If } x < 2, \text{ then } x \text{ is a perfect square}$$

It is easy to show that $p \Rightarrow q$ because $p \rightarrow q$ is a true statement in all (four) possible cases, namely:

If $1 < 2$, then 1 is a perfect square.
If $2 < 2$, then 2 is a perfect square.
If $3 < 2$, then 3 is a perfect square.
If $4 < 2$, then 4 is a perfect square.

In mathematics one often makes assertions of the form "$p \rightarrow q$ *and* $q \rightarrow p$." This logical form, called the *biconditional* of p and q, is usually symbolized as $p \leftrightarrow q$ (with superposition of an arrow pointing from p toward q on one pointing from q toward p). It can be expressed verbally as p *if and only if* q, abbreviated by mathematicians to p *iff* q. From what we have just stated,

$$p \leftrightarrow q = [(p \rightarrow q) \wedge (q \rightarrow p)]$$

We shall make use of this definition in deriving the following truth table for the biconditional.

Truth Table for $[(p \to q) \wedge (q \to p)]$ *[The Biconditional]*

p	q	$p \to q$	$q \to p$	$[(p \to q) \wedge (q \to p)]$
1	1	1	1	1
1	0	0	1	0
0	1	1	0	0
0	0	1	1	1

In the above truth table, the third and fourth columns were obtained from the first two by application of the truth table for the conditional, and the last column was obtained by applying the truth table for conjunction to the third and fourth columns. Since the symbol at the top of the last column is identical in meaning with the biconditional, $p \leftrightarrow q$, we can abbreviate the above table as follows.

Truth Table for the Biconditional

p	q	$p \leftrightarrow q$
1	1	1
1	0	0
0	1	0
0	0	1

Those logicians who read "$p \to q$" as "p *implies q materially*" read "$p \leftrightarrow q$" as "p is *materially equivalent* to q." Thus the first and last rows of the truth table for *material equivalence* (the biconditional) show that two statements are *truly* equivalent in material meaning if they are both true or both false.

Again, there is the question of *logical equivalence*, $p \Leftrightarrow q$, which occurs in relation to special subjects or special problems that limit the domains of p and q in such a way that the logical form $p \leftrightarrow q$ *always* yields a true material equivalence when permissible substitutions (propositions in the prescribed domain) are substituted for p and q. Thus, we leave as exercises for the reader the task of proving

(1) $(x$ is a perfect square$) \Leftrightarrow (x < 2)$
   when the domain of x is $I = \{0, 1, 2\}$;
(2) $(x$ is equilateral$) \Leftrightarrow (x$ is equiangular$)$
   when the domain of x is the set of all triangles in the Euclidean plane.

The reader will observe in both exercises that substitution of a value of x from the prescribed domain yields two true statements on both sides of the equivalence symbol or else two false statements so that, in every case, the material equivalence is true.

Earlier in this chapter we asserted that logical forms are "equivalent" if they have the same truth tables. Now we can reveal that *logical equivalence* was meant. The forms considered were composites of the same propositional variables.

Given any two such forms, say, $f(p, q, r, \ldots)$ and $g(p, q, r, \ldots)$, substitution of certain truth values for $p, q, r, \ldots$ may cause f to have truth value 0. But in that case, g will also have truth value 0 because the truth tables for f and g are the same. Substitution of other truth values for $p, q, r, \ldots$ may give f truth value 1, and hence g truth value 1. But this means that f and g are both true or both false for every possible choice of truth values for p, q, r, etc. Then, in accordance with the truth table for the biconditional, f and g will be materially equivalent in every possible instance, and hence $f \Leftrightarrow g$, that is, f is *logically equivalent* to g.

The concept of logical equivalence is important because one of the logical rules of proof is the *rule of replacement for sentences*. That rule states that any logical form may be replaced by a logically equivalent form. It is often very helpful, in the course of a proof or logical argument, to be able to make such a replacement.

The form which one may wish to replace by a logical equivalent is very often a conditional sentence. Thus, in plane geometry, when it is required to prove that if alternate interior angles are equal, then lines are parallel, it is easier to demonstrate that if lines are not parallel, that is, if they intersect, then they form unequal alternate interior angles. The latter conditional statement is called the *contrapositive* of the original one and can be shown to be *logically equivalent* to it.

The *contrapositive* of $p \to q$ is defined to be $(\sim q) \to (\sim p)$. To prove the logical equivalence of the two forms, the reader can derive the following truth table for the contrapositive and verify that it is identical with that for the conditional.

Truth Table for $(\sim q) \to (\sim p)$

p	q	$\sim q$	$\sim p$	$(\sim q) \to (\sim p)$
1	1	0	0	1
1	0	1	0	0
0	1	0	1	1
0	0	1	1	1

Here once more the reader might prefer to use Boolean manipulations. The class corresponding to $p \to q$ can be symbolized by $Q \oplus P'$ and that corresponding to $(\sim q) \to (\sim p)$ would be $P' \oplus (Q')'$. But, since $(Q')' = Q$, the latter class is identical with the former. Hence the two logical forms are equivalent since they correspond to the same class.

The equivalence of the conditional and its contrapositive carries with it the fact that $p \Rightarrow q$ and $(\sim q) \Rightarrow (\sim p)$ are also equivalent.

If in his school days the reader did not make the acquaintance of the contrapositive form of a conditional proposition, $p \to q$, he surely encountered the *converse*, $q \to p$. He can now employ truth-table analysis to show that the conditional and its converse are *not* equivalent. Thus the truth or falsity of a conditional statement does not necessitate the respective truth or falsity of its converse. Hence, if a conditional sentence is a *logical implication*, this is not necessarily the case for its converse. For example, we have the logical implication: If a man lives in Chicago, then he lives in the United States. The converse would be: If a man lives in the United States, then he lives in Chicago. The truth of the latter conditional sentence is contingent on the man selected; its antecedent may be true (for example, when Mr. Jones actually lives in the United States) and its consequent false (Mr. Jones lives in Boston). Hence the converse sentence is *not* a logical implication.

On the other hand, consider: If the opposite sides of a quadrilateral are parallel, then it is a parallelogram—whose converse is: If a quadrilateral is a parallelogram, then its opposite sides are parallel. Here both conditional sentences are logical implications, that is, they are laws in the universe of plane quadrilaterals. In the particular illustration, the two implications provide the customary *definition* of a parallelogram. In fact, every definition involves a logical implication and its converse, which must also be a logical implication. In other cases, however, when it is given as the premise of an argument that some conditional sentence is a logical implication, it *cannot* be assumed that the converse is a logical implication, and reasoning is required to demonstrate whether or not this is so.

Suppose that the conditional $p \to q$ is transformed in a different way so as to form its *inverse* or *obverse*, $(\sim p) \to (\sim q)$. The contrapositive of this last conditional sentence must be logically equivalent to it, and this contrapositive is $q \to p$. But $q \to p$ is the *converse* of the original conditional sentence, and hence the inverse and the converse of any conditional are logically equivalent forms.

A common type of fallacy in everyday thought results from failure to realize that the converse or inverse of a logical implication and not the implication itself is being applied. This is the case when someone says, "There is no fracture because the X-ray was negative." Now if the X-ray reveals a break, then a fracture is present, but the inverse is false. It is not true that if the X-ray does not reveal a break, then no fracture is present. Or again, there is a comment like "He's so absolutely mad that he must be a genius." Now, granting (from a conventional point of view) the law that if a person is a genius, then he must be mad, the converse need not be a universal truth. It is not true that if a man is mad, he is necessarily a genius.

The rule of replacement for sentences makes it possible to substitute one logical form for another providing the two forms are *equivalent*. But the rules of proof of the propositional calculus permit a second type of substitution. The *rule of substitution for variables* states: In any tautology, it is permissible to substitute for all occurrences of a particular propositional variable any logical form whatsoever, and the result of such a substitution will be a tautology.

Thus, different substitutions for p in $p \lor \sim p$, the law of the excluded middle, would lead to the tautologies

$$(q \land r) \lor \sim (q \land r)$$
$$(q \to r) \lor \sim (q \to r)$$

Again, consider the tautology (page 121)

$$(r \wedge s) \Rightarrow (r \vee s)$$

(Since this is a tautology, it is a *logical implication* and hence we have used the symbol ⇒). If we replace r by $p \vee q$ and s by $t \to v$, we arrive at the more complicated tautology,

$$[(p \vee q) \wedge (t \to v)] \Rightarrow [(p \vee q) \vee (t \to v)]$$

an implication true whatever the truth values of the *four* components, p, q, t, v.

In recent years there has been great interest in *logical machines*, that is, electronic calculators that will solve problems in formal logic. In the past, there were some purely mechanical machines like the "logical piano" designed by the logician William Stanley Jevons (1835–1882), but none of those machines was as versatile as the modern electronic "brains." It was Claude E. Shannon who, in a 1938 research paper, presented his discovery of the remarkable analogy between the truth values of propositions and the states of switches and relays in an electric circuit. Shannon showed, in fact, that "switching algebra" is a concrete interpretation of the abstract binary Boolean algebra and hence is isomorphic to the truth-table algebra of propositions. After Shannon's discovery, Boolean algebra became important in the design of switching circuits for all sorts of purposes—telephone networks, computers for general arithmetic calculation, computers to test the syllogisms of classical logic or the logical forms of modern symbolic logic, etc.

In Shannon's interpretation, a closed switch or a closed circuit (one in which current flows) can be considered to be in "state 1" and an open switch or an open circuit (one in which no current flows) can be described to be in "state 0." Then combinations of open or closed switches in a circuit will control the current, cutting it off or permitting it to flow, that is, yielding state 0 or state 1. We shall picture some circuits in which A, B, C, . . . will be used to represent switches. Switches in a circuit may operate so that two or more open and close simultaneously. In that case, our diagrams will label all such switches with the same letter, A, say. Again, A and A' will refer to different switches that operate simultaneously but are controlled so as to be in opposite states. In other words, when A is in state 1, A' is in state 0, and when A is in state 0, A' is in state 1.

In Figure 6.4 the switches A and B are connected in "series," a combination represented by $A \otimes B$, or by AB. Current will flow between the terminals if and only if both switches are closed, and this corresponds to the binary Boolean product

Figure 6.4   *A* ⊗ *B* interpreted by switches in series

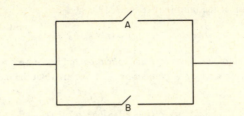

Figure 6.5 $A \oplus B$ interpreted by switches in parallel

$1 \otimes 1 = 1$. The possibilities where current does *not* flow (Figure 6.4) correspond to the other three entries in the multiplication table of the binary Boolean algebra, that is, to $0 \otimes 0$, $0 \otimes 1$, $1 \otimes 0$, products which are all equal to 0.

Again, in Figure 6.5 the switches A and B are connected in "parallel," a combination represented by $A \oplus B$. Current will flow between the terminals if A is closed, or if B is closed, or if both A and B are closed, states corresponding to the three entries $1 \oplus 0$, $0 \oplus 1$, $1 \oplus 1$ in the binary Boolean addition table. All three sums are equal to 1. If A and B are both open, one has $0 \oplus 0 = 0$, which signifies that no current flows.

No matter how circuits are constructed with switches in series or in parallel, the total picture will be represented by an expression involving "sums" and "products" of $A, B, C, \ldots, A', B', C', \ldots$, that is, by a "Boolean polynomial." In turn, the *state* of the entire network will be determined by appropriate substitution of 1's and 0's and computation of the corresponding binary Boolean sums and products. The kind of mechanical manipulation one carries on when literal symbols stand for subsets of a universal set may be carried out when the symbols represent switches. In that way, simplification of a Boolean polynomial may make it possible to replace a complicated switching network by a simpler arrangement that is

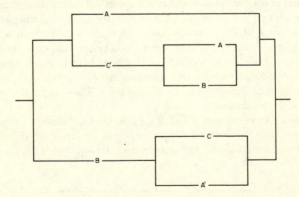

Figure 6.6 Switching circuit for $A \oplus C' (A \oplus B) \oplus B (C \oplus A')$

equivalent to it. For example, consider the circuit diagrammed in Figure 6.6. With Shannon's interpretation, the circuit can be represented by the polynomial

$$A \oplus C'(A \oplus B) \oplus B(C \oplus A')$$

which, by application of the first distributive law (and the commutativity of $\otimes$), becomes

$$A \oplus AC' \oplus BC' \oplus BC \oplus A'B$$

The first two terms have sum A by virtue of the second law of absorption. The sum of the third and fourth terms can be expressed as $B(C' \oplus C) = BI = B$. Hence the polynomial simplifies to

$$A \oplus B \oplus A'B$$

Applying the second law of absorption to the last two terms gives the sum B, and hence the original polynomial reduces to

$$A \oplus B$$

Thus the complicated network of Figure 6.6 is equivalent to the simple circuit of two switches in parallel pictured in Figure 6.5.

Our brief excursion into the algebra of switching circuits concludes our discussion of the elements of the first two divisions of symbolic logic (in Russell's partition). But there are several final points which require emphasis. The first of these is that we have actually treated some aspects of what logicians call quantification. When we found truth sets for open sentences (in the present chapter and also in Chapter 4), we observed that the sentences were true of *all, some,* or *no* values of the variables. To express such facts, we used quantifying phrases like "for all x," "for some x," thereby converting an open sentence into a proposition (page 69). When the sentence was true for *no* values, that is, when the truth set was $\emptyset$, we merely *negated* the sentence and found that the negation was true "for all x" (page 114). The only thing we failed to do was to introduce symbols for the quantifying phrases. Let us repair that omission by revealing that $\forall_x$ (read "for all x") is the symbol for the *universal quantifier* and that $\exists_x$ (read "for some x" or "there exists an x such that") is the symbol for the *existential quantifier.*

We remark that the symbolic treatment of quantification lends itself to mechanization of logical procedures involving quantifiers. However, the elementary logical material we have been discussing does not call for any elaborate manipulation of quantifier symbols. Only when quantification concerns two or more variables is it worthwhile to formalize the handling of the inverted A's and reversed E's, Such *multiple quantification* was involved implicitly when various algebraic postulates were stated, for example, "*For all x* and *for all y, xy = yx.*" This law can now be symbolized as follows:

$$\forall_x \forall_y \ (xy = yx)$$

Again, let us symbolize two quantified statements which were made (page 69) in connection with the solution of open sentences in the variables x and y, with domain $\{1, 2, 3, 4, 5\}$, namely,

$$\forall_x \forall_y \ (x + y > 1)$$
$$\exists_x \exists_y \ (x + y = 1)$$

The first statement is *true* and the second is *false*.

It would not have mattered if we had reversed the order of the quantifier symbols in the three instances just cited. But such symbols are *not* commutative in every case. Thus, if the domain of x and y is the set of integers $\{\ldots, -3, -2, -1, 0, 1, 2, 3, \ldots\}$, the first of the following statements is *true* and the second, formed by *reversing the order of the quantifiers*, is *false*:

$$\forall_x \exists_y \ (y < x)$$
$$\exists_y \forall_x \ (y < x)$$

The first statement can be read as "For each and every x, there exists a y such that $y < x$." The reader can convince himself of the truth of the statement by choosing $y = x - 1$, that is, the pairs $x = 3, y = 2$; $x = 2, y = 1$; $x = 0, y = -1$; etc. Or he can choose $y = x - 4$, with pairs like $x = 3, y = -1$ and $x = 2, y = -2$, etc.

The second statement above can be read as "There exists a y such that, for every x, $y < x$." In other words, "There exists an integer y smaller than every integer." This is false, since the scale of integers continues downward forever and ever. It never stops at some smallest integer. Even if we were to imagine such a smallest integer, it could not qualify as y in our statement, since such a y would be *equal* to one integer (the hypothetically smallest one) and hence would *not* be less than *every* integer. Thus we see that, in general,

$$\forall_x \exists_y \ p(x, y) \neq \exists_y \forall_x \ p(x, y)$$

where $p(x, y)$ is an open sentence or propositional form in two variables, x and y.

The algebra of multiple quantification is part of a general topic which will be treated more fully in our next chapter, namely, the logic of relations. In anticipation, however, we shall give a single example in which the concepts of "inverse" relation and multiple quantification make possible a certain kind of *non*syllogistic reasoning with which classical logic cannot cope. Thus if one accepts the premise that "someone loves everybody," Aristotelian logic does not permit a conclusion which seems obvious but can be rigorously validated by the modern algebra of relations, namely, that "everybody is loved by someone."

Finally, in accordance with our spiral scheme of organization, let us reveal that even after the treatment of relations in our next chapter, the story of modern symbolic logic will not be complete. The Boolean algebra of classes and the two-valued truth-table method were considered basic in logic until 1921, when two climactic research papers appeared, one by the Polish mathematician Jan Lukasiewicz on three-valued logic, and one by the American Emil L. Post (1897–1954) on general n-valued systems, where n is any integer greater than 1. One example of a three-valued truth table is given below.

Truth Table for p → q in Three-Valued Logic

p	q	p → q
1	1	1
1	2	2
1	0	0
2	1	1
2	2	2
2	0	2
0	1	1
0	2	1
0	0	1

In the table, truth values 0, 1, and 2 have been assigned, but for the sake of clarity, the reader might replace 0 by F = "false," 1 by T = "true," and 2 by M = "possible" (that is, not known to be either true or false). Examination of the above table shows that the first, third, seventh, and ninth rows represent the truth table for the conditional of two-valued logic. Also, the new table, even though more general, preserves the essential postulate for the truth value of conditional statements, namely, the assumption that such a statement is false if and only if its antecedent is true and its consequent false.

7

Forefathers of Modern Mathematics and Their Legacy

In evaluating contributions to mathematics during any period of its history, the conceptions which must be judged most profound are those that unify what has gone before and, at the same time, open up wide avenues for future generalization. One such integrating, trail-blazing idea is to be found in the *method* of analytic geometry, as formulated by René Descartes (1596–1650) and Pierre de Fermat (1601–1665).

The two forefathers, as we have named them, arrived at their ideas independently and almost simultaneously. Fermat's thinking on the subject seems to have been carried on around 1629. In 1636 he communicated his thoughts through correspondence with other mathematicians, in particular in a special letter to the geometer and physicist, Gilles Persone de Roberval (1602–1675). Another of Fermat's letters (to Pascal) contributed to the beginnings of a different subject, probability theory (Chapter 13). But the full details of his conception of analytic geometry were published only posthumously in a paper entitled *Isogoge ad locus planos et solidos*.

Descartes arrived at his formulation after 1629, but he holds priority over Fermat in the matter of publication date. Descartes' scientific masterpiece, the *Discours de la Méthode pour bien conduire sa raison et chercher la vérité dans les sciences*, was published in 1637. The last of three appendices to this work was entitled *La Géométrie*. In Descartes' final afterthought, analytic geometry was first given to the world at large.

Descartes and Fermat, both French, both living in the same era, both geniuses interested in some of the same mathematical questions, were both unaware that their geometric method was a major scientific bequest to posterity, something too important to be relegated to an appendix or to be confided casually in an epistle. In all other respects the two men were very different.

Descartes, forefather of modern mathematics, has also been called the father of modern philosophy. He was creative in other fields as well: physics, cosmology, chemistry, physiology, and psychology. The progenitor of so much science came of a noble family. His father was in comfortable circumstances and expected his son to lead the life of a French gentleman. René was a delicate child and did not start his formal schooling until the age of eight. At the Jesuit college at La Flèche the rector, Father Charlet, observed the boy's physical frailty and advised him to lie in bed as late as he pleased in the mornings. It was thus that Descartes formed the lifelong habit of spending his mornings in bed whenever he wished to think.

At an early age he entered the army, a strange thing for a man to do who expressed his personal philosophy in the words, "I desire only tranquillity and repose." As it happens, there was a brief interval of tranquillity and repose during the wars of William the Silent, and at that time Descartes began to think about analytic geometry. Then, at the age of twenty-five, he left the army for good. Later, in 1628, he retired to Holland for twenty years of research in philosophy and mathematics, He did not neglect other scientific studies, and much of his time was given to the preparation of an imposing treatise, *Le Monde*. This book presented a scientific rationalization of the Book of Genesis, and gave Descartes' physical doctrine of the universe. He determined to preserve all his work for posthumous publication as his legacy to science, but in 1637 his friends persuaded him to release his greatest work, the *Discours de la Méthode*.

In 1649 Descartes decided to vary his existence by accepting an invitation to the court of Queen Christina of Sweden. This headstrong, masculine girl, whose tyrannic behavior is historical, was Descartes' undoing. The poor scholar, who could not endure the harsh climate of Sweden or the difficult routine set by the twenty-three-year-old queen, fell ill and died of pneumonia within a few months of his arrival in Stockholm.

Words are an easy form of penance. The eccentric tyrant wrote to a friend: "The greatest of philosophers has just died. If I were superstitious, I should weep like a child over his death, and I should bitterly repent having drawn this bright star from its course. His death depresses me; it will always fill me with justified but useless regret."

The facts of Fermat's life are known, but there is not too much to tell. He was born in 1601 at Beaumont-de-Lomagne, near Toulouse, in southwestern France. His father was a leather merchant and his mother came from a family of parliamentary jurists. He received his early education at home and subsequently studied at Toulouse, where, in 1631, he became commissioner of requests. Seventeen years later he was promoted to the position of councillor in the parliament at Toulouse, and this job filled the next seventeen years, right up to the day of his death at Castres, where he was trying a case. In his official capacity he was noted for remarkable legal knowledge and strict integrity. His mathematical activities were carried on as a hobby.

A month after he had been appointed to his first position, he married Louise de Long, his second cousin. There were five children of this marriage, two daughters and three sons. Both girls became nuns, and Fermat's son Samuel carried on the family legal tradition by writing a number of books on law. He also edited his father's work and translated several Greek scientific and mathematical works.

Fermat was a "co-inventor" in relation to analytic geometry, probability theory, differential calculus. In a broad sense, he may also be said to have helped the launching of the "calculus of variations" and modern theories of "optimization." In the higher arithmetic or theory of numbers, however, he has few peers in mathematical history. We shall have more to say about his contributions to that subject later (Chapter 21), when we explain why "Fermat's last theorem" is still an enigma that challenges the best mathematical minds.

Nowadays historians of mathematics point out that Fermat and Descartes should not be given complete credit for the creation of analytic geometry, because some of its ideas were anticipated by others. The notion of coordinates is fundamental in Cartesian geometry, and that subject is in fact often called *coordinate geometry*. In this matter, however, ancient Egyptian surveyors were forerunners of Descartes and Fermat. Also, Hipparchus (*ca.* 140 B.C.), the greatest astronomer of antiquity, used longitude and latitude as the coordinates of a point on the earth's surface. Then, too, analytic geometry associates algebra and geometry. But Greek geometers, as we have already seen, stated fundamental algebraic laws in geometric terms, solved quadratic equations by the geometric method of "completing a square," and carried out a great quantity of painstaking geometric algebra. Again, the major content of traditional college textbooks in coordinate geometry involves properties of curves described as *conic sections*. Emphasis on this material goes back to Menaechmus. Historians indicate that an indubitable source of inspiration was the profound *Conic Sections* of Apollonius of Perga (*ca.* 225 B.C.).

Actually none of these early contributions employed the essential methodology of Descartes and Fermat, but in the opinion of the author of the present book, thoughts that came close to their ideas must be credited to Nicole Oresme (1323–1382). A native of Normandy, he taught at the University of Paris, became headmaster of the College of Navarre in 1356, and was ultimately named bishop of Lisieux. One of his mathematical treatises, written about 1370, uses "Cartesian" coordinates (more than 260 years in advance of our "forefathers"). He went beyond the ancients by considering not only coordinates but *relations* between them. Thus he derived the equation of a straight line, that is, the *relation* between the coordinates of any point on the line. He also considered the graphic representation of certain scientific formulas. He gave some thought to "solid" or three-dimensional analytic geometry and even attempted to formulate something analogous to a modern four-dimensional space. What prevented Oresme from getting beyond first steps was the lack of a good algebraic symbolism. That missing link was gradually made available, notably in the work of François Viète (1540–1603) and in the contributions of Descartes himself.

Our brief historic résumé actually puts the cart before the horse. For how can one render critical judgment about who invented something without knowing just what the invention is? Hence the present chapter will consider the question: What constitutes analytic geometry and what makes it important? We see the subject not as a geographic scheme of location or an exercise in blending algebra and geometry or a painless path to Euclidean and more advanced geometries, but as a means of expressing relationships (*relations*, in mathematical terminology) of all kinds and making important deductions about them. This point of view makes analytic geometry the very core of mathematics, since many mathematical subjects —for example, trigonometry, calculus, "complex analysis," probability, mathematical statistics, and even one essential part of logic—are theories of general or special relations.

It is often said that Descartes and Fermat wedded algebra and geometry (after some preliminary matchmaking by predecessors). But today the key to that romance is found in other matings. A one-to-one correspondence can pair the real numbers with the points of a straight line and also match every geometric configuration (point set) on the line with some aggregate of real numbers (Chapter 2). The same sort of correspondence can mate the figures of plane geometry with sets whose members are *ordered pairs* of real numbers. The first of these "marriages" produces a "one-dimensional" analytic geometry. From the second, the plane analytic geometry of Descartes and Fermat can be derived.

For some purposes one might wish to develop analytic geometry on a straight line, and we shall do this to some extent in connection with relativity (Chapter 18) and other advanced geometries (Chapter 17). But a "one-dimensional" universe does not provide opportunities to indicate the full flavor and scope of analytic geometry. Descartes and Fermat developed their initial ideas with reference to a *plane* geometry, and we shall do likewise.

In coordinate geometry, as we have said, geometric configurations in the plane are matched or identified with sets containing *ordered pairs of real numbers*. For present purposes the concept of real number needs no elaboration beyond the material provided in our examination of the points of view of Eudoxus, Dedekind, and Cantor (Chapter 2). But we have not yet probed the general notion of *ordered pair*, and hence that idea must now be examined.

It is said that the novelist Charles Dickens kept a file of peculiar names so that he might combine them, as the need arose, to form appellations that would "sound" like the characters he created. Although we shall not list Micawbers or Pecksniffs, let us consider the following sets of names:

$$S = \{\text{Martin, Randolph, Spencer}\}$$
$$T = \{\text{Joyce, Leslie}\}$$

We can indicate the idea of an *ordered pair* by picking a name from S followed by one from T, and then reversing the process, for example, Randolph Joyce and Joyce Randolph. Here we have the same pair of names in each case but two *different* ordered pairs, since Randolph is the given name in the first pair and the surname in the second. Also, we might consider Randolph Joyce as the name of a man and Joyce Randolph as that of a woman.

If we form all possible ordered pairs resulting from the choice of a given name from S and a surname from T, we arrive at an aggregate that is called the *Cartesian product* of S by T, symbolized as $S \times T$ (read "S cross T"). Thus $S \times T$ will contain *six* full names like Martin Joyce, Randolph Joyce, etc.

Now $T \times S$ ("T cross S") is a different set, containing the six ordered pairs formed by choosing given names from T and surnames from S. Although the six full names in $T \times S$ are not required to be listed in alphabetical or any other order, we shall present them in a pattern that can be associated with other concepts. Thus

$$T \times S = \begin{cases} \text{Joyce Spencer} & \text{Leslie Spencer} \\ \text{Joyce Randolph} & \text{Leslie Randolph} \\ \text{Joyce Martin} & \text{Leslie Martin} \end{cases}$$

Here we see that the ordered pairs or full names in $T \times S$ differ from those in $S \times T$. Hence the Cartesian product defines a *noncommutative* multiplication of sets.

In passing, we remark that the pattern of $T \times S$ matches Figure 2.2a, the Pythagorean figurate polygon for the numerical product, 2×3. In Chapter 2 we defined the multiplication of cardinal numbers as repeated addition. We stated that 2×3 was the result of adding 3 two times, that is, the cardinal number of the rectangular pattern formed by 2 columns each containing 3 elements. But, having defined the meaning of Cartesian product, we can now give a new definition of the multiplication of cardinals, one that matches a numerical product with a product of sets. Now 2×3 is defined as the cardinal number of the *Cartesian product* of a set containing 2 elements by a set containing 3 elements. Then either Figure 2.2a or our $T \times S$ would be such a product. The cardinal number of either set is 6, and hence $2 \times 3 = 6$. The new definition of multiplication is considered superior to the old not only because it matches two species of product but because it can be generalized to apply to *infinite* cardinal numbers.

If we give a slightly different interpretation to the sets S and T above, it will enable us to illustrate what is meant by a *binary relation*, a relation associating *two* elements, one in a set S and a second in a set T. Let us then think of the elements of S and T as the *given names* of five distinct individuals, three men and two women. Then Martin is the name of a man who may possibly be related to Joyce or Leslie. For example, he might be the father or the brother of either or both women. Then each of the statements, "Martin is the father of Joyce" and "Martin is the father of Leslie" associates *two* people, one in S and one in T. The other men might be cousins of the women, and one or more of the men might possibly be "related" to Joyce or Leslie in the mathematical sense of being taller than she is, or living within a mile of her home, or belonging to the same church as she does, etc. In mathematics, "is the father of," "is the brother of," "is taller than," "belongs to the same church as," etc. would all be described as *binary relations* on or from S to T. To assist us in formulating the meaning of such a relation more exactly, we now list the ordered pairs in $S \times T$ (once again in a pattern that can be linked with other ideas). Thus

$$S \times T = \begin{cases} \text{(Martin, Leslie)} & \text{(Randolph, Leslie)} & \text{(Spencer, Leslie)} \\ \text{(Martin, Joyce)} & \text{(Randolph, Joyce)} & \text{(Spencer, Joyce)} \end{cases}$$

To avoid any ambiguity that may arise from verbal descriptions, we next introduce variables s and t with domains (replacement sets) S and T, respectively, so
that we may provide exact formal definitions. Then the open sentence "s is the son
of t" would be said to express a *binary relation* on or from S to T. Also, the truth
or solution set of the open sentence is called the *graph* of the relation (because, so
often, one provides a pictorial representation).

To obtain the graph of the above relation one can substitute in turn each
ordered pair of $S \times T$ in the defining open sentence. Those pairs that lead to true
statements form the graph. We must, of course, have such personal information
about the five parties as will enable us to say what is true and what is false. At
any rate, not all six statements resulting from the substitution of pairs can be true.
Thus, if it is true that Martin is the son of Joyce, then he is surely not the son of
Leslie. Suppose then that the facts are such as to yield

$$\{(\text{Martin, Joyce}), (\text{Spencer, Joyce})\}$$

as the graph of the binary relation. We observe that this graph is a *subset* of the
Cartesian product, $S \times T$.

If the reader wishes to see the aptness of our terminology, he can picture $S \times T$
in the Pythagorean figurate form as a rectangular universe of 6 points, as in
Figure 7.1. Then the two enclosed points [representing (Martin, Joyce), (Spencer,
Joyce)] constitute a "geometric" graph of the relation on S to T, "is the son of."

Figure 7.1   Graph of the relation " is the son of "

We might find graphs for other binary relations on S to T, and, in every case,
the graph would be a subset of $S \times T$. Thus the graph of "is the brother of"
might be $\{(\text{Randolph, Joyce}), (\text{Randolph, Leslie})\}$, that is, the two points in the
second column of Figure 7.1.

Also, if we were to reverse the order of elements of the pairs in the two relations whose graphs we have just considered, we would obtain the graphs of the
respective *inverse* relations, "is the mother of" and "is the sister of," binary relations on T to S. Thus, in general, binary relations on S to T will possess inverses
whose graphs are subsets of $T \times S$.

As additional examples of binary relations on S to T, let us consider "is 300
years older than." Since we imagine S and T to contain names of living people, no
member of S can possibly be 300 years older than a member of T, and the graph
of the relation in question is $\varnothing$, the null or empty set (which is considered to be a
subset not only of $S \times T$ but of every set). The relation "is 300 years older than"
can be described as *nullary*.

Now suppose that all five persons named in S and T happen to live in the same
town. Then the graph of the relation "lives in the same town as" would be our entire

universe, $S \times T$, an improper subset of itself. Here we are dealing with a *universal* relation.

Because the graph of a binary relation on S to T, where S and T are *any* sets, is always a collection of ordered pairs forming a subset of $S \times T$, it is customary, even though an "abuse of language" is entailed, to identify relation and graph. Then a *binary relation* on S to T is *defined* as any subset of $S \times T$, that is, as *any set of ordered pairs* where each pair is a member of $S \times T$. We shall adhere to this definition from now on and reserve the term *graph* for a pictorial representation of a relation.

Then, in the illustration we have been considering, the reader can form different subsets of $S \times T$, thereby defining different relations on S to T. Thus he might form the subset

{(Randolph, Joyce), (Spencer, Joyce), (Spencer, Leslie)}

This subset of three ordered pairs is a binary relation, by *definition*. One is not required to connect the members of each pair by an assertion like "*s* is taller than *t*" or "*s* presented a Christmas gift to *t*." The binary relation illustrated does not associate all members of S with elements of T. Hence, to indicate what elements are actually associated in a binary relation, that is, in an aggregate of ordered pairs, one calls the set of *first* elements of the pairs the *domain*, the set of *second* elements the *range* of the relation. In the relation above, the domain is {Randolph, Spencer} and the range is {Joyce, Leslie}. Thus the relation restricts the variable *s* to a smaller domain than its original replacement set.

If $\emptyset$ and the entire Cartesian product $S \times T$ are exceptional by being nullary and universal relations on S to T, some of the other binary relations are special in other ways. Thus, "husband of" (in a monogamous society) and "son of" are binary relations which a member of S can have to *at most one* member of T. On the other hand, a man in a set S could be "brother of," "cousin of," "taller than," more than one woman in T. The special nature of the former type of relation is indicated graphically in Figure 7.1. There the graph of "son of" exhibits *one point* in the first and third "vertical" columns and no points in the second column, that is, *at most one* point in any vertical column. When we defined the relation "brother of" as the aggregate {(Randolph, Joyce), (Randolph, Leslie)} with geometric graph the *two* points of the second vertical column of Figure 7.1, we were illustrating a less specialized type of relation.

A different choice of S and T would bring out more clearly the specialization we mean to indicate. If S is a list of items on sale in a supermarket and T is a list of prices, then an item in S can be paired with only one price in T. There might, of course, be a number of items with the same price. The relation might thus be considered a "many to one" correspondence, where each item *s* "determines" a unique price *t*. The same fact is often emphasized by calling *s* the *independent* variable and *t* the *dependent* variable. The inverse relation would be a "one to many" correspondence and would *not* be "determinate," because naming a price would not fix a unique article having that price.

The formula, $s^3 = t$, where s is the length of the edge of a cube and t is its volume, would define a relation on S to T, where S and T are certain sets of numbers. Here is a relation even more specialized than the price list, since a length "determines" a unique volume and inversely, for physical or geometric cubes. This binary relation is a *one-to-one correspondence* (Chapter 1).

Since, in many scientific situations, determinate or unique results are absolutely essential, the special type of binary relation we have illustrated is exceedingly important. It is called a *function* (a term due to Leibniz) or a *mapping*. To provide a formal definition, we now assert that a binary relation on S to T is called a *function* or a *mapping* if and only if it is a set of ordered pairs in $S \times T$ no two of which have the same first element. The set of *first* elements of the pairs is called the *domain* of the function, and the set of *second* elements is called the *range* of the function.

The use of the term "mapping" as a synonym for function derives from geography, where, for example, New York City, Boston, and Philadelphia may be represented by unique points, n, b, p, on some map. If S is the set whose elements are the three cities named, T is the set of *all* points representing cities on the map, and $W = \{n, b, p\}$, a subset of T, the most recent usage in advanced mathematics would describe the situation as

$$f : S \longrightarrow T$$

which is read as "f is a mapping of S into T." Or, alternatively, there is

$$f : S \xrightarrow{\text{onto}} W$$

which is read as "f is a mapping of S onto W." In either case, f is defined by the set of ordered pairs, $\{(\text{N.Y.C.}, n), (\text{Boston}, b), (\text{Phila.}, p)\}$, and is a function or mapping because each city is paired with *only one* point of the map.

To provide illustrations of binary relations more like those of plane analytic geometry, we must choose $S = T$, where S is a set of *numbers*. Hence, let us consider $S = \{2, 5\}$ and form the Cartesian product $S \times S = S^2$. Then

$$S^2 = \{(2, 2), (2, 5), (5, 2), (5, 5)\}$$

We now study binary relations on S to S, or, in briefer terminology, binary relations on S, that is, subsets of S^2.

Thus the subset $\{(2, 2), (5, 5)\}$ is a binary relation on S. Moreover, it is a *function* for which S is both domain and range. It is not necessary to express this function by sentence or formula. But if the reader wishes to do so, he can use (x, y) to symbolize an ordered pair in S^2 and then describe the above relation by $x = y$. The sentences $x^2 = y^2$, $x^3 = y^3$, etc. will all define this same relation (if we assume that substitutions like 2^2, 5^3, etc. are meaningful). Although the relation is a specific one, it may be characterized in these many varied ways. In fact, any set whatsoever has many different defining properties (Chapter 1).

As another example, $\{(2, 5), (5, 2), (5, 5)\}$ is a binary relation on S, but not a function. Although a defining sentence is not required, the reader may wish to formulate the relation as "x and y are not both equal to 2" or as "$x = 5$ and/or $x \neq y$" (or in some other way).

With $S = \{2, 5\}$ there are 16 different subsets of S^2 (including $\emptyset$ and S^2) and hence 16 distinct binary relations on S. We leave to the reader the task of listing them. He can do this by starting with $\emptyset$ and S^2, continuing with all subsets containing one ordered pair, all subsets containing two ordered pairs, etc.

The concept of Cartesian product can be generalized to 3, 4, . . . , n, any finite number of given sets. If the sets are S_1, S_2, S_3, . . ., S_n, the Cartesian product, $S_1 \times S_2 \times S_3 \times \ldots \times S_n$, is defined to be the set of all ordered n-tuples, $(s_1, s_2, s_3, \ldots, s_n)$, where $s_1, s_2, s_3, \ldots, s_n$ are variables whose domains are $S_1, S_2, S_3, \ldots, S_n$, respectively. Then, as a special case, $S_1 = S_2 = S_3 = \ldots = S_n = S$, and one has the Cartesian products S^2, S^3, S^4, . . ., S^n.

If $S = \{2, 5\}$, we have already formed S^2, and now we can consider the *eight* ordered triples in S^3. Thus the reader can complete the roster for

$$S^3 = \{(2, 2, 2), (2, 2, 5), (2, 5, 2), (2, 5, 5), \ldots\}$$

In this "three-dimensional" universe one can form subsets of S^3, that is, *ternary* relations on S. There are 256 distinct ternary relations (including $\emptyset$ and S^3).

The reader can complete the roster of 16 ordered quadruples in

$$S^4 = \{(2, 2, 2, 2), (2, 2, 2, 5), (2, 2, 5, 2), \ldots\}$$

He can also list some subsets of S^4, that is, some quaternary relations on S. By carrying out the exercise we have just suggested, the reader will be constructing a finite four-dimensional (sic) Cartesian "geometry."

Our illustrations have shown that the number and variety of binary, ternary, quaternary, etc. relations on S is dependent on the choice of S. Hence the essential difference between our finite "geometries" and the analytic geometry of Descartes and Fermat lies in the fundamental set on which relations are defined. In contrast to our petite $S = \{2, 5\}$, the basis of Cartesian geometry is R, the real number continuum (Chapter 2), the *infinite* aggregate of all real numbers arranged in order of size. This fact makes the geometry *analytic* because, crudely speaking, *analysis* is the part of mathematics which deals with variables which have *continuous* domains.

The founding "forefathers" were interested in binary relations on R, that is, subsets of R^2 and, to a lesser extent, with ternary relations on R (subsets of R^3). But, as our examples have indicated, modern analytic geometry would extend to subsets of R^4, R^5, . . ., R^n. Now R and all its powers—R^2, R^3, etc.—are infinite sets (Chapter 24) and contain an infinite number of different subsets. This means that the number of distinct relations on R is infinite. Hence it would not be possible to provide a complete roster of binary, ternary, etc. relations on R in the manner we have employed for finite geometries. Thus one cannot examine *all* relations on R since their number is infinite, and therefore analytic geometry concentrates on certain types which are of special interest for practical or theoretical purposes.

In their formulation of plane analytic geometry, Descartes and Fermat arithmetized Euclidean plane geometry by matching or virtually identifying each point of
the plane with an ordered pair of real numbers called the *coordinates* of the point.
In this way every geometric figure in the plane was identified with a set of ordered
pairs of real numbers, that is, with a binary relation. In other words, plane configurations like the vertices of a decagon, a straight line, or a continuous curve
are merely binary relations. Definition by roster would be impossible for the infinite
aggregate of points on a continuous curve. Hence Euclid, Descartes, Fermat, and
all geometers made use of formulation by defining properties. Now, as we have
pointed out, sets of ordered pairs, like sets in general, have many defining characteristics. But a major feature of analytic geometry is that defining properties are expressed *algebraically*, by equations and inequalities containing variables whose
replacement set is R. Then such equations and inequalities are subjected to manipulations, and this makes *proof* a mere matter of *al-jabr w'almuqabalah*. The logic
of the situation is inherent in the fundamental laws of algebra.

As we have stated, the essence of plane analytic geometry lies in the matching
of ordered pairs of real numbers with points of a plane. Such a one-to-one correspondence can be established in many ways, but it seems appropriate to begin with
an examination of Descartes' technique. The Cartesian coordinate system, which
is a commonplace of school mathematics and everyday graphic representation,
generalizes the coordinate system on a line (Chapter 2). In the plane, the Cartesian
frame of reference consists of two perpendicular lines or coordinate axes with the
same origin and the same unit of length (Figure 7.2). We can imagine the Cartesian
scheme as resembling a network of streets and avenues, where streets run north
and south, and avenues east and west. The X-axis is "Main Avenue" and the
Y-axis is "Main Street." Streets east of Main Street are given positive numbers;
those west receive negative numbers. Avenues north and south of Main Avenue are

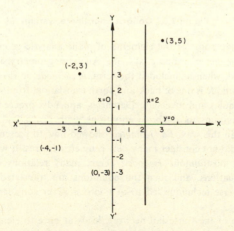

Figure 7.2

positive and negative, respectively. The *Cartesian coordinates* of any point in the plane are the street and the avenue (in that order) on which the point is located, for example, $(3, 5)$, $(-2, 3)$, $(-4, -1)$, $(0, -3)$. (See Figure 7.2.)

In the pure mathematical scheme, the "streets" and "avenues" are assumed to be capable of ever-so-fine subdivision, so that the plane is covered by a sort of gossamer network. This makes it possible to conceive of points like $(-4.00058, \sqrt[3]{7})$ or $(6\sqrt[6]{3}, -2\pi)$; that is, in the pure conception, there is a "$6\sqrt[6]{3}$ Street" and a "-2π Avenue" which intersect at a point of the plane.

In summary, every point in the plane has a pair of coordinates, (x, y) where the x-coordinate or *abscissa* is the directed (signed) distance from the Y-axis and the y-coordinate or *ordinate* is the directed distance from the X-axis (distance being measured along a perpendicular). Conversely, every ordered pair of real numbers can be interpreted as Cartesian coordinates and applied to locating a point in the plane. Our summary establishes the oft-repeated claim of one-to-one correspondence between points of the plane and ordered pairs of real numbers.

But the one-to-one correspondence can be established in many ways. One can, for example, vary the position of the coordinate axes or use different units of length on the two axes or even use axes that meet obliquely instead of at right angles (Figure 7.3). Every point will have a pair of real coordinates, but the coordinates will vary with the frame of reference.

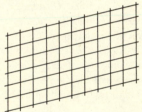

Figure 7.3 Oblique coordinate system

Let us develop some of the elements of plane analytic geometry by using a Cartesian *rectangular* coordinate system. Analytic geometry is a *method,* as Descartes implied, when he included the term, *la méthode,* in the title of his great treatise. The technique is one of back-and-forth translation, from verbal statements to algebraic formulas and inversely. Descartes' appendix proceeded, in the main, from verbal definitions to algebraic equations, with just a few instances of the inverse routine. In this way, *La Géométrie* showed how to present Euclid in algebraic dress, but did not consider many new geometric figures. It was just the opposite in Fermat's posthumous *Isogoge.* There, many relations are first defined by algebraic equations, and then the equations are translated into geometric graphs. This inverse technique led to new curves never considered by Euclid or Apollonius.

Observation of the Cartesian network leads at once to elementary algebraic formulations. For example, in Figure 7.2, points of the Y-axis are seen to have coordinates like $(0, -3)$, $(0, -1)$, $(0, 1)$, $(0, 2.6)$, etc. The Y-axis, then, is identical with the set of all such ordered pairs in which the x-coordinate is equal to zero. Moreover, no point outside the Y-axis will be on "0 Street," that is, have abscissa equal to zero. Hence the equation $x = 0$ expresses a defining property for the set of ordered pairs, that is, for the binary relation with which the Y-axis is identified. In the language of analytic geometry, one says that $x = 0$ is the *equation* of the Y-axis.

The reader will see that "2 Street" contains points like $(2, -5)$, $(2, 0)$, $(2, 6)$, etc. and that points not located on this street cannot have abscissa equal to 2. Hence $x = 2$ is the equation of "2 Street," and, more generally, $x = a$ (where a represents any real number) is the formula for a "street," that is, for a line parallel to the Y-axis. Similarly, $y = b$ is the formula for an "avenue," that is, for a parallel to the X-axis. If $b = 0$, $y = 0$ is the equation of the X-axis itself.

In the equations for parallels to the coordinate axes, only one variable, either x or y, is named and there is no restriction on the missing variable. This thought will assist us if we ask, in the spirit of Fermat: Given the inequality $x < 2$, what is its geometric graph? To answer, we observe that the sentence $x < 2$ places no limitation on y. Then its geometric graph must contain all points whose abscissa is less than 2 and whose ordinate is any real number whatsoever. For example, *all* points of "1 Street," "1/2 Street," "0 Street," "−3 Street," etc. will belong to the graph. In short, the picture will contain all points of all streets to the left of "2 Street," that is, all points to the left of the line, $x = 2$. The graph of $x < 2$ is indicated by the shaded area in Figure 7.4*a*. In the diagram, $x = 2$ is dotted in order to indicate that it is not part of the graph. If that line were included, the graph would illustrate the relation $x \leq 2$ (x is equal to or less than 2). Figure 7.4*b* pictures the graph of $y > 1$, and Figure 7.4*c* indicates the graph of the compound sentence "$x \leq 2$ and $y \geq 1$." The reader will readily see that of all the graphs considered thus far, only the parallels to the X-axis, that is, the lines $y = b$, represent *functions*, since, in all other cases, there may be more than one value of y corresponding to a

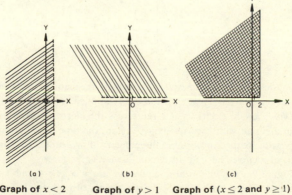

(a)   (b)   (c)

Graph of $x < 2$ Graph of $y > 1$ Graph of ($x \leq 2$ and $y \geq 1$)

Figure 7.4

given value of x. The domain of the function, $y = b$, is the entire X-axis or real number continuum R, and the range is the single number b. One can therefore understand why $y = b$ is called a *constant* function.

A straight line need not be parallel to one of the coordinate axes, and we shall have to consider lines in more general positions in the plane. We now assert (leaving proof for the next chapter): The equation of *any* straight line is a first-degree equation in x and y, that is, an equation of the form

$$ax + by + c = 0$$

where a, b, and c are real numbers. Conversely (if we rule out the cases $a = b = 0$ and $a = b = c = 0$), the Cartesian graph of any such equation must be a straight line.

It is at once apparent that if $a = 0$ and $b \neq 0$ in the above formula, one obtains the parallels to the X-axis, and if $b = 0$, $a \neq 0$, one has the parallels to the Y-axis. Suppose now that $c = 0$, $a \neq 0$, $b \neq 0$. Then $ax + by = 0$ or, after some algebraic manipulation, $y = -(b/a)x$, which we shall write as $y = kx$. Let $k = 3$ and consider the straight line graph of $y = 3x$. Euclid tells us that there is one and only one straight line through two distinct points of the plane. Hence we need only plot two points of $y = 3x$. It is readily seen that $(0, 0)$ and $(1, 3)$ check in the given equation. Its picture is shown in Figure 7.5. In general, $y = kx$ is a straight line through the origin. The equation and the graph formulate a type of function that is said to represent a *direct variation*.

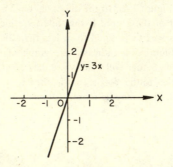

Figure 7.5

In physics, $y = kx$ becomes *Hooke's law* of mechanics if x is interpreted as the elongation of a spring when a force y is applied. Again, the same equation defines *Ohm's law* in electricity if y represents the electromotive force and x the current in a circuit with fixed resistance. Again, the type of variation represented by $y = kx$ is of frequent occurrence in geometry. For example, x might stand for the radius of a circle and y for its circumference.

In Figure 7.6 the lines whose equations are $y = 2x + 4$ and $x + y = 7$ have been graphed. There, once more, only two points of each line were plotted, and then those points were joined by means of a straight-edge. One is free to graph any

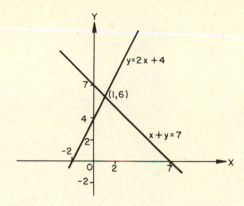

Figure 7.6

two points of the line, but very often one selects the points where it cuts the axes. In the case of $y = 2x + 4$, we can say that

for $x = 0$, $y = 2(0) + 4$; $y = 4$ (this is called the *Y-intercept*)
for $y = 0$, $0 = 2x + 4$; $x = -2$ (this is called the *X-intercept*)

The substitutions show that $(0, 4)$ and $(-2, 0)$ are points of the line. These points are plotted and joined in Figure 7.6. It is readily seen that for $x + y = 7$, both the *X*-intercept and the *Y*-intercept are equal to 7. The intercepts are graphed and connected by a line in Figure 7.6.

Now, if one is asked to solve the compound sentence "$y = 2x + 4$ *and* $x + y = 7$," he can do so graphically by consulting Figure 7.6. It is seen that the only point on both lines is $(1, 6)$. Therefore $\{(1, 6)\}$ is the solution set of the above compound sentence, that is, of the *system* of *simultaneous linear equations*,

$$y = 2x + 4$$
$$x + y = 7$$

We have just associated Cartesian graphs with the *solution* of *algebraic equations* and, in fact, some of the uses of analytic geometry are to clarify meanings in classical algebra and to provide *approximate* solutions when the technique of exact solution is difficult or impossible in finite terms.

If we refer to the equation $y = 2x + 4$, its form can assist us in the graphing of the inequality $y > 2x + 4$. In the equation, substitution of $x = 0$ yields $y = 4$. For the inequality, if $x = 0$, $y > 4$. Therefore $(0, 4.1)$, $(0, 5)$, $(0, 6)$ are all points in the graph of the inequality. These points all lie on the *Y*-axis *above* the *Y*-intercept of the line, $y = 2x + 4$. We can reason similarly that $x = 1$, $y = 6$ is a point of the line, but that, for the inequality, when $x = 1$, $y > 6$. Then the graph of the inequality must contain all points above $(1, 6)$ on the "vertical" line through $(1, 6)$. By continuing with such reasoning, we can understand why the shaded region in Figure

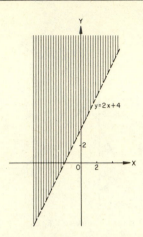

Figure 7.7 Graph of $y > 2x + 4$

7.7 represents the graph of $y > 2x + 4$, since that graph must contain *all* points above the line $y = 2x + 4$. The unshaded region below that line represents the graph of $y < 2x + 4$.

To provide algebraic definition for a straight-line *segment* (for example, the portion of the line $y = 2x + 4$ indicated in Figure 7.8) one may employ a compound sentence like "$y = 2x + 4$ *and* $-1 \leq x \leq 2$." The effect of the second clause is to restrict the domain of x (and hence the range of y). The effect is to cut the line down to the segment between the points $(-1, 2)$ and $(2, 8)$ in Figure 7.8.

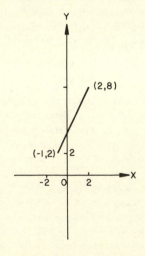

Figure 7.8   Graph of $(y = 2x + 4$ **and** $-1 \leq x \leq 2)$

The description of the Cartesian coordinate system applied various Euclidean concepts such as parallelism or perpendicularity of lines. Our discussion of graphs has also used Euclidean ideas, and we have translated some of them into algebraic equivalents. Now much of Euclidean geometry is *metric*, that is, concerned with *measurement*, and determination of a length or *distance* between two points is the basic procedure. Special theorems then make it possible to measure certain areas by determining two distances, and certain volumes by measuring three lengths. Hence, if analytic geometry is to be metric, one must express length or distance in algebraic form.

A distance formula for the Cartesian plane is obtained by translation of the Pythagorean theorem into an algebraic formula involving coordinates. In Figure

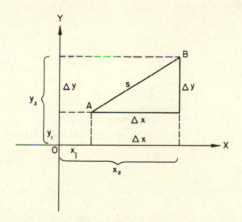

Figure 7.9

7.9 the distance between A and B is the length AB, the hypotenuse of a right triangle. Figure 7.9 indicates that

$$s^2 = (x_2 - x_1)^2 + (y_2 - y_1)^2$$

In this equation, $x_2 - x_1$ is a change in the variable x or a difference between two of its values. A similar statement can be made about $y_2 - y_1$. It is customary to use the Greek letter Δ (delta) corresponding to the Roman D, the initial letter of *Difference*, to symbolize such changes. One writes Δx (read "delta x") for $x_2 - x_1$ and Δy for $y_2 - y_1$. Then the above equation becomes

$$s^2 = (\Delta x)^2 + (\Delta y)^2$$

and the formula for distance in the Euclidean plane is

$$s = \sqrt{(\Delta x)^2 + (\Delta y)^2}$$

We observe that if, in Figure 7.9, we reversed directions and proceeded from Q to P, then Δx would symbolize $x_1 - x_2$ and Δy would be equal to $y_1 - y_2$. It would not matter in the formula, because in it the differences are squared. Thus, if in one case $\Delta x = 3$ and in the other $\Delta x = -3$, then in both cases $(\Delta x)^2 = 9$.

Without drawing a diagram, we can apply our formula to find the distance between $(3, -2)$ and $(1, 5)$. Here, $\Delta x = 1 - 3 = -2$ and $\Delta y = 5 - (-2) = 7$. Hence

$$s = \sqrt{(-2)^2 + 7^2} = \sqrt{4 + 49} = \sqrt{53}$$

The line segment between the two points has an irrational length, that is, one incommensurable with our unit of length.

Let us now use our distance formula to translate the Euclidean definition of a circle into an algebraic equation. Euclid defines a circle as the set of all points at a given distance (called the radius) from a given point (called the center). If we choose $(2, -1)$ as the center of a circle of radius 4, and let (x, y) represent any point on the circle, then $\Delta x = x - 2$, $\Delta y = y + 1$, and $s = 4$ (the radius of the circle). Substitution in

$$(\Delta x)^2 + (\Delta y)^2 = s^2$$

yields for the equation of the circle

$$(x - 2)^2 + (y + 1)^2 = 16$$

To obtain a formula for the equation of *any* circle in the plane, we assume a pair of real numbers, (a, b), as the coordinates of its center and r, a positive real number, as its radius. Then

$$(x - a)^2 + (y - b)^2 = r^2$$

is the equation of the circle. If the center is at the origin, the equation takes the simple form $x^2 + y^2 = r^2$.

We observe that, for a rigorous proof, one ought to indicate that the solution set of the equation derived does *not* include any points not on the circle. If some

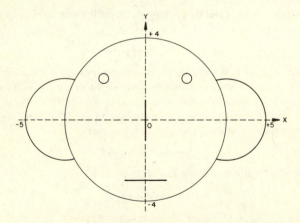

Figure 7.10   Cartesian Cherub

ordered pair which we shall symbolize by (u, w) is part of the solution set, then it must check in the equation, that is

$$(u - a)^2 + (w - b)^2 = r^2$$

which means that the distance of (u, w) from the center (a, b) is equal to r. Therefore the point (u, w) must be on the given circle and nowhere else, in accordance with the Euclidean definition of a circle.

The reader can apply the analytic geometry developed up to this point by verifying that the lines, circles, or segments thereof, in the cartoon of the *Cartesian Cherub* (Figure 7.10) are geometric graphs of the following relations.

Face	$x^2 + y^2 = 16$
Right eye	$(x - 2)^2 + (y - 2)^2 = \dfrac{1}{4}$
Left eye	$(x + 2)^2 + (y - 2)^2 = \dfrac{1}{4}$
Mouth	$y = -3$ and $-1 \leq x \leq 1$
Nose	$x = 0$ and $-1 \leq y \leq 1$
Right ear	$(x - 4)^2 + y^2 = 1$ and $x > \dfrac{31}{8}$
Left ear	$(x + 4)^2 + y^2 = 1$ and $x < -\dfrac{31}{8}$

In spite of the apparent complexity of formulas for our Cartesian Cherub, he is nevertheless a kindergarten creature of analytic geometry, since he is composed of linear and circular segments, defined by first- and second-degree algebraic equations and inequalities. He does not even illustrate *all* types of curve that can correspond to *quadratic* equations in x and y. We now claim that all such curves are *conic sections*. The name is associated with the original definition of such curves in Hellenic geometry. If a right circular *cone* (hourglass figure like two ice cream cones placed end to end) whose curved surface extends indefinitely up and down is cut

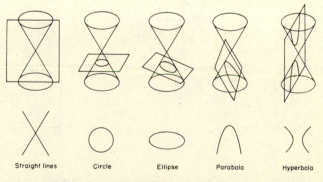

Straight lines   Circle   Ellipse   Parabola   Hyperbola

Figure 7.11   The conic sections

by a plane, the curve of intersection is called a conic section or merely a *conic*. In Figure 7.11 we have illustrated the cone and the five typical conics: circle, ellipse, parabola, hyperbola (a curve of two branches), and two intersecting straight lines.

The geometric properties of the conics were worked out in thorough, beautiful, but not simple algebraic form by Apollonius. Without his work, the great astronomer Kepler (who lived *before* Descartes and Fermat) might never have been able to discover the laws of planetary motion (1609), and we might not as yet be living in an age of man-made satellites. One of Kepler's laws states that the path of each planet is an ellipse. Kepler had to create a sort of "integral calculus" to study planetary motion along the elliptic path (page 169). Newton went further with that idea but, in any event, without Kepler's laws he would probably have been unable to formulate his theory of universal gravitation, and Einstein would not have had the occasion to challenge that theory by inventing a new one. Hence some of the greatest moments in science go back to Apollonius' original inspiration.

Nature is very fond of the ellipse and almost equally partial to the other conic sections, as we shall illustrate. Now it is very much easier to derive important scientific properties of these curves by algebraic manipulation of their Cartesian equations than by using the methods of Apollonius. We shall not, however, carry out the translation from Apollonian properties to algebraic formulas, but shall merely give the results. For example, the equation of an ellipse in the standard position illustrated in Figure 7.12a is

$$\frac{x^2}{a^2} + \frac{y^2}{b^2} = 1$$

where a and b are the lengths of the semiaxes. We can give a rationale for this equation by observing that some ellipses, for example, the planetary paths, are almost circular. If, in fact, $b = a$ in the above equation, and one multiplies both

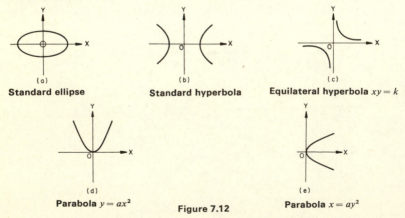

(a)
Standard ellipse

(b)
Standard hyperbola

(c)
Equilateral hyperbola $xy = k$

(d)
Parabola $y = ax^2$

Figure 7.12

(e)
Parabola $x = ay^2$

sides by a^2, the result is $x^2 + y^2 = a^2$, the equation of a circle. Thus the ellipse might seem something like a distorted circle.

If one changes the sum in the above equation to a difference, the result is

$$\frac{x^2}{a^2} - \frac{y^2}{b^2} = 1$$

the equation of the standard hyperbola of Figure 7.12b. In this equation, if $b = a$, one arrives at a so-called *equilateral* hyperbola, not essentially different in appearance from the one in Figure 7.12b. But it is sometimes a convenience to rotate (Chapter 17) such a hyperbola to a position like that in Figure 7.12c. Then its equation becomes

$$xy = k$$

where k is any real number other than zero. The equation (or its graph) is said to express the *inverse variation* of the variables x and y. As an instance of such a relation, there is Boyle's law, $xy = k$, where $k > 0$, and x represents the volume of some gas when it is subjected to pressure y but kept at constant temperature. From a hyperbola like that of Figure 7.12c one can read volume, given pressure, and conversely.

Nature requires that the path of a bullet or baseball and the trajectory of a comet be (approximately) *parabolic*. The equations of the parabolas in the standard positions of Figures 7.12d and 7.12e have the form $y = ax^2$ and $x = ay^2$ respectively.

To derive the equations of the conics, one usually employs definitions that are more suited to algebraic handling than the descriptions as particular cross-sections of a cone. Thus an ellipse is described as the set of all points (in a plane) for which the sum of distances from two fixed points is constant. (We note that this is a metric definition, which mentions distances.) Each fixed point is called a *focus* of the ellipse. Figure 7.13 shows how to draw an ellipse mechanically from this description. According to Kepler's laws, the sun is at one focus of each planetary orbit. Since the paths of the planets are almost circular, the foci of the ellipse are close together, and the position of the sun is almost central.

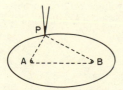

Figure 7.13   A loop of fixed length is allowed to slide around the foci while the tracing pencil holds it taut, tracing the curve. Because the length between the foci will always be the same, what remains of the cord (sum of distances) is also constant.

In analytic geometry it can be demonstrated algebraically that the ellipse has a property important for acoustical or optical applications. It is shown that if light or sound is emitted from one focus of an ellipse, it must be reflected to the other. This fact explains the phenomenon of whispering galleries with elliptical cross-sections. A sound uttered at one focus will be heard distinctly at the other even when foci are widely separated.

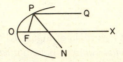

Figure 7.14

A similar property of the parabola is used in automobile headlights and reflecting telescopes. A parabola has a single focus which suggests that it is a sort of infinite ellipse, with a second focus "at infinity." If a light is placed at the focus F (Figure 7.14), all rays hitting the parabolic surface will be sent out in a parallel beam, that is, reflected to the imagined second focus at infinity. In reverse, parallel rays from a star ("the focus at infinity") are all reflected to form a sharp image of the star at the focus of a parabolic reflecting telescope.

To avoid technicalities, we have, in many cases, given the results of the analytic method and have passed over details of the proofs. Let us then give some specific examples to indicate the clarity and effectiveness of Cartesian techniques. We shall explain just how Menaechmus and Omar Khayyám solved some interesting cubic equations by the use of conic sections. What they did was highly original and, from their viewpoint, *difficult* because all their reasoning was geometric, with the conics of the hourglass picture. For us everything will merely be a case of the most elementary sort of analytic geometry.

It is traditional to refer to certain questions as the "famous problems of antiquity." One of these is the "duplication of the cube" or the *Delian problem*. The story goes that Apollo, being angered, visited a plague on the people of Delos. They appealed to the oracle, who commanded them to double the size of the altar but to keep its shape the same. The form of the altar was cubic and, hence Apollo's command was to duplicate the cube, that is, to construct a cube with volume double that of the original.

Let us represent the side of the initial altar by 1 (linear unit). Then its volume was $1 \cdot 1 \cdot 1 = 1$ cubic unit. The new altar was to have a volume of 2 cubic units. Calling the length of its edge x, its volume would be x^3, and the condition of the problem would be expressed by the cubic equation

$$x^3 = 2$$

Today this cubic is no more difficult than $x^3 = 64$, for which we can give $x = 4$ as one root. The side of the new altar should be the irrational number or incommensurable length,

$$x = \sqrt[3]{2}$$

and if we desire a *rational* estimate, there are the Babylonian cube root tables, or our own. But we are considering Greek geometric algebra, and according to the rules that Plato had formulated, geometers were restricted to the use of straight-edge and compasses, that is, straight lines and circles, in performing constructions. Therefore the problem of the Greeks was to construct, with straight-edge and compasses, a line segment of length $\sqrt[3]{2}$ units. Modern algebra has proved that this is an *impossible* task, and that the only irrationals the use of circles and straight lines will produce are square roots and combinations or iterations of such roots. Nevertheless, around 350 B.C. Menaechmus was able to "duplicate the cube," that is, to solve $x^3 = 2$ *geometrically*. But he had to defy Plato's stricture on tools. Instead of drawing circles and straight lines, Menaechmus drew a *parabola* and a *hyperbola*. The intersection of these curves provided the desired solution.

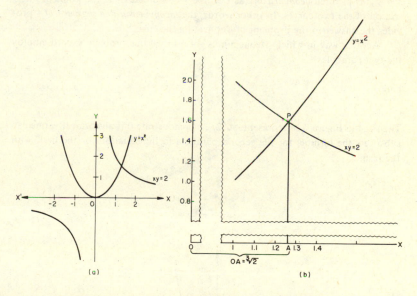

Figure 7.15

From the Cartesian point of view, Menaechmus' hyperbola can be considered to be the one in Figure 7.15a, that is, the hyperbola whose equation is $xy = 2$. Likewise, the equation of the parabola he used (Figure 7.15a) is $y = x^2$. These curves intersect (Figure 7.15b) at a point whose coordinates are approximately (1.26, 1.59). To do justice to Menaechmus, he actually stated that OA (enlarged to a suitable scale) would provide the *exact* length for the altar to appease Apollo. To prove that he was right, one can obtain the coordinates exactly by switching from graphic to algebraic considerations. The point (x, y) whose coordinates are sought lies on both curves or, algebraically, the values of x and y must meet the requirements of *both* the above formulas for the curves. Combining these formulas by substituting the second in the first,

$$x(x^2) = 2$$

or

$$x^3 = 2$$

and

$$x = \sqrt[3]{2}$$

Since $y = x^2$,

$$y = \sqrt[3]{4}$$

The equation $x^3 = 2$ is the cubic associated with the duplication of the cube, and $x = \sqrt[3]{2}$ represents the *irrational* number which answers this problem. In the graph, $x = \sqrt[3]{2}$ is pictured *geometrically* as OA, the x-coordinate of P, the point of intersection of the two curves. In other words, the *incommensurable* segment OA provides the answer to the problem of duplicating the cube.

       Another way in which Menaechmus solved the same problem was to employ the two parabolas

$$y^2 = 2x$$

and

$$x^2 = y$$

Then OA in Figure 7.16 represents $\sqrt[3]{2}$. To demonstrate this algebraically, one can substitute the formula for the second parabola in the equation for the first, with the result

$$x^4 = 2x$$

or

$$x^4 - 2x = 0$$
$$x(x^3 - 2) = 0$$

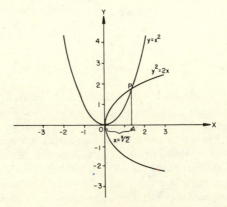

Figure 7.16

so that

$$x = 0 \quad \text{and} \quad x = \sqrt[3]{2}$$

These are the two real roots of the equation, and are the x-coordinates of the points of intersection of the two parabolas.

Now let us present an example of Omar Khayyám's methodology. As we explained in a previous chapter, he was seeking to solve cubic equations. Consider then the cubic

$$x^3 - 3x^2 + x + 1 = 0$$

The intersections of a circle and the upper branch of a hyperbola (Figure 7.17) were used by Omar to give the roots $x = 1$, $x = 2.4$ (approximately). The latter root is irrational, and Omar would have given the exact answer as the (incommensurable) length of segment OA in Figure 7.17.

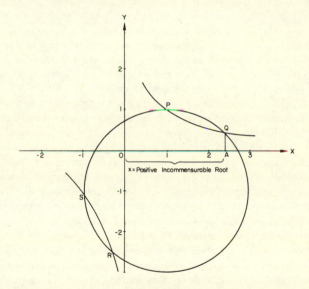

Figure 7.17

Omar was, as has been emphasized, a geometric algebraist like his Greek predecessors. Therefore it might be difficult to follow the reasoning by which he proved his method to be correct. Modern algebraic symbolism will clarify it, however, and provide

$$xy = 1 \quad (\text{or} \quad y = \frac{1}{x}, \quad \text{if } x \neq 0)$$

and

$$x^2 - 2x + y^2 + 2y - 2 = 0$$

as the Cartesian equations of hyperbola and circle, respectively.

To obtain the coordinates of the points of intersection of the graphs, one proceeds as in the previous examples, by substituting the first equation in the second. The result is

$$x^2 - 2x + \left(\frac{1}{x}\right)^2 + 2\left(\frac{1}{x}\right) - 2 = 0$$

or

$$x^4 - 2x^3 + 1 + 2x - 2x^2 = 0$$

or

$$x^4 - 2x^3 - 2x^2 + 2x + 1 = 0$$

This equation can be factored into

$$(x + 1)(x^3 - 3x^2 + x + 1) = 0$$

The first factor leads to the answer $x = -1$, which corresponds to point S in Figure 7.17. Since Omar did not consider negative answers, this factor did not appear in his reasoning. The second factor leads to the given cubic equation. Thus it has been proved that three points of intersection of the two graphs—namely, P, Q, and R—will provide the roots of this cubic. Again, R corresponds to a negative root, and therefore Omar used only one branch of the hyperbola to obtain the positive answers associated with points P and Q.

The examples of Menaechmus' duplication of the cube and Omar's solution of cubics were selected in order to suggest how and why analytic geometry is such a powerful tool. In our illustration we indicated certain conic sections by their equations and carried out algebraic processes with these equations, without any need for diagrams. When the algebra was complete, the final results were connected with a geometric picture. But in the interim it was not necessary to draw in points, lines, circles, or any other curves indicated by the algebraic work. Similarly, in all examples of analytic geometry, an advantage accrues because thought is not obstructed by complicated diagrams and proof is almost entirely a matter of mechanical algebraic manipulation.

Moreover, there is a further and greater advantage in the analytic method. Greek plane geometry defined and made deductions about geometric figures having counterparts in the physical world. But the abstract concepts of analytic geometry permit much greater freedom. Write any open sentence in x and y, that is, specify any subset of R^2. Then consider its geometric picture and possible applications. Will it provide approximate solutions for some problem? Will it be useful as a mechanical device? Will it describe the motion of some natural object? Since one is at liberty to consider *any* subset of R^2, infinitely more binary relations are available for study by the method of Descartes than by the geometry of the Greeks.

Straight lines and conic sections are amenable to algebraic representation in terms of Cartesian coordinates. But in order to give simple expression to other relations, different frames of reference may be preferable. In one of these, "avenues" meet "streets" obliquely—somewhat in the fashion that Broadway cuts

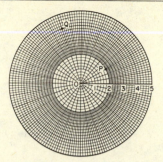

Figure 7.18   Polar coordinate system

Forty-Second Street in New York (see Figure 7.3). In another layout, avenues are concentric circles and streets are lines radiating from the center of the circles (Figure 7.18). Until surrealism hits the city planners, such a scheme will probably not be adopted. Still the Paris boulevards suggest the circular scheme, and the layout of Washington illustrates a radial arrangement of streets.

In the latter scheme, called a *polar coordinate system*, the coordinates of a point (boulevard and street) would be the *distance from the origin* and the *angle* formed with Main Street. In Figure 7.18 we have indicated origin and initial line, point P with coordinates $(2, 30°)$ and Q, with coordinates $(4, 120°)$. In this scheme, the equations of certain important curves are especially simple. A circle with center at the origin and radius k would have the equation $r = k$. For the *spiral of Archi-*

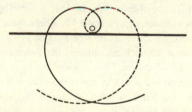

Figure 7.19   Spiral of Archimedes, $r = \theta$. (Dotted line corresponds to negative values of θ.)

medes in Figure 7.19 there is $r = \theta$; for the *cardioid* of Figure 7.20 there is $r = 1 + \cos\theta$; for the *four-leaved rose* of Figure 7.21 there is $r = \sin 2\theta$. And mathematicians have other city layouts to which they can resort to simplify the algebraic representation of other important curves. To understand the last two equations above may seem to require a knowledge of elementary trigonometry, but the reader can appreciate the simplicity of their form if we state that the corresponding Cartesian equations would be

$$(x^2 + y^2 - x)^2 = x^2 + y^2 \quad \text{and} \quad (x^2 + y^2)^3 = 4x^2 y^2$$

which are anything but simple, especially if the problem is to proceed from equation to geometric graph.

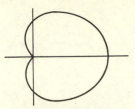

Figure 7.20 Cardioid

Figure 7.21 Four-leaved rose

If one sort of generalization of the Cartesian coordinate system is to be found in the use of different frames of reference, another is furnished by the extension of Cartesian coordinates to space, as in Figure 4.3, where three mutually perpendicular number lines or coordinate axes are illustrated. Just as a 2-dimensional Cartesian coordinate system uses a network of squares, the 3-dimensional framework can be pictured as an array of cubes. Then various surfaces become the geometric graphs of ternary relations expressed by equations in x, y, z, where (x, y, z) are the Cartesian coordinates of a point in space. As for 4-, 5-, or n-dimensional coordinate systems, we need not picture them—the algebraic work will take care of all the facts, and the geometric counterparts are superfluous. An algebraist does not balk at manipulating equations and inequalities involving many variables. On the other hand, a Greek geometer was decidedly limited by what he could see. An algebraist can create a fourth dimension merely by a stroke of his pen, and although he will not try to visualize the result, geometric language is a great convenience. To say that a relation like

$$x^2 + y^2 + z^2 + w^2 = 9$$

is a *hypersphere* with radius 3 is so much easier than to state that the relation is the set of all ordered quadruples of real numbers such that the sum of the squares of these four numbers is always 9.

We have featured analytic geometry as an important general method. We have also emphasized its service in representing binary and ternary relations graphically, and in providing terminology and techniques for the study of relations of higher order. Since a *function* is the most important type of binary relation, we shall close this chapter by discussing traditional and modern vocabulary associated with the function concept, and by presenting some additional and more general examples which will provide a link with the chapters to follow. The reader must become acquainted with the older modes of expression, because they are the ones he is likely to encounter in reading elementary mathematical literature which is not completely "new."

A function, as stated earlier, is a set of ordered pairs in which no two pairs contain the same first member. In other words, to each element of the domain there corresponds *only one* element in the range. Our illustrations have specified particular functions in various ways, namely, by tabulation, graph, defining sentence, or defining formula. The last method is associated with some special vocabulary, in which, traditionally, the value of a variable y is considered to be "dependent on" or "determined by" the value of the variable x.

Then if, for example, R, the real number continuum, is fixed as the replacement set for x and y, a particular function f might be specified as the set of all (x, y) for which $y = x^2 - x + 4$. In connection with the Cartesian graph of this function, we may wish to plot some ordered pairs, and hence to substitute $x = 1, 2, 3, -1, -2, -3$, etc. in the above defining formula in order to find the corresponding values of y. For this purpose, and also to show the "dependence" of y on x, one convenient notation is $y(x)$, which is read as "y of x." Another is $f(x)$, which is read as "f of x." In our example,

$$y(x) = x^2 - x + 4$$

or

$$f(x) = x^2 - x + 4$$

Then

$$f(1) = 1^2 - 1 + 4 = 4$$
$$f(2) = 2^2 - 2 + 4 = 6$$
$$f(-1) = 1^2 + 1 + 4 = 6$$

and therefore $(1, 4)$, $(2, 6)$, $(-1, 6)$ are ordered pairs belonging to the function, or coordinate pairs for points in its graph. All the symbols, $f(1)$, $f(2)$, $f(-1)$, are values in the range of the function, and hence $f(x)$ is the general formula for such a value.

For practice in the particular notation (and not for plotting the graph of the function) the reader should verify that

$$f(a) = a^2 - a + 4$$
$$f(a^3) = (a^3)^2 - a^3 + 4 = a^6 - a^3 + 4$$
$$f(x^2 + 1) = (x^2 + 1)^2 - (x^2 + 1) + 4$$
$$= x^4 + x^2 + 4$$

and that if

$$g(x) = x^2 + 1$$
$$f(g(x)) = f(x^2 + 1) = x^4 + x^2 + 4$$
$$g(f(x)) = g(x^2 - x + 4) = (x^2 - x + 4)^2 + 1$$
$$= x^4 - 2x^3 + 9x^2 - 8x + 17$$

where the last examples indicate that, in general, $f(g(x)) \neq g(f(x))$.

As far as the function originally defined is concerned, the new notation merely indicates that this function is the set of ordered pairs $(x, f(x))$ where $f(x) = x^2 - x + 4$, or it is the set (x, y), where $y = f(x)$. The meaning of $f(1)$ or "f of 1" is the value of y corresponding to $x = 1$, and the meaning of $f(x)$ or "f of x" is the value of the variable y corresponding to a given value of the variable x.

Now in some of the older usage there may be reference to "a function $f(x)$," which, taken literally, is confusing, since $f(x)$ is a value of the variable y, and not a set of ordered pairs. But if one accepts "a function $f(x)$" as an idiom, and reads it as "a function f of x," then the sentence $y = f(x)$ (which does define a function) will tend to be read as "y is a function f of x" or simply as "y is a function of x," which, again, is not literally true since y is a single variable, and not a pair of related variables defining a set of ordered pairs. The same idiom is applied when, in referring to the formulas $A = \pi r^2$ and $d = kt$, it is claimed that the area of a circle is a function of its radius, and that for a body moving uniformly (or, in fact, for any moving body), distance is a function of time.

Again, the reader will find $f(2)$ described as the value of the function at 2 (instead of the value of y corresponding to $x = 2$). The idiom makes $f(x)$ "the value of the function at x" (instead of the value of y corresponding to a given value of x). Also, such examples as $f(g(x))$ and $g(f(x))$ which we treated earlier may be called "composite functions."

The common usage which we have described may confuse because it appears to identify the function with the values in its range. Since such pains are taken to define a function exactly, the reader may wonder why the exact formulations are "more honored in the breach than in the observance." All we can say in explanation is that the function concept has a long history, rooted in antiquity (*vide* the many *tabulations* of the Babylonians), and that members of the modern mathematical world are like Shakespearean actors who have read and recited and studied Elizabethan English to such an extent that they slip into the outmoded idioms even in ordinary conversation, especially when they feel that the expressions have special aptness or beauty.

If then, we give in to tradition, we shall consider the symbol $f(x, y)$, read as "f of x and y" to *mean* a "function of x and y." But can it possibly be meaningful to say, for example, that $z = f(x, y)$ is a function defined by $f(x, y) = x^2 + y^2$? The equation $z = x^2 + y^2$ defines a *ternary relation*. Hence how can we speak of a *function*, which is a special *binary* relation?

Let us then justify the usage. If

$$f(x, y) = x^2 + y^2$$

we can find

$$f(1, 1) = 1^2 + 1^2 = 2$$
$$f(1, 2) = 1^2 + 2^2 = 5$$
$$f(1, 3) = 1^2 + 3^2 = 10$$

so that $(1, 1, 2)$, $(1, 2, 5)$, and $(1, 3, 10)$ are ordered triples belonging to the *ternary relation* defined. But let us interpret things somewhat differently by examining $z = f(x, y)$ once more, and also its Cartesian graph (Figure 7.22), which is a surface

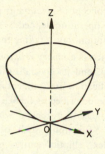

Figure 7.22   Surface $z = x^2 + y^2$

in 3-dimensional space. If, now, we think of $(1, 1)$ or $(1, 2)$ or $(1, 3)$, etc. as a *point* or *vector* (Chapter 4) in the XY-plane, then each one of these ordered pairs can be conceived as a *single entity*. Then the above equation pairs each of these entities with a number. Thus $(1, 2)$ is paired with $f(1, 2)$ or 5, and $(1, 3)$ is paired with $f(1, 3)$ or 10, etc. Therefore, in the present instance a *binary relation* is formulated by the set of ordered pairs, $((x, y), z)$ or $((x, y), f(x, y))$. Furthermore, the binary relation is a *function* because $f(x, y)$ has a *unique* value when $(1, 1)$ or $(1, 2)$ or any other real number pair is substituted for (x, y). This uniqueness can be seen from the diagram of Figure 7.22, where a "vertical" line through $(1, 1)$ or $(1, 2)$ or any point of the XY-plane meets the surface $z = x^2 + y^2$ in *only one* point.

In similar fashion, a quaternary relation, for example, the one we termed a hypersphere (page 160), can be converted into a *binary relation*, $((x, y, z), w)$. If we substitute $(1, 0, 2)$ for (x, y, z) in the equation of the hypersphere, the result is

$$1^2 + 0^2 + 2^2 + w^2 = 9$$
$$w^2 = 4$$

Therefore

$$w = +2 \quad \text{or} \quad w = -2$$

Here w is *not* determined uniquely, that is, the ordered pairs, $((1, 0, 2), 2)$ and $((1, 0, 2), -2)$, having the *same first member*, both belong to the binary relation, which is therefore *not a function*. But if $w = f(x, y, z)$ where

$$f(x, y, z) = x^2 + 2y - z$$

then $((x, y, z), f(x, y, z))$ is seen to be a function. In our examples, $f(x, y)$ and $f(x, y, z)$ symbolize values in the range of a function but, in common usage, may be referred to as *functions* (of x and y, of x, y, and z, respectively), and there are "functions" $f(x_1, x_2, x_3, \ldots, x_n)$.

By special interpretation, we have converted ternary and quaternary relations into binary relations, and we could do the same for n-ary relations, if $n > 4$. Here we have converted a relation into one of lower order. But we can reverse the procedure if it should suit our purposes. Perhaps we might prefer, in arithmetic or algebra, to base ideas on the concept of *relation* in preference to *operation*. Then, for example, addition, a binary operation defined by $z = x + y$, could be considered a *ternary relation*. A *unary* operation, say, "taking the square root of *one* number," can be defined by $y = \sqrt{x}$, which is a *binary relation*.

The concept of a relation, and its specialization, the notion of a function, form a central theme for classification and unification in mathematics. Many subdivisions of mathematics can be described as theories associated with special types of function or relation. The index to a general work on mathematics may well reveal as subheadings under "function" such qualifying terms as algebraic, analytic, Bessel, beta, complex, elliptic, entire, gamma, harmonic, Lagrange, Legendre, logarithmic, orthogonal, periodic, propositional, real, trigonometric, vector, etc. Some of these types have already appeared in our story, and others will arise as we continue. For example, open sentences or propositional forms were formerly called "propositional functions," that is, in common usage, functions of the variables contained in the sentences. *Periodic* functions are the mathematical tools for dealing with natural and man-made periodicity—the recurrent changes in tides, the orbiting of planets and satellites, business cycles, alternating current oscillations, mechanical vibrations, the waves and vibrations associated with physical theories of light and sound.

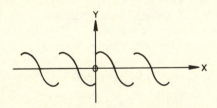

Figure 7.23   A periodic function

As usual, Cartesian graphic representation will clarify a concept. Figure 7.23 pictures a periodic function, so called because the same pattern is repeated again and again forever. If such a pattern is formulated by $y = f(x)$, then the fact that the height of the "wave" is periodically the same leads to the definition of a periodic function as one for which

$$f(x + p) = f(x)$$

If *p* is the least number for which the statement is true, it is called the *fundamental period* of the function.

Trigonometry, which some school students consider a glorified geometry with superimposed computational torture, has a more important aspect, which makes it the ABC of periodicity. Thus a thumbnail sketch would say that trigonometry is the study of the graph of the function $y = $ sine x (Figure 7.24). This graph shows that the fundamental period is 360° or 2π radians, since the pattern for x between 0° and 360° is repeated again and again, to left and right, forever. (Other periods are 720°, 1080°, etc.)

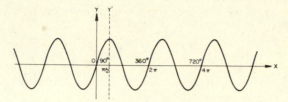

Figure 7.24   Graph of $y = \sin x$

The cosine curve, that is, the graph of the function $y = \cos x$ has the same shape as that for $y = \sin x$ and can be obtained either by shifting the sine curve through a distance $\frac{1}{2}\pi$ to the left or by sliding the coordinate axes through a distance of $\frac{1}{2}\pi$ to the right (the dotted position in the diagram) so that the reading for cos 0° will be the same as that for sin 90° and cos 30° = sin 120°, etc. The *tangent* function of trigonometry is defined by $y = \sin x/\cos x$, where the domain of the function is the

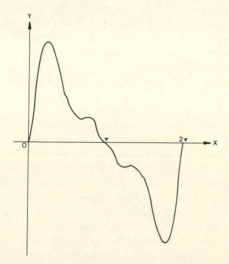

Figure 7.25   Pattern for $y = 5 \sin x + 3 \sin 2x + 2 \sin 3x$

set of all real numbers except those for which cos $x = 0$. The cosine graph will tell which values of x are excluded from the domain.

Adjustment of the scales on the X and Y axes in Figure 7.24 will provide the graphs of $y = \sin 2x$, $y = \sin 3x$, $y = 5 \sin 7x$, etc. The sum of a number of such sine functions, for example,

$$y = 5 \sin x + 3 \sin 2x + 2 \sin 3x$$

will have an irregular wave pattern akin to Figure 7.25. The French mathematical physicist J. B. J. Fourier (1768–1830) showed that, conversely, any well-determined graph, subject to a few restrictions unimportant in science, can be resolved into a sum of sine curves. It is a remarkable fact that you might scribble a curve thus:

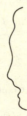

and it would nevertheless have a sine equation. Of course, three or four sine curves might be too few. You might need a dozen or even a hundred for a good approximation. Since such a graph represents a function, Fourier's discovery means that all the elementary relations of science are expressible in terms of sines. In a sense, then, trigonometry is at the base of all scientific formulation. The idea of periodicity dominates portions of pure mathematics as well as applied science. There is, for example, the theory of *elliptic functions*, a sort of higher trigonometry dealing with "doubly periodic" functions. Mathematicians and physicists alike hardly associate the importance of trigonometry with its initial elementary use in the fields of surveying and astronomy.

There are broader classifications of functions which include trigonometric and elliptic functions merely as subcategories. For example, one overall way of describing functions or relations is by means of their domains. Thus mathematics contains "a theory of functions of a real variable" and "a theory of functions of a complex variable" signifying the study of $f(x)$ and $f(z)$, respectively, where the domain of x is R, the real continuum, and the domain of z is the aggregate of complex numbers. These descriptions employ the traditional idioms associated with functionality. Actually, general relations and not merely functions are studied in the subjects named. Also, modern pure mathematicians do not like the "function of" mode of expression, and therefore the older terms are now replaced by descriptions like "Real Analysis," "Complex Analysis," "Real Functions," etc. This nomenclature provides the transition to our next chapters, in which we shall be discussing the great names and the fundamental notions associated with the beginnings of "real analysis" or "functions of a real variable."

8

A Calculus for Heaven and Earth

The place is Prague, where a young scientist in search of a job had brought his family. For a talented, genial, pious, imaginative, and romantic personality, who possessed a willingness to carry out the most laborious tasks at a salary of 500 florins less per year than the usual rate, employment was not hard to find. Thus, in 1601, our protagonist became court mathematician to Emperor Rudolph II. If he found the task of compiling the *Rudolphine* tables an arduous one, other duties must have been more pleasurable—casting the horoscope of the emperor, of Wallenstein, and of other political magnates, and writing mystical interpretations of the triple conjunction of Mars, Jupiter, Saturn. It is strange to recount that these were major activities of Johannes Kepler (1571–1630), the founder of modern astronomy. To quote Henri Poincaré, leading mathematician of the early twentieth century:

> Even in its primitive, pseudoscientific phases, astronomy has been a boon to humanity, for it was astrology that enabled men like Tycho Brahe and Kepler to earn a living, by predicting the fate of naïve kings from the conjunctions of heavenly bodies. If these monarchs had not been so credulous, the great astronomers might have been too poor to engage in scientific research and we might still wallow in ignorance and believe that nature is governed by whims.*

During the years 1601–1609, Kepler's real scientific work was carried on as a sideline, but finally he completed his "great Martian labor" and, as he said, was able to "lead the captive planet to the foot of the imperial throne." It is our present purpose to examine the nature of his triumph over the planet Mars. Another decade was needed before he vanquished the solar system completely, and this second ten-year period followed the pattern of his earlier existence—financial difficulties, personal bereavement, laborious tabulations for all the states of upper Austria, and much astrology paralleling the development of the greatest satellite

---

* Henri Poincaré, *La Valeur de la Science,* Flammarion, Paris, 1913, p. 157. (Translation by the Author.)

theory of all time. None of these hardships diminished his mysticism or crushed his romanticism.

In the *Harmonices Mundi* of 1619 he presented his final planetary law, but he actually devoted more space to the celestial harmonies to which the "spirit of the sun" was the sole auditor (except for Kepler himself). One need only examine the 1619 treatise to find exact musical notes for earth's song of sorrow, the dull moan of Venus, the staccato rhythms of Mercury, and the contrapuntal role of the other planets. When the whole strange mélange of science and fantasy was complete, Kepler dedicated it to James I of England, in whom it must have struck a sympathetic chord, for the musical astronomer soon received an invitation to England's court.

Kepler demonstrated his extravagance in voluminous correspondence. In a typical letter addressed to Baron Peter Heinrich von Strahlendorf in 1613 he describes the courtship preceding his second marriage; he weighs the merits of eleven candidates for his hand; he writes:

I had been waiting impatiently for the visit of the wife of Herr Helmhard, wondering what she would have to say about the third candidate, and whether her words would sway me in favor of this third lady instead of her two predecessors. But when I had at last heard from Frau Helmhard, I made up my mind to accept the fourth candidate on my list. But I still felt bad that the fifth possible choice had in the interim withdrawn from the picture. Just then fate stepped in: The fourth lady became aware of some reluctance on my part and gave her word to someone else, a man who had wooed her persistently for a long time and had painted a glowing picture of their future together. Now I was as much annoyed about losing her as I had been at the withdrawal of the fifth. There must be something wrong with my emotions, which seem to be stimulated by any hesitation on my part, any weighing of pros and cons. From what I have learned subsequently, it was just as well that I had no success with the fourth. As far as the fifth is concerned, there is still the question of why, though she was destined for me, God permitted her to have six rivals in the course of one year. The name of my final choice is *Susanna*. If my bride lacks wealth and family background, she nevertheless has all the simpler virtues. Her father was a carpenter by profession. She received her education, which must take the place of a dowry, at the home of the Stahrembergs. She is not young, but she is modest and unpretentious. I feel that she will be highly efficient in running my household. The wedding will take place on the day of the eclipse of the moon, when the astronomical spirit is in hiding, as I want to rejoice in the festival day. . . .*

The celebration took place at Linz, and Kepler, with mathematical issues always in the back of his mind, became interested in finding the correct volume of the wine barrels associated with the festivities. He may have imbibed rather freely of the contents he was shortly to measure, for the problem was inspired by a heated argument with the wine merchant, whom Kepler accused of making errors of cubature in his own favor. The net result was the appearance in 1615 of the great astronomer's *Nova stereometria doliorum vinariorum* (New Solid Mensuration of

---

* This is a free translation by the author of pp. 25–27, Vol. 2, of M. Caspar and W. von Dyck, *Johannes Kepler in seinen Briefen*, Verlag Oldenbourg, Munich and Berlin, 1930.

Wine Casks). The mathematics of the wine barrels and that of the conquest of Mars can both be classified as a crude type of *integral calculus*.

When Kepler first came to Prague, remarkably accurate observations of the heavens were available. These were a legacy from his predecessor, the Danish astronomer whom he called the *phoenix*, Tycho Brahe. But years more of observation and computation were required before Kepler could state the first of his famous planetary laws: *The orbits of the planets are ellipses with the sun at one focus*. This discovery was the result of imaginative genius combined with prosaic *triangulation*, a type of mathematics that can be carried out by scale drawing or trigonometry. Kepler's second law states: *The focal radius joining a planet to the sun sweeps out equal areas in equal times*. We shall now examine the reasoning on which Kepler based his second law, in order to show that his technique helped to launch the subject we call integral calculus.

In the first place, Kepler chose the planet Mars for prolonged observation. This was a lucky choice, since Mars' orbit has greater "eccentricity," or differs more from a circle than all other planetary ellipses except that of Mercury. Therefore, even with the optical instruments available in his day, Kepler was able to observe the eccentricity of Mars' orbit. Having charted Mars' elliptic trajectory, Kepler made a more profound study of just how the planet moves in the course of its celestial journey. He concentrated on points like *A, B, C, D, E, F* in Figure 8.1*a*, which represent positions of Mars such that the time required for the planet's progress from *A* to *B* is the same as that from *C* to *D* or from *E* to *F*. Drawing the focal radii *SA, SB, SC*, etc., Kepler computed the areas of sectors *ASB, CSD, ESF* and found them equal. He did this by a method of approximation that used elementary geometry. First, he subdivided sector *ASB* into very thin sectors, then sector *CSD* into very thin sectors, etc. In each of the small sectors there is little difference between the curved arc of the ellipse and its chord, so that for approximate purposes it suffices to measure the area of the triangle inscribed in the elliptic

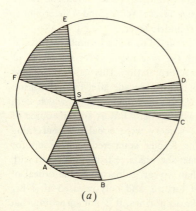

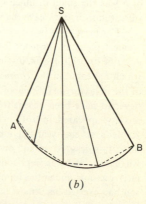

(*a*)   (*b*)

Figure 8.1   Kepler's second law

sector, and this is found by the elementary Euclidean formula $\frac{1}{2}bh$. Then the area of sector ASB can be found by considering it to be the sum of numerous small triangles (Figure 8.1b), and in the same way, areas can be computed for sectors CSD and ESF. Kepler found the three large sectors to be equal in area, and equal to other sectors formed by S (the sun) and focal radii to orbital points whose separation in time was the same as that from A to B or C to D, etc. These were the numerical results which he summarized in the second of his two laws, enunciated in 1609.

In the summation of numerous small triangles to estimate the area of an elliptic sector, Kepler may have felt that he was merely using common sense, but we would say that he had solved a problem in *integral calculus*, and that he had used approximate or *numerical integration*. His stereometry of wine casks comes under the same heading. In that instance, he visualized the volume of a barrel as made up of numerous shallow circular cylinders somewhat like those in Figure 8.2. The

Figure 8.2   The sum of cylinders approximates the volume of the upper half of a barrel.

elementary formula, $\pi r^2 h$, for the volume of a circular cylinder, was then applicable to each thin layer of the barrel. The indefatigable Kepler measured the circular periphery of the cask at each level, and substituting this length for C in $C = 2\pi r$, computed r for the circular bases of the shallow cylinders at each level. The value, of h was the small distance between successive layers. With r and h determined, substitution in $\pi r^2 h$ gave the volume of the cylindric layer.

For more than two decades Kepler had struggled to find a formula that would connect planetary *distance* from the sun with the *time* required for the planet to describe its orbit. He finally succeeded in such a formulation. This was his third law, which appeared in the *Harmonices Mundi* of 1619, a decade after the publication of the first two laws. Some of the delay was caused by family troubles, which seem never to have been absent from the noted astronomer's life. A special source of anxiety appeared in 1615 when rumors of sorcery were circulated concerning Kepler's seventy-year-old mother. Catherine Kepler countered with a suit for libel; this dragged on and on until 1620, when she was arrested and charged with witchcraft. She was imprisoned for more than a year, subjected to the ordeal of interrogation under the imminent threat of torture, and only released after her son had exerted his utmost influence with the authorities. It is not surprising that her death followed shortly thereafter.

The "satellite" law that was released in the midst of all these troubles expresses a relation between the *time* for one complete satellite trip and the (average) *distance* of the satellite from the heavenly body about which it is revolving. Kepler found that, for every planet, the quotient T^2/r^3 has the same numerical value, where T is its *period*, that is, the time for a complete circuit, and r is its mean or average distance from the sun. In other words $T^2/r^3 = k$, where k is a numerical constant, or $T^2 = kr^3$. This last formulation is the one usually given: The square of the time which any planet takes to complete its orbit is proportional to the cube of its mean distance from the sun. Kepler considered this discovery his greatest. Is it any wonder, then, that his comments were even more extravagant than usual?

> What I predicted twenty-two years ago ... and believed in long before that ... what I revealed to my friends ... and sixteen years ago urged them to accept as valid ... that for which I joined Tycho Brahe, for which I settled in Prague, the problem which inspired my lifelong devotion to astronomy ... I have solved at last. ... The book (containing it) may be read now or by posterity, it does not matter. If it must wait a century to be read, I care not, since God himself has had to wait six thousand years for a proper observer of His handiwork.*

It is true that Kepler's laws terminated millennia of geometrical astronomy based on the empirical study of the heavens. But his inductions were not the last mathematical word on the subject. Galileo's mechanics and Newton's universal gravitation were yet to come, and to provide the foundations for modern dynamical astronomy. By making certain assumptions about the gravitational attraction between two bodies, Newton was able to provide a *deductive* proof for Kepler's laws. Between Kepler's assertion of these rules to which all satellites must conform and the logical demonstration by Newton, about seventy years were to elapse. During this interval the mathematical tools required for Newton's proof were forged. The whole modern trend in mathematics was begun when analytic geometry was crystallized by Descartes and Fermat, and the calculus was given definitive form by Newton and Leibniz.

But did Newton or did Leibniz create the calculus? Such a question is pertinent only if one believes in the type of popular folklore which demands that all concepts be considered as originating during a specific period of history in the mind of a single mathematician. One result of this unfortunately typical attitude is that mathematical development has been accompanied by numerous priority disputes, often provoked by rival groups of scientists rather than by the mathematicians themselves. The calculus, with which this chapter is concerned, gave rise to the most prolonged and bitter of all controversies of this sort, the "two-hundred-year war" between British and Continental scientists concerning the question of whether Newton or Leibniz was the inventor of the subject. It is naïve to give complete credit to either of these renowned mathematicians, since the roll call of forerunners is lengthy and distinguished. The most serious consequence of the

---

* M. Caspar and W. von Dyck, *loc. cit.*

calculus dispute was its damaging effect on British mathematics. The late Norbert Wiener* described this sorry situation:

> We have not much doubt that Leibniz' work was somewhat later but independent (of Newton's), and that Leibniz' notation was far superior to Newton's. . . . It was not long before patriotic and misguidedly loyal colleagues of both discoverers instigated a quarrel, the effects of which have scarcely yet (1949) died out. In particular, it became an act of faith and of patriotic loyalty for the British mathematicians to use the less flexible Newtonian notation and to affect to look down on the new work done by the Leibnizian school on the Continent. . . . When the great continental school of the Bernoullis and Euler arose (not to mention Lagrange and Laplace who came later) there were no men of comparable calibre north of the Channel to compete with them on anything like a plane of equality. . . .
>
> Not until the nineteenth century was well under way . . . was there awareness of what . . . Laplace and Lagrange had done in mathematics. Even then mathematical education at the English universities was devoted to (preparation for) the passing of severe examinations like the Cambridge Tripos, rather than to the development of original mathematical workers. . . . Mathematical talent in the British Isles went instead to the formation of a great school of mathematical physicists. . . . At about the turn of the century an awareness of the great work of the continental mathematicians smuggled itself into England by non-academic bypaths. . . . G. H. Hardy (1877–1947) and his associates . . . represent the first generation (of British mathematicians) to have had contact from the beginning of their training with modern continental analysis, and the first, except for William Henry Young (1863–1942) (Chapter 27) to have familiar personal contacts with all the leaders of their work on the Continent and to be regarded by the latter as friendly equals.

To avoid any continuation of the ugly controversy whose effects Wiener describes, Newton and Leibniz will *not* here be named as sole creators of the calculus, but rather as the mathematicians who made the major contributions to a subject that had existed since antiquity, first by establishing its "fundamental theorem," and second by providing a definite algebraic symbolism and a systematic set of rules for performing operations. These two elements are the essence of elementary calculus as it is studied by college students today, and if this is what any one means by "calculus," then he would be correct in attributing the source of such subject matter to Newton and Leibniz.

The true story of the creation of the calculus is the history of the gradual evolution of its various concepts. Where should one begin? If a logical, philosophical point of view is adopted in preference to the algebraic-manipulative stress of elementary calculus courses, then, as we have seen (Chapter 2), one might begin with the Pythagorean discovery of the $\sqrt{2}$ and Eudoxus' theory of the irrationals. These bring into focus the real number system, the need for "infinite" processes, the meaning of geometric continuity, all of which are essential to the *theoretical* foundations of calculus and the general (infinitesimal) analysis that includes it. If, on the other hand, one desires a mathematical tool related to the physical

---

* Norbert Wiener, "Godfrey Harold Hardy," *Bulletin of the Amer. Math. Society,* Vol. 55, No. 1, Part 1 (January 1949), pp. 73–74.

measurement of lengths, areas, volumes, weights, pressures, velocities, accelerations, etc., and this was indeed the viewpoint of Newton and Leibniz, one should start the story of the calculus with Archimedes.

Archimedes' ideas were imitated in Kepler's measurement of the areas of elliptic sectors and the volumes of wine casks. As an additional exercise in numerical integration in the spirit of Kepler, one might approximate the area within the curve of Figure 8.3. There are 161 *complete* small squares within the boundary,

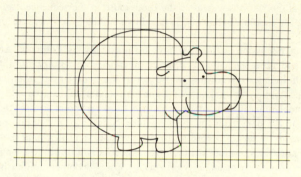

Figure 8.3

and if each square represents a square centimeter, say, then the desired area is somewhat in excess of 161 sq. cm. Counting each peripheral portion of a square as if it were entire, one obtained 214 sq. cm., an estimate exceeding the area sought. The area itself lies somewhere between the two estimates. "Common sense" suggests that the *mean* of the two approximations, namely, 187.5 sq. cm., would be a better estimate of the area than either previous figure. In similar fashion, one could approximate the volume of a solid object by counting the number of very tiny cubes which could be packed within its surface.

A proper scientific attitude would call for some standardization in the techniques of approximate integration to replace rough and ready methods like the above, and also to decrease experimental errors and estimate their magnitude. In the sort of measurement used by Kepler, it is evident that finer cross-section paper, smaller cubes, "narrower" triangles, and thinner cylindrical layers would yield better estimates. If a first approximation of some area seems too crude, one can decrease the size of the squares or triangles, etc., and obtain a second approximation. If this again seems poor, one can use still smaller meshes in a third approximation and, in *theory*, if not in practice, such improvement in approximation could be continued *ad infinitum*.

This was the thought behind the Greek *method of exhaustion* initiated by Antiphon the Sophist (*ca.* 430 B.C.), perfected by Eudoxus, and put to such excellent use by Archimedes that, if popular demand for a unique originator must be satisfied, he, rather than Newton or Leibniz, should be credited with the invention of integral calculus. The method of exhaustion is still indicated, after a fashion, to school students of geometry in connection with deriving the area of a circle. First the pupils may inscribe a regular hexagon in a circle, then double the number of sides to obtain a regular dodecagon, double the number again to obtain a regular polygon of 24 sides, etc. As pupils study the sequence of inscribed regular polygons with 6, 12, 24, 48, 96, 192, 384, . . . sides, they are led to an intuitive justification of the claim that ultimately the polygonal area *exhausts* the area within the circumference. In more modern terminology, we would say that there is a *limiting* value to which all polygonal areas after the thousandth in the sequence, say, are very, very close. If such a *limit* exists, it is said to be the *exact* measure of the area sought.

In our discussion so far, we have assumed, just as Eudoxus, Archimedes, Newton, and all mathematicians prior to the nineteenth century did, that lengths, areas, and volumes are entities whose nature we comprehend intuitively. Thus, given a description or diagram of some object, our problem would become a computational one. But the modern mathematical viewpoint stresses that reliance on intuition is a dangerous logical procedure. Nowadays an "infinite process" of approximation is set up according to definite rules, and if a *limit* exists for the approximations in question, the limiting number is said to define a length or area or volume or pressure or quantity of work or any of the other entities that integral calculus serves to measure. Thus, if we are to advance beyond a naïve idea of area or volume, it is the concepts of *infinite process* and *limit* that we must master, and these are the basic notions of the calculus.

Archimedes, in his form of integral calculus, first considered various geometric figures as if they were material bodies to which those laws of mechanics which he had discovered were applicable. Having obtained formulas for certain lengths, areas, volumes by means of mechanical principles, he provided rigorous logical proofs that involved the method of exhaustion. In this way he was able to find areas of ellipses, parabolic segments, sectors of a spiral, certain volumes, and also the centers of gravity of segments of a parabola, cone, sphere.

After Archimedes' contribution, little was added to integral calculus until Kepler's day, when a number of mathematicians revived the Archimedean techniques and attempted to improve them. Among these, in addition to Kepler himself, were the Flemish mathematical physicist Simon Stevin (1548–1620) and Bonaventura Cavalieri (1598–1647), a pupil of Galileo and professor at Bologna. *Cavalieri's principle* makes an "integral calculus" of sorts possible in school mathematics today.

Cavalieri stated: Two solids (lying between parallel planes) have the same volume if cross-sections at equal heights have equal areas. Cavalieri conceived of a solid as if it were built of a stack of cards. Thus he pictured a box as if built of ordinary rectangular cards, all the same size, a square pyramid as if built of square cards of varying sizes, a triangular pyramid as if it were a stack of triangular cards diminishing in size toward the vertex, circular cylinders and cones as constructed of circular cards (Figure 8.4). Then, if the circular base of a cylinder and the rectangular base of a box are equal in area, and if the two solids have the same height

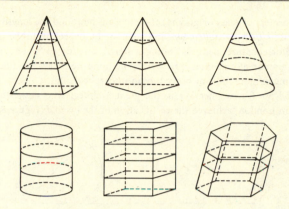

Figure 8.4   Cavalieri's principle

(Figure 8.4) their volumes must be equal, in accordance with Cavalieri's principle. One might conceive of the box as a stack of 100 ordinary (rectangular) cards, say, and the cylinder as a stack of 100 circular cards, where a circular card face has the same area as the rectangular card face in the other stack. The cards are not mere surfaces, however, but solids, since each has slight thickness. Then every card in either pack has the same small *volume*. If this volume is represented by x, the total volume of the box is $100x$, and likewise for the cylinder. But it is an easy matter to obtain the volume of a box if its dimensions are known. This volume is just the product of these dimensions, or the product of the base area by the height. The base area is just that of the rectangular cards (or the circular cards). Then the volume of the cylinder is the same as that of the box and hence is the product of the area of its (circular) base and its height. Thus the formula for the volume of a cylinder is $V = \pi r^2 h$. If a cylinder is not circular, that is, if it is built of elliptical cards or cards of some other shape, if in fact it is built of pentagonal or hexagonal or octagonal cards (in which case it is usually called a *prism*) (Figure 8.4), Cavalieri's principle is still applicable as long as the cards employed have face areas identical with those of the rectangular cards used for the box. In every case the volume formula will be $V = Bh$, where B is the area of the cylinder base = box base and h is the height of the cylinder (or box).

In addition to Kepler's approximative species and Cavalieri's intuitive type of integral calculus, there were many other pre-Newtonian contributions to the same subject. Fermat, whose role in analytic geometry has already been discussed, created integral calculus methods to find areas like that in Figure 8.16, where the boundary consists of three straight-line segments (the X-axis and two ordinates) and an arc of a "general parabola," $y = x^3$. In fact, he developed a formula for such an area when $y = x^n$ (where n is *any* positive integer), also for the case $y = x^{m/n}$ (m and n are positive integers), and also where the curve is a "general hyperbola," $y = 1/x^n$ (where n is a positive integer greater than 1). He found still more general areas, determined the volume of a paraboloid of revolution, and located various centroids (centers of mass or gravity). In addition, he found the lengths of certain curves. This is called *rectification*, and one might expect it to be easier than quadrature or cubature, because finding a length sounds so much simpler than determining an area or a volume. However, the opposite is true; rectification is, in general, a much more difficult procedure in integral calculus.

In 1658 Sir Christopher Wren, the great architect who designed St. Paul's Cathedral in London, achieved the rectification of the *cycloid* (Figure 8.5), a curve

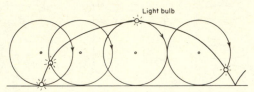

Light bulb

Figure 8.5   **A light attached to one point of the rim of a wagon wheel in motion generates a cycloid.**

with so many beautiful properties that it is called the "Helen of Geometry"; he also determined the centroid of the area under a cycloidal arch. Blaise Pascal, whose correspondence with Fermat on probability problems provides one of the favorite anecdotes in mathematical history (Chapter 13), was also deeply interested in the cycloid, and in fact specialized in some of the more arduous issues of quadrature and cubature—for example, where the bounding curves or surfaces have trigonometric formulas.

Still another forerunner of Newton and Leibniz was Christiaan Huygens (1629–1695). Surprisingly, his integral calculus techniques were, in the main, the classical methods of Archimedes. Perhaps his greatest service to the subject was the mathematical education which he gave to Gottfried Wilhelm Leibniz (1646–1716). Up to the age of twenty-six, the "co-inventor" of the calculus knew almost nothing about the more modern aspects of mathematics, even though he had formulated the basic notions of modern symbolic logic while he was still in his teens (Chapter 6). He had taken his bachelor's degree in philosophy at seventeen, and his doctorate in law at twenty. He then became a diplomatic assistant to the Elector of Mainz, and a particular diplomatic mission brought him to Paris in 1672. (Half a dozen years earlier, Newton had already formulated but not published his first thoughts on the calculus.) There Leibniz met Huygens and begged him for tutoring in mathematics. The Dutch mathematical physicist accepted the young diplomat as a pupil, since he recognized his genius almost immediately. Without any special instruction, Leibniz had already worked out details of his logic, and constructed his remarkable calculating machine, which was far superior to the first calculating machine in history, the one Pascal had invented at the age of nineteen. Whereas Pascal's machine could add and subtract, Leibniz' device could also multiply, divide and extract roots. But just now the point of emphasis is Huygen's influence on the integral calculus, directly, through his own discoveries, and indirectly, through his influence on Leibniz.

Another who can be mentioned for both his direct and his indirect influence is the Scotsman James Gregory (1638–1675). Like Pascal, he was short-lived, and like him, he solved some of the more difficult types of problem in integral calculus. His most important achievement was the theory of infinite series (Chapter 22), and here he and other British contemporaries influenced Leibniz. In fact, Leibniz is sometimes erroneously credited with discovering the "Gregory series."

But one must also mention the influences on Newton. The most immediate of these was John Wallis (1616–1703), British analyst and mathematics professor at Oxford for fifty-four years. It was a reading of his treatise, the *Arithmetica Infinitorum*, that precipitated Newton's considerations of a decade later. But the last and best pre-Newtonian ideas on integral calculus were provided by Pietro Mengoli (1626–1686) of Bologna in his *Geometria Speciosa* (1659). In this work he represented certain areas as *limits* of sums of rectangles, and in doing so, provided a *theoretical* definition for an "integral." After Augustin-Louis Cauchy (1789–1857), the first of France's great modern analysts, had placed calculus on a sound logical foundation, the "limit of a sum" definition of areas, volumes, or other entities to be measured came to be known as the *Mengoli-Cauchy integral*. This completes the story of integral calculus prior to Newton and Leibniz. Since their major contribution was to the *differential* calculus, we shall pause here for further biographical facts about the two men prior to introducing that subject.

Isaac Newton (1642–1727), often cited as the greatest genius the race has ever produced, showed no signs of promise in his early years. Like many talented children, he found school a complete bore. In lieu of studying, he devoted himself to the construction of kites, water wheels, tops, mechanical toys, sundials, ingenious clocks, and all sorts of amazing gadgets. His academic record was so poor that his mother decided farming would be the ideal vocational choice for Isaac. His father, who had died shortly before the boy's birth, had been a Lincolnshire farmer, and it seemed natural to have the son continue in the same work. Newton's uncle had a better understanding of the boy's talents. One day he discovered Isaac reading under a hedge instead of performing his agrarian duties. That clinched matters, and he persuaded Newton's mother to send the boy to Trinity College, Cambridge.

In the years 1665–1666, when the Great Plague was raging in England, Newton went back to the farm and in spare moments started his work on gravitation, optics, and calculus. In connection with the third subject, he developed the *binomial theorem*, which was no mean stunt for a young man of twenty-three.

When Newton was twenty-six years of age, his teacher, Isaac Barrow, himself a mathematician of note, resigned from the Lucasian professorship at Trinity to devote himself to theology. He named his pupil as his successor. Newton's first work at Cambridge was in the field of optics. He built his own reflecting telescope, which won him a membership in the Royal Society. He advanced the corpuscular theory of light, which later scientists rejected in favor of the wave theory. Recent developments in atomic physics and quantum mechanics have caused scientists to reconsider Newton's notions about the nature of light, and the present opinion is best described by saying that for some phenomena it is convenient to accept

Newton's view. Light in these cases is considered as particles of energy called quanta or *photons*. In other situations, theories are simplified by picturing light as one type of electromagnetic wave.

Newton presented the fruits of his most profound thought in the immortal *Principia* (1687). A brief summary can hardly convey a true impression of the monumental nature of this work. It contained, among other revolutionary scientific material, Newton's dynamics, his law of universal gravitation, and his "system of the world." The *Principia* seems to have ended the most important part of Newton's scientific career, and it was certainly a grand finale. After the publication of his work, Newton started life anew in a political capacity. A chance incident was the cause of it. In 1687 James II sought to impose his will on Cambridge University. Newton led the group that resisted and, as a reward for his courage, was elected to represent the University in Parliament the following year. This "opportunity" was fatal. He was later (1696) appointed Warden of the Mint and finally Master of the Mint (1699). He was one of the best Masters the Mint ever had.

Some mathematicians feel regret that a "high priest of science" should have wasted his genius in public office. Others cite Newton as a shining example of a scholar who deserted his ivory tower. Still others feel that Newton had spent himself early, that he realized his power for scientific thought was waning and hence sought other occupations. Some biographers assert that Newton's financial need was the cause of his action. We must remark that, whatever the evaluation of Newton's latter-day intellectual power, he was mathematically active to the day of his death at the age of eighty-five, although he did not produce any other work comparable to the *Principia*.

As for critical estimates of Newton's work, we first quote the generous tribute of his rival Leibniz, who said that, taking mathematics from the beginning of the world to the time when Newton lived, what Newton did was much the better half. Many still agree with Leibniz on this point. Lagrange, whom Frederick the Great named the "greatest mathematician in Europe," once remarked humorously that Newton was the greatest genius that ever lived and the most fortunate, since only once can the system of the universe be established.

Newton's modesty is revealed in his own appraisal of his work: "I do not know what I may appear to the world; but to myself I seem to have been only like a boy playing on the seashore, and diverting myself in now and then finding a smoother pebble or a prettier shell than ordinary, whilst the great ocean of truth lay all undiscovered before me."

Newton died in 1727 and was buried in Westminster Abbey. Voltaire attended the funeral, and it is said that, at a later date, his eye would grow bright and his cheek flush when he mentioned that a professor of mathematics, only because he was great in his vocation, had been buried like a king who had done good to his subjects.

Earlier in this book we discussed certain aspects of Leibniz' life in relation to his most important contribution to mathematics, the initiation of symbolic logic. Here we remark that if he was not, like Newton, the greatest genius of all time, he was not very far from the highest rank. His talents were all-round, for in addition to mathematics and logic, he contributed ideas in such varied areas as law, religion, history, economics, and metaphysics. But his intellectual powers were not given the recognition they deserved during his lifetime. In fact, from the scientist's point of view, Leibniz' active life ceased at thirty, for he spent the last forty years of his life as historian of the Brunswick family. His last days were unhappy ones. In 1714 his patron, Elector George Louis of Brunswick, left for London to become George I of England. Leibniz would have liked to go with him for one last round with the English mathematicians and Newton, Master of the Mint. However, George left him at Hanover. Two years later Leibniz died and was buried in an obscure grave. Let us forget this unhappy ending and return to our theme—the branch of mathematics that he helped to develop for the proper handling of a world in flux.

We are now ready to talk about the evolution of the *differential calculus*. If we seek sources in antiquity, we find only one slight suggestion in the work of Archimedes. In fact, there was no real development of the subject before the seventeenth century. Like the *integral* calculus, the *differential* calculus is also concerned with infinite processes and associated limits. Its most beautiful and useful theorem provides, for many problems of area, volume, etc., a simple, truly easy alternative to the Mengoli-Cauchy "limit of the sum" procedure. For this reason, beginners in the calculus find it easier to master the concepts of the newer branch of calculus first, and then establish its connection with the older integral calculus. The names of those who paved the way for Newton and Leibniz in *differential* calculus will be revealed a little later on, since an appreciation of their contributions can be increased by a preliminary short journey along the "easier path" to the calculus.

"Wages have gone up 20 per cent this year." "And what about prices?" you ask before rejoicing at the approach of an era of prosperity. Johnny has gained fifteen pounds this year, but still wears that lean and hungry look, for he has also grown three inches. The ocean has made an inroad of five miles on this coastline. "What a disaster!" you feel, until you learn that it has taken a million years for this change to take place. *A comparison of two related changes* is called for in each case. We compare increases in wages with rise in prices before concluding that times are improving, increase in weight with growth in height before prescribing for Johnny, change in distance with change in time to realize that the speed of shoreline advance is slight.

Consider the comparison of change in *amount* with the corresponding change in *time* for a simple interest problem in which the principal is $100 and the rate 4 per cent. Initially the amount is $100; at the end of each year the amount is increased by the annual interest of $4, so that the amounts are

$$100 + 4 \cdot 1 \text{ at end of first year}$$
$$100 + 4 \cdot 2 \text{ at end of second year}$$
$$100 + 4 \cdot 3 \text{ at end of third year}$$
$$\text{etc.}$$

The formula connecting amount and time is obviously

$$A = 100 + 4t$$

where A is the amount and t the number of years. Newton referred to a variable like A, that is dependent on (is a function of) time, as a *fluent*.

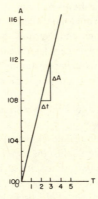

Figure 8.6

Note that this formula and its graph (Figure 8.6) are linear. The tabulation below indicates changes in time and the corresponding changes in amount.

Change in Time	Years t	\$ A	Change in Amount
	0	100	
1	1	104	4
1	2	108	4
1	3	112	4
1	4	116	4

The quotient,

$$\frac{\text{change in amount}}{\text{change in time}} = \frac{4}{1} = \frac{8}{2} = \frac{12}{3}$$

$$= \$4 \text{ per year}$$

The quotient derived from the table represents the change in amount for one year, or the *rate of change* of amount. In other words, the annual interest is the rate of change in amount with respect to time.

In the previous chapter, Δ (delta), the usual symbol for *change* or *difference*, was introduced. Applying this symbol, we assert that $\Delta A = 4$ and $\Delta t = 1$ between each pair of successive points in our table and graph. Then the

$$\textit{difference quotient} = \frac{\Delta A}{\Delta t} = \frac{4}{1} = 4 = \text{rate of change}$$

The fact that this difference quotient is the same between any pair of points is characteristic of a straight line graph; when a graph is not linear, the difference quotient varies, as will be indicated shortly. As another linear function, consider

$$y = \frac{2x}{3} + 5$$

If the values 0, 3, 6, 9, etc. are substituted for x and the corresponding values of y are found, the coordinates of points on the straight line graph are obtained (Figure 8.7). The changes Δy and Δx are tabulated below and lead to the

$$difference\ quotient = \frac{\Delta y}{\Delta x} = \frac{2}{3}\ \text{for this line}$$

Δx	x	y	Δy
	0	5	
3	3	7	2
3	6	9	2
3	9	11	2
3	12	13	2

In other words, whatever the meaning of y and x, the rate of change in y is 2/3 of a unit for each unit change in x.

The rate of change $\Delta y/\Delta x$ is also termed the *slope of the line*. It is evident that this rate is actually connected with the ordinary "slope" of an incline. Here Δy represents the *vertical* rise from point to point, Δx the *horizontal* progress. The ratio of the two lengths tells what the rise in height is for one unit of horizontal advance. The greater the ratio, the steeper the slope. In our first illustration, the slope 4 gives a steeper line than the slope 2/3 in our second illustration.

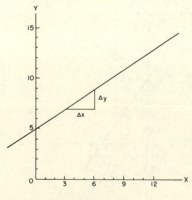

Figure 8.7

By an examination of

$$y = 4x + 100$$

and

$$y = \frac{2x}{3} + 5$$

we see that the slopes 4 and 2/3 appear as the coefficient of x in the equation. We shall now prove that this is generally true. When a linear equation (for a non-vertical line) is put into the general form

$$y = ax + b$$

we can consider any two points satisfying this relation. Let the x-coordinates of these points be c and d, respectively. Then, by substitution in this equation, the corresponding y-coordinates are $ac + b$ and $ad + b$. Studying changes, we have

$$
\begin{array}{c|c}
x & y \\
\hline
c & ac + b \\
d & ad + b
\end{array}
$$

$$\Delta x = d - c \qquad \Delta y = ad - ac = a(d - c)$$

and slope

$$\frac{\Delta y}{\Delta x} = a$$

since the factor $d - c$ divides out. Then the slope of

$$y = -\frac{3x}{2} + 5$$

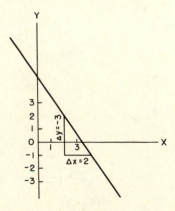

Figure 8.8

is $-3/2$. In other words,

$$\frac{\Delta y}{\Delta x} = -\frac{3}{2} = \frac{-3}{2} = \frac{3}{-2}$$

This means that when x increases 2, y decreases 3 (see Figure 8.8), or when x decreases 2, y increases 3.

Let us apply the notion of rate of change to some concrete situations. The following is a timetable showing the time of arrival and departure of a train at various stations. We have filled in the columns t (time), s (distance), Δs, Δt, and $\Delta s/\Delta t$ and plotted the data in Figure 8.9.

Δs (miles)	s (distance in miles)	Town	s (time in hours)	Δt (hours)	$\frac{\Delta s}{\Delta t}$ (miles per hour)
	0	dep. A	9:00		
25				$\frac{1}{2}$	50
	25	arr. B	9:30		
0				$\frac{1}{6}$	0
	25	dep. B	9:40		
20				$\frac{1}{2}$	40
	45	arr. C	10:10		
0				$\frac{1}{6}$	0
	45	dep. C	10:20		
15				$\frac{1}{4}$	60
	60	arr. D	10:35		

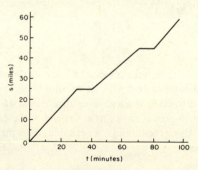

Figure 8.9

In the picture we have five different line segments representing the various portions of the trip. The five values of $\Delta s/\Delta t$ indicate the slopes of these segments, and also the velocity of the train from point to point. From A to B this velocity was 50 miles an hour; during the first stop the velocity was naturally zero; then the train proceeded at 40 miles an hour, etc.

To get the average rate for the entire trip, we seek the constant rate the train would have followed if it had proceeded uniformly from A to D without stopping. The picture of its motion would have been a straight-line segment from the first to the last points of the graph. The slope of this line would have been the uniform velocity followed. In other words, this slope represents the average rate from A to D. Reading from either the graph or the table, we see that, from the first to the last points,

$$\Delta s = 60$$

$$\Delta t = 1\frac{7}{12}$$

$$\frac{\Delta s}{\Delta t} = \frac{60}{19/12} = 37.9 \text{ mi. per hr.}$$

This exercise illustrates how the quotient $\Delta y/\Delta x$ (above $\Delta s/\Delta t$) is useful even when the picture of a situation is not a single straight line. In our example the picture was a broken line, and in other cases it will be a curved line. We see that in such situations $\Delta y/\Delta x$ will give an average rate of change between two points.

As a further illustration let us now consider the famous apple whose fall is supposed to have initiated the theory of universal gravitation. It is shown in this theory that the formula

$$s = 16t^2 \qquad (t \text{ seconds, } s \text{ feet})$$

governs the motion of the apple.* Let us suppose that it took the apple exactly 1 second to fall. By the formula, we see that it fell 16 feet. Then

Δt	t	s	Δs
	0	0	
1			16
	1	16	

$\dfrac{\Delta s}{\Delta t} = \dfrac{16}{1} = 16$ ft. per sec. (*average velocity during the second*)

Was the apple moving 16 feet per second when it hit the ground? Evidently not for, as we know, the apple started with no velocity at all and increased its speed under gravitational attraction. Hence, since it averaged 16 feet per second, its velocity must have been greater than this at the end of the first second to balance its slow initial speed. Using the formula, we can find where the apple was at the end of half a second. Then

Δt	t	s	Δs
	1/2	4	
1/2			12
	1	16	

Thus $\dfrac{\Delta s}{\Delta t} = \dfrac{12}{1/2} = 24$ ft. per sec. (*average velocity during the half-second preceding impact*)

---

* Strictly correct only if the apple falls in a vacuum. The factor 16 is approximate also.

Similarly,

Δt	t	s	Δs
	3/4	9	
1/4			7
	1	16	

$\dfrac{\Delta s}{\Delta t} = \dfrac{7}{1/4} = 28$ ft. per sec. (*average velocity during the last quarter-second*)

Again,

Δt	t	s	Δs
	7/8	49/4	
1/8			15/4
	1	16	

$\dfrac{\Delta s}{\Delta t} = \dfrac{15/4}{1/8} = 30$ ft. per sec. (*average velocity during the last eighth-second*)

Finally,

Δt	t	s	Δs
	0.99	15.6816	
0.01			0.3184
	1	16	

$\dfrac{\Delta s}{\Delta t} = \dfrac{0.3184}{0.01} = 31.84$ ft. per sec. (*average velocity during the last hundredth of a second*)

It requires no Sherlock Holmes to sense that this last approximation is the best indication of the velocity of the apple when it hits the ground.

Now this simple fact is at the basis of the differential calculus. To get the *instantaneous rate*, in this case, we let Δt get smaller and smaller. Notice that Δs gets smaller too. The *limiting value* of $\Delta s/\Delta t$ is called the *instantaneous rate of change* or *derivative* of s with respect to t. In our particular problem, we have taken Δt as small as 0.01. We could take Δt as small as 0.001, 0.0001, etc. If we did so, we should find $\Delta s/\Delta t$ to have a value closer and closer to 32. We could get a value as close to 32 as we pleased by taking Δt small enough. This value, 32 feet per second, is the limiting value of $\Delta s/\Delta t$ as Δt approaches zero. Better and better average rates are obtained by taking smaller and smaller intervals, but there is no "best" average rate, since if the rate for any interval, however small, is figured, you can always furnish a still better rate by decreasing the interval further. The *limit* does the trick because it is a number to which all "good" rates are close.

Newton used the term *fluent* for the dependent variable in a function of time. Thus he would have called s a fluent in the function, $s = 16t^2$, which we have been discussing. He used the word *fluxion* for the derivative or instantaneous rate of change of a fluent, and referred to calculus as *fluxions*. But we shall not always be handling fluents and fluxions. Therefore let us define the instantaneous rate of change or derivative for *any* functional relation associating two variables, y and x, in such a way that the value of y is determined by the value of x. If there is a limiting value of $\Delta y/\Delta x$, as Δx approaches zero, we shall call it the instantaneous rate of change of y with respect to x. The process of finding the derivative is called *differentiation*, and hence the term *differential calculus*.

Let us illustrate differentiation, that is, the finding of an instantaneous rate. Suppose that a 2-inch metal cube is heated. Just how fast will it expand? The formula for the volume of a cube is

$$V = x^3$$

By substituting values of the edge close to 2 and using this formula to find V, we shall be able to obtain good *average rates of expansion*. Using the accompanying table, we see that, if the edge were to increase

x (inches)	2	2.01	2.1	2.5	3	4
V (cubic inches)	8	8.1204	9.3	15.6	27	64

from 2 to 2.5 inches, then

$$\text{change in } x = \Delta x = 0.5$$
$$\text{change in } V = \Delta V = 7.6$$

and

$$\frac{\Delta V}{\Delta x} = \frac{7.6}{0.5} = 15.2 \text{ cu. in. per in.}$$

If the edge increases from 2 to 2.1,

$$\frac{\Delta V}{\Delta x} = \frac{1.3}{0.1} = 13 \text{ cu. in. per in. (see table above)}$$

If the edge increases from 2 to 2.01,

$$\frac{\Delta V}{\Delta x} = \frac{0.1204}{0.01} = 12.04 \text{ cu. in. per in.}$$

The rates 15.2, 13, 12.04 are evidently getting closer and closer to 12 cu. in. per in. By taking $\Delta x = 0.001, 0.0001$, etc. the average rate would come still closer to 12, which is the *limit* of the series of numbers, and the instantaneous rate of expansion of the cube with respect to its edge is 12 cu. in. per in. when $x = 2$.

Naturally, serious workers in the calculus do not use the excessive computation we have employed for determining a limit. They have exact meanings and systematic methods. We shall shortly indicate the nature of such techniques.

Since graphs have proved themselves so revealing on previous occasions, let us see what they will do for our concept of the derivative. Let us graph the formula $V = x^3$, which we have just used, for those values of the edge selected in the table above. In Figure 8.10 we have plotted points and joined them by a smooth curve. Notice the relative positions of the lines *PA*, *PB*, *PC*, and *PD* with relation to the curve of the formula. They cut across it, and hence are called *secants*. But *PC* and

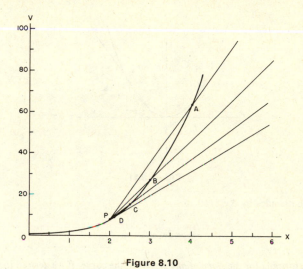

Figure 8.10

PD cut off much less of the curve than *PA* and *PB*. The secant *PC* cuts off a very small segment, and *PD* barely grazes the curve. The line *PC* has the slope

$$\frac{\Delta V}{\Delta x} = \frac{7.6}{0.5} = 15.2$$

and *PD* has the slope 13. The *limiting* line, which all these secants approach, is called the *tangent* to the curve at *P*. It will touch the curve at *P* only. Thus the geometric meaning of a derivative, or instantaneous rate, is the *slope of a tangent to a curve*.

Let us now investigate more efficient ways of finding derivatives. Let us take the function $y = x^2$ and obtain a formula for the derivative for any value of x, that is, a formula for the slope of any tangent to the parabola $y = x^2$. In Figure 8.11 let *P* with coordinates (x, y) be any point on the curve. Then the coordinates of an adjacent point *Q* will be $(x + \Delta x, y + \Delta y)$.

The slope of the secant *PQ* is $\Delta y/\Delta x$. To find a formula for $\Delta y/\Delta x$, let us take cognizance of the fact that *Q* and *P* are points on the curve and hence their coordinates must satisfy the condition $y = x^2$, that is, for *Q*,

$$y + \Delta y = (x + \Delta x)^2$$

or

$$y + \Delta y = x^2 + 2x\Delta x + (\Delta x)^2$$

and for *P*,

$$y = x^2$$

Subtracting the second equation from the first,

$$\Delta y = 2x\Delta x + (\Delta x)^2$$

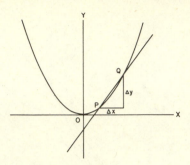

Figure 8.11

Dividing both sides by Δx,

$$\frac{\Delta y}{\Delta x} = 2x + \Delta x$$

This is the formula for the *slope of any secant* to the curve. If we desired the slope of the secant between the points whose x-coordinates are 3 and 5, respectively, we should have

$$x = 3 \quad \Delta x = 2$$

and

$$\frac{\Delta y}{\Delta x} = 2x + \Delta x = 6 + 2 = 8$$

To find the derivative or slope of the tangent, we want the limit of the slope of the secant as two points approach coincidence. If, in the formula just applied, namely,

$$\text{slope of secant} = 2x + \Delta x$$

we let Δx approach zero, that is, become smaller and smaller, the limit approached will obviously be $2x$, and this will be the formula for the derivative or slope of the tangent.

A symbol usually used for the derivative is dy/dx. Then, in this case,

$$\frac{dy}{dx} = 2x$$

Applying this formula, the slope of the tangent at the point where $x = 3$ is

$$\frac{dy}{dx} = 2x = 2 \cdot 3 = 6$$

and the slope of the tangent at any other point is readily found.

We might use this result a little differently. If we think of $y = x^2$ as the formula for the area of a square, $dy/dx = 2x$ is the formula for instantaneous rate of change of area relative to length of a side. Then if a square is expanding (as a result of temperature change, for example), its rate of expansion at the instant when its side is 3 inches is 6 square inches per inch.

The geometric interpretation of the derivative of a function as the slope of the tangent to the curve representing the function, is a good point of departure for discussing anticipators of Newton and Leibniz in the matter of the differential calculus. These forerunners were all interested in methods of constructing tangents to special curves, or to curves in general. This was, then, equivalent to finding derivatives for specific functions, or for all functions. Fermat's consideration of tangents was, except for symbolism, virtually that explained above. He studied them in connection with maximum and minimum problems, handling these questions much as we shall later on. But then he went further and considered the more general idea that economy is not limited to specific problems but is an intrinsic characteristic of natural phenomena. He enunciated a "principle of least time" and from it deduced the laws of geometrical optics. Later such principles became the typical issues of the calculus of variations, an advanced calculus treating very general maxima and minima.

Archimedes had considered only one special case of tangent construction (or finding a derivative), namely, that connected with his favorite curve, the one we call the spiral of Archimedes. This is the faint inkling of differential calculus to which we have alluded. Nothing else was contributed until Fermat's time. He came closest to the ideas of the differential calculus, but there were others who contributed to the development by considering special cases or particular formulas. They were Descartes; Evangelista Torricelli (1608–1647), the mathematical physicist connected with the theory of the barometer; Isaac Barrow; Gilles Persone de Roberval (1602–1675); Johannes Hudde (1633–1704), mayor of Amsterdam. René François Walther (1622–1685), Baron de Sluse and Canon of Liège, studied Fermat's method and improved on it. There is no doubt that Newton's method of constructing tangents was identical with the De Sluse procedure, and in 1673, Newton actually gave priority to De Sluse, and at another time gave credit to Fermat and Barrow.

In the study of differential calculus, formulas for the derivatives of many types of function are established. In a fashion similar to that used for showing that the derivative of $y = x^2$ is $dy/dx = 2x$, it can be shown that the derivative for any power of x,

$$y = x^n$$

is

$$\frac{dy}{dx} = nx^{n-1}$$

Putting the symbols into words, the derivative is equal to the *exponent multiplied by the base with exponent reduced by* 1. In formal calculus this rule is proved to hold where n is *any* rational number. Hence n may be positive or negative or zero, integral or fractional.

Thus the derivative of
$$y = x^4$$
is
$$\frac{dy}{dx} = 4x^3$$
and the derivative of
$$y = x^7$$
is
$$\frac{dy}{dx} = 7x^6$$
and the derivative of
$$y = x \text{ (that is, } y = x^1)$$
is
$$\frac{dy}{dx} = 1x^0 = 1$$
If a coefficient is present, then the derivative of
$$y = ax^n$$
is
$$\frac{dy}{dx} = nax^{n-1}$$

The fact that the coefficient is carried along may be justified by a specific instance. Let us refer to page 188, where we derived the fact that if
$$y = x^2$$
then
$$\frac{dy}{dx} = 2x$$
If we go through a similar derivation for
$$y = 5x^2$$
then
$$y + \Delta y = 5(x + \Delta x)^2$$
or
$$y + \Delta y = 5x^2 + 10x\Delta x + 5(\Delta x)^2$$
But
$$y \quad\quad = 5x^2$$
Hence, by subtraction,
$$\Delta y = 10x\Delta x + 5(\Delta x)^2$$
and
$$\frac{\Delta y}{\Delta x} = 10x + 5\Delta x$$

Let Δx approach zero. Then, in the limit,
$$\frac{dy}{dx} = 10x$$

Now compare this procedure with that on page 187, and notice that the effect of the coefficient 5 is merely that all terms in the right member of each equation are multiplied by 5. Hence the net result is a derivative 5 times as great. Thus the derivative of

$$y = 3x^5$$

is

$$\frac{dy}{dx} = 15x^4$$

and the derivative of

$$y = 3x$$

is

$$\frac{dy}{dx} = 3x^0 = 3$$

and the derivative of

$$y = 4, \text{ (that is, } 4x^0)$$

is

$$\frac{dy}{dx} = 0 \cdot 4x^{-1} = 0$$

(if we assume that the fundamental law holds for $n = 0$).

The derivative of any constant is zero, for if

$$y = a \text{ (that is, } ax^0)$$

then

$$\frac{dy}{dx} = 0 \cdot ax^{-1} = 0$$

The derivative of a sum is equal to the sum of the derivatives of the individual terms. The reason this law holds is that a derivative is a limit, and it is a fundamental fact that the limit of a sum is equal to the sum of the limits of terms. As an illustration of this statement, let x, y, z, and w represent four variables and S their sum, so that

$$S = x + y + z + w$$

Let us suppose that x, y, z, and w approach the limits 2, 4, 5, 10, respectively. We claim that S will approach the limit $2 + 4 + 5 + 10 = 21$, for this reason: At some stage of their variation, x, y, z, and w will be within 0.001 of their goals. Suppose that at this stage each is 0.001 less than its limit. Then

$$S = 1.999 + 3.999 + 4.999 + 9.999 = 20.996$$

We see that S will be within 0.004 of 21. If the variables x, y, z, and w are within 0.000001 of their limits, then, at worst, S will be within 0.000004 of 21, so that S evidently approaches 21 as a limit. In fact, to repeat, it is generally true that the limit of a sum is equal to the sum of the limits of its terms.

Thus if

$$y = 2x^4 - 3x^3 + 7x^2 - 4x + 5$$
$$\frac{dy}{dx} = 8x^3 - 9x^2 + 14x - 4$$

and the derivative of

$$y = 3x^5 - 4x^3 + 2x - 7$$

is

$$\frac{dy}{dx} = 15x^4 - 12x^2 + 2$$

If the slope of the tangent to the curve

$$y = 3x^5 - 4x^3 + 2x - 7$$

at the point where $x = 1$ is desired, we merely substitute $x = 1$ in the formula for the derivative, thus obtaining

$$\text{slope} = \frac{dy}{dx} = 15 - 12 + 2 = 5$$

The fact that the derivative of

$$y = 3x$$

is

$$\frac{dy}{dx} = 3$$

that is, the derivative is a *constant*, can be readily interpreted graphically; $y = 3x$ can be pictured as a *line*. The tangent at any point is the line itself, and therefore its slope is always the same.

Since the slope of any line is a constant, the derivative of a linear expression should always be a constant. We see that this is actually the case. The derivative for any line

$$y = ax + b$$

is

$$\frac{dy}{dx} = a \quad \text{(a constant)}$$

The fact that the derivative of

$$y = 4$$

is

$$\frac{dy}{dx} = 0$$

can be interpreted by plotting the line $y = 4$ (see table and Figure 8.12). This is a line parallel to the X-axis. Its slope is actually zero.

x	y
0	4
1	4
2	4
3	4

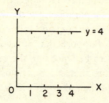

Figure 8.12

Let us apply our rules to a few practical problems. The distance s for stopping an automobile under normal conditions is given by the formula $s = 0.097V^2$, where s is measured in feet and V in miles per hour. Find how fast s increases with V when $V = 20$. First, differentiating in the formula, we have

$$\frac{ds}{dV} = 0.194V$$

Substituting $V = 20$,

$$\frac{ds}{dV} = 3.88 \text{ ft. per mi. per hr.}$$

that is, at the velocity under consideration (20 mi. per hr.), *the distance for stopping was increasing about 3.9 feet for each mile per hour increase in velocity.*

The arch of a certain bridge is parabolic, and the height (y ft.) above the water at any horizontal distance (x ft.) from the center is $y = 80 - 0.005x^2$. Find the slope of the bridge at any point (that is, the slope of the tangent) and, in particular, the slope 40 ft. from the center.

$$y = 80 - 0.005x^2$$
$$\frac{dy}{dx} = -0.01x$$
$$= (-0.01)(40) = -0.4 \quad \text{when} \quad x = 40$$

This means that for a horizontal advance of 10 ft., the tangent drops 4 ft. (see Figure 8.13).

To put the derivative to other uses, we remark that graphs of parabolas like those in Figure 8.13 and Figure 7.12d exhibit *maximum* or *minimum* points. Notice

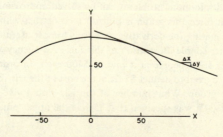

Figure 8.13

that at these highest and lowest points the tangent is horizontal, that is, its slope $dy/dx = 0$. Even if a graph is not a parabola, the derivative will be zero for *extrema*.

The fact that $dy/dx = 0$ at maximum and minimum points is useful in solving important practical problems. Suppose that a rectangular pasture, one side of which is bounded by a river, is to be fenced on the other three sides. What are the dimensions of the largest pasture that can be enclosed with 4000 ft. of fence? In Figure 8.14 we have represented the unknown width of the pasture by x. Two sides

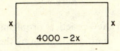

Figure 8.14

of the fence are x ft. in length, and the third side must be $4000 - 2x$. Let us call the area y. Then, since the area of a rectangle is equal to the product of the length by the width,

$$y = x(4000 - 2x)$$
$$y = 4000x - 2x^2$$

Now, it is just this quantity y, or $4000x - 2x^2$, which we desire to be a maximum. We could, of course, graph this equation for different values of x (different widths), and then, from the graph, estimate which value of x will produce the greatest area. Incidentally, this graph will be a quadratic parabola with a maximum point. If we recall, however, that the slope of the tangent is zero at a maximum point, we can avoid the work of graphing.

$$y = 4000x - 2x^2$$
$$\frac{dy}{dx} = 4000 - 4x$$
$$4000 - 4x = 0 \quad \text{at a maximum point}$$
$$\therefore \ 4x = 4000$$
$$x = 1000$$
$$4000 - 2x = 2000$$

The dimensions of the rectangle should be 1000 ft. by 2000 ft.

Maximum and minimum problems can be solved approximately by the reading of a graph, but often this graph is difficult to construct with accuracy. Hence the technique of equating the derivative to zero is a most useful one.

To give another simple illustration of this fact: An agency agreed to conduct a tour for a group of 50 people at a rate of \$400 each. In order to secure more tourists the agency agreed to deduct \$5 from the cost of the trip for each additional person joining the group. What number of tourists would give the agency maximum gross receipts? (It was specified that 75 was the upper practical limit for the size of the group.)

If 6 people were to join the group, the reduction in the cost of the tour would be $30 per person. If 10 joined, it would be $50, etc. If we represent by x the unknown number of additional tourists, the reduction will be $5x$ dollars, and

$$\text{cost of tour} = 400 - 5x$$
$$\text{number of tourists} = 50 + x$$

If we multiply the number of tourists by the cost of the tour, we shall obtain the gross receipts of the company. Let y symbolize these gross receipts. Then

$$y = (400 - 5x)(50 + x)$$

or

$$y = 2000 + 150x - 5x^2$$

To find the maximum gross receipts, we use the technique of finding the derivative and equating the result to zero.

$$\frac{dy}{dx} = 150 - 10x$$
$$150 - 10x = 0$$
$$10x = 150$$
$$x = 15$$

If there are 15 additional tourists for maximum receipts, the number in the group will be 65.

To return momentarily to the process of differentiation, if

$$y = 2x^4 - 5x^3 + x^2 - 7x + 3$$

is a special formula or the equation of a curve or the expression of some physical law, then

$$\frac{dy}{dx} = 8x^3 - 15x^2 + 2x - 7$$

is its derivative, slope, or rate of change. Now one may find the derivative of a derivative, or the *second derivative*, symbolized by d^2y/dx^2. For the formula above,

$$\frac{d^2y}{dx^2} = 24x^2 - 30x + 2$$

Continuing, one may find the third, fourth, fifth, etc. derivatives. For the formula above,

$$\frac{d^3y}{dx^3} = 48x - 30$$
$$\frac{d^4y}{dx^4} = 48$$
$$\frac{d^5y}{dx^5} = 0$$

The most important derivatives in physical applications are the first and second, and these have various special meanings. For example, if x represents time and y distance, then dy/dx represents *velocity*. In this case d^2y/dx^2, the rate of change in velocity, is called the *acceleration*. The second derivative has other special interpretations, depending on the meaning of the related variables x and y. When the relation between x and y is graphed, then, in one interpretation, d^2y/dx^2 is associated with the *curvature* of the graph.

The meaning and use of higher derivatives are particular questions of the differential calculus. But in some applications of the calculus the situation is reversed—the derivative is known and the formula is to be determined. Suppose, for example, it is known that a ball is rolling down an incline, and that its velocity is changing at the constant rate of 10 ft. per sec. Rate of change of velocity is termed *acceleration*. Here we know that the acceleration

$$\frac{dV}{dt} = 10$$

It is not difficult to see that

$$V = 10t + c$$

for, if a derivative is constant, it must come from a linear function. In the equation, c is a numerical quantity or constant, to be determined by known physical conditions. For example, if we know that the ball was started down the incline with a velocity of 30 ft. per sec., that is, for

$$t = 0, \quad V = 30$$

we can substitute this information in

$$V = 10t + c$$

and get

$$30 = 10 \cdot 0 + c$$
$$c = 30$$

thus determining the constant in question. Then the formula for the velocity of the body at any time is

$$V = 10t + 30$$

We can use this formula to obtain another for the position of the ball at any time. *Velocity* is another name for *instantaneous rate of change of distance*, so that

$$V = 10t + 30$$

is equivalent to

$$\frac{ds}{dt} = 10t + 30$$

where s represents distance.

It is a little harder to guess this time, but not too difficult to see that

$$s = 5t^2 + 30t + k$$

where k is once again a constant to be determined by known physical conditions.

If we know that the ball started at a position 20 ft. from the top of the incline, that is, for

$$t = 0, \quad s = 20$$

we can substitute this information in the formula for s in order to obtain k.

$$20 = 5 \cdot 0 + 30 \cdot 0 + k$$
$$k = 20$$

and

$$s = 5t^2 + 30t + 20$$

In the problem just illustrated, ds/dt was given as a function of t, and then we found s as a function of t. This is the reverse of the first calculus process discussed, and hence it is called *antidifferentiation* or the finding of an *antiderivative*. Incidentally, the relation between ds/dt and t is described as a *differential equation*. If, then, we are given the differential equation

$$\frac{dy}{dx} = 3x^2 + 4x - 5$$

it is easy to guess that

$$y = x^3 + 2x^2 - 5x + c$$

and to sense the rule: *In each term raise the power by one and divide by the new exponent.*

It is not always easy to guess the formula for finding an antiderivative. Calculus has its set of rules and regulations for doing this. Whereas most of the functions that occur in elementary mathematics and science can be differentiated, even the most innocent-looking expressions may fail to be the derivatives of elementary functions. Then, too, it is evident that there will always be an unknown constant in the answer. Unless some additional information is given, it will not be possible to determine this constant.

In spite of these difficulties, antidifferentiation will handle many vital problems. Chief among the geometric types are the calculations of the *length* of a curve, or the *area* enclosed by a curve, or the *volume* bounded by a surface. Since Euclidean methods can handle only a limited number of curves and solids, the calculus is a far superior metric tool.

We shall give one illustration of how antidifferentiation will determine an area. Let us say that the shaded area, marked A in Figure 8.15, is a growing quantity, for our point of view demands that we consider all things in a state of flux. The small amount by which it grows when $P(x, y)$ moves to $Q(x + \Delta x, y + \Delta y)$ is marked ΔA. It is bounded by the curve, the X-axis, and the ordinates of P and Q. The area ΔA is made up of a rectangle, surmounted by a small triangular area, PQR, one side of which is the curve. We shall treat this little triangle as if it were a right triangle. If Δx, and hence Δy and ΔA, were very small, the error in this assumption would be negligible. Then

$$\Delta A = \text{rectangle} + \text{triangle}$$
$$\Delta A = y\Delta x + \tfrac{1}{2}\Delta x \Delta y$$

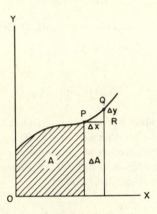

Figure 8.15

since the area of a triangle is one-half the product of base and height. Dividing both sides of this equation by Δx, we have

$$\frac{\Delta A}{\Delta x} = y + \tfrac{1}{2}\Delta y$$

Now if Δx approaches zero (see Figure 8.15) Δy and ΔA will do likewise. The limit of $\Delta A/\Delta x$ in this case is dA/dx and

$$\frac{dA}{dx} = y$$

that is, *the rate at which the area is changing at any point of a curve is equal to the height of the curve at that point.*

This proof lacks the mathematical rigor of twentieth-century standards, but even the great inventors of the calculus depended on intuition rather than logic at many points. Emulating them, we have given a demonstration sufficient for our present purpose.

To illustrate the use of the formula

$$\frac{dA}{dx} = y$$

let us find the area (Figure 8.16) under the curve $y = x^3$, and bounded by the curve, the X-axis, and the ordinates at $x = 1$ and $x = 4$.

$$\frac{dA}{dx} = y$$

or

$$\frac{dA}{dx} = x^3$$

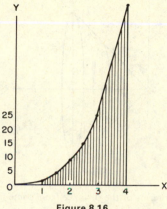

Figure 8.16

is the differential equation in this case. We readily guess that

$$A = \tfrac{1}{4}x^4 + c$$

is the *general solution* of the differential equation.

Now in the special case of a growing area that starts at $x = 1$ and ends at $x = 4$, we know that $A = 0$ when $x = 1$; that is, initially there was no area. Substituting this *initial condition* in the general solution

$$A = \tfrac{1}{4}x^4 + c$$

we have

$$0 = \tfrac{1}{4} \cdot 1 + c$$
$$-\tfrac{1}{4} = c$$

and

$$A = \tfrac{1}{4}x^4 - \tfrac{1}{4}$$

is a particular solution, the specific formula for the growing area.

We want the value of this area when it has grown to the position $x = 4$. Substituting this value, we have

$$A = \tfrac{1}{4} \cdot 4^4 - \tfrac{1}{4} = 63\tfrac{3}{4} \text{ square units}$$

This is an area Euclid could not measure—in fact, would not have considered —since the curve $y = x^3$ was unknown to him. Thus a little simple algebra will measure an infinite variety of areas unknown to Greek geometers.

An extension to three dimensions of the work involved in obtaining $dA/dx = y$ will give $dV/dx = A$. By increasing the dimension of each quantity by one, we can get this formula by analogy. Thus A (area) becomes V (volume); y (length) becomes A (area). In the first of these formulas y is the length of a line segment that varies its position, that is, moves so as to generate the required area. Then by analogy A is a generating area forming a volume as it moves. Therefore we can find the volume of those solids for which it is possible to express the area of a

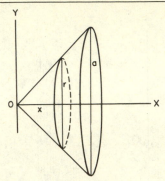

Figure 8.17

moving cross-section in terms of the distance from some fixed point, providing we can find an antiderivative for the expression thus obtained.

Suppose that we desire the volume of the cone of Figure 8.17, which is such that the radius of the circular cross-section is always equal to the distance from the origin, in other words, $r = x$. Then

$$\frac{dV}{dx} = A$$

leads to the differential equation

$$\frac{dV}{dx} = \pi x^2$$

whose general solution is

$$V = \tfrac{1}{3}\pi x^3 + c$$

But the growing volume was just starting when $x = 0$; that is, there is the initial condition, $V = 0$ when $x = 0$. Substituting this initial condition in the general solution,

$$0 = \tfrac{1}{3}\pi \cdot 0 + c$$
$$0 = c$$

and

$$V = \tfrac{1}{3}\pi x^3$$

is the particular solution, giving a formula for the growing volume. If the total height of the cone is a, and accordingly the radius of the base is also a, then the volume $V = \tfrac{1}{3}\pi x^3$ will grow until $x = a$ and

$$V = \tfrac{1}{3}\pi a^3$$

Again, suppose that the volume of the horn in Figure 8.18 is desired, and we are told that the radius of a cross-section is given by the formula

$$r = 0.04x^2$$

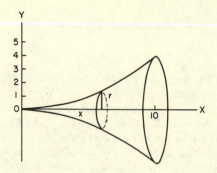

Figure 8.18

and the depth of the horn is 10. Then

$$\frac{dV}{dx} = A$$

leads to a differential equation

$$\frac{dV}{dx} = 0.0016\ \pi x^4$$

$$V = 0.00032\ \pi x^5 + c$$

Since

$$V = 0 \quad \text{when} \quad x = 0$$
$$0 = 0 + c$$
$$c = 0$$
$$V = 0.00032\ \pi x^5$$

Since the depth is 10,

$$V = 0.00032\ \pi (10)^5$$
$$V = 320\ \pi \text{ cubic units}$$

Numerous physical formulas can be found by antidifferentiation—for example, *the work done by a variable force*. Thus

$$\frac{dW}{dx} = F$$

or the *instantaneous rate of change of work with respect to distance is equal to the force*.

If the force required to stretch a certain spring x inches is $F = 20x$, how much work will be done in elongating the spring 5 inches? We have the differential equation

$$\frac{dW}{dx} = 20x$$

whose general solution is

$$W = 10x^2 + c$$

Since

$$W = 0 \text{ when } x = 0, \ c = 0$$

and

$$W = 10x^2 = 250, \text{ for } x = 5$$

We have seen how areas and volumes can be found by antidifferentiation. An analogous geometric problem is that of finding the *length* of a curve, *rectification*, as it is called. One might think that it would be easier to find a length by anti-differentiation than an area or a volume, but this is *not* the case, as we have already pointed out (page 176). Therefore, instead of going into the algebraic intricacies of rectification, it will be treated theoretically, and in the process, there will be a return to the questions which launched the present discussion of the calculus. From the original point of view, that of the early numerical integration, the approximate length of a curve PQ (Figure 8.19) can be found by measuring the length of the

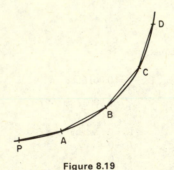

Figure 8.19

numerous small straight chords PA, AB, BC, etc. The closer together A, B, C, etc. are, and the shorter the chords, the better they will approximate the arcs PA, AB, BC, etc. The sum

$$PA + AB + BC + \ldots$$

is the approximate length of the curve. The actual length of the curve is defined to be the *limit* of this sum as the chords become smaller and smaller but more and more numerous. Finding this limit, if it exists, is described as *integration*.

Again, the area under the curve in Figure 8.20 can be approximated by finding the sum of the numerous small rectangles. The more numerous and the narrower these rectangles, the better the approximation. The limit of the sum as the rectangles grow more and more numerous, and their widths tend toward zero, is defined to be the required area. Once again, finding this limit, if it exists, is called *integration*, and one would say that the area as well as the length above are *integrals* in the sense of Mengoli-Cauchy (page 199). But we have already found areas like that in Figure 8.20 by the process of *antidifferentiation*. The fact that, in most instances, antidifferentiation may be substituted for integration, or that integration is the process *inverse* to differentiation is known as the *fundamental theorem of the calculus*. A formal proof of this theorem will not be given. It has, however, been applied

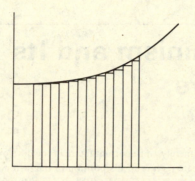

Figure 8.20

right along when the relatively simple reversal of differentiation was performed in preference to a consideration of limits of sets of approximating sums. The notation used by Leibniz for the *integral sign* is a medieval S, standing for *summa*, and the area of Figure 8.16 can be represented by

$$A = \int_1^4 x^3 dx$$

Nevertheless the actual algebraic procedure for finding the numerical value of this area is the *same*, whether one uses the Leibnizian symbolism with its underlying notion of the limit of a variable, or employs the antidifferentiation process. That this is so in the vast majority of problems where a geometric or physical measurement is required was proved by Newton and Leibniz, a theorem which was their most vital theoretical contribution to that subject for whose complete creation their followers were much too eager to give them full credit.

9

Determinism and Its Creators

To consider Newton merely an important figure in the development of the calculus is to give an inadequate picture of his position in the history of mathematics and science. Therefore we must now examine the *raison d'être* for his formulation of the "fluxion of a fluent," namely, the Newtonian *dynamics* which related forces to motions, terrestrial and celestial, and which culminated in his cosmological theory of universal gravitation. Our purpose is not only to consider Newton as a physicist, but also to establish a link with mathematical giants still to come—Euler, Lagrange, Laplace, Hamilton, and Jacobi—who developed Newton's mechanics to greater and greater heights in accordance with a principle that has come to be labeled "determinism."

The doctrine of scientific determinism, or the principle of causality, as it is sometimes called, originated in the fact that the motion of a particle in the mechanics of Newton, or of a system of particles in the analytic mechanics of Euler and Lagrange, or of a planet in the celestial mechanics of Laplace, is completely and unambiguously *determined* for all the future by the knowledge of the position and velocity of the particle or set of particles or celestial body at a single instant of time. This is the result of the fundamental postulates of each type of mechanics, in which laws are all formulated in a certain way. To quote Poincaré once more: "For Newton a physical law was a relation between the present state of the world and its condition immediately after, or, in other words, physical laws are differential equations."

Now in the previous chapter we solved some very elementary differential equations. In one case, a pair of such equations described a particular motion (page 196). In that example the given initial position and velocity enabled us to obtain definite algebraic formulas *determining* the position and the velocity of an object at all future times. Therefore this sort of physical determinism can come as no surprise to us. It seems a mere question of seeking *exact* formulation.

So much for determinism in the best sense of the word. But the virtue of the principle is destroyed if one extrapolates its application to nonscientific situations or converts it into a dogma for physics by insisting that a law of physics *must* be a rigorous rule that determines the evolution of a physical "system" uniquely through all of time when the state of the system at a single instant is known.

Physicists did consider the doctrine a "must" up to the day of James Clerk Maxwell (1831–1879). Then his ideas and, later, modern quantum theory with its non-determinate probabilistic laws showed the value of "may" instead of the "must" for determinism. Before that time, however, certain philosophers seized on the principle of causality with great avidity and proceeded to envision all of nature and human affairs in a mechanistic light, with no "free will" or the like.

We could give an account of the lives and works of the mathematicians who will be mentioned in this chapter without ever mentioning determinism, but we have discussed the idea in order to unify the mathematical activity of the eighteenth and early nineteenth centuries, and also so that we may be able to contrast the deterministic viewpoint with that of the probabilistic (statistical) mathematics and physics to be discussed in the next few chapters.

To start the story of determinism from the beginning, we must consider young Newton's thoughts at the time when the theory of universal gravitation was germinating in his fertile mind. Although the *Principia*, in which his classic laws of motion are stated, did not appear until 1687, he conceived the postulates of dynamics during the most creative period of his life, the years 1665–1667. Posterity clings to the legend of the falling apple as the inspiration for universal gravitation, but if Newton considered the motion of any terrestrial object at all, the thought was a momentary one, for he was more interested in cosmic motions and theories.

From antiquity up to the period just before Galileo (1564–1642) and Descartes, the explanation of celestial motion was almost as inadequate as having spirits, genii, or demons (figuratively, if not literally) shove the moon and the planets around their curved orbits in much the same way that humans or animals push or pull in order to support mundane motions. Even Kepler believed that planets held to their orbits or that, in general, objects remained in motion because they were sustained by some supporting force. With Galileo, the earlier theories of motion were completely altered. It is true that he had some precursors, for example, the Venetian, Giovanni Battista Benedetti, who, in his *Diversarum Speculationum Mathematicarum et Physicarum* (1585), set down ideas on accelerated motion that were akin to those which Galileo was soon to demonstrate experimentally at Pisa.

Galileo's concept, incorporated in the first two of Newton's laws of motion, was that *no force at all* is required to sustain motion, and that a body moving under no force enjoys constant speed in a constant direction. (If this speed is zero, the body is at rest.) If any force acts upon a body, Galileo, and then Newton, assumed that the speed or the direction or both must vary. They indicated that experimental data can merely *approximate* predictions made from their assumptions, since one cannot realize physical objects moving under no force at all. Galileo gave formulas for the distances, velocities, and accelerations of bodies falling near the earth's surface. We believe that young Newton, sent home from Cambridge to Woolsthorpe, Lincolnshire, during an epidemic of bubonic plague, stated to himself, if not to the world, the two laws that generalized Galileo's results:

Every body will continue in a state of rest or of uniform motion (constant velocity) in a straight line unless it is compelled by some external force to change that state.

If there is an impressed force, the time rate of change of momentum is proportional to this force and takes place in the direction in which the force acts.

In the second law of motion, *momentum* signifies the product of mass and velocity. Then the force is proportional to the *derivative* of *mv* with respect to *t* (time), or since mass was considered constant in classic mechanics, the second law states that

$$F = km \frac{dv}{dt}$$

where k is a proportionality factor and dv/dt, the exact rate of change of velocity, is the instantaneous *acceleration*. Symbolizing that by a, we have $F = kma$ as a concise formulation of Newton's second law. A still simpler form,

$$F = ma$$

is possible providing $k = 1$. This can be accomplished if suitable units of measurement are used, for example, if time is measured in *seconds*, distance in *feet*, force (hence weight) in *pounds*, and mass in *slugs*. For a reader not familiar with the *slug*, let it be stated that a *slug* is the mass of a body weighing about 32.2 pounds at the earth's surface.

In classical physics the mass of a body is assumed to be constant or unchanging, in accordance with the law of conservation of mass. The weight of a body, on the other hand, varies with its position, being more at the poles of the earth than at the equator, and much less on the moon than on the earth. But, as Newton wrote, "By experiments made with the greatest accuracy, I have always found the quantity of matter in bodies to be proportional to their weight." Thus, if m and w represent mass and weight, respectively, then, at sea level, $w = 32.2m$ or $m = w/32.2$ (approximately) if m is measured in slugs and w in pounds.

What was the thread of thought leading from the laws of motion to universal gravitation, where Newton first assumed that heavenly bodies exert forces on one another, and then went further in seeing such forces as a universal property of all the matter in the universe? The ultimate Newtonian postulate assigned a specific numerical measure to the gravitational attraction between any two particles of matter in the physical world. If the theory actually was initiated in the apple orchard of a Lincolnshire hamlet, the first question could have been: If the earth's gravitational force is responsible for the fall of the apple, how far does such gravitational influence extend—to the highest mountains, to the moon, or beyond?

In the orchard or in his study, somewhere, sometime in 1666, the Cambridge student might have continued: Since the moon is not at rest and does not move with constant speed in a straight line, there must be an external force pulling on it at every instant. Is this force due to the gravitational influence of the earth? If so, this influence is not merely a terrestrial phenomenon, confined to Woolsthorpe apples, or objects dropped from the leaning tower of Pisa, but it extends at least as far as the moon. Next—none of the planets fly off on tangents. What forces deflect them from straight-line paths and keep them in orbits around the sun? Perhaps the sun acts on them in the way the earth seems to draw the apple and moon toward its center. Then the sun must exert its own particular gravitational influence that attracts all the other objects in the solar system. Finally, the pulling of one heavenly body on another exists throughout the *universe* and is not a phenomenon limited to the force that makes nearby objects fall to the earth's surface.

Granting that this or any hypothesis is a good one, it must be given exact mathematical formulation so that it will lead to numerical results that can be compared with observations in order to provide a test of the theory in question. Thus what Newton required was a formula for the universal gravitational force that he imagined—that is, a function expressing the dependence of gravitational force on other variables. He assumed that the gravitational pull, influence, or attraction, exerted by a body would depend on its mass, and that a huge mass like the sun would be able to pull much harder than a relatively smaller planetary mass. But in spite of the tug of the earth on the moon, the latter does not fall to the earth, whereas the apple does. Again, the gravitational pull of the huge solar mass is just sufficient to divert the planets from linear to elliptic orbits, but not large enough to make them fall into the central fire. Therefore, Newton reasoned, gravitational force must diminish with distance. The simplest law he might have assumed would have been that the force varies inversely with the distance, that is, $F = k/d$, where F represents force, d distance, and k is a constant that contains other factors like the masses which are pulling on one another and the adjustments for particular units of measure. This would mean that when distance is doubled, gravitational force would be half as great, when it is trebled, the force would be one-third as great, etc. But Newton felt that gravitational attraction must fall off more rapidly than inverse variation permits. Hence he decided tentatively that this force might vary inversely as the *square* of the distance, that is, that

$$F = \frac{k}{d^2}$$

This implies that when distance is doubled, other factors remaining constant, the gravitational tug is one-fourth as great; when distance is trebled, the gravitational pull is one-ninth as great, etc. When, *ultimately*, Newton expressed k in terms of the masses, M and m, of the two bodies pulling on one another, his law became

$$F = \frac{GMm}{d^2}$$

where G is the "gravitational constant." By suitable choice of the units of mass and distance, one can obtain $G = 1$, and the force of attraction is just Mm/d^2. Also, in its initial form, the principle was applied only to bodies that were mass "particles." In its final form it became the *law of universal gravitation*: Any two particles of matter in the universe attract one another with a force directly proportional to the product of their masses and inversely proportional to the square of the distance between them.

But in 1666, all of this was *conjecture* on Newton's part, and he wished to provide a deductive *proof* that gravitational force actually obeys the inverse-square law, $F = k/d^2$. He was able to do so by assuming the validity of Kepler's third law. In later years Newton reversed the process by assuming the inverse-square law and deducing Kepler's formula from it. In a logical sense, one would say that the two laws are equivalent, or that, more generally, if the truth of proposition p implies the truth of proposition q and, conversely, q implies p, then p and q are equivalent propositions (Chapter 6).

Whether Kepler's third law is assumed and Newton's inverse-square law is deduced, or vice versa, would not matter in a pure (abstract) mathematical science. But in physical science, postulates are supposed to correspond approximately to reality. Therefore Newton devised a test in which he carried out certain calculations based on the actual motion of the moon. In his own words, "I compared the force requisite to keep the Moon in her Orb with the force of gravity at the surface of the Earth, and found the answer pretty nearly."

The "pretty nearly" may explain why he did not publicize the inverse-square law in 1666 when he first deduced it, but postponed the release for sixteen years. In his test, he had argued as follows: The distance of the moon from the earth's center is about 60 times the radius of the earth, and therefore, *if the inverse-square law is true* (in the physical universe), the gravitational effect of the earth on a unit of mass near its surface must be 3600 times its effect on such a mass located on the moon. But the "effect" for a body on the earth's surface is the weight of the body, given by the formula $w = 32.2m$. Therefore, if $m = 1$, $w = 32.2$ lb. Dividing this by 3600 gives 0.00895 lb. as the tug of the earth on a unit mass on the moon. But Newton's computation from the moon's motion (*without assuming the inverse-square law*) was 0.00775 lb., and this may have made him doubt the inverse-square law, for his "pretty nearly" signified a 15 per cent difference between his test result and that predicted by the inverse-square law. But what had happened was that he had used 3440 miles as an estimate of the earth's radius, based on his assumption that a degree on a meridian measures 60 miles.

Although Newton is supposed to have been a typically "absent-minded professor," it is hard to reconcile the triviality of such a mistake with the magnitude of the idea being tested. Pemberton, a friend, said that "being absent from books" at Woolsthorpe, Newton confused English miles with nautical miles, or that he took the wrong value from an unreliable text he had with him at the farm. But it appears that although Picard made accurate measurements of the earth's radius in 1672, and gave 69.1 miles as the length of a meridional degree, Newton apparently did not become aware of this good approximation until 1682. The reader may wonder why Newton's tardiness matters at all except as an illustration of the remoteness of genius from all ordinary activity or the inability of great mathematicians to carry on everyday arithmetic. But the upshot of his failure to release the inverse-square law at the very beginning was an ugly priority dispute with the physicist Robert Hooke, with whom he had already begun to quarrel around 1672 concerning the nature of light. When this was combined with the controversy with Leibniz on the invention of the calculus, it cast a long shadow on Newton's personal life.

There are alternative explanations to the error-in-arithmetic supposition. Some feel that Newton was a perfectionist and would not accept the figure 32.2 pounds, which was based on Galileo's experiments, and that he wished to derive this result theoretically. This he was not prepared to do in 1666, for he had not advanced sufficiently far in the integral calculus. What he required was a theorem he finally established in 1685: The gravitational attraction between two homogeneous spheres acts as if their total masses are concentrated at their centers. The proof of this proposition and its importance in the theory of universal gravitation were revealed in the *System of the World*, the last and crowning portion of Newton's *Principia*. The theorem enabled him to treat heavenly bodies as if they were *particles* to which the universal law $F = GMm/d^2$ applies.

In 1684 Newton had released some of his ideas in a short treatise entitled *De Motu Corporum* (On the Motion of Bodies). This work, revised and enlarged, was to become Book I of the *Principia*. Neither the *De Motu* nor Newton's *chef-d'oeuvre* would ever have been issued if not for the persuasive powers of Edmond Halley (1656–1742), the great astronomer. He coaxed Newton into publication. The result was that the manuscript of *Philosophiae Naturalis Principia Mathematica* was presented to the British Royal Society on April 28, 1686. On May 19 the Society's council decided that "Mr. Newton's work should be printed forthwith." But the Society's treasury was completely depleted by the previous expenditure for the publication of a book on the habits of fish. Halley stepped into the breach, thereby rendering an immeasurable service to Newton, to the Royal Society, and to all of scientific posterity. The astronomer, who was in no sense a wealthy man, published the *Principia* at his own expense. Thus, in July 1687 there appeared the greatest work of science ever written, one small quarto volume of five-hundred pages, selling at nine shillings.

Some of the ideas in Book III of the *Principia* will eternally link Newton and Halley in men's minds. In this part of his work, Newton set down his notions on comets. Their paths had been considered *parabolic*. Hence a comet would travel on an open curve and never return to a previous position. Newton showed, however, that the parabolic path is just an approximation, that the orbits are actually elongated ellipses, and hence the motion of comets is periodic and they must return after a very long time. Halley applied Newton's comet theory to calculate the orbits of all comets for which he had any records of observations. He found that at intervals of approximately 75 years brilliant comets had been seen, specifically in the years 1531, 1607, and 1682, this last appearance having been observed by Halley himself. The Newtonian computations indicated *all three* comet orbits to be the same elongated celestial ellipse; that is, in reality there was *only one* comet, with a period of about 75 years!

The deterministic laws of universal gravitation predicted that *Halley's comet* would reappear at the end of 1758 or the beginning of 1759. Although neither Newton nor Halley was alive to witness the fact, the comet returned within a month of the time they had estimated. Moreover, it returned after another 76 years, in 1835, and then, after a 75-year interval, in 1910. Readers may question grandparents and great-grandparents about that last appearance. Even in 1910 some of the older superstitions about comets were still circulating. Halley's comet was substituted for the "bogeyman," and children were told that if they did not behave, the world would end when it passed through the tail of the comet.

To leave superstition and return to the *Principia*, we can quote Einstein's evaluation: "Nature was an open book to Newton." To give a brief description of the contents of the work which Laplace, the "Newton of France," termed the "most outstanding production of human genius in all of time," one should start with the introduction. There Newton discusses his space-time notions, acknowledges his debt to Galileo, and deduces from the latter's experimental results the three famous laws of motion. In Book I of the *Principia* there is the first *printed* statement by Newton concerning the discovery of the method of fluxions, also the deduction of Kepler's law from the inverse-square postulate, and the general "two-body problem." In that problem the orbits of earth and moon, sun and a single planet, etc. are deduced under the assumption that the gravitational tug of other celestial bodies is negligible. In the same portion of the *Principia* Newton deals with laws of force other than the inverse square. Such laws are applicable to the mechanics of rigid bodies. Newton shows that they lead to oscillatory motions, paths that are not closed, and so on. Book I concludes with material that seems irrelevant—Newton's thoughts on the "corpuscular" theory of light.

The second book of the *Principia* deals with motion in a resisting medium. Today these ideas would apply to the movement of ships, underwater missiles, bullets, rockets, and planes in the air. Newton assumed that for certain speeds, $F = kv$, or that the resistance of the medium is proportional to the velocity of the moving object, and that for higher speeds $F = kv^2$. The paths deduced from such axioms were suitable for ballistics and aerodynamics until modern times, when speeds far beyond any in Newton's experience became available.

Book II also contains the fundamentals of the science of hydrodynamics, the first printed mathematical discussion of wave motion, and a germ of the "calculus of variations" (Chapter 23). Newton discusses the optical "diffraction" phenomena which the Italian Grimaldi had first discovered. These are best explained by considering light to be made up of waves rather than corpuscles. Among the multitude of other physical concepts in Book II is Newton's notion of sound. He related how he had once determined the velocity of sound approximately by listening to the echoes in the cloisters of Neville's Court Walk at Trinity College. Therefore it has become traditional for Cambridge students to test the quadruple echo that Newton claimed to have heard. At another point of Book II Newton neatly demolishes

Descartes' "vortex theory." The formulator of analytic geometry had explained planetary motion as the result of vortices or whirlpools sweeping the planets around the sun. Newton says, "The vortex theory is in complete conflict with astronomical observations, and instead of explaining celestial motions, merely confuses our ideas about them."

The *De Mundi Systemate*, the System of the World, which forms Book III of the *Principia*, contains the full development of the theory of universal gravitation. In it the orbits of the planets and their satellites are specified, comets are discussed, numerous difficulties with the moon's orbit are handled, a theory of tides is furnished, and the "precession of the equinoxes" is explained for the first time in scientific history. Astronomical observations, even in earlier eras, had revealed certain irregularities in the motion of the moon. Subsequently, similar orbit irregularities had been observed for the satellites of Jupiter and Saturn. Newton explains these phenomena as due to the "perturbative" effect of the sun. Here is the beginning of the famous "three-body problem." Kepler's ellipses are only valid when a planet moves around the sun, or a satellite orbits around a planet, *without* the presence of a third body nearby in the heavens. When such a body is present, its gravitational pull will draw a planet away from its elliptic course. In Book III of the *Principia* the numerical effect of "perturbations" is considered, and these correspond to the observed irregularities in satellite orbits. When, in mid-nineteenth century, the orbit of Uranus appeared to be "perturbed," the astronomers Adams and Leverrier independently came to the conclusion that the perturbation must be due to a planet as yet unknown to astronomers. They determined its position in the heavens on the basis of Newton's theory. Shortly thereafter the new planet was actually observed very near the point predicted, and it was named Neptune. Similar considerations led to the twentieth-century discovery of Pluto.

The source of these astronomic discoveries is, as we have emphasized, the Newtonian *System of the World*. Among the many consequences of universal gravitation treated in that part of the *Principia* is the fact that the earth must be flattened at the poles, that is, its shape is that of an oblate spheroid. In France, Giovanni Domenico Cassini (1625–1712), director of the Paris Observatory, and his son, Jacques (1677–1756), who succeeded him in the position, were misled by erroneous geodetic measurements and held to the theory of a prolate spheroid for the earth's shape, that is, to the idea of *elongation* rather than flattening at the poles. The French Academy of Sciences decided that a test must be made to decide between Newton and the Cassinis. Accordingly, expeditions were sent out to measure arcs on the earth's surface at widely separated points. In 1735 (eight years after Newton's death) an expedition under La Condamine was sent to Ecuador and another under Maupertuis was despatched to Lapland. Voltaire, who disliked Maupertuis, and also felt that Newton must have been correct, called the director of the Lapland expedition "Marquess of the Arctic Circle," "dear flattener of the world and of Cassini" and "Sir Isaac Maupertuis." He also addressed to La Condamine the rhyme

Vous avez confirmé dans les lieux pleins d'ennui
Ce que Newton connut sans sortir de chez lui.

(You have confirmed in dreary far-off lands
What Newton knew without e'er leaving home.)*

No other physical scientist has ever been able to create anything on a par with Newton's *Principia*, which is the source from which all subsequent (scientific) blessings flowed. Even when relativity and quantum theory made their appearance, Newtonian physics was the starting point—if only to challenge some of its postulates or to illustrate a deterministic pattern with which to disagree. But Newton had worthy successors who, in successive generations, modified and generalized his ideas. Nevertheless, on the whole, they followed the master by specializing in analysis, by applying that subject to the formulation of deterministic laws for physics, and by earning each in turn, the title of "foremost mathematician in Europe." Thus, the calculus became "analysis incarnate" in the hands of Léonard (Leonhard) Euler (1707–1783). Joseph-Louis Lagrange (1736–1813) produced the *Mécanique Analytique*, as profound an accomplishment in general mechanics as universal gravitation had been in celestial mechanics, and one of a sequence of events indicating that mechanics, perhaps even more than astronomy, is a source of inspiration for pure mathematical theory. This significant role of mechanics is evident in much of the work of Euler, Lagrange, and their mathematical disciples and contemporaries. But astronomy was to the fore in the research of Pierre-Simon Laplace (1749–1827). During the years 1799–1825 he released volume after volume (five in all) of his *Mécanique Céleste*. Many subjects were enriched by his mathematical fertility, but his devotion to celestial mechanics was almost an obsession; every concept he created was related directly or indirectly to the field in which he out-Newtoned Newton.

His production was small, however, compared to that of Euler, whose collected writings fill some eighty large volumes, an all-time record among mathematicians. There was quality as well as quantity in Euler's works, and among the many topics which claim space in the vast collection are number theory, artillery, northern lights, sound, the tides, navigation, ship-building, astronomy, hydrodynamics, magnetism, light, telescopic design, canal construction, annuities, lotteries. All such writing was outside his fields of major achievement. He completed and expounded all the formalisms of classic analysis in his *Introductio in Analysin Infinitorum*, which the mathematical historian Carl B. Boyer calls the "foremost textbook of modern times."† As if all this were not enough, Switzerland's greatest mathematician can be considered one of the founders of the *calculus of variations* and that exceedingly important research subject of today, *combinatorial topology* (Chapter 25).

---

* Translation by the author.

† See his article bearing this title in the *American Mathematical Monthly*, April 1951, pp. 223–226.

Not only was Euler the most prolific mathematician in all of history, but he appears likewise to have given rise to the greatest number of legends. It is said that he could repeat the *Aeneid* from beginning to end. He was capable of remarkable concentration and would write his research papers with a child on each knee, while the rest of his youngsters raised uninhibited pandemonium all about him (he was the father of thirteen children in all). Euler, like most mathematicians, had a keen interest in music, but his book on *New Musical Theories* was said to contain "too much mathematics for musicians and too much music for mathematicians."

Frederick the Great was one of Euler's patrons. As a result, the mathematician found himself conducting a correspondence course in scientific subjects for Frederick's niece, Princess Philippine von Schwedt, to keep her busy while she (and the rest of the royal court) were interned at Magdeburg during the critical phases of the Seven Years' War. In a letter of August 1760 Euler was explaining to the princess the use of the level, a surveying instrument. As an exercise he asked, "Would a straight line drawn from your new home, Magdeburg, to your old home, Berlin, be a *horizontal* one?" Then he answered his own question in the negative, stating that Berlin is higher than Magdeburg, because the former city lies on the Spree, the latter on the Elbe, and the Spree empties into the Havel, which in turn flows into the Elbe. Euler's justification was wrong! The elevations of Berlin and Magdeburg are 33 and 41 meters, respectively. The fallacy in his argument lies in the fact that the junction of the Havel and Elbe does not take place at Magdeburg but at a point far below! "Even Homer was known to nod!"

The setting for Euler's more serious scientific activities was a world which had just witnessed the most bitter of all priority disputes, the Newton-Leibniz controversy concerning the invention of the calculus. Professor Rudolph Langer has described this situation and has given an account of Euler's career.*

The public condemnation of Leibniz by the British Royal Society took place in 1713, and three years later, before he had completed his defense, Leibniz died. The English stood to a man by Newton. Germany, on the other hand, was at this time in a state of utter political turmoil and exhaustion, and so it happened that no countryman of Leibniz was at hand to come to his defense or even to understand his work. It was fortunate for posterity, therefore, that there was, nevertheless, one man on the European continent who could champion his cause and who was equipped by ability and temperament to do so effectively. This man was Jean Bernoulli, of the famous Bernoulli family that was to produce eight mathematicians in three generations.

Jacques Bernoulli (1654–1705) was professor of mathematics at Basel, Switzerland, from 1687 until his death. Jean, his brother, thirteen years younger, was at first located in Holland, but succeeded upon the death of Jacques to the professorship at Basel. These two men were among the few to whom the publications of Leibniz were a revelation. They were fired by the importance of the invention of the calculus, and from its very birth took so decisive a hand in its development that Leibniz is known to have pronounced it as much theirs as his. The death of Jacques Bernoulli preceded that of Leibniz, and hence when Leibniz died Jean looked upon himself as the logical scientific successor and proceeded to champion the cause which had become his. A man of great genius, he was in

---

* R. E. Langer, "The Life of Léonard Euler," *Scripta Mathematica*, Vol. 3 (1935), pp. 61 ff.

his personal associations a game cock or a dragon, as the occasion required. It became his mission in life to defeat single-handed the whole tribe of the English. Unmerciful in his ridicule, of the greatest incisiveness as a critic and writer, pitiless of anyone who dared take issue with him, he turned out a veritable flood of publications which were filled alike with his own great works and discoveries and with invective for those (the English) who followed as disciples of a different system. Thus he helped to save the calculus of Leibniz from oblivion and, by applying the impetus of his own genius, he so accelerated its development as to insure for it subsequently an almost complete triumph over the fluxions of Newton. Bernoulli was active without interruption at Basel for forty-two years, and with ever growing fame, acknowledged as the greatest living mathematician, he drew to his circle, like an oracle, mathematicians from all the countries of continental Europe.

This was the environment Euler entered as a young man. He was born in 1707 not far from the city of Basel. His father, a country preacher, was fond of mathematics, had studied under Jacques Bernoulli, and was the young Euler's first teacher in this subject. An apt pupil in everything, Euler entered the university and obtained his baccalaureate at the age of fifteen. It was his father's wish that he turn to theology, but the boy's love for mathematics and the proximity of the compelling personality of Jean Bernoulli were not to be resisted. So Euler sought instruction from Bernoulli, and was given books and papers to read, with Bernoulli's promise to be available once each week to answer questions and clear up difficult points. It was during this period that Euler became a friend of Bernoulli's sons, Nicolas, Daniel, and Jean II, all mathematicians of high rank, the first two being twelve and seven years Euler's senior, respectively, and Jean II three years his junior.

Euler's first paper was published at the age of eighteen. In the succeeding year he wrote on the propagation of sound. When the Paris Academy offered a prize for an essay on the masting of ships, the young Euler, at the age of twenty and devoid of any practical experience with ships, wrote and won the prize. He remarked in the conclusion of his essay that he had not considered it necessary to check his results by experiment, for *since they were deduced from the surest foundations in mechanics, their truth or correctness could not be questioned.* This was characteristic of Euler's lifelong attitude. He never ceased to regard the deductive power of the mind as being of unquestionable supremacy.

Peter the Great of Russia had at the instigation of Leibniz conceived the plan of an academy at St. Petersburg, and his successor, the Empress Catherine the First, put this plan into effect. Thus it occurred that Nicolas and Daniel Bernoulli were called from Basel. At parting they promised Euler to seek an opening for him, and in the following year they wrote to him about a position in physiology, with the recommendation that he make haste to learn some medicine and anatomy and apply for it. This Euler proceeded to do, and in his twentieth year set out from his native city for St. Petersburg. Upon his arrival he was made an associate in mathematics—the matter of physiology apparently having been forgotten.

The academy consisted of twenty professors representing the various fields of learning and was an able, inspiring, and argumentative group of men. It had been stipulated in the contracts of appointment that each member was to bring with him two students, and when the academy was ready to function as a university the wisdom of this provision became evident. There was no other student body. In this circumstance the members found complete freedom for scientific work, and Euler immediately began voluminous publication in the academy's proceedings.

Soon, however, the academy fell upon evil days. On the very day Euler entered Russia, Empress Catherine died. The ensuing regency considered the academy a futile source of expense, and after three years Euler was driven to the verge of accepting a lieutenancy in the Russian navy. The sudden death of the boy Czar and the accession of the Empress Anna, however, brought a change for the better. Meanwhile Nicolas Bernoulli had died, and Euler had assumed the professorship of natural philosophy, a chair which he later resigned for the professorship in mathematics when the latter was vacated by Daniel Bernoulli's return to his native country. The betterment in Euler's financial circumstances was soon followed by his marriage to the daughter of a Swiss painter resident in St. Petersburg. He had formed many friendships. All along, his mathematical production proceeded at a fabulous pace.

The year following his marriage marked the first major misfortune in Euler's life. Because of illness, which legend ascribed to overexertion in a feat of astronomical calculation, he lost the sight of his right eye—in fact, the eye itself.

The great number and significance of Euler's works had meanwhile attracted serious attention throughout Europe. The extraordinary activity of Euler may be indicated by the fact that at the age of thirty-three he had produced a total of eighty papers and books in pure science alone, although he had all the while been engaged in an astounding number of other ways. Thus he is to be found as supervisor of the Russian government department of geography and as commissioner of weights and measures, giving much of his time to details and writing essays on the construction and testing of scales and other apparatus of measurement. He wrote the elementary mathematical textbooks for use in the Russian schools, was called upon to solve the problem of raising and hanging the great bell of Moscow, and so on. In 1741 Euler received a call to go to Berlin. The administrative state of the St. Petersburg Academy during the preceding years had not been an entirely happy one, and as the death of the Empress Anna was anticipated with foreboding, he welcomed and heeded the call. He had been active for fourteen years at St. Petersburg and had reached the age of thirty-four.

Frederick the Great became King of Prussia in 1740. The scientific society at Berlin which had been founded by Leibniz forty years earlier was in a state of decay, and the new king became bent upon invigorating it and reorganizing it into a great academy. To this end he brought to Berlin as an organizer the mathematician Maupertuis. Maupertuis was a scientist and an adventurer—a man of brilliant personality, a gifted conversationalist, an enthusiast for science, and a mathematician and philosopher of respectable accomplishment. He had studied under Jean Bernoulli on the Continent, but had also been in England, where he had developed a strong liking for the Newtonian methods. Mathematicians of acknowledged greatness were essential to add luster to Frederick's new academy, and strong efforts were immediately made to draw the Bernoullis—Jean, now seventy-three years old, and his sons, Daniel and Jean II from Switzerland, and Euler from St. Petersburg. The Bernoullis were not to be had, but Euler came to Berlin to join a brilliant association of scholars.

It was customary during this period for the Paris Academy to pose for each biennium a subject for a prize dissertation, and Euler often entered the competition. He won with such regularity that it has been said he must have almost come to regard the prize as a regular subsidy to his salary. It is interesting to note that this Parisian prize was carried off twice by Jean Bernoulli, four times by his son Jean II, ten times by Daniel Bernoulli, and twelve times by Euler. In all, the prize thus went twenty-eight times to these men who were all natives of the same city, Basel, Switzerland.

The epoch in which these events took place was marked by a number of intense scientific disputes which were widely participated in throughout Europe. The questions generally hinged on philosophical points, and Euler, who entered the discussions with vigor, soon showed himself to be a mediocre, or poorer, philosopher. It became evident to many through these controversies that the great mathematician was quite mortal in other fields, and Daniel Bernoulli is known to have advised Euler to stay out of print in matters nonmathematical. Euler, far from being daunted, continued, curiously, often on the wrong side of the argument. Such a turn of events was an occasion of much joy to

Voltaire, who at the time was also at the court of Frederick and an aspirant for the royal favor; Euler was not spared the sting of his satire.

Voltaire ferreted out a paper written by Euler in his youth, in which he had treated the hypothetical example of a shaft through the center of the earth with a heavy particle moving in it under the earth's attraction. His formulas showed that, on reaching the earth's center, the particle would abruptly turn and retrace its course rather than continue through. This paradoxical result arose from an unpermissible interchange of the order of two infinite processes, a matter which at that time was not properly understood. Voltaire's attacks were enjoyed by Euler as much as by anyone. Incidentally, they never deterred him from engaging in philosophical disputes with gusto.

While Euler was at Berlin, the mathematician Jean le Rond d'Alembert (1717–1783) was active in Paris. He had attained acknowledged prominence and worked on many of the problems upon which Euler was also engaged. The correspondence between the two men was frequent and cordial until about 1757, when a difference of opinion concerning the solution of the problem of the vibrating string caused the association to lapse. D'Alembert was a man of fine culture. Besides his mathematical pursuits he was active as a philosopher and as a writer. It is a curious fact that his talent in letters never extended to mathematical exposition. In this of all subjects his writings were often so obscure and difficult to read that many of his results, in fact, became generally known only through Euler's presentations of them.

Despite his great scientific success, Euler was not spared frustration and discouragements in Berlin, mainly because he had never won real favor with the king. The management of the Berlin Academy had gradually fallen, in fact though not in name, into Euler's hands, and Euler was widely recommended to Frederick for the position of nominal as well as actual head of the organization. It was Frederick's wish, however, to have as the leader of his academy not just a scientist but also a man of the world, a man of social grace and brilliance to dine at his board and through his appearance to add luster to the institution. This was a role for which Euler, the typical German burgher, was not adapted. Accordingly, Euler found himself seeking in vain for even the most trivial royal favors, such as suitable appointments for his sons, and in Frederick's correspondence one may find frequent sarcastic and disparaging references to Euler, his lack of polish, his missing eye, the great cyclops of a geometer, etc. This irritation of the king's no doubt had, in part, another cause. Although mathematical accomplishments were regarded in highest terms at the time, Euler's most serious efforts had never taken him beyond the elementary stages of the subject, and this limitation had engendered a natural distaste and distrust. He had repeatedly, and always unsuccessfully, tried to wring from men of learning an admission that mathematics after all was of no great account. Now Euler was a mathematician and a mathematician alone. Frederick would have none of the suggestion to make him president of the academy. In this matter his choice fell upon D'Alembert, much to Euler's consternation. Despite fabulous offers, however, D'Alembert refused to exchange his position at Paris for a residence in Frederick's palace. He visited Berlin at the insistence of the king, allayed Euler's fears by the great cordiality which he brought to the meeting—and by urging upon the king the unsuitableness of placing any living person in a position of academic superiority over so great a man.

D'Alembert's refusal served only to heighten Euler's disfavor in Frederick's eyes. After many irritating incidents, none of major importance, matters came to a climax in 1766, and Euler, now fifty-nine years old, felt compelled to resign his position and leave the circle in which he had lived and worked for twenty-four years. During all these years he had maintained the friendliest of relations with the academy at St. Petersburg. He had published regularly in its proceedings and had taken students coming from there into his home. This cordiality was reciprocated to such a degree that during a Russian invasion of Berlin two Russian soldiers were especially assigned to guard Euler and his household, and when his country home was inadvertently pillaged, he was personally indemnified by the Russian Empress.

His resignation was finally accepted in Berlin, and Euler set out for St. Petersburg with his entire household of eighteen persons. The Empress Catherine (the Great) was elated at winning so great a man for her academy, and Euler was entertained *en route* by the Prince of Courland, by the City of Riga, and for many days in Warsaw by the King of Poland. On his arrival in Russia, the Empress presented him with a house completely furnished, and with one of the royal cooks. His sons were given positions of importance, and in every way Euler felt the greatest gratification. Frederick meanwhile had made the best of his situation, and had filled Euler's place with the one living man worthy to succeed him, namely, Lagrange, then thirty years old. Euler's relations with Lagrange were of the warmest, the older man being among the first and the most willing to admit and laud the greatness of the younger.

Euler's life had not been free from tragedy. At Berlin he had seen the death of eight of his children. Now, soon after his return to St. Petersburg, a calamity of another sort occurred—he lost the sight in his one remaining eye. The severity of such a blow to a man of Euler's profession and activity hardly requires description. Fortunately the calamity failed to have the anticipated effect. It served only to reveal completely the afflicted man's genius. With the stimulus of outer distractions removed, the play of Euler's imagination became greater than ever. His most phenomenal memory, which could instantly recall anything he had ever seen or heard, even to the minutiae of the uncountable formulas he had dealt with, allowed him to continue his work despite his blindness. Indeed, the rate of his publications increased, so that his papers written during his seventeen years of blindness number almost four hundred, not to mention a respectable number of complete books.

The mechanics by which this process of production was achieved seems roughly to have been somewhat as follows. A large table covered with a slate top to form a blackboard was placed in the middle of his large study. With younger men as his assistants, he dictated the text of his works and at the appropriate points he would write with chalk upon the table the formulas to be used. The increase in the rate of his papers was evidently due to the relief from the care for editorial details.

In the year 1771 there occurred a great fire in St. Petersburg which almost ended Euler's career. His house was engulfed by the flames and only through a servant's heroism was the blind man brought out to safety. His home and many manuscripts were destroyed, but the financial loss was made good to him by the Empress. Euler happened to be engaged at this time upon a theory of the motion of the moon, a work which involved computations of stupendous magnitude and complexity. It seems incredible that such a work should have been done in blindness and under such devastating distractions as that of the fire. Yet it was carried to so successful a conclusion that it won him two prizes at the Academy of Paris. The year of the fire also marked for Euler the birth of a great hope which was to end tragically. A surgical operation having been undertaken upon his eye, his sight was restored, to his unbounded joy. But not for long. Failure was somehow involved, and amidst agonies described as horrible, he was plunged again, and finally, into complete blindness.

Euler's blindness had raised his reputation to legendary proportions. His home became a scientific shrine visited by a steady stream of pilgrims, including even the heir to the Prussian throne. They would generally find Euler, to their surprise, as enthusiastic as in his youth for the discovery of new mathematical facts. The thrill had never worn off. And nonmathematicians were astounded at his fund of information on all conceivable subjects. He was conversant with the literature of the ancients, knew the histories of all nations in all periods, was versed in medicine and chemistry, and so on.

The final years of Euler's life were marked by the loss of his wife and his two daughters. His hearing became impaired, and as a result he took a less active part than formerly in the affairs of the academy. But his mathematical work did not diminish, nor did he lose the great imagination which had led him so well throughout his life. Thus in his papers on hydrostatics and hydrodynamics he considered the effects of varying density and temperature on the equilibrium of the air. From his results he read the general cause of winds and applied it to the trade winds and to the monsoons of the Indian Ocean. He determined the shape of the earth and its surrounding media, saw in his formulas the description not only of the tides but of the motions of fluids in vases, pumps, pipes, etc., derived from them the velocity of the propagation of sound, which in turn yields him the formulas for the notes of a flute, and so on.

According to legend, Euler's last mathematical activity was a prophetic one. On September 18, 1783, the greatest Swiss mathematician passed the afternoon in thoughts concerning the orbits of *satellites*. In his accustomed work habits since he had become completely blind fifteen years earlier, he outlined in large characters on a slate the calculation of the orbit of Uranus, the planet which Herschel had just discovered. After he had dined with friends, he resumed the calculations while he was drinking his tea. Next he asked for his grandson and played with the child for a few minutes. Suddenly he turned to the slate and inscribed "I die" beneath the satellite computations. The chalk dropped from his hand and, in the words of Condorcet "*il cessa de calculer et de vivre.*"

Among Euler's younger contemporaries were France's leading mathematicians of the era, Joseph-Louis Lagrange and Pierre-Simon Laplace. When one considers how many scientists, authors, and artists achieved fame only posthumously, these mathematicians were indeed extremely fortunate to receive in their lifetime the recognition and rewards they so richly merited. Napoleon, who was a great admirer of men of science, honored Lagrange and Laplace by making them Senators, Counts of the Empire, and Grand Officers of the Legion of Honor. Through the influence of Euler, who recognized the youthful genius of Lagrange, the latter was made an Associate of the Berlin Adademy when he was only twenty-three, and somewhat later was made director of the mathematics division of the academy when the great Swiss analyst departed for St. Petersburg. And so in 1766, at the age of thirty, Lagrange went to Berlin, where for twenty years he served as court mathematician to Frederick the Great. In the words of that ruler, "the greatest monarch in Europe deserves the greatest mathematician." During the first few years in Berlin, he received several of the highest awards of the French Academy of Sciences, which also made him one of its associate members when he was only twenty-four. By 1785 he attained the lofty heights of full membership in this academy. In 1816 he was elected to membership in the academy of immortals, the French Academy, a distinction accorded only to the greatest French scholars.

Although Lagrange and Laplace worked in related fields of mathematics and might have been considered rivals, there is no evidence of disputes or harsh feelings between them. Temperamentally they were quite different. Lagrange was modest, retiring, shy, and the soul of tact. Most of his life he suffered from what was then called "bilious hypochondria," a malady whose symptoms were recurrent periods of mental depression. It is related that when Lagrange's masterpiece, the *Mécanique*

Analytique, came from the printer, it remained unopened on his desk for two years. By contrast, Laplace was ambitious, outgoing, shrewd, and aggressive. Although he was on occasion severely criticized for failing to give proper credit to other mathematicians, it is remarkable that he appears to have made no enemies. In fact, he was considered a good politician because he was able to carry on his work through different regimes without any loss of position or presitge, and eventually he was made a marquis by Louis XVIII.

Lagrange and Laplace were far apart in their attitudes toward mathematics. The former revered complete abstraction, and his analytical mechanics is not physics but "pure" mathematics. Laplace, although capable of the highest type of abstract thinking, was more interested in putting theories to practical use, and hence is considered mainly a physicist, astronomer, and "applied" mathematician.

In their mathematical specialties both men carried forward the mechanics of Newton. Lagrange completed and generalized the mechanics of rigid bodies. To emphasize the fact that the *Mécanique Analytique* was "pure" mathematics, no diagrams were included. This great treatise was called by Hamilton (the successor of Lagrange in the field of mechanics) "a scientific poem by the Shakespeare of mathematics." Laplace continued the mathematics associated with universal gravitation, that is, the *celestial* mechanics of Newton. The title of Laplace's greatest book is *Mécanique Céleste*. Deterministic contributions of Lagrange and Laplace to mechanics will be discussed in the present chapter. It must be emphasized, however, that the two French leaders made important contributions to other subjects as well. Lagrange played an essential role in the advancement of almost every branch of pure mathematics, especially in the calculus of variations, the theory of numbers, Diophantine analysis, "invariants," and algebraic solvability (see Chapter 16). Laplace's researches, in addition to those in celestial mechanics, concerned probability theory, potential theory, partial differential equations, and various exceedingly useful "operational" methods like the *Laplace transform* of modern mathematical physics and engineering.

Joseph-Louis Lagrange was born in 1736 in Turin, which was part of the Kingdom of Sardinia at the time. His grandfather, who was of French noble birth, had entered the service of the King of Sardinia and had settled in Turin, where he married into a prominent Italian family. Lagrange's father, an erstwhile wealthy man who ultimately lost his entire fortune through unwise speculation, was war treasurer of Sardinia. Young Joseph-Louis' extraordinary mathematical talent displayed itself very early. Hence he was able at the age of eighteen to become a professor of mathematics at the Artillery School in Turin. When he was nineteen he began a correspondence with Euler, who was amazed at the young man's mathematical creativity. Some four years later, Lagrange sent Euler a solution to a problem that the latter had been working on without success for a long time. It is to Euler's credit that he did not appropriate the method as his own but instead allowed Lagrange to publish first and to obtain full credit as the discoverer of the calculus of variations.

While he was at Frederick's court, Lagrange sent for one of his female relatives in Turin and married her. This was no romantic marriage. In fact, some of his correspondence states that he had no taste for marriage but merely wished to be cared for. Nevertheless he developed a genuine affection for his wife, and when she developed a fatal illness, he nursed her himself. Her death left him grief-stricken and precipitated one of his periods of depression.

In 1787 Lagrange was invited to Paris by Louis XVI, who offered him French nationality and made him a "veteran pensioner" of the French Academy of Sciences. When the French Revolution broke out two years later, friends of the mathematician urged him to return to Berlin, but he decided to remain in Paris, a decision he had occasion to regret when he witnessed the excesses of the Reign of Terror. It is recorded that he ran the great risk of expressing himself concerning the execution of Lavoisier, France's greatest chemist. He stated: "It took them only a moment to cause his head to fall, and a hundred years perhaps will not suffice to produce its like."

In 1792, at the age of fifty-six, Lagrange made a second marriage, this time to the sixteen-year-old daughter of Lemonnier, an astronomer who was his colleague in the Academy of Sciences. Despite the great difference in their ages, the marriage turned out to be a very successful one.

When the École Polytechnique was founded in 1797, Lagrange became its first professor of mathematics. It is interesting to note that Laplace, who was one of the chief organizers of the École, offered the position to Lagrange. It was also during this period that both mathematicians played leading roles on the committee that perfected the metric system of weights and measures.

During the last years of his life Lagrange devoted his energies to a revision and extension of his *Mécanique Analytique*. Unfortunately, in 1813, death intervened before the task was completed. Lagrange was aware that the end was approaching, and in the presence of friends he summarized his life thus: "Death is not to be dreaded and when it comes without pain, it is a last function which is not unpleasant. I have had my career; I have gained some celebrity in mathematics. I never hated anyone, I have done nothing bad, and it would be well to end."

When Lagrange was a schoolboy of thirteen in Turin, Laplace was an infant in a little Norman village a thousand miles away. He was born on March 23, 1749. Concerning his early years, the mathematical physicist Sir Edmund Whittaker (1873–1956) said that many statements by historians are erroneous. For example, Laplace's father was *not* a poor peasant, as usually claimed, but the owner of a small estate and a member of the bourgeoisie. One uncle was a surgeon and another was a priest who taught at the Benedictine priory at Beaumont. It is believed that the latter stimulated Laplace's interest in mathematics even before his admission to the University of Caen at the age of sixteen. While he was still a student at the university he wrote his first paper, which came to the attention of Lagrange just as, earlier, Lagrange's genius had been demonstrated to Euler.

At the age of eighteen Laplace decided to go to Paris to make a career for himself. Provided with a letter of recommendation, he sought an interview with

D'Alembert to no avail. Later he tried a different approach by incorporating his ideas on mechanics in a letter. He got an immediate reply; D'Alembert wrote: "You needed no special introduction; you have recommended yourself, and my support is your due." Shortly thereafter Laplace was appointed a professor of mathematics at the École Militaire of Paris. This marked the beginning of the career which was to earn for him the title of the "Newton of France." He did, in fact, extend the implications of universal gravitation to the entire solar system. All his ideas in this area were included in his five-volume treatise, the *Mécanique Céleste* (1799–1825).

In 1784 Laplace received a lucrative position as "examiner to the royal artillery." In this capacity he had occasion to test the sixteen-year-old Napoleon when the latter applied for admission to the École Militaire. What might the course of world history have been if Napoleon had been failed by Laplace? Or, if he had been admitted over a protest by Laplace, what might Laplace's history have been? What it actually was like in some important respects is related by Whittaker.*

When Laplace began his active career as a mathematician, more than eighty years had elapsed since the publication of Newton's *Principia*. For long after its first appearance, that greatest of all works of science had met with considerable opposition. The most eminent mathematicians of the end of the seventeenth century—Huygens, Leibniz, John Bernoulli, Cassini—declared against the Newtonian theory of gravitation; this occurred even in Cambridge, Newton's own University. It was not until 1745 that the theory of the motions of the heavenly bodies began to be carried beyond the stage it had reached in the *Principia*; and for forty years after that, attention was focused almost exclusively on certain phenomena, which were well attested by astronomical observation, and which seemed to be irreconcilable with the Newtonian theory.

The most striking of these was what was called the great inequality of Jupiter and Saturn. From a comparison of ancient and current observations, it was found that for many centuries the mean motion, or average angular velocity round the sun, of Jupiter, had been continually increasing, while that of Saturn had been continually decreasing. Hence it could be inferred, by Kepler's third law, that the mean distance of Jupiter from the sun was always decreasing, and that of Saturn increasing; so the ultimate fate of Jupiter would be to fall into the sun, while Saturn would wander into space and be lost altogether to the solar system. This of course, assumed that the phenomenon was truly secular, that is to say, that it proceeded always in the same direction, with a cumulative effect. . . . Attempts to bring the great inequality of Jupiter and Saturn within the compass of the Newtonian theory of gravitation were made by Euler in 1748 and 1752 and by Lagrange in 1763, but without success. In 1773 Laplace, then aged twenty-four, took the matter up and . . . finally arrived at a complete explanation of the great inequality; it was not secular, but was periodic, of very long period; and it was due to the fact that the mean motions of the two planets were nearly commensurable. The comparison of his theory with observation left no doubt of its truth.

Laplace's explanation of the great inequality of Jupiter and Saturn was the first of a long series of triumphs, which are recorded in the two thousand pages of his *Mécanique Céleste*, and which achieved the complete justification of the Newtonian theory. He also made many important discoveries in theoretical physics, and indeed he was interested in

* Sir Edmund Whittaker, "Laplace," *American Mathematical Monthly*, Vol. 56, No. 6 (June–July 1949), pp. 369 ff.

everything that helped to interpret nature; pure mathematics for its own sake, however, did not greatly appeal to him, and his contributions to pure mathematics were mostly thrown off as mere by-products of his great works in natural philosophy. Yet there are several cases where Laplace's papers have become an important branch of pure mathematics. When in the course of his researches he comes to a situation where a heavy piece of pure mathematical working is needed, he often says, "*Il est facile de voir*," and gives the result without saying how he got it. His power of solving problems in pure mathematics has perhaps never been equaled, but he seems to have thought nothing of it, and to have assumed that it was possessed by all the readers of his works.

Lagrange's mechanics and Laplace's celestial mechanical discoveries involved the solution of differential equations, the essential feature of scientific determinism. Such equations involve not only variables like the x, y, z's of algebra but also derivatives, that is, rates of change, which, in physical applications, may be velocities, accelerations, or the like. To illustrate the most elementary sort of differential equation arising in mechanics, let us consider the case of a bullet that is fired straight upward from a point 5 feet above the ground with an initial speed of 1200 feet per second. Its future velocity and position are completely *determined* by the knowledge of its speed and position at the instant of firing. The effect of gravity will retard its motion and its speed will decrease until it comes to rest at the highest point reached. Then, reversing its direction, it will fall toward the ground with gradually increasing speed. This motion results from a negative acceleration and, according to Galileo and Newton, the acceleration does not vary as time goes on but is constant and approximately equal to -32 feet per second each second. (Since no account is taken of air resistance, this assumption is strictly true only for objects falling in a vacuum.) The expression of the physical law by a *differential equation* is

$$\frac{dv}{dt} = -32$$

By antidifferentiation, one obtains

$$v = -32t + c_1$$

Substituting the given fact,

$$v = 1200 \quad \text{when} \quad t = 0$$

(that is, the *initial* or *boundary* condition)* leads to

$$1200 = 0 + c_1$$

or

$$c_1 = 1200$$

and the formula that *determines* the future velocity of the bullet is

$$v = -32t + 1200$$

We can ascertain that after half a minute (30 seconds),

$$v = -32(30) + 1200 = 240$$

---

* Such particular requirements may be called *boundary conditions*, although conditions specified at $t = 0$ (or at any designated *time*) are usually called *initial conditions*.

so that the speed has slowed down to 240 ft. per sec. To find the time when the bullet begins to fall, $v = 0$ is substituted in the above formula:

$$0 = -32t + 1200$$

$$t = 37.5 \text{ sec.}$$

To find the height of the bullet at this instant or at any other time, we recall that $v = dx/dt$, and substituting this derivative in place of v leads to a second differential equation,

$$\frac{dx}{dt} = -32t + 1200$$

After antidifferentiation,

$$x = -16t^2 + 1200t + c_2$$

The knowledge that $x = 5$ when $t = 0$ indicates that $c_2 = 5$, and the deterministic formula for future position is

$$x = -16t^2 + 1200t + 5$$

The bullet reaches its maximum height when $t = 37.5$, and if we substitute this value, we find that the greatest height is $x = 22,505$ ft. We might wish to know when the bullet will return to its initial position, $x = 5$ ft. (the height at which it was fired). Substituting 5 for x yields

$$5 = -16t^2 + 1200t + 5$$

$$-16t^2 + 1200t = 0$$

$$16t(-t + 75) = 0$$

$$16t = 0 \quad \text{or} \quad -t + 75 = 0$$

$$t = 0 \qquad\qquad t = 75$$

The second answer indicates that the bullet returns to the position at which it was fired after 75 seconds, or that it rises for as many seconds as it falls.

The formulation of a physical law as a differential equation is rarely as simple as $dv/dt = -32$. For example, there is

$$\frac{d^4y}{dx^4} + 2k^2 \frac{d^2y}{dx^2} + k^4y = 0$$

a differential equation associated with the bending of an elastic plate. This equation involves an independent variable x, a dependent variable y, as well as the *second* and *fourth* derivatives of y with respect to x. (The letter k is a constant whose numerical value is determined by physical facts associated with a particular problem.) The *order* of the equation is said to be equal to 4, because that is the order of d^4y/dx^4, the highest-order derivative appearing in the equation.

To *solve* the equation means to determine a relation between y and x satisfying the equation. If y should be expressible in terms of x, then substitution of the expression for y as well as its second and fourth derivatives with respect to x in the left member of the above differential equation should yield the result 0. Inspection of that equation indicates that no mere antidifferentiation will solve it, that is, will lead to a relation of the desired type. In the case of the above bullet problem which can be formulated either as

$$\frac{dv}{dt} = -32$$

or as

$$\frac{d^2x}{dt^2} = -32$$

and also in an equation like

$$\frac{d^5y}{dx^5} = 12x^2 - 6x + 3$$

a derivative of some order is expressed as a function of the independent variable. Solution merely requires finding a sequence of antiderivatives. Thus, in the last illustration, one obtains

$$\frac{d^4y}{dx^4} = 4x^3 - 3x^2 + 3x + c_1$$

$$\frac{d^3y}{dx^3} = x^4 - x^3 + \frac{3}{2}x^2 + c_1x + c_2$$

$$\cdot\ \cdot\ \cdot$$
$$\cdot\ \cdot\ \cdot$$

$$y = \frac{1}{210}x^7 - \frac{1}{120}x^6 + \frac{1}{40}x^5 + \frac{c_1}{24}x^4 + \frac{c_2}{6}x^3 + \frac{c_3}{2}x^2 + c_4x + c_5$$

This result, called the *general solution*, contains *five* arbitrary constants, corresponding to the fact that the differential equation is of *fifth* order. It is always true that the number of arbitrary constants in the general solution is equal to the order of the differential equation. If a sufficient number of initial or boundary conditions is given, the numerical values of the constants can be determined and a *particular solution* is obtained.

But it must be emphasized once again that the easy mode of solution just illustrated is not usually possible. The reader may inspect the following formulations of physical laws as differential equations to observe that in no case is the equation expressed so as to make a sequence of antidifferentiations possible.

$$\frac{dI}{dt} + 12I = 40 \sin 120\pi t$$

(*I* is the current at any time in a 20-volt AC circuit with a 6-ohm resistor and a 0.5-henry inductance)

$$\frac{d^2Q}{dt^2} + 12\frac{dQ}{dt} + 100Q = 40 \sin 120\pi t$$ (If a condenser with capacity 0.02 farad is placed in the above circuit, Q is the charge on this condenser at any time)

$$\frac{d^2y}{dx^2} = k\sqrt{1 + \left(\frac{dy}{dx}\right)^2}$$ (Curve of suspension of an inextensible flexible cable)

$$\frac{d^2v}{dx^2} - 2v\frac{dv}{dx} + 6v = 0$$ (Oscillator in quantum mechanics)

$$\frac{d^4y}{dx^4} = k^4y$$ (Curve of rotating shaft in rapid motion)

$$\frac{dy}{dt} = ky$$ (Law of growth if k is a positive physical constant. Law of decay of radium, uranium, etc. if k is negative)

If one returns once more to the elementary equation,

$$\frac{dv}{dt} = -32$$

then the general solution is

$$v = -32t + c_1$$

This solution would apply not only to the bullet in the original problem, but also to any object (*in vacuo*) subject to the effect of gravity. The arbitrary constant c_1 is the value of v when $t = 0$; that is, c_1 is the initial velocity with which the object is launched. Since c_1 can, in theory, be equal to any real number whatsoever, the general solution incorporates an infinite number of formulas, one for each value of c_1. This is shown in Figure 9.1, which pictures the general solution as an infinite

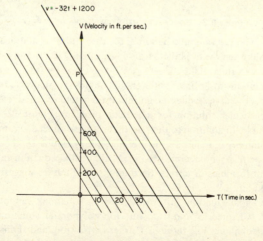

Figure 9.1 **General solution of** $dv/dt = -32$ **and particular solution for a special initial condition**

set of parallel lines with *slope* $dv/dt = -32$. The particular solution for the initial condition, $v = 1200$ when $t = 0$, is singled out from the set as the line through the point P (0, 1200).

But Figure 9.1 provides only the general solution for the velocity v. To obtain that for the distance x, one can start anew with

$$\frac{d^2x}{dt^2} = -32$$

to obtain

$$\frac{dx}{dt} = -32t + c_1$$

and

$$x = -16t^2 + c_1 t + c_2$$

This includes an infinite number of distance formulas for different initial velocities and different points of fire. Figure 9.2 reveals this general solution as an infinite set

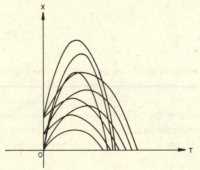

Figure 9.2   General solution of $d^2x/dt^2 = -32$

of parabolas. The curve corresponding to the initial conditions, $x = 5$ and $v = dx/dt = 1200$ when $t = 0$, is the parabola through Q (0, 5) and tangent at this point to a line with slope 1200.

General solutions to ordinary differential equations can always be pictured as such infinite sets of curves, where given initial or boundary conditions select one curve as the particular solution for a specific problem. Figure 9.3 provides an additional example. In addition to their aesthetic appeal, such diagrams are helpful in clarifying a number of issues. In the first place, they emphasize the advantage of describing a type of phenomenon by a *single* differential equation rather than by an infinite set of formulas or curves. Secondly, sets of curves suggest certain procedures for solution.

Most differential equations, like most algebraic equations, cannot be solved in "finite terms" (Chapters 5 and 16). But for polynomial equations, there are approximative techniques like those of Horner or Newton, and similarly there are graphic and numerical methods for obtaining approximate solutions to differential equations. The approximation is a curve or a set of curves very close to the *exact*

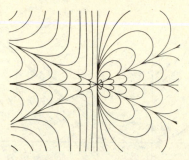

**Figure 9.3   General solution of a differential
equation of the first order**

forms. The analogy with algebraic equations will be continued in order to provide further insight into the meaning of a differential equation and also to furnish a crude idea of graphic or numerical approximative procedures. Given an algebraic equation like

$$y = x^2 - 3x - 10$$

or

$$4x^2 + y^2 = 36$$

one can plot it graphically as a curve. A knowledge of the elements of analytic geometry would enable identification of the above equations as representing a parabola and an ellipse, respectively. But a student of school algebra will painstakingly assign a set of varied values to x and then compute the corresponding values of y so as to obtain the coordinates of points on the curve. The reader may place himself *in loco studentis* as far as differential equations go. Thus, if he is asked to obtain a graphic solution of

$$\frac{dy}{dx} = x - y$$

(an equation not solvable by antidifferentiation), he can substitute various pairs of values for the variables x and y in the right member, and then compute the corresponding values of dy/dx. What he will obtain if he substitutes $x = 2$, $y = 1$ is the slope $dy/dx = 2 - 1 = 1$, and hence the inclination $= 45°$ for the tangent to the curve through the point $(2, 1)$. If a large number of such substitutions are made and the corresponding tangent directions determined, these can be plotted as in Figure 9.4, where the *direction field* suggests the set of curves forming the general solution. The heavy line in the diagram is the curve through the point $(0, 0)$, that is, the particular solution corresponding to the initial or boundary condition, $y = 0$ when $x = 0$.

Thus a crude graphic method for obtaining a particular solution of a *first-order* differential equation is as follows. One starts at the point corresponding to the

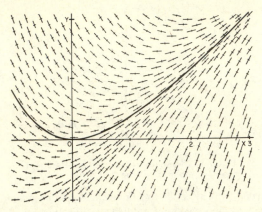

Figure 9.4   Direction field for $dy/dx = x - y$

boundary condition, say, $y = -1$ when $x = 4$. Then one substitutes these values in the differential equation to find the numerical value of dy/dx, the slope, and finally one plots the point $(4, -1)$ and a small arrow through this point in the direction specified by the value of the slope. Now one computes or else estimates from the graph the coordinates of the tip of the arrow. Suppose that they are $(4.2, -0.7)$. Next these values are substituted for x and y in the differential equation to find the numerical value of the slope dy/dx. A small arrow with this slope is drawn through the above point, the coordinates of the tip of this arrow are computed or read from the graph, and the whole process is repeated, and so on.

Sometimes one can obtain a solution or, at any rate, get some idea of its nature, by interpreting the differential equation geometrically (graphically) without carrying out all the numerical work involved in plotting its direction field. Thus in the case of the first-order equation,

$$\frac{dy}{dx} = \frac{y}{x}$$

which is not solvable directly by antidifferentiation, the slope dy/dx at each point of the direction field is equal to the direction of the line joining that point to the origin (Figure 9.5). Then the lines through the origin (Figure 9.5) form the general solution, which must therefore have the form

$$y = cx$$

Note that the origin $(0, 0)$ is an exceptional or "singular point," because substitution of $x = 0$, $y = 0$ in the differential equation yields $dy/dx = 0/0$, which is *indeterminate*. The graph of Figure 9.5 indicates why this is so, for instead of having a unique solution through the origin, there is an infinite number of solutions (straight lines).

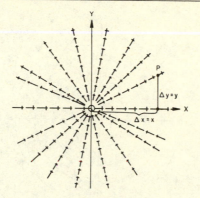

Figure 9.5   Direction field for $dy/dx = y/x$

Again, to solve

$$\frac{dy}{dx} = -\frac{x}{y}$$

through geometric considerations, one must use a fact of analytic geometry, namely that if m and m' are the slopes of two lines, then $mm' = -1$ is the necessary and sufficient condition for the lines to be perpendicular. Now if a line is drawn from the origin to the point (x, y), its slope is

$$\frac{\Delta y}{\Delta x} = \frac{y-0}{x-0} = \frac{y}{x}$$

which makes the line so drawn perpendicular to any line with slope $-x/y$. Hence, for the given differential equation, the direction field at every point is perpendicular to the line joining that point to the origin (Figure 9.6). The field is suggested in Figure 9.6, and indicates that the general solution is the set of circles

$$x^2 + y^2 = c^2$$

To obtain exact solutions of equations of first and higher orders when no fortuitous geometric property is present requires considerable theory and technique. Fortunately, the major theme of this chapter requires only special differential equations arising in mechanics, and there is no need to develop a detailed analysis of other types.

The study of the differential equations of *rational* or *analytic mechanics* was initiated by Euler and brought to fruition by Lagrange. Ultimately, Hamilton added a final generalization, one that is suitable even in today's quantum mechanics. Analytic mechanics provides a pure mathematical treatment of *dynamics*, the study of the *motion* of material bodies subject to forces. Lagrange's abstract "systems" idealize the corresponding physical realities. The "state" of such systems at a particular time is described by two or more real numbers. (Above, two numbers furnished the *position* and *velocity* of the bullet.) As time goes on the two, three, . . . ,

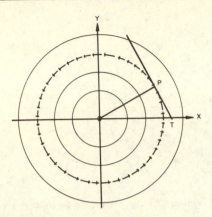

Figure 9.6   Direction field for $dy/dx = -x/y$

n, real numbers describing the state of a dynamical system will in general vary, and if we name these real variables $x_1, x_2, \ldots, x_n$, a dynamical system is such that the rates of change, $dx_1/dt, dx_2/dt, \ldots, dx_n/dt$, depend merely on the values of the variables themselves. Hence the laws of motion can be expressed by n differential equations which give formulas for $dx_1/dt, dx_2/dt, \ldots, dx_n/dt$ in terms of constants and one or more of the variables $x_1, x_2, \ldots, x_n$. In the example of the motion of the bullet, we may think of x, v as x_1, x_2 and then

$$\frac{dx_1}{dt} = x_2$$

$$\frac{dx_2}{dt} = -32$$

A pair of dynamical equations that cannot be solved directly by antidifferentiation in spite of simplicity of appearance is the following:

$$\frac{dx_1}{dt} = x_2$$

$$\frac{dx_2}{dt} = -kx_1$$

where k is a numerical constant. One of the first dynamical problems that gave rise to equations of this form arose in connection with clock-making. In 1656 Christiaan Huygens, Dutch mathematician and physicist, constructed the first pendulum clock. If the arc through which a pendulum swings is *small*, then the above are the differential equations for its motion, where x_1 and x_2 represent the position and the velocity, respectively, of the pendulum bob on this arc.

Equations of the same form express the motion of a stretched spring, according to a law discovered by Robert Hooke, that contentious contemporary of Newton. *Hooke's law* states that the distance by which a force will extend an elastic spring varies directly as this force. This means that if this force is doubled, trebled, quadrupled, etc., the same is true of the displacement of the spring. Algebraically, $F = Kx_1$, where K is a constant depending on the particular spring, F and x_1 are force and distance, respectively.

If the spring is released, there will be a reaction, a restoring force, in accordance with Newton's third law: *To every action there is an equal and opposite reaction.* Then the restoring force can be represented by $-Kx_1$. Newton's second law can be formulated as

Force = Mass × Acceleration or, if x_2 represents velocity,

$$F = m\frac{dx_2}{dt}$$

(when suitable units of measure are used), and Hooke's law states that

$$F = -Kx_1$$

Equating the two different expressions for the force,

$$m\frac{dx_2}{dt} = -Kx_1$$

and

$$\frac{dx_2}{dt} = -\frac{K}{m}x_1$$

or

$$\frac{dx_2}{dt} = -kx_1$$

where we have replaced the constant K/m by the single constant k. In this way Hooke's law leads to a differential equation like the second one of the original pair. The first equation of the pair, $dx_1/dt = x_2$, merely states the familiar definition of velocity as rate of change of distance with respect to time.

Once more, a knowledge of the position and velocity at a single instant will determine the motion for all the future. For example, suppose that $K = 10$ for a certain spring on which a 5-pound weight is suspended. The weight is pulled 3 inches or $\frac{1}{4}$ foot below the equilibrium point, and then released. By Hooke's law, the restoring force is

$$F = -10x_1$$

Since a weight of 5 pounds has a mass of 5/32 slug (approximately), Newton's second law implies that

$$F = \frac{5}{32}\left(\frac{dx_2}{dt}\right)$$

Equating the two expressions for the force, one obtains

$$\frac{5}{32}\left(\frac{dx_2}{dt}\right) = -10x_1$$

or

$$\frac{dx_2}{dt} = -64x_1$$

as a differential equation to be satisfied by the motion of the spring. Mathematical theory supplies the general solution

$$x_1 = c_1 \sin 8t + c_2 \cos 8t$$

where c_1 and c_2 are to be determined from the initial conditions. At the beginning, or when $t = 0$, $x_1 = \frac{1}{4}$. Substituting,

$$\tfrac{1}{4} = c_1 \sin 0 + c_2 \cos 0$$

or

$$\tfrac{1}{4} = c_1 \cdot 0 + c_2 \cdot 1$$

and

$$c_2 = \tfrac{1}{4}$$

Therefore

$$x_1 = c_1 \sin 8t + \tfrac{1}{4} \cos 8t$$

In elementary calculus it is proved that the derivatives (with respect to t) of $\sin at$ and $\cos at$ (a, a constant) are $a \cos at$ and $-a \sin at$, respectively. Hence, differentiating with respect to t in the last equation above,

$$\frac{dx_1}{dt} = x_2 = 8c_1 \cos 8t - 2 \sin 8t$$

Initially the weight was *released*, that is, $x_2 = 0$ when $t = 0$. Substituting these values yields

$$0 = 8c_1 \cos 0 - 2 \sin 0$$

or

$$0 = 8c_1 \cdot 1 - 2 \cdot 0$$

and

$$0 = c_1$$

Therefore the motion is described by the pair of equations

$$x_1 = \tfrac{1}{4} \cos 8t$$

$$x_2 = -2 \sin 8t$$

The reader can check that these equations (as well as the general solution) satisfy the pair of differential equations

$$\frac{dx_1}{dt} = x_2$$

$$\frac{dx_2}{dt} = -64x_1$$

The deterministic formulas for x_1 and x_2 are the equations for a *simple harmonic motion*. The equation for x_1 indicates that, in theory, the weight will vibrate up and down forever between positions $\frac{1}{4}$ ft. or 3 in. above and the same distance below the equilibrium point. The period of the vibration is about 0.8 sec., that is, this will be the interval of time between an instant when the weight is at the lowest position and the instant when it is next at this position. Let us see what effect a change in initial conditions has on the motion. Suppose that when $t = 0$, $x_1 = 0$ and $x_2 = 3$, that is, the weight, initially in the equilibrium position, is struck so as to give it a downward velocity of 3 ft. per sec. Then substituting these conditions, one obtains $c_2 = 0$ and $c_1 = 3/8$ so that the equations for the simple harmonic motion in this instance are

$$x_1 = 3/8 \sin 8t$$

$$x_2 = 3 \cos 8t$$

The period is the same as in the previous case, but the oscillations take place between points 3/8 ft. or $4\frac{1}{2}$ in. above and below the equilibrium point.

Since the small oscillations of a simple pendulum are described by a pair of differential equations of the same sort as those just considered, the pendulum will also have a simple harmonic motion. With our electric clocks and time-telling gadgets of today, we scarcely realize the epochal nature of Galileo's and Huygen's discoveries of the pendulum laws. Until these were known, it was not the common man, but the astronomer who suffered most, because it was not possible to measure small fractions of a day with any real precision. One day in 1583 the nineteen-year-old Galileo, while praying in the cathedral at Pisa, chanced to observe the swinging of the great lamp which had just been lighted. Using the beating of his own heart as a clock, he timed the oscillations and found the time to be about the same for every vibration, even after the amplitude of the swinging had greatly diminished.

Thus Galileo discovered that the motion of a pendulum is approximately *isochronous*; that is, the time of return to its initial position is about the same whether it makes a large movement at high velocity under a strong restoring force, or a small movement at low velocity with slight restoring force. Galileo used this fact to invent an instrument by which a doctor could take a patient's pulse, and then forgot about it until late in life, when suddenly he realized that the isochronism of the pendulum might make it possible to construct clocks that would constitute a great improvement over the crude timekeepers then in use. The pure mathematical pendulum of the differential equations above maintains its motion forever, but in clockmaking or other applications, some external source of power is required to prevent the pendulum from coming gradually to rest. Galileo never completed a mechanical design for supplying such power, and hence it was not he but Huygens who constructed the first pendulum clock. In such a clock the pendulum is kept in motion by a slight force applied as it passes its lowest point. Each time it swings it causes a spring or weight to move a wheel through a certain specified distance, and this motion is transmitted to the clock hands by means of gears.

Springs and clocks were some of the earliest practical applications of the Galileo-Newtonian mechanics, but in the strictest sense the classic laws of motion were not entirely suited to the physical problems involved. Newton's laws apply only to particles, that is, masses so tiny that they can be considered to be concentrated at points, and therefore have a position defined by one Cartesian coordinate x if located on a line, by two coordinates (x, y) or three coordinates (x, y, z) if located in a plane or in space, respectively. But if the motion under consideration is that of an entire physical body, and not merely that of a single particle, the mechanics of Newton must be generalized and the mathematics becomes tremendously more difficult. The generalized physical laws and the associated mathematical techniques were contributed by Euler, Lagrange, and Hamilton.

One far-reaching principle of mechanics was due to Euler's younger contemporary, D'Alembert, who is probably best known to general readers for his association with Diderot in the preparation of the *Encyclopedia*. In our discussion of the mechanics of rigid bodies, we have alluded to *external* or impressed, rather than internal, forces as the primary cause of motion. It is *D'Alembert's principle* that furnishes the reason for this fact. The internal forces of a rigid body are gravitational attractions, molecular reactions, various interactions among its numerous particles. External or impressed forces are those exerted by other bodies. The weight of a rigid body is an external force because it is exerted by another body, the earth. When men pull on a tow rope in removing a wrecked car from the road, the force exerted is external to the car.

Newton's second law, $F = ma$, applies only to a single particle. Here F represents the resultant or vectorial sum of all the forces, external and internal, acting on such a particle. It is this resultant force that is effective in producing the actual motion of the particle, and because this force is equal to the product ma, it is customary to call ma the *effective force* on the particle. In 1742 D'Alembert generalized Newton's $F = ma$ by the assertion that if ma, the effective force, is computed for each particle of a rigid body or general mechanical system and all these effective forces are totaled, this sum is equal to the total of *external* forces acting on the system. Such is *D'Alembert's principle*. An alternative statement of this law emphasized that the internal forces of a system have no effect on its motion and that it is only external forces that can cause any movement: *The internal actions and reactions of any system of bodies in motion are in equilibrium* (that is, they cancel one another and their total effect is nil).

After enunciating his principle, D'Alembert applied it to hydromechanics with fruitful results, and from that time on his researches in both pure and applied mathematics were of such outstanding quality as to win recognition not only from his native Academy of Sciences but also, as mentioned earlier, from the Berlin Academy of Frederick the Great. In his later years he acted as a sort of big brother to Lagrange, his successor in the field of mechanics. In the course of a lengthy correspondence he encouraged the young man's scientific efforts, counseled him to take better care of his health, and roused him from his periodic spells of despondency.

Lagrange, possibly the greatest figure in the history of mechanics, considered very general dynamical systems. In 1834 Hamilton generalized Lagrange's methods still further. Hamilton expressed mechanical laws in the form of "canonical equations," which are the utmost in generality. Both these and the less general Lagrange equations resemble the differential equations expressing Newton's second law. If the acceleration of a body varies, then $F = ma$, or

$$F = m \frac{dv}{dt}$$

Since mass was considered to be constant in classic mechanics,

$$F = \frac{d(mv)}{dt}$$

Now mv, the product of mass and velocity, is called *momentum*, and if we examine the last equation above, we can see why Newton's second law is often stated as: *The rate of change of momentum* (of a particle) *is equal to the force.*

In the Hamilton differential equations the law expressed is: The rate of change of the *generalized* momentum is equal to the *generalized* force. The Hamiltonian equations are the most powerful generalizations of classical mechanics and are adequate to deal with the most complicated systems imaginable.

Although the basic postulates of quantum mechanics differ from those of the traditional subject in very fundamental aspects, it is a remarkable fact that the enormous generality of the Hamilton equations make them suitable, *mutatis mutandis*, even in the new, radically different mechanics. The same cannot be said of Lagrange's equations, even though in classical mechanics these imply the Hamiltonian equations and conversely. It is hard to explain this paradox without going into the details of quantum mechanics, but the essential point is that in Lagrange's mechanics the function defining the energy of the system is expressed in terms of generalized *velocities*, whereas the *Hamiltonian* is a function of generalized *momenta*. Now a momentum has meaning in quantum mechanics but a velocity has not. One might argue that, after all, momentum is equal to mv, the product of mass and velocity, so that a little algebraic manipulation can convert an expression involving velocities into one with momenta and vice versa. This is actually the classical procedure for deducing the Hamiltonian equations from the Lagrangian and conversely, but analogous reasoning would *not* hold in quantum mechanics, for there one starts with generalized momenta, the concept of momentum being autonomous. A momentum is a *matrix* (Chapter 10), an entity dealing with a whole set *en masse*, and is *not* the product of the mass and velocity of the subatomic particles that are the object of study in quantum mechanics. In fact, the nature of quantum theory does not permit a well-defined velocity concept, and it would be illogical to base other notions on an entity whose measurement is ambiguous. The indeterminism of velocity is part of the famous Heisenberg "principle of indeterminacy." In 1927 Heisenberg advanced the theory that it is impossible to fix both the position and the velocity of an electron with perfect precision, that if we increase the accuracy of one of these measurements, it automatically decreases the precision of the other. If our measuring tools were perfect, or nearly so, this would not be the case. Our instruments are part of the universe we are studying, and they share its characteristics, one of which, according to the quantum theory, is the *discreteness* or lack of continuity of matter and energy. This makes ever-so-fine subdivision of units of measure not only a practical but a theoretical impossibility. The smallest possible subdivision of mass is that of an electron, and the smallest unit of energy the "quantum." Such are the notions from which Heisenberg deduced the principle establishing the indeterminacy of velocity. This issue does *not* arise in classical mechanics because the phenomena considered are macrocosmic and it is not a question of motion in a subatomic world where the inevitable coarseness of our tools is important. The older mechanics is valid up to the boundaries of the atom, and in fact, for the outer confines of the atom the quantum equations become identical with the classical ones.

All our illustrations have exhibited "ordinary" differential equations, but the vast majority of physical laws are couched as "partial" differential equations. Therefore, to appreciate the significance of major physical laws, it is necessary to know what is meant by a partial differential equation. Thus we shall now introduce that concept.

In the discussion of calculus in the previous chapter, only functions of a single variable were considered, that is, there was an independent variable (x), and a dependent variable (y). But the formulation of natural phenomena cannot be limited to relations containing only two variables, but may involve many variables. Even in elementary analytic geometry three variables are usually required for the equation of a surface in space. Thus $z = 2x + 3y + 1$ is the equation of a certain plane and $z = x^2 + 4y$ is the equation of a certain "quadric" surface.

In the equation for a plane, if one were to consider only those points for which $y = 1$, or those for which $y = 4$, or those for which $y = b$ (any constant value), the above equation would become $z = 2x + 3 + 1$ or $z = 2x + 12 + 1$ or $z = 2x + 3b + 1$, that is, a straight line in the given plane, and then the elementary calculus of the previous chapter would be applicable. The slope of each one of the straight lines would be $dz/dx = 2$. Reasoning similarly with the quadric surface, that is, considering only those points of the surface where $y = b$, the equation for such points becomes $z = x^2 + 4b$ (a *parabola* on the surface) and the derivative (slope of the tangent to the parabola) is $dz/dx = 2x$. In both examples treated, one is dealing merely with a *part* of the surface, a line or curve within it. This results from having y remain constant in value while only x and z change. To indicate this phenomenon, it is customary to refer to the derivatives just found as *partial* derivatives and to employ a special symbolism, namely $\partial z/\partial x$.

In analogous fashion one may consider only those parts of surfaces where x has a specified value, $x = a$. Then substitution in the equations above would yield $z = 2a + 3y + 1$ and $z = a^2 + 4y$, respectively, both straight lines, with $\partial z/\partial y = 4$ as the slope of the first line and $\partial z/\partial y = 4$ as the slope of the second.

It is thus possible to interpret partial derivatives, like ordinary derivatives, as slopes, rates of change, etc. As another example, consider $A = 2\pi rh + 2\pi r^2$, the formula for the total surface area of a circular cylinder, where r is the radius and h the altitude. Here $\partial A/\partial h = 2\pi r$ gives the rate of change of surface area when the radius is held constant while the height is permitted to increase or decrease so that the cylinder gets taller or shorter. If $r = 10$, then $\partial A/\partial h = 20\pi = 62.8$ (approximately) and the area is increasing or decreasing at the rate of roughly 63 sq. in. per in. (of height). If, on the other hand, the altitude is held constant and the radius varies, $\partial A/\partial r = 2\pi h + 4\pi r$. If $r = 10$ and $h = 5$ when the radius begins to increase (or decrease), $\partial A/\partial r = 10\pi + 40\pi = 50\pi = 157$ (approximately), so that the area is changing at the rate of 157 sq. in. per in.

A simple example of the use of partial derivatives in physics is furnished by the thermodynamic law of Charles (1746–1823) and Gay-Lussac (1778–1850). That law, which expresses the relation of volume, pressure, and absolute temperature in an ideal gas, has the algebraic form $V = kTP^{-1}$. Therefore $\partial V/\partial T = kP^{-1}$ gives the rate of change in the volume of any gas when its pressure is held constant but its temperature is allowed to vary, and $\partial V/\partial P = -kTP^{-2}$ gives the rate of change in volume when temperature is held constant while pressure varies.

Partial derivatives, like ordinary derivatives, can also be extended to second or higher orders. Consider the surface

$$z = 2x^3 + 5xy^2 - y^3$$

If the reader will proceed as above, first taking $y = b$ (any constant), and then $x = a$ (any constant), he will find that $\partial z/\partial x = 6x^2 + 5y^2$ and $\partial z/\partial y = 10xy - 3y^2$. (Actually he will obtain $6x^2 + 5b^2$ and $10ay - 3y^2$. But this is understood in the formulas. In $\partial z/\partial x$ only x and z vary. Hence y must be a constant. Likewise x must be a constant in $\partial z/\partial y$. In a more general case, if u depends on x, y, z, t, w, etc. and $\partial u/\partial w$ is found, it is understood that only u and w are variables, and if other letters—x, y, z, t, etc.—appear in the formula for $\partial u/\partial w$, it is known that they are to be held constant.)

To obtain a *second* partial derivative, let $y = l$, a constant, in

$$\frac{\partial z}{\partial x} = 6x^2 + 5y^2$$

Then

$$\frac{\partial z}{\partial x} = 6x^2 + 5l^2$$

and

$$\frac{\partial^2 z}{\partial x^2} = 12x$$

For a different second derivative, let $x = k$, a constant, in $\partial z/\partial x$. Then

$$\frac{\partial z}{\partial x} = 6k^2 + 5y^2$$

and

$$\frac{\partial^2 z}{\partial y \partial x} = 10y$$

Proceeding in similar fashion with the expression for $\partial z/\partial y$, one finds that

$$\frac{\partial^2 z}{\partial y^2} = 10x - 6y$$

and

$$\frac{\partial^2 z}{\partial x \partial y} = 10y$$

We observe that, in this case,

$$\frac{\partial^2 z}{\partial y \partial x} = \frac{\partial^2 z}{\partial x \partial y} = 10y$$

In many other cases (but not in all) it will be true that

$$\frac{\partial^2 z}{\partial y \partial x} = \frac{\partial^2 z}{\partial x \partial y}$$

that is, in such special instances, differentiation with respect to x and y is *commutative*, or if one differentiates first with respect to x and then with respect to y, the result will be the same as if one proceeds in the opposite way, varying y first while holding x constant, then varying x while y is constant. It can be proved that if the two partial derivatives above are continuous (for a pair of values of x and y), then these derivatives are equal (for the particular values of x and y) and there is "commutativity."

To reverse the procedure above, let us carry out some antidifferentiations with partial derivatives and, in the process, solve some *partial differential equations*. Consider first the ordinary differential equation

$$\frac{dy}{dx} = 2x$$

Then the general solution is

$$y = x^2 + c$$

Now if z is an unknown function of two variables, x and y, and if from the conditions of some problem it is known that

$$\frac{\partial z}{\partial x} = 2x$$

the solution of this partial differential equation can be found by antidifferentiation in a manner suggested above. The solution is

$$z = x^2 + g(y)$$

where $g(y)$ is an *arbitrary function* of y (instead of an arbitrary constant as in the case of the ordinary differential equation). To check, the reader may substitute any function of y for $g(y)$. Thus let

$$g(y) = y^5 - 17y^2 + 3$$

Then
$$z = x^2 + y^5 - 17y^2 + 3$$
and
$$\frac{\partial z}{\partial x} = 2x$$

so that the given condition is satisfied.

For practice the reader may solve

$$\frac{\partial z}{\partial x} = y^3$$

to obtain the answer

$$z = xy^3 + g(y)$$

Returning momentarily to ordinary differential equations, let us find the general solution of

$$\frac{d^2y}{dx^2} = 0$$

If we let

$$w = \frac{dy}{dx}$$

then the original equation is seen to be equivalent to

$$\frac{dw}{dx} = 0 \qquad \text{and} \qquad w = c_1$$

or

$$\frac{dy}{dx} = c_1 \qquad \text{and} \qquad y = c_1 x + c_2$$

Now if z is an unknown function of x and y, let us solve

$$\frac{\partial^2 z}{\partial x^2} = 0$$

Let

$$w = \frac{\partial z}{\partial x}$$

so that

$$\frac{\partial w}{\partial x} = 0 \qquad \text{and} \qquad w = g(y)$$

or

$$\frac{\partial z}{\partial x} = g(y) \qquad \text{and} \qquad z = xg(y) + h(y)$$

This general solution contains two arbitrary functions, just as the solution of the corresponding ordinary differential equation contained two arbitrary constants.

To connect the question of partial derivatives with that of physical laws, the partial differential equations of mathematical physics are generally of the "second order," which signifies that they contain no derivatives higher than the second. For example, there are generalizations of the differential equation $\partial^2 z / \partial x^2 = 0$, which was just solved. These are

$$\frac{\partial^2 z}{\partial x^2} + \frac{\partial^2 z}{\partial y^2} = 0 \quad \text{and} \quad \frac{\partial^2 u}{\partial x^2} + \frac{\partial^2 u}{\partial y^2} + \frac{\partial^2 u}{\partial z^2} = 0$$

They are called the *Laplace equations* in two and three dimensions, respectively, and are generally considered to express the *most important laws* of mathematical physics. It is easy to furnish particular examples of functions satisfying Laplace's equation, since the variety of such functions is infinite. Thus let us show that

$$\frac{\partial^2 z}{\partial x^2} + \frac{\partial^2 z}{\partial y^2} = 0$$

is satisfied by

$$z = x^3 - 3xy^2$$

For this function

$$\frac{\partial z}{\partial x} = 3x^2 \qquad \frac{\partial z}{\partial y} = -6xy$$

$$\frac{\partial^2 z}{\partial x^2} = 6x \qquad \frac{\partial^2 z}{\partial y^2} = -6x$$

and therefore

$$\frac{\partial^2 z}{\partial x^2} + \frac{\partial^2 z}{\partial y^2} = 6x - 6x = 0$$

In ordinary differential equations the problem is usually to find a general solution, that is, to find *all* functions meeting the requirements expressed by the equation. Then if a particular solution is desired, initial conditions are substituted and from these the arbitrary constants in the general solution are determined. In the case of partial differential equations, it is usually neither feasible nor desirable to seek general solutions, although we have done so for some special cases. Instead, one makes use, almost from the outset, of the special restrictions furnished by the conditions of the particular problem. Thus a vibrating string can be shown to satisfy

$$\frac{\partial^2 y}{\partial t^2} = c^2 \frac{\partial^2 y}{\partial x^2}$$

where (x, y) are the coordinates of a point on the string, and t represents time. The constant c is determined from the physical condition of the string. Thus, if the tension is constant throughout the string,

$$c^2 = \frac{\text{tension}}{\text{mass density}}$$

Suppose, for example, that the length of the string is 4 ft., its weight is 2 oz., and it is stretched so that the tension throughout is 4 lb. Then mass (in slugs) is $1/8 \div 32 = 1/256$, and mass density is $1/256 \div 4 = 1/1024$. Hence

$$c^2 = \frac{4}{\dfrac{1}{1024}} = 64^2$$

Therefore, under these conditions, the equation becomes

$$\frac{\partial^2 y}{\partial t^2} = (64)^2 \frac{\partial^2 y}{\partial x^2}$$

Suppose that a solution is required which will satisfy the following conditions. First, there is the *boundary condition* specifying that the ends of the string are attached to the X-axis at the points $(0, 0)$ and $(4, 0)$. Next, there is the initial condition stating that when $t = 0$, the string is stretched slightly and drawn away from its straight-line position so that its shape is given by $y = 1/4 \sin \pi x/4$ (Figure

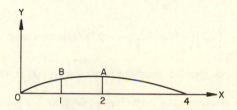

Figure 9.7   String stretched to shape $y = \tfrac{1}{4} \sin \pi x/4$

9.7), and then it is released. The last fact signifies that initially the points of the string will have zero velocity in a vertical direction, that is, when $t = 0$, $\partial y/\partial t = 0$. The particular solution satisfying all these conditions can be shown to be $y = 1/4 \sin \pi x/4 \cos 16\pi t$. This formula will give the position of any point on the string at any time. Thus, if one is interested in following the progress of the midpoint of the string, this can be done by substituting $x = 2$ (since the string is 4 ft. long) in the solution, to obtain

$$y = \frac{1}{4} \sin\left(\frac{\pi}{4} \cdot 2\right) \cos 16\pi t$$

Now $\sin \pi/2$ (radians) $= \sin 90° = 1$, and the result is therefore

$$y = \frac{1}{4} \cos 16\pi t$$

Where is point A after 1/32 sec., or after 1/16 sec.? This calls for substituting $t = 1/32$ and $t = 1/16$ to obtain

$$y = \frac{1}{4} \cos\left(16\pi \cdot \frac{1}{32}\right) = \frac{1}{4} \cos \frac{\pi}{2}$$

and

$$y = \frac{1}{4} \cos \left(16\pi \cdot \frac{1}{16} \right) = \frac{1}{4} \cos \pi$$

Now π radians $= 180°$, $\pi/2 = 90°$, and $\cos 90° = 0$, $\cos 180° = -1$. The results are therefore $y = 0$ and $y = -1/4$, or point A is on the X-axis (at $x = 2$) after 1/32 sec., and is 1/4 ft. or 3 in. below it after 1/16 sec.

In the same way, it is possible to tell the position of any other point of the string at any time. For example, where is the point B in Figure 9.7 after 3 1/16 seconds? If we assume $x = 1$ for this point and substitute in the solution of the differential equation, then

$$y = \frac{1}{4} \sin \left(\frac{\pi}{4} \cdot 1 \right) \cos \left(16\pi \cdot \frac{49}{16} \right)$$

$$= \frac{1}{4} \sin \frac{\pi}{4} \cos 49\pi$$

$$= \frac{1}{4} \left(\frac{\sqrt{2}}{2} \right) (-1) = -\frac{\sqrt{2}}{8} = -0.18 \text{ (approximately)}$$

Therefore, at the time specified, the point in question is 0.18 ft. or about 2.16 in. below the X-axis.

The question of whether any mathematical solution to a physical problem or to the corresponding partial differential equation (with arbitrary boundary conditions) can be found and the mathematical devices for obtaining such solutions (Fourier series, for example) are all subjected to study in the modern theory of partial differential equations and their application to physical problems. In a certain sense, one might consider deterministic mathematical physics to be the theory of boundary value problems. Therefore, to discuss all the possible physical situations described by partial differential equations and associated boundary conditions would require a profound study of advanced mathematics and physics. Nevertheless it is possible to give a few elementary illustrations.

Suppose, for example, that the periphery of a thin metal plate, with insulated faces, is maintained at a constant temperature, or more generally that each point of the periphery is kept at a constant temperature which may, however, vary from point to point. Thus, in a circular plate half the circumference might be kept at one temperature, and the other half at a different temperature. Then after a while the distribution of temperatures in the plate will reach a "steady state," a permanent condition where the temperature at each particular point remains the same as time goes on. The specific temperatures at each point will then depend on the size and shape of the metal sheet and the temperatures maintained at the boundary points. It can be proved that the formula for the steady state distribution of temperatures must satisfy *Laplace's differential equation* in two dimensions, that is, if z represents the temperature at a point P of the plate with Cartesian coordinates (x, y),

$$\frac{\partial^2 z}{\partial x^2} + \frac{\partial^2 z}{\partial y^2} = 0$$

As to boundary conditions, it might be known, for example, that the boundary is a circle and the temperature at each point of the circumference might be specified. The solution would be different if the shape were rectangular rather than circular, and would also depend on just how the peripheral temperatures on the rectangle are assigned.

To give another example of a physical interpretation of Laplace's equation, one can assert that the possible distributions for the Newtonian "gravitational potential" in the region of space surrounding the sun, or any other material body, are governed by the differential equation in question. At a point (x, y, z) *outside* matter (or in empty space) the "potential" u (a term due to *Lagrange*) must satisfy

$$\frac{\partial^2 u}{\partial x^2} + \frac{\partial^2 u}{\partial y^2} + \frac{\partial^2 u}{\partial z^2} = 0$$

and at a point *inside* matter, the corresponding physical law is expressed by a slightly different partial differential equation, that of Poisson (1781–1840), namely,

$$\frac{\partial^2 u}{\partial x^2} + \frac{\partial^2 u}{\partial y^2} + \frac{\partial^2 u}{\partial z^2} = -4\pi\rho$$

where ρ is the density of mass at the point (x, y, z).

Without going too deeply into the meaning of the "potential," it can be described as a continuous numerical function $u(x, y, z)$ from which the gravitational or electrical or magnetic force acting at the point (x, y, z) can be derived. In classical physics the potential at this point can be interpreted as the *work* that must be done to bring a unit of mass or a unit of electric charge from a very great distance to the point in question. A "velocity potential" is one from which velocities can be derived. In an ideal, incompressible fluid where there are no vortices about which the fluid will tend to rotate, and where there is no change in the total amount of fluid (that is, no sources of increase or "sinks" causing decrease), the velocity potential must satisfy Laplace's equation. This fact is important in hydrodynamics and aerodynamics.

The partial differential equations mentioned thus far have not involved time as a variable, because the phenomena referred to have been permanent or static—the temperatures have been those for the "steady state," or the masses, charges, velocities have remained invariant as time goes on, although the physical entities involved have varied from point to point. If, however, one considers Poincaré's "present state of the world and condition immediately after," partial derivatives with respect to time must be involved. Most physical systems do, in fact, change with time, and t (time) will appear in equations and solutions. Also there will be *initial* ($t = 0$) as well as boundary conditions to exert restrictions on the possible answers to particular problems. Thus, in heat flow one may be concerned with the temperature distribution prior to the steady state, or such a condition may never be reached. For example, if z represents temperature, its distribution in a thin insulated rod at *any* time must satisfy

$$\frac{\partial^2 z}{\partial x^2} = k \frac{\partial z}{\partial t}$$

whereas the distribution prior to the steady state in an insulated plate must meet the requirement

$$\frac{\partial^2 z}{\partial x^2} + \frac{\partial^2 z}{\partial y^2} = k \frac{\partial z}{\partial t}$$

In these instances, the particular solutions will have to satisfy not only boundary conditions involving the shape and size of the object and its peripheral temperatures, but also initial conditions like the following: When $t = 0$ the temperature $z = 30°C$ throughout the rod or metallic plate.

The partial differential equation

$$\frac{\partial^2 y}{\partial t^2} = c^2 \frac{\partial^2 y}{\partial x^2}$$

has been related to the phenomenon of a vibrating string, but it is possible to give it a more general interpretation by which it is termed the one-dimensional *wave equation*, where c is equal to the velocity of propagation of the waves. A generalization is the three-dimensional wave equation

$$\frac{\partial^2 u}{\partial t^2} = c^2 \left(\frac{\partial^2 u}{\partial x^2} + \frac{\partial^2 u}{\partial y^2} + \frac{\partial^2 u}{\partial z^2} \right)$$

where c is the speed of propagation of the waves. This equation applies to all types of electromagnetic waves, or to waves in any homogeneous elastic medium. In mathematical physics, the electromagnetic condition in free space is said to be determined by two vectors, an electric force and a magnetic force. That the components of these forces satisfy the wave equation above is a consequence of *Maxwell's laws*, which have been the starting point of all electromagnetic and optical theories from Maxwell's day to the present time. These famous principles are expressed by a set of partial differential equations. Instead of recording the abstract mathematical form of Maxwell's equations, one might give their physical interpretation somewhat as follows:

(1) The electric flux across a closed surface is zero.
(2) The magnetic flux across a closed surface is zero.
(3) A variable magnetic field generates an electric field. (Faraday's law of induction).
(4) A variable electric field generates a magnetic field [Maxwell's hypothesis formulated on the grounds of symmetry with (3)].

In his general theory of relativity Einstein expressed by means of certain celebrated "field equations" the laws to which his four-dimensional space-time continuum was to conform. These were differential equations controlling the possible distributions and changes of the "gravitational field" and were the gravitational analogues of the Maxwell equations we have just discussed. When general relativity and the differential-geometric concepts fundamental to that subject are discussed later in this book, a reaffirmation of the theme of the present chapter will be found in the words of the famous Norwegian mathematician, Marius Sophus Lie (1842–1899), who stated as his belief, "The theory of differential equations is the most important branch of modern (nineteenth-century) mathematics."

But for a proper appreciation of *nondeterministic* physics—for example, quantum mechanics—it will be necessary to have some understanding of probability theory and statistics. These subjects will be discussed next, not only for the sake of application to quantum mechanics, but because they play an essential role in many other modern phases of mathematics.

10

The Elements of Strategy in War and Peace

In tracing the development of arithmetic or geometry or other mathematical subjects, it is customary to point to motivations arising from the practical needs of advancing civilization. But the origins and the evolution of probability theory furnish a counterexample to the traditional evaluation of mathematics as a sort of virtuous handmaiden ever ready to serve worthwhile scientific purposes. The historic fact is that church and state have had to wage an age-long battle against the very games of chance which motivated a probability calculus, encouraged its development, and provided the simplest empirical counterparts for its theoretical results. To indicate that lowly gaming has continued to inspire lofty mathematics up to the present day, one need only reveal that the game of poker played a germinal role in the recent theory of games formulated by John von Neumann (1903–1957). The nature and significance of this theory are described in a statement made by Harold W. Kuhn and A. W. Tucker,* two of von Neumann's friends who are specialists in the field.

Of the many areas of mathematics shaped by his genius, none shows more clearly the influence of John von Neumann than the theory of games. This modern approach to problems of competition and cooperation was given a broad foundation in his superlative 1928 paper, *Zur Theorie der Gesellschaftsspiele* (Theory of Parlor Games). A decade later when the Austrian economist Oskar Morgenstern came to Princeton, von Neumann's interest in the theory was reawakened. The result of active and intensive collaboration was the von Neumann-Morgenstern treatise, *Theory of Games and Economic Behavior* (1944). Together, the paper and treatise contain a remarkably complete outline of the subject as we know it today.

Some of the basic concepts of the von Neumann theory will be discussed in the present work. The history of the subject shows that somewhat earlier (1921) than the 1928 von Neumann research paper, the great French analyst Émile Borel (1871–1956) had also attempted to mathematicize the concept of strategy. He considered only the simplest examples, however, and provided no proof of the basic theorem which is the crux of von Neumann's theory. A third leading

---

* H. W. Kuhn and A. W. Tucker, "John von Neumann's Work in the Theory of Games and Mathematical Economics," *Bulletin of the American Mathematical Society*, Vol. 64, No. 3, Part 2 (May 1958), pp. 100 ff.

mathematician of the recent period must also be brought into the story, namely, Abraham Wald (1902–1950), whose theory of *statistical decision functions* is intimately related to the theory of games but is somewhat broader in its scope and applicability. Professor J. Wolfowitz, a leading mathematical statistician of our day and a collaborator in some of Wald's important discoveries, has stated that Wald's decision theory includes almost all problems which are the *raison d'être* of mathematical statistics.

The overall purposes of Chapters 10–14 are, first, to survey the simplest features of the von Neumann and Wald theories; second, to present the most elementary aspects of the mathematical statistics which preceded Wald's decision theory and was embraced by it; and third, to discuss those probabilistic concepts required for the formulation or solution of problems of strategy, decision, or statistical inference. The treatment of probability will be integrated with the other subjects of discussion, and therefore in subsequent chapters will be related both to games of chance and to the Borel-von Neumann strategic games.

Before proceeding to the mathematics of strategy, however, it would seem fitting to say something about the mathematicians who contributed the content of recent theories. Because the creation of game theory was only a single phase of von Neumann's career, his name is bound to arise in connection with most of the important fields of modern mathematical research. It would be futile to list all the areas in which he made notable discoveries, partly because the list would be so lengthy, but chiefly because the mere names of the subjects do not usually convey meaning to nonspecialists. Perhaps, in addition to game theory, the only other mathematical activity of von Neumann to attract popular attention was that connected with his theory of automatic large-scale computers and machines that are self-repairing, self-correcting, and ultimately will be self-reproducing. His final efforts were made in this direction, and shortly before his death in 1957, he prepared (but was unable to deliver) a lecture comparing the human brain with a "mechanical brain" or computer.

Concerning von Neumann's individual "gray matter," Hans Bethe, a leading nuclear physicist, once said: "I wonder whether a brain like von Neumann's does not indicate a species superior to that of man." Those scientists who admired that brain went further—they worshiped it, and held the view that when the mathematical history of the present era is written, von Neumann's name will be placed first. Even without such forecasts, one must be interested in the biography of a contemporary hero.

John von Neumann's native city was Budapest, where his father was a wealthy banker. The boy's genius manifested itself early, and by the time he entered the *gymnasium* at the age of ten, his mathematical knowledge was so great that instead of attending regular classes he read advanced mathematics under the direction of leading Hungarian mathematicians. At the age of twenty-one he acquired two degrees, one in chemical engineering at Zurich, and the other a Ph.D. in mathematics from the University of Budapest. Three years later he became a *privat dozent* at the University of Berlin, and then, in 1930, accepted a visiting professorship at Princeton University. In 1933 he and Einstein were among the first full professors to be appointed to the newly organized Institute for Advanced Study. Von Neumann thus became the youngest member of its permanent faculty.

During World War II, he was consultant at Los Alamos, and his research speeded up the making of the A-bomb. Later he became chief adviser on nuclear weapons to the United States Air Force. He was also instrumental in the government decision to accelerate the intercontinental ballistic missile program. In October 1954, President Eisenhower appointed him to the Atomic Energy Commission; but after von Neumann had devoted his "giant brain" to this task for a brief six months, X-ray examinations revealed the source of the prolonged and cruel illness that was to terminate his career. Although he knew the diagnosis was cancer, he continued his heavy schedule on the AEC, and even when he became invalided, he maintained communication through a telephone connected directly with the AEC office.

There are, in addition to the facts of his accomplishment, many tales connected with his feats of memory and the rapidity of his thought. In the latter category is a story of how he obtained within a few minutes the answer to a problem that had required two years for solution by the research group known as Project Rand. As for memory feats, he could recite the smallest details in Gibbon's *Decline and Fall* or recount the minutest features of the battles in the American War between the States. It is said that during the last period of his fatal illness, his brother would read to him in German from Goethe's *Faust*, and that each time a page was turned, von Neumann would recite from memory the continuation of the passage on the following page. Dying at fifty-three, he was not able to finish his greatest mathematical "poems" in the leisurely way that the eighty-year-old Goethe had been able to complete *Faust*.

It is surprising to see in how many respects the life of Émile Borel was similar to that of von Neumann, with one strong exception. Borel lived to the ripe old age of eighty-five, dying peacefully just a year before the American game-theoretician succumbed to his tragic illness. Moreover, the age at which Borel *began* his consideration of games of strategy was approximately the same as that at which von Neumann gave his final thought to the subject. Probability theory and statistics are specialties to which Borel devoted his later years. He is more renowned for earlier discoveries which make him one of the founders of the twentieth-century phase of mathematical analysis. Although von Neumann's genius was far more universal, nevertheless as far as probability and analysis are concerned, both men were interested in the same problems. Both combined careers in mathematics with distinguished public service.

Émile Borel was born in 1871 in Saint-Affrique in south-central France. His father, a Protestant minister, was also the principal of the elementary school which Émile attended. The younger Borel's subsequent academic life was to be associated again and again with the great École Normale Supérieure of Paris, first as a student, later as an instructor, and in the decade from 1910 to 1920 as assistant director.

Thereafter Borel's scientific reputation was widespread. In 1921, he was elected to the French Academy of Sciences, and very soon scientific societies outside his native country recognized his merit. He received the Grand Cross of the Legion of Honor and was a member of the council of that order. In 1955, he was the first recipient of the Gold Medal of the French National Center for Scientific Research.

Although Borel was first and foremost a mathematician, he lent his ability to a variety of causes. For example, in 1926, when the Rockefeller Foundation offered to provide half the funds necessary for the creation of a French center of research in mathematics and mathematical physics, it was Borel who set up a suitable plan and who succeeded in getting M. de Rothschild to match the Rockefeller contribution. Within two years, the institute was inaugurated and named after Henri Poincaré, the renowned mathematical leader of the early twentieth century.

Émile Borel also displayed his versatility by participation in military and political activities. His services during World War I earned him the *Croix de Guerre*. In 1924, he started his political career as mayor of his home town. Then he was elected to the Chamber of Deputies and remained a deputy for twelve years. His abiding love for science is shown in his successful campaign for a law requiring industry to support pure research. In 1925 he became Minister of the Navy. Finally, he retired from politics in 1936, and returned to his first love, mathematics. During the German occupation in 1941, he was imprisoned in Fresnes because of his aid to the Resistance movement, for which he was later awarded the Medal of the Resistance. After World War II, his age did not deter him from writing a goodly number of semipopular mathematical monographs; he continued at this task right up to the time of his death in 1956.

More than thirty years earlier, he had written in a treatise on probability: "Probabilistic problems concerning military or economic or financial matters are not without analogy to questions about games. Their solution requires that mathematics be supplemented by *strategy*."

Borel's statement as well as the *date* of publication (1944) and *title* of the von Neumann-Morgenstern treatise on games of strategy indicate why, in the beginning, examples of such games were chosen from military or economic situations. Thus one was asked to picture the case of two planes in combat, or the arrival at a price in an economic exchange relation between buyer and seller. These are both approximated by the situation in a duel.

After World War II, game theorists offered less violent examples of strategic problems. Thus there is the "engagement game," where a girl believes that certain young men in her social circle are potential husbands. We shall imagine that the number of prospects is eleven. Since the girl is not too well acquainted with the young men, any one of them seems as appealing as another. She decides that she will pick one of them at random, attract his interest, and become engaged to him. As part of her plan, she expects to break her engagement if, after a while, she should feel that any one of the remaining ten men might be a more suitable husband. In that case, she will make a random selection from the ten, become engaged to this second man, etc. The question is: At what point should she marry? Should the second, third, fourth, etc. engagement be the final one? If she marries one of the earlier candidates, there may possibly be a more desirable man among those left. But if she delays, she will be getting older, may tire of the game, may have doubts that the few remaining men will be superior, and, in fact, may not find the same candidates left, since they may marry in the interim or depart from the locality. Therefore, she must decide on the best strategic choice for the number of engagements. Her decision will depend on a variety of special personal factors.

The "engagement problem" cannot be taken too seriously as far as *real* young men and women are concerned. If, however, the situation seems too frivolous or too extravagant, let us remind the reader that this very game was played more than three hundred years ago by the great Johannes Kepler who (as he himself recounts) considered in turn, each of eleven potential candidates for his hand (Chapter 8). It is evident that Kepler did not treat romantic situations in scientific fashion, and hence did not analyze facts in advance and then decide what the *optimum* number of engagements should be. Young ladies and gentlemen today will have greater faith in the von Neumann theory if we state that the "engagement problem" and the "duel problem" are both difficult to solve.

A duel would be called a *two-person game* because there are two players. Chess is described as a two-person game, even when the opponents are opposing teams. But the von Neumann theory also deals with three-person, four-person, . . ., n-person games, where $n > 2$, in which there are 3, 4, . . ., n players, respectively. Some of these games are *cooperative*, so called because the players form coalitions before the game starts in order to coordinate the strategies they will use. Solitaire, certain puzzles, Robinson Crusoe on a desert island, are all examples of *one-person games*, the simple random games of probability theory. Since no issues of strategy are involved, one-person games did not require the special scrutiny of Borel and von Neumann.

In most ordinary parlor games that are played for money, wealth is neither created nor destroyed. In technical language, these are described as *zero-sum games*, since at each move the sum of the gains of the players is zero. Matching pennies, for example, is a zero-sum game, because in each move one player wins what the other loses, and hence the sum of the gains is zero. But *non-zero-sum games* are also important in the general theory because they serve as models for economic processes in which the wealth of participants is increased or decreased. In the present work, however, discussion will be limited to zero-sum, two-person games because these are the model for Wald's decision theory and can therefore be applied to situations far more vital than parlor games or gamblers' activities.

The zero-sum, two-person games are called *rectangular* or *matrix* games because an overall picture of the possible outcomes of any particular move can be furnished by listing the potential gains of one player (and corresponding losses of the other) in the form of a rectangular array or *matrix*, to use the usual mathematical term. An example of such a game is one in which each of the two players writes a number on a slip of paper without permitting his opponent to see his choice. The first player is permitted to select 1 or 2 or 3, and the second player may choose 1 or 2 or 3 or 4. After the choices have been made, the two players match the slips of paper and the second player is required to pay the first an amount listed in the following *payoff matrix*. We shall name the players R and C as a mnemonic device for recalling that their choices involve, respectively, the *R*ows and *C*olumns of the matrix.

		C chooses		
	1	2	3	4
1	2	1	4	2
R chooses 2	-3	2	5	0
3	3	-1	-5	-1

From this point on, symbols like R_2 and C_3 will have a double meaning: R_2 will signify that R has written 2 on his slip of paper; C_3 means C has written 3; R_2 will also mean Row 2; C_3 will also mean Column 3. Then the entry in the second row, third column of the matrix, namely 5, indicates that C must pay five units of money to R. If the respective choices are R_3 and C_2, then the entry in the third row, second column, namely -1, indicates that C must pay -1 to R, that is, R must pay one unit of money to C. Thus R sustains a gain or loss according to whether the payoff is positive or negative. Thus the pair of choices (R_2, C_1) would result in a loss of 3 for R.

The rules of the game we have been considering are formulated by a "3 by 4" (symbolized as 3×4) payoff matrix, so called because there are three rows and four columns. In game theory it is customary merely to indicate the matrix without the marginal entries. This permits a variety of interpretations for the same payoff matrix. In the present instance, as a trivial variant, the players could point fingers (1 or 2 or 3 for R and 1 or 2 or 3 or 4 for C) in the fashion used in the Italian game of Morra. In a more subtle interpretation, R and C might be competing organizations where R must choose from three detailed courses of action (symbolized by 1, 2, 3), and C from four (1, 2, 3, 4). But the *matrix* itself is the game from the point of view of the von Neumann theory, where the word "game" means the set of rules, and where the word "play" is substituted for "game" in statements like "We engaged in two games of chess," which becomes "We engaged in two plays of chess."

Next, let us consider the game with the following 3×3 payoff matrix.

$$\begin{pmatrix} 3 & 1 & 2 \\ 6 & 0 & -3 \\ -5 & -1 & 4 \end{pmatrix}$$

The reader may, if he finds it helpful, visualize the game as consisting of two steps. First, there is the selection of one of the numbers 1, 2, 3 by both R and C; next, they match their choices, with the resulting payoff as indicated in the foregoing matrix. Let us suppose for the sake of simplicity that there is to be only a single "play." If the unit of money is worth \$1,000,000, a great deal is at stake in this one play, for an examination of the matrix indicates that the play (R_2, C_1) will require C to pay 6 units or \$6,000,000 to R, while if the play is (R_3, C_1), R will lose \$5,000,000 to C.

The problem then is as follows: Are there optimum "strategies" for R and C, that is, can a row and column be specified to assure least risk and maximum gain under the circumstances? The answer is in the affirmative, as will now be demonstrated. First, consider R's line of reasoning. The possibility of winning the \$6,000,000 indicated in position (R_2, C_1) of the matrix is tempting. If he is to aim for this, his selection must be R_2, the second row. But suppose C does *not* play C_1, but plays C_2 or C_3. Then the entries at (R_2, C_2) and (R_2, C_3) of the matrix, namely, 0 and -3, indicate that R's choice of R_2 may win nothing for him or else may cause him to lose \$3,000,000 to C. The entry -3 is the worst that can happen, the *minimum* payoff for the second row. R feels that if he selects that row, -3 is almost sure to be his lot. For, he reasons, C knows the rules, since the matrix is also available to him. Therefore he will *not* select C_1. Instead, he will surely choose C_3 so that R will have to pay him \$3,000,000.

Now R decides to examine the other rows. Perhaps I can win the \$4,000,000 in (R_3, C_3), he thinks. But, again, if I pick the third row, what's the worst that can happen? He sees that the entry -5 in (R_3, C_1) is the *minimum* for the third row, or that the worst that can happen from choosing this row is that he will have to pay \$5,000,000 to C. R_3 seems an even more risky choice than R_2. Finally, R examines the first row and sees that the worst that can happen is that he will win \$1,000,000. Hence he reasons that it is safest to select R_1 in anticipation of receiving the maximum of the minima, termed the *maximin* of the row payoffs. His rationale is that in playing a competitive game a person always considers his opponent's potential behavior and that C will therefore put himself in R's place, duplicating the complete argument outlined above. As a result, he will realize that R dare not select R_2 or R_3 and must choose R_1. That being the case, C will feel that he himself should select C_2 in order to cut R's winnings to a *minimum*. Then from R's point of view, the solution of the game must be the pair of "strategies," R_1 and C_2, with a *value* of \$1,000,000 to him.

But is this the best solution for C, who stands to lose \$1,000,000 by it? Let us see how the matrix appears to him. He would like to aim at the -5 in (R_3, C_1). But, he thinks, if I pick the first column, R could defeat me either by an accidental selection of the second row or by a choice of this row because he suspects my purpose. Then instead of winning \$5,000,000 I would have to pay him \$6,000,000, the *maximum* payoff for the first column. No, I cannot risk playing C_1 *if* a better course is open to me. Perhaps I should aim for the -3 (gain of \$3,000,000 for C) in position (R_2, C_3) of the matrix. But if I play C_3, R may play R_3. I would then stand to lose the \$4,000,000 corresponding to (R_3, C_3). Finally, what would happen if I were to play C_2? I would be hoping for (R_3, C_2) where I would win \$1,000,000, but the worst that could happen is indicated by the *maximum* payoff in this column, namely, \$1,000,000 at (R_1, C_2). This is the *minimax* of the column payoffs, that is, the minimum of the three column maxima, which are 6, 1, 4 for C_1, C_2, C_3,

respectively. Hence, I shall select the strategy C_2, knowing that R will be able to duplicate my line of reasoning, and accordingly will pick R_1. It appears to me that the
solution of the game must be the pair of strategies (R_1, C_2) where I lose $1,000,000
to R. Therefore, I consider the proposed game unfair!

Since the reasoning of both R and C leads to the same pair of strategies and
to the same entry in the game matrix, the Borel-von Neumann theory would say
that the game in question is *determined* or solvable. In general, a matrix game is
said to have a solution in the form of a pair of *pure* (row-column) strategies if

$$maximum \ (minimum \ of \ a \ row) =$$
$$minimum \ (maximum \ of \ a \ column)$$

For this reason, von Neumann used the term "*minimax* solution." The matrix
entry corresponding to C's minimax and R's maximin is called the *value* of the game
(to R). The reasoning of both players forces the value to be the lowest payoff in its
row and the highest in its column. Hence its position in the matrix is often called a
saddle point by geometric analogy (see Figure 10.1), even though positions in a

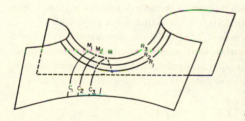

Figure 10.1 The surface in the
diagram is shaped like a *saddle*.
On it the curves R_1, R_2, R_3 are
analogous to matrix rows, and
C_1, C_2, C_3 are analogous to matrix
columns. The points that are "row"
minima and "column" maxima are
indicated. Point *M*, which is both
the greatest of the minima and the
least of the maxima, is called a
saddle point.

matrix lack the continuity of those on a surface.

Let us apply the minimax criterion to the game with the following payoff
matrix.

$$\begin{pmatrix} 2 & -2 & -3 \\ 1 & 0 & 2 \\ -1 & -1 & 3 \end{pmatrix}$$

Here R must examine the row minima, which are $-3, 0, -1$, and find

$$\text{max (min of row)} = 0$$

C must consider the column maxima, which are $2, 0, 3$, and find

$$\text{min (max of column)} = 0$$

Since R's *maximin* $= C$'s *minimax*, the game has a solution, namely, (R_2, C_2). The
value of the game is *zero*, and hence it is *fair* because neither player will gain at
the other's expense if the game is played in minimax fashion.

The reader can verify that the game with payoff matrix

$$\begin{pmatrix} -4 & 3 & 0 & 15 \\ 5 & 6 & 4 & 5 \\ 8 & 0 & 2 & 6 \end{pmatrix}$$

has value 4 (to R) with optimal strategies R_2 and C_3, and that the game whose payoff matrix is

$$\begin{pmatrix} 3 & 1 \\ 4 & 1 \end{pmatrix}$$

has two solutions, both with value 1, namely, (R_1, C_2) and (R_2, C_2). Evidently there would be something inconsistent about several optimal solutions if they led to different values; and it can be proved, in fact, that where more than one solution exists, the value is the same for all solutions.

In some of the matrix games, a preliminary consideration of the matrix may eliminate certain choices immediately and thus simplify the process of locating a saddle point. In the 3 × 4 rectangular matrix above, for example, the fourth column can be eliminated as a choice for C. By comparing its entries with those in the third column, C would observe that each number in the fourth column is greater than the corresponding one in the third. Hence, no matter what row R chooses, C would be worse off by playing C_4 than C_3. Therefore his optimum strategy should be obtainable by removing the fourth column and considering only the first three. Also, since R can readily figure out what his opponent's line of reasoning must be, *both* players will consider the first three columns only, that is, the 3 × 3 matrix

$$\begin{pmatrix} -4 & 3 & 0 \\ 5 & 6 & 4 \\ 8 & 0 & 2 \end{pmatrix}$$

Now from R's point of view, the first row of this matrix should never be selected as a strategy, since the payoffs in the second row would always be greater, no matter what column C chooses. Hence R's strategies should be chosen from

$$\begin{pmatrix} 5 & 6 & 4 \\ 8 & 0 & 2 \end{pmatrix}$$

and C will figure out that R must limit himself to the above 2 × 3 matrix. In that matrix the first column would be eliminated as a strategy when C compares it with the third, and only

$$\begin{pmatrix} 6 & 4 \\ 0 & 2 \end{pmatrix}$$

would remain. Here

$$\text{max (min of row)} = \text{min (max of column)} = 4$$

which is the value of the original game to R. To find the optimum strategies one must locate the entry 4 in the original matrix, and so arrive at the solution (R_2, C_3). (Observe that one could carry the process of elimination still further by eliminating the second row of the 2 × 2 matrix.)

The reader can apply the procedure followed above and thus remove strategies that would be inadmissible for R or C in the following 4×3 matrix game:

$$\begin{pmatrix} 0 & 3 & 12 \\ 2 & 3 & -7 \\ 8 & 2 & 5 \\ 6 & 3 & 8 \end{pmatrix}$$

The solution arrived at should be (R_4, C_2) with value 3 to R.

Let us next see whether the minimax criterion will provide a solution for the familiar game of matching pennies. In this game, each player places a penny in his hand with heads or tails up, without letting his opponent see his choice. Next, the players match the coins. If both are heads, or both are tails, C gives his penny to R. If the coins are different, R gives his penny to C. Hence the payoff matrix is

$$\begin{pmatrix} 1 & -1 \\ -1 & 1 \end{pmatrix}$$

If R examines this for row minima, he obtains -1, -1, and hence

$$\text{max (min of row)} = -1$$

If C examines this for column maxima, he obtains 1, 1, and

$$\text{min (max of column)} = 1$$

Since the maximin and minimax are not equal, there are no optimal pure strategies for a single play. In other words, the game has no solution of the type we have considered thus far.

But almost every one has engaged in matching pennies and tried to choose heads and tails in about equal numbers in *repeated plays*, being careful not to follow any regular pattern which can be observed by his opponent. The very symmetry of the payoff matrix would suggest that this is a good strategy. The strategies for R and C are the same, and can be achieved practically by having each player toss his penny at every play with the result that, in the long run, the relative frequency for each player's choice of heads (or tails) will be $\frac{1}{2}$.

The solution in the game of matching pennies is typical, for in those rectangular games where a saddle point is lacking, the von Neumann solution always consists of varying choices in random fashion from play to play in a way that will produce the optimal long-run results. The player is then said to employ a *randomized* or *mixed* strategy in contrast with the *pure* strategy where the same row or column must be selected no matter how many times the play is repeated. In general, the problem of solving a matrix game consists in determining the optimal relative frequencies for the selection of different rows and columns. In the case of matching pennies, the choice is $(\frac{1}{2}, \frac{1}{2})$ for R and likewise for C. Pure strategies can, in fact, be classified as special cases of mixed strategies. Thus, instead of saying that the matrix game originally considered has as solution the pure strategies R_1 and C_2, one can say that its solution is the pair of *randomized* or *mixed* strategies, $(1, 0, 0)$ for R and $(0, 1, 0, 0)$ for C.

Will intuition provide a solution to the game with the following payoff matrix?

$$\begin{pmatrix} 2 & 1 \\ 0 & 3 \end{pmatrix}$$

Since

$$\max \ (\min \ of \ row) = 1$$

and

$$\min \ (\max \ of \ column) = 2$$

there is no saddle point and one cannot pick an optimal pair of pure strategies. Hence, if there is a best policy, it must involve randomized strategies. From R's point of view it might seem desirable to select the first row, because he would be sure to win something no matter what strategy C chooses. But there are good arguments for selecting the second row in some plays, because there is always the chance of winning 3. Moreover, R should not risk the pure strategy of always playing the first row, for then C would become aware of what is happening, and after a few plays would always select the second column, and in successive plays R's winnings would always be limited to 1. Therefore it would be sensible for R to vary his choices in different plays but in accordance with a relative frequency favoring the selection of the first row. Now C's reasoning would lead him, like R, to use a mixed strategy which varies the choice of C_1 or C_2 in random fashion from play to play.

In the next chapter it will be *proved* that the *optimum* mixed strategies for R and C are $(\frac{3}{4}, \frac{1}{4})$ and $(\frac{1}{2}, \frac{1}{2})$, respectively, which means that the random mixing of choices should be such that R chooses the first row three times as often as the second in the long run, and C selects both columns equally frequently. To achieve randomness in successive plays, C might toss a coin and R might draw a marble from a sack containing thirty blue and ten red marbles, selecting the first row whenever a blue marble is drawn, the second row whenever the marble is red. (The marble must be replaced and the sack thoroughly shaken after each drawing.)

Although our present interest concerns games of strategy, the above considerations involve concepts associated with games of chance. The relative frequencies which describe a mixed strategy correspond in fact to mathematical *probabilities*. If we have had no previous experience with the random devices in the above examples, we may reason *a priori* that when a well-balanced coin is tossed, the chances for "heads" are 1 in 2, that the likelihood of drawing a blue marble from the sack described above is 3/4. The expectation is that the ratios in question will be achieved approximately in very numerous plays.

The intuitive meaning of the term "probability" derives from just such situations. In general, it is assumed that if a random experiment (like those involved in achieving mixed strategies) is repeated many times under essentially stable conditions, the relative frequency of a particular result will ultimately fluctuate only slightly, that is, will tend to a *limit* as the number of repetitions of the experiment is made greater and greater. This limiting value of the relative frequency is considered the "true" probability that the event in question will occur. (This is not a limit in the sense of the calculus, because the relative frequency is not a definite, determined function of the number of repetitions.)

The notion of a probability as the limit of a relative frequency is at the basis of the *empirical* or *statistical* approximations so common in life insurance and other areas of vital statistics. For example, a department store can afford to double a layette gratis on the advent of twins, since the empirical probability of double births is slight. Again, when premiums are to be computed, mortality table ratios based on 100,000 cases may be considered a sufficiently good approximation to the "true" probabilities involved.

Thus for a proper understanding and handling of randomized strategies, we must make a short detour into probability theory. We shall require, in particular, one special concept from that theory if we are to formulate the Borel-von Neumann-Wald criterion for solving a rectangular game whose matrix lacks a saddle point.

11

Probabilistic Models, Great Expectations, and Randomized Strategies

To fulfill the promise made at the close of the previous chapter, we shall now present briefly certain aspects of the theory of probability prior to resuming our discussion of games of strategy. A practical issue associated with such games was the motivation for formulating a concept of probability that is useful to the applied mathematician. But the pure mathematician sees in probability theory a methodology for the construction of "mathematical models" for a certain type of physical experiment.

A *mathematical model* (page 47) is an abstract idealization of various features of a real situation in the same sense that pure Euclidean plane geometry is the abstract counterpart of the surveyor's or the engineer's conception of physical points, lines, polygons, circles, etc. and their properties. In every model the totality of things under discussion is called the *universal set* or *universe of discourse* (Chapter 6). In a geometric model the universe of discourse may be a line or a plane or a "space," but, in any case, it is a *set of points*. In a probabilistic model the universe of discourse is also conceived to be a set of "points," where a "point" corresponds to an outcome of some "random" experiment.

Tossing a coin, rolling a die, playing bridge are all examples of what is meant by a *random experiment*. A telephone interview in connection with the rating of television programs is another example of a random experiment. Although such an experiment can be repeated many times under essentially the same conditions, the outcome will vary and no specific result can be predicted with certainty. Nevertheless, the set of possible outcomes is known in advance. Thus one can say that these are {heads, tails} for the tossing of a coin and {1, 2, 3, 4, 5, 6} for the rolling of a die. If a telephone interview is an inquiry as to which TV channel is being observed at the time, the possible responses in a certain part of the United States are {II, IV, V, VII, IX, XI, XIII}. Such outcome sets are, of course, convenient idealizations, since, for example, a coin might land on its side, a die might bounce out of the window, and the answer to the telephone query might be, "I have no TV set."

Varied interests and purposes may lead to essentially different sets of elements for the description of the potential outcomes of one and the same random experiment. Thus, although we have used an aggregate of *six* numbers to represent the possible results of tossing a die, a gambler betting on "ace" might claim that, as far as he is concerned, the set of *two* elements {ace, not-ace} suffices. He might point out that no matter how the die falls, the outcome can be placed in one of his two categories. Again, in a situation where bets are being placed on {less than 4, equal to 4, greater than 4}, the collection of *three* categories might be considered an adequate description of all eventualities. The reader will be able to provide still other aggregates corresponding to the same totality of outcomes. One might, of course, insist on the original set of six elements by arguing that it is fundamental, since no outcome listed can be decomposed further, whereas in the other descriptions, categories like "not-ace," "less than 4," "greater than 4" are really composite and can be broken down into finer classifications.

Each aggregate describing all possible outcomes of tossing a die would be called a *sample space* of the die-tossing experiment. The terminology derives from mathematical statistics, a subject in which a random experiment is often a sampling procedure and hence an experimental outcome is a "sample." The set of possible samples is thus the sample set, or sample space. In general, a *sample space* of a random experiment is a set of elements such that any outcome of the experiment is represented by one, and only one, element of the set. Each of these elements is termed a *point* of the sample space.

In constructing a mathematical model of some random experiment, the first step is specification of the particular sample space of interest, which then becomes the *universe of discourse* or *universal set* for the model. Next, assumptions related to this abstract set of points can be formulated and deductions made. The model will be practically useful if its postulates, definitions, and theorems approximate physical realities. Since our objective is to build a *probabilistic* mathematical model, we shall begin by assigning "probabilities" on the basis of intuition or observation, with the idea of idealizing "common sense" in the axioms and definitions laid down for the abstract probabilities of the model.

Suppose, then, that the random experiment consists of tossing a single perfectly symmetric die, and that the universe of discourse is the sample space $S = \{1, 2, 3, 4, 5, 6\}$. Although a gambler does not usually define what he means by the word "probability," he might say that the probability of "ace" is 1/6, that the probability of "deuce" is 1/6, likewise for "trey," etc. If he has had no previous experience in tossing the particular die involved, it seems natural to assume that one outcome is as likely as another, that is, to assign the same "weight" to each element or point of the universe of discourse. To find the probability that the die will turn up either "5" or "6" a gambler might compute the proportion of "favorable" outcomes in the universe. Since there are six outcomes in the sample space S, only two of which favor the gambler's bet, the probability in question is 2/6. In the same way, the probability that the outcome will be less than "5" would be figured as 4/6, because there are four "favorable" outcomes, namely, $\{1, 2, 3, 4\}$. Since $2/6 = 1/6 + 1/6$, and $4/6 = 1/6 + 1/6 + 1/6 + 1/6$, one might say that the probability for the occurrence of some event is a number obtained by adding the "weights" of all the "favorable" outcomes.

We remark in passing that if our hypothetical gambler has not defined the word "probability," he has nevertheless indicated how he thinks his chances of winning should be measured. But if he holds with the statistician and conceives of a probability as the limit of a relative frequency in a long sequence of trials (page 257), he must expect that about one-third of such trials with a die would lead to "5" or "6," and about two-thirds to an outcome less than "5."

The experimental results or events to which reference has just been made can be defined by {5, 6} and {1, 2, 3, 4}, respectively, that is, by collections of outcomes in the sample space of interest. In an abstract probability model, an *event* is defined as a set of points, one which is a *subset* of the universe of discourse (the sample space selected). An event can therefore be specified by listing the elements (outcomes) which belong to the associated subset when these are finite in number or, alternatively (and in the infinite case), by stating some property which characterizes the component outcomes.

In the die-tossing experiment it was assumed that outcomes in a certain space were equally likely and hence should be given equal weight in computing probabilities for various events. In figuring probabilities in this way, we were in fact applying the *classic* point of view of Jacques Bernoulli and Pierre-Simon Laplace. Their definition of probability stated: In the case of a random experiment with a *finite* number of possible outcomes which, by agreement, are considered *equally likely*, the probability, $P(E)$, that some event E will occur is the ratio

$$P(E) = \frac{m}{n}$$

where n is the total number of possible outcomes, and m is the number of outcomes in the subset defining the event E.

Let us apply the Bernoulli-Laplace definition once more to the same die-tossing experiment and universe of discourse considered above. If we ask, "What is the probability that the outcome will be an odd number?" the event in question can be defined by the subset {1, 3, 5}, and the probability of this event is 3/6. Similarly, the event "greater than 1" signifies any outcome or element in the subset {2, 3, 4, 5, 6}, and the required probability is 5/6.

Events that may seem special are "greater than 5," "a positive integer less than 7," "the number 17." The subset for the first of these is {6}, a singleton set or elementary event, which has probability 1/6. Since "a positive integer less than 7" means any integer in the collection {1, 2, 3, 4, 5, 6}, the "subset" in this case is the entire universe of discourse. The whole set is called an "improper" subset of itself, while other subsets are called "proper." At any rate, the probability of a positive integer less than 7 is 6/6 = 1. When a die is tossed, it is *certain* that the result will be a positive integer less than 7. Here we have an illustration of the fact that if an event is certain to occur, its probability is 1. In the same way, if one asked for the probability that the die will not turn up 75, the favorable subset would once again be the entire sample space, and the probability would be 1.

For the probability that the die will turn up 17, common sense gives *zero* as the answer. Here one would say there are no favorable outcomes, and hence the re-required ratio is $0/6 = 0$. But if every event, without exception, is to define a subset of the fundamental set, we shall require a subset for the event in question, or for any *impossible* event. Mathematics has available for this purpose the *null* or *empty* set, symbolized by $\emptyset$ (Chapters 1 and 6). The null set is a subset of every aggregate and hence is useful in connection with many other issues. At this point, $\emptyset$ is the subset corresponding to the event, "die will turn up 17," and since there are zero outcomes in $\emptyset$, the probability is figured to be $0/6 = 0$. Thus an impossible event has probability zero.

Even in those simplest cases we have considered, various laws of probability theory are illustrated—namely, that probabilities are real numbers between 0 and 1 inclusive, the former figure being assigned to impossibility, the latter to certainty, that is, to the empty set and the entire universe, respectively. Again, the probability that an event will occur is the sum of the weights of the outcomes (elements) in the subset which defines this event.

Another fundamental law will be illustrated if we ask, in relation to the die-tossing experiment, "What is the probability of either 'ace' or an even number?" The event in question is the subset $E = \{1, 2, 4, 6\}$, and hence its probability is $4/6$. But someone may point out that the verbal description of the event indicates that it is actually composed of *two* other events, namely, "ace" and "even," which are *mutually exclusive*, since they cannot both occur. Or, in terms of sets of outcomes, $E = \{1, 2, 4, 6\}$ is the fusion or *union* of the two *disjoint* (nonoverlapping) sets, $E_1 = \{1\}$ and $E_2 = \{2, 4, 6\}$. Symbolically, $E = E_1 \cup E_2$, where $E_1 \cap E_2 = \emptyset$. Thus our first interpretation called for $P(E)$, and the new point of view demands $P(E_1$ or $E_2)$, that is, $P(E_1 \cup E_2)$. The answer must be the same whether or not E is subdivided into components. Therefore, $P(E_1$ or $E_2)$ must also be equal to $4/6$. Now

$$P(\text{ace}) = P(E_1) = \frac{1}{6}$$

and

$$P(\text{even}) = P(E_2) = \frac{3}{6}$$

Therefore

$$P(E_1 \text{ or } E_2) = P(E_1) + P(E_2)$$

which is a general law of probability for a pair of mutually exclusive events.

The law illustrated is readily extended to the case of 3, 4, ..., n, any finite number, of disjoint or mutually exclusive events, so that if $E_1, E_2, E_3, \ldots, E_n$ are disjoint, that is, if $E_1 \cap E_2 = \emptyset$, $E_1 \cap E_3 = \emptyset$, $E_2 \cap E_3 = \emptyset$, etc.

$$P(E_1 \text{ or } E_2 \text{ or } E_3 \text{ or } \ldots \text{ or } E_n) = P(E_1) + P(E_2) + P(E_3) + \ldots + P(E_n)$$

In all the examples thus far, the Bernoulli-Laplace definition of probability, with its assumption of equal likelihood, has been applied because the classic criterion seems natural for assigning probabilities a priori (before any experimentation is carried out) in games of chance and in the case of random experiments resembling such games. But whenever the classic definition is used, it is essential to verify that equal likelihood is a realistic assumption. Thus, let us ask, "If a fair coin is tossed twice, what is the probability that it will turn up 'heads' exactly once in the two throws?" To assist us we can select as universe of discourse the sample space $S = \{HH, HT, TH, TT\}$, where TH, for example, symbolizes "tails" in the first throw, "heads" in the second. The subset describing "heads exactly once" is $\{HT, TH\}$. Therefore $m = 2$, $n = 4$, and the required probability is 2/4 or 1/2.

Now it is not unusual for a student to offer a different analysis. He may say that the question concerns the number of "heads," and therefore he selects as his universe of discourse the sample space $U = \{0, 1, 2\}$. Then he assumes equal likelihood for each of the three outcomes with the result that the probability of $\{1\}$ or "exactly one head" is figured as 1/3. We tell him that his assumption is not realistic because his $\{1\}$ actually includes the *two* outcomes we labeled as HT and TH in our own analysis, whereas his $\{0\}$ and $\{2\}$ are not further decomposable. If the student insists on using $U = \{0, 1, 2\}$ as the set of possible outcomes, he must not assign equal weights. Instead, he must give twice as much weight to "heads exactly once" as to either of the other outcomes. Then he must choose the weights $\{1/4, 1/2, 1/4\}$ for the respective outcomes in $U = \{0, 1, 2\}$. Observe that whatever weights are assigned to outcomes in the universal set, the sum of such weights must be equal to 1, since that set represents an event that is *certain* and hence must have probability 1.

In the above instance it was easy to justify equal likelihood for elements of one sample space, and to demonstrate that outcomes in another space were *not* equally likely. But in the case of intricate games or complicated random experiments, a formidable logical argument may be necessary. For this reason, equal likelihood is a weakness in the classic definition. The practical difficulty of reasoning about equally likely cases is not, however, the only defect in the Bernoulli-Laplace definition. Let us return, for example, to the same die-tossing experiment and the same sample space we have considered repeatedly. If we are informed that the die is loaded, how should probabilities be assigned? Thus, what would be the probability of "ace," that is, of $\{1\}$? Obviously, one can no longer apply the classic definition. It is impossible, in fact, to assign a realistic probability to $\{1\}$ or to any event related to tossing a biased die if the assignment must be made a priori. But if one accepts the point of view that a probability is the limit of an empirical relative frequency in many trials, one can estimate the probability of $\{1\}$ a posteriori. Thus, if "ace" should occur 29 times in 100 tosses, we might take 0.29 as a crude approximation of the ideal or "true" probability, and use this estimate for future tosses.

Again, there is the case mentioned earlier where the random experiment consists of a telephone inquiry and the sample space is $S = \{II, IV, V, VII, IX, XI, XIII\}$, the Roman numerals indicating TV channels. It would not be sensible to assign a weight of 1/7 to each response, for if viewers were as likely to be observing one TV program as another offered at the same time, what would be the purpose of TV rating schemes? Granted that the seven weights should not all be equal, it is still not possible by logical argument alone to determine the weights a priori. But if one has available considerable data from previous polls, and these show a relative frequency of 0.2 for the response "Channel II," a relative frequency of 0.15 for "Channel V," etc., such empirical results can guide the assignment of weights. Suppose, then, that the weights based on past experience are {0.2, 0.2, 0.15, 0.15, 0.1, 0.1, 0.1}. Now one might wish to predict the likelihood of certain outcomes when a new poll is taken, for example, to ask, "What is the probability that the response will be IV or V?" Then $P(E)$ where $E = \{IV, V\}$ is obtained by adding the weights for IV and V. Hence $P(E) = 0.2 + 0.15 = 0.35$. Similarly, if $E = \{II, VII, XIII\}$, $P(E) = 0.2 + 0.15 + 0.1 = 0.45$. The probability is zero that the response will be "I am watching Channel XXV" or "I am watching all the channels (on my one TV set) at this very instant." These two events are equivalent since both responses define $\emptyset$, the null set.

When modern probability theory is treated in formal axiomatic fashion as a mathematical science based on a system of postulates, some of the "laws" to which we have alluded can be taken as postulates (assumptions) and others can be deduced as theorems. Thus one might conceive of a probability model as having three components, namely, (1) S, the sample space selected as universe of discourse; (2) an aggregate of events, that is, of subsets of S; (3) the assignment of a real number to each event E in (2). This real number, $P(E)$, called the probability of the corresponding event, must satisfy the following axioms:

(1) $0 \leq P(E) \leq 1$. (A probability must be a positive number between 0 and 1 inclusive.)

(2) $P(S) = 1$ (If an event is *certain*, that is, describable by S, then its probability is 1.)

(3) If E_1 and E_2 are mutually exclusive events, that is, disjoint subsets,

$$P(E_1 \text{ or } E_2) = P(E_1 \cup E_2) = P(E_1) + P(E_2)$$

From these postulates one can deduce the other probabilistic laws we have mentioned. For example, to prove that an *impossible* event, that is, one definable by $\emptyset$, must have probability 0, we observe first that, obviously,

$$\emptyset \cup S = S$$

Therefore

$$P(\emptyset \cup S) = P(S) = 1, \quad \text{by axiom 2}$$

But, by axiom 3,

$$P(\emptyset \cup S) = P(\emptyset) + P(S) = P(\emptyset) + 1$$

Since the two results obtained for $P(\emptyset \cup S)$ must be equal,

$$P(\emptyset) + 1 = 1$$

and hence

$$P(\emptyset) = 0$$

It will be observed that in modern probability theory there is no postulate prescribing a specific method for assigning probabilities in a particular model. This fact makes the pure theory very general, because it permits great freedom of choice for the probabilities related to conceptual random experiments. The modern theory does *not* postulate equal likelihood, but it also does *not* forbid its use in appropriate situations. Then, if the sample space selected as universe is a *finite set*, a convenient method for specifying probabilities is as follows. First, assign positive real weights (equal *or* unequal) to the elements (points) in the universal set so that the sum of these weights is 1. Next, define the probability of any event (subset) as the sum of the weights of the elements (points) that make up this event (subset).

We digress to indicate that the above technique for assigning probabilities can be said to involve *two* functions, a *point function* for the specification of weights and a *set function* for the assignment of probabilities. As an illustration, suppose that the universe of discourse is a sample space containing *three* elements (points). Then the tabulation below represents one possible function for the assignment of weights. Its *domain* is made up of all *points* of the *universe* (hence the name "*point* function"), and its *range* is the collection of *weights*.

Point or Element	e_1	e_2	e_3
Weight	0.1	0.6	0.3

The space $S = \{e_1, e_2, e_3\}$ contains eight subsets, defining eight essentially different events. These subsets are named in the first row of the tabulation below, and the second row lists the corresponding probabilities, computed by adding weights of component elements. The table represents a function whose *domain* is a collection of *sets* (hence the term "*set* function") and whose range is an aggregate of positive real numbers between 0 and 1 inclusive.

E	$\emptyset$	S	$\{e_1\}$	$\{e_2\}$	$\{e_3\}$	$\{e_1,e_2\}$	$\{e_1,e_3\}$	$\{e_2,e_3\}$
$P(E)$	0	1	0.1	0.6	0.3	0.7	0.4	0.9

For the sake of simplicity we have restricted ourselves thus far to a consideration of *finite* sample spaces. The classic definition of probability is applicable only to such spaces, but if *unequal* positive weights are permissible, they can be applied to a sample space containing an *infinite sequence* of points. To obtain an example of such a space, let us suppose that the random experiment consists of selecting any positive integer at random, or observing the record of a Geiger counter which is counting cosmic rays. In either case, an associated sample space is the sequence of natural numbers, that is, the set $S = \{1, 2, 3, \ldots\}$.

There are, however, sample spaces corresponding to real experiments in which it is impossible to use the technique of assigning weights to individual points. For example, suppose that we are told that the length of an object lies between 2 and 3 cm., and we are asked to guess the "true length" without examining the object. A sample space for such a guess is the infinite set of all real numbers between 2 and 3. This space can be represented on a Cartesian graph (Figure 11.1)

Figure 11.1   Sample space: all points of the interval (2, 3)

as the line segment between the points 2 and 3 of the *X*-axis. The geometric picture will be helpful when ultimately (Chapter 14) probabilities are assigned to events in such a sample space.

In accordance with probabilistic laws, impossibility, that is, *Ø*, and certainty, *S*, *always* have probabilities of 0 and 1, respectively. In Chapter 14 we shall see that the converse statements are *false* when the universal set is a sample space like the one we have just illustrated (a *continuum*). Thus in the case of a random choice of any real number between 2 and 3, theory gives *zero* as the probability that 2.40173, say, will be selected, although it is conceivable that just that choice might be made. In this instance, zero probability does *not* imply impossibility, *Ø*.

Again, *any* subset of a *finite* sample space defines an *event*. But for certain sample spaces containing an infinite number of outcomes, it may be necessary to exclude some subsets from consideration as events because one cannot assign probabilities to them in consistent fashion. The mathematician is forced to limit himself to "measurable" subsets. That the "measure theory" suitable for infinite sample spaces is due in considerable part to Borel will not surprise readers. What will seem strange, however, is that Borel created the appropriate analytic concepts in his youth, many years before he decided to write on probability.

After the above detour into advanced probability theory, we shall now revert to elementary instances in order to obtain a simple formulation for a special concept due to Pascal, who was an important figure in the history of probability. In a fair lottery, it is reasonable to assume that one number is as likely to be drawn as another. Therefore, if a thousand tickets are sold and you hold just one, the probability that you will obtain the grand prize is 1/1000. If the grand prize is $500, Pascal would have said that your mathematical *expectation* is 1/1000 of $500, that is, $0.50. In a sense, that is the true value of your ticket, for it represents the amount you would get if the $500 were divided equally among the thousand ticket-holders. If you held six hundred tickets in the lottery, your expectation would be six hundred times as much as in the case of a single ticket, that is,

$$\text{expectation} = (600)\,(1/1000)\,(\$500)$$

$$= (600/1000)\,(\$500)$$

$$= (\text{probability of winning prize})(\text{amount of prize})$$

In general, expectation is *defined* in this way; in other words,

$$\text{expectation} = pA$$

where *A* is the prize or payoff and *p* is the probability of winning the payoff.

In the illustration above, it was assumed that you held 600 out of 1000 lottery tickets and therefore your probability of winning the grand prize was 3/5. That probability was figured on the basis of the classic definition. If we were to reinterpret it as a *statistical* probability, it would signify a relative frequency approximated in many repetitions of the same experiment. Let us then imagine 5000 successive lotteries with a winning ticket in 3000 of the lotteries, and a losing ticket in the remaining 2000. The relative frequency of success would then be 3/5. If the grand prize is $10, the *expectation* would be $6, in accordance with the above formula. Now the *mean* or "average" payoff in the 5000 games would be

$$\frac{\overbrace{0 + 0 + \ldots\ldots + 0}^{2000\ times} \quad + \quad \overbrace{10 + 10 + \ldots\ldots + 10}^{3000\ times}}{5000}$$

that is,

$$\frac{(2000)\ (0) + (3000)\ (10)}{5000}$$

which is equal to

$$(2/5)(0) + (3/5)(10) = 6$$

and this is the value of the expectation obtained by using the formula. In other words, expectation can be interpreted as the *mean or average* payoff in a very long series of trials.

How can you compute your expectation in a more complicated gambling situation? Suppose that you are betting on the outcome of tossing a die, and if it turns up "ace," your opponent must pay you $10; if it turns up 2 or 3 or 4, you must pay him $2; otherwise there is no exchange of money. Here the probabilities of winning $10, –$2, $0 are 1/6, 3/6, 2/6, respectively. If we consider these as relative frequencies in a long series of tosses of a die, the following tabulation would approximate the results of six thousand tosses.

Payoff	Frequency
10	1000
−2	3000
0	2000

Now interpreting the expectation as the mean or average in these very numerous trials, we have

$$\text{expectation} = \frac{(1000)\ (10) + (3000)\ (-2) + (2000)\ (0)}{6000}$$

$$= (1/6)\ (10) + (3/6)\ (-2) + (2/6)\ (0) = 2/3$$

The expected or average winning is $2/3 per play. Now observe that each of the three terms in the sum above is itself an expectation. Thus (1/6) (10) is the expectation from a lottery with prize $10 where your chances of winning are 1 in 6. Or it is your expectation if you bet that the die will turn up "ace." Similarly, the second term in the sum above, $(3/6)(-2)$, is your expectation if you bet that the die will turn up 2 or 3 or 4, etc. Such a breakdown of the total expectation into component expectations is typical. In general, if a sample space of a random experiment contains n possible outcomes, occurring with probabilities $p_1, p_2, \ldots, p_n$, and associated with payoffs $a_1, a_2, \ldots, a_n$, then $p_1 a_1$ is the expectation based on the first outcome, $p_2 a_2$ the expectation based on the second, etc. The total expectation is then defined as

$$\text{expectation} = p_1 a_1 + p_2 a_2 + \ldots + p_n a_n$$

This formula for the expectation is basic in the theory of strategic games as well as in the Wald decision theory, and will therefore be applied repeatedly. At this point it will enable us to present a criterion for the selection of optimum randomized strategies in matrix games.

If the matrix of a rectangular game has a saddle point, a solution in terms of pure strategies exists; that is, there is a best choice of row and column for a single play, and the value of the game to R is the corresponding matrix entry. The solution consists of *randomized* strategies, however, when the matrix lacks a saddle point. In that case, R and C must decide on the optimum relative frequencies for the variation in choice of rows and columns from play to play.

When the choices are varied or "mixed" in repeated plays, it would seem sensible for R and C to consider *average* payoffs. Their objectives would, of course, be diametrically opposed. R would try to maximize the average, and C would aim to make it as small as possible. Therefore, the mean payoff in a long series of plays, that is, R's *expectation*, is said to be the *value* of the game to him.

In the previous chapter, it was stated (without proof) that the matrix game

$$\begin{pmatrix} 2 & 1 \\ 0 & 3 \end{pmatrix}$$

has as solution the pair of mixed strategies $(3/4, 1/4)$ and $(1/2, 1/2)$ for R and C, respectively. We are now ready to indicate why that choice of relative frequencies is optimum. If R selects the first row in 3/4 of all plays and C chooses the first column 1/2 the time, the payoff listed in (R_1, C_1) of the matrix will accrue to R in 3/4 of 1/2, or 3/8, of all plays. Similarly, the payoff in (R_2, C_1) will result in 1/4 of 1/2, or 1/8, of all plays, and the relative frequency or probability matrix corresponding to the given mixed strategies is

$$\begin{pmatrix} 3/8 & 3/8 \\ 1/8 & 1/8 \end{pmatrix}$$

Then R's expectation is $(3/8)(2)$ from those plays in which (R_1, C_1) is the choice, since 2 is the gain for the corresponding position of the payoff matrix. Likewise, his expectation is $(3/8)(1)$ from the plays in which (R_1, C_2) is the selection, etc. By what has been stated earlier, his overall expectation or average winning in the long run, that is, the value v of the game is

$$v = (3/8)(2) + (3/8)(1) + (1/8)(0) + (1/8)(3) = 3/2$$

In order to show that the mixed strategies above are truly optimum, one must consider what would happen if the relative frequencies were selected differently. If x is the long-run relative frequency (probability) for the choice of R_1, and y is that for C_1, the probability matrix for the mixed strategies is

$$\begin{pmatrix} xy & x(1-y) \\ y(1-x) & (1-x)(1-y) \end{pmatrix}$$

and the overall expectation or long-run average gain is a function of x and y which will be symbolized by $E(x, y)$. Then

$$E(x, y) = 2xy + x(1-y) + 0y(1-x) + 3(1-x)(1-y)$$

or

$$E(x, y) = 4xy - 2x - 3y + 3$$

By algebraic manipulation one can obtain the equivalent form

$$E(x, y) = 4(x - 3/4)(y - 1/2) + 3/2$$

(The reader can check this by carrying out the multiplication in the second expression.)

If $x = 3/4$ is substituted in the alternative form, the result is

$$E(3/4, y) = 4(0)(y - 1/2) + 3/2 = 3/2$$

which indicates that R's expectation from the mixed strategy (3/4, 1/4) is 3/2 *no matter what strategy* $(y, 1 - y)$ is used by C. Likewise, if C plays the mixed strategy (1/2, 1/2), his expected loss to R is 3/2, *no matter what strategy* R uses, that is,

$$E(x, 1/2) = 4(x - 3/4)(0) + 3/2 = 3/2$$

What must now be shown is that R's average winning would be *less* and C's average loss would be *greater* from a choice of strategies other than those prescribed. To see this, one should reexamine

$$E(x, y) = 4(x - 3/4)(y - 1/2) + 3/2$$

and consider the effect of substituting for x and y relative frequencies other than 3/4 and 1/2. If R chooses x greater than 3/4 with the thought of possibly *adding* something to the 3/2 of the second term, C can (if he observes that R is choosing the first row very often) counter by selecting y less than 1/2. Then the first term above would be *negative*, and therefore the new result would be an average winning less than 3/2 for R. (For any choice where $x > 3/4$, C's selection of $y = 0$, that is, the pure strategy (0, 1), would make R's expectation the *minimum* possible for his strategy.)

Suppose, instead, that R chooses a value of x less than 3/4. Either by chance or because he foresees R's plan, C can counter by playing y greater than 1/2, thus making the first term negative and diminishing the expectation to a value below 3/2. (For any choice in which $x < 3/4$, C's selection of $y = 1$, that is, the pure strategy (1, 0), would make R's expectation the minimum for his strategy.)

Since R runs the risk of an average winning of *less* than 3/2 in any strategy where $x > 3/4$ or $x < 3/4$, his best choice is $x = 3/4$, which assures him an expectation of 3/2, no matter what C's strategy is. An analogy can be formed between R's method of selecting a mixed strategy here and his reasoning in deciding on a pure strategy in the simpler games considered in the previous chapter. There he selected a strategy that would guarantee him the *maximin* payoff. Here he selects a mixed strategy that will guarantee him the *maximin expectation* or *maximin average* winning in the long run. This was indicated earlier where, for *every* x or mixed strategy of R, it was seen that C would pick y so as to give R the *minimum* expectation for his strategy. Of all the minima for different values of x, the greatest is 3/2, corresponding to $x = 3/4$. Hence, (3/4, 1/4) is R's optimum strategy.

Thus far, only R's optimum mixed strategy has been determined, and it remains to select a strategy for C. If the expression for $E(x, y)$ is examined once again, it will be seen that any attempt on the part of C to diminish his loss to R below 3/2 can be countered by R. If C selects $y < 1/2$, R can choose $x < 3/4$, and thereby increase C's average loss. (By selecting $x = 0$, that is, by playing the pure strategy (0, 1), R can make this average loss a *maximum*.) If C selects $y > 1/2$, R can choose $x > 3/4$ and in fact can choose $x = 1$, that is, the pure strategy (1, 0), which will make C's average loss a *maximum*. But if C chooses $y = 1/2$, no strategy of R can increase C's average loss beyond 3/2, which is thus the *minimum* of the many maxima for different values of y. Then $y = 1/2$, or the mixed strategy (1/2, 1/2) is optimum for C, since it corresponds to 3/2, the *minimax* expected payoff to R.

To obtain some practice in solving a game where randomized strategies are involved, the reader can duplicate both the technique and the argument outlined above in order to show that in the matrix game

$$\begin{pmatrix} 1 & 3 \\ 2 & -1 \end{pmatrix}$$

the optimum strategies for R and C are (3/5, 2/5) and (4/5, 1/5), respectively, with a value of 7/5 for R. It is suggested also that the reader return to the game of matching pennies and provide *proof* that it is fair and that the common-sense strategies are actually optimum.

It was pointed out in the previous chapter that pure strategies are merely special cases of randomized strategies. Thus in the case of a 2×2 matrix game, R's possible pure strategies could be symbolized as the "mixed" strategies (1, 0) and (0, 1). The same is true for C. Therefore, although the rationale of the procedure in the case of mixed strategies may seem more difficult to follow, the facts in both types of game are strictly analogous. In both cases R must choose a strategy which will assure him a certain *maximin*, whereas C must direct his behavior toward a *minimax*. The equality of the maximin and the minimax makes it possible to arrive at a solution. In the more general case, the equal minimax and maximin were associated with the expectation function $E(x, y)$. But to continue the analogy further, one can say that the solution corresponds to a *saddle point* of this function. Thus in a 2×2 game, if the optimum strategies for R and C are symbolized by

$(x^*, 1 - x^*)$ and $(y^*, 1 - y^*)$, the reader may imagine the surface pictured in Figure 10.1 as the graph of the function $E(x, y)$, and the point with coordinates (x^*, y^*) as the saddle point indicated in the diagram. Finally, the value of the game is an *expectation* even in the case of a solution by pure strategies; for if it corresponds to some matrix entry, this *same* payoff will be obtained in all repetitions of the game, however numerous, and will therefore be equal to the long-run average.

Most of the game-theory concepts discussed thus far were mentioned by Borel in his 1921 paper and in his subsequent publications on probability theory. But it is in the matter of general *proofs* that von Neumann went far beyond him and laid the foundation for a true mathematical theory. Von Neumann proved that *every matrix game can be solved*. This fundamental proposition is called the *minimax theorem* because the existence of a solution results from his proof that in every case

$$\text{maximum} \begin{bmatrix} \text{minimum } E(x, y) \\ \text{(for all } y) \end{bmatrix}$$
$$\text{(for all } x)$$

and

$$\text{minimum} \begin{bmatrix} \text{maximum } E(x, y) \\ \text{(for all } x) \end{bmatrix}$$
$$\text{(for all } y)$$

exist and are equal. The maximin or minimax is the value v of the game. The x and y corresponding to this value are such that if R selects x, he can expect to win at least an amount v; and if C selects y, he can expect to lose no more than v. This wording is in terms of 2×2 matrix games but can be extended to matrix games of *all orders*. That Borel did little more than initiate concepts is indicated by the fact that he conjectured the minimax theorem to be false for the higher-order matrix games.

The reason Borel may originally have entertained some doubts about the truth of a *general* minimax theorem is that intuitive evidence was lacking. Clues to the fact would naturally lie in finding randomized strategies for specific games of higher order. But the task of computing mixed strategies becomes more arduous as the number of rows and columns of the game matrix increases. In a proof of the minimax theorem, whether it is von Neumann's or one of the alternative demonstrations provided later on by other mathematicians, the existence of a solution is deduced from advanced mathematical theorems concerned with special types of aggregate and special kinds of function.

To obtain some idea of the larger problem, let us advance a small step further by solving the following 2×3 game:

$$\begin{pmatrix} 3 & 2 & 1 \\ -1 & 0 & 3 \end{pmatrix}$$

Examination of this matrix indicates that it has no saddle point. Therefore there is no solution prescribing pure strategies for both R and C. The minimax theorem assures us that there is a solution of some sort. Let us represent R's strategy by $(x, 1 - x)$. Now suppose, for the moment, that C does not reason about his opponent's possible moves but merely decides to select column C_1, that is, the strategy $(1, 0, 0)$. Then the probability matrix will be

$$\begin{pmatrix} 1 \cdot x & 0 \cdot x & 0 \cdot x \\ 1(1-x) & 0(1-x) & 0(1-x) \end{pmatrix} = \begin{pmatrix} x & 0 & 0 \\ 1-x & 0 & 0 \end{pmatrix}$$

and R's expectation will be

$$3x - 1(1-x) = 4x - 1$$

Now the reader can show in the same way that if C selects the pure strategy C_2, that is, the strategy $(0, 1, 0)$, R's expectation will be $2x$, and that if C selects C_3, R's expectations will be $-2x + 3$. We shall represent the three expectation functions by

$$E_1(x) = 4x - 1$$
$$E_2(x) = 2x$$
$$E_3(x) = -2x + 3$$

where x, the relative frequency with which R chooses R_1, may have any value between 0 and 1, inclusive. The three linear expectation functions are graphed in Figure 11.2. Now suppose that R chooses the value of x corresponding to point A in the diagram. Then AB, the ordinate to the lowest intersection with the three expectation lines, represents the minimum amount which R can expect if C limits himself to pure strategies. Therefore, the heavy broken line in Figure 11.2 indicates

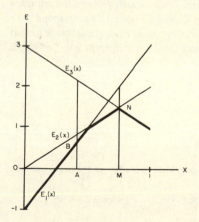

Figure 11.2 Solution of the matrix game:
$$\begin{pmatrix} 3 & 2 & 1 \\ -1 & 0 & 3 \end{pmatrix}$$

the minimum expectations for all choices of x if C chooses pure strategies. Now if R chooses $x^* = 3/4$, represented by M in the diagram, his expectation will be $MN = 3/2$, the maximum of the minimum expectations for C's pure strategies. In addition, the diagram indicates that whether C selects C_2 or C_3, R's expectation will still be $3/2$, and if C should select C_1, R may expect a greater amount. Moreover, we shall show that even if C chooses a mixed strategy involving C_2 and C_3, R's expectation will still be 3/2. Suppose, then, that C's strategy is $(0, y_2, y_3)$ where $y_2 + y_3 = 1$. Then the probability matrix when R's strategy is (3/4, 1/4) will be

$$\begin{pmatrix} 0 & 3y_2/4 & 3y_3/4 \\ 0 & y_2/4 & y_3/4 \end{pmatrix}$$

and $E(x^*, y_2, y_3) = 3y_2/2 + 3y_3/4 + 3y_3/4 = 3(y_2 + y_3)/2 = 3/2$. One can show in the same way that if C uses C_1 in a mixed strategy, R's expectation will be greater than 3/2. Therefore, whatever C's optimum strategy is, it cannot include C_1. Moreover, $x^* = 3/4$, that is, R's mixed strategy (3/4, 1/4), has the characteristic property of an optimum strategy since, whatever strategy C uses, pure or mixed, he cannot diminish R's expectation. Hence we conclude that $x^* = 3/4$ is R's optimum strategy. As far as C is concerned, the graph and the above computation indicate that C_1 must be ruled out, so that the problem reduces to solving the game

$$\begin{pmatrix} 2 & 1 \\ 0 & 3 \end{pmatrix}$$

This was the 2×2 game solved earlier in this chapter, with the result (3/4, 1/4) for R, which checks with the graphic solution, and (1/2, 1/2) for C, or, in terms of the 2×3 matrix, (0, 1/2, 1/2).

The illustrative example is typical of any $2 \times n$ game. A graph like Figure 11.2 will determine R's optimum strategy and eliminate all but two of C's pure strategies. Then, solution of a 2×2 game will determine C's optimum strategy. Sometimes a $2 \times n$ game will present special features. For example, Figure 11.3 corresponds to a fair game in which R has an infinite number of optimum strategies, strange as that may seem, namely, all those where x^* has any value in the interval $2/9 \leq x^* \leq 3/5$.

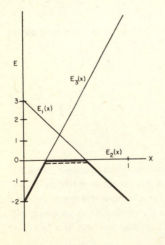

Figure 11.3   Solution of the matrix game:
$$\begin{pmatrix} -2 & 0 & 7 \\ 3 & 0 & -2 \end{pmatrix}$$
R: $(x^*, 1 - x^*)$ where $2/9 \leq x^* \leq 3/5$
C: (0, 1, 0)
$v = 0$

Next, consider the game with the 3×2 matrix

$$\begin{pmatrix} 2 & -3 \\ 0 & 0 \\ -7 & 2 \end{pmatrix}$$

The entries in the first column represent C's potential losses if he chooses C_1. But instead of speaking of losses of 2, 0, —7, one might equally well talk of *gains* of —2, 0, 7. Similarly, the second column could be conceived as containing potential *gains* for C of 3, 0, —2. Therefore, the game in question is equivalent to the 2×3 game

$$\begin{pmatrix} -2 & 0 & 7 \\ 3 & 0 & -2 \end{pmatrix}$$

This game, already solved in Figure 11.3, will provide the solution to the present game if the roles of R and C are interchanged. Thus, in the present game, it will be C who has the infinite number of optimum strategies and R who has R_2 (of the 3×2 matrix) as optimum strategy. In general, $m \times 2$ games can be solved by converting them into $2 \times m$ games in the manner illustrated.

Now to advance a step further, let us see how a *maiko* (a youthful geisha-in-training) transforms baseball into a 3×3 matrix game. Japan is a popular country for tourists nowadays, especially those from the United States. One of the festive tours includes a dinner party where several *maikos*, supervised by a senior geisha, entertain—serving *sukiyaki*, heating and pouring *sake*, playing musical instruments, singing and engaging in pantomime dances. In one of these dances, each *maiko* selects a male tourist as a partner, and the pantomime is "playing baseball." The winning "team" is determined at the end of the dance by having each partner gesture with fist, palm of the hand, or two fingers; it is simply the children's game "scissors, paper, stone" which the fourteen-year-old *maikos* are young enough to enjoy. Fist = stone, palm = paper, fingers = scissors. Since stone sharpens scissors, scissors cut paper, and paper covers (can be wrapped around) stone, the inferior object loses and must pay off to the superior. Hence, a possible payoff matrix is

	Scissors	Paper	Stone
Scissors	0	1	−1
Paper	−1	0	1
Stone	1	−1	0

This geisha game is obviously "symmetric" because the *maiko* and her opponent, the American tourist, have the same set of pure strategies, with the same rewards attached to each choice. (The matrix is described as "skew-symmetric.") Intuition tells us that such a symmetric game must be fair, and hence $v = 0$. The fact that all penalties are equal in magnitude suggests that the optimum strategy for R or for C should be (1/3, 1/3, 1/3). But how can one *prove* that this is so, and how can one solve more general 3×3 games?

If (x_1, x_2, x_3) and (y_1, y_2, y_3) represent the strategies of R and C in a 3×3 game, then $x_i \geq 0$, $y_j \geq 0$, where $i, j = 1, 2, 3$, and

$$x_1 + x_2 + x_3 = 1$$
$$y_1 + y_2 + y_3 = 1$$

In the matrix for "scissors, paper, stone," let us suppose that C plays the pure strategy C_1. Then the probability matrix will be

$$\begin{pmatrix} x_1 & 0 & 0 \\ x_2 & 0 & 0 \\ x_3 & 0 & 0 \end{pmatrix}$$

and R's expectation will be

$$E_1(x_1, x_2, x_3) = -x_2 + x_3$$

Similarly, if C should play C_2 or C_3, one would have

$$E_2(x_1, x_2, x_3) = \quad x_1 - x_3$$
$$E_3(x_1, x_2, x_3) = -x_1 + x_2$$

Now, if R's strategy is *optimum*, one of these expectations may possibly be equal to the value of the game. On the other hand, since C is selecting a pure strategy, his choice may *not* be optimum, and hence R's various expectations above may be greater than what he could expect if C used his best strategy. Therefore, $(x_1{}^*, x_2{}^*, x_3{}^*)$, R's optimum strategy, satisfies the *inequalities*

$$-x_2 + x_3 \geq v$$
$$x_1 - x_3 \geq v$$
$$-x_1 + x_2 \geq v$$

By reversing the behavior of R and C, one reaches the conclusion that $(y_1{}^*, y_2{}^*, y_3{}^*)$, C's optimum strategy, satisfies the inequalities

$$y_2 - y_3 \leq v$$
$$-y_1 + y_3 \leq v$$
$$y_1 - y_2 \leq v$$

Originally we had two equations and six inequalities which the x^*'s and y^*'s must satisfy, and now we have six more, involving the additional unknown v.

In the present specially simple example, it is easy to verify that all equations and inequalities are satisfied by the solution which intuition suggested, but matters are not so simple with nonsymmetric 3×3 matrix games, or with games whose matrices are of higher order. In such advanced cases, solutions can be provided by *linear programming*, which is a technique for finding the maximum or minimum value of a linear function subject to various constraints in the form of linear inequalities. But we shall not discuss this special numerical discipline, since it would involve considerable computational detail. Our objectives will take us elsewhere—to strategic games that are more general in scope and applicability.

12

General Games and Statistical Decision Theory

While the von Neumann minimax theorem establishes the solvability of all two-person rectangular games, even those in which, say, a 1,000,000 × 2,000,000 matrix is involved, there still remains the task of considering more general games—in which there may be more than two players and each player may have several moves, etc. However, it turns out that there is no new theoretical difficulty, providing a general game is *finite*, that is, involves only a finite number of moves with a finite number of alternatives at each move. A finite game, as we shall illustrate shortly, can always be "normalized," that is, converted into an equivalent matrix game. Hence the minimax theorem and the method of solution we have discussed apply to all finite games, even the most general.

To furnish a very simple example, let us consider a game with three moves where R and C play alternately. Specifically, suppose that R makes the first move and that each move consists of a player choosing a number from the set $\{1, 2\}$, and that players are always aware of the choices made in previous moves. Thus a possible "play" of the game would be to have R choose 2, then C choose 1, then R choose 1. We shall symbolize this way of playing the game by (2, 1, 1). There are other possible plays, for example, (1, 1, 1), (1, 1, 2), (1, 2, 1), etc.

The various plays can be represented by a game *tree* (Figure 12.1), a helpful graphic device. Any complete play is pictured as a path proceeding from the root to the top of the tree. There are as many different paths as there are terminal points at the top, and it is easier to count the latter. In Figure 12.1 there are eight points at the top and therefore eight possible plays of the game. For each such play, the rules of the game specify the amount that C must pay to R. If (x, y, z) symbolizes a sequence of choices made by R and C in the three moves, that is, a path in the

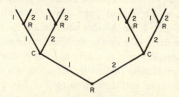

Figure 12.1   Game tree for game with three moves

tree, then the function $M(x, y, z)$, tabulated below, is called the *payoff function* for the game.

$M(x, y, z)$

x	y	z	M
1	1	1	−4
1	1	2	−2
1	2	1	6
1	2	2	−8
2	1	1	10
2	1	2	4
2	2	1	4
2	2	2	12

The domain of this function consists of the eight possible paths (or sequences or ordered number triples) specified in the table, and the range consists of the eight corresponding monetary payoffs.

What meaning can be assigned to *strategy* in the game we have formulated? A pure strategy for a specified player (R or C) is conceived as a complete set of directions telling exactly what choices should be made at each of his moves under all possible circumstances that may arise, that is, for all possible sets of information about previous moves. The instructions are of the sort that could be followed by a substitute player or could be fed into a machine which would play the game in automatic fashion. In the present game, a strategy for C would concern only his move, the second, but would require a *pair* of directions—what to do if R has chosen 1 in the first move, and what to do if R has chosen 2. Thus C might plan the following strategy: If R has selected 1, choose a *D*ifferent number; if he has selected 2, pick the *S*ame number. If he were to code this strategy for convenience, he might symbolize it by *DS*. In effect, his strategy would select the number 2 no matter what R's first move has been. Another strategy for C would be *SS*, signifying that if R has chosen 1 in the first move, C will choose this *S*ame number in the second move, and that if R has chosen 2, C will choose that *S*ame number. Thus there would be *four* different pure strategies open to C, those coded as *SS*, *SD*, *DD*, and *DS*.

R's strategies are a little harder to symbolize. He might plan to select 1 in the first move and then code his contemplated behavior on the third move by using one of the four pairs above. Thus, four pure strategies available to him are 1–*SS*, 1–*SD*, 1–*DD*, and 1–*DS*. If he decides on strategy 1–*DD*, this means that he plans to select 1 in the first move and in the third move will apply DD on the basis of *C's* choice in the second move. If C has selected 1, R will choose *D*ifferently; if C has selected 2, R will choose *D*ifferently. In effect, R's total strategy consists of choosing 1 in the *first* move, and then selecting 2 or 1 in the *third* move according to whether C has picked 1 or 2 in the second move. Four additional pure strategies are available to R, namely, those where he plays 2 in the first move. In all, then, R has *eight* possible pure strategies from which to choose.

Since R and C have eight and four possible pure strategies, respectively, the game under consideration can be described by the following 8×4 matrix, and the methods of solution for matrix games can be applied. The *normalization* of the game has been accomplished.

	SS	SD	DD	DS
1–SS	− 4	− 4	− 8	− 8
1–SD	− 4	− 4	6	6
1–DD	− 2	− 2	6	6
1–DS	− 2	− 2	− 8	− 8
2–SS	12	10	10	12
2–SD	4	10	10	4
2–DD	4	4	4	4
2–DS	12	4	4	12

The payoffs in the above matrix were computed with the use of the payoff matrix $M(x, y, z)$. For example, the entry at (R_3, C_2), that is, in the third row, second column, must correspond to the strategy pair, 1–DD for R and SD for C. This means that R chooses 1, C chooses the *Same* number (hence 1), then R chooses a *Different* number from the one C picked so that the sequence of choices is $(1, 1, 2)$ with payoff $M(1, 1, 2) = -2$. Again, (R_6, C_4) implies the sequence $(2, 2, 1)$ with payoff $M(2, 2, 1) = 4$.

Examination of the rows of the matrix shows that R's *maximin* is a payoff of 10 occurring in the fifth row, and C's *minimax* is also 10, occurring in the second and third columns. Thus the *value* of the game to R is 10, and there are two saddle points, hence two solutions involving a pure strategy pair, namely, (R_5, C_2) and (R_5, C_3). In terms of the game, this signifies that R has one optimal strategy—to select 2 in the first move and then in the third move to choose the same number that C has picked in the second move. C, on the other hand, has two equally good strategies available. He can play SD, which amounts to choosing 1 in the move assigned to him, or else he can decide to choose the number differing from that in R's first move. Because R is planning to select 2 in the first move, either of C's strategies will compel him to select 1, and as a result, R's strategy will require him to pick 1 on the third move. Thus, although there are two solutions to the normalized game, they both result in the same optimal sequence or "path to the top," namely, $(2, 1, 1)$, and hence to the same payoff, 10.

It will be observed that after the general game considered above was normalized, the solution was effected more easily than in some of the more elementary games we have illustrated. This was because the strategy matrix of the more complicated game had saddle points, and hence had solutions in terms of pure strategies. This relative ease of solution will always occur in any game having "perfect information," which means that at any move a player has complete knowledge of the choices made in all previous moves. A special theorem of game theory establishes the fact that in all games with perfect information the normalized form, that is, the *strategy matrix*, will have at least one saddle point and hence a solution in terms of pure strategies.

By contrast, the uncomplicated pastime of matching pennies is not a game with perfect information. If it were, the matrix would have a saddle point, and we saw this was not the case. One can think of matching pennies as a game with two moves, where R plays the first move and C the second move (or vice versa). But one must supply an additional condition if this new form of matching pennies is used—namely, that the player in the second move is ignorant of the choice made in the first move. This additional fact is indicated in the game tree for matching pennies (Figure 12.2)

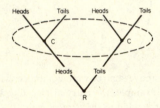

Figure 12.2   Game tree for matching pennies

by encircling the two vertices corresponding to the second move in order to indicate that C, if he is the player for that move, does *not* know at which of the vertices he is located.

Those games of several moves that we have considered may appear somewhat trivial, but after the connection between game theory and Wald's statistical decision theory has been made clear, it will be possible to provide more challenging instances of general games and also to show that the common link between the two theories is the same minimax principle which has been applied to all the strategic games illustrated up to this point.

To create an analogy with zero-sum, two-person games, Wald cast the fundamental decision process in the form of a "game against nature," in which player R is the statistician or any individual who must *decide* among alternative actions—R_1, R_2, etc.—because he is faced with an uncertain prospect whose outcome will be influenced by chance factors and may turn out to be C_1, C_2, etc., with varying consequences, monetary or otherwise, for R. In a specific example, a man is faced with the prospect that the winter ahead may be $C_1 =$ mild, $C_2 =$ average, or $C_3 =$ severe. Although he is not gambling with a real opponent, it is convenient to use the fiction of an adversary C, the one Wald called "nature," whose "states," or pure strategies, are C_1, C_2, and C_3. Suppose, then, that your decision problem, the game in which you oppose "nature," involves the choice of a strategy for purchasing and storing your winter supply of coal. Suppose that if you place your order during July and August the price is $15 per ton, but that coal ordered later costs $22

per ton. From previous experience, you know that it requires five, six, or eight tons
to heat your home, depending on whether the winter is mild, average, or severe.
Moreover, any coal that remains after the winter ahead will be a total loss to you
because you are selling your home to people who plan to install an oil burner. If
your pure strategies are orders of five, or six, or eight tons during July and August,
the payoff matrix is readily computed as

$$
\begin{array}{c}
\\
\text{(Order 5)} \\
\text{(Order 6)} \\
\text{(Order 8)}
\end{array}
\begin{array}{ccc}
\text{Mild} & \text{Average} & \text{Severe} \\
\begin{pmatrix} -75 & -97 & -141 \\ -90 & -90 & -134 \\ -120 & -120 & -120 \end{pmatrix}
\end{array}
$$

Inspection of this matrix shows that the payoff, -120, for (R_3, C_3) is a maximin for rows (R's pure strategies) as well as a minimax for columns (nature's pure
strategies). Therefore (R_3, C_3) is the minimax solution. This means that your best
bet is to spend \$120 for eight tons of coal. But it would seem to be pushing fiction
too far to state that nature's optimum strategy will be to provide a severe winter,
or that, in decision problems in general, she "plays" so as to give you the minimum
possible or extract the maximum from you. Nevertheless, as Wald pointed out, if
nature's true state is unknown to you, it is not unreasonable for you to behave *as if*
she were malevolent and wished to oppose you with a minimax strategy. Put otherwise, if you are unwilling to take chances in an uncertain situation, you had better
picture the worst eventuality and act accordingly.

But some theoreticians think that a decision-maker should be more optimistic.[*]
One way he can accomplish this is to argue against the notion of complete uncertainty and say, just as Bernoulli and Laplace did, that in the absence of observational data, one must assume that nature is as likely to do one thing as another.
Therefore, in the problem under consideration one would assign the a priori probabilities $(1/3, 1/3, 1/3)$, respectively, to nature's three possible states. The assumption
involved is a powerful one, for it postulates that nature is *not* maneuvering against
the decision-maker, since her strategy is the *known* mixed strategy $(1/3, 1/3, 1/3)$.
The decision situation is thus reduced from a two-person game of strategy to a mere
one-person game of chance (page 250). To solve this new game, the decision-maker
would make use of the known (assumed) probabilities, and reason as follows: If
pure strategy R_1 (five tons) is selected, the expectation is

$$
\frac{1}{3}(-75) + \frac{1}{3}(-97) + \frac{1}{3}(-141) = -\frac{313}{3}
$$

Similarly, the expectations if R_2 and R_3 are chosen can be computed to be $-314/3$
and $-360/3$, both lower than the expectation if R_1 is picked, and therefore R_1 is the
best strategy. Thus, postulating equal probabilities for nature's states leads to the

---

[*] See Appendix to this chapter.

decision to order five tons of coal instead of the eight tons prescribed by the more conservative Wald solution.*

But Wald would not have approved of the arbitrary assumption of equal likelihood. He would not, however, have been averse to assuming probabilities based on experimental data compiled in the past. Thus, careful observation of many *previous* winters in the particular geographical vicinity, a long-range meteorological forecast, etc., might be combined to yield relative frequencies that would approximate the chances for the winter ahead. Then past experience might lead to approximations like (0.1, 0.6, 0.3) for the probabilities of nature's states. If these probabilities are postulated, then the expectations under R_1, R_2, R_3 are $-\$108$, $-\$103.20$, $-\$120$, respectively. Therefore, R_2 is the optimum strategy, an answer differing from the two previous solutions of the same problem.

The reader may be troubled by the fact that we have apparently obtained three different answers to the same problem. In fact, however, three different problems were solved. In the first instance, it was assumed that nothing whatsoever could be predicted about the winter ahead. In the second and third problems, one still had to take chances with the winter ahead but the probabilities were known, since in one case they were assigned on the basis of classic probability theory, and in the other instance they were estimated from empirical data.

In trying to make the optimum decision about ordering a coal supply for the winter ahead, the decision-maker did not carry out any special experimentation. In one solution it was assumed that he used empirical data, but the meteorological observations involved were made by others in the past and were not designed to shed light on his particular decision problem. It is much more typical of the Wald *statistical* decision procedure to design some special experiment in order to "spy" on nature and thus obtain some revealing evidence of what her true state is likely to be.

As an example, there is the sort of statistical decision process which might be used to control the quality of manufactured articles. Suppose that the statistician advises that before any lot is shipped, three items should be drawn at random and inspected, the lot to be accepted or rejected according to the outcome of this random experiment. In addition to specifying the experiment, the statistician may advise the quality-control inspector to make his decision in accordance with the following *statistical decision function*, to use Wald's terminology.

Outcome (Number of Defectives)	0	1	2	3
Action or Decision	a_1	a_1	a_2	a_2

where a_1 symbolizes "accept" and a_2 stands for "reject."

---

* The Bernoulli-Laplace assumption is only one of several alternatives to Wald's methodology. Again, see Appendix to this chapter.

We observe that the *domain* of this statistical decision function is the sample space $S = \{0, 1, 2, 3\}$, and that the *range* is $A = \{a_1, a_2\}$, the aggregate of contemplated actions. The illustrative function would require rejection of the entire lot if the sample should contain two or three defective items. But the statistician might advise a different strategy by prescribing the following statistical decision function:

Outcome (Number of Defectives)	0	1	2	3
Action or Decision	a_1	a_1	a_1	a_2

which would permit more leniency in the control of quality. Or, if stringency is desired, he might suggest that the lot be passed only if every item in the sample is satisfactory, a policy described by the function

Outcome (Number of Defectives)	0	1	2	3
Action or Decision	a_1	a_2	a_2	a_2

The preceding example is typical, because it illustrates the fundamental role of the *statistical decision function*. Such a function can be defined as a set of ordered pairs of the form (z, a), where z is the outcome of a random experiment and a is the action or decision corresponding to this outcome.

As a further illustration, suppose that in some decision problem the statistician suggests a random experiment whose sample space $S = \{z_1, z_2, \ldots, z_5\}$ contains five possible outcomes. Suppose also that the choice of actions must be made from a set $A = \{a_1, a_2, a_3\}$. Then, in theory, there are very many different statistical decision functions which could be applied to the problem. A few of these functions are indicated by listing some of the ways actions might be assigned to outcomes:

	Outcome				
	z_1	z_2	z_3	z_4	z_5
	a_1	a_1	a_1	a_2	a_3
Action	a_1	a_1	a_1	a_3	a_2
or	a_2	a_3	a_1	a_1	a_1
Decision	a_3	a_1	a_1	a_1	a_2
	a_2	a_1	a_3	a_2	a_2

If the situation is regarded abstractly, without giving concrete meanings to the z's and the a's, there will be many other decision functions which can be applied. In fact, it can be shown that the potentially available decision functions number $3^5 = 243$ (page 295). Of course, when there are specific interpretations, like $a_1 = $ *accept*, $a_2 = $ *reject*, $a_3 = $ *inspect more items*, common sense would immediately eliminate some functions, for example,

Outcome	z_1	z_2	z_3	z_4	z_5
Decision	a_1	a_1	a_1	a_1	a_1

which would prescribe acceptance of a lot no matter what the results of sampling inspection should be. The number of available decision functions can be increased even beyond 243 by considering varied ways of spying on nature, that is, other random experiments and the (numerous) decision functions corresponding to each one.

With such a plethora of choice even in an apparently simple situation where the experimental outcomes and possible actions are small in number, how can a statistician advise an optimum procedure? Wald showed that once again it is a matter of playing a strategic game against nature, a somewhat subtler game than in the case of decision without experimentation. There the decision-maker's pure strategies were various courses of action. In *statistical decision*, as we have labeled the type in which experimentation may be used, it is the statistician who plays the game against nature, and his pure strategies are the (numerous) possible decision rules. Such a game can be pictured by a tree like that in Figure 12.3 where any path from the root to the top represents a possible mode of play.

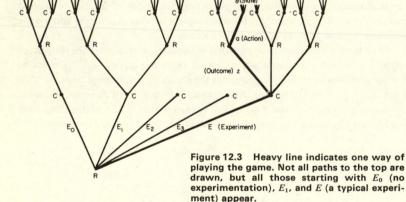

Figure 12.3   Heavy line indicates one way of playing the game. Not all paths to the top are drawn, but all those starting with E_0 (no experimentation), E_1, and E (a typical experiment) appear.

In a general game with several moves, a pure strategy always signifies a complete set of directions prepared *prior* to playing the game. Then nature's pure strategy would consist of instructions for the second and fourth moves (see Figure 12.3), that is, for the experimental outcome and her actual "state."

Since the statistician (or the individual whom he is advising) plays the first and third moves (Figure 12.3), his pure strategy must specify in advance *both* the experiment to be performed in the first move and the statistical decision function which is to guide action in the third move. Hereafter we shall refer to this dual specification as a *decision rule*.

In Figure 12.3 we observe that one of the lowest branches of the tree is marked E_0. This represents a dummy experiment, that is, no experiment at all, with a dummy outcome, z_o, for the second move. The statistician may wish to start the path to the top along branch E_o, which signifies that, in actuality, only the third and fourth moves are made. In other words, he may advise a decision rule involving no special experimentation. Perhaps in certain situations it would be too costly or it might delay action unduly. Thus statistical decision includes as a special case the simple decision game without experimentation.

Although it is the *definition* of pure strategy that requires the decision rule to be selected *prior* to play, this regulation is also good common sense. If the decision-maker were not compelled to agree irrevocably to act in a specified fashion, he might change his mind if an experimental result appeared discouraging. Such behavior would be unscientific and might actually work to his disadvantage, because he would be acting contrary to what the statistician advises as the optimum procedure on the basis of the Wald theory.

Earlier it was seen that a finite game can always be normalized or converted into a matrix game in which the pure strategies of the two players correspond to rows and columns, respectively. Therefore, when any game of statistical decision is normalized, the rows represent the (numerous) decision rules—R_1, R_2, R_3, etc.,— and the columns correspond to nature's possible strategies—C_1, C_2, etc.

To illustrate such a normalized game, let us consider the following situation, where facts and figures are simplified for the sake of comprehension, and where it is assumed that preliminary screening has eliminated a great many of the possible decision rules. Suppose, then, that in the production of a certain complex object, different manufacturers hold contracts for supplying the various parts. One particular manufacturer is to produce parts of a specified type. The parts are to be packed in pairs, since two will always be needed in the construction of the object. The cost of producing one of these parts is $1, and if it is satisfactory, the manufacturer will be paid $4. If it is defective, he will be fined $2, because imperfections cause damage, delay, etc. Also, the contract calls for the scrapping of parts not shipped (to prevent the manufacturer from selling them elsewhere). Finally, if the manufacturer wishes a part to be tested for quality, this can only be done by taking it apart completely, which renders it worthless.

The manufacturer asks his statistician for advice on some sort of scheme that will insure reasonable quality in the parts shipped but will not be too costly. A statistician would actually consider a large variety of decision rules for the handling of a pair of parts, but for the sake of simplicity we assume that only four rules remain after preliminary screening, namely,

Rule 1.   Ship both parts (a reasonable rule if there has been previous evidence that quality is good).

Rule 2.   Select one part at random and pronounce it satisfactory. The part that remains must be scrapped, in accordance with the contract.

Rule 3.   Select one part at random and test it. Judge the quality of the other part to be the same. Thus, if the part tested is good, the other part is considered to be good, too. If the test shows the part chosen to be defective, the other part is assumed to be defective and is scrapped.

Rule 4.   Select one of the two parts at random and test it. Assume the other part to be opposite in quality to the one tested. Thus, if the tested part is satisfactory, the other part is assumed to be defective and is scrapped, whereas if the test shows a defective part, the other part is assumed to be satisfactory.

These decision rules are then considered as four pure strategies open to the statistician R in his game against nature C. There are three pure strategies possible for nature, namely, to have 0, 1, or 2 defectives in the package containing a pair of parts. The payoff matrix for this game is

		Nature (C)		
		0 Defective	1 Defective	2 Defectives
	Rule 1	6	0	−6
(R)	Rule 2	2	−1	−4
	Rule 3	2	−3	−2
	Rule 4	−2	0	−4

The reader can readily check the entries in the payoff matrix. Those in the *first column*, for example, correspond to nature's kindest strategy, "no defective parts." Rule 1 will have the manufacturer ship both and hence receive $8. But he has spent $2 to produce the parts, and therefore his gain or payoff is $6. Under Rule 2, only one part will be shipped and the other destroyed. Hence the gain will be $4 − 2 = \$2$. Under Rule 3 the part selected will be satisfactory, and hence the other will be judged correctly as satisfactory, and shipped with a gain of $4 − 2 = \$2$. Under Rule 4 the part tested would be satisfactory, and hence the other would be judged defective and scrapped. Both parts would thus be a total loss, and the payoff therefore − $2.

The entries in the third column can be computed similarly. Some of those in the second column may require some additional explanation. For example, the entry at position (R_2, C_2) is − 1, and this is figured as follows: Under Rule 2, one of the parts is selected at random. Therefore the probability of selecting the good part is 1/2, and likewise for the defective part. In the former case, there would be a gain of $4 − 2 = 2$, in the latter a loss of $−2 − 2 = −4$. For the payoff listed at (R_2, C_2), we have computed the *expectation* or *long-run average*.

$$(\tfrac{1}{2})(2) + (\tfrac{1}{2})(-4) = 1 - 2 = -1$$

Now for the solution of the game, the first column can be removed as a strategy for nature, because the corresponding entries in the last column would always be more profitable for her. There remains the matrix

$$\begin{pmatrix} 0 & -6 \\ -1 & -4 \\ -3 & -2 \\ 0 & -4 \end{pmatrix}$$

The first row of this matrix can be removed as a strategy for R, since the last row would be a better strategy. Likewise, the second row can be removed for the same reason. This leaves

$$\begin{pmatrix} -3 & -2 \\ 0 & -4 \end{pmatrix}$$

There is no saddle point and hence no solution in terms of pure strategies, because max (min of row) $= -3$ and min (max of column) $= -2$ and these numbers are not the same. The reader may use the methods of the previous chapter to show that the solution is the pair of mixed strategies $(4/5, 1/5)$, $(2/5, 3/5)$ for R and C, respectively, or, in the original matrix, $(0, 0, 4/5, 1/5)$ and $(0, 2/5, 3/5)$, with the value of the game an expected loss of $2.40 for R. This means that as pairs of parts are submitted to the statistician (or the quality-control inspector following his advice), he should always select one part at random and test it. Sometimes he will judge the other part to be of the same quality as the one tested, and sometimes to be of the opposite quality. He uses a random device (like a sack containing forty blue and ten red marbles) to achieve the relative frequencies $(4/5, 1/5)$. The solution seems to be in accord with common sense, because it prescribes a *statistical* decision procedure where the experimentation is the sort of sampling inspection called for by either Rule 3 or Rule 4. Also, the more frequent use of Rule 3 seems rational if one thinks of the common notion of judging an entire group by means of a sample. Thus the whole group of two articles should more often than not be like the one tested.

Again, some statisticians would disagree with the preceding solution, which is based on the assumed malevolence of nature. (Thus, since the mixed strategy $(0, 2/5, 3/5)$ signifies that nature will never select C_1, one must play the game as if nature would never produce a pair of parts where both are satisfactory.) The classic viewpoint would have the statisticians use Rule 1 all the time because the first row of the matrix has the maximum total, namely zero; and hence if one were to follow Bernoulli and Laplace, he would optimistically ship both parts each time without hesitation or inspection. Another criterion (see Appendix) would lead to the exclusive use of Rule 3 where the statistician always tests one part at random and judges the quality of the other to be the same.

Still another model for this same situation will be provided in the next chapter where, after a brief return to the simpler games of chance, new concepts will present nature as a lady of infinite variety, and more general methods will enable us to adjust our behavior as best we can to her manifold moods.

The decision problems we have considered, like the games treated somewhat earlier, were presented to illustrate fundamental concepts of the von Neumann and Wald theories rather than to indicate significant applications in realistic situations. Hence one might ask whether the ideas we have discussed suffice for such applications. The answer is that the *techniques* of solution, as we have explained them, are adequate, but that issues arising in the behavioral sciences may call for something additional. The statistician or decision-maker in every case is nevertheless like the "R" of our problems, in that he must measure as best he can the relative merits of future eventualities in which chance elements will play an important role. But his evaluation of prospects may *not* be entirely a question of dollars and cents, pounds and pence, for he may wish to give weight to nonmonetary factors as well—his moral convictions, his emotional stability when risks are involved, and a host of other subjective intangibles.

Even a decision involving the simplest sort of gamble may involve subjective factors. If, for example, a friend offered the following bet, would you accept or refuse? A fair coin is to be tossed, and if it turns up "heads," he will pay you $1200. In case the result is "tails," however, you must pay him $1000. If your liquid assets are under $1000 and/or you do not like to take chances, you will doubtless refuse the bet, even though there is a chance of winning $1200 and your expectation (the criterion in the von Neumann and Wald theories) is

$$(\tfrac{1}{2}) (1200) + (\tfrac{1}{2}) (-1000) = \$100$$

On the other hand, a gambling addict who is already in debt would doubtless accept the wager no matter what logical arguments against such behavior were presented. A wealthy man who enjoys gambling might accept, unless the stakes seemed too small to challenge his interest. The man who considers all forms of gambling immoral would, of course, reject the wager. The sum total of the situation is that when people are faced with a gamble, their behavior is not necessarily governed by expected monetary payoff.

Decisions seem to be determined by the "true" values for an individual of the prospects or gambles which he must face. But is there any way of measuring these "true" values? Von Neumann and Morgenstern dealt with this very issue in the later editions of the *Theory of Games and Economic Behavior*. They showed that if an individual has preferences that satisfy certain assumptions, there exists a measure of (true) value, which they called *utility*.

In spite of von Neumann's proof of the existence of a utility measure, the practical task of setting up a utility scale to fit a specific situation may be an exceedingly difficult one. But suppose that in a particular problem the decision-maker has been able to assign numerical utilities to each pure strategy pair (R_i, C_j). Then the fundamental matrix becomes a *utility* payoff matrix, where the gains for R are measured in *utiles*, the units of value measurement in his special scale. The matrix governs the game, and except for the fact that its entries are to be read as utiles and not as dollars, one proceeds in exactly the fashion we have already explained.

The von Neumann utility measure provides an answer to the question of subjective values and, in addition, resolves the issue of whether or not expectation, the average payoff in very numerous trials, is a suitable criterion for a decision that is to be made just once. If the expectation formula (page 267) is applied to a matrix row of *utility* payoffs, it turns out that the result need *not* be conceived as an average based on many trials. Instead, the expectation formula is merely a prescription for figuring the "true" value or *utility* of a gamble more complicated than the basic prospects to which the decision-maker originally assigned measures. In other words, "expected" utility is just utility. If after computation of "expected," that is, *actual* utility for various gambles, the decision-maker selects what he considers an optimum strategy and claims that it is the best decision because it offers maximum utility, he will seem to be engaging in circular reasoning. As indicated earlier, his ideas of what prospects are good, better, best are supposed to precede consideration of a decision problem. Subsequently, a utility scale must be derived from these preferences. If originally he had been able to assign utility measures to *all* possible gambles that might arise in the particular problem, he would have known which eventuality he considered best. But when prospects are numerous and complicated, it is easier to rate only a few simple cases at the beginning and then to derive utilities of other prospects if and when such utility measurements are needed. Hence, in choosing a prospect after the expectation formula reveals maximum utility of the particular gamble, an individual is not reasoning in a circle but merely acting in consistent agreement with his original preferences.

As a final word, we might point to the fact that certain researchers in the behavioral sciences feel that, in addition to utility measure, *subjective* probabilities should be introduced. Since the classic and the statistical concepts of probability have already given rise to certain logical difficulties, we shall not introduce new troubles by dealing with the issue of subjective probabilities. In summary, then, utility theory will not modify our treatment of decision problems. Illustrative examples will, in general, be such that monetary payoffs are actually a suitable measure of utility. If not, it will be assumed that payoffs are in utiles, and that the term "expectation" merely signifies the utility of a mixed gamble.

In the chapters that follow we shall return to a more elementary level and consider some of the special types of decision problem that characterized mathematical statistics prior to the promulgation of Wald's theory. Before looking backward, however, it seems appropriate to conclude the present chapter with a biography of the late Abraham Wald. Two of the sources for our account are eulogies written by Morgenstern and Wolfowitz.*

Abraham Wald was born in Cluj, Rumania, in 1902. His grandfather was a famous rabbi, his father a small businessman with many intellectual interests. There were five other children in the family. One brother, Martin, considered as gifted as Abraham, was an electrical engineer with numerous inventions to his credit.

Wald's early education did not follow the usual pattern. He was not admitted to the local *gymnasium* because, as the son of an Orthodox Jew, he would not attend school on Saturdays. Therefore he studied by himself and passed the entrance examinations of the University of Cluj. After graduation, he experienced considerable difficulty in entering the University of Vienna because of religious restrictions. Thus he spent a year at the engineering school in Vienna before he was finally permitted to study pure mathematics at the university.

---

* Oskar Morgenstern, "Abraham Wald," *Econometrica,* Vol. 19, No. 4 (October 1951), pp. 361 ff.; Jacob Wolfowitz, "Abraham Wald," *Annals of Mathematical Statistics,* Vol. 23 (1952), pp. 1–13.

In spite of his exceptional ability and the numerous research papers he wrote in the years 1931 to 1936, his religion made an academic post in Austria an impossibility. In the course of his quest for a source of income and an opportunity to work in applied mathematics, Wald met Oskar Morgenstern, who was then director of the Institute for Business Cycle Research. Morgenstern appreciated Wald's talents, employed him at the Institute, and became his lifelong friend.

Wald's research in mathematical statistics gained such renown that he was invited in 1937 to become a staff member of the United States Cowles Commission, and in 1938 to serve as a research associate at Columbia University. His decision to accept these offers was a wise one since, in the interim, the Nazis had taken control of Austria. Columbia University established a special department of mathematical statistics and named Wald as its first chairman. Students from all over the world flocked to his courses, and some of them are today's leading statisticians.

As we have already indicated. Wald's greatest achievement was the theory of statistical decision functions. But his total accomplishment included research involving more than ninety publications—both books and papers. In these many reports of his discoveries are contained some of the most decisive new ideas of modern mathematical statistics. A few of the new concepts arose from his research activities during World War II, when he was a member of the Statistical Research Group at Columbia University. One of the problems he had to solve led him to his now-famous *sequential analysis* (page 323), which was immediately applied in the war industries and saved untold millions of dollars. In fact, a visit to many American factories today will show immediately the now-familiar charts of sequential testing as indispensable tools in mass production. But sequential analysis is only a special part of Wald's broader decision theory.

Wald did not escape tragedy. At the end of the war, he learned that eight members of his immediate family, among them his parents, had been murdered by the Nazis. Even this cruel blow failed to embitter him, although a certain sadness could be felt to be with him for the rest of his life. He succeeded in bringing the sole survivor, his brother Hermann, to the United States.

In 1950 Wald received an invitation from the Indian Government to lecture at Indian universities and research centers. He accepted eagerly, and in November of that year he and his wife set out. En route, he lectured in Paris and Rome. He delivered several of his scheduled lectures in India, but the tour was not to be completed. On December 13, 1950, an airplane, lost in a fog, crashed into a peak of the Nilgiris, killing all aboard, among them Abraham Wald and his wife.

The influence of his great ideas remains, and their importance increases. Some of them have already been discussed in order to acquaint readers with the ultimate in mathematical statistics and also because the author holds Wald in particular reverence. She worked with him during World War II as a member of the Columbia University Statistical Research Group, and witnessed the development of his sequential analysis and his initial formulations in statistical decision theory.

Appendix

In addition to the Bernoulli-Laplace method and the Wald minimax principle, two other criteria have been proposed for solving games against nature. When the four criteria are applied to the game whose matrix is given below, the solutions prescribe four different strategies or decisions for R. The reader will see at once that the classic criterion would pick R_4 whereas Wald's method would select R_1.

$$\begin{pmatrix} 2 & 1 & 1 & 1 \\ 0 & 6 & 0 & 0 \\ 1 & 4 & 0 & 0 \\ 3 & 2 & 0 & 2 \end{pmatrix}$$

Professor Leonid Hurwicz has formulated a criterion whose purpose is to modify the extreme pessimism of Wald's assumption that nature will do her worst. Hurwicz suggests that R's optimism be assessed by a real number α, where $0 \le \alpha \le 1$. Then R must compute, for each row, the quantity $\alpha M + (1 - \alpha)m$, where M and m are the maximum and minimum payoffs in the row. Finally, R must select the row for which this quantity is greatest. We observe that if R is least optimistic or most pessimistic, $\alpha = 0$, and the quantity computed for each row is $\alpha M + (1 - \alpha)m = m$, the minimum. Thus, R will be selecting the row containing the *maximin*, and hence for $\alpha = 0$, Hurwicz's criterion reduces to Wald's. At the opposite extreme, $\alpha = 1$ when R is most optimistic. In that case, Hurwicz's index is $1M + 0m = M$, the maximum of a row. Then R will select the strategy corresponding to the maximum of row maxima if he is extremely optimistic.

In the foregoing payoff matrix, let us suppose that α is *not* extreme but has some value as yet unspecified. Then, for the first row,

$$\alpha M + (1 - \alpha)m = 2\alpha + 1(1 - \alpha) = \alpha + 1$$

Since $m = 0$ in the other rows, the Hurwicz index is seen to be 6α, 4α, and 3α, respectively. For *any* possible value of α greater than zero, 6α is the maximum of these three. It remains to compare 6α, the index for the second row, with $\alpha + 1$, the index for the first. The two strategies will be equally good if $6\alpha = \alpha + 1$, that is, if R's optimism measures $\alpha = 0.2$. If $6\alpha < \alpha + 1$, that is, if $\alpha < 0.2$, then R_1 is better than R_2, and R_1, the Wald solution, is the best of the four strategies (which seems sensible because α is close to zero). But if $6\alpha > \alpha + 1$, that is, if $\alpha > 0.2$, R_2 is the best strategy.

The fourth criterion in current use is due to Professor Leonard J. Savage, who suggests that a *regret* matrix be computed first. R's regret is defined to be the difference between the actual payoff resulting from his selection of some pure strategy and the payoff he might have received had he known what C's strategy would be. Thus if R chooses R_1 and nature chooses C_2 in the foregoing matrix, the payoff to R is 1. But if he had known nature would choose the second column, he would have played R_2 so as to obtain 6, the *maximum* payoff in that column. His *regret* is therefore $1 - 6 = -5$. In general, if R plays R_i, the ith row, and nature plays C_j the jth column, R's *regret* is the difference between the resulting payoff and the maximum payoff in the jth column. The regret matrix corresponding to the given payoff matrix is therefore

$$\begin{pmatrix} -1 & -5 & 0 & -1 \\ -3 & 0 & -1 & -2 \\ -2 & -2 & -1 & -2 \\ 0 & -4 & -1 & 0 \end{pmatrix}$$

The Savage criterion prescribes that R pick the row corresponding to the *maximin* regret. The respective minima for rows are -5, -3, -2, and -4. Therefore -2 is the maximin and R_3 the best strategy. We observe that regrets are either zero or negative in value, and therefore the minima actually correspond to the regret of *greatest* magnitude attached to each strategy. When R seeks the maximum of the minimum regrets, he is in fact trying to diminish the magnitude of his regret by bringing it as close to zero (no regret) as possible.

Here R's use of the maximin strategy for the regret matrix has him proceed as if he were applying the Wald criterion to this matrix. There is one difference, however. The choice of R_3 in the above regret matrix does *not* signify that there is a saddle point in the third row. The minimum regret in the third row, namely, -2, occurs in three different positions. In no one of these positions is it true that the entry -2 is also the maximum of its column. Naturally, -2 cannot be the maximum of a column since that must always be zero (no regret). Thus, the Wald criterion would have called for a solution in terms of mixed strategies. Savage, however, does not accept the fundamental assumption used by Wald, namely, that R must proceed as if nature were actually maneuvering against him. Savage sees nature as an opponent with no known objectives or prejudices. Thus, in the present problem all that can be assumed is that nature's strategy is unknown, and therefore R must act so as to have least regret, no matter what she does.

In summary, the four different criteria lead to four different solutions:

Bernoulli-Laplace	R_4
Wald	R_1
Hurwicz	R_2 if $\alpha > 0.2$
Savage	R_3

This situation places the problem squarely in the hands of the practicing statistician. In the case of a particular problem, he will have to use the criterion which seems most suitable for the given facts.

If the Bernoulli-Laplace assumption of equal likelihood is gratuitous and the Wald principle is pessimistic, mathematicians are able to point out that the Hurwicz and Savage philosophies for solution of a decision problem also have shortcomings. Hence, at the present time no criterion has all the characteristics that will satisfy both the pure mathematician and the practical decision-maker.

13

From Dice to Quantum Theory and Quality Control

Statistical decision theory as formulated by Wald might be described as mathematical statistics considered from an advanced point of view. Since one of our objectives is to present mathematical subject matter as it exists today, it has seemed best to present the all-encompassing picture first and fill in special technical details later on. As a result, it has not been possible to provide challenging problems of statistical decision, because they usually require special concepts from probability theory or from the earlier less general species of mathematical statistics. For the sake of significant applications, we must now backtrack. Therefore, the present chapter will resume the discussion of probability theory initiated in Chapter 11, and Chapter 14 will treat some of the basic notions of the twentieth-century mathematical statistics which Wald incorporated into his theory. In both chapters the historic evolution of ideas will be presented, and concepts will be related both to theoretical science and to practical problems of statistical decision.

For the reasons cited, it will be necessary to start the story of probability anew. How did it all begin? It is conjectured that the infinitesimal germs of those notions that were to blossom into the work of Borel, von Neumann, Morgenstern, and Wald may have existed even 40,000 years ago. It now appears that games of chance were already in vogue in that early era and that *astragali*, the knucklebones (hucklebones) of animals, served as dice. The *astragalus* is a small bone in the ankle, just under the heel bone or *talus*. The matter of ancient dice-rolling and the entire history of games of chance have been investigated thoroughly by Dr. F. N. David of the University College, London, who has presented the results in a most interesting and significant paper. She tells us:*

The astragali of animals with hooves are different from those with feet, such as man, dog, and cat. From the comparison in Figure 13.1, we note how in the case of the dog the astragalus is developed on one side to allow for the support of the bones of the feet. The astragalus of the hooved animal is almost symmetrical about a longitudinal axis, and it is a pleasant toy to play with. In France and Greece, children still play games with astragali, and it is possible to buy pieces of metal fashioned into idealized shapes but still recognizable as astragali.

---

* F. N. David, "Studies in the History of Probability and Statistics I," *Biometrika*, Vol. 42 (1955), pp. 1–15. See also, F. N. David, *Games, Gods, and Gambling*, Hafner, New York, 1962.

Sheep   Dog

Figure 13.1   Drawings of the astragali in sheep and dog

Dr. David states that many early games of chance probably originated in Egypt and did not start in Greece as is often claimed. She writes: "However, Herodotus, the first Greek historian, like his present-day counterparts, was willing to believe that the Greeks (or allied peoples) had invented nearly everything."

She quotes Herodotus' account of the famine in Lydia (*ca.* 1500 B.C.):

For some time the Lydians bore the affliction patiently, but finding that it did not pass away, they set to work to devise remedies for the evil. Various expedients were discovered by various persons; dice and hucklebones (i.e., astragali) and ball and all such games were invented, except tables (backgammon), the invention of which they do not claim as theirs. The plan adopted against famine was to engage in games on one day so entirely as not to feel any craving for food, and the next day to eat and abstain from games. In this way passed eighteen years.

It is Dr. David's opinion that some sort of theory of probability might have developed early in history if dice had not been associated with divination rather than with mathematics. This was true in classical Greece and Rome, and still holds among the Buddhists in Tibet. Even as late as 1737, John Wesley decided by the drawing of lots whether to marry or not.

Allusions to gambling with knucklebones and dice appear in the literature of all lands throughout the ages. The existence of various gamblers' handbooks, and the evidence of the constant warfare of the Christian Church against the vices associated with gaming, are proof that variants of the early games of chance continued on and on. Be that as it may, there was no genuine mathematical analysis of such games until Cardano's *Liber de Ludo Aleae* (Manual on Games of Chance), written *ca.* 1526, but not published until 1663, almost a century after the algebraist's death.

Cardano's distinction as a physician and the contributions to algebra that made him the leading mathematician of his day are emphasized in histories of science, but he is rarely given the credit he deserves for founding a *theory* of probability. The great Galileo added a few additional theoretical notions in his own work on dice, *Sopra le Scoperte de i Dadi* (On Discoveries About Dice), first published in 1718, although written much earlier. Galileo's manual is remarkably lucid in style, with explanations so thorough and simple as to read like those offered to high school students today. But Cardano's major role and Galileo's interest in probability have suffered neglect by the historians, and it has become traditional to attribute the creation of probability theory to Pascal and Fermat.

It is true that the two French mathematicians contributed to the theory in the course of a correspondence initiated by a gambling problem of Antoine Gombaud, Chevalier de Méré (1607–1684). In 1654 Pascal wrote to Fermat: "Monsieur le Chevalier de Méré is very bright, but he is no mathematician, and that, as you know, is a very grave defect."

De Méré (as he is usually called) knew that the odds favored the appearance of at least one "6" in 4 throws of a single die. His analysis alluded to 36 possible outcomes of rolling a pair of dice, or six times as many as in the case of a single die. Hence he felt that six times as many tosses, that is, 24, should be carried out with a *pair* of dice, and then "*double* 6" would be favored. When he discovered, however, that this was *not* the case, he declaimed loudly that "mathematical theorems are not always true and mathematics is self-contradictory." He put the problem to Pascal and asked for an explanation. Pascal wrote to Fermat about the matter, and in the ensuing correspondence both mathematicians demonstrated, each in a different fashion, that in 4 throws with one die, the odds are 671 to 625 in favor of having at least one "6" turn up, but that the probability of a "double 6" at least once in 24 throws is only about 0.49. A method of arriving at this conclusion will be indicated shortly.

It must be remarked that with Pascal, Fermat, and others who contributed to the development of probability theory, such work was merely incidental to broader scientific activity. So it was with Christiaan Huygens who, as a young man of twenty-seven, not yet the monumental figure who ranked close to Newton in the quality of his discoveries in mechanics, optics, astronomy, and mathematics, wrote a small tract, *De Ratiociniis in Aleae Ludo* (On Reasoning in the Game of Dice). This appeared in 1657 as an appendix to Frans van Schooten's *Exercitationes Mathematicae*. The date was prior to that of publication of Cardano's and Galileo's texts. Hence Huygens' was the *first* printed work on the theory of games of chance, and in this sense he must be considered a founder of probability theory.

Since, by 1500, the human race has been indulging in games of chance for at least four thousand years and possibly for much longer, why was there no probability *theory* until Cardano made a small beginning in his gamblers' handbook *ca.* 1526? One explanation, and it may indicate the major factor, is to be found in religious beliefs and ethical principles. If gambling is considered immoral, then it is also sinful to use games of chance as models of scientific situations.

But after Cardano, Galileo, Fermat, Pascal, and Huygens had made successive contributions to a theory of probability, why was progress still slow? The answer is that general issues and worthwhile applications could not be treated until there was a fully formulated methodology for determining, without resort to enumeration, the number of elements in the large (but finite) aggregates to which probabilistic reasoning may apply. At certain points in previous discussions, the reader must have been aware of this problem. For example, when there were three possible actions, $\{a_1, a_2, a_3\}$, from which to choose in formulating a decision function associated with five possible outcomes, it was asserted that there would be 243 different ways of assigning actions to outcomes. But the only verification suggested was enumeration of all the varied possibilities—$a_1a_1a_1a_1a_1$, $a_2a_2a_2a_2a_2$, $a_1a_1a_1a_2a_3$, $a_2a_1a_3a_1a_3$, etc.

The reader is thus somewhat in the same position as the early probabilists and must therefore consider the elements of the *combinatorial* algebra or combinatorial analysis they began to develop, that is, he must study principles which substitute reasoning for enumeration. As a very simple example, suppose that a man plans to travel from Chicago to London, stopping at New York City en route. He can journey from Chicago to New York via one of five convenient flights, $\{A, B, C, D, E\}$. Next he has three choices—the *Queen Elizabeth 2*, the *France*, or a transoceanic jet. Let us symbolize these possibilities as $\{1, 2, 3\}$. In how many different ways can he plan his trip? One might list all the possible plans—$A1$, $A2$, $A3$, $B1$,... etc., and then count them. But in order to arrive at a general principle, let us consider the planning of the trip as if it were a game with two moves, the first a choice of *letter* and the second a selection of *number*. This "game" can be pictured as a tree in which each path from root to top is a possible plan. Since there are 15 paths (Figure 13.2), as many as terminal vertices at the top of the tree, there are 15 possible

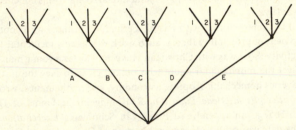

Figure 13.2   Tree for possible trips from Chicago to London via New York

plans. But to help us obtain a general rule, we observe the pattern of the tree. Each lower branch is topped by three branches, so that the paths subdivide into sets of three. Thus if it is planned to take flight A, the paths or possible journeys to London are $\{A1, A2, A3\}$. Similarly, if one were to take flight B, the paths or possible trips would be $\{B1, B2, B3,\}$ etc. There would be 5 such sets of three because there are 5 ways of starting out. Hence there are $5 \times 3 = 15$ paths or possible trips.

Now suppose that there had been 7 suitable flights and 4 choices for transoceanic travel. Then the tree for this situation would exhibit paths in sets of *four* like $\{A1, A2, A3, A4\}$, $\{B1, B2, B3, B4\}$, etc., and since there would be 7 ways of starting the journey, there would be 7 such sets, and hence $7 \times 4 = 28$ paths or possible trips. And, in general, if there were m ways of starting out and n ways of continuing, there would be mn possibilities for the entire journey.

Suppose now that the original travel plan is extended so that following a stay in England, our voyager will go on to the Continent. He considers the following countries as possibilities for his next stop—Belgium, Holland, France, Switzerland. In how many ways can he plan the new three-stage journey? To answer, one need only permit the original tree to grow taller. Let us label the countries listed as

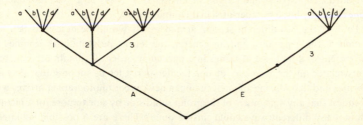

Figure 13.3   Part of tree for three-stage trip

$\{a, b, c, d\}$. In Figure 13.3 part of the taller tree is pictured, and the diagram indicates that the paths to the top (trips from Chicago to the Continent via New York and London) fall into sets of *four*. Each path in the original tree (trip to London) is topped by four new branches. Because trip $A1$ can be completed in four different ways, one has the four paths to the top, $\{A1a, A1b, A1c, A1d\}$. Similarly, journey $A2$ can be rounded out into the trips $\{A2a, A2b, A2c, A2d\}$, etc. Finally, $E3$ can be completed in one of the four ways $\{E3a, E3b, E3c, E3d\}$. There will be $5 \times 3 = 15$ such sets of four, because each set corresponds to a path of the original tree (possible way of traveling to London), and it was shown that there are 15 such paths. Hence, the paths of the taller tree, the three-branch paths, will number $5 \times 3 \times 4 = 60$, that is, there are 60 possible ways of making the three-stage journey. In general, if there are n_1 choices for the first lap of the journey, n_2 for the second, and n_3 for the third, then the trip to London can be planned in $n_1 n_2$ ways, and the entire journey can be planned in $n_1 n_2 n_3$ ways. One can generalize this further by picturing a trip with four or five or six, etc., successive stages. And one can go still further by considering other sequences of actions, in no way connected with planning a trip.

Therefore, by analogy with the cases illustrated, a completely *general* principle for counting is as follows: If one thing can be done in n_1 different ways, and after this is done, a second thing can be done in n_2 ways, and subsequently, a third thing can be done in n_3 ways, a fourth thing in n_4 ways, etc., then the number of ways in which the sequence of things can be done is the *product*, $n_1 n_2 n_3 n_4 \ldots$.

Let is apply the general principle to the problem previously considered, where one had to assign an action or decision to each of five outcomes, and each action had to be chosen from the set $\{a_1, a_2, a_3\}$. To find the number of possible assignments, we can say that $n_1 = 3$, because the action corresponding to the first outcome can be chosen to be a_1 or a_2 or a_3. Next, $n_2 = 3$, because the same choice is available. Then $n_3 = 3$, $n_4 = 3$, $n_5 = 3$. Hence the five choices can be made in $n_1 n_2 n_3 n_4 n_5 = 3 \cdot 3 \cdot 3 \cdot 3 \cdot 3 = 3^5 = 243$ ways. This was the claim made previously, but never proved.

To provide another example, suppose that 6 persons in a theater party have consecutive seats in the same row. In how many different ways can they arrange themselves? Since the first to arrive has a choice of 6 seats, the second a choice of

5, the third a choice of 4, and so on, the answer is $6 \cdot 5 \cdot 4 \cdot 3 \cdot 2 \cdot 1 = 720$, offering an unexpectedly large number of social complications.

Each of the 720 different seating arrangements would be described as a *permutation* of the 6 persons, and, in general, the term "permutation" signifies an arrangement of a finite set of things in a particular order. To obtain other permutations related to the members of our theater party, let us suppose that they are celebrities and that, during an intermission, a newspaper photographer arrives with the request that any 4 members of the group stand side by side to pose for a picture. In how many different ways could this be done? There are 6 possible volunteers for the first position in the lineup, and after someone has filled this position, there are 5 possibilities for the second position, thereafter 4 possibilities for the third position, etc. Then there are $6 \cdot 5 \cdot 4 \cdot 3 = 360$ different ways of satisfying the photographer's request. Each of these represents a permutation, that is, an ordered arrangement of *four* members of the group. If we symbolize the members as $\{A_1, A_2, A_3, A_4, A_5, A_6\}$, then the 360 permutations or possible photographs include arrangements like

$$A_1 A_2 A_3 A_4, \ A_2 A_1 A_3 A_4, \ A_4 A_1 A_3 A_2, \ldots$$
$$A_2 A_1 A_5 A_6, \ A_6 A_2 A_5 A_1, \ A_5 A_2 A_6 A_1, \ldots$$
$$A_2 A_3 A_4 A_6, \ A_4 A_3 A_6 A_2, \text{ etc.}$$

We have applied a fundamental combinatorial principle in order to avoid the need for such enumeration, but nevertheless have listed a few of the possible permutations in order to indicate that *any* subset of size 4 might be drawn from the entire set of 6 persons and that the individuals in such a subset could be arranged in any conceivable order. Therefore, the figure 360 gives the totality of all permutations of all subsets of size 4 which can be formed using members of the entire set of six.

There is a convenient notation for representing answers to some permutation problems. Instead of saying that there are $6 \cdot 5 \cdot 4 \cdot 3 \cdot 2 \cdot 1$ permutations of members of the theater party, one would describe the number as *factorial* 6, symbolized by 6!. Thus, 6! signifies the product of the first six positive integers. If members $\{A_1, A_2, A_3, A_4\}$ volunteer for the requested photograph, then the photographer can permute this group of 4 persons in $4 \cdot 3 \cdot 2 \cdot 1 = 4!$, that is, *factorial* 4 ways. The reasoning which led to 6! and 4! shows that for any set of 6 distinguishable elements (4 elements), whether these be persons, pictures, coins, books, symbols, etc., the number of permutations is 6! (4!). If 10 distinguishable elements are to be permuted, there would be 10 choices for the first position, 9 for the second, 8 for the third, etc., and hence 10! permutations. In general, if there are n distinguishable elements in a set, the number of permutations of these elements is $n!$.

Instead of considering a set of 6 persons, let us return again to a sample space for the rolling of a die, $S = \{1, 2, 3, 4, 5, 6\}$, in order to explain the answers given by Pascal and Fermat to the classic questions raised by the Chevalier de Méré (page 293). At first the Chevalier was concerned with tossing a single fair die 4 times. Let us describe the possible outcomes by sequences like 1-2-6-4, 2-6-6-6,

4-5-2-2, 3-3-3-3, 6-1-6-4, 6-6-6-6, 4-3-5-1, etc. Since De Méré was betting on the occurrence of a "6" at least once in 4 throws, the favorable sequences in the above list would be 1-2-6-4, 2-6-6-6, 6-1-6-4, 6-6-6-6. What is the total number of sequences, and how many favor De Méré's bet? There are 6 possibilities for the first toss of the die (first number in the sequence), 6 for the second toss, etc., so that there would be $6 \cdot 6 \cdot 6 \cdot 6 = 1296$ possible sequences, all considered *equally likely* in classic probability theory. To find how many of these are "favorable," it is easier to compute the number of *unfavorable* outcomes and subtract from the total. For each toss (position in the sequence) there are 5 unfavorable possibilities, namely any of the numbers $\{1, 2, 3, 4, 5\}$. So failure for a "6" to appear could occur in $5 \cdot 5 \cdot 5 \cdot 5 = 625$ different ways. Since $1296 - 625 = 671$, a "6" would appear once or oftener in 671 games out of 1296, so that the probability is $671 \div 1296 = 0.52$, and the odds are 671 to 625 in favor of having a "6" turn up at least once.

Now the second of De Méré's wagers concerned a pair of dice. Let us describe the possible outcomes as 1-1, 1-2, 1-3, 1-4, 1-5, 1-6, 2-1, 2-2, 2-3, etc. If the reader wishes to distinguish between outcomes like 1-2 and 2-1, for example, he can think of rolling a pair of dice of different colors, say, black and white. Then "black 1, white 2" is a different outcome from "black 2, white 1." Since there are 6 different ways in which the black die can turn up and the same is true of the white die, there should be $6 \times 6 = 36$ terms in the above list of potential outcomes. But the pair of dice is to be rolled 24 times, Since there are 36 possibilities for the first toss, 36 for the second, 36 for the third, etc., the total number of possible sequences for the 24 tosses is $(36)^{24}$. Because De Méré's bet was on "double 6," that is, 6-6, the other 35 ordered number pairs can be considered unfavorable. Then there are $(35)^{24}$ unfavorable results in 24 tosses of a pair of dice, and the probability that "double 6" will never appear in the course of these 24 trials is $(35)^{24}/(36)^{24} = 0.51$. (This approximation can be obtained by using logarithms.) Therefore the probability that "double 6" will appear one or more times is 0.49, the fact that seemed so utterly puzzling to De Méré.

De Méré's difficulties may have stimulated Pascal and Fermat to formulate combinatorial principles, but a much more innocent situation will help to explain an important probabilistic concept which will, in turn, make it possible to prove a fundamental law. Let us, then, return to consideration of the random device described at the end of Chapter 10, namely, a sack containing 30 blue and 10 red marbles. If a marble is drawn at random, the probabilities are P (blue) = 3/4 and P (red) = 1/4. If we are now told that 32 of the marbles are small and 8 large, we can compute P (small) = 4/5 and P (large) = 1/5. Suppose that the complete classification according to color and size is

	Blue	Red	Total
Small	27	5	32
Large	3	5	8
Total	30	10	40

This tabulation makes it possible to compute P (small and blue) $= 27/40$. P (large and red) $= 5/40$, etc.

Imagine that someone draws a marble from the sack, informs you that it is *large*, and asks you to make a bet about its color. To figure your chances, you wish to assign a value to P (blue | large), which is read as "the probability of *blue*, given *large*." Then you must consider only the 8 large marbles to which the second row of the tabulation refers. Since 3 of those marbles are blue, P (blue | large) $= 3/8$ and P (red | large) $= 5/8$.

Each of these figures is described as a *conditional* probability, in contrast with the original "unconditional" or "absolute" probabilities. We observe in the example above that the conditional probabilities for *blue* and *red* (3/8, 5/8) are less and greater, respectively, than the corresponding unconditional probabilities (3/4, 1/4). Thus the concept of conditional probability makes a species of "statistical inference" possible, because the revision of the original probability estimates is a conclusion based on experimental evidence (in the illustration, the fact that the marble drawn is large).

The example considered also shows that a conditional probability results from a reduction in the original sample space. The absolute probabilities for *blue* and *red* were based on a sample space of 40 elements, whereas the conditional probabilities referred to a set of only 8 elements, which was a *proper subset* of the original sample space.

To make sure that the reader understands the concept of conditional probability, he should proceed as follows.

(1) Interpret symbols like P (red | small) $=$ "the probability of *red*, given *small*."
(2) Evaluate some conditional probabilities like P (red | small), P (large | blue), P (small | red). (The answers are 5/32, 1/10, 1/2.)
(3) Observe that the reduced sample spaces in (2) contain 32, 30, 10 elements, respectively.
(4) Compare the answers in (2) with the corresponding unconditional probabilities, 1/4, 1/5, 4/5.

Let us now make a more detailed analysis of the computation of a conditional probability. For example,

$$P \text{ (blue | large)} = \frac{3}{8} = \frac{3/40}{8/40} = \frac{P \text{ (large and blue)}}{P \text{ (large)}}$$

This equation expresses a conditional probability as the quotient of two unconditional probabilities and suggests the general definition: The conditional probability of E_2, given E_1, is symbolized by $P(E_2 | E_1)$ and is expressed by the formula

$$P(E_2 | E_1) = \frac{P(E_1 \text{ and } E_2)}{P(E_1)}$$

providing $P(E_1) \neq 0$. (In those cases where E_1 has probability zero, a conditional probability cannot be defined.) But if $P(E_1) \neq 0$, we can multiply both sides of the above equation by this probability to obtain

$$P(E_1 \text{ and } E_2) = P(E_1) \cdot P(E_2 \mid E_1)$$

a theorem expressing the important *multiplication law* of probability theory. Although we have proved the multiplication formula only for the case where $P(E_1) \neq 0$, it is valid even when $P(E_1) = 0$, if it is agreed that the product in the right member of the formula is zero whatever the value of $P(E_2 \mid E_1)$.

In Chapter 11 it was emphasized that an event can be defined by a set of points which is a subset of the sample space of interest. Then if two events, E_1 and E_2, are defined by such point sets, the event "E_1 and E_2" is the *intersection* of the two sets. This intersection is customarily symbolized by $E_1 \cap E_2$. The intersection of two sets consists of all points (elements) common to those two sets. But the present discussion will employ the terminology "E_1 and E_2" rather than the symbolism $E_1 \cap E_2$, partly because events will usually be defined by verbal statements and partly because we wish to avoid confusion with $E_1 \cup E_2$, the *union* of two sets (pages 20 and 261), which corresponds to "E_1 and/or E_2."

The multiplication law given above is a formula for the probability of $E_1 \cap E_2$. It is useful for assigning probabilities in problems like: A sack contains 10 blue and 15 red marbles. One marble is drawn at random, and then a second marble is drawn at random from the remaining marbles. What is the probability that the first marble is red and the second blue? Here $E_1 = $ "first red" and $E_2 = $ "second blue." Then

$$P \text{ (first red and second blue)} = P \text{ (first red) } P \text{ (second blue} \mid \text{first red)}$$

$$= \left(\frac{15}{25}\right)\left(\frac{10}{24}\right) = \frac{1}{4}$$

In our illustrations the conditional probabilities have all differed from the corresponding absolute or unconditional probabilities. But a conditional probability may in some instances be equal to the original "absolute" probability. For example, consider the case where a coin is tossed and then a die is rolled. If the possible outcomes are described by the sample space $S = \{H1, H2, \ldots, H6, T1, \ldots, T6\}$, the probability that the die will turn up deuce, that is, the probability of the event $\{H2, T2\}$, is $P \text{ (deuce)} = 2/12 = 1/6$. Suppose that the experimenter tells us that the coin has turned up "heads." This information leads to the reduced sample space $S_1 = \{H1, H2, \ldots, H6\}$, containing only 6 outcomes, and in this space the event, "die will turn up deuce" is described by $\{H2\}$. Hence the *conditional probability* $P \text{ (deuce} \mid \text{heads)} = 1/6$, which is equal to the original "unconditional" probability for deuce. This result is not surprising, because the additional information that the coin has turned up "heads" seems to have no bearing on the behavior of the die. One would say that the event "die turns up deuce" is *independent* of the event "coin turns up heads," a fact borne out by the equality of the conditional and "absolute" probabilities computed above.

In general, if $P(E_2 | E_1) = P(E_2)$, event E_2 is said to be *independent* of E_1. In that case, substitution of $P(E_2)$ for $P(E_2 | E_1)$ in the *multiplication law* gives

$$P(E_1 \text{ and } E_2) = P(E_1)P(E_2)$$

as the condition for E_2 to be independent of E_1. Now, if $P(E_1) = 0$, the conditional probability $P(E_2 | E_1)$ is undefined, as explained earlier. Hence we could not compare it with the unconditional probability $P(E_2)$ in order to test for independence, nor could we carry out a substitution in the multiplication law, as specified above. Nevertheless, it is customary to accept the above equation as a criterion of independence even when $P(E_1) = 0$.

In summary, then, if

$$P(E_1 \text{ and } E_2) = P(E_1) \cdot P(E_2)$$

E_2 is said to be independent of E_1. Observe that this equation can also be written in the form

$$P(E_2 \text{ and } E_1) = P(E_2) \cdot P(E_1)$$

because the order of factors in the right member is immaterial, and because the event $(E_2 \text{ and } E_1)$ is obviously the same as $(E_1 \text{ and } E_2)$. But the alternative form of the equation expresses the condition for E_1 to be independent of E_2. Hence, if E_2 is independent of E_1, then E_1 is also independent of E_2. Thus, independence is a symmetrical relation which makes it possible to speak of a *pair* of independent events.

The equation formulated repeatedly above therefore provides a *definition* for two independent events, E_1 and E_2. In other words, two events are said to be independent if, and only if, the probability of their joint occurrence is equal to the product of their absolute probabilities.

To apply this definition, consider the experiment where one card is drawn at random from a bridge deck. Are the events E_1, a spade, and E_2, a picture card, independent? We see that $P(E_1) = 13/52 = 1/4$ and $P(E_2) = 12/52 = 3/13$. Now the event "spade and picture" is $(E_1 \text{ and } E_2) = \{$king of spades, queen of spades, jack of spades$\}$, and thus $P(E_1 \text{ and } E_2) = 3/52$. Since

$$\frac{3}{52} = \left(\frac{1}{4}\right)\left(\frac{3}{13}\right)$$
$$P(E_1 \text{ and } E_2) = P(E_1) \cdot P(E_2)$$

and therefore events E_1 and E_2 are independent. One might have anticipated this result on intuitive grounds, because the fact that a card is a spade gives no clue whatsoever as to whether or not it is a picture card, and vice versa.

With reference to the same random experiment, what is the probability that a card will be either a spade or a picture card, or both? What we seek, then, is $P(E_1 \text{ and/or } E_2)$, that is, $P(E_1 \cup E_2)$. To count favorable cases, one might compute

$$13 \text{ spades} + 12 \text{ picture cards}$$

But this would include the king, queen, and jack of spades twice. Therefore one must subtract the three cases that are *both* spades and picture cards, so that

$$P\,(E_1 \text{ and/or } E_2) = \frac{13 + 12 - 3}{52}$$

$$= \frac{13}{52} + \frac{12}{52} - \frac{3}{52}$$

From the results of the previous example, we recognize that the three fractions in the right member are $P(E_1)$, $P(E_2)$, $P(E_1 \text{ and } E_2)$, respectively, so that the equation expresses the fact that

$$P\,(E_1 \text{ and/or } E_2) = P\,(E_1) + P\,(E_2) - P\,(E_1 \text{ and } E_2)$$

The answer to the problem is $22/52 = 11/26$, but we are much more interested in the fact that the last equation above expresses a general law, one which includes as a special case the axiom applicable to mutually exclusive events (page 263). Thus, if one uses the general formula to compute the probability that a card drawn from a deck is E_1, a spade, or E_2, the queen of diamonds, then $P\,(E_1 \text{ and } E_2) = 0$, since a card cannot be both a spade and a diamond. Therefore the required probability is

$$\frac{13}{52} + \frac{1}{52} = \frac{14}{52} = \frac{7}{26}$$

If E_1 and E_2 cannot both occur, that is, if $P\,(E_1 \text{ and } E_2) = 0$, the general formula reduces to that for mutually exclusive events, namely,

$$P\,(E_1 \text{ or } E_2) = P\,(E_1) + P\,(E_2)$$

To apply some of the above probabilistic principles to a problem related to statistical decision, we shall consider an example similar to a hypothetical situation described in the previous chapter. There we pictured that weather data would yield empirical probabilities for the occurrence of a mild, average, or severe winter, or else that, lacking such data, one might possibly assume equal likelihood for the occurrence of each state of nature in the winter ahead. In either case the probabilities would be assigned *prior* to the experimentation associated with the statistical decision problem. The incorporation into the formula for conditional probability of some of the probabilistic laws discussed above leads to an important principle of statistical inference, called *Bayes' rule*, which provides a method of reassessing probabilities *after* the experimentation associated with statistical decisions is carried out. We remark in passing that the principle was formulated in 1763 by the Reverend Thomas Bayes, a British clergyman with a strong bent for mathematics.

For a specific example of the Bayes technique, let us consider a question that would arise in a statistical decision problem related to production in a small factory having only two machines A and B. Suppose that past experience indicates that A produces 30 per cent of the articles with about 1 per cent defective, and that among the 70 per cent produced by B, about 3 per cent are defective. The experimentation to be carried out consists in selecting a package of items at random, and inspecting one article drawn at random from it. If the article turns out to be defective, what is the probability that it was manufactured by A?

What is sought is $P\,(A\,|\,\text{defective})$, and by the definition of conditional probability,

$$P\,(A\,|\,\text{defective}) = \frac{P\,(A \text{ and defective})}{P\,(\text{defective})}$$

Now the multiplication law can be applied to the numerator of the fraction on the right to yield

$$P\,(A \text{ and defective}) = P\,(A)\,P\,(\text{defective}\,|\,A)$$
$$= (0.3)(0.01) = 0.003$$

But the denominator, $P\,(\text{defective})$, concerns an event that can occur in two *mutually exclusive* ways, namely, as the result of manufacture by A or by B. Therefore,

$$P\,(\text{defective}) = P\,(A \text{ and defective}) + P\,(B \text{ and defective})$$
$$= 0.003 + (0.7)(0.03)$$
$$= 0.003 + 0.021 = 0.024$$

Substituting the numerical values just obtained, we have

$$P(A\,|\,\text{defective}) = \frac{0.003}{0.024} = \frac{1}{8}$$

Thus the probabilities for manufacture by A or B are 1/8 and 7/8, respectively. If we imagine that the "states of nature" in the decision problem refer to manufacture by A or by B, then we have here a reassessment of the probabilities for these states of nature, since they were given a priori as 0.3 and 0.7, respectively.

The above computation of revised probability estimates involved substitution (with the aid of two fundamental principles) in the formula for conditional probability. Now Bayes' rule is nothing but a formulation of this procedure. To use the customary symbolism for the above problem, it is usual to say that one has two states of nature or two "hypotheses"—H_1 and H_2—and that the experiment has resulted in the event E. Then the computations carried out above can be formulated as

$$P\,(H_1\,|\,E) = \frac{P\,(H_1)\,P\,(E\,|\,H_1)}{P\,(H_1)\,P\,(E\,|\,H_1) + P\,(H_2)\,P\,(E\,|\,H_2)}$$

with a similar formula for $P\,(H_2\,|\,E)$. The pair of formulas constitutes Bayes' rule when there are two states of nature. The expression of Bayes' rule where there are n, any finite number of states, consists of n formulas like the above, with n terms in the denominator of each. This rule is considered the first instance in mathematical history of a specific method of statistical inference.

Let us now leave the macrocosm within which Bayes operated in order to examine further details of the microcosm of combinatorial analysis. A little earlier, we spoke of a theater party of 6 persons and showed that they could take their seats in 6! ways. During the first intermission at the theater, 4 persons were to volunteer for a photograph, and we saw that this gave rise to $6 \cdot 5 \cdot 4 \cdot 3 = 360$ possibilities. During the second intermission, which is lengthy, members of the group have time to talk about all sorts of things—impressions of the play, personal matters, etc. Someone mentions that an organization to which they all belong has asked for a committee of 4 volunteers to serve in a fund-raising drive. "Since A_1, A_2, A_3, and A_4 will appear in a newspaper picture, they will become better known and would therefore be a good committee for the special purpose," A_5 comments. But the others disagree. Hence a natural question is: In how many different ways could a committee of 4 be selected from the 6 members of the theater party?

The number of potential committees is *not* 360, since the issue is *not* a question of permutation. The 4 committee members need *not* be named in one particular order. Members A_1, A_2, A_3, A_4 may have posed for the picture in the order named or in any one of the 4! different orders in which these 4 individuals can be lined up. They would nevertheless constitute only a *single committee*. The same fact could be asserted about $\{A_3, A_4, A_5, A_6\}$, or about $\{A_1, A_3, A_4, A_5\}$, or any subset of 4 members. Each such subset would constitute one committee but would provide 4! or 24 permutations. Therefore there are 24 times as many permutations of 4 members as there are committees, and the number of different committees of four is $360 \div 24 = 15$. In general, if a set contains 6 elements of any nature, there are 15 different *unordered* subsets containing 4 elements.

The symbol $\binom{6}{4}$ is often used to represent the number of unordered subsets of 4 elements that can be selected from a set of 6 elements. Such unordered subsets were formerly called *combinations* to distinguish them from permutations, which are *ordered* subsets. The name "combination" is the origin of the term *combinatorial analysis (algebra)*.

If the committee or subset selected is $\{A_1, A_2, A_3, A_4\}$, there is a subset containing the *two* members omitted, namely, $\{A_5, A_6\}$. Again, if $\{A_1, A_3, A_5, A_6\}$ is the subset picked, then the subset $\{A_2, A_4\}$ contains the *two* members not selected. Therefore each selection of a committee or subset of 4 members is inevitably accompanied by the simultaneous designation of a residual subset of 2 members. One could reverse the procedure and thus observe that selecting a committee of 2 members involves the simultaneous choice of a subset of 4. Therefore

$$\binom{6}{4} = \binom{6}{2} = 15$$

One can verify this last result by computing $\binom{6}{2}$ directly.

Instead of saying that a selection of a subset of 4 elements from a set of 6 leaves a residual subset of 2 elements, one can say that the act of choice effects a *partition* of the entire set into subsets containing 4 and 2 elements, respectively. The number of partitions of this type is symbolized by

$$\binom{6}{4,\ 2}$$

and therefore

$$\binom{6}{4,\ 2} = \binom{6}{4} = \binom{6}{2} = 15$$

To obtain some practice in evaluating and interpreting "combinations," consider

$$\binom{9}{6} = \binom{9}{3} = \binom{9}{6,\ 3}$$

These symbols represent the number of subsets of size 6 in a set of size 9, the number of subsets of size 3, and the number of partitions into subsets containing 6 and 3, respectively. These numbers, moreover, are equal to one another. Let us evaluate $\binom{9}{3}$ by imitating the procedure followed in the earlier illustration. Then we must first treat the problem as if it were a question of finding permutations of 3 elements chosen from the entire set of 9 elements. There would be $9 \times 8 \times 7$ such permutations. But a subset like $\{A_1, A_2, A_3\}$ or any other subset of 3 elements would give rise to 3! permutations. Therefore the number of permutations is 6 times as large as the number of combinations and

$$\binom{9}{3} = \frac{9 \cdot 8 \cdot 7}{3!} = \frac{9 \cdot 8 \cdot 7}{1 \cdot 2 \cdot 3} = 84$$

Because any selection of a subset of size 3 leaves a residual subset of size 6, and vice versa,

$$\binom{9}{6} = \binom{9}{3} = 84$$

The concepts of partition and combination are useful in those areas of modern quantum mechanics which require that the elementary particles of a mechanical system be assigned at random to small "cells" or regions of an (abstract) space. In issues involving electrons, protons, neutrons, and all particles with "fractional spin," the *Fermi-Dirac statistical mechanics*, named after Enrico Fermi (1901–1955) and Professor Paul Dirac of Cambridge University, is a suitable model. In that subject the particles are considered indistinguishable, and only one may be assigned to a cell. Particles thus restricted are called *fermions*, a term used in combinatorial algebra for indistinguishable elements of *any* sort when these are to be placed in various categories, with a maximum of one element in any category. Suppose that there are 6 fermions and 20 cells. Then it is a matter of choosing from the entire set of 20 cells the subset of 6 cells that is to be occupied by the fermions. Reasoning in the fashion explained above,

$$\binom{20}{6} = \frac{20 \cdot 19 \cdot 18 \cdot 17 \cdot 16 \cdot 15}{1 \cdot 2 \cdot 3 \cdot 4 \cdot 5 \cdot 6} = 38,760$$

possible ways of assigning the fermions.

In the quantum mechanics of those particles which have "whole number spin," namely, photons, pions, and K-mesons, the *Bose-Einstein statistical mechanics* applies. The *bosons*, so named after the Indian statistician S. N. Bose, are like fermions in being indistinguishable, but differ from the latter in that any number may occupy a cell. Again, in combinatorial problems, the term *boson* is applied to *any* indistinguishable elements which are to be distributed into categories with no limitations on the number per category. Thus, if there are 3 bosons and 6 cells, one possible interpretation is the problem of how to present 3 identical coins to 6 children, and another is the question of what stops may occur when 3 passengers are to be discharged from an elevator which can stop at 6 floors. Some of the possible distributions of 3 bosons into 6 cells are

```
 b  |     |     |  b  |     |  b
bb  |     |     |     |     |  b
    |  b  |     |  bb |     |
    |     |     |     |     |  bbb
    |     | bbb |     |     |
```

From these it is seen that 8 positions are involved, namely, those of the 3 bosons and the 5 bars separating the 6 cells. Then the problem becomes that of selecting from 8 positions the 3 positions in which the bosons should be placed. If it were a matter of permutation, there would be 8 choices of position for the first boson, 7 choices for the second, 6 for the third, so that there would be $8 \times 7 \times 6$ arrangements possible. But the principle of combinations rules that this figure must be divided by $3!$. Therefore the number of different groups of 3 positions, or the number of ways the bosons can be distributed, is

$$\frac{8 \cdot 7 \cdot 6}{1 \cdot 2 \cdot 3} = 56$$

This problem might have been interpreted in terms of the dividing bars: From 8 possible positions, how many different ways can a combination of 5 be selected for the 5 dividing bars? The answer would be $\binom{8}{5} = 56$, which must, of course, be the same as the previous result. Moreover, if the problem were to place 5 bosons in 4 cells, there would still be 56 possible ways of accomplishing this, since there would again be 8 positions, this time for 5 bosons and 3 bars dividing the 4 cells, and one would have to choose either the 5 boson positions or the 3 for bars so that the number of placements would be figured either as $\binom{8}{5}$ or $\binom{8}{3}$.

Because boson analysis is not limited to quantum mechanics, one of the preceding results says that there are 56 ways of distributing 3 identical coins among 6 children (if a child may be given one or more coins) and 56 ways an elevator may discharge 3 passengers when stops on 6 floors are possible. Again, one can apply the same analysis to dice-rolling. If 3 dice are tossed, there are $6 \times 6 \times 6 = 216$ possible permutations. If the 3 dice were different in color or size, these 216 permutations would be distinguishable. Thus {red 1, black 2, white 6} would appear as a different occurrence from {red 2, black 1, white 6} or {red 6, black 1, white 2}. If the dice are alike, however, these occurrences are indistinguishable, and gamblers eventually discovered that there are only 56 distinguishable tosses. To reason this out combinatorially instead of by enumeration, one can think of 6 cells (or *categories* corresponding to the 6 ways any die can turn up), and then consider the 3 dice as bosons. This is permissible since they are indistinguishable, and several of them may turn up the same way, that is, be in the same category or "cell." Then, since there are 3 bosons and 6 cells, this is the same instance already considered and hence it can be figured, as previously, that there are 56 distinguishable occurrences.

In the early gambling era of probability theory, this figure was computed and applied to matters very different from Bose-Einstein statistical mechanics. In fact, around 1000 A.D., a bishop, Wibold of Cambray, applied the problem of three dice to clerical purposes. A description is given by the British mathematical statistician M. G. Kendall.*

Wibold enumerated 56 virtues—one corresponding to each of the ways in which three dice can be thrown, irrespective of order. Apparently a monk threw a die three times, or threw three dice, and hence chose a virtue which he was to practice during the next twenty-four hours. It does not sound like much of a game, but the important point is that the falls of dice were correctly counted. There was no attempt at assessing relative probabilities.

The use of dice for the purpose of choosing among a number of possibilities may well be much older than Wibold and continued for long after his time. There exist several medieval poems in English, setting out the interpretations to be placed on the throws of three dice. The best known is the "Chaunce of the Dyse" which is in rhyme royal; one verse for each of the 56 possible throws of three dice.

The earliest approach to the counting of the number of ways in which three dice can fall (permutations included) appears to occur in a Latin poem *De Vetula*. This remarkable work was regarded as Ovid's for some time, and is included among certain medieval editions of his poems. It is, however, suppositious, and several candidates have been proposed for authorship. The relevant passage may be briefly and freely construed as follows:

---

* M. G. Kendall, "Studies in the History of Probability and Statistics II," *Biometrika*, Vol. 43 (1956), pp. 1–14.

If all three numbers are alike, there are six possibilities; if two are alike and the other different, there are 30 cases, because the pair can be chosen in 6 ways and the other in 5; and if all three are different, there are 20 ways, because $30 \times 4 = 120$, but each possibility arises in 6 ways. There are 56 possibilities.

Perhaps the reader may find this explanation in *De Vetula* easier to understand than the boson method used earlier. At any rate, before we leave the question, its relevance to probability theory must be pointed out. The assumption in rolling fair dice is that each of the 216 *permutations* is equally likely to occur, or that each has the probability of occurrence 1/216. This reasoning treats the permutations as if they were distinguishable. Thus a throw of $\{1, 3, 5\}$ can occur in 6 ways {red 1, black 3, white 5}, {red 1, white 5, black 3}, etc., so that the probability of such a throw is $6/216 = 1/36$ and a throw of $\{1, 1, 3\} = $ {red 1, black 1, white 3}, {red 1, black 3, white 1}, {red 3, black 1, white 1} can occur in 3 ways, so that the probability of such a throw is $3/216 = 1/72$. In the "classical" Maxwell-Boltzmann-Gibbs statistical mechanics (Chapter 15), devised for thermodynamic application, the assumptions made concerning equal likelihood are similar to those in dice-rolling.

But the Fermi-Dirac and Bose-Einstein types of statistical mechanics are models based on experimentation related to the various particles of modern quantum theory, and empirical results justify the assumption of equal likelihood *not* for the permutations, but for the various *distinguishable* groupings of fermions or bosons. Thus, since there are 56 distinguishable distributions of 3 bosons over 6 cells, any one of these would be assumed to have a probability of 1/56. Referring to the case of 3 dice where there are also 56 distinguishable cases, the probability of $\{1, 3, 5\}$ was seen to be 1/36. The toss $\{1, 3, 5\}$ can be interpreted as a die in category or "cell" 1, another in "cell" 3, another in "cell" 5. But the probability that the quantum bosons will occupy cells 1, 3, 5 is equal to 1/56. Here we see a difference between the model for fair dice and the Bose-Einstein model for certain types of particles. Again, take the dice throw $\{1, 1, 1\}$, whose probability is 1/216. For the quantum bosons, this would mean that all three are in the first cell, an occurrence with probability 1/56. To see the difference among the models for dice (Maxwell-Boltzmann-Gibbs), bosons, and fermions, consider the distribution of 3 fermions over 6 cells. Then the dice throw or boson placement $\{1, 1, 1\}$ is impossible for fermions; that is, such placement has probability 0 for fermions, because one fermion at most may occupy a cell, and the presence of all three in the first cell cannot occur. Thus three different probabilities for the same event, namely, 1/216, 1/56, and 0, have been obtained in the three different models. If this fact appears puzzling, one must recall that the theorems of a mathematical model, whether it is geometry or statistical mechanics, are deduced from the axioms or assumptions. In the three kinds of statistical mechanics, in order to describe properly the elements involved, there is a difference in the selection of the postulate concerning equal likelihood. This is logically permissible, just as, in the case of Euclidean, Lobachevskian, and Riemannian geometries, there are different parallel postulates.

As a somewhat different application of combinatorial analysis, consider a "single sampling inspection" scheme where the manufacturer ships items in lots of

50 and tests quality by selecting 2 items at random from each lot. He applies a *decision function* prescribing that if both articles are defective, the lot is to be rejected, otherwise it is to be accepted. In the technical language of quality control, the *acceptance number* is 1, which means that if the number of defectives does not exceed 1, the lot is to be accepted. One might ask: What is the probability that a lot 30 per cent defective, that is, containining 15 defective items, will be accepted?

First, one can readily figure the probability that the lot will be rejected. By combinatorial algebra, there are $(50 \cdot 49)/(1 \cdot 2) = 1225$ different possible ways a sample of size 2 can be selected, and among these there will be $(15 \cdot 14)/(1 \cdot 2) = 105$ samples in which both items are defective. Hence, the probability of rejection is $105/1225 = 3/35$, and therefore the probability of acceptance is 32/35, that is, the chances are better than 9 in 10 that the lot will be accepted.

In the study of the different decision functions associated with various sampling schemes, an *operating characteristic* function may be helpful. This function indicates how the probability of acceptance depends on the *quality* of the lot, that is, the proportion of defective items. If p represents the proportion defective in the lot, then, as in the preceding example, the number defective is $50p$; the probability of rejection is

$$\frac{50p\,(50p - 1)}{50 \cdot 49}$$

and if y represents the probability of acceptance,

$$y = 1 - \frac{50p\,(50p - 1)}{50 \cdot 49}$$

or

$$y = 1 + \frac{p}{49} - \frac{50p^2}{49}$$

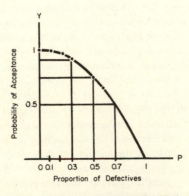

Figure 13.4 Operating characteristic (*OC*) curve for sampling inspection scheme

This equation is called the *operating characteristic* of the particular sampling scheme and its graph (Figure 13.4) is called the corresponding *OC* curve.

There may be 1 or 2 or 3 or . . . 50 defectives in a lot, and hence *p* may have the value 0.02 or 0.04 or 0.06 or . . . 1. Hence, although the *OC* curve appears to be continuous, strictly speaking, only those 50 points which correspond to the above values of *p* have meaning for sampling inspection of lots of size 50. In most industrial situations there are many more items in a lot—500 or 1000, say—and therefore 500 or 1000 pertinent points, very close to one another on the *OC* curve. For the sake of convenience, this curve is graphed as a continuum and not as a set of 50 or 500 or 1000 "discrete" points.

The *OC* curve of Figure 13.4 shows that if quality is standardized at $p = 0.3$, that is, if all lots with *p* less than 0.3 are considered "good" and all others "bad," then the chance of acceptance of any "good" lot is greater than 0.9. But "bad" lots, unless they are very bad, have a good chance of passing inspection. For example, the *OC* curve indicates that lots that are 50 per cent defective have a probability of 0.75, or 3 chances in 4, of being accepted, and even those that are 70 per cent defective have a 50-50 chance of acceptance.

Figure 13.5 shows the *ideal* form for the *OC* curve of a sampling scheme if

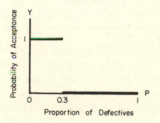

Figure 13.5   Ideal *OC* curve if quality is standardized at $p = 0.3$

quality is standardized at $p = 0.3$. This ideal *OC* curve would guarantee acceptance for "good" lots (probability 1) and assure rejection (probability of acceptance = 0) for "bad" lots. No decision function associated with a sampling plan will lead to an *OC* curve exactly like that of Figure 13.5, but the more closely the curve approximates the ideal, the better the decision rule. The *OC* curve of Figure 13.4 approximates the ideal up to $p = 0.3$, and therefore the sampling plan it represents is a satisfactory one, providing the quality of production is good. But this *OC* curve is very far from the ideal for values of *p* between 0.3 and 1. Figure 13.6*a* and 13.6*b* show the effect on an *OC* curve of some single sampling scheme if the acceptance number is decreased or the sample size is increased. Common sense, as well as an examination of Figure 13.6*b*, would lead to an increase in the size of the sample inspected as a remedy for the specific weakness which Figure 13.4 indicated in the original decision rule.

By taking some clues from the comparison of *OC* curves, we can furnish a more realistic solution to a statistical decision problem considered in the preceding chapter (pages 283–285). There we treated the question as a two-person game in

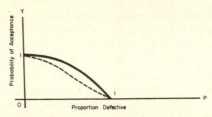

**Figure 13.6(a)   Effect on OC curve of decrease in acceptance
number (--- represents OC curve corresponding to smaller acceptance number)**

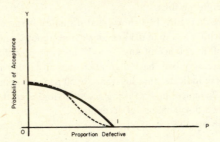

**Figure 13.6(b)   Effect on OC curve of increase in sample size
(--- represents OC curve corresponding to larger sample)**

which there is a theoretical conflict between the statistician and nature. But instead of imagining that nature is maneuvering against the statistician, one might merely assume that she has selected a definite strategy which is either completely unknown to the statistician or which, on the other hand, he may be able to estimate by means of observation. If the latter situation obtains, then nature's strategy becomes known, and the problem is reduced to a one-person game, that is, a simple game of chance. An *OC* function makes the merit of a particular decision rule dependent on the relative frequency of defective items in the manufacturing process. Using this same idea, we shall now consider nature's pure strategy to be p, the proportion of defective items she "chooses" to produce.

Up to this point, we have followed the pattern used in formulating an *OC* function. Now we depart in order to include a factor which the *OC* function neglects, namely, the question of gain or loss, an important feature of game theory and the Wald decision process. For each possible strategy of nature, that is, for each value of p, a decision rule will assign a specific expected payoff or gain, and thus an *expected gain function* is defined. Figure 13.7 gives the four gain functions for the four decision rules of the previous chapter. The combinatorial analysis involved in deriving these gain functions will be explained subsequently. At this point, Figure 13.7 will be applied to the selection of optimum decision rules.

First, let us imagine a manufacturing process which is in its early stages, so that not much empirical data is available for the estimation of p. Since p (in the mathematical model) can be any real number between 0 and 1, inclusive, nature's available pure strategies are infinite in number and, in theory, form a *continuum*.

Thus we picture the statistician R as playing a two-person game of strategy against nature, C. Then in Figure 13.7 the graph for each decision rule is analogous to a matrix row, except that the number of payoffs in a "row" cannot, in the present case, be read off one after the other, since they constitute a continuum. For the same reason, one cannot draw all the matrix "columns" in Figure 13.7, but whenever we wish to indicate a particular pure strategy of nature, that is, a specific value of p, we shall indicate the "column" in question by drawing the corresponding line parallel to the Y-axis.

It will now be seen that the infinite-columned "matrix" has a saddle point, that is, a point representing R's maximin and C's minimax. Examining each graph of Figure 13.7 *in toto*, the minimum expected gains for Rules 1, 2, 3, and 4 are -6, -4, $-13/6$, -4, respectively. The maximum of the four minima, that is, R's maximin, is $-13/6$, corresponding to Rule 3. The broken line in Figure 13.7 represents the maximum "payoff" that nature would have to accord the statistician for each value of p. Her minimax is $-13/6$, corresponding to a strategy which can be read as $p = 0.83$ (approximately) in Figure 13.7. The exact value which can be computed from the equations to be formulated later (page 315) is $p = 5/6$. At any rate, since $-13/6$ is both R's maximin and C's minimax, the game has the solution (Rule 3,

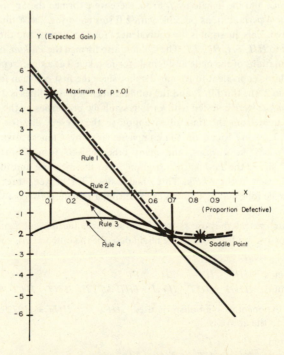

Figure 13.7   Expected gain functions for the four decision rules in the statistical decision game of Chapter 12

$p = 5/6$). Nature's "choice" of $p = 5/6$ is a pessimistic fiction, but in accordance with the Wald theory, Rule 3 is the optimum selection for the statistician.

If the manufacturing process is in a state of *statistical control* so that quality has been stabilized except for small fluctuations, and if enough previous sampling has been done to show that p is approximately 0.1, then nature's strategy is uniquely specified and R plays a one-person game by examining Figure 13.7 and observing that Rule 1 is the best, since it would provide the maximum expected gain (4.8) for the given value of p. Suppose, on the other hand, that the manufacturing process has just begun and it is felt that quality is exceedingly poor, the proportion of defectives being about 70 per cent. Then the diagram indicates that it would make little difference which decision rule is used.

In order to see just how the gain functions of Figure 13.7 have been computed, it will be necessary to return to combinatorial analysis once again, and it is inevitable that games of chance should provide helpful idealizations or models. In the present case, the simplest of all games of chance, coin-tossing, will serve. It is worth noting that problems concerned with the tossing of several coins were considered in the Orient 3000 years ago. In the *I-king*, one of the oldest Chinese mathematical classics, whose probable date is about 1150 B.C., there appear the "two principles": the male *yang*, and the female *ying*. From these were formed the *Sz'Siang* or "four figures," the 4 permutations possible with 2 forms choosing 2 at a time, repetition being allowed. This question is the equivalent of tossing 2 coins, and the four figures correspond to *HH, TH, HT, TT*. The Chinese also formed the *Pa-kua*, or 8 trigrams, the 8 permutations of the male and female forms taken 3 at a time, repetition being allowed. This is equivalent to tossing 3 coins, since the first coin can fall in 2 ways, the second in 2, the third in 2, and the total number of possibilities is $2 \times 2 \times 2 = 8$.

The *Pa-kua* were invested with various symbolic properties by the Chinese and have been used from the time of invention to the present day for purposes of divination. They are found on the compasses used by Chinese diviners and on amulets, charms, fans, vases, and many other objects from China and India. Figure 13.8 shows the *Pa-kua* or 8 trigrams, also the meaning and direction connected with each in the *I-king*. Three males stood for heaven, three females for earth, a female followed by two males for steam, a female between two males for fire, etc.

But our present purpose is merely to associate the ancient figures with the results of tossing modern coins. One might describe the outcomes of coin-tossing as

1	coin	H	T					
2	coins	*HH*	*HT*	*TH*	*TT*			
3	coins	*HHH*	*HHT*	*HTH*	*THH*	*TTH*	*THT*	*HTT*	*TTT*

If the exponential symbolism is used, $HH = H^2$, $HHT = H^2T$, etc. Then an alternative to the above is

1	coin	H	T					
2	coins	H^2	HT	TH	T^2			
3	coins	H^3	H^2T	TH^2	TH^2	T^2H	T^2H	HT^2	T^3

| heaven | steam | fire | thunder | wind | water | mountain | earth |
| S. | SE. | E. | N.E. | S.W. | W. | N.W. | N. |

Figure 13.8   The Pa-kua or eight trigrams

As far as most games of chance are concerned, $HT = TH$, $HHT = HTH$, etc., for the person making a wager on the coins is not usually interested in the arrangement but merely in the actual number of heads or tails. Therefore a third listing of the same facts is

$$
\begin{array}{llll}
\text{1 coin} & H & T & \\
\text{2 coins} & H^2 & 2HT & T^2 \\
\text{3 coins} & H^3 & 3H^2T & 3HT^2 \quad T^3
\end{array}
$$

These are readily recognized as the terms of the algebraic expansions of $(H + T)^1$, $(H + T)^2$, and $(H + T)^3$, respectively. Therefore it may seem reasonable to conjecture that the situation for 4 coins will be symbolized by the terms of $(H + T)^4$, and for any number n of coins by $(H + T)^n$. This fact can actually be proved by the method of "mathematical induction." If the reader will apply the "boson technique" to coin-tossing, he will see that the number of distinguishable occurrences with 3 coins is a matter of distributing 3 bosons (the coins) into 2 categories or "cells" (H or T). This calls for 3 bosons and 1 dividing bar or 4 positions from which 3 (or 1) must be chosen. Therefore, 4 distinguishable occurrences are possible, namely, H^3, H^2T, HT^2, T^3. Similarly with n coins, there are $n + 1$ possibilities. Hence $(H + T)^n$ must have $n + 1$ terms.

In connection with this area of probability theory, Pascal made considerable use of the famous *arithmetic triangle*, pictured below.

$$
\begin{array}{ccccccccccc}
 & & & & & 1 & & & & & \\
 & & & & 1 & & 1 & & & & \\
 & & & 1 & & 2 & & 1 & & & \\
 & & 1 & & 3 & & 3 & & 1 & & \\
 & 1 & & 4 & & 6 & & 4 & & 1 & \\
1 & & 5 & & 10 & & 10 & & 5 & & 1
\end{array}
$$

etc.

For this reason, he is often credited with originating the device, but it now appears that Tartaglia (Chapter 5) and subsequently Cardano were among the first Europeans to consider its properties. (It was known to the Chinese much earlier.) In the triangle, the numbers in any row after the first two are obtained from those in the preceding row by copying the terminal 1's and adding together the successive pairs of numbers from left to right to give the new row. Thus to get the fourth row, put a 1 at each end, and, looking at the third row, say $1 + 2 = 3$, $2 + 1 = 3$ to get the other numbers; for the fifth row, place a 1 at each end and, to obtain the other numbers put $1 + 3 = 4$, $3 + 3 = 6$, $3 + 1 = 4$. Moreover, the first row represents $(H + T)^0$, the second row gives the coefficients for $(H + T)^1$, the third for $(H + T)^2$, the fourth for $(H + T)^3$, etc. Using the fifth line, we can write $(H + T)^4 = H^4 + 4H^3T + 6H^2T^2 + 4HT^3 + T^4$ and get the probabilities for tossing 4 coins. The expansion shows 1 case of 4 heads, 4 cases of 3 heads and 1 tail, 6 cases of 2 heads and 2 tails, etc., out of a total of 16 outcomes considered equally likely in classic probability theory. Thus one obtains the (relative) *frequency function* for the number of heads when 4 fair coins are tossed (or one fair coin is tossed four times). The tabulation of this frequency function is

Number of Heads	0	1	2	3	4
(Long-Run) Relative Frequency or Probability	1/16	4/16	6/16	4/16	1/16

The probabilities in this table could have been obtained merely by substituting $H = 1/2$, $T = 1/2$ in the successive terms of the expansion of $(H + T)^4$, and in the same way the probabilities for the number of heads in tossing a fair coin n times can be obtained from $(H + T)^n$.

The binomial expansion can furnish the appropriate frequency function even when the coin is not fair. Suppose, for example, that the coin is heavily loaded to favor tails and the probability of heads on a single toss is 0.2 (and hence the probability of tails 0.8); the substitution of $H = 0.2$, $T = 0.8$ in $H^2 + 2HT + T^2 = (0.2)^2 + 2(0.2)(0.8) + (0.8)^2$ will give the *frequency function* for the number of heads in 2 tosses of the unfair coin, as

Number of Heads	0	1	2
(Long-Run) Relative Frequency or Probability	0.64	0.32	0.04

This frequency function can be applied to the inspection sampling of 2 items from a manufacturing process leading to 20 per cent defective and 80 per cent satisfactory merely by substituting "number of defectives" for "number of heads" in the preceding tabulation. In the more general case mentioned earlier, where the relative frequency of defectives is unknown because it cannot be specified exactly, the probability that an article will be defective can be represented by p, and the probability that it will be satisfactory by $1 - p$. Then $H^2 + 2HT + T^2 = p^2 + 2p(1 - p) + (1 - p)^2$, which provides the frequency function for Rule 1 (page 284). This frequency function and the payoff or gain associated with each number of defectives in accordance with the rule is

Number of Defectives	0	1	2
(Long-Run) Relative Frequency or Probability	$(1-p)^2$	$2p(1-p)$	p^2
Payoff or Gain (in dollars)	6	0	−6

For the graphs of Figure 13.7, the expected or long-run average gain was used, and in the present case the expected gain for Rule 1 is $6(1 - p)^2 + 0 \cdot 2p(1 - p) - 6p^2 = 6 - 12p$. Therefore the gain function for Rule 1 appears as the straight-line graph $E_1(p) = 6 - 12p$ in Figure 13.7. The gain functions for the other three rules are found in similar fashion. For example, the payoffs for Rule 3, as listed in the matrix for the decision game (page 284), are 2, -3, -2, respectively. Therefore

$$E_3(p) = 2(1 - p)^2 - 3 \cdot 2p(1 - p) - 2p^2$$
$$= 2 - 10p + 6p^2$$

In summary, the four expected gain functions are

$$E_1(p) = \quad 6 - 12p$$
$$E_2(p) = \quad 2 - 6p$$
$$E_3(p) = \quad 2 - 10p + 6p^2$$
$$E_4(p) = -2 + 4p - 6p^2$$

14

Realm of Random Variables

"What is geometry?" This was a question put to Oswald Veblen (1880–1960), the renowned American geometer. "Geometry is what geometers do," he responded in all seriousness. If, by analogy, modern mathematical statistics is what a modern statistician does, then a discussion of Wald's very general statistical theory does not give an adequate picture of the facts. One must fill in the details of the broad outline by explaining the elements of the early twentieth-century statistics developed by W. S. Gosset (1876–1947), Jerzy Neyman, Egon Pearson and, above all, by that world leader in mathematical statistics, Sir Ronald Aylmer Fisher (1890–1962).

Although the roots of probability theory go back to primitive man, the origins of modern mathematical statistics are more recent. The subject was fostered by applied mathematicians like the Belgian L. A. J. Quételet (1796–1874), director of the royal observatory at Brussels, who in 1829 designed and analyzed the first Belgian census. Next came Sir Francis Galton (1822–1911), a cousin of Darwin and, like him, a student of heredity. Modern mathematical statistics took its first giant stride in the contributions of Karl Pearson (1857–1936), British mathematician and biometrician.

Statisticians from Karl Pearson to R. A. Fisher have considered mathematical statistics as a sort of adjunct to the traditional "scientific method." Francis Bacon broke that technique down into a series of steps among which were experimentation, the drawing of tentative conclusions from the observations obtained (formation of hypotheses), and the determination whether the conclusions agree with observed facts. In connection with the first step, the modern statistician has a theory of the proper *design of experiments*. The second and third steps involve the entire issue of *statistical inference*, a subject concerned with the major difficulty in all inductive reasoning, namely, the uncertainty attendant upon forming generalizations on the basis of a limited number of observations.

Of course, one may be able to collect *all* the information about the phenomenon in which one is interested—the ages of all workers in a certain factory, the number of children in every family in a particular small town, the grades of all freshman students in the final examination in mathematics at a certain college. The data might be tabulated, graphed, and summarized by giving the mean or median figure and the range. Such procedures are referred to as *descriptive statistics*. Any inductions like "The men in the factory range from twenty-one to fifty-two years of age" or "All freshman final grades in mathematics were lower than 95 per cent" are trivial in the sense that they are merely summaries.

Descriptive facts become important only if the aggregate being described is *not* the entire set associated with a phenomenon but a random *sample* (proper subset) of a larger *population*. There is inductive inference if one concludes that various attributes of the sample also apply to the population as a whole. Or one engages in *inductive behavior*, Professor Neyman's term, when one takes an action (makes a decision) which a Wald decision function associates with the particular sample observation. As we have said, one purpose of the theory of statistical inference is to *measure* the uncertainty attached to inductive generalization or inductive behavior. Thus, once again there is the matter of assigning appropriate probabilities.

We are now ready to tell what a modern practical statistician does. He may design random experiments and/or he may carry out such experiments, draw conclusions from the results, and measure the fallibility of these conclusions. In particular, he tests statistical hypotheses and estimates population "parameters" on the basis of sample observations.

There are certain basic theoretical concepts associated with all activities of the statistician. First and foremost is the notion of a *random variable*. The number of heads obtained in coin-tossing, the number of defectives in a sample, the number of dollars of payoff or gain in the tabulation on page 314, and the number of aces in repeated throws of a single die are all examples of what the statistician calls a *random variable*. It is evident that throughout our discussion of probability and game theory we made use of random variables without mentioning the fact. Whenever a unique number x is associated with each possible outcome of a random experiment, the number x is called a random variable. Let us illustrate different random variables which might be associated with the same simple random experiment, namely, the one where a coin is tossed twice. An associated sample space is $S = \{HH, HT, TH, TT\}$. If we are interested in the number of heads, the values of the relevant random variable x are

Outcome	*HH*	*HT*	*TH*	*TT*
Value of x	2	1	1	0

If we were interested in the number of tails, the respective values of the random variable would be 0, 1, 1, 2. Suppose that we are playing a game against nature and are to receive 5 cents if the outcome is *HH*, pay the bank 10 cents if the outcome is *TT*, and otherwise incur neither a gain nor a loss. In that case, the random variable x, whose values are tabulated below, is the sort of payoff function which appeared repeatedly in our earlier discussion of strategic games and decisions.

Outcome	*HH*	*HT*	*TH*	*TT*
Value of x	5	0	0	− 10

The term "function" was used to describe this example of a random variable, but the other random variables illustrated were also functions. In each case, a sample space formed the *domain* of the function and the set of values of x constituted the *range*. In general, then, a random variable is a *function* whose *domain* is a sample space of some random experiment and whose *range* is a set of real numbers. In general, too, different random variables can be associated with the same experiment. In tossing a single die five times, for example, the number of aces is one random variable, the number of 6's is another random variable, the number of times a throw results in less than 5 another, the number of times the result of a throw is greater than 2 another; then a gambler's gain in betting connected with each of these random variables is also a random variable. Thus, if the outcome of five tosses of a die is the sequence 2—1—6—1—5, the respective values of the first four random variables described above would be 2, 1, 3, 2.

What the random variable accomplishes, in effect, is the substitution of a *number* for the verbal description (possibly lengthy) of an outcome of some random experiment. Then, instead of having to deal with a sample space or aggregate of possible outcomes, one can handle a *set of numbers*, namely the *range of the random variable*. In the previous chapters, there was much discussion of how to assign weights to the outcomes in a sample space—whether to use empirical relative frequencies, whether to assume equal likelihood, etc. Such weights will now be associated with the values of the random variable which replace the basic outcomes. The result will be a *frequency function* like those used in the previous chapter (pages 314–315).

For an immediate example of such a function, let us picture a game in which a marble is to be drawn from a bag containing 3 red, 4 white, 1 blue, and 2 black marbles. Let us suppose that the following payoffs are the values of the random variable of interest:

Outcome	Red	White	Blue	Black
Value of x	-2	0	1	2

Since the probabilities of drawing marbles of the various colors are 0.3, 0.4, 0.1, and 0.2, respectively, the *frequency function* for the random variable is

x	-2	0	1	2
$f(x)$	0.3	0.4	0.1	0.2

There is a frequency function associated with every random variable; this is the most important aspect of such a variable, since it is this function which provides the probabilities required by the statistician. The frequency function, also known as the *probability density function* (because one physical interpretation links it with density of mass), has the set of possible values of the random variable as its domain and the corresponding probabilities as its range.

Thus the number of heads in coin-tossing is a random variable with a *binomial* density function, so called because the probabilities are given by the terms of a binomial expansion (page 313). The number of defectives in sampling from a large population of items also has a binomial frequency function and, in general, the same is true of the number of "successes" in a sequence of *independent* trials where the probability of occurrence of some event is *constant* from trial to trial. Because Bernoulli made a profound study of such sequences, it is customary to call them *Bernoulli trials* and also to describe a binomial frequency function as Bernoullian. The frequency functions considered in the previous chapter were all of the Bernoulli type.

In the present chapter, two different random variables were associated with tossing a coin twice. In both cases, probabilities would be given by the terms of the binomial expansion, $(H + T)^2 = H^2 + 2HT + T^2$. If the coin is fair and the random variable represents the number of heads, the frequency function would be

x	0	1	2
$f(x)$	$\frac{1}{4}$	$\frac{1}{2}$	$\frac{1}{4}$

In the case where the values of the random variable are payoffs of -10, 0, 5, respectively, the frequency function would be

x	-10	0	5
$f(x)$	$\frac{1}{4}$	$\frac{1}{2}$	$\frac{1}{4}$

If the coin is unfair, and the probability of "heads" is 0.6, the values of $f(x)$ in both frequency functions above would be replaced by 0.16, 0.48, 0.36.

The binomial probability density function is one of the fundamental types considered in mathematical statistics. But in *theory*, if not in practice, there can be infinite variety among probability density functions. For example, the following tabulation would qualify.

x	-4	-3	-2	0	2
$f(x)$	0.43	0.18	0.14	0.21	0.04

We leave to the reader the task of devising a random experiment or a game with cards, marbles, etc., for which the preceding tabulation would be a suitable frequency function. We shall discuss more significant frequency functions later in the chapter but, in theory, a tabulation like the last one is always a possibility, providing it has two major properties which will be seen to exist in all the illustrative examples we have furnished. First, since the values of $f(x)$ are *probabilities*, they must be numbers between 0 and 1, inclusive. Second, since some outcome or other in the sample space is *certain* to occur, some one of the values of x within its range must be observed, and therefore the probabilities represented by $f(x)$ must have a total sum equal to 1. This is in accordance with probability theory, since events that are certain to occur correspond to S, the sample space selected as the universal set, and $P(S) = 1$.

But now we must see how a mathematical statistician *applies* frequency functions to problems of inductive inference. As we have said, one part of Bacon's scientific method consisted of formulating tentative hypotheses and observing how well their consequences would agree with subsequent observations. In mathematical statistics, however, one does not deal with general hypotheses such as "Cancer is caused by a virus" or "There is life on Mars." A *statistical hypothesis* is an assumption about the *frequency function* of a random variable. For example, there is the following: In a sample survey, the number of people who will favor candidate Jones is a random variable with a *binomial frequency function* in which the probability of success is 0.7. Or again: The number of cases in which Brand A toothpaste will prevent cavities is a binomial variable with $p = 0.5$.

During the years 1928–1933, Professor Jerzy Neyman, in collaboration with Karl Pearson's son, Professor Egon S. Pearson, developed a theory for testing statistical hypotheses. The two statisticians devised a method which was later seen to be a special case of Wald's decision process. To explain the elements of the Neyman-Pearson theory, we shall once more draw examples from games of chance. Therefore, let us now imagine that a gambler has observed a certain coin-tossing game and suspects that the coin being used is unfair and is, in fact, strongly biased in favor of "heads." The bank thinks otherwise, maintaining in effect the truth of the *statistical hypothesis*: The frequency function for the number of "heads" is *binomial*, with $p = \frac{1}{2}$. The gambler proposes to test this hypothesis by having the particular coin tossed 10 times in succession. If it turns up "heads" each time, the hypothesis is to be rejected. Now even if the hypothesis is true, that is, even if the coin is perfectly fair, it may happen *by chance* that 10 "heads" in a row will result. The probability of that event is $H^{10} = (\frac{1}{2})^{10} = 1/1024$. What this signifies is that if many such tests with fair coins were to be carried out, about once in every 1000 tests the freak phenomenon would be observed and the hypothesis would be unjustly rejected.

Another gambler, who has observed even more coin-tossing games, feels very sure that the bias exists. He says that the hypothesis should be rejected even if there are only 8 or 9 "heads" among the 10 tosses. What would be the risk of faulty judgment in that case? The coin is to be pronounced unfair if 8 or 9 or 10 "heads" occur. In case the coin is in actuality absolutely fair, the probability that one of those outcomes will result is

$$H^{10} + 10H^9T + 45H^8T^2 = \left(\frac{1}{2}\right)^{10} + 10\left(\frac{1}{2}\right)^{10} + 45\left(\frac{1}{2}\right)^{10} = \frac{56}{1024}$$

$$= 0.05 \text{ (approximately)}$$

One might then say that the probability of an erroneous judgment is 0.05.

Now a third gambler appears on the scene. He believes the coin is unfair because he has heard rumors to that effect. However, he is not certain whether the bias is in favor of "heads" or "tails." Therefore, he suggests rejection of the hypothesis if there are extremes in either direction—9 or 10 "heads," 9 or 10 "tails." In that case, the probability of a chance occurrence of the phenomenon is obtained by substitution in $H^{10} + 10H^9T + 10HT^9 + T^{10}$ and is readily computed to be about 0.02. Again, this figure measures the risk of erroneous rejection of the "fairness" hypothesis.

Even when more serious statistical hypotheses are subjected to test, it is because there is some uncertainty about the frequency function of a random variable, just as in the case of the tosses of a coin. Also, there will be the same sort of risk in pronouncing judgment on the hypothesis at the conclusion of an experiment designed to test it. The usual situation fits into the mold of decision theory. One must choose between two alternative hypotheses or else weigh a specific hypothesis against a whole set of possible alternatives.

To furnish an illustration analogous to that of coin-tossing, suppose that past experience with the use of a certain drug for a particular malady has shown that it is effective in about 50 per cent of all cases. A pharmaceutical firm offers a new drug for the same ailment with the claim that sufficient experimentation has been carried out to prove it will effect cures in 70 per cent of all cases. In one hospital it is decided to subject the new drug to test prior to considering it as a possible replacement for the older palliative. The doctors formulate a "null" hypothesis, as is usually the procedure in the Neyman-Pearson theory: The new drug is *no better* than the old; that is, the probability that it will cure is merely 0.5, and not 0.7 as claimed. There is good reason for framing statistical hypotheses in this negative form. In discussing the limitations of inductive reasoning, it is customary to point out that millions, even billions, of repetitions of a phenomenon cannot constitute an absolute proof. But to show that some claim is false, only a single counterexample is needed. Hence it is customary to devise a test that may lead to the rejection of some "null" hypothesis and thus lend credence to the acceptance of some alternative hypothesis, the latter being one for which researchers claim to have substantiating evidence.

To be specific on the Neyman-Pearson technique, suppose that in the example under consideration the alternative assumptions and the test procedure agreed upon in advance by the hospital staff are

Null hypothesis	$p = 0.5$ for new drug (see above)
Alternative hypothesis	$p = 0.7$ for new drug (as claimed)
Decision rule	

(a) *Experimental test*: Give new drug to 10 patients selected at random from those suffering from the particular ailment.

(b) *Decision function*:

Number of Cures	0–9	10
Decision or Action	Accept null hypothesis	Reject null hypothesis and accept alternative (as a new working hypothesis)

When the new drug is given to the patients, the cure of all 10 may result even though the drug is no more effective than the old, for a "freak" once-in-a-thousand occurrence is always possible. But the doctors have designed the experiment in advance with the understanding that they are willing to risk such a chance, and hence should the entire group of patients be cured, they will reject the null hypothesis and accept its alternative, in this case the assertion that the probability of a cure with the new drug is 0.7.

The Neyman-Pearson theory, like the more general Wald theory, always weighs the risks involved in testing a statistical hypothesis. As indicated, one kind of error may be committed by rejecting a (null) hypothesis when in actuality it is true. Neyman and Pearson call this a Type I error. If on the other hand the (null) hypothesis is false (and some alternative true), a Type II error is committed if the (null) hypothesis is accepted. What is called for in mathematical statistics is to give a numerical measure of the risk of committing either kind of error. In the foregoing illustration, the probability of occurrence of a Type I error is the probability of the "freak" event, and this measures 0.001 (approximately). Usually the design of an experiment includes the maximum risk that will be tolerated for a Type I error, and this is called the *level of significance*. In the present case this level = 0.001.

To compute the chance of a Type II error in the example under consideration, one must find the probability that if the alternative hypothesis ($p = 0.7$) is true and the null hypothesis ($p = 0.5$) false, the latter will nevertheless be accepted. This type of error will occur if, although $p = 0.7$, fewer than 10 cures occur. If $p = 0.7$, the probability of 10 cures would be $(0.7)^{10} = 0.028$. Therefore the probability of fewer than 10 cures is $1 - 0.028 = 0.972$, which is the chance of a Type II error. About 97 chances in 100 seems like a large risk to take. Can it be diminished?

The desired decrease in risk can be accomplished by altering the decision function associated with the experiment. Thus, suppose that the level of significance is set at 0.05, which means that the probability of a Type I error must not exceed 0.05. It might then be specified at the outset that the null hypothesis will be rejected if the new drug should cure 8 or 9 or 10 of the group of 10 patients. To show that this meets the prescribed level of significance, one can substitute $H = 0.5$, $T = 0.5$ in $H^{10} + 10H^9T + 45H^8T^2$. (See the eleventh line of the "arithmetical triangle" (page 313) for the coefficients in this expansion.) The result is

$$(0.5)^{10} + 10 (0.5)^9(0.5) + 45 (0.5)^8(0.5)^2 = 0.05 \text{ (approximately)}$$

If one substitutes $H = 0.7$, $T = 0.3$ instead, with the result 0.382, this figure is the probability of 8 or 9 or 10 heads if the alternative hypothesis is true. But the Type II error would then occur should that hypothesis be rejected, that is, should the number of cures be fewer than 8. The probability of that occurrence is $1 - 0.382 = 0.618$. A risk of 62 in 100 is considerably less than the 97 in 100 of the original design. Hence by running a greater risk of a Type I error (0.05 rather than 0.001), the risk of a Type II error has been diminished.

The situation just illustrated is typical. If the size of the sample group subjected to a test is kept the same, the probability of a Type II error can only be decreased by increasing the risk of committing a Type I error, and the experimenter who is limited in the matter of sample size must weigh the risks in accordance with the situation. In the example considered, doctors eager to retain the old drug and dubious about the new one should design a test in which the chance of a Type I error is slight. If on the other hand they are not satisfied with the old drug and are eager to replace it by one that may possibly be better, they should diminish the probability of a Type II error and run a greater risk of the opposite sort of mistake.

If the size of the sample subjected to the test can be enlarged, then *both* risks can be diminished. For example, if the new drug can be administered to 100 patients suffering from the malady it purports to cure, and if the doctors decide in advance to reject the null hypothesis if 60 or more cures should be effected, the probabilities of Type I and Type II errors can both be diminished to about 0.02 each. (The computation involved in proving this would be arduous for the reader. It is usually carried out by a process of approximation which will be treated later in the present chapter.)

In the preceding example a choice was to be made between two specific statistical hypotheses, but in actual practice one may have a whole set of possible alternatives to the null hypothesis. Thus the hypothesis that $p = 0.5$ for the new drug might be compared with the possibility that p has *any* value greater than 0.5. Then the null hypothesis would still be the assumption that the number of cures is a binomial variable with $p = 0.5$, and the alternative would be that the number of cures is a binomial variable where p has some value greater than 0.5 (and, of course, equal to or less than 1).

The testing of a statistical hypothesis illustrates a very special type of statistical decision function, one where the range of actions contains *only two* values, namely, "accept" or "reject." The Neyman-Pearson technique is also restricted by the fact that implicitly the decision rules are such that the cost of experimentation is a certain *fixed* amount, whether this be measured in dollars or utiles. Wald indicated another reason the traditional testing of statistical hypotheses is less general than his own decision procedure, namely, that related experimentation must be carried out in a single stage. He implied that use of his notion of *sequential analysis* would make the Neyman-Pearson procedure less specialized.

Thus, in the illustration where a new drug was under examination, a single experiment was to be carried out, and the size of the sample to be tested was fixed at ten *prior* to the performance of the experiment. Setting the size of a sample in advance was standard statistical procedure until Wald proposed a modification. In his sequential technique, size is made dependent upon the course of the test, and criteria are determined by the observations as they occur. The sequential method can be described thus: A rule is given for making one of the following three decisions at any stage of an experiment: (a) to take action 1; (b) to take action 2; (c) to continue the experiment by making an additional observation. Wald proved that observations will not have to continue indefinitely, but that, on the contrary, the size of the sample required in order to arrive at a terminal decision is, as a rule, far smaller than where the standard technique of fixed sample size is employed.

At any rate, whatever sampling technique is followed, whatever the problem of statistical decision, one must specify the frequency function of some random variable. In the tests considered earlier, we knew that the associated random variable had a binomial frequency function, but our information was incomplete because we did not have the exact value of the basic *parameter p*. In the examples given, the tests weighed one value of p against another. In mathematical statistics, there are other decision procedures for handling the same problem. One of them will give a specific *estimate* of p on the basis of observed values of the random variable. For example, if a coin is tossed 100 times and it turns up "heads" 47 times, one might give 0.47 as an estimate of the true value of p, the probability of heads. Thus, if one uses the relative frequency as the estimate, the associated decision function assigns a specific estimate to each possible experimental outcome. Increasing or decreasing the number of observations would vary the decision rule. Or one might decide to estimate p by using some aspect of the observations other than the relative frequency of "heads." Choosing the *optimum* method of estimation would once again be a case of choice among decision rules, that is, a typical Wald game of statistical decision would be called for. Much more statistical theory, in particular much more discussion of the frequency functions of random variables, is needed before one can play this last sophisticated game.

Still another method of solving the same problem is to obtain an *interval* estimate of p, for example, (0.42, 0.62). In such a case experimental data might enable us to state that our "confidence" is 0.9 that the true value of p lies within the interval named. Or the observed values of the random variable might lead us to state, with "confidence" 0.8, that p must have a value in the interval (0.47, 0.57). Just what "confidence" means in this connection and just how the intervals are obtained call for the same sort of statistical theory as in the case of making specific numerical estimates of the parameter p. Again, we have instances of the decision process—decision functions which assign interval estimates to experimental outcomes, and general decision "games" where one selects the optimum rule for obtaining the interval estimate.

The testing of statistical hypotheses as well as "point" and interval estimation represented the most typical instances of statistical decision prior to Wald's theory. They remain important and, as we have stated, we cannot handle them without a more varied knowledge of frequency functions and their attributes.

Just how its frequency function characterizes a random variable will become apparent gradually. In the case of a specific binomial variable, for example, the frequency function provides the *probabilities* for various events. It is not necessary, therefore, to limit a discussion of coin-tossing to absolutely fair coins for which the outcomes are equally likely. The assumption that a particular binomial frequency function fits the situation where coins are unfair (or where the probability of a defective item is *not* 1/2, but 0.03, say) will enable us to assign probabilities to the different events in which we may be interested. As we study other types of random variable, we shall find that, in general, it is the frequency function which determines the probabilities. Postulating a particular frequency function to fit a special situation is a *modern replacement* for assumptions like *equal likelihood.*

In statistical problems, we may be interested not only in probabilities but also in other aspects of a random variable. Thus, although the frequency function may tell us that some values of a random variable are more likely to occur than others, we might like an overall picture. In games of chance and games of strategy, the *expectation* or long-run *mean* provides part of such a picture. For a random variable with a finite range, the formula for the mean μ is the same as that for the expectation (page 267), namely,

$$\mu = p_1 x_1 + p_2 x_2 + \cdots + p_n x_n$$

Let us apply this formula to find the mean or expected value of a binomial variable with $p = 1/2$ and $n = 3$. Substitution in $(H + T)^3$ leads to the frequency function

Number of Heads (x)	0	1	2	3
Probability	1/8	3/8	3/8	1/8

Then the mean is

$$\mu = p_1 x_1 + p_2 x_2 + p_3 x_3 + p_4 x_4$$

$$= \left(\frac{1}{8}\right)(0) + \left(\frac{3}{8}\right)(1) + \left(\frac{3}{8}\right)(2) + \left(\frac{1}{8}\right)(3)$$

$$= \frac{12}{8} = \frac{3}{2} = 1.5$$

The mean $\mu = 1.5$ gives an overall picture by telling us that although in many repetitions of a certain coin-tossing experiment, results will vary at random between 0 and 3, respectively, they can be characterized as 1.5 on the average.

It will be observed that, in the example, $1.5 = (3)(1/2) = np$. It can be proved that, in general, the mean or expectation for a binomial variable is given by the formula

$$\mu = np$$

Hence in a manufacturing process where the proportion of defective items is $p = 0.04$ and articles are shipped in lots of fifty ($n = 50$), the average or expected number of defectives per lot is

$$\mu = np = (50)(0.04) = 2$$

But one may desire something more than the range and the mean to give an overall picture of how the values of the random variable will be distributed about the mean in repeated experimentation. Using the preceding figures, the mean number of defectives per lot is 2. But the range of the random variable is 0, 1, 2, 3,

4, ..., 50. Thus, although the average is 2, it is possible that the number of defectives in some lots might be high. Then the manufacturer would want to know whether the average 2 is the result of many occurrences of, say, 0, 1, 2, 3, 4, or whether it comes about because many lots with 20 defectives, say, are balanced by an enormous number with no defectives at all. Envisioning the latter situation might cause the manufacturer some anxiety, because he would have trouble with the many customers receiving the unsatisfactory lots.

To see how one might measure the dispersion or spread of values, imagine that the following values of a random variable (arranged in order of magnitude) were obtained in 10 repetitions of a random experiment: {4, 6, 8, 8, 8, 8, 9, 9, 10, 10}. This aggregate is a *sample* taken from the *population* of values which one would obtain by repeating the same experiment again and again indefinitely. The mean of the sample, $m = 8$, is readily computed. (It will be noted that m, the sample mean, is to be distinguished from μ, the population mean or expectation of the random variable.) Now we wish to study how the sample values are dispersed about the sample mean. One way of indicating this scatter would be to state that the sample values range from 4 to 10. Another would be to list the size of the deviation of each value from the sample mean, namely, {4, 2, 0, 0, 0, 0, 1, 1, 2, 2}, and then average these deviations. The average, 1.2, would be called the "mean deviation from the mean." In actuality, the first two of the above deviations were negative, but the mean deviation uses only the *magnitude* and not the direction of the deviations.

Another way of avoiding negative numbers is to *square* the deviations, and this is done in defining the *variance*: *The variance is the mean square deviation from the mean.* The variance of the preceding sample, symbolized by s^2, would therefore be equal to

$$s^2 = \left(\frac{1}{10}\right)(16 + 4 + 0 + 0 + 0 + 0 + 1 + 1 + 4 + 4) = 3$$

If the values of the random variable were lengths expressed in inches, then the deviations would be measured in inches, the squares of the deviations and hence the variance in square inches. Therefore, if one seeks a measure of variability with the same dimensionality as the values of the random variable, one can take the square root of the variance to obtain the *standard deviation*, sometimes called by the more descriptive name *root mean square* (deviation from the mean). In the present instance,

$$\text{standard deviation} = s = \sqrt{3} = 1.7 \text{ (approximately)}$$

The values of the random variable (or any linear function of them) can be compared directly with the standard deviation, which thus forms a sort of "standard" or unit of linear measurement for a specific random variable. For example, in the present case, one might state that the highest value in the sample is roughly $6s$, or that the mean deviation from the mean is about $0.7s$.

The equation which gave the value of s^2 could be expressed in the form

$$s^2 = \left(\frac{1}{10}\right)(4)^2 + \left(\frac{1}{10}\right)(2)^2 + \left(\frac{4}{10}\right)(0)^2 + \left(\frac{2}{10}\right)(1)^2 + \left(\frac{2}{10}\right)(2)^2$$

Here the fractional coefficients are the relative frequencies of the sample values. Each relative frequency is multiplied by the square of a deviation from the sample mean. If a sample is very large, that is, if the randon experiment is repeated many times, the relative frequencies approximate true probabilities. This fact suggests why the definition of the population variance, σ^2, is analogous in form to the preceding equation. In fact,

$$\sigma^2 = p_1 d_1{}^2 + p_2 d_2{}^2 + \cdots + p_n d_n{}^2$$

where $p_1 p_2, \ldots, p_n$ are the probabilities which the frequency function assigns to the occurrence of $\{x_1\}$, $\{x_2\}$, ..., $\{x_n\}$, respectively, and $d_1, d_2, \ldots, d_n$ are the corresponding deviations from the population mean or expectation,

$$\mu = p_1 x_1 + p_2 x_2 + \cdots + p_n x_n$$

The reader will observe the similarity in form in the formulas for μ and σ^2. This occurs because both measures are *expectations* or long-run averages, that is,

$$\mu = E(x)$$

$$\sigma^2 = E(d^2) = E(x - \mu)^2$$

The reader can think of μ and σ^2 as if they were computed in the same way as m and s^2, except that the relative frequencies are obtained from an enormous number of experimental repetitions.

The definitions given for population mean and variance apply to the case where the range of the random variable contains only a *finite* number of values. When the number of possible values is infinite, questions of mathematical analysis are involved—the convergence of a series or the existence and evaluation of a definite integral. But in every case the mean is a measure of "location" or "central tendency," and the variance or the standard deviation is a measure of dispersion. Special properties of certain types of random variable will be related to these measures. Thus, let us mention in passing that in the case of a *binomial* variable, it can be proved that

$$\sigma^2 = np\,(1 - p)$$

where n is the number of trials and p is the probability of "success" at each trial.

The binomial random variable will now lead us to an even more important species. Returning once again to the tossing of a fair coin, let us consider the frequency function for the number of heads when such a coin is tossed 20 times. Figure 14.1, which represents this frequency function, is usually described as a *probability* or *frequency polygon*. Note the resemblance of the polygon to a smooth

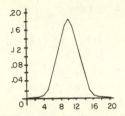

Figure 14.1   Probability or frequency polygon of $(H + T)^{20}$

curve. If the number of trials with the coin is made greater, this resemblance to a smooth curve becomes more marked. In fact, as the number of tosses of the coin becomes greater and greater, the corresponding frequency polygon approaches a certain curve as a *limit*. This curve (Figure 14.2), bell-shaped in appearance, is

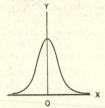

Figure 14.2   Normal probability curve

called the *normal probability curve*, and its equation (in standard form) is

$$y = \frac{1}{\sqrt{2\pi}} e^{-x^2/2}$$

where e is the natural logarithmic base.

The approach of the binomial frequency polygon to the normal curve was discovered by Abraham de Moivre (1667–1754). French by birth, De Moivre emigrated to England after the revocation of the Edict of Nantes had made France less tolerant toward Protestants. He spent part of his life in England among the gamblers of a London coffeehouse and, like Cardano, wrote a gambler's manual, *The Doctrine of Chances*. The limit theorem which he proved became the prototype for other limit theorems which are at the heart of probability theory and mathematical statistics today. Until recently, Laplace, and not De Moivre, was credited with being the first to relate the binomial and the normal frequency functions. Laplace did contribute notably to the theory of the normal probability curve, and later Gauss, the mathematical giant of the nineteenth century, made a thorough study of its properties. For this reason, one often finds this important frequency function called the law of Laplace, or the Gaussian curve.

De Moivre's limit theorem indicates that the normal curve, or more practically, tabulations of the normal frequency function, can be used to provide *approximate* probabilities when a fair coin is tossed a very large number of times. In fact, the

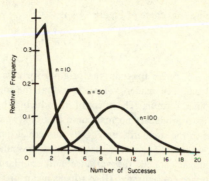

Figure 14.3   Frequency polygon of $(0.1 + 0.9)^n$ for increasing values of n

normal law is a good approximation for the probabilities of "success" in the case of *any* Bernoulli sequence with very numerous trials. This is suggested by Figure 14.3, which represents $(H + T)^n$ with $H = 0.1$, $T = 0.9$, for increasing values of n. The diagram shows how the frequency polygon for tossing an *unfair* coin approaches a normal density function as the number of trials increases. In many different situations which have nothing to do with coin-tossing, empirical frequency functions approximate a normal density. This is the case with vital statistics, anthropometry, psychological or educational tests, and the like. Figure 14.4 furnishes an example. The reason that numerical measurements from so many varied fields conform to the frequency function of one particular type of random variable and that the normal frequency function plays a major role in theoretical statistics will appear presently. At this point, it should be remarked that a minor algebraic modification of the standard normal formula makes it more suitable for specific statistical problems. Thus, although the standard curve has the equation

$$y = \frac{1}{\sqrt{2\pi}} e^{-x^2/2}$$

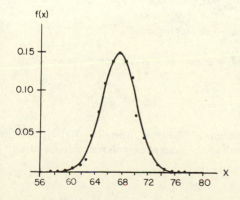

Figure 14.4 Fitting of normal curve to the empirical frequency function of heights of adult males in a large random sample. Mean = 67.5 in. Standard deviation = 2.6 in.

the formula for the normal curve of Figure 14.4 is

$$y = \frac{1}{2.6\sqrt{2\pi}}\, e^{-\frac{1}{2}\left(\frac{x-67.5}{2.6}\right)^2}$$

This substitutes $(x - 67.5)/2.6$ for x in the original formula and introduces the additional factor $1/2.6$. The reader will observe that the mean height, 67.5 in., appears in the modified formula. The effect of introducing this mean value in the foregoing manner is to produce a normal curve that is symmetric about $x = 67.5$ instead of $x = 0$, as in the standard curve. The numerical quantity, 2.6 in., is the *standard deviation*, whereas $(2.6)^2$ is the *variance* of the population of heights.

As we have already stated, the variance and standard deviation measure whether the heights are concentrated close to the mean or widely spread out to right and left. A smaller standard deviation would have indicated a greater concentration of the heights represented in a diagram like Figure 14.4, and a larger standard deviation would have indicated a greater dispersion above and below the mean height. Thus statistical applications make use of a whole *family* of normal curves,

$$y = \frac{1}{\sigma\sqrt{2\pi}}\, e^{-\frac{1}{2}\left(\frac{x-\mu}{\sigma}\right)^2}$$

one curve for each pair (μ, σ^2).

As an example, let us consider the member of this family which would approximate the frequency polygon graphed in Figure 14.1. The polygon corresponds to the experiment of tossing a fair coin 20 times. Hence, $p = \frac{1}{2}$, $n = 20$, and by the formulas for the mean and variance of a binomial variable, $\mu = np = (20)(\frac{1}{2}) = 10$, and $\sigma^2 = np(1 - p) = (20)(\frac{1}{2})(\frac{1}{2}) = 5$. Substitution in the formula for the normal family yields the normal frequency function

$$f(x) = \frac{1}{\sqrt{10\pi}}\, e^{-\frac{(x-10)^2}{10}}$$

The reader can check, by substituting various values of x, that this curve is a good fit for the polygon of Figure 14.1. For example, if $x = 10$,

$$\frac{1}{\sqrt{10\pi}}\, e^{\circ} = \frac{1}{\sqrt{10\pi}} = 0.18 \text{ (approximately)}$$

which is close to the reading for $f(10)$ in Figure 14.1. Again, substitution of either $x = 5$ or $x = 15$ yields

$$f(5) = f(15) = \frac{1}{\sqrt{10\pi}}\, e^{-2.5} = 0.18e^{-2.5}$$

Numerical tables give the approximation $e^{-2.5} = 0.082$. Therefore, $f(5) = f(15) = 0.015$ (approximately), which is just about the value of the ordinates corresponding to $x = 5$ and $x = 15$ in Figure 14.1.

The typical decision procedures that were discussed in the case of a binomial frequency function apply to normal densities as well. One can test a statistical hypothesis where the assumed frequency function is normal. If the value of μ is known, one may use "point" or interval estimation to approximate σ^2. Or one may use these types of estimation to approximate both μ and σ^2. And again more theory would be needed to explain our measure of "confidence" in interval estimates, or our selection of an optimum decision rule for point estimates.

Many special properties of a normal variable were studied by Laplace and Gauss. For example, they showed that the probability is approximately 2/3 that a normal variable will not differ from its mean by more than σ. In the population of heights previously considered, this would signify that about 2/3 of the men must have heights between $67.5 - 2.6 = 64.9$ in. and $67.5 + 2.6 = 70.1$ in. Furthermore, in a normal population about 95 per cent of all cases must lie in the range ($\mu - 2\sigma$, $\mu + 2\sigma$) and practically all cases in the range ($\mu - 3\sigma$, $\mu + 3\sigma$). All these properties of a normal variable are approximately true for a binomial variable when n is large, since the frequency function of the latter approximates a normal curve.

If we consider a binomial variable where $p = 0.1$ and $n = 100$, its frequency function will be described by one of the frequency polygons in Figure 14.3. For this case,

$$\mu = np = 10$$
$$\sigma^2 = np(1 - p) = 9$$
$$\sigma = 3$$

Then, in accordance with what was stated earlier, practically all experimental results should lie between $10 - 3\sigma$ and $10 + 3\sigma$, that is, between 1 and 19. The diagram confirms this fact by indicating negligible relative frequencies for $x > 19$. A variety of interpretations can be provided. In the first case, if we think of an experiment where 100 unfair coins with $H = 0.1$ are tossed, numerous repetitions of this experiment will practically never lead to more than 19 "heads" or fewer than 1. If we think of a manufacturing process where 10 per cent of all items produced have some minor defects and articles are packed in lots of 100, there will hardly ever be more than 19 defectives in a lot even though the quality of production is so poor. Again, if it is known from past experience that about 10 per cent of the people who reserve theater tickets over the telephone do not call for them at the box office, a policy where 100 such reservations are accepted for each performance entails, at worst, a possible loss on 19 sets of tickets for some (few) performances.

Since so much has been said about binomial and normal frequency functions, it is necessary to emphasize that there are many other types of probability density function. It would not be feasible to discuss the latter in great detail, but it will be worthwhile to make some general statements about random variables and then to give examples that are neither binomial nor normal.

In the first place, there is a distinction among random variables that was implied in our initial discussion of probability theory when we illustrated both finite and infinite sample spaces. Just as *cardinal numbers* like 1 or 2 or 3 or 4 or 5, etc., tell how many outcomes are contained in a finite sample space, there are *transfinite* cardinal numbers for infinite sets. The theory of such numbers was developed by a leading mathematician of the modern era, Georg Cantor. Instead of saying that the aggregate of natural numbers is *countably infinite* (see Chapter 24), one can use Cantor's symbolism and state that its cardinal number is $\aleph_0$ (read *aleph null*). A higher transfinite cardinal, which Cantor symbolized by C, is assigned to an infinite set which is a *continuum*, an aggregate like the totality of points in a continuous line segment of any length—an interval on the X-axis, say, or the entire X-axis.

Now the range of a random variable may also be a finite set, a countably infinite set, or a continuum.* For example, in those cases where a different value of the random variable is assigned to each outcome in a sample space, there will be as many values as outcomes, and hence the range of the random variable will be a finite aggregate, a countably infinite set, or a continuum according as the sample space is one of these three types. In statistics a random variable is described as *discrete* (Chapter 2) if its range is finite or countably infinite, since in either case the values in the range can be arranged as a *sequence*. A *continuous* random variable is one whose range is a continuum. A *binomial* random variable is *discrete* because its range is finite. A *normal* variable is *continuous* because its range is the set of all real numbers, that is, the whole X-axis, which is a continuum.

We have not as yet illustrated a discrete random variable with a countably infinite range. Let us then consider the discrete random variable whose range is the set of natural numbers and whose frequency function is

$$f(x) = \frac{1}{2^x}$$

which can be tabulated

Value of x	1	2	3	4	...	n	...
$f(x)$	$\frac{1}{2}$	$\frac{1}{2^2}$	$\frac{1}{2^3}$	$\frac{1}{2^4}$	...	$\frac{1}{2^n}$	...

In this instance, one possible interpretation is that the random experiment consists of tossing a fair coin until it turns up "heads" for the first time. Then x is the number of tosses required, and could therefore have one of the values 1, 2, 3, 4, 5, etc. In theory, one might have to toss the coin many times before obtaining "heads." That this does not happen often in practice is indicated by the probabilities which the frequency function assigns. Thus the probability that a dozen tosses will be needed ($x = 12$) is $1/2^{12} = 1/4096$. In other words, in very numerous repetitions

---

* In advanced statistics, the range can also be mixed in type, containing both a continuum and a set of isolated points; for example, the values might be 0, 1, 2, and any real number in the interval from 4 to 5 on the X-axis.

of a random experiment, only once in about 4000 times would a dozen tosses be required before the appearance of "heads." If x is, say 1,000,000 or more, the probability is so very small that it is extremely unlikely (even if possible) that such a large number of tosses would be needed.

The illustration can be used to review probabilistic issues treated in Chapter 11. There we indicated that if an event is certain to occur, its probability must be equal to 1 whether the sample space is finite or infinite. Let us use the preceding function to compute the probability that "heads" will appear on some toss, whether it is the first or the second or the third or the billionth, etc., and see whether the answer is equal to 1, as it should be. Originally, an event was defined as a subset (proper or improper) of the sample space. But such a subset of the *domain* of the random variable corresponds to a subset of the *range*, and the frequency function assigns probabilities to the latter. Hence the event "heads in some toss" can be described by the *improper* subset of the range $E = \{1, 2, 3, 4, \ldots \text{ forever}\}$; that is, the event E corresponds to the entire range, and $P(E)$ is the sum of the respective weights assigned to the values 1, 2, 3, 4, etc., in accordance with another fundamental law of probability theory. Then

$$P(E) = f(1) + f(2) + f(3) + \cdots + f(n) + \cdots$$

and

$$P(E) = \frac{1}{2} + \frac{1}{2^2} + \frac{1}{2^3} + \cdots + \frac{1}{2^n} + \cdots$$

The infinite series in the right member is *convergent*. It is, in fact, a *geometric* series, and the reader will have no difficulty in showing that its sum is equal to 1, which is consistent with probabilistic law.

What we observe is that, in general, for any random variable with a countable infinity of values, it is necessary that the probabilities assigned by the frequency function form a *convergent* infinite series whose sum is 1. We can see why the classic assumption of equal likelihood cannot apply to a countably infinite sample space, for if one assigns the same probability p, however small, to each outcome or to the corresponding value of the random variable, then the infinite series, $p + p + p + p + \ldots$, would diverge, because the sum of the first n terms would be np, which could be made to exceed all bounds by making n sufficiently large.

It was Bernoulli who studied the most important type of random variable with a *finite* discrete range, namely, the binomial. De Moivre and Laplace provided the properties of the frequency function par excellence for continuous random variables, namely, the normal. And now the French mathematician and mathematical physicist Siméon Denis Poisson (1781–1840) must be credited with an analogous contribution for random variables whose range is countably infinite. The three fundamental types of random variable are related, moreover, since both the normal and the Poisson frequency functions are special limiting cases of the frequency function for a binomial variable. A Poisson probability density is a good approximation to the binomial when p, the probability of "success," is small and n, the number of trials, is large, but the product np is moderate in size. The formula for a Poisson frequency function is

$$f(x) = \frac{e^{-\mu}\mu^x}{x!}$$

where x, the random variable, has the countably infinite range {0, 1, 2, 3, 4, ... forever}, e is the base of natural logarithms, and $\mu = np$, n and p being the number of trials and the probability of "success" for the related binomial variable. Figure 14.5 indicates how good an approximation a Poisson density can furnish even when p is not so very small or n so very large.

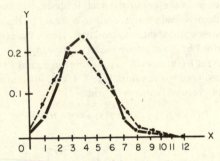

Figure 14.5   Binomial (——) and Poisson (- - -) frequency polygons for $n = 12$, $p = 1/3$, and $\mu = 4$

To see how the Poisson approximation can make the computation of probabilities less arduous, suppose that for a binomial variable, $p = 0.02$ and $n = 200$. Then $np = 4$, which seems "moderate" in size. These figures might be interpreted as giving $p = 0.02$, the proportion of defective items in a manufacturing process under control where articles are packed in lots of 200 (with a mean or expectation of $np = 4$ defective items). Then, if one wished to know the probabilities that a lot would contain 0 or 1 or 2 or 3 or 4, etc., defective items, one would evaluate successive terms of the binomial expansion

$$[(1 - p) + p]^n = [(0.98) + (0.02)]^{200}$$

a lengthy and tedious computational procedure. Instead, let us use the Poisson formula to *approximate* the probability that the number of defectives will be less than 4. We must find the probabilities for the outcomes 0, 1, 2, 3, and add them. We seek

$$\frac{e^{-4}4^0}{0!} + \frac{e^{-4}4^1}{1!} + \frac{e^{-4}4^2}{2!} + \frac{e^{-4}4^3}{3!} = e^{-4}\left(1 + 4 + 8 + \frac{32}{3}\right)$$

where $0! = 1$ in accordance with a mathematical convention. From numerical tables, one can obtain $e^{-4} = 0.0183$. Hence the required probability is $(0.0183)\ (71/3) = 0.433$. The more arduous computation using the binomial expansion gives the result 0.437, showing once more the closeness of the Poisson approximation.

But the Poisson frequency function is *not* limited to the purpose of approximating binomial densities. It exists in its own right and, like the normal density, seems to be applicable to numerous and varied situations in the real world. As an example, we shall indicate how well a Poisson density will fit the empirical frequency function for the occurrence of vacancies in the United States Supreme Court during the years 1837–1932.

Number of Vacancies per Year (x)	0	1	2	3
Relative Frequency (empirical probability)	59/96 (0.61)	27/96 (0.28)	9/96 (0.09)	1/96 (0.01)

The mean number of vacancies per year would be

$$\left(\frac{59}{96}\right)(0) + \left(\frac{27}{96}\right)(1) + \left(\frac{9}{96}\right)(2) + \left(\frac{1}{96}\right)(3) = \left(\frac{48}{96}\right) = 0.5$$

The corresponding Poisson frequency function is

$$f(x) = \frac{e^{-0.5}(0.5)^x}{x!}$$

To show that it fits the preceding tabulation *approximately*, one can substitute $x = 0$ and obtain $f(0) = e^{-0.5} = 0.61$ (from numerical tables), which agrees with the frequency for $x = 0$ in the tabulation; if $x = 1$, $f(1) = e^{-0.5} (0.5)/1 = 0.30$, which approximates the tabulated value, 0.28; if $x = 2$, $f(2) = 0.61 (0.5)^2/2! = 0.08$ (the tabulation has 0.09); $f(3) = 0.61 (0.5)^3/3! = 0.01$; $f(4) = 0.001$; and the probabilities for $x = 5$, $x = 6$, etc. are still smaller. It will be observed that the empirical tabulation has a zero relative frequency for $x > 3$ so that the probabilities of 0.001 or less *approximate* the tabular values. A *theoretical* Poisson variable can, of course, assume any one of the infinite set of values 0, 1, 2, 3, 4, ... forever.

As a final point, let us show that the sum of the weights assigned by a Poisson frequency function is equal to 1. Once again this is necessary because it is *certain* that $x = 0$ or 1 or 2 or 3 or some other positive integer, and probabilistic law requires that such an event have probability 1. Under the Poisson formula the probability of $E = \{0, 1, 2, 3, 4, \ldots,$ forever$\}$ is

$$\frac{\mu^0 e^{-\mu}}{0!} + \frac{\mu^1 e^{-\mu}}{1!} + \frac{\mu^2 e^{-\mu}}{2!} + \frac{\mu^3 e^{-\mu}}{3!} + \cdots = e^{-\mu}\left(1 + \mu + \frac{\mu^2}{2!} + \frac{\mu^3}{3!} + \cdots\right)$$

Now it is shown in the theory of infinite series that the expression within parentheses is equal to e^{μ}. Therefore the required probability is $e^{-\mu}e^{\mu} = e^0 = 1$, which is the desired result.

By this stage of our story it is becoming increasingly evident that the statistician assigns probabilities by postulating suitable frequency functions. Usually it is past experience or present observation which provides the justification for the assumption. Thus, alternative frequency functions might be tested by the Neyman-Pearson theory. But how can probabilities be assigned when empirical evidence is lacking? In the case of a *finite* sample space, one can always revert to classic theory and assign equal likelihood to outcomes. There is something analogous for a *continuous* random variable whose range is a finite interval. Suppose, for example, that a telephone answering service has failed to record the exact time for some incoming call and that the operator knows only that she received it at some time between noon and 2 P.M. What is the probability that the call came at 1:15 P.M., that is, between 1:15 and 1:16 P.M.? Complete ignorance compels us to assume that any one-minute interval is as likely as another, and since there are 120 one-minute intervals between noon and 2 P.M., the answer is 1/120. On the same basis, the probability that the call came between 12:30 and 1 P.M. would be 30/120 = 1/4. If the timing of a call could be carried out with a high-precision instrument, the exact moment might be measured as 18.347 minutes after noon, for example. Therefore an idealized model for the situation would treat x, the number of minutes after noon, as a *continuous* random variable whose range is the continuous interval (0, 120) and whose frequency function is $f(x) = 1/120$ (Figure 14.6). Such a fre-

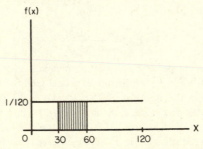

Figure 14.6   Uniform or rectangular frequency function
$f(x) = 1/120$ **in the interval** (0, 120)

quency function is described as *uniform* or *rectangular*. In Figure 14.6, the shaded rectangular area represents the probability that the call came in between 12:30 and 1 P.M. The probability that it came in during the 1/4-minute interval after 1 P.M. would be represented by a rectangle with base 1/4, height 1/120, and area 1/480. In general, in the case of *continuous* random variables, *events* are pictured as *intervals* or *sets of intervals* within the range of the variable, and the probabilities can be pictured by areas, as in the present case. It will be observed that we did *not* ask for the probability that the call would arrive at *exactly* 1 P.M., but phrased the question in terms of a short interval thereafter. To make the time closer to 1 P.M., one might consider the length of the interval (base of the rectangle) as, say, only 0.01 minute. Then the corresponding probability would be only 1/12,000. The rectangle which measures probability would have very little area. If one insisted, even though the concept of an event forbids it, on the probability of a call at exactly 1 P.M., there would be an interval of zero length, and the area of the probability rectangle would be (1/120)(0) = 0. The probability of occurrence on the dot, whether it is

1 P.M. or 12:15 P.M. or any other instant, is *zero*. Although it is unlikely or physically inconceivable that calls would arrive at *exactly* these moments (if we could measure the time very precisely), it is *not* abstractly impossible. This illustrates a statement we made in our first treatment of probabilistic concepts, that in the case of infinite continuous sample spaces an impossible event must have probability zero, whereas the converse is false. Zero probability does *not* imply absolute impossibility. As the complementary fact, a probability measuring 1 does not imply certainty.

In general, the formula for a uniform or rectangular frequency function is

$$f(x) = \frac{1}{c}$$

where the range of the random variable x is the interval $(0, c)$. This density function assumes "equal likelihood," as it were, except that the random variable associated with outcomes has a continuous range because its values are distances, intervals of time, or the like. Thus the foregoing formula would provide the probabilities in roulette if the wheel is fair and has circumference c. In that case the random variable x would be the distance at which the wheel would stop.

It has been stressed repeatedly that the properties of a random variable are related to its frequency function. But it is necessary to mention that the *cumulative distribution function* (or briefly, the *distribution function*), is more suitable to the purposes of advanced probability and statistics than the frequency function. The term referred originally to the distribution of a total probability of 1 over the range of the random variable. Associated with the following empirical frequency function (first tabulation) for the intelligence quotients of a group of 200 students, there is (second tabulation) a *cumulative* relative frequency function (the empirical distribution function).

Intelligence Quotient	88–92	93–97	98–102	103–107	108–112	113–117	118–122	123–127	128–132	133–137	138–142	143–147
Relative Frequency	0.01	0.03	0.065	0.13	0.16	0.20	0.16	0.12	0.07	0.025	0.02	0.01

Intelligence Quotient	87.5	92.5	97.5	102.5	107.5	112.5	117.5	122.5	127.5	132.5	137.5	142.5	147.5
Cumulative Relative Frequency	0	0.01	0.04	0.105	0.235	0.395	0.595	0.755	0.875	0.945	0.97	0.99	1.00

The cumulative table refers to the proportion of students whose I.Q. does not exceed a specific I.Q. Thus 23.5 per cent of the students have I.Q.'s not exceeding 107.5, and 97 per cent have I.Q.'s not exceeding 137.5. In accordance with mathematical usage, an I.Q. of 92 signifies a *measurement* made to the nearest integer, which means that it is not exactly 92 but nearer to 92 than to 91 or 93, that is, a measurement equal to or greater than 91.5 but less than 92.5. This explains the

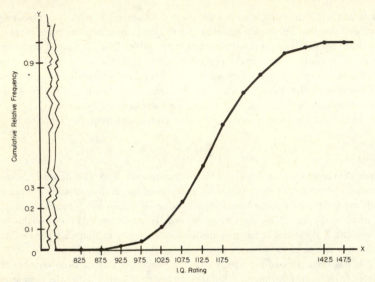

Figure 14.7   Cumulative frequency polygon for an empirical distribution of I.Q.'s

entries for intelligence quotients in the cumulative table. Thus 0.04 or 4 per cent of
the I.Q.'s are 97 or less, and since an I.Q. of 97 actually signifies a measurement equal
to or greater than 96.5 but less than 97.5, the cumulative table more properly indicates that 97.5 is an upper bound for 4 per cent of the I.Q.'s. Figure 14.7 is the
cumulative frequency polygon corresponding to the foregoing tabulation. Figure
14.8 is the cumulative normal distribution graph. Figure 14.9 is the distribution
graph corresponding to a uniform frequency function, and the step graph of Figure
14.10 represents the cumulative distribution for the tossing of a single die. A step
function is typical for discrete random variables. Apropos of the present subject,
the reader must be warned that the term "distribution" is used rather loosely in
mathematical writing and may refer either to the frequency function or the cumulative distribution function.

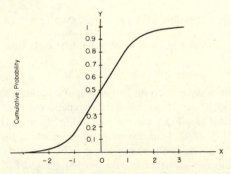

Figure 14.8   Cumulative distribution function for a normal variable

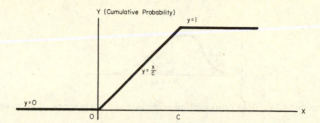

Figure 14.9   **Cumulative distribution function corresponding to a uniform frequency function**

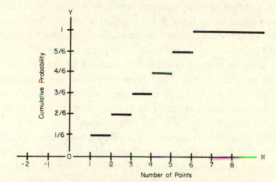

Figure 14.10   Cumulative distribution function for the tossing of a fair die

To give one instance of how a distribution function may solve a problem more easily than a frequency function, we return to the case of a continuous random variable where an event, as we have stated, is represented by an interval or a set of intervals on the X-axis, and the probability that this event will occur is represented by a geometric area like that in Figure 14.11, an area under the curve representing the frequency function of the random variable. In general, there would therefore be a question concerning the existence and evaluation of a definite integral. The case previously considered, where the frequency function was uniform (Figure 14.6), was specially simple because the required area was rectangular. But suppose that one is given a *distribution function* of some continuous random variable, for example, $F(x) = \sin x$ where the range of x is $(0, \pi/2)$, and one seeks the probability of the event represented by the interval $(\pi/6, \pi/2)$. Since $F(\pi/2)$ is the probability that $x \leq \pi/2$ and $F(\pi/6)$ is the probability that $x \leq \pi/6$, therefore $F(\pi/2) - F(\pi/6) = \sin \pi/2 - \sin \pi/6 = 1 - 1/2 = 1/2$ is the probability that x lies in the interval $(\pi/6, \pi/2)$. Thus it was possible to obtain the probability of the event in question without requiring the reader to have a knowledge of integral calculus. Probabilities can be derived from the distribution function by simple substitution.

Although the statistician is free to assume any sort of frequency or distribution function to fit a particular situation, it must be stated that the normal frequency and distribution functions (and other functions derived from them) are the most

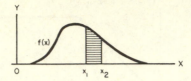

Figure 14.11   Area representing probability of occurrence of an event

important types for the *theoretical* purposes of mathematical statistics. The reason this is so will now be explained. That so many empirical phenomena fit into the normal mold, and that so many different species of random variable have distributions that are normal in the limit, is the result of a modern *limit theorem*, a major generalization of De Moivre's limit theorem. The conceptual background for the more inclusive proposition is to be found in the point of view of Laplace and Gauss, who believed that the normal law must of necessity be almost universal for data describing natural phenomena. Their prescription for any departures from the standard frequency function was to gather more and more data, because adequate observation was supposed inevitably to banish irregularities and smooth figures into the normal pattern.

To Laplace and Gauss, normal curves were *curves of error* like those obtained from tabulations of fine measurements of some physical entity, where any random deviation from the true measurement is an error. In other kinds of statistical data, the "errors" were taken to be the differences between the values of the random variable and its mean or average value. Thus the amounts by which the heights represented in Figure 14.4 differ from 5 ft. 7.5 in. can be interpreted as nature's "errors." It is assumed that gross or coarse errors as well as systematic errors have been removed from the results of measurement and that only random effects remain. Such random errors can be considered as the total effect of numerous smaller components—slight defects in different parts of the apparatus, small variations in temperature and pressure, imperfections in the vision of the observer, random factors affecting the manual dexterity of the observer and his general alertness, unconscious bias of the observer, and so on. Each of these components is itself a random variable which does not usually influence the other components, and the total "error" is the sum of numerous random variables, usually statistically independent of one another.

The *central limit theorem* of probability theory asserts that a random variable which is such a sum must have a *normal* frequency function. A rigorous statement of this basic theorem or a discussion of its proof would take the reader into advanced issues. Suffice it to say that in very broad terms the theorem in its most general form signifies that any random variable whose numerical value is influenced by a large number of independent causes, each contributing a small "impulse" to the total effect, must be approximately normal. This is the logical justification for the frequent occurrence of normal variables in statistical applications to so many different fields.

The central limit theorem explains the use of the normal frequency function in what statisticians call *large sample theory*. If, for example, one is drawing a large sample at random from a very large set of people (the "population") and tabulating heights in order to obtain a frequency table like that graphed in Figure 14.4, the height of the first person whose name is drawn is a random variable. This means that if a sample of size 100 is drawn, and then another such sample is drawn again and again, the *first* height obtained will vary in random fashion from sample to sample and there will be a frequency function for the set of all possible first heights. The same is true of the *second* height, the *third* height, etc. Naming these random variables $x_1, x_2, x_3, \ldots, x_{100}$, respectively, the central limit theorem demonstrates that y, the sum of these variables, will have a frequency function that is approximately normal. Thus the normal variable is

$$y = x_1 + \cdots + x_{100}$$

Dividing both sides by 100,

$$\frac{y}{100} = \frac{x_1 + x_2 + \cdots + x_{100}}{100}$$

yields the *sample* mean, and since it is one-hundredth of the variable that is normally distributed, its values from sample to sample will be proportional to those of the normal variable y. Hence, on intuitive grounds, the sample mean should have a normal frequency function.

As an example of the application of the properties of the normal distribution to large sample theory, suppose that measurements (idealized here for the sake of simplicity) taken over a long period show that the mean gross weight of individuals departing from a particular airport is 165 lb., and that the distribution of weights is approximately normal, with a standard deviation equal to 23.3 lb. Then very few (in fact less than 0.3 per cent) of the passengers would have had weights differing from the mean weight by more than three times the standard deviation, that is, 70 lb. Thus, almost all passengers would have had gross weights between 95 lb. and 235 lb. Now the central limit theorem implies that the mean weights for all flights carrying a full passenger load of, say, 80, would fluctuate normally about the "grand mean" of 165 lb. with a standard deviation of $23.3/\sqrt{80}$ and that practically all the mean weights would differ from this grand mean by less than $70/\sqrt{80} = 7.8$ lb. (For flights of size n, the standard deviation would be $23.3/\sqrt{n}$ and the maximum difference from the grand mean $70/\sqrt{n}$ lb.) Therefore it could be safely assumed that the mean weight of passengers from flight to flight in 80-passenger planes would rarely exceed 173 lb., and the total passenger weight would hardly ever be more than 13,840 lb., or about 7 tons.

Even if the gross weights recorded over a long period were not to follow a normal distribution, the central limit theorem implies that the means from flight to flight would nevertheless be distributed normally about the grand mean in all cases where n, the number of passengers, is sufficiently large. The standard deviation of the population of passenger weights is substituted in the formula $\sigma/\sqrt{n}$, and $3\sigma/\sqrt{n}$ is subtracted from and added to the grand mean in a way analogous to the foregoing in order to estimate the range in mean weight from flight to flight.

As a second application, let us use large sample theory to obtain a *confidence interval* for the mean of a population. Suppose that a random sample of 100 observations is drawn from a population with unknown mean μ and known standard deviation $\sigma = 3$. By virtue of the central limit theorem, the sample means of all the possible samples of size 100 will be distributed normally about the population mean with a standard deviation equal to $\sigma/\sqrt{n} = 3/\sqrt{100} = 0.3$. As stated earlier, about 95 per cent of the observed values of a normal variable will differ from the population mean by less than 2 standard deviations. Therefore, in the present case, 95 per cent of all sample means will lie in the interval $(\mu - 0.6, \mu + 0.6)$. Suppose that the mean in the particular sample observed is 14.2. We might imagine that this mean value falls in the above interval, but to be conservative, we can picture the observed mean at one of the ends of the interval so that it deviates from the true mean by the maximum amount of any number in the interval. In that case, either

$$\mu - 0.6 = 14.2 \quad \text{and} \quad \mu = 14.8$$

or

$$\mu + 0.6 = 14.2 \quad \text{and} \quad \mu = 13.6$$

The two extremes lead us to say that (13.6, 14.8) is a 95 per cent *confidence interval* for μ, or that our confidence is 0.95 that the true value of μ will lie in this interval. The exact probabilistic meaning of the statement is: In repeated sampling from the specified population, the different sample means will lead to various confidence intervals computed in the preceding manner. About 95 per cent of the confidence intervals so obtained will contain the true value of μ, the population mean.

In sampling situations where sample size is small and in other statistical problems where the random variable is not normally distributed, nonnormal

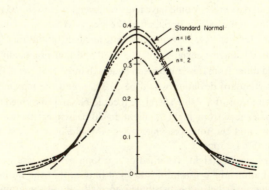

Figure 14.12   Student t-distributions (frequency functions) compared with standard normal density

distributions come into use. Figure 14.12 shows the "Student"* t-distribution (t-frequency function) used for a sample of size 16. Its appearance is not too different from that of the normal curve. The other "Student" curves in Figure 14.12 are for smaller samples. The theory of the t-distribution makes it possible to use a small sample to obtain a confidence interval for the mean of a normal population when neither this mean nor the standard deviation of the population is known.

Random variables and their frequency functions have been related to various problems of statistical inference, and these in turn have been considered as special instances of Wald's general decision process. But the particular objectives and the limited size of the present work have made it necessary to restrict discussion to the elementary aspects of mathematical statistics and the inclusive theory of decision. Thus there are *multivariate* problems where more than one random variable is associated with an experiment, as we shall illustrate in our next chapter. Then one would obtain probabilities from frequency functions like $f(x, y)$ or, say, $f(x_1, x_2, x_3, x_4)$, or distribution functions like $F(x, y)$ or $F(x_1, x_2, x_3, x_4)$. Although the underlying notions remain the same, *mutatis mutandis*, some new issues arise; for example, the question whether the random variables are statistically independent of one another. Then there is an area of inference involving functional relationships between random and nonrandom (controlled) variables. There is also the matter of designing the experiments associated with the subtler problems of statistical decisions. A new methodology being developed at the present time is called *distribution-free* or *nonparametric inference*. Its purpose is to avoid the traditional pattern explained in this chapter, where inference always involves the assumption that the random variable has a specific type of frequency function—binomial, normal, Poisson, etc. In spite of all such unexplored territory, the author believes that the reader has been offered a maximum minimum of those fundamental concepts which are invariant in the more extended theories and which can therefore provide an avenue of approach to their comprehension.

---

* "Student" was the pseudonym of William Sealy Gosset, who discovered the relationship of the t-curves to *small* sample theory, a field he investigated as early as 1908. He was a statistical consultant to the Guiness brewery in Dublin. A ruling of that firm forbade its employees to publish the results of research, but the regulation was relaxed to permit Gosset to write under a pen name.

15

Demons, Energy, Maxwell, and Gibbs

Just as regular as the appearance of angle-trisectors on the mathematical horizon are the periodic attempts of someone in the world of letters to prove that Shakespeare was Bacon, Marlowe, Raleigh, Jonson, or even Queen Elizabeth I. To the sophisticated it does not matter whether the Bard who gave the world its greatest poetry and drama is a legendary figure or someone who can be labeled with a name, date, profession, geographical habitat, scholastic degree. But, we assure ourselves, it couldn't happen now. A creator of immortal work can no longer remain unknown to his contemporaries. Yet let any "inquiring reporter" quiz the man on the street or, for that matter, the man on the campus, on the identity of Josiah Willard Gibbs (1839–1903). It is a safe conjecture that few will recognize the name of the foremost native American mathematical physicist.

Gibbs's greatest discovery, *statistical mechanics*, links his research with the subject matter of the preceding chapters, where the fundamentals of probability and statistics were presented. Why and how did Gibbs's selection of one particular branch of mathematical physics come about, and why did that subject require the use of modern probabilistic concepts? To answer these questions, part of the present chapter must be concerned with some pre-Gibbsian concepts and also with the ideas of James Clerk Maxwell (1831–1879), the renowned Scottish mathematical physicist who terminated one classical line of thought in physics and founded the new era of statistical physics. He was a contemporary of Gibbs, and one of the few leading scientists of the day to understand the latter's ideas and to appreciate the genius they disclosed.

Maxwell is best known for the electromagnetic "field equations," the set of differential equations governing electrodynamics (Chapter 9) in the same way that Hamilton's canonical equations rule ordinary dynamics. But Maxwell was also interested in the kinetic theory of gases, and he found in that subject a possible source of conflict with traditional thermodynamic principles, namely, with ideas originally set forth by Sadi Carnot (1796–1832), the founder of modern thermodynamics.

Carnot had assumed that heat energy cannot "spontaneously," "naturally," of its own accord, flow from a colder to a hotter body. In modern refrigerators, heat energy does pass from the colder interior to the warmer kitchen, but this is *not* spontaneous, since there is an additional external supply of energy, namely electrical. If one disconnects this external factor, the refrigerator defrosts, with the natural passage of heat energy from the hotter room to the cooler icebox.

Maxwell lived long before the electric comforts that were direct or indirect consequences of his electromagnetic equations, but he was aware of many examples confirming Carnot's postulate. However, one *counterexample* that Maxwell offered has become famous. If his predecessors had their assumptions, he saw no reason why he could not himself postulate a situation where *intelligence* could contradict Carnot's principle. In Maxwell's illustration an ideal gas fills a vessel that is divided into two compartments, the temperature of the left compartment being higher than that of the right. According to the kinetic theory, temperature is an average effect, so that the very, very numerous individual gas molecules in both compartments vary in speed and direction of their motions, the average effect of those in the left compartment being higher than the average in the right. Now Maxwell imagined a highly intelligent, skillful, "sorting demon" to be placed at a tiny trapdoor between the compartments. At the approach of a very slow-moving particle in the left chamber, the demon opened the door to permit passage of the particle to the right chamber; whenever an exceptionally speedy molecule in the right chamber came near, the little "devil" placed the shutter the other way so that the rapid molecule would pass into the left compartment. Moreover, this was done effortlessly, that is, without the expenditure of energy. By such repeated sorting and sifting, a great many slow particles were concentrated in the right, and a large proportion of the rapid molecules in the left compartment. Heat energy of a gas is measured by the mean kinetic energy ($\frac{1}{2}mv^2$) of the gas molecules. If two molecules with speeds of 800 and 200 meters per second are exchanged, an exchange in kinetic energy proportional to $640,000 - 40,000 = 600,000$ is effected, and thus there is an increase in the mean kinetic energy of the left chamber, with a corresponding decrease on the right. Such exchanges would increase and decrease the left and right average effects, respectively, so that Maxwell's sorting demon could, by means of numerous exchanges, transfer considerable heat energy from the cooler right to the warmer left. Maxwell stated that it is *highly probable* that heat energy will pass from a hotter to a colder body, but that the reverse is *not* impossible, merely unlikely, and that if such a phenomenon has never been observed, it may be because the probability of its occurrence is so small. Perhaps it can occur only once in a million years.

The above statements suggest that a thermodynamic law may be a probabilistic rather than an absolute principle. In fact, in 1859 Maxwell asserted in precise form the first major natural law that is *statistical*, and not absolute, deterministic, causal. It was connected with the kinetic theory of gases. In that subject it is assumed that, in a fixed volume of gas in thermal equilibrium (at constant temperature), the individual molecules will be fairly uniformly distributed throughout the containing vessel and will be moving in all possible directions with different speeds. Maxwell's statistical law is a frequency or density function of a random variable (Chapter 14), namely, molecular speed. Therefore the law enables one to find the proportion or

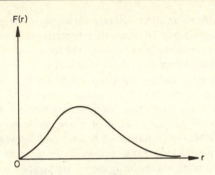

Figure 15.1   Maxwell-Boltzmann distribution of molecular speeds

relative frequency of molecules moving at different speeds—close to zero, at about 500 meters per second, or much higher, etc. Figure 15.1 represents this frequency function. In the diagram, we see that relative frequencies or probabilities are small for low or high speeds. Stripped of all technicalities, Maxwell's formula is a *frequency function*,

$$f(r) = ar^2 e^{-b^2 r^2}$$

Here r represents molecular speed. The physical constants, a and b, depend on the particular gas and its absolute temperature. Subsequently we shall see how Maxwell derived his law.

By making assumptions concerning molecular collisions and applying certain other postulates, the Austrian Ludwig Boltzmann (1844–1906) was able to improve the mathematical rigor in Maxwell's derivations, and subsequently it became the custom to speak of the above law as the Maxwell-Boltzmann distribution of molecular speeds.* Émile Borel, that leader in analysis, pure mathematical probability, and game theory, whose role was discussed in Chapter 10, called this law "one of the most beautiful and fruitful of all applications of probability theory to physics." The Maxwell-Boltzmann distribution was a purely theoretical construction, but around 1930 apparatus became available for obtaining actual experimental measurements of molecular speeds, and the resulting figures were in close agreement with the hypothetical ones obtained from the Maxwell-Boltzmann law.

Up to this point we have emphasized only molecular *speed*, that is, the magnitude of molecular *velocity*, without considering the *direction* of molecular motion. A velocity is a *vector*, however, having both direction and size. In space, such a vector (Figure 4.5) can be considered as the resultant or "sum" of three components parallel to the three coordinate axes, respectively (Chapter 4). Maxwell's line

---

* As pointed out in Chapter 14, physicists sometimes use the term *distribution* where mathematicians prefer *frequency* or *density* function, reserving the term "distribution" for the *cumulative* frequency function.

of thought led him to the assumption that the x-component of the velocity would have a normal frequency function,

$$f(x) = \frac{1}{\sigma\sqrt{2\pi}} \, e^{-x^2/2\sigma^2}$$

and the y and z components would also be distributed by the same normal frequency function, that is,

$$f(y) = \frac{1}{\sigma\sqrt{2\pi}} \, e^{-y^2/2\sigma^2}, \qquad f(z) = \frac{1}{\sigma\sqrt{2\pi}} \, e^{-z^2/2\sigma^2}$$

Here $\sigma^2 = kT/m$, where T represents the absolute temperature of the particular volume of gas, m is the mass of a molecule of this gas, and the proportionality factor k is an exceedingly important physical constant, called *Boltzmann's constant*.

The value of the variance σ^2 indicates whether the set of velocity components is concentrated fairly close to the mean or is spread out. The mean of each of the above distributions is *zero*. Therefore a small value for σ^2 (and hence a low absolute temperature, since σ^2 is proportional to T) signifies that the x-components of molecular velocity will, on the whole, be close to zero, that is, very small. The same will be true of y and z components, so that the resultant velocities may have *any* direction but will, on the whole, be small in magnitude at low temperatures and much more varied in size at high temperatures.

Maxwell's logic in probabilistic proofs was brilliantly intuitive but lacking in mathematical rigor. For example, he assumed that molecular velocity components are statistically independent of one another, and in this case, probability theory permits the multiplication of respective probabilities to obtain the joint frequency function,

$$f(x, y, z) = \left(\frac{1}{\sigma\sqrt{2\pi}} \, e^{-x^2/2\sigma^2}\right) \left(\frac{1}{\sigma\sqrt{2\pi}} \, e^{-y^2/2\sigma^2}\right) \left(\frac{1}{\sigma\sqrt{2\pi}} \, e^{-z^2/2\sigma^2}\right)$$

$$= \frac{1}{2\pi\sigma^3\sqrt{2\pi}} \, e^{-(x^2 + y^2 + z^2)/2\sigma^2}$$

The above function can be applied to answer questions like the following: In a certain volume of nitrogen gas at 0°C (or absolute temperature $T = 273°$), what proportion of molecules would have *velocities* with components $x = 200$ (meters per second), $y = 300$, $z = 250$?* Since T, m (mass of a nitrogen molecule) and k (Boltzmann's constant) are known, $\sigma^2 = kT/m$ can be computed. Then the values of σ^2, x, y, and z are substituted in the formula above. Numerical tables can be used to evaluate the right member of this formula, and thus the desired probability or relative frequency can be determined.

The first statistical concept illustrated by the density function $f(x, y, z)$ is that of a *multivariate* frequency or density function, that is, one where two or more

---

* Since the frequency function is of the *continuous* type, the question is interpreted to mean that x is in the *interval* (199.5, 200.5), y in the interval (299.5, 300.5), z in the interval (249.5, 250.5).

random variables are involved. The distribution above is one type of multivariate *normal* function. There are three random variables, namely, x, y, and z, the velocity components. But interest may center in the *speed*, that is, the magnitude of velocity, (without regard for the direction of motion). For studying speeds one can take the multivariate distribution of velocities and from it derive the Maxwell-Boltzmann univariate distribution of molecular speeds. The procedure would be as follows.

Banishing technicalities, one can consider the above multivariate formula as

$$f(x, y, z) = Ae^{-b^2r^2}$$

where $r^2 = x^2 + y^2 + z^2$ (see Figure 4.5), and the positive value of r is the molecular *speed* (not velocity). But then one must be careful to interpret this formula correctly. For example, in the problem about a small volume of nitrogen gas, $r^2 = 192,500$, or the speed, $r = 439$ meters per second (approximately) would be substituted. Now one could use the very same substitution to answer a great many different questions, for example, to obtain the proportion of molecules whose velocity components are $x = -200, y = 300, z = 250$, or the proportion of those for which $x = 200, y = -300$, $z = -250$, etc. In the last two instances the velocity is different from that of the original problem, but the speed is the same; the difference in velocity is due to a difference in the *direction* of motion. Again, the substitution, and hence the answer, would be the same because r, the speed, is the same, but the velocity would be different if $x = -300, y = 250, z = -200$, or $x = -250, y = -300, z = -200$, and it would be approximately the same if $x = 248, y = -305, z = 200$, etc. There are countless sets of different components, that is, different velocities all corresponding to the *same* speed, about 439 meters per second. The original multivariate density function gives the same answer for any single velocity in this infinite set.

If one were to draw a sphere with center at the origin and radius equal to 439, all the vectors drawn from the origin to the surface of this sphere would represent all the different velocities with this same speed. Then, if one were not interested in velocities, but merely in speed, one might ask the question: What proportion of the nitrogen molecules have speeds approximately equal to 439? This would require a summation of all the relative frequencies corresponding to the vectors drawn to all points of the sphere. The number of relative frequencies (or vectors) is infinite, and the summation is carried out by integral calculus. In general, one would be performing a summation of frequencies corresponding to all vectors whose end-points are on a sphere of radius r. This is similar to the summation used in obtaining the area of the sphere, and the result in the latter case is $4\pi r^2$. Therefore it seems logical that, if the individual frequency corresponding to one vector or one point of the sphere is $Ae^{-b^2r^2}$, the total frequency should be $4\pi r^2$ times as great, and hence that the frequency function for speeds should be

$$f(r) = 4\pi r^2 \cdot \frac{1}{2\pi\sigma^3\sqrt{2\pi}} e^{-b^2r^2} = \frac{1}{\sigma^3}\sqrt{\frac{2}{\pi}} r^2 e^{-b^2r^2}$$

or

$$f(r) = ar^2 e^{-b^2 r^2}$$

This is exactly the formula for the Maxwell-Boltzmann distribution of molecular speeds (page 346).

The thermodynamics of Maxwell and Boltzmann, with its three random variables, extends the concept of univariate distribution emphasized in the previous chapters. But more profound thermodynamic issues require much further generalization, since a study of molecular motion may involve as many as 10^{24} variables, or even more, as the present chapter will indicate. When the number of variables is so enormous, it would be wise to offer some simpler method for presenting thermodynamic theory. A more elementary approach was sought and ultimately achieved in the statistical mechanics of Josiah Willard Gibbs.

Each year since 1923, when the American Mathematical Society established the Josiah Willard Gibbs Lectureship, there is simultaneous recognition of the work of Gibbs and that of the distinguished mathematician who is invited to deliver a memorial address on a mathematical topic of current interest. In 1930 Edwin Bidwell Wilson (1879–1964) was asked to talk about Gibbs himself, and subsequently we shall quote from Wilson's "Reminiscences of Gibbs by a Student and Colleague." In 1934 it was Einstein and a decade thereafter John von Neumann who were Gibbs lecturers, and others whose mathematical research has been or will be discussed in the present book gave the special addresses in other years: James Pierpont, G. H. Hardy, Hermann Weyl, Norbert Wiener, Kurt Gödel, Marston Morse, and Eugene P. Wigner.

Not enough of this sort of reverence was accorded to Gibbs during his lifetime (1839–1903), most of which was spent at Yale University (first as son of a professor of divinity, then as a student, then as a professor of mathematical physics). The *vector analysis* that is indispensable to every serious student of physics today owes its form to the Yale mathematician. From the point of view of the physicist, the Gibbs vector calculus was a major simplification and improvement of Hamilton's quaternions. Yet it was received at first with a hostility akin to the initial responses to Beethoven's masterpieces, Puccini's melodies, and Picasso's abstractions. The British physicist P. G. Tait, in his 1890 text on quaternions, called the Gibbs technique a "hermaphrodite monster, compounded of the notations of Hamilton and Grassmann." Three years later, a letter in the journal *Nature* contained the New England professor's gentle rebuttal. The final sentence of Gibbs's epistle stated that "the world is too large, and the current of modern thought is too broad, to be confined to the *ipse dixit* even of a Hamilton."

By the turn of the century Gibbs had won the battle as far as the British world of science was concerned. The Royal Society of London awarded him the Copley Medal in 1901 for his thermodynamical discoveries. He was still a prophet without sufficient honor in his own land, however. When an American college president was seeking talent at Cambridge, England, he asked J. J. Thomson to recommend a molecular physicist. The latter's response was, "You needn't come to Europe for that, the best man in the field is your countryman, Willard Gibbs." "You mean

Wolcott Gibbs (a chemist)," the president suggested. Thomson reaffirmed the name of *Willard* Gibbs and enlarged on the latter's contributions in the field of molecular physics. But the college president was not interested; he had already decided that a man unknown to him must lack personality and would not do for the job.

And so Gibbs remained at Yale, developing the vector analysis that was only a small preliminary to his most outstanding achievement, the creation of *statistical mechanics*, and anticipating the physico-chemical *phase rule* which he was to deduce as one consequence. Edwin Bidwell Wilson's reminiscences of Gibbs describe how feverishly Gibbs worked during the last period of his life in order to complete within a brief eight or nine months in 1901 and 1902 the entire manuscript of his *Elementary Principles of Statistical Mechanics*. In that work Gibbs developed a theory that not only provided a rigorous foundation for thermodynamics but was, in addition, of more general applicability and not limited to phenomena associated with heat energy. When some of Gibbs's assumptions are modified, his mechanics becomes suitable for various types of particles in the quantum theory today. One alteration leads to the so-called Bose-Einstein statistics, another to the Fermi-Dirac variety (Chapter 13). But the fundamental point of view remains that of Gibbs. As one of his admirers, the British physical chemist F. G. Donnan, has put it: "Of such work it is, and always will be true that the eager hand of Time may add something, but can take nothing away."

Gibbs could not have known that his ideas would involve him (posthumously) in a major crisis in modern physics and philosophy, namely, the conflict between the believers in deterministic (strictly causal) natural laws and the adherents of the doctrine that indeterminism, with its reliance on probabilistic prediction, is unavoidable in describing the physical world. In his later years Einstein placed himself firmly in the former camp. "I cannot believe God plays dice with the universe," he said. This was a reversal of an earlier attitude during the period when he had made notable contributions to statistical physics by solving the problem of "Brownian motion," or again by developing the quantum-mechanical properties of "bosons," the subatomic particles to which the Bose-Einstein statistical mechanics applies (Chapter 13).

As for Brownian motion, it is so named after the English botanist Robert Brown, who, a little over a century ago, observed the erratic paths followed by grains of pollen or tiny plant spores suspended in water. That type of motion is characteristic of all sufficiently small particles suspended in any kind of fluid—for example, the smoke and dust in our atmosphere. Such movements, which are readily visible through a microscope, were explained by Einstein as resulting from the innumerable kicks, shoves, and tosses inflicted by agitated water or air molecules in the course of their thermal motion. We cannot see this motion, but we can judge its nature indirectly from the measurement of the visible Brownian motion. Einstein established statistical (not causal) laws for this phenomenon, and subsequent experiments by the French physicist Perrin provided verification of the Einstein hypotheses. Nobel prize winner Max Born (1882–1970) says of Einstein's paper on

Brownian motion. " By its simplicity and clarity, this paper is a classic of our science."
Of Einstein's continued research in the same field, Born writes:*

> These investigations . . . have done more than any other work to convince physicists
> of the reality of atoms and molecules, of the kinetic theory of heat and of the *fundamental
> part of probability in natural laws*. At that time the statistical aspect of physics was pre-
> ponderant in Einstein's mind; yet at the same time he worked on relativity where rigorous
> causality reigned. His conviction seems always to have been that probability is used to
> cover our ignorance and that only the vastness of this ignorance pushes statistics into the
> forefront.

This attitude is not shared by other great physicists or philosophers today, and
it seems to them that Einstein's dislike of statistical physics acquired exaggerated
magnitude as he grew older. He expressed it in correspondence with Born, in papers
attacking specific results of quantum mechanics where probabilistic methodology is
used, and in many other ways. Since it has become the fashion to offer psychoana-
lytic interpretations of behavior (as seen in recent studies of Lewis Carroll, of Freud
himself) and these are often couched in terms of "traumas" received in early life,
we present the following fact for what it is worth in this connection and for its
coincidental association with Willard Gibbs.

In 1901, when Gibbs was completing the manuscript of his *Statistical Mechanics*, Einstein, unaware, was starting a paper which was to duplicate all the essential
features of Gibbs's *chef-d'oeuvre*. This occurrence provides one of the many instances of the phenomenon of simultaneous discovery, so often repeated in the
history of science. When a mathematical or physical subject has evolved to the ripe
stage that demands an all-embracing theory or a rigorous logical foundation, the
foremost thinkers in the particular field—Newton and Leibniz, or Einstein and
Gibbs, as the case may be—will arrive independently at the same theoretical
results. The American Mathematical Society memorializes Gibbs each year because
his theory encompassed and superseded all the thermodynamic theories formulated
in the two millennia preceding his day. Hence anyone can realize what profound
thought Einstein must have given to the simultaneous discovery embodied in his
Kinetic Theory of Thermal Equilibrium and the Second Law of Thermodynamics
(1902) and *Theoretical Foundations of Thermodynamics* (1903). Is it not possible that
the youth (Einstein was in his twenties at the time) who was to create the greatest of all
all causal theories experienced some unconscious shock, some major frustration, in
failing to be the *first* to discover the methodology leading to the greatest of all
theories in statistical physics? Was it reasoning or intuition or such a trauma as we
have imagined that brought about his complete withdrawal from the statistical
school and his pointed prejudice against the use of probabilistic logic in physics?
Since our subject is mathematics and not psychology, we shall not attempt to
answer the question. In the domain of fact, not speculation, Einstein must be
credited with original ideas in *both* relativity and quantum mechanics. He carried

---

* Max Born, "Einstein's Statistical Theories," *Library of Living Philosophers*, Vol. 7,
Einstein: Philosopher-Scientist, 1949.

one theory to a profound ultimate conclusion, but we shall never know whether he might not have done the same with those early thermodynamic hypotheses whose methodology he renounced.

What was the actual content of the 1901–1903 statistical mechanical theory of Gibbs? As far as physics is concerned, one might say that Gibbs added little to the mere physical facts of the Maxwell-Boltzmann theory. His methods, however, are vastly superior, for he constructed a *pure mathematical subject* of very great generality, one which gave impetus to the development of the modern theory of *stochastic* or *random processes*. On the basis of a small number of postulates or assumptions, he *deduced* theorems of wide applicability to a *variety* of physical situations.

One feature of major importance in Gibbs's statistical mechanics and also in Einstein's study of Brownian motion is that they were among the very first mathematical contributions to introduce ideas and issues related to *random* or *stochastic processes*. (The adjective "stochastic" derives from the Greek *stochos*, which means "guess.") We must therefore explain what is meant by such a process. But before doing so, we remark that random processes are of theoretical and practical importance today not only in statistical mechanics but also in astronomy, biology, epidemiology, economics, communication and control theory, industrial engineering, oceanography, and other subjects.

In the broadest sense, the theory of stochastic processes is the *dynamic* part of mathematical statistics. If the random variables of interest, that is, the frequency and distribution functions, change as *time* passes, the whole set of random variables is called a stochastic process. To start with a very elementary and hence somewhat trivial example, we can imagine a coin to be tossed in very many repeated random experiments so that as time goes on, wear and tear change p, the probability of "heads." Let us consider plausible values of p initially, after a million tosses, after two million tosses, etc. and label the corresponding "instants" of time as

$$t = 0, 1, 2, 3, 4, \ldots$$

with

$$p = 0.50, 0.49, 0.48, 0.40, 0.50, \ldots$$

Then if we consider a random variable with values 0 and 1 corresponding to the outcomes "tails" and "heads," respectively, the initial and the fifth frequency functions are both as follows:

x	0	1
$f(x)$	0.50	0.50

whereas the second and fourth frequency functions are given by the tabulation

x	0	1
$f(x)$	0.51	0.49

The reader will readily formulate the third frequency function and make conjectures about the functions corresponding to $t = 5, 6, 7, \ldots$ The whole *sequence* of random variables or frequency functions is an example of a *discrete parameter stochastic process*, a finite process if the sequence terminates, otherwise an infinite one.

Observe that even if the above sequence forms an infinite process, it may be one that acquires a certain stability after a while. For, if we imagine that the coin-tossing apparatus is repaired so as to make the fall of the coin almost frictionless, then ultimately we might (in theory) arrive at a situation where the terms of the sequence (rounded off to hundredths) become

$$0.52, 0.52, 0.52, \ldots \text{ forever}$$

We have reached an *equilibrium* frequency function where the probabilities of "heads" and "tails" are 0.52 and 0.48, respectively. In other words, we might have a sequence which has a *limit*.

Since games of chance always provide the simplest illustrations for probabilistic issues, let us rely on coin-tossing experiments once more to give an example of a *finite* stochastic process. In Figure 12.1 there is a diagram of a "game tree" which was related to a strategic game described in that chapter. Let us now use the same diagram but have R and C play a game of chance instead of a game of strategy. Let there be three moves once again, with R and C playing alternately, choosing a number from the set $\{1, 2\}$. But now R tosses a penny in the first move and a nickel in the third, and C tosses a dime in the second move. Either player selects the number 1 only when the coin turns up "heads." Suppose that the probability of "heads" is 0.4, 0.6, 0.3 for penny, dime, and nickel, respectively. Then we have a sequence of three different frequency functions. If there were betting associated with the game and if it were played repeatedly, we might be interested in the expectations or means. The reader can verify that they form the sequence $\{1.6, 1.4, 1.7\}$. If one is to answer probabilistic questions like "What is the likelihood that 2 will be selected only once in the three moves?" or if, as in Chapter 12, there are payoffs at the *end* of the game, then it will be advisable to consider the process as a whole and to formulate a frequency function in which the values of the random variable are the eight paths from the base to the top of the tree (the eight ordered triples of number selections). Since the three coin-tossing experiments are *independent* trials. it is just a matter of multiplying three probabilities to obtain the probability for a particular path. Thus the frequency function for the whole process will assign to the respective paths the probabilities 0.072, 0.168, 0.048, 0.112, etc., with 0.252 the probability for the most likely path, the one corresponding to (2, 1, 2).

We might have made the above illustration of a stochastic process more general by requiring the second and third moves of the game to be dependent on the results of the previous move. Thus, in the second move, C might toss one of two different coins (with different probabilities for "heads") according to whether R chooses 1 or 2 in the first move, and likewise the nature of R's random experiment in the third move might depend on C's choice in the second move. But, in any case, the simplicity of our example would make it artificial rather than realistic. It was selected only to anticipate in a very elementary way the picture of Gibbs's stochastic process, where there are an infinite number of paths or "trajectories" about whose frequency function Gibbs theorized.

An *infinite* stochastic process need *not* be of the discrete parameter type, that is, the instants of time in the "index set" need *not* be a sequence. Instead the index set might contain the numbers in a *continuum*; for example, we might have as index set all values of t in the interval $(0, 1)$, or the index set might be the continuum $t \geq 0$, the set of all nonnegative real numbers, arranged in order of size. In either case we would have a *continuous parameter stochastic process*. Gibbs's stochastic process was of the continuous parameter type.

In summary, a *stochastic process* is usually described as a *family of random variables*. Observe that a "family" is not always identical in meaning with a "set." In one "family" illustrated above, the numbers 0.50 and 0.49 appear repeatedly in correspondence with different index numbers. If we were considering the *set* of distinct numerical values, the numbers in question would appear only once. Another point is that, with our emphasis on the "dynamic," we have considered time to be the parameter whose values are given in the index set. But in various applications the parameter may have some other concrete meaning, whereas in the general pure mathematical stochastic process, the parameter is an abstraction with no specific interpretation.

Before we proceed to Gibbs's theory itself, we must remark that, in one respect, it is simpler than the general stochastic process, since only a single random experiment is involved, namely, the initial one of selecting the initial conditions for a mechanical system. The random variables and their frequency functions do vary in time, however, both as a result of the indeterminacy of the initial conditions and the strictly deterministic mechanical laws (which force the paths or trajectories to follow certain routes). The reader can obtain one type of picture of the Gibbs process by imagining it to be a tremendous generalization of the tree diagram of Figure 12.1. The root of the tree in the Gibbs process can be located at any one of a continuum of points. Then each one of the infinity of trees branches out in a continuously infinite number of directions. The position of the root and the direction of branching depend on the outcome of the initial random experiment. But there are no further random experiments; the initial branches just grow up and out forever, in directions prescribed by the laws of mechanics. (Our own picture, later on, will be somewhat different—in order to prevent the growing branches from getting in each other's way.)

What then were the basic axioms of Gibbs's statistical mechanics? In the first place, he postulated that certain "systems" obey the most general laws of dynamics as set down by Hamilton. These systems are mathematical abstractions, but the physical counterparts are finite portions of matter—solid, liquid, or gaseous. Each system is a mechanical one, whose state at a particular time can be specified by a certain number of generalized coordinates and an equal number of generalized momenta, these coordinates and momenta being governed by the canonical equations of Hamilton (Chapter 9). A most elementary physical realization of such a system is furnished by a single particle constrained to move in a straight line. Its position at any time can be described by giving its distance from a fixed point or origin of coordinates on the line, but this single numerical fact merely gives its "mechanical configuration" at the time in question. For its complete "state," its velocity or momentum at this instant must also be given. In elementary physics, momentum is just the product of mass and velocity, and since mass is constant (at low speeds), momentum is proportional to velocity and it matters little which of the two is used. However, it must be emphasized that the Hamiltonian equations make use of generalized or "conjugate" momenta that are very different from the mass-velocity products of elementary mechanics. From the Hamiltonian equations and a theorem of Joseph Liouville (1809–1882), the great French analyst, Gibbs was able to deduce two important "conservation" laws. We shall call the Gibbs deductions the *conservation of volume* and the *conservation of probability density*, the latter term suggesting the statistical aspect of his mechanics. Actually his "conservation" laws are just part of the integral calculus required for the proper formulation of the stochastic process he had in mind, that is, for giving algebraic expression to the continuous family of probability density functions and the ultimate "equilibrium" density function. In summary, by starting with the "canonical" equations of Hamilton, Gibbs arrived at principles essential for his own probabilistic physical laws.

A mechanical system slightly more advanced than a particle moving in a straight line is a monatomic molecule moving in space. Here *three* position coordinates and *three* components of momentum, that is, *six* numerical facts, will be required to fix the state of the system at a particular instant. In a system of n molecules, $6n$ numbers will suffice. Suppose that the system is a "gram-molecular mass" or "mole" of gas, the standard volume* used in chemistry. Then $n = Avogadro's number$, that is, the number of molecules in a mole of any gas. Experiments have shown that $n = 6.06 \times 10^{23}$ (approximately). Hence $6n = 36.36 \times 10^{23}$ or about 3.64×10^{24}. Thus $6n$, the number of generalized coordinates required to describe the state of a mole of gas at a particular instant, exceeds 10^{24}. Even though the gaseous form of matter is the simplest of the three forms, and a mole is a relatively small volume, the recording of the coordinates at a single instant would be a formidable task, requiring more than 1,000,000,000,000,000,000,000,000 entries! But as soon as the gas molecules move, some or all of these numbers change—within a thousandth of a second, and change again within the next thousandth of a second! Even if the facts were observable directly or indirectly, recording them would be a hopeless undertaking.

But returning momentarily to the single particle moving on a straight line, the state at any time is the pair of values of

$$(q, p) = (\text{position, momentum})$$

* This is the volume of 2 grams of hydrogen at standard temperature and pressure.

determined by the differential equations of dynamics (Newtonian in this elementary case, Hamiltonian in the general case). As the point moves on the straight line, the values of the number pair (q, p) will change. To facilitate the study of the changes in state as time goes on, any number pair can be plotted as the rectangular coordinates of a point in a Cartesian graph. There will be different points for different states, and as the point moves on the straight line, its evolution through successive (q, p) states will be a curve in the Cartesian plane.

Gibbs would have referred to the state of a system as a *phase*, to the representative point as a *phase point*, and to the Cartesian plane as the *phase space*. In an earlier example, the case of a bullet moving in a vertical straight line (page 222), we considered just such a system. Let us take an arbitrary system of mass measure in which the bullet has unit mass. Hence $m = 1$, $p = mv = v$ in this elementary instance, and the formulas previously obtained give

$$q = -16t^2 + 1200t + 5$$
$$p = -32t + 1200$$

When $t = 0$, $q = 5$, $p = 1200$; when $t = 10$, $q = 10,405$, $p = 880$; when $t = 30$, $q = 21,605$, $p = 240$. These three states or phases are represented by points A, B, C in Figure 15.2.

If we were completely ignorant of the initial conditions, that is, the position and momentum when the bullet was fired, then *any point* in the phase space (Cartesian plane) might represent some possible phase or state, or *any* one of a whole family of curves (parabolas) might represent the history of successive states. But our illustration is *deterministic*, for we know both the proper differential equations and the initial conditions. The evolutionary curve is a portion of a specific parabola. By algebraic elimination of t from the above formulas for q and p, the reader can obtain

$$q = 22505 - \frac{1}{64} p^2$$

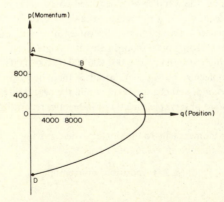

Figure 15.2   Phase-space trajectory for the firing of a bullet

as the equation of the parabola. When the bullet returns after 75 seconds, let us terminate the story. At that instant, $q = 5$, $p = -1200$ (phase point D). We might continue until the bullet hits the ground at $t = 75.01$ when $q = 0, p = -1200.32$, but this adds little to the picture.

For the sake of clarification, one usually resorts to simple instances but they often lack some of the features of the general case and may lead to some wrong impressions. In the example above, the evolution of states and the curve representing the history came to an abrupt end. This is *not* typical of the systems of statistical mechanics, where the succession of states usually goes on and on; so does the curve giving the dynamic history.

Let us consider a slightly more advanced example. A *simple harmonic vibration* will not reveal all general features but will contain aspects not indicated in the bullet example. We shall introduce the procedures of advanced mechanics and form the *Hamiltonian energy function*, which is the sum of the *kinetic* and *potential* energies of the vibrating particle. In elementary physics one learns that kinetic energy is expressed by $\frac{1}{2}mv^2 = m^2v^2/2m = p^2/2m$. It can be shown that the potential energy of the oscillating particle is proportional to $\frac{1}{2}q^2$, hence equal to $kq^2/2$, where k is a physical constant. For convenience, let us consider a particle of unit mass. Then $m = 1$ and the total energy H can be expressed as

$$H = \frac{kq^2}{2} + \frac{p^2}{2}$$

If there is no friction (only possible in a pure mathematical vibration), the total energy H will remain constant, that is, will be *conserved* as time goes on, and the particle will constitute a *conservative system*. The constancy of the total energy is expressed by the following equation.

$$\frac{kq^2}{2} + \frac{p^2}{2} = c$$

This is a relation between q and p, namely, the equation of the curve representing the succession of states (phases) of the vibrating system. The equation represents an ellipse (Figure 15.3) with semiaxes equal to $\sqrt{2c/k}$ and $\sqrt{2c}$. If c, the total energy, is fixed, one of the axes—and hence the shape of the ellipse—depends on the physical constant k. Any point on the ellipse represents the phase (q, p) at some time,

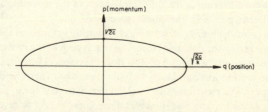

Figure 15.3   Phase-space trajectory for a harmonic oscillator

and the ellipse is described over and over again forever, corresponding to the fact that the particle is vibrating periodically in a straight line, repeating the same positions and momenta forever.

It should be emphasized that in both of the above illustrations the *actual motion* of the system takes place in a straight line. The parabola in one case, and the ellipse in the other, is the progress chart, the *dynamical history* of the system, *not* the path of the moving particle. Second, both examples are atypical because of the simplicity of the systems—the evolution of the bullet because it terminates, the phase curve of the oscillator because it is periodic.

A *general* phase curve usually goes on and on into different parts of the phase space, visiting all regions consistent with the energy of the particular system. Finally, in the actual systems with which statistical mechanics is concerned there would almost never be anything as elementary as the motion of a single particle in a straight line, and the phase space in which the evolutionary curve is described would not be a plane but an abstract higher-dimensional space. In general, then, one deals with abstract paths in an abstract space, but as in other parts of mathematics, quasi-geometric terminology makes for concise, lucid expression. From what we have already explained, more than 10^{24} coordinates are needed for the specification of the state of a gram-molecular mass of gas, which is a most elementary sort of system in statistical mechanics, and hence the phase space would, even in a simple case, have a very large number of dimensions.

Thus far we have mentioned only the *mechanical* foundations of Gibbs's theory, but the huge number 10^{24} is a good motivation and point of departure for the *statistical* side of the subject. Gibbs imagined not one system with 10^{24} phase variables, but a huge set of such systems, an *ensemble*, as he called it, a "population" of systems, if we use the statistical terminology of Chapter 14. He conceived of the systems of the ensemble as macroscopically identical, but microscopically different, just as, from a crude statistical point of view, men in a population are all alike in general physical makeup but differ individually in height, weight, age, life expectancy, etc. To return to Gibbs's ensembles for an example, suppose that the system of interest, the one whose physical properties are of concern, is a container filled with gas; then Gibbs's mental construction forms an enormous aggregate of duplicate containers filled with the same gas in which, however, the phases (states)—the positions and momenta of the molecules at a particular instant—vary from one container to another. Then the values of $(q_1, q_2, \ldots, q_n, p_1, p_2, \ldots, p_n)$ at that instant are different for the different systems of the ensemble and are represented by different points in the phase space. As time goes on, these different representative points will move, describing curves or trajectories, as they are usually called, each one giving the dynamic history of one system of the ensemble. One can imagine the phase points as a cloud of dust (in a hyperspace) whose individual dust particles are in motion in accordance with dynamic laws. Then the *density* of this cloud may appear, at a particular time, to be different from place to place, or might possibly vary at each particular place as time goes on. As a result of certain *dynamical characteristics*, the dust particles (phase points) for the numerous systems never collide, and if they do not go round and round in tornado fashion, that is, if they are not periodic, each one ultimately visits every part of the phase space.

Gibbs's task, as we see it today, was to formulate the family of probability density functions for the states (phases) of the systems of his ensemble at different times. His methodology contrasts with the formulation of the single Maxwell-Boltzmann frequency function for a particular system at a particular time. If one is really ignorant of certain physical characteristics of the particular system, then Gibbs's approach permits finding statistical averages over a large number of similar systems (at a given time). If, on the other hand, one has some knowledge of a particular system (or a particular type of system) but feels that conditions are changing or evolving with time, then Gibbs's point of view permits averages for a particular system over a period of time.

From the above statements, the reader may gain the impression that the Maxwell-Boltzmann distribution is erroneous because it would change to a different one within a billionth of a second and to still another one in the next billionth of a second, etc. But Maxwell and Boltzmann *assumed* that, after a long time, a certain distribution of the speeds would be reached and *maintained* (as in the hypothetical coin-tossing sequence on page 353). At that stage an ideal gas is said to have a constant absolute temperature and to be in *thermal equilibrium*. If there is any preliminary variation in frequency functions, the ultimate distribution is that of Maxwell and Boltzmann and this *defines* thermal equilibrium of a gas as a state where the speeds of the molecules obey the Maxwell-Boltzmann law.

In certain parts of statistical physics, the whole family of random variables and their distributions as they vary in time must be considered, but the construction of theories based on a unique equilibrium distribution is a great simplification. In this respect Gibbs followed the example of Maxwell and Boltzmann, but he generalized their single system to an ensemble of systems. Moreover, he soon abandoned one of their physical postulates, namely, that of an isolated system not subject to energy exchanges with the surrounding medium. That assumption was a simplifying one because it made the total energy constant, that is, made the system *conservative*. But Gibbs did not overlook the advantage of conservative systems and, in fact, he imagined an ensemble of such systems. Now, however isolated a thermodynamic system may be, there is always some energy exchange with surroundings, and it is unrealistic to make contrary assumptions. Therefore Gibbs postulated an ensemble where statistical equilibrium is attained *not* by isolation but by free interaction. This is analogous to the thermal equilibrium described above, and if we wish, we can imagine the one real mechanical system of interest as immersed in a sort of "ice-water bath" made up of the enormous number of other systems of the ensemble. Since such an ensemble in equilibrium involves simpler statistical theory, Gibbs considered it first. For the case where the systems of the ensemble are conservative and one knows their energy, but nothing else about them, Gibbs made the simplest of all possible assumptions, namely, that for a given fixed value of the energy, the phase points are uniformly distributed throughout a "cloud" (in a hyperspace) so that the density is the *same everywhere*. This corresponds to the uniform or rectangular frequency function of Chapter 14 and, as indicated in that chapter, is the analogue for continuous distributions of Laplace's assumption of equal likelihood in the case of a finite sample space. Here it would signify that any state of a system is as likely as any other and would represent complete ignorance, as it were, of any determinate factors. The uniformity of density (for a fixed value of the energy) is what Gibbs labeled a *microcanonical distribution*.

In statistical mechanics, the frequency and distribution functions are not usually expressed directly in terms of the phase variables $q_1, q_2, \ldots, q_n, p_1, p_2, \ldots, p_n$, but are given as expressions involving H, the *Hamiltonian*, which is a function of the q's and p's. The Hamiltonian represents the total energy of a system, and in the so-called conservative systems its numerical value remains constant with time. (This does not mean that the q's and p's are always the same. They vary, but whenever a particular set is substituted in the H-function, the resulting value of that function is the same.)

To make analytic handling easier, it is customary to think of such a conservative system as having a slight range of variation in energy between H and $H + \Delta H$ where ΔH is very small. Then Gibbs's microcanonical distribution can be pictured as a uniform very thin cloud in phase space. The microcanonical distribution is sometimes called an *energy shell* or surface distribution. To see why, we refer to Figure 15.3, in which the ellipse represents the evolution of one conservative system with total energy $H = c$. If that system is just one member of an ensemble whose other systems may have the energy $H = c$ or an energy value slightly greater— that is, where the energy range lies between H and $H + \Delta H$ (ΔH a small positive number)—then in Figure 15.4 any points on or between the two ellipses may be

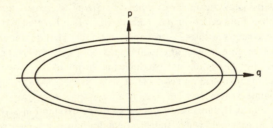

Figure 15.4.   Energy shell

phase points, and the thin elliptic cloud is the energy shell. If we conceive of ΔH as getting smaller and smaller, the energy shell will become just an elliptic curve in the limit. In general phase space, the energy shell between two surfaces (actually "hypersurfaces") would become just a surface as $\Delta H \to 0$.

But as we have indicated, both the empirical physicist and the theoretician found the conservatism of the microcanonical ensemble lacking in generality.

Therefore the major theme of Gibbs's *chef-d'oeuvre* was the postulation of a *canonical distribution*, whose frequency function is

$$f(H) = ce^{-H/\theta}$$

where $f(H)$ is the probability or density in the neighborhood of a cloud in phase space, e is the natural logarithmic base, H (the Hamiltonian) is the random variable, c and θ are positive numerical parameters. (Observe that if H, the total energy, is fixed, $f(H)$ takes a constant value, and the canonical distribution reduces to the microcanonical one.)

The canonical distribution gives the probability that a system selected at random from the Gibbs ensemble will have energy that approximates an H which we may specify by particular numerical values of the q's and p's. (These values determine H). As for the parameters in Gibbs's formula, θ, which he called the *modulus* of the ensemble, is proportional to the variance, being small if the distribution is concentrated, that is, if most systems of the ensemble have values of H very close to its mean. Now the range of energy values in a canonical ensemble is small if one excludes the few extreme cases. It would not be mathematically correct to drop these extreme states, but from their very tiny relative frequencies one can pronounce the phases in question very unlikely to occur. The parameter c varies inversely as $\sqrt{\theta}$ and also contains a "normalizing" factor which can be adjusted so that the canonical formula gives either the actual frequency (number of systems) or the relative frequency approximating a selected state ($q_1, q_2, \ldots, p_n$), that is, a specified value of H.

In the special case where the ensemble systems are portions of an ideal gas,

$$\theta = kT$$

where T is the absolute temperature of the gas and k is Boltzmann's constant (1.38×10^{-16}). This exceedingly small number makes θ tiny, and thus the distribution is concentrated. When this special meaning is assigned to θ in Gibbs's formula, it reduces to the Maxwell-Boltzmann distribution. But, as emphasized originally, the Gibbs ensembles need not be composed of portions of a gas, and therefore his canonical distribution is of more general applicability. Thus, in relation to that distribution, Gibbs found various averages and then gave them physical interpretations. He called one of these averages the *mean index of phase probability*. He was able to show that except for a change in algebraic sign, the average in question corresponds to what physicists call *entropy*, one of the most fundamental of all thermodynamic concepts.

The use of averages in statistical mechanics raised certain challenging issues that were to provide a stimulus to pure mathematics. Gibbs himself had avoided some logical difficulties by basing his mechanics on classic probability theory, Liouville's theorem, and various bold assumptions of his own. But Maxwell and Boltzmann had felt less secure than Gibbs. They were much concerned in their kinetic theory of gases with the paradoxical situation where the deterministic behavior of individual gas molecules had to be reconciled with the indeterminate statistical laws governing aggregates of those molecules. Thus the Maxwell-Boltzmann distribution leads to certain averages connected with the enormous aggregate of molecular velocities, those overall magnitudes being considered to correspond to physical entities like pressure and temperature.

For a *direct* experimental check on the theoretical averages of Maxwell and Boltzmann one would want to measure simultaneously, at some particular instant of time, the velocities of more than 10^{24} gas molecules. To carry out such an experiment would be a sheer impossibility, and hence one might next consider whether it is humanly possible to concentrate on a single one of the vast assemblage of molecules and measure its velocity repeatedly as time goes on. It would seem that averages obtained from such figures should check with those for many similar molecules at one instant, just as tossing a single coin a million times should lead to approximately the same proportion of heads as one throw with a million coins (identical with the original one).

Following the progress of just one molecule sounds simple enough, but it is evident that this would be no more possible empirically than the instantaneous measurement of the velocity of each of 10^{24} molecules. However, we can follow the single molecule in *theory* by means of the dynamical differential equations, which give a unique position and a unique velocity of the molecule at each instant of time, providing we know its position and velocity at just a single particular instant. Then, after we have obtained a quadrillion velocities at different instants, it is possible to find their mean, median, mode, variance, or any desired average. The figures can then be compared with the corresponding averages for the equilibrium aggregate of the Maxwell-Boltzmann distribution.

But now another problem arises: Which one of the 10^{24} molecules should one select for the tabulation of velocities at a quadrillion instants of time? In terms of solving the dynamical differential equations, what initial conditions, that is, what initial molecular position and velocity, should we assign? It would be fortunate if it did not matter, that is, if the long-time behavior of one molecule were the same as that of every other molecule. Then the averages would be the same no matter what initial position and velocity were selected, as long as the numbers did not contradict given physical facts. First Maxwell and then Boltzmann postulated that this would be the case and that, moreover, the *time average*, that is, the deterministic figure obtained by averaging positions, velocities, or other variables for a single molecule over a very long period of time, would be equal to the corresponding statistical average for the aggregate of molecules at one instant of time. Maxwell called this axiom the principle of "continuity of path" and Boltzmann named it the *ergodic hypothesis*, the term we use today. The problem of the equivalence of the two types of average is not limited to kinetic theory but extends to general stochastic processes. In the case of statistical mechanics, we must ask: Is the time average of some variable associated with the single system of interest in agreement with the ensemble average computed *at any time* for the Gibbsian ensemble of "duplicates" of this system?

With all these thoughts in mind, Maxwell and Boltzmann conceived their *ergodic hypothesis* thus: The trajectory (in phase space) representing the evolution of any particular system of a microcanonical ensemble will, in the course of time, pass through *every* point of the corresponding energy surface (hypersurface). If the hypothesis is assumed, then one can deduce as a theorem: The *length of time* during which a particular system exhibits a given set of phases is proportional to the relative frequency of the latter; that is, *every* trajectory in phase space spends in a given region of that space a length of time proportional to the extent (area, volume, *hypervolume*) of the region.

No sooner had Maxwell and Boltzmann advanced this hypothesis than the physicist Lord Kelvin (1824–1907) indicated that it seemed unreasonable to him because special periodic motions must contradict it. For example, if rigid particles of some gas are bouncing up and down or to right or left between parallel faces of a rectangular container, the orderliness of the motion will prevent the frequent collisions that produce randomness in position and velocity as time goes on. As a result, the phases for such a system will not vary greatly, and the corresponding trajectory will not visit every part of phase space. Therefore, important physical quantities—time averages like pressure on the walls of the container, for example, would be very different from the corresponding statistical averages over the microcanonical ensemble. Kelvin arrived at this conclusion intuitively, but Poincaré later gave rigorous mathematical proof in a thorough study of periodic motions. Thus the ergodic hypothesis was shown to be *false*.

Although the ergodic hypothesis in its original form is untenable, various modifications, so-called *quasi-ergodic* hypotheses, have been shown to be *true*. In the applications of stochastic processes it is most desirable to be able to use time and ensemble averages interchangeably and hence to know under what conditions such equivalence is valid. The most important proof of an ergodic theorem was provided in 1931 by George David Birkhoff (1884–1944). According to Marston Morse, professor at the Institute for Advanced Study, Birkhoff was "during the major part of his life . . . the acknowledged leader of American mathematics," and of all his mathematical achievements, "popular opinion focuses attention on his proof of Poincaré's Last Theorem and the *Ergodic Theorem*" so that "much of his other work is obscured."[*]

According to Paul R. Halmos, "*Modern* ergodic theory started early in 1931 with a most significant observation made by B. O. Koopman." This thought occurred in a short paper delivered before the National Academy of Sciences. Halmos tells how Koopman inspired John von Neumann to prove a *mean ergodic theorem*. "Shortly after he proved it he discussed it with G. D. Birkhoff and shortly after that Birkhoff proved the individual ergodic theorem—the priority explained in a subsequent note of Birkhoff and Koopman."[†]

---

[*] Marston Morse, "George David Birkhoff and His Mathematical Work," *Bulletin of the American Mathematical Society*, Vol. 52, No. 5, Part 1 (May 1956), pp. 359 and 389.

[†] Paul R. Halmos, "Von Neumann on Measure and Ergodic Theory," *Bulletin of the American Mathematical Society*, Vol. 64, No. 3. Part 2 (May 1958), p. 91.

To compare the two ergodic theorems, let us state that Birkhoff's is a much more general proposition. The conditions prescribed in von Neumann's theorem do *not* lead to complete agreement in time and ensemble averages, but there is equality in the case of the mean and the variance, the most important averages in physical applications. Von Neumann's theorem shows that the ergodic hypothesis is *approximately* true and gives a way of measuring the precision of the approximation; Birkhoff's theorem establishes that the hypothesis is practically a universal truth.

A whole ergodic theory of related subject matter has grown up since 1931, always to the accompaniment of great names, and is still expanding. By 1931, however, quantum mechanics had deprived the ergodic hypothesis of most of its physical significance, but that did not diminish the importance of ergodic *theorems* in pure mathematics or their value in applied fields other than statistical mechanics.

The deductive demonstration of ergodic theorems is too advanced for presentation in a work for the general reader. But we emphasize that those propositions provide a proper theoretical basis for the computation of statistical mechanical averages. One such mean, mentioned earlier, is named *entropy*. In popularizations of science, entropy is variously described as a measure of randomness, an index of disorder, an amount of uncertainty, "Time's Arrow," etc. The famous second thermodynamic law, often called the Law of Entropy, has as one corollary the ultimate "heat death" of our universe. This is a lugubrious picture, and it is interesting that some of the apparent "refutations" of the pessimistic Second Law have come from the field of biology, where biophysicists and biochemists every now and then cite living processes involving energy transformations that provide counterexamples to the Law of Entropy.

One of Gibbs's foremost students and colleagues, Edwin Bidwell Wilson, was to devote many years to a different aspect of biological processes in his career as professor of vital statistics at Harvard's School of Public Health. His work in biological statistics was not, however, the first evidence of Gibbs's influence, for Wilson, who prepared the *Vector Analysis* textbook based on the Yale professor's superior substitute for the Hamiltonian quaternions, taught mathematics at Yale after his teacher's death, then mathematical physics at the Massachusetts Institute of Technology, where he became head of the physics department. Twenty-seven years after Gibbs's death, his illustrious pupil was asked to deliver the eighth Josiah Willard Gibbs Lecture before the American Mathematical Society and the American Association for the Advancement of Science.

The biography of a mathematician or artist or writer is essentially nothing more than the list of his creative works; legendary or fictional incidents and anecdotes arise from attempts to force "human interest" details. Therefore the *factual* reminiscences of Gibbs in Wilson's 1930 lecture are exceptional biographical material, and we shall quote some parts of that lecture.

Wilson, who was graduated from Harvard in 1899 at the head of his class, with highest honors in mathematics, had been advised by William Fogg Osgood (1864–1943), the renowned Harvard mathematician, to get a change in point of view by taking graduate work at Yale. He followed this advice and thus became Gibbs's student and disciple. We continue the story in Wilson's own words.*

Gibbs was born on February 11, 1839. He prepared for college at the Hopkins Grammar School. He was graduated from Yale in the class of 1858 at the age of nineteen. In college his interests appear to have been Latin and mathematics as he took prizes in each in more than one year of his course. He took the Bristed Scholarship of $95 for the best examination in Greek, Latin and Mathematics. He won the Latin Oration in both Junior and Senior years. He was awarded the Clark Scholarship of $120 for the best examination in the studies of the college course which was conferred subject to the condition that the recipient continue as a graduate for one or two years pursuing non-professional studies. He did so continue and in 1863 got his Ph.D. degree with a thesis: "On the form of the teeth of wheels in spur gearing." In the Yale catalogs of 1863–64 and 1864–65 he is listed as Tutor in Latin; in that of 1865–66 he appears as Tutor in Natural Philosophy. Afterwards he went abroad to study. In the catalog of 1871–72 he reappears as Professor of Mathematical Physics and so continues until his death. Except for his periods as Tutor he taught only graduate work, although particularly competent undergraduates might be admitted to his courses, especially the Vector Analysis.

Gibbs lectured without notes and what specific preparation he generally made I do not know. It was almost always some very simple affair on which he would go astray rather than something recondite. The year I took thermodynamics he could not make his Carnot engine run right. There was a tradition, perhaps unwarranted, that the Carnot engine was apt to trouble him. Sometimes he would unravel his difficulty before the end of the hour and it was then an especial treat to see his mind work; sometimes the end of the hour would come sooner and he would have to leave the matter over until the next time when he would appear with a sheet of paper containing the demonstration.

By far the longest conversation I ever had with him, and of course the last, took place in June 1902 when I was leaving for Paris for a year's study. He said that he did not wish to determine my line of future interest but that he hoped I would consider taking some work in applied mathematics in Paris in addition to any I might take in pure mathematics. He ventured the opinion that one good use to which anybody might put a superior training in pure mathematics was to the study of the problems set us by nature. He remarked that in the thirty years of his professorship of mathematical physics he had had but a half-dozen students adequately prepared to follow his lectures. He did me the honor to include me in the list, though I myself never felt that my preparation in physics had been adequate. I asked why he had given exclusively such advanced courses, why he had not offered some more elementary work to prepare his students. He replied that he had not felt called upon to do so but that if I were willing he would be glad to have me look forward to giving upon my return a general introductory course on mathematical physics, and at any rate he would be happy if I would bear the possibility in mind while abroad. He then went on to say that if I should choose to occupy myself somewhat seriously with mathematical physics he had a considerable number of problems on which he thought I could make progress and that he would be glad to talk about them on my return. How much I have regretted that he did not talk of them at the time, but he gave no inkling of them.

Finally he proceeded to say something of his own plans for the future. He remarked that if he could depend on living to be as old as Methuselah he would continue to study for several hundred years yet, but that as he could not expect any such span of years he had decided to set about preparing some matters for publication. There were three lines of activity he desired to pursue: (1) The revision and extension of his work on thermodynamics, to which he said he had some additions to make covering more recently discovered experimental facts not yet adequately incorporated into the theory and other additions of theory apparently not yet exemplified in experiment. (2) A contribution to multiple algebra* on which he said he had some ideas he thought worth while even though the subject appeared at the time not to be of much interest to mathematicians, most of whom were devoting their attention to analysis. (3) A revision of his method of computing orbits which should certainly be revised now that it had recently been printed verbatim by Buchholz in the third edition of Klinkerfues' Astronomy when certain important improvements were only too obvious. He asked what I should think he had best first undertake, but without waiting for reply answered that the astronomers were conservative and unlikely to be appreciative of improvements in his methods for orbits, that the mathematicians were not impatient to learn of his ideas in multiple algebra and that on the whole he felt it was more important to set about the work on thermodynamics to which he had made no published contribution of significance for about twenty-five years.

He wrote the *Statistical Mechanics* in something like nine months while he carried on his regular teaching. All through the winter and spring of 1900–1901 he worked not only by day, the light in his study in Sloane could be seen burning at night. The manuscript was finished in the summer at Intervale, N.H. After Gibbs died in April, 1903, A. W. Phillips of Yale told me that it was this severe work that killed him. He said that they had gone together to the express office to dispatch the copy to Scribners, that up to that time Gibbs had been quite himself but that from the time they turned away from the office he slumped, the elasticity was gone from his gait, he was a worn out old man, and never fully came back.

This is a thrilling story but sad. However, it may not be true. I communicated it to my old friend Ralph Van Name, nephew of Willard Gibbs, who writes: "This may be true, but it was not apparent to his family," and later, "my comment on the incident of the delivery of the manuscript of the *Statistical Mechanics* was not made in a spirit of criticism, but merely as a statement of my recollection, and of that of my sister, whom I had consulted about it. Though both of us were in Europe at the time of Willard Gibbs's death, I did not leave New Haven until June 1902, and she not until March 1903. It is unquestionably true that my uncle worked to the limit of his strength in trying to get the volume finished on time, and that he did not get over the effects for a good while. But both of us have the impression that he seemed to be in practically normal health and spirits by the autumn of 1901 . . . . My uncle's final illness was a sudden and acute attack of a nature which has no obvious connection with his overwork two years before—it was an intestinal obstruction which the doctors were unable to relieve." I may say that all through the academic year 1901–1902 Gibbs seemed to me to be in normal condition, and in his conversation of June 1902 of which I have given so long an account seemed to be looking cheerfully and healthily ahead with real pleasure in the prospect which he was outlining and with no discernible feeling that it might not be finished—indeed he spoke as one surely counting on being active on my return fifteen months later.

If I have gone at such length into this story I have done so chiefly because it so well illustrates stories which come with the best intention of truth from persons near to Gibbs, with just as high desires to tell the truth and nothing but the truth as I have on this occasion, but which none the less cannot be wholly credited, quite as I do not wholly credit as fact my own statements. There is the story that at home, where he lived all his life with his sister who had married his friend and classmate Addison Van Name, he always

---

* See Chapters 4 and 28, where the concept of a "multiple algebra" or algebra of hypercomplex numbers is developed.

insisted on mixing the salad on the ground that he was a better authority than the others on the equilibrium of heterogeneous substances. A very pretty conceit, and one vouched for by a colleague much closer to Gibbs than I, but I daresay both the fact and the statement of reason for the fact would not be substantiated by the family. Another story refers to his letter to *Nature* in comment on and disproof of Lord Kelvin's proposed experiment to determine the velocity of longitudinal waves in the ether. It is said that when a colleague told him that he had just seen the letter in print Gibbs blushed and said that he could not believe the Editor of *Nature* would print it. That illustrates his modest and retiring disposition, which was a conspicuous trait, but seems hardly credible.

One often hears lament at Yale and elsewhere that Gibbs's colleagues did not capitalize his great discoveries in physical chemistry by developing the subject experimentally and intensively in New Haven from 1876 on. The comment often takes the turn of wondering how much greater role American science would have played in the growth of physical chemistry if Gibbs had accepted the offer to go to the Johns Hopkins University instead of remaining at Yale. How much difference would it have made? Perhaps very little. What efforts Gibbs made to develop physical chemistry at Yale I do not know; perhaps none. That he knew his thermodynamic work was important and knew so when he printed it I have no doubt; but I have noted above that he appeared not to have lectured upon it in his cycle of courses for about 15 years after its completion, preferring for some reason to teach other subjects, and the subject matter of the memoir is not such as would be likely to diffuse around any university without exposition by the master unless by chance there were at hand some almost equally competent person who very much needed the work as a basis for his own, and knew that he needed it. Gibbs was not an advertiser for personal renown nor a propagandist for science; he was a scholar, scion of an old scholarly family, living before the days when research had become *research*. Probably he had faith that when the time was ripe for his thermodynamics, the doctrine would spread.

Another beautiful legend is that Gibbs was not appreciated in this country or at Yale during his life. It is probably true that his name was not well known to the ordinary Yale alumnus before the recent time when his photograph and some eulogy of him were widely circulated to the alumni during a drive for funds. But the efforts which were made to arrange for printing his long and costly paper in the Transactions of the Connecticut Academy in 1876–78 were a high testimonial to the faith of his local contemporaries in his work. He was elected to the National Academy at 40, the average age of election being 50, and only the year after the appearance of the second half of the thermodynamic memoir. In 1881 he received the Rumford Medal from the American Academy of Arts and Sciences which means that a group of his contemporaries in Boston appreciated promptly and highly his contributions in the field of heat. Of course he did not have the notice which Einstein receives today; he had no press agents and surely wanted none. There seems to be every evidence that he received the type of recognition to be expected.

There is no problem requiring more brains, sounder judgment, better total adjustment internal and external, than that of uniting the logical and operational techniques of pure mathematics with the infinite variety of observable fact which Nature offers to our contemplation with a ringing challenge to our best abilities. It was in this field that Gibbs was supreme. He had studied with Weierstrass and was not unmindful of mathematical rigor; in the paper in which he pointed out that phenomenon of the convergence of Fourier series which has come to be known as the Gibbs phenomenon he showed his appreciation of mathematical precision as he did on other occasions. But fundamentally he was not interested in rigor for itself, he was inspired by the greater problem of the union between reflective analytical thought and the world of fact.

16

Sweet Manuscript of Youth

Yet ah that Spring should vanish with the Rose
That Youth's sweet-scented manuscript should close!
Omar Khayyám

In Galois' mind the fever of life that he had lived recently became a present spectacle; the prison walls seemed once more about him. He stood up. The few moments of liberty he had had were only those after his release from the hospital, where he had spent the last month of his prison term. Even those few had been cut short by the two strangers who had met him on the street corner that very morning and forced him to accept their challenge to a duel scheduled for the next day.

This was how everything was ending. He could have died happily side by side with the engineers during the glorious July days in 1830. He could have died happily in another revolution to wipe out the reactionary Louis Philippe. And now he might die over some wretched girl completely unworthy of the sacrifice of his life. But he must not reason himself out of the duel like a coward. If he had not the courage to face the consequences of his own stupidity, he had no place among the brave friends in the revolution to come. And he realized, with a smile of misery, the irony of the situation, for if he did have the courage to go through with the duel, it might end fatally, and there would still be no place for him in the revolution.

And what of his plans to complete what so many other mathematicians had left undone? Like his plans for revolution these would die with him—except for the little that he had already accomplished, and what he could do during this night.

In prison the nights had been long; so long that he sometimes felt that a man could live a whole life in a single night. And now he was beginning a night in which he must accomplish what others had had generations to work out; already the night seemed waning, it seemed to be the moment before dawn.

He must waste no more time in revery. On the table were blank sheets he had made ready for the night's labor. He sat down to the table, took up the pen, and in a moment was lost in a fever of mathematical theory.

Galois pressed a hand to his aching forehead. Two pages gone, two hours, and all the field of thought lay untouched before him. He laid his head on his arms and wept. "I have no time," "I have no time," he scrawled in the margins. Galois glanced over the pages. What he was writing would keep the Academicians busy for years to come—just this bit!—and how much more could he have accomplished with a lifetime before him! He clenched his fists in an agony of hatred and despair at the thought of how Cauchy had lost his memoir—if it had been seen by the Academy, he might never have come to this death. How fate—or man—had conspired to break him! Fourier was to present another paper of Galois' to the Academy—and had died. The manuscript was lost. Only a few months ago Poisson had promised to read still another paper of his to the Academy, and because the fool could not understand it, he had returned it unread.

Fate was in league with a kingdom of fools to destroy him, as they had driven his father to suicide two years ago. What else could have kept him out of the École Polytechnique? He recalled the second time he had gone to take the entrance examinations. It was a bitter memory. Why were all the things he wanted in the hands of fools? He would have been among people like himself at the Polytechnique; but he had been forced to return to Louis-le-Grand, and after plaguing that faculty with his outspoken radicalism he had been expelled even from there.

The night was wearing on. He looked up wearily from his work. It was useless to go on. He could never do in this brief space the many things he had left to do; the pages he had written were his scientific will and he knew that what he was leaving as a boy was far more than most men leave in their old age; but the tragic waste was there and ate into his heart; when a bullet found its mark in the morning, a machine would run down almost before it had really begun to operate. Hastily he gathered together the papers and letters written all during the night and stuffed them into his pockets. He went up to the window for his coat. A chill moist draft of air fell upon him from the window. Very shortly he was due at the field.

On the field it was cold, with a thin morning mist. Dimly Galois saw, as though from a great distance, the figure of his opponent. Galois sucked in the cold air, trying to bring a feeling of reality into his consciousness, but the dizzying dream sensation persisted; he saw the raised arm of Dubois, the referee, and heard the voice cutting the air. "Ready" and "Aim"; then it was the dream and a sudden stab and burning in the abdomen and . . .

Was it the dream? The hospital walls again and the stiff white linen; the hard cot and the loneliness—it could not be—the door was opening softly and a figure from the past slipped in and was by his side—why should he not weep if all this was true?

"You will be well, Évariste," the boy said—it was his brother—"you will be well," and he wept and took Évariste's hand.

This was not the way, Galois thought, and he whispered "Don't cry, I need all my courage to die at twenty."

If there is anything to tell of the brief tragic life of Évariste Galois (1811–1832), we have touched on it in our vignette. Let us not dwell on the sordid details of the wretchedly unhappy existence of our "Keats," but study instead the facts contained in the mathematical testament which the twenty-year-old youth wrote on the eve of the fatal duel.

Galois' will and the research papers he had prepared earlier in his life (starting at the age of sixteen) were eventually published and recognized as the work of one of the greatest geniuses of all time. Fourteen years after the boy's death, his testament and early memoirs were placed in the hands of the great French analyst Joseph Liouville, whom we have mentioned in connection with Gibbs's statistical mechanics. The Galois papers were then published in the *Journal de Mathématiques pures et appliquées*, of which Liouville was the editor. In his introduction, Liouville implied that Galois' material and style were beyond the capabilities of the "referees" of the Academy of Sciences, who had rejected or lost the youth's writings.

Posterity credits Galois perhaps even more than Hamilton with the founding of modern "abstract algebra." Whereas Hamilton's quaternions, designed with physics in mind, led him to break ground for modern axiomatics, Galois' point of view emphasized far-reaching general theorems associated with abstract structures like groups and fields. He gave tremendous impetus to the development of the theory of abstract groups, a subject which has a unifying effect within mathematics itself and also has important applications in modern physics. We shall consider Galois' contribution in the present chapter and provide examples of unification and physical application in the two chapters that follow.

Earlier in this book (Chapter 4) we alluded to *Galois fields* and told how Galois' great theorem spelled the end of classical algebra. Although the statement of the epochal proposition has already been given (page 98), details of the proof and the related concepts created by Galois were postponed, lest our readers, like the Academy's "referees," find Galois' notions incomprehensible. But now that we have gradually built up some background in modern advanced mathematical ideas, we feel that we can safely wind around a wider loop of the original "spiral" of thought. This will require that we emphasize once again just what Galois meant by the term *group*.

In playing "Hamilton's game" of gradual removal of the postulates for a field, we finally arrived at the simplest of all structures, a *groupoid*, which is a system, {S, ∘ }, where S is a set of elements and ∘ symbolizes a binary operation on S. Of course, the structureless system consisting of the set S alone, with no operations or relations whatsoever, is even simpler.

From now on we shall consider "operation on a set S" to signify that the result of the operation is a unique member within S. In other words, the term "operation" will hereafter imply *uniqueness* and *closure*. Therefore, "square root" will *not* be considered a unary operation on the set of real numbers since, for example, the square root of 4 is *not* unique. The result is either $+2$ or -2. Again, subtraction will *not* be considered a binary operation on the set of natural numbers, because closure is lacking, since $2-5$, for example, is not defined within the natural number aggregate. In summary, we shall hereafter say that ∘ is a binary operation on S if and only if $a \circ b$ is a unique member of S, when a and b are any elements of S.

We shall, in the spirit of Galois, now consider *abstract* groupoids in which the members of the fundamental set S are *undefined* elements which need not be numbers but may be capable of varied interpretations in particular situations. We must try to think of the binary operation $\circ$ as an abstraction, too, and not as a sort of addition or multiplication, but merely as a way of combining pairs of elements in S. For this purpose, we review the comments at the end of Chapter 7, where it was stated that any *binary operation* on S can be expressed as a *ternary relation* on S, that is, as a subset of S^3. Suppose then that we consider a groupoid $\{S, \circ\}$ where $S = \{a, b\}$ and $a \neq b$. To define a specific *binary* operation on S there must be a unique answer within S for $a \circ a, a \circ b, b \circ a, b \circ b$. There are two possible choices for the answer in each of the four cases, since that answer must be either a or b. Hence, applying the fundamental combinatorial principle of Chapter 13, there are $2 \cdot 2 \cdot 2 \cdot 2 = 16$ permissible ways of assigning a set of answers to the four combinations and hence 16 potential binary operations on S. By making any one of these choices, we define a specific binary operation on S. From the abstract point of view, we are free to postulate that, for example, $a \circ a = b$, $a \circ b = b$, $b \circ a = a$, $b \circ b = a$. This is equivalent to the following ternary relation or subset of S^3:

$$\{(a, a, b), (a, b, b), (b, a, a), (b, b, a)\}$$

The reader can, if he wishes, list some of the other fifteen ternary relations that define binary operations. Thus, from the combinatorial point of view, there are 16 possibilities for groupoids when the fundamental set contains 2 elements. (Some of these, however, are isomorphic, that is, abstractly identical, as we shall see.) If we rewrite the example above as

$$((a, a), b), ((a, b), b), ((b, a), a), ((b, b), a)$$

it will illustrate that a binary operation is a *special type* of ternary relation on S, namely, a *function* whose domain is S^2 and whose range is a subset (proper or improper) of S. So much for the purely logical meaning of binary operations on a set.

Let us now reverse Hamilton's game by *adding* postulates instead of deleting them. In other words, instead of starting with a field and proceding to more general structures, let us start with a groupoid and advance in the direction of specialization. Thus a *special* groupoid in which the binary operation is *associative* is called a *semigroup*. For purposes of illustration, let us choose $S = \{a, b\}$. Then we shall be able to see that not all sixteen groupoids $\{S, \circ\}$ are semigroups. Let us show that the groupoid of our previous example is *not* a semigroup. For convenience in the present instance, we shall rewrite our previous definition of the particular binary operation in the form of a "multiplication table," a format introduced by Cayley, and used by us on page 78 to define multiplication of quaternions. Thus we say that we have the groupoid, $\{S, \circ\}$, where $S = \{a, b\}$ and $\circ$ is defined by

$\circ$	a	b
a	b	b
b	a	a

One uses this table by reading $a \circ b$, for example, as the entry in "Row a" and "Column b." Thus $a \circ b = b$. To show that $\circ$ is *not* an associative binary operation on S, we need only a single counterexample. Hence we now indicate that

$$(a \circ b) \circ a \neq a \circ (b \circ a)$$

For, using our "multiplication table," we obtain

$$(a \circ b) \circ a = b \circ a = a$$

and

$$a \circ (b \circ a) = a \circ a = b$$

But some of the other fifteen possibilities for groupoids with $S = \{a, b\}$ as fundamental set are semigroups. For example, if $\circ$ is the binary operation with "multiplication table"

$\circ$	a	b
a	a	a
b	b	b

it is not difficult to prove that $\circ$ is associative. The proof would involve demonstrating the truth of eight statements like

$$(a \circ a) \circ a = a \circ (a \circ a)$$
$$(a \circ b) \circ a = a \circ (b \circ a)$$
$$(a \circ b) \circ b = a \circ (b \circ b)$$

etc. We shall leave the task to the reader but, as a sample, we shall show that

$$(b \circ a) \circ b = b \circ (a \circ b)$$

This is true because, consulting the above table, we have

$$(b \circ a) \circ b = b \circ b = b$$

and

$$b \circ (a \circ b) = b \circ a = b$$

When all eight statements are proved to be true, it is established that $\circ$, as defined by the above table, is associative, and $\{S, \circ\}$ is a semigroup.

The reader will see that the semigroup illustrated is not a structure permitting interpretation in terms of ordinary addition or multiplication of numbers. For, if a and b are numbers and if $\circ$ is interpreted as $+$, the table says that $a + a = a$ which would mean that $a = 0$. But the table tells us that $a + b = a$ and hence, if $a = 0$, $0 + b = 0$ or $b = 0$, which is impossible, since a and b are distinct elements. If we interpret $\circ$ as multiplication, the table says that $a \circ a = a$, which now means

$a^2 = a$ or $a^2 - a = 0$, $a(a - 1) = 0$ so that $a = 0$ or $a = 1$. Suppose that we choose $a = 0$. The table says $b \circ a = b$, which now means $ba = b$ and $b \circ 0 = b$, or $b = 0$, which is impossible since a and b are distinct elements. Suppose that $a = 1$. The table says that $ab = a$, which would mean $b = 1$, which is once again impossible since a and b are distinct elements.

Since there is apparently no concrete interpretation of the semigroup in terms of ordinary addition or multiplication of numbers, perhaps we can suggest a hypothetical application and the reader can then find other interpretations of his own. Suppose that a psychologist experiments with the sensation of taste by having subjects taste things that are either markedly "sweet" or "salty." He claims (hypothetically) that the subjects have an extreme prejudice in favor of the first sensaton experienced, whether "sweet" or "salty." If we interpret a as "sweet," b as "salty," and $\circ$ as "is followed by," then $a \circ a = a$ and $a \circ b = a$ signify that tasting a sweet thing followed by tasting something salty produces a sweet sensation just as much as the succession of two "sweets." Similarly, $b \circ a = b$ and $b \circ b = b$ signify that "salty followed by sweet" produces a salty sensation just as much as tasting two salty things in a row. The reader will observe that associativity permits us to carry out extended products like $a \circ b \circ b \circ a \circ a \circ b$, etc., and since in the present instance the first "factor" determines the binary "product," it also determines the extended product. Decidedly hypothetically, the sweet or salty sensation continues no matter how many successive things are tasted.

The reader can establish that the system $\{S, \circ \}$ is a semigroup if $S = \{a, b\}$ and $\circ$ is defined by the "multiplication table"

$\circ$	a	b
a	b	b
b	b	b

All he needs to do is to prove that $\circ$ is associative, a fact that is practically obvious since all binary "products" yield the same result. Perhaps one might provide an interpretation of the semigroup in terms of a paranoid personality who makes a distinction between $a = $ "good" and $b = $ "bad" if only a single event of any importance occurs in a particular period. If one event is followed by another (or still others), he pronounces every composite *bad*. In this interpretation, $\circ = $ "is followed by."

Next, if $\circ$ is a binary operation defined by the table

$\circ$	a	b
a	a	a
b	a	a

one obviously obtains a semigroup. If *a* and *b* are interpreted as "good" and "bad," one might consider the semigroup a model for the behavior of the super-optimist who can distinguish between good and bad for single events but pronounces all composites *good*.

The reader can see that there is similarity in the hypothetical patterns of behavior of the paranoid and the optimist. In other words, although the multiplication tables apparently define different binary operations, the two semigroups are *isomorphic* (Chapter 1) or structurally identical. To make this exact, one should show that in the two semigroups there is matching of elements, operations, and the results of operations. This is evident since the correspondence is:

Semigroup I   Semigroup II

$$a \longleftrightarrow b$$
$$b \longleftrightarrow a$$
$$\circ \longleftrightarrow \circ$$

and also

$$a \circ a \longleftrightarrow b \circ b$$
$$a \circ b \longleftrightarrow b \circ a$$
$$b \circ a \longleftrightarrow a \circ b$$
$$b \circ b \longleftrightarrow a \circ a$$

The reader can give still other definitions (among the sixteen possible) for the binary operation $\circ$ on the set $S = \{a, b\}$ and test for associativity, that is, see whether or not the groupoid defined is a semigroup. He can show, for example, that multiplication is *not* associative for the first table below and hence that the groupoid it defines is not a semigroup. Thus the table indicates that $(ba)b$ is *not* equal to $b(ab)$. Next he can demonstrate that the groupoid defined by the second table is isomorphic to the groupoid of the first. It follows that associativity of multiplication and the property of being a semigroup are lacking in the second groupoid as well as in the first.

$\circ$	a	b
a	a	b
b	a	a

$\circ$	a	b
a	b	b
b	a	b

Our interpretations suggest that if we are to obtain nontrivial applications, we may have to consider a larger fundamental set or else continue the reversal of Hamilton's game in order to obtain specialization of a semigroup.

A special semigroup such that S contains an *identity* or *unit element* and also an *inverse* for every element is called a *group*. A special group in which the binary operation is *commutative* is called an *abelian group* (Chapter 4). We shall not continue the process of specialization any further, but before proceeding we shall summarize by means of the definition: A *group* is a system, $\{S, \circ\}$, that is, a *set S* and a *binary operation* $\circ$ on S such that the following three postulates are satisfied.

(1) *Associativity*: If a, b, and c are any elements in S, then

$$(a \circ b) \circ c = a \circ (b \circ c)$$

(2) *Identity*: There is in S a unique element I (called the *identity* or *unit element*) such that, for any element a in S,

$$a \circ I = I \circ a = a$$

(3) *Inverses*: For any element a of S, there exists a unique element a^{-1}, called the *inverse* of a, such that

$$a \circ a^{-1} = a^{-1} \circ a = I$$

If, once more, our universe is the set $S = \{a, b\}$ and we wish to define a binary operation $\circ$ so that $\{S, \circ\}$ will be a group, it can be shown that of the sixteen multiplication tables possible for a groupoid, only two will define a *group* and the two groups so defined are isomorphic or structurally identical. In other words, there is essentially only one group of *order* 2, that is, for a system of two elements. The two group multiplication tables are

$\circ$	a	b		$\circ$	a	b
a	a	b	and	a	b	a
b	b	a		b	a	b

In the first case, a is the *identity* or *unit element*, and in the second case, b is this element. This is so because the first table indicates that the "product" of a with any element of S leaves the element unchanged, for example, $a \circ a = a$, $b \circ a = b$, $a \circ b = b$. The second table indicates that when any element is combined with b that element is unchanged. In both tables each element of S is its own inverse since $a \circ a$ and $b \circ b$ are both equal to the unit element (a in the first table, b in the second). We observe also that, in either table, the group defined is commutative or abelian, since $a \circ b = b \circ a$. The reader can prove that the groups defined are isomorphic or abstractly identical by matching $a \leftrightarrow b$, $b \leftrightarrow a$, and using the tables to see that $a \circ a \leftrightarrow b \circ b$, $a \circ b \leftrightarrow b \circ a$, etc. Or one can say that both tables refer to an identity element, which can be labeled I ($a = I$ in the first table, $b = I$ in the second table), and another element, called X ($b = X$ in the first table, $a = X$ in the second table). With this replacement, the tables become

$\circ$	I	X		$\circ$	X	I
I	I	X		X	I	X
X	X	I		I	X	I

These "multiplication tables" are identical and differ only in the order of the rows and columns labeled I and X.

What are some of the concrete interpretations that can be given to the single abstract group of order 2? One example is furnished by taking ∘ to be ordinary multiplication and $I = +1$, $X = -1$. Then the above tables become

×	+1	−1
+1	+1	−1
−1	−1	+1

which obviously gives the correct products for the numbers listed.

Another interpretation of the same abstract group is provided by taking $I =$ "even," $X =$ "odd," that is,

$$I = \{\ldots, -2, 0, +2, +4, \ldots\}$$
$$X = \{\ldots, -1, +1, +3, \ldots\}$$

If we take ∘ = ⊕, where ⊕ is the kind of "addition" in which each number in one set is added (in the ordinary sense) to every number of the other, then it can be seen that $\{S, \oplus\}$ is a group. To show that "even" ⊕ "odd" = "odd," that is, that $I \oplus X = X$, we form

$$I \oplus X = \{\ldots, -2-1, -2+1, -2+3, \ldots, 0-1, 0+1, 0+3, \ldots, +2-1, \ldots\}$$
$$= \{\ldots, -3, -1, +1, +3, \ldots\} = X$$

(where we list the odd integers that occur repeatedly only once). The illustration we are considering is a concrete interpretation of the one and only *finite* abstract group with two elements, but it will be observed that each of the two elements is an *infinite set*.

There is another way of describing the interpretation just given. The alternative description will now be revealed because it avoids the necessity of alluding to "infinite" sets before we discuss them (Chapter 24); also it may seem simpler to the reader; furthermore it will be helpful in later applications. To begin, we recall that in previous chapters we stated repeatedly that a single idea may have many different names. Thus "cardinal 3" may be named *drei* or III or $2 + 1$ or $8 - 5$, etc. and all such names are considered equivalent. Again, the ratio $1 : 2$ or the fraction $1/2$ has many equivalent names like $4/8$ or 0.5, etc. Then, in the present instance, we might consider all even numbers as "congruent" or equivalent (in the sense of "evenness" or exact divisibility by 2) and take a single even integer 0 as the representative of I, the class of even numbers. We could then say that the symbol 0 represents the idea of evenness, and -2 or $+2$ or $+4$, etc. are just other names for the same idea. In the same way, the integer 1 can be taken as the representative of the set of odd numbers, all of which are equivalent in the sense of leaving the *remainder* or "residue" 1 when divided by 2. Then we can say that the symbol 1 represents "oddness," that is, the property described, and -1, $+3$, $+5$, etc. are just other names for the same idea. Then instead of saying "odd" $\oplus$ "even" = $X \oplus I = X$, we shall say $1 \oplus 0 = 1$, which will seem satisfactory to the reader, as will $0 \oplus 0 = 0$, and $0 \oplus 1 = 1$. Instead of saying "odd" $\oplus$ "odd" = "even," we now say $1 \oplus 1 = 0$, which may trouble the reader unless he realizes that it merely means that the ordinary sum of *any pair* of odd integers is even (or that in true statements like $3 + 5 = 8$, $7 + 9 = 16$, etc., each number on the left side of the equations is equivalent to 1 and the numbers on the right are equivalent to 0 in the sense just indicated). In the present instance, $\oplus$ is called *addition modulo 2*, and the symbols 0, 1 are often said to represent the *residue classes I* and *X* (whose members leave "residues" 0 and 1 when divided by 2, the "modulus"). Then $\{S, \circ\}$ is a group in which S consists of the residue classes 0 and 1, $\circ = \oplus$, which is *addition modulo* 2 as defined by the table

$\oplus$	0	1
0	0	1
1	1	0

Since the "new" school mathematics includes *addition modulo* 2 in its courses of study, that type of addition (with modulus 2 and other moduli) is often called "clock addition." Thus, in an ordinary clock the *modulus* would be 12, and the "residues" after division by 12 would be $0, 1, 2, \ldots, 11$. If it is now 9 o'clock and we ask what time it will be in 5 hours, the answer is 2 o'clock. If $\oplus$ is addition modulo 12, $9 \oplus 5 = 2$, and similarly $8 \oplus 7 = 3$. If it is now 9 o'clock and we ask what time it will be 29 hours from now, then starting at 9, we have to go around the clock twice and then 5 hours more. This shows that 29 is just another name for 5 in the modulo 12 system, that is, in the sense that it leaves a "residue" 5 when divided by 12. Hence $9 \oplus 5 = 2$ o'clock gives the answer to the question raised. If we take a clock which (Figure 16.1) has just two markings, 0 and 1, we can use it to picture the binary operation of our previous illustration, namely, *addition modulo* 2. In particular, if the clock of Figure 16.1 now reads 1 o'clock and we ask what time it will be one "interval" from now, the answer is 0 o'clock, so that we have a picture of $1 \oplus 1 = 0$ in the modulo 2 system.

An interpretation similar to modulo 2 addition can be obtained if in the multiplication table for the abstract group of order 2 we consider *I* and *X* as two positions

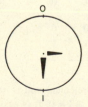

Figure 16.1 The time is "half past zero" on a modulo 2 clock.

of a wheel, I as its original position and X its position after rotation through 180°. But the wheel would also be in the original position I after rotation (in either direction) through 360° or 720°, etc. and would be in position X after rotation (in either direction) through 540° or 900°, etc. Thus there is an "equivalence class" of rotations for each of the two positions, and we have a group of order 2 in which the elements are the two sets of rotations

$$I = \{0°, \pm 360°, \pm 720°, \ldots, \pm 360n°, \ldots\}$$
$$X = \{\pm 180°, \pm 540°, \pm 900°, \ldots, \pm(180 + 360n)°, \ldots\}$$

and here $\circ$ = "is followed by," an operation signifying that one rotation is to be succeeded by another rotation.

We could give still other interpretations of the single abstract group of order 2, but the whole point is that all such concrete groups are isomorphic and thus abstractly identical. If we increase the number of elements in the fundamental set S, there is a special theorem of group theory establishing the fact that there is one and only one abstract group for the orders $n = 2, 3, 5, 7, 11, 13, 17, 19, \ldots$ or any *prime number*. A prime number is a natural number whose only exact divisors are itself and 1. The theorem states that there is one and only one abstract group for any prime number. Then if we take $n = 3$ or $S = \{a, b, c\}$ where a, b, and c are distinct elements, there is one abstract group $\{S, \circ\}$. The reader can establish this fact by bearing the group postulates in mind and then deriving the multiplication table by a process of trial with groupoids as we did in the case of the groupoid of order 2.

But let us accept the fact that all groups of order 3 are isomorphic. Then we can derive the general multiplication table from a specific instance. Let us show that if $\circ = \oplus = addition\ modulo\ 3$, and $S = \{0,1,2\}$ where 0, 1, and 2 are the "residues" or "residue classes" modulo 3, then $\{S, \circ\} = \{S, \oplus\}$ is a group. The reader will have no trouble with addition modulo 3 except possibly with the combinations, $1 \oplus 2 = 2 \oplus 1 = 0$, $2 \oplus 2 = 1$. He can carry this out on a clock with only 3 "hours" labeled 0, 1, 2 by means of questions like: If it is 1 o'clock now, what time will it be 2 "hours" from now? Or he can carry out the ordinary additions $1 + 2 = 3$, $2 + 2 = 4$ and say that 3 and 4 are equivalent (modulo 3) to 0 and 1, respectively. The modulo 3 addition table is

$\oplus$	0	1	2
0	0	1	2
1	1	2	0
2	2	0	1

Here 0 is obviously the identity element and is its own inverse. Also 1 and 2 are inverse elements since $1 \oplus 2 = 2 \oplus 1 = 0$. It is easy to show that $\oplus$ is associative, and this would complete the proof that the system $\{S = \{0, 1, 2\}, \oplus\}$ is a group.

Since there is only one abstract group of order 3, we can represent it by replacing $S = \{0, 1, 2\}$ in the above table by $S = \{I, a, b\}$. Then $\{S, \circ\}$ is a group in which $\circ$ is defined by

$\circ$	I	a	b
I	I	a	b
a	a	b	I
b	b	I	a

An interpretation of the group of order 3 in terms of rotations of a wheel would take I as the original position of the wheel, a and b as positions after rotations through 120° and 240°, respectively. Here $\circ$ = "is followed by" and rotations of 0°, $\pm 360°$, $\pm 720°$, ..., $\pm 360n°$, where n is any integer, are equivalent, as are rotations through 120°, 480°, $-240°$, ..., $120° \pm 360n°$, etc.

The table defining a binary operation on a set is customarily called the "multiplication table," and results of operations are usually called products. The reader will therefore understand such terminology when we use it. Also, instead of $a \circ b$, $a \circ b \circ c$, etc., it is customary to employ the usual product symbolism of elementary algebra and write ab, abc, etc. In fact, $a \circ a \circ a \circ a$ would not be written $aaaa$ but more compactly as a^4.

The use of such symbolism will indicate a special property of the abstract group of order 3. Consulting the multiplication table, we see that $aa = b$, that is, $a^2 = b$. From the same table, $ba = I$. Substituting a^2 for b in this last statement, we have $a^2a = I$ or $a^3 = I$. Thus the three group elements, a, b, I, can be symbolized as a, a^2, a^3, that is, as "powers" of one of them.

If all the elements of a group can be expressed as powers of one of them, the group is described as *cyclic*. Thus the group of order 3 is cyclic. If we examine the multiplication table of the abstract group of order 2 (page 375), we see that $X \circ X = XX = X^2 = I$. Hence the two group elements are X and X^2. Therefore the group of order 2 is cyclic. In group theory it is proved that *all groups of prime order are cyclic*. Thus if the order of a group is 7, its elements are $a, a^2, ..., a^6$, $a^7 = I$ and it is easy to "multiply" any pair of elements.

Since the group of order 2 is cyclic, we can represent its elements by $a, a^2 = I$. This explains one of the interpretations we have already given. If $\circ$ = multiplication of *numbers* and I is a *multiplicative* identity, then $I = 1$. Hence $a^2 = 1$ and $a = +1$ or $a = -1$. Since a must differ from I, $a = -1$. Therefore our group is $\{S = \{+1, -1\}, \times\}$, as defined by the table on page 376.

The abstract group of order 3 is cyclic, and its elements are $a, a^2, a^3 = I$. Once again if we interpret $\circ$ as multiplication of numbers and I as the multiplicative identity 1, $a^3 = 1$. We cannot have $a = 1$ since a and I are distinct elements. We can solve $a^3 = 1$ by writing it as $a^3 - 1 = 0$ or

$$(a - 1)(a^2 + a + 1) = 0$$

The first factor gives $a = 1$ and application of the quadratic formula (page 90) gives $a = -1/2 + i/2\sqrt{3}$, $a = -1/2 - i/2\sqrt{3}$. Since these answers satisfy the equation $a^3 = 1$, they are called *cube roots of* 1. Thus 1 has the real cube root 1, and two complex cube roots. A little manipulative algebra shows that either complex root is the square of the other. The first complex root above is often symbolized by the Greek letter omega, ω, and hence the three cube roots of 1 are ω, ω^2, $\omega^3 = 1$. Therefore an interpretation of the abstract group of order 3 is given by

$$\{S = \{\omega, \omega^2, \omega^3 = 1\}, \times\}$$

Thus far our study of finite groups has led only to abelian groups. To obtain a noncommutative group, we refer to the multiplication table for quaternions in Chapter 4. If the quaternion multiplication is to define a binary operation on a finite set, the table will have to be enlarged so as to provide *closure*. For example, in quaternion multiplication, $i^2 = -1$ and $kj = -i$. Hence -1 and $-i$ will have to be included in S. If $S = \{1, i, j, k, -1, -i, -j, -k\}$ and $\circ =$ quaternion multiplication, $\{S, \circ\}$ is the *quaternion group* of order 8, which is not commutative. By using Hamilton's assumption (Chapter 4) that $q_1 q_2 = -q_2 q_1$ where q_1 and q_2 are any two quaternions, the reader will be able to state that $ij = -ji$ and hence, from the table in Chapter 4, $-ji = -k$. From such considerations he can construct the multiplication table (with 8 rows and 8 columns) corresponding to the set S, and then show that $\{S, \circ\}$ is a group.

One can obtain a finite *abstract* group of any order merely by deciding to postulate a cyclic group. Thus, although the quaternion group is one abstract group of order 8, there is also the abstract cyclic group of order 8 whose elements are $\{a, a^2, \ldots, a^7, a^8 = I\}$. This is a commutative (abelian) group and hence certainly *not* isomorphic to the quaternion group. As a numerical interpretation there is the multiplicative group whose elements are the 8 solutions (in the complex domain) of the polynomial equation, $x^8 = 1$, that is, the eight eighth-roots of 1, two of which are the real numbers $+1$ and -1, the other six being complex. There are three more abstract groups of order 8, but we shall not discuss them here. Instead, we shall summarize the situation by saying that there are 5 nonisomorphic abstract groups of order 8, of which 3 are abelian and 2 nonabelian.

The reader may think that mathematicians are able to make statements like those in the preceding paragraph about finite groups of any given order, however large. But that is not the case. There is at least one abstract group for every finite order, namely, the cyclic group. If the order is prime, the cyclic group is the only one. But there is as yet no universal formula that will tell the number and nature of abstract groups for all composite (nonprime) orders. The way this unsolved problem is being tackled today derives directly from the thought of the boy Galois.

All the groups discussed in the present chapter have been of *finite* order. But we have illustrated infinite groups earlier in this book. Thus if S is the set of integers and $\circ$ is ordinary addition, $\{S, \circ\}$ is an abelian group. Its identity element is 0, and there is an inverse, $-a$, for every element a. We observe that the set of natural numbers, $S = \{1, 2, 3, \ldots\}$ is *not* a group with respect to addition because there is no identity element (and this also makes it impossible to define additive inverses). If S is the set of rational numbers (or the set of real numbers) with zero excluded and $\circ$ is ordinary multiplication, then $\{S, \circ\}$ is an abelian group. The unit element is 1 and inverses are reciprocals. If $\circ$ is multiplication and S is the set of natural numbers, $\{S, \circ\}$ is not a group because inverses are lacking. In our next chapter we shall illustrate infinite groups whose elements are not numbers.

Cauchy and Galois considered finite groups in which the elements were *not* points or numbers (complex or hypercomplex) or qualities but *mappings* (functions) of a certain kind. In Chapter 7 a mapping (function) was defined as a binary relation or set of ordered pairs in which every element of the domain is matched with a unique element of the range. We gave as an example (page 141) the mapping of a set of three cities *into* a set of points, and *onto* a proper subset of that set. Thus, as another example, if $S = \{a, b, c, d\}$ and $T = \{e, f, g\}$, then (using the format we shall employ in the present chapter)

$$\begin{pmatrix} a & b & c & d \\ f & g & f & f \end{pmatrix}$$

represents a mapping whose domain is S and whose range is $\{f, g\}$. It is a mapping of S into T, or onto $\{f, g\}$, a proper subset of T.

The ordered pairs in the mapping illustrated are (a, f), (b, g), (c, f), and (d, f). If we reverse the order of elements in each pair, we obtain the *inverse relation*, $\{(f, a), (g, b), (f, c), (f, d)\}$. This inverse relation is *not a mapping* (*function*) because there are three pairs containing the same first element, f. Corresponding to f, an element of the domain, there are *three* elements of the range. Another way of pointing up the same fact is to state that, in general, a mapping may be a "many-to-one correspondence." In the original mapping, for example, the "many" elements, a, c, and d, all correspond to f. Then the inverse of such a mapping will be a "one-to-many correspondence" and is therefore, in general, *not* a mapping.

By contrast, there are special cases like the mapping

$$\begin{pmatrix} a & b & c \\ d & e & f \end{pmatrix}$$

which is a *one-to-one correspondence* and has an inverse that is a mapping, namely

$$\begin{pmatrix} d & e & f \\ a & b & c \end{pmatrix}$$

Cauchy and Galois considered a still further specialization—a function that is not only a *one-to-one* correspondence, but also maps the fundamental set *onto itself*. For example, if $S = \{a, b, c, d\}$, then

$$\begin{pmatrix} a & b & c & d \\ d & a & c & b \end{pmatrix}$$

is a *one-to-one mapping* of S *onto itself*. Here S is both domain and range. The effect of the mapping is merely to rearrange or *permute* the elements of the set. Hence today we call it a *permutation*, a usage akin to that in the combinatorial algebra of Chapter 13.

We shall now be considering *groups* of permutations not only because they are applied in Galois' theory but also because such groups can be used to represent *all finite groups*, since it can be proved that *every finite group is isomorphic to some group of permutations*.

In the case of a *finite* set, it is tautological to refer to a *one-to-one* mapping of the set *onto* itself, since any mapping onto itself must be one-to-one. For if it were many-to-one, several elements of the domain would be mapped onto a single element of the range. Hence there would be fewer elements in the range than in the domain and therefore the latter could not be mapped *onto* the former, as required. However, in the case of an infinite set, as the next chapter will show, a mapping of such a set *onto* itself need *not* necessarily be one-to-one (and a one-to-one mapping can be *into* instead of *onto* itself).

But we are now discussing finite sets and finite groups. Hence we can state that a *permutation is a mapping of a finite set onto itself*. As an example, from everyday life, suppose that three objects, a paperweight, a pencil, and an eraser, are located in certain positions on top of a desk. For convenience, let us attach the labels 1, 2, 3, respectively, to these objects. Then, the symbolism

$$\begin{pmatrix} 1 & 2 & 3 \\ 3 & 1 & 2 \end{pmatrix}$$

will be used to signify that objects are to be shifted so that those named on the lower line replace or are *substituted* for the corresponding ones on the upper line. This suggests why Cayley and Galois called permutations *substitutions*. In the present example, eraser is *substituted* for paperweight, paperweight replaces pencil, pencil replaces eraser. The following array of numbers represents the very same permutation or substitution of objects (except that the original location of objects on the desk is different, being pencil, paperweight, and eraser, in that order):

$$\begin{pmatrix} 2 & 1 & 3 \\ 1 & 3 & 2 \end{pmatrix}$$

Likewise, there are four other formats for this same permutation, namely,

$$\begin{pmatrix} 1 & 3 & 2 \\ 3 & 2 & 1 \end{pmatrix}, \quad \begin{pmatrix} 2 & 3 & 1 \\ 1 & 2 & 3 \end{pmatrix}, \quad \begin{pmatrix} 3 & 1 & 2 \\ 2 & 3 & 1 \end{pmatrix}, \quad \begin{pmatrix} 3 & 2 & 1 \\ 2 & 1 & 3 \end{pmatrix}$$

Again, the rearrangement keeping the paperweight wherever it happens to be, but interchanging positions of pencil and eraser is a *single* permutation or substitution with any one of the six formats:

$$\begin{pmatrix} 1 & 2 & 3 \\ 1 & 3 & 2 \end{pmatrix}, \quad \begin{pmatrix} 1 & 3 & 2 \\ 1 & 2 & 3 \end{pmatrix}, \quad \begin{pmatrix} 2 & 1 & 3 \\ 3 & 1 & 2 \end{pmatrix}, \quad \begin{pmatrix} 2 & 3 & 1 \\ 3 & 2 & 1 \end{pmatrix}, \quad \begin{pmatrix} 3 & 1 & 2 \\ 2 & 3 & 1 \end{pmatrix}, \quad \begin{pmatrix} 3 & 2 & 1 \\ 2 & 3 & 1 \end{pmatrix}$$

Presently we shall want to show that certain sets of permutations constitute groups with respect to ∘ = "is followed by" or "succession." Just what will happen, then, if one permutation follows another? If we label the two illustrative permutations above A and B respectively, what is the meaning of $A \circ B = AB = $ "A followed by B"? Picturing AB in terms of the objects involved, we have

First, $A = $ $= \begin{pmatrix} 1 & 2 & 3 \\ 3 & 1 & 2 \end{pmatrix}$

Then, $B = $ $= \begin{pmatrix} 3 & 1 & 2 \\ 2 & 1 & 3 \end{pmatrix}$

The net result is that the eraser is in its original position on the desk, but the pencil and paperweight have been interchanged, that is,

$$AB = \begin{pmatrix} 1 & 2 & 3 \\ 2 & 1 & 3 \end{pmatrix}$$

an array which could have been obtained merely by deleting the common row (or intermediate arrangement of objects), 3 1 2, in the above arrays for A and B.

Let us reverse the order of succession, and find the meaning of BA. If we express B in the form

$$B = \begin{pmatrix} 1 & 2 & 3 \\ 1 & 3 & 2 \end{pmatrix}$$

then we shall want to use that one of the six forms for A that has 1 3 2 as the top row (so that B and A will have a common row that can be deleted). Hence we have

$$BA = \begin{pmatrix} 1 & 2 & 3 \\ -1--3--2- \end{pmatrix} \begin{pmatrix} -1--3--2- \\ 3 & 2 & 1 \end{pmatrix} = \begin{pmatrix} 1 & 2 & 3 \\ 3 & 2 & 1 \end{pmatrix}$$

which is *not* the same permutation as AB. In the case of BA, the pencil is in its original position whereas the other two objects are interchanged. Thus the "multiplication" of A and B is *not* commutative.

Our examples illustrate that the succession of two permutations of certain objects, that is, the succession of two mappings of a finite set onto itself, results in some permutation (rearrangement) of the objects (some mapping of the set onto itself). Then if we are given an aggregate or class whose elements are *all* possible permutations of a particular finite set (that is, all possible mappings of that set onto itself), it will be *closed* under the operation of succession. In other words, "is followed by" or "succession" will truly be a *binary operation* on the class of *all* possible permutations of a particular finite set.

If we consider the same example of three objects on a desk (or any three distinguishable objects), we know (from Chapter 13) that 3! or 6 different arrangements are possible. Hence we can define 6 distinct permutations or substitutions involving any three objects. We shall now label these substitutions $S_1, S_2, \ldots, S_6$ and show that, given the class, $C = \{I = S_1, S_2, \ldots, S_6\}$, and $\circ$ = "succession," the system $\{C, \circ\}$ is a *group*. As we have just explained, $\circ$ will truly be a binary operation on C because C includes *all* possible permutations of the particular set of objects. Now the six permutations or substitutions that are members of C are

$$I = S_1 = \begin{pmatrix} 1 & 2 & 3 \\ 1 & 2 & 3 \end{pmatrix} \quad \text{(the \textit{identity})}$$

$$S_2 = \begin{pmatrix} 1 & 2 & 3 \\ 2 & 3 & 1 \end{pmatrix}$$

$$S_3 = \begin{pmatrix} 1 & 2 & 3 \\ 3 & 1 & 2 \end{pmatrix}$$

$$S_4 = \begin{pmatrix} 1 & 2 & 3 \\ 1 & 3 & 2 \end{pmatrix} \quad \text{(1 is left fixed)}$$

$$S_5 = \begin{pmatrix} 1 & 2 & 3 \\ 3 & 2 & 1 \end{pmatrix} \quad \text{(2 is left fixed)}$$

$$S_6 = \begin{pmatrix} 1 & 2 & 3 \\ 2 & 1 & 3 \end{pmatrix} \quad \text{(3 is left fixed)}$$

To prove that $\{C, \circ\}$ is a group, one must show (1) that $\circ$ is *associative*, (2) that there is a unit or identity element in C, and (3) that each one of the six elements of C has an inverse in C.

Although the operation $\circ$ = "succession" need not be commutative, it is is always *associative* for a class of permutations, as we shall indicate by a general argument below. But if the reader wishes to do so, he can establish associativity in the present instance for all triples of permutations formed from substitutions in C; that is, he can find the result, $(S_2 S_3) S_4$, for example, and show that it is the same as $S_2 (S_3 S_4)$.

We have already labeled S_1 as I, the identity element, but we should show that it really is a unit element, by proving $I S_2 = S_2 I = S_2$, $I S_3 = S_3 I = S_3$, etc. This presents no problem, since, for example,

$$I S_4 = \begin{pmatrix} 1 & 2 & 3 \\ 1 & 2 & 3 \end{pmatrix} \begin{pmatrix} 1 & 2 & 3 \\ 1 & 3 & 2 \end{pmatrix} = \begin{pmatrix} 1 & 2 & 3 \\ 1 & 3 & 2 \end{pmatrix} = S_4$$

$$S_4 I = \begin{pmatrix} 1 & 2 & 3 \\ 1 & 3 & 2 \end{pmatrix} \begin{pmatrix} 1 & 3 & 2 \\ 1 & 3 & 2 \end{pmatrix} = \begin{pmatrix} 1 & 2 & 3 \\ 1 & 3 & 2 \end{pmatrix} = S_4$$

As for inverses, the reader can show that each of the permutations S_1, S_4, S_5, and S_6 is its own inverse. The argument in the case of S_6, that is, the proof that $S_6 S_6 = S_6{}^2 = I$ is as follows:

$$S_6 S_6 = \begin{pmatrix} 1 & 2 & 3 \\ -2--1--3- \end{pmatrix} \begin{pmatrix} -2--1--3- \\ 1 & 2 & 3 \end{pmatrix} = \begin{pmatrix} 1 & 2 & 3 \\ 1 & 2 & 3 \end{pmatrix} = I$$

It is also easy to show that S_2 and S_3 are inverse elements. The argument proving that $S_2 S_3 = S_3 S_2 = I$ is as follows:

$$S_2 S_3 = \begin{pmatrix} 1 & 2 & 3 \\ -2--3--1- \end{pmatrix} \begin{pmatrix} -2--3--1- \\ 1 & 2 & 3 \end{pmatrix} = \begin{pmatrix} 1 & 2 & 3 \\ 1 & 2 & 3 \end{pmatrix} = I$$

$$S_3 S_2 = \begin{pmatrix} 1 & 2 & 3 \\ -3--1--2- \end{pmatrix} \begin{pmatrix} -3--1--2- \\ 1 & 2 & 3 \end{pmatrix} = \begin{pmatrix} 1 & 2 & 3 \\ 1 & 2 & 3 \end{pmatrix} = I$$

Since all three requirements of the definition of a group are met, the class of 6 permutations of 3 objects constitutes a group with respect to "succession." The same is true for the 2! or 2 permutations of 2 objects, the 4! or 24 permutations of 4 objects, the 5! or 120 permutations of 5 objects, etc. Hence if ∘ = succession, the $n!$ permutation of n objects constitute a group, called the *symmetric group* of *degree n*. The nature of the symmetry will appear presently. In Galois theory, the 2, 3, 4, 5, . . . , n objects that are permuted are not pencils or paperweights but the 2, 3, 4, 5, . . . , n roots of a quadratic, cubic, quartic, quintic, . . . , nth degree polynomial equation.

Let us now show, as we promised to do, that ∘ = "succession of permutations" is an associative operation. If P_1, P_2, and P_3 are *any three* permutations of n objects, we must now indicate that

$$(P_1 P_2)P_3 = P_1(P_2 P_3)$$

The reader will see why this statement is true if he considers P_1, P_2, and P_3 as shuffles of a single suit of cards. Let him concentrate on what happens to the top card in each shuffle. A similar chain of reasoning will then apply to the bottom card or to a card in any position. Let us suppose that the top card is the ace and that shuffles have the following effect:

$$P_1 = \begin{pmatrix} ace --- \\ queen --- \end{pmatrix}, \quad P_2 = \begin{pmatrix} queen --- \\ king --- \end{pmatrix}, \quad P_3 = \begin{pmatrix} king --- \\ deuce --- \end{pmatrix}$$

Then the effect of P_1 followed by P_2 would be to replace ace by king. Thus

$$P_1 P_2 = \begin{pmatrix} ace --- \\ king --- \end{pmatrix} \quad \text{and} \quad P_3 = \begin{pmatrix} king --- \\ deuce --- \end{pmatrix}$$

Hence $(P_1 P_2)P_3$ would replace ace by deuce.

Let us now consider $P_1(P_2 P_3)$, the right member of the above equation. We see from the effects stated above that since P_2 replaces queen by king, and P_3 replaces king by deuce, P_2 followed by P_3 would have the effect of replacing queen by deuce, that is,

$$P_2 P_3 = \begin{pmatrix} queen --- \\ deuce --- \end{pmatrix}$$

Then $P_1 (P_2P_3)$ means

$$P_1 = \begin{pmatrix} \text{ace} - - - \\ \text{queen} - - - \end{pmatrix} \quad \text{followed by} \quad P_2P_3 = \begin{pmatrix} \text{queen} - - - \\ \text{deuce} - - - \end{pmatrix}$$

and $P_1(P_2P_3)$ replaces ace by deuce, which is the same as the effect of $(P_1P_2)P_3$.

In summary, succession of permutations (or, in general, of *mappings of any set onto itself*) is associative because three shuffles in succession have the same effect whether you are like the "player on the right" who examines the effect on a typical position in the deck after the first shuffle (and the last) or like the "player on the left" who makes his examinations after the second and third shuffles.

If all the elements of one group are also elements of another group, then the former is called a *subgroup* of the latter. We note that any group is thus a subgroup of itself. If we refer to this special case as an "improper subgroup," all subgroups of a group except the group itself can be termed *proper*. For example, in the illustration considered on page 384, namely, the group of six permutations of three objects (the symmetric group of "degree" 3), it can be proved that the group has subgroups with the following rosters:

$$A = \{S_1, S_2, S_3\}$$
$$G_1 = \{S_1, S_4\}$$
$$G_2 = \{S_1, S_5\}$$
$$G_3 = \{S_1, S_5\}$$
$$J = \{S_1\}$$

The reader will find it easy to prove that each one of these aggregates qualifies as a group under the operation of "succession," but we shall nevertheless give him some help by indicating how to prove that the system $\{A, \circ\}$ is a group. In the first place, we must be sure that $\circ = $ "succession" is truly a binary operation on the set of elements in A. We cannot use our previous general argument concerning closure because the class A does *not* include *all* permutations of three objects. If we are to show that $\circ$ is a binary operation on A, that is, possesses closure on A, we must show that we can construct a "multiplication table" giving S_1 or S_2 or S_3 as the answer to any of the nine products S_1S_1, S_1S_2, ..., S_2S_1, S_2S_2, etc. Since S_1 is the element we previously labeled I, we know that $S_1S_2 = IS_2 = S_2$, etc., and the entries in the first row (and first column) of the multiplication table will therefore be S_1, S_2, S_3, respectively. We also showed previously that $S_2S_3 = S_3S_2 = I$, and this will provide for two more entries. Hence we need only find the meaning of S_2S_2 and S_3S_3. We see that

$$S_2S_2 = \begin{pmatrix} 1 & 2 & 3 \\ -2--3--1- \end{pmatrix} \begin{pmatrix} -2--3--1- \\ 3 & 1 & 2 \end{pmatrix} = \begin{pmatrix} 1 & 2 & 3 \\ 3 & 1 & 2 \end{pmatrix} = S_3$$

$$S_3S_3 = \begin{pmatrix} 1 & 2 & 3 \\ -3--1--2- \end{pmatrix} \begin{pmatrix} -3--1--2- \\ 2 & 3 & 1 \end{pmatrix} = \begin{pmatrix} 1 & 2 & 3 \\ 2 & 3 & 1 \end{pmatrix} = S_2$$

Now we can fill in the multiplication table with entries that are either S_1 or S_2 or S_3. Hence "succession" is truly a binary operation on the class A, and the multiplication table is as follows:

∘	S_1	S_2	S_3
S_1	S_1	S_2	S_3
S_2	S_2	S_3	S_1
S_3	S_3	S_1	S_2

The general argument that "succession" is associative applies in the present case. $S_1 = I$ is obviously an identity or unit element in A. Once again S_1 is its own inverse and S_2 is the inverse of S_3, whereas S_3 is the inverse of S_2, because $S_2 S_3 = S_3 S_2 = S_1 = I$, and this completes the proof that $\{A, \circ\}$ is a group.

Recalling that the number of elements in a group is called the *order* of the group, we can state that A, G_1, G_2, G_3, J have orders 3, 2, 2, 2, 1, respectively, and the entire symmetric group of third degree has order 6. Note that the orders of the subgroups are exact divisors of the order of the entire group. Lagrange established, in fact, one of the fundamental theorems of finite group theory when he proved that the order of a subgroup of a finite group must be an exact divisor of the order of the entire group.

A consequence of the Lagrange theorem is that the ratio of the order of a group to the order of a subgroup must be an integer. This integer is called the *index* of the subgroup in the larger group. Thus, for the subgroups tabulated above, the index of A in the symmetric group is 2, and the index for G_1 is 3; the indices of G_2, G_3, J in the symmetric group are 3, 3, 6, respectively; the index of J in G_1 is 2.

As has just been indicated, the group A has index 2 in the symmetric group because it contains just half as many elements as the entire symmetric group of degree 3. Now it can be proved that for the symmetric group of *any* degree there is always a unique subgroup containing half as many elements as the entire group. In other words, the symmetric group contains $n!$ permutations, or is of order $n!$, and the subgroup in question has order $n!/2$, with index 2 in the larger group. This special subgroup is called the *alternating* group of degree n. In the case of degrees 4, 5, 6, for example, the symmetric groups have orders 24, 120, 720 and the corresponding alternating groups have orders 12, 60, 360, respectively.

Forming subgroups of a given finite group may be a fruitful way of creating new groups, but Galois' great success came about through specializing the subgroups formed. He created the concept of a *normal subgroup* which he described (for a special reason which will appear shortly) as *invariant* or *self-conjugate*. A normal subgroup may be defined in various ways, but a general reader would probably find the following definition the easiest to understand: Given that K

is a subgroup of G, then it is a *normal subgroup* if and only if it *commutes* with every element of G.

The meaning of the definition just stated can best be clarified through an example. Let us then show that in the case of the symmetric group of degree 3, the alternating group is not only a subgroup but is also a *normal* subgroup. Using the notation previously employed, the rosters of the group and subgroup are

$$G = \{I, S_2, S_3, S_4, S_5, S_6\}$$
$$A = \{I, \ S_2, \ S_3\}$$

To prove that A is *normal*, we must show that $Ax = xA$ where x is any one of the six members of G. Thus let us choose $x = S_4$, for example, and show that $AS_4 = S_4A$. First, AS_4 signifies that *each element* of A, the alternating group, is to be "multiplied" by S_4 to form the set

$$\{S_4, S_2S_4, S_3S_4\}$$

The reader can use the definitions of S_2, S_3, and S_4 on page 384 to show that $S_2S_4 = S_5$ and $S_3S_4 = S_6$ so that

$$AS_4 = \{S_4, S_5, S_6\}$$

Sometimes AS_4 is described as a *right coset* of the subgroup A. Let us now "multiply" S_4 by each element of A to form the *left coset*,

$$S_4A = \{S_4, S_4S_2, S_4S_3\}$$

The reader can show that $S_4S_2 = S_6$ and $S_4S_3 = S_5$ so that

$$S_4A = \{S_4, S_6, S_5\}$$

which is the *same aggregate* as AS_4 even though the elements are listed in different order. We observe that the commutativity of A with S_4 is a matter of proving that a right coset of A is identical with the corresponding left coset.

The reader can imitate the above example in order to prove that $AS_5 = S_5A$ and $AS_6 = S_6A$. It is obvious that $AI = IA$ for, if each element of A is "multiplied" either on the right or on the left by I, it is unchanged. Hence $AI = A$ and $IA = A$. It is less obvious that AS_2, S_2A, AS_3, S_3A are all identical with A, but if we interpret any one of these symbols, it signifies that the members of A are all "multiplied" by a member of A (S_2 or S_3). Since A is a subgroup, the "products" are all members of A. Because A commutes with all six members of G, A is a *normal* subgroup of G.

Now let us explain Galois' use of the adjectives "invariant" and "self-conjugate" in referring to a normal subgroup. We can continue with the above example, in which the condition for the normality of A was expressed by

$$Ax = xA$$

where x is any element of G. Since G is a group, every element must have an inverse. Let us symbolize the inverse of x in the usual way by x^{-1}, and then "multiply" or follow x^{-1} by either side of the above equation to obtain

$$x^{-1}Ax = x^{-1}xA$$

Since x and x^{-1} are inverse elements, $x^{-1}x = I$, and hence the right member of the equation becomes IA, or A. Therefore

$$x^{-1}Ax = A$$

The last statement indicates that the normal subgroup A (or any normal subgroup) will be *invariant* under a certain procedure, called "conjugation" of its elements. Therefore a normal subgroup is *invariant* or *self-conjugate*. The last equation above can be interpreted to mean that if x is any element whatsoever in G, then

$$\{x^{-1}Ix, \, x^{-1}S_1x, \, x^{-1}S_2x\}$$

is merely a listing of the membership of the normal subgroup A. Here $x^{-1}S_1x$ is called the conjugate of S_1 with respect to x and $x^{-1}S_2x$ is the conjugate of S_2 with respect to x. Observe that $x^{-1}Ix = I$, and hence I is its own conjugate (is self-conjugate) with respect to any element of G.

To obtain some practice with the concept of normality, the reader can prove that $H = \{I, a^2, a^4\}$ and $K = \{I, a^3\}$ are both normal subgroups of the cyclic group, $G = \{a, a^2, a^3, \ldots, a^6 = I\}$. He can verify that K has *two* right cosets which are the same as the corresponding left cosets, respectively. As an additional exercise, he can prove that all subgroups of an abelian group must be normal. He can also investigate subgroups of the noncommutative quaternion group and prove that they are all normal. At the other extreme from such groups there are the *simple* groups *none* of whose proper subgroups (except the identity) is normal.

Galois applied not only the concept of *normal* subgroup but that of *maximal* normal subgroup. If G is a given group and G_1 is a proper normal subgroup of G, then G_1 is called a *maximal* normal subgroup of G if no other proper normal subgroup of G contains G_1. This definition permits a group to have more than one maximal normal subgroup, and such subgroups need *not* be of the same order.

Galois' procedure for determining the "algebraic" solvability of a given polynomial equation of degree n starts by associating with the equation a certain permutation group. Today we call that group the *Galois group* of the equation. Galois defined the "group of an equation" in such a way that, for specified polynomial equations of degree n, the group must be either the symmetric group of degree n or a proper subgroup of that group. The first step in the Galois process then is to determine the Galois group of the given equation.

The second step is to choose a maximal normal subgroup of the Galois group. (There may be several such subgroups, and any one of them may be selected.) The next step is to choose a maximal normal subgroup of the maximal normal subgroup chosen in the second step, etc. In successive steps there is a selection of a maximal normal subgroup of the maximal normal subgroup chosen in the previous step. The process will terminate (and this might happen even as early as the second or third step) when the only possible choice for a maximal normal subgroup is $\{I\}$, where I is the identity element of the Galois group. The process must come to an end in a finite number of steps because the Galois group is always of *finite* order and a maximal normal subgroup is *proper* and hence of lower order. Thus the sequence of choices must lead to smaller and smaller subgroups, inevitably terminating with $\{I\}$.

If G is the Galois group, then the sequence

$$G, G_1, G_2, \ldots, G_r, \{I\}$$

where each $G_1, G_2, \ldots, \{I\}$ is a maximal normal subgroup of the preceding one, is called a *composition series*. If the orders of the groups in the composition series are respectively,

$$g, g_1, g_2, \ldots, g_r, 1$$

we can compute the *index* of each normal subgroup in its predecessor as

$$\frac{g}{g_1}, \frac{g_1}{g_2}, \ldots, \frac{g_r}{1}$$

We shall shortly see that these *composition indices* are of the utmost importance.

As we have already stated, the Galois group for the *general* polynomial equation of degree n is always the symmetric group of degree n. It can be proved that the symmetric group has *only one* maximal normal subgroup, namely, the *alternating* group of degree n. Therefore, in the general case, the first of the composition indices is always 2 because the order of the symmetric group is $n!$ and that of the alternating group is $\frac{1}{2}n!$. The composition indices for various composition series starting with the symmetric group for $n = 2, 3, 4, 5, 6, 7$ are listed below:

n	Composition Indices
2	2
3	2, 3
4	2, 3, 2, 2
5	2, 60
6	2, 360
7	2, 2520

The composition indices listed for $n = 4$ were obtained from choices of maximal normal subgroups as follows:

		Order	Index
S:	Symmetric group	24	
			$\dfrac{24}{12} = 2$
A:	Alternating group	12	
			$\dfrac{12}{4} = 3$
H:	Maximal normal subgroup of alternating group	4	
			$\dfrac{4}{2} = 2$
K:	Maximal normal subgroup of H	2	
			$\dfrac{2}{1} = 2$
J:	{Identity permutation} = maximal normal subgroup of K	1	

We have indicated that when a maximal normal subgroup is to be selected, there may possibly be more than one maximal normal subgroup from which to choose. If the possible choices are subgroups of different orders, different indices will result. Then the continuation of the procedure thereafter will vary with the selection made, since the membership—and hence the order and the index—of the next maximal normal subgroup selected depends on the previous choice. Long after Galois, the group-theorists Camille Jordan (1838–1922) and Otto Hölder (1859–1937) proved that the very same set of composition indices will result, no matter what choice of maximal normal subgroup is made at any point of the process. The index numbers may appear in different orders, however, depending on the normal subgroup selection at each step.

And now it is possible to explain how Galois defined a *solvable group*. Such a group is one in which the composition indices are all *prime* numbers. The above tabulations indicate, therefore, that the symmetric groups for $n = 2$, 3, and 4 are solvable; the others are *not* solvable. Galois proved that if $n > 4$, the sequence of maximal normal subgroups is: Symmetric group, alternating group, identity, and hence that the composition indices are 2, $n!/2$. Since $n!/2$ is not a prime number (but is composite) for $n > 4$, the symmetric groups of degree higher than 4 are *not* solvable groups.

Galois' criterion for solvability is: *A polynomial equation is algebraically solvable if and only if its group is solvable.* Since the symmetric group is the "group of the equation" if the equation is perfectly general, and since the symmetric group is *not* a solvable group for $n > 4$, it follows that it is impossible to solve "algebraically" the general polynomial equation of degree greater than four.

This is the theorem to which we alluded in Chapter 5, where we explained why Galois' discovery provided a terminus for traditional algebra.

Several points must be emphasized in connection with this great theorem. The first is the significance of the words "to solve algebraically." By an "algebraic solution" of the general polynomial equation of the nth degree,

$$a_0 x^n + a_1 x^{n-1} + a_2 x^{n-2} + \cdots + a_n = 0$$

is meant a formula expressed in terms of the coefficients a_0, a_1, a_2, ..., a_n, and using only a *finite* number of additions, subtractions, multiplications, divisions, and root extractions. Because the root extractions are expressed by radical signs, algebraic solution is often called *solution by radicals*.

The impossibility of solution by radicals of polynomial equations of degree higher than 4 applies only to general polynomial equations. Many special equations of higher degree, that is, equations in which the literal coefficients are not arbitrary but are related in some way, or else have particular numerical values, can be solved by radicals, and the Galois criterion indicates that this is so. In applying the Galois theory in such cases, it turns out that the group of the particular equation is *not* the symmetric group, but some *proper subgroup* of the symmetric group. Then the search for maximal normal subgroups starts with the group of the equation, which is a smaller group than the symmetric group. Since the starting point is different, so are the possibilities for maximal normal subgroups at each step, and if $n > 4$, the set of index numbers may, in special cases, consist only of *primes*, thereby indicating that the particular equation has a solvable group, and hence that this equation is algebraically solvable.

Another point to be emphasized is that "algebraic" solution requires expression in a *finite* number of arithmetic steps. Solution of general equations of degree higher than 4 is possible if an *infinite* number of steps is permitted. But such solutions are nonalgebraic, and are sometimes expressed in terms of special nonalgebraic (*transcendental*) functions. The reader may think of such functions as formulated by the *infinite series* so important in analysis, and in this way realize that an infinite number of arithmetic steps is involved. Now trigonometric functions (which are nonalgebraic or transcendental functions) are effective in obtaining solutions when the cubic formula yields irreducible results (page 96). Therefore, mathematicians after Galois' day conceived the idea that the *elliptic functions*, which generalize ordinary trigonometric functions, might offer a means of expressing solutions of some higher-degree equations that are not solvable algebraically. Thus, Charles Hermite (1822–1905) succeeded in solving the general quintic equation ($n = 5$) in terms of *elliptic modular functions*. Felix Klein (1849–1925) was also interested in the general quintic, and geometrized the problem of its solution by relating it to rotations of the regular icosahedron, a polyhedron whose surface is made up of 20 equilateral triangles (Chapter 25). He suggested similar geometrizations for equations of still higher degrees. Hilbert indicated the types of transcendental function required for expressing the solutions of ninth-degree equations, and Poincaré discussed the application of his *automorphic functions* (generalizations of elliptic functions) to equations that are not solvable algebraically.

To round out the discussion of Galois' theory, one should explain exactly what is meant by the "group of an equation," provide some instances where it is not the symmetric group but a subgroup of the latter, and examine how this may affect the possibility of algebraic solvability or the nature of the solutions. A problem of Galois' theory then begins *not* with a general equation whose coefficients are completely arbitrary, but with an equation where there is some restriction on the nature of the coefficients. If these coefficients are given specific numerical values, the equation is a particular one. Even when the coefficients are literal, they are considered to be *rationally known*, or "known," in the following sense. The values they may assume are limited to a number *field F*. In Chapter 4 we listed the postulates for a *field*. Among them were two requiring closure with respect to addition and multiplication. But the existence of additive and multiplicative inverses provides closure also with respect to subtraction and division (except by zero). Hence a field is closed under all four "rational" operations.

A problem of Galois theory always begins with a given polynomial equation and a given field of numbers *F*, to which the coefficients of the given equation are required to belong. For example, the given equation might be

$$x^7 - a = 0$$

and the given field *F* might be the set of rational numbers. This means that *a* must be a rational number; it cannot have the value $\sqrt{3}$, for example.

Given an equation and a number field *F*, the first question asked is: Are the solutions of the equation "known," that is, are they numbers in the field *F*? For example, given the equation

$$x^2 - 2 = 0$$

with *F*, the field of rational numbers. The coefficient, -2, belongs to this field, as required. The theory of equations indicates that if this equation has rational solutions, they must be exact divisors of -2. Then $+1$, -1, $+2$, -2 are the only rational numbers that could possibly be solutions. Substitution in the equation shows that they are *not* solutions. Therefore the answer to the question raised above is no. The solutions do *not* belong to *F*, the rational field. Given the same equation with *F* specified as the field of numbers, $a + b\sqrt{2}$, where *a* and *b* are any rational numbers (page 80), one has an entirely different problem. Here the answer to the crucial question is yes, since $+\sqrt{2}$ and $-\sqrt{2}$ check in the equation and are members of the specified field.

As one more example, let *F* be the field of real numbers and let the given equation be

$$x^2 + 1 = 0$$

The solutions cannot be real numbers because $x^2 = -1$ and there is no real number whose square is -1. Therefore the solutions are *not* "known," that is, are not quantities in the given field. If, however, the given field had been the set of complex numbers, the equation would have been solvable in that field, since $x = +i$ and $x = -i$ are solutions.

The equations $x^2 - 2 = 0$ and $x^2 + 1 = 0$ illustrate a special point, namely, that if an equation is not solvable for a particular field, it may become solvable when the field is embedded in a larger field. Thus, in the first case there was "extension" from the field of numbers a, where a is rational, to the field $a + b\sqrt{2}$ (a and b rational). The second equation above is not solvable in the field of numbers a, where a is real, but is solvable in the field $a + b\sqrt{-1}$, (a and b real), an "extension" of the original field.

Presented with F, the field of rational numbers, and a quintic equation like

$$x^5 + 6x^4 - x^3 - 19x^2 + 103x - 1 = 0$$

with coefficients in F, what sort of information will help in the determination of its Galois group (and thus ultimately make it possible to decide on its solvability by radicals)? An elementary fact of the theory of equations indicates that the only possible solutions in the rational field F must be exact divisors of the constant term, -1, of this equation. Therefore the only rational possibilities are $+1$ or -1. Substitution of these values indicates that they do not satisfy the equation. Although this equation has no rational solutions, certain functions of the solutions are rationally "known," that is, are numbers in F, as we shall now indicate. We know that the polynomial in the left member can be factored as follows (Chapter 5):

$$(x - x_1)(x - x_2) \ldots (x - x_5)$$

to indicate solutions $x_1, x_2, \ldots, x_5$, real or complex. Now the sum and product of the solutions are "known." Specifically,

$$x_1 + x_2 + x_3 + x_4 + x_5 = -6$$
$$x_1 x_2 x_3 x_4 x_5 = 1$$

To see how the numerical values -6 and 1 are obtained, let us state that it is merely a case of generalizing certain elementary facts of school algebra. Thus the general quadratic equation

$$x^2 + bx + c = 0$$

(whose coefficients are arbitrary) can be expressed as

$$(x - x_1)(x - x_2) = 0$$

where x_1 and x_2 are the solutions (distinct or identical). Since the two forms of the quadratic must be identical,

$$(x - x_1)(x - x_2) = x^2 + bx + c$$

and expanding the left member,

$$x^2 - (x_1 + x_2)x + x_1 x_2 = x^2 + bx + c$$

Therefore

$$x_1 + x_2 = -b$$

and

$$x_1 x_2 = c$$

The reader may recall the derivation of these formulas from school algebra, or their statement as a theorem: In a quadratic equation, if the coefficient of x^2 is 1, then the sum of the solutions is equal to the negative of the coefficient of x, and the product of the roots is equal to the constant term.

From the more advanced point of view of the present chapter, interest in the formulas lies in the fact that they express the functions $x_1 + x_2$ and $x_1 x_2$ in terms of the coefficients of the quadratic equation, so that these functions are "known," that is, belong to any field F prescribed for the coefficients b and c. In passing, it should be mentioned that the *symmetric group* of degree 2 contains only the permutations $I = \begin{pmatrix} 1 & 2 \\ 1 & 2 \end{pmatrix}$, the identity, and $P = \begin{pmatrix} 1 & 2 \\ 2 & 1 \end{pmatrix}$, the interchange of the two roots. Therefore, $x_1 + x_2$ and $x_1 x_2$ are described as *symmetric functions* because they are unaltered by I and changed to the equivalent forms, $x_2 + x_1$ and $x_2 x_1$, by P; that is, they are *invariant* under all permutations of the symmetric group of degree 2.

In the case of a particular quadratic like

$$x^2 + 2x - 4 = 0$$
$$b = 2, \qquad c = -4$$

and the formulas reveal that

$$x_1 + x_2 = -2$$
$$x_1 x_2 = -4$$

The solutions of the particular quadratic were found (Chapter 5) to be

$$x_1 = -1 + \sqrt{5}, \qquad x_2 = -1 - \sqrt{5}$$

If the given field F is that of the rational numbers, one observes that while the sum and product of the solutions are in F, the solutions themselves do not belong to this field, and the equation is not solvable within F.

Returning to the general quadratic,

$$x^2 + bx + c = 0$$

it will now be indicated that symmetric functions other than $x_1 + x_2$ and $x_1 x_2$ can be expressed in terms of these two elementary functions, and therefore in terms of b and c, the coefficients of the quadratic. Thus

$$x_1{}^5 x_2{}^5 = (x_1 x_2)^5 = c^5$$
$$x_1{}^2 + x_2{}^2 = (x_1 + x_2)^2 - 2x_1 x_2 = b^2 - 2c$$
$$x_1{}^3 - 6x_1 x_2 + x_2{}^3 = x_1{}^3 - 6c + x_2{}^3$$
$$\qquad = -6c + (x_1 + x_2)^3 - 3x_1{}^2 x_2 - 3x_1 x_2{}^2$$
$$\qquad = -6c - b^3 - 3x_1 x_2 (x_1 + x_2)$$
$$\qquad = -6c - b^3 + 3bc$$
$$\frac{x_1{}^2 + x_2{}^2}{3x_1{}^2 x_2{}^2} = \frac{b^2 - 2c}{3c^2}$$

Since the expressions in terms of the coefficients involve only the rational operations—addition, subtraction, multiplication, and division—on these coefficients, the symmetric functions are "known," that is, are equal to quantities in the field of the coefficients. For the particular quadratic above, where $b = 2$, $c = -4$, the four functions have the rational numerical values -1024, 12, -8, and $1/4$, respectively.

The facts illustrated for the quadratic equation generalize to equations of higher degree. For the general cubic

$$x^3 + ax^2 + bx + c = 0$$

with roots x_1, x_2, x_3, the expression of the elementary symmetric functions (those *invariant* under the *symmetric group* of degree 3) in terms of the coefficients is

$$x_1 + x_2 + x_3 = -a$$
$$x_1x_2 + x_2x_3 + x_3x_1 = b$$
$$x_1x_2x_3 = -c$$

For the general equation of nth degree,

$$x^n + a_1x^{n-1} + a_2x^{n-2} + a_3x^{n-3} + \cdots + a_{n-1}x + a_n = 0$$

with roots x_1, x_2, x_3, ..., x_n,

$$x_1 + x_2 + x_3 + \cdots + x_n = -a_1$$
$$x_1x_2 + x_2x_3 + \cdots + x_{n-1}x_n = a_2$$
$$x_1x_2x_3 + x_2x_3x_4 + \cdots + x_{n-2}x_{n-1}x_n = -a_3$$
$$\cdots \quad \cdots \quad \cdots \quad \cdots \quad \cdots \quad \cdots \quad \cdots \quad \cdots \quad \cdots$$
$$x_1x_2x_3 \cdots x_{n-1}x_n = a_n \quad \text{or} \quad -a_n$$

according to whether n is even or odd.

In equations of higher degree, just as in the case of the quadratic, more complicated symmetric functions are expressed in terms of the above elementary symmetric polynomials and ultimately in terms of the coefficients. In fact, not only can symmetric *polynomials* be so expressed, but also *rational* symmetric functions, which are defined as quotients of symmetric polynomials, for example,

$$\frac{4x_1{}^3x_2{}^3x_3{}^3 - 3x_1x_2 - 3x_2x_3 - 3x_3x_1}{x_1{}^2 + x_2{}^2 + x_3{}^2} = \frac{-4c^3 - 3b}{a^2 - 2b}$$

where x_1, x_2, x_3 are the solutions of the general cubic equation. There is a fundamental theorem of great importance: Every rational symmetric function (with coefficients in F) of the solutions of an algebraic equation can be expressed rationally in terms of the coefficients, and therefore is "known" or equal to a quantity in F, the field of the coefficients. This fundamental theorem can be reworded as follows: If, for a given algebraic equation and a given field F, a rational function (with coefficients in F) of the solutions is invariant under all permutations of the symmetric group, then this function is "known" or equal to a quantity in the given field F.

If the equation is *general* and the field is that formed by rational operations (addition, multiplication, etc.) on the coefficients, the converse of the above theorem is true. When a proposition and its converse are both true, they provide "necessary and sufficient conditions" that can be used to furnish a *definition* (see Chapter 6). Thus the above theorem and its converse might be worded so as to provide a definition of the symmetric group.

A slight modification of the above conditions leads directly to a definition of the *group of an equation*, that is, its *Galois group*: The group G of a given equation for a given field F is a group possessing the two characteristics:

(1) If a rational function (with coefficients in F) of the solutions of the equation remains numerically unchanged by all the permutations of G, it is equal to a quantity in F.

(2) Conversely, if a rational function (with coefficients in F) of the solutions of the equation is equal to a quantity in F, it remains numerically unchanged by all the permutations of G.

Applying this definition, let us determine the Galois group, or group of the equation,

$$x^3 - 3x + 1 = 0$$

for the field F of all rational numbers. Its group is either the entire *symmetric group* of degree 3 (page 384) or one of the *proper subgroups* of this group (page 386). Now

$$(x_1 - x_2)^2(x_1 - x_3)^2(x_2 - x_3)^2$$

is a symmetric function, since it is unchanged by all permutations of the symmetric group. Therefore, by the fundamental theorem, it can be expressed rationally in terms of the coefficients, or is in the field F. In other words, its numerical value must be a rational number. By the type of manipulation previously used, it can be shown that for the general cubic

$$x^3 + bx + c = 0$$

the above symmetric function of the roots is equal to $-4b^3 - 27c^2$. For the particular case above, $b = -3$, $c = 1$, and the function has the value $(-4)(-27) - 27 = 81$. Since

$$(x_1 - x_2)^2(x_1 - x_3)^2(x_2 - x_3)^2 = 81$$

the rational function $(x_1 - x_2)(x_1 - x_3)(x_2 - x_3)$ must be equal to either $+9$ or -9. The latter function is equal to a rational number, a number of the field F. Therefore this function should be numerically invariant under all permutations of the Galois group for the field F. Then this group cannot be the symmetric group, because the permutation $S_4 = \begin{pmatrix} 1 & 2 & 3 \\ 1 & 3 & 2 \end{pmatrix}$, for example, changes the function into

$$(x_1 - x_3)(x_1 - x_2)(x_3 - x_2)$$

which would have the value -9 if the original function is equal to $+9$, or the value $+9$ if the original function is equal to -9. Only three permutations of the symmetric group leave the rational function invariant, namely, the identity and the permutations $\begin{pmatrix} 1\ 2\ 3 \\ 2\ 3\ 1 \end{pmatrix}$ and $\begin{pmatrix} 1\ 2\ 3 \\ 3\ 1\ 2 \end{pmatrix}$. These form the alternating group A (page 387). Since the Galois group cannot be the symmetric group, it must then be the alternating group, or else a proper subgroup of this group. The only proper subgroup of A is the identity group. Then A or $\{I\}$ are the two possibilities for the Galois group. Now the rational function x_1 (consisting of just one of the solutions) is numerically unaltered by the identity permutation, the only permutation in $\{I\}$. Therefore, if $\{I\}$ should be the Galois group, x_1 would have to be "known," that is, it would have to be a rational number in the field F. But a theorem from the elementary theory of equations indicates that if the equation

$$x^3 - 3x + 1 = 0$$

had a rational solution, it would have to be an exact divisor of the constant term $+1$. Hence it would be $+1$ or -1. When we substitute these values in the equation, it is evident that they do not satisfy it. Therefore it has *no rational root*, and x_1 cannot possibly be a rational number. Thus $\{I\}$ cannot be the Galois group, and therefore the Galois group for the particular equation and the rational field F is

$$A: \left\{ S_1 = \begin{pmatrix} 1\ 2\ 3 \\ 1\ 2\ 3 \end{pmatrix}, \ S_2 = \begin{pmatrix} 1\ 2\ 3 \\ 2\ 3\ 1 \end{pmatrix}, \ S_3 = \begin{pmatrix} 1\ 2\ 3 \\ 3\ 1\ 2 \end{pmatrix} \right\}$$

It is often exceedingly difficult in practice to determine the Galois group for a given equation and a given field F. Various ways of approaching the problem are based on different (but equivalent) definitions of the group of an equation, or on special theorems. At any rate, when the Galois group for the field F has been found, the Galois criterion will indicate whether or not the group is "solvable." If it is a solvable group, the (theoretical) procedure is as follows: There is an auxiliary or *resolvent* equation (defined in a special way) which *can actually be solved by radicals*. Its degree is the first index number in the series of prime index numbers (page 391). Then the field F is extended to a field F' by "adjunction" of the radicals appearing in the solution of the auxiliary equation. Let us explain what is meant by *adjunction*.

If, for example, $\sqrt{2}$ is to be adjoined to the field of rational numbers, one must form all rational combinations (sums, differences, products, quotients) involving $\sqrt{2}$ and the rational numbers. Thus $5 + \sqrt{2}$, $\frac{1}{2} - \sqrt{2}$, $7\sqrt{2}$, $(2 + 6\sqrt{2}) \div (-1 - 3\sqrt{2})$ would all be numbers in the extension field and, in fact, that field would contain all numbers $a + b\sqrt{2}$ where a and b are rational numbers. The rational field corresponds to $b = 0$ and is a *subfield* of the extension. Again, if $\sqrt[3]{2}$ is adjoined to the rational field, the extension field will have to contain $(\sqrt[3]{2})(\sqrt[3]{2}) = \sqrt[3]{4}$, $(2 - 3\sqrt[3]{4}) \div (5 + 7\sqrt[3]{2})$, etc.

When the field F' has been specified, the Galois process must start anew with the determination of the Galois group G' for the field F', the extension of F. This new Galois group, G', is a maximal normal subgroup of G, the original Galois group. Then a second auxiliary or resolvent equation is solved. Its degree is that of the second index number in the series of composition indices. The field F' is extended to F'' by adjunction of radicals appearing in the solutions of the second auxiliary equation.

The procedure is repeated by finding the Galois group for the field F'', solving a third auxiliary equation, adjoining its radical solutions to F'' to yield the extension F''', etc. Since a composition series is of finite length, iteration of the procedure described will yield, for some extension field $F^{(m)}$, a Galois group equal to $\{I\}$, the group containing only the identity element. Since the solutions x_1, x_2, x_3, ..., x_n, are all functions invariant under the identity permutation, they must (by the definition of the Galois group) be equal to quantities in the final extension field. Since the only adjunctions to the original field are radicals, the equation can thus be solved in terms of radicals. In addition, since the nature of these radicals is known (whether they are square roots or cube roots, etc.) the character of the solutions is fully determined by the Galois procedure, even if it is not always the simplest practical way to solve an equation.

In closing the present discussion, we shall give historic bearings for the ideas involved. Lagrange's contributions to the theory of equations were doubtless the most potent anticipations of Galois' own ideas. In a 1770–1771 memoir, Lagrange attempted to find a uniform procedure for solving equations of all degrees. He analyzed the methods that had yielded general solutions for degrees 2, 3, 4, and found that in each case the technique involved the use of a *resolvent* equation. Although the latter was of lower degree than the original for $n = 2, 3, 4$, Lagrange discovered that application of the previously successful pattern to the quintic ($n = 5$) led to an irreducible sextic ($n = 6$), and the problem became more difficult instead of being resolved.

He then suggested timidly that perhaps the general quintic cannot be solved by the apparently "universal" method. In other words, he hinted at the impossibility of solution by radicals, and let the matter drop. Perhaps he was discouraged by the reception accorded to his great 1770–1771 memoir. The historian of the French Academy, for example, wrote: "Monsieur de la Grange (Lagrange) seems convinced that it will be necessary to replace the customary procedure for solving equations, while Monsieur Vandermonde (an algebraist, contemporary with Lagrange) is inclined to believe the latter still may succeed. As for me, I favor Monsieur Vandermonde's opinion because it is less disheartening."

But Paolo Ruffini (1765–1822), an Italian physician who taught mathematics as well as medicine at the University of Modena, continued the thread of Lagrange's thoughts. In 1799 Ruffini wrote a book on the theory of equations, in which he gave a proof that virtually established the unsolvability of the quintic by radicals. Ruffini's demonstration was improved by Abel, and in 1822 the young Norwegian mathematician proved conclusively that the general equation of degree $n > 4$ cannot be solved algebraically. There were some minor flaws in Abel's proof, but it inspired Galois to seek a fundamental reason for unsolvability or, more positively stated, a basic criterion for algebraic solvability.

The present chapter has dealt with Galois' resolution of the deeper issues of solvability. His group-theoretic approach provided a simpler proof of Abel's theorem and, in general, replaced the algebraic theories of Lagrange, Ruffini, and Abel. Concerning the mathematical testament which Galois wrote on the eve of his death and in which he incorporated his theory, Hermann Weyl, a leading twentieth-century mathematician, had this to say: "If judged by the novelty and profundity of ideas it contains, it is perhaps the most substantial piece of writing in the whole literature of mankind."*

The effect that this great document had upon modern abstract algebra has been noted and will be emphasized again later. Chapter 17 will indicate how the group concept can be used to unify various types of geometry, and Chapter 18 will illustrate an application to modern mathematical physics. There it will be revealed that the special theory of relativity is, in abstract content, nothing more than the geometry of one special group, and is therefore another important instance of the influence of

> ... the marvelous boy,
> The sleepless soul that perished in his pride.†

---

* Hermann Weyl, *Symmetry*, Princeton University Press, Princeton N.J., 1952, p. 138.

† William Wordsworth, "Resolution and Independence," Stanza 7.

17

The Unification of Geometry

Yet may we not entirely overlook
The pleasures gathered from . . . geometric science . . .
With . . . awe and wonder,
. . . meditate.
On the relations those abstractions bear
To Nature's laws.

Wordsworth

At Düsseldorf, on the night of April 25th, 1849, there was anxiety in the house of the secretary to the Regierungspräsident. Without, the cannon thundered on the barricades raised by the insurgent Rhinelanders against their hated Prussian rulers. Within, although all had been prepared for flight, there was no thought of departure; on that night was born a son to the stern Prussian secretary. That son was Felix Klein. His birth was marked by the final crushing of the revolution of 1848; his life measured the domination of Prussia over Germany, and typifies all that was best and noblest in that domination; with his last illness came the consummation of its downfall.

Few mathematicians have left such ample material for forming an opinion of their life and work as Felix Klein (1849–1925). We have his life, written by his own hand two years before his death. We have his Collected Mathematical Papers, in three volumes thoroughly revised by himself, and interspersed with supplementary notes and introductory articles of an autobiographical character. We have the greater part of his mathematical lectures in print, lectures which had for many years enjoyed a considerable publicity in lithographed form; we have even a faithful record of lectures given by him in the years just preceding his death, carefully annotated by his colleague and successor, Professor Richard Courant. Klein's personal influence was as great, or greater even outside his own country, than that, perhaps, of any mathematician of modern times. He owed this to his forceful and attractive personality, to his wide mathematical outlook and to his objective openmindedness, only occasionally tinged by the play of personal feeling, for Klein had by no means the cold, calculating nature, supposedly typical of the mathematician. He owed it also partly to that intangible agent, so often and so callously cited "Luck," which afforded him, by a succession of unforeseen external events, the opportunity without which a man of the highest intellectual and moral endowments may remain mute inglorious. He owed it partly to his success in interesting his audiences, even in branches of mathematics of which his own command was comparatively slight. And he owed it to his untiring devotion to the cause of education.

Klein's genius had been precocious. At seventeen he was chosen by Julius Plücker (1801–1868) as his assistant in his physics laboratory at Bonn; that laboratory where Plücker had invented what today we call the Geissler tube. Plücker had reverted in his later years to his early interest in geometry. When he died in 1868, he left an unfinished manuscript, entitled "New Geometry of Space, founded on the straight line as element." The task of completing the work and issuing the second half of the book was entrusted to Plücker's young assistant, Felix Klein.

Shortly after he began his work on the "New Geometry" Klein became affiliated with the University of Göttingèn. He was there for a short time in 1869, for a second stay from 1871 to 1876, and his final migration thither was in 1886. Between his first and second periods at Göttingen there were three important events in his life—a sojourn in Berlin, a trip to Paris, and the Franco-German war, in which he took part in the ambulance corps, and from which he was sent home invalided with typhoid fever.

From August 1869 to March 1870 Klein was at the University of Berlin where he hoped to profit from personal contact with Weierstrass (Chapter 22). But in the latter's mind a veritable antagonism to Klein seems to have arisen. When Klein spoke at Weierstrass' Seminar and made the suggestion that there was a connection between non-Euclidean geometry and Cayley's Theory of the Absolute, Weierstrass absolutely rejected the notion. Later Klein developed his idea in a series of papers extending over several years [page 424].

At the early age of twenty-three, Klein obtained the full chair of mathematics at the University of Erlangen. There he met and married the beautiful, cultured Anna Hegel, daughter of an Erlangen professor and granddaughter of the philosopher Hegel. In mathematics he produced what he himself regarded as his most notable achievement, the so-called Erlanger Programme. It would be no exaggeration to say that it has revolutionized the treatment of Geometry. Only in details has exception been taken to any part of it, which is the more remarkable when we consider the age of the author and the state of mathematics at the time when it was composed. The notes which Klein himself has added, and the account which he has given of its production, have greatly increased its interest. The new idea which lies at its basis was that all the various species of Geometries which, during the 19th century in particular, had multiplied exceedingly (metrical geometry, projective geometry, line geometry, etc.) could be regarded from a single standpoint, that of the Theory of Groups, each different geometry being conceived as the theory of the invariants of an appropriate group. This idea, which as then formulated, seemed "very astonishing," was an abiding one in the mind of Klein, who thought he foresaw its extension to regions of mathematics other than geometry. In particular, the development by Einstein of the Theory of Relativity revived in his later years Klein's interest in his Erlanger Programme, and it was with some mortification that he found among his physicist friends no interest whatever in the possibility of a physical interpretation of the most general group of conformal transformations of four dimensional space, as leaving invariant the Maxwell-Lorentz equations of relativity.

The above is an excerpt from a biography of Felix Klein by the British mathematician William Henry Young (1863–1942),[*] one of the leading analysts of the twentieth century and a personal friend of the German mathematician. This friendship had come about through Young's wife, Grace Chisholm Young (1868–1944), Klein's "favorite pupil," an excellent mathematician in her own right. She was the first woman to receive a German Ph.D. (1895) based on regular examinations. More will be said later about the Youngs, husband and wife, and their specific discoveries in the field of modern analysis (Chapter 27). We have paused at the above point in Young's biography of Klein in order to examine the background, content, and application of the mathematical contribution which Klein (according to Young) regarded as his greatest achievement, namely, the Erlanger Program.

[*] Royal Society of London, *Proceedings*, Series A, Vol. 121 (November–December 1928), pp. 1 ff.

In the Program, Klein advanced the idea that any species of geometry is simply the study of *invariants* associated with a particular *group of transformations*. If we explain that the term "transformation" (favored in geometry) is merely a synonym for function or mapping and that the Klein transformations are, like the Cauchy-Galois permutations, *one-to-one* mappings of a set *onto* itself, the reader will see the kinship of Klein's notions and those of Galois. The great difference arises from the fact that algebra is concerned with *finite* sets like the roots of a polynomial equation, whereas traditional geometries apply to "spaces" like the line or plane or 3-dimensional Euclidean space, etc., and these are the *infinite* continuous sets of "points" we called R, R^2, R^3, etc. in Chapter 7. In the Klein geometries the points of such a space may be shifted to new positions, but the space or set of points as a whole remains intact.

To connect invariance with geometry, one need only return to Euclid. In the physical geometry that predated his *Elements*, it was taken for granted that measuring tools were *rigid bodies*, that is, that they remained *invariant* in size and shape as they were moved about in the course of use. Modern surveyors modify this point of view when they make corrections for temperature, tension, and sag of tape. Nevertheless their work is still based on Euclidean geometry, with the tacit assumption: A figure can be moved about freely in space without changing its size or shape.

In Euclid, congruent figures are defined as those that can be made to coincide. If one figure is already superposed on another, an immediate verdict of congruence can be rendered. But if figures in coincidence were the customary situation. Euclid would hardly have troubled to define congruence or to prove so many special congruence theorems. In practical affairs, the tool must usually be brought into contact with the object to be measured, whereas the Euclidean congruence proofs generally start by moving a triangle through a sequence of positions. Then deductive reasoning is used to demonstrate that the final position is one of complete coincidence with a second triangle.

Suppose that we are given that triangle $A'B'C'$ of Figure 17.1 agrees with triangle ABC in the sides AB, AC and the angle A, and that we are required to demonstrate the congruence of the two triangles. We must prove that the triangles can be made to coincide. Let us slide triangle ABC so that every point of it moves

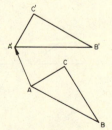

Figure 17.1

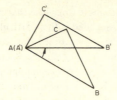

Figure 17.2

in the direction AA' until A falls on A' (Figure 17.1). Such a motion is called a *translation*. Next, using A of Figure 17.2 as a pivot, we carry out a counterclockwise *rotation* of ABC through an angle sufficient to bring the triangle into superposition with $A'B'C'$.

We have given the Euclidean picture of translation and rotation of a particular triangle to a new position. Klein extended the motion of the triangle to a *transformation* of the entire plane, a mapping of that set of points onto itself. Thus, not only points of the triangle but all points of the plane slide in the direction of AA' (Figure 17.1) through a distance equal to AA'. Again, all points of the plane describe a circular motion about A as pivot (Figure 17.2).

If the original position of triangle ABC is somewhat different (Figure 17.3), a translation-rotation motion will not suffice to bring about its coincidence with $A'B'C'$. After translation in the direction AA' and rotation through an angle sufficient to accomplish the superposition of AB on $A'B'$ (Figure 17.3), we might use AB as a hinge and revolve ABC as if it were a door, until it returns to the plane of $A'B'C'$, at which point coincidence is achieved. But if we were creatures confined to a plane, that is, if for us a physical third dimension did not exist, the final door-like motion of ABC would be a physical impossibility, We might, instead, think of AB as a mirror with ABC transformed into $A'B'C'$ by *reflection*. Then the triangles of Figure 17.3 are mirror images of one another, and in this sense identical, even if not physically superposable.

The *proper* congruence motions, that is, the displacements or *rigid motions* that can be carried out within the plane itself, are thus translations, rotations, or combinations of translations and rotations. A *reflection* like the one described above, or the combination of a reflection with a proper displacement, is an

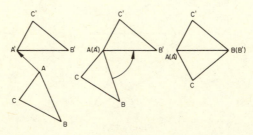

Figure 17.3

improper rigid motion, one that would be a physical impossibility within the plane itself. Later we shall show that the set of *all* rigid motions,* proper and improper, is a group which has the set of proper rigid motions as a (proper) subgroup. Klein asserted that plane Euclidean *metric* geometry can be described as the study of properties unchanged by the group of all rigid motions. As illustrations of such metric properties, there are equality of the radii of a circle, equality of sides and angles in regular polygons, equality and parallelism of opposite sides in a parallelogram, perpendicularity of adjacent sides in a rectangle, area formulas. All of these characteristics are dependent on two *fundamental invariants* in terms of which they can be expressed, namely, *distance* and *angle*.

Euclid's pure geometry of measurement assumed that the effects of motion must leave size *invariant*. The issue of whether this is a good assumption for motion in the real world arises in twentieth-century physics. We have all heard of the "Lorentz-Fitzgerald contractions" of early relativity. To resolve some of the paradoxes in physics, the Euclidean assumption of rigidity was abandoned in cases where relative speed of motion is very large, that is, commensurate with the velocity of light. Felix Klein was concerned with this very question later in his career and did relate it to the Erlanger Program. Although the Dutch physicist H. A. Lorentz and the Irish scientist G. F. Fitzgerald advanced the contraction hypothesis some twenty years after Klein formulated his Program, the Euclidean assumption of free mobility bears closely on relativity. In fact the postulate that displacement does not affect the geometric properties of a figure is often called the *principle of relativity for Euclidean geometry*. The classic axiom contains implications concerning the kind of space to which Euclidean geometry applies. If free mobility is permissible, no *point* and no *direction* in space are preferable to other points and directions. In relativity language, the fact that all points are on an equal footing becomes: Euclidean space is *homogeneous*. That there is no special significance for measurement of a line segment in whether its direction is north or east or south-southwest, etc. is stated: Euclidean space is *isotropic*.

Our presentation thus far has emphasized *measurement* as the fundamental question of geometry. But draftsmen make scale drawings; architects construct models; photographers prepare enlargements. Hence properties of *similar* figures as well as of congruent ones are the concern of both physical and pure Euclidean geometry. Congruent figures are considered equivalent because they are exact duplicates. But for a draftsman a scale drawing is equivalent to the actuality it represents. Except for aesthetic effect, a colossal statue of a human figure is a work of sculpture geometrically equivalent to a life-size model. We can define this new type of equivalence by means of a generalization of the idea of displacement. In similarity geometry, two figures are "equivalent" if they can be brought into coincidence by means of a rigid motion combined with an enlargement or contraction. Since the scale ratio is permitted to have the special value 1:1, congruent figures are included in the new definition, that is, congruent figures are similar, but not conversely. Similarity geometry is the study of properties unchanged by all

---

* These are also called *isometries*, signifying "same measures."

possible combinations of rigid motion with change of scale (enlargement or contraction). These properties are fewer in number than in the metric case. Distances are no longer invariant, since enlargement or contraction affects them. Hence metric theorems involving lengths are not part of the new geometry, but angles are still invariant.

We are now fully launched on the Klein technique of classification. His Erlanger Program takes the various types of geometry existing in 1872 and makes each one after the first a generalization of some preceding one. As we have indicated, fewer properties are invariant in a generalization, and for this reason only part of the subject matter of the preceding geometry is pertinent. The rigid motions of Euclidean metric geometry have already been extended so as to include change of scale. Now they can be generalized still further to define figures which, for the purposes of some special geometry, can be considered equivalent. The generalizations are transformations that tend to distort figures, for example, the *affine* transformations. In the corresponding *affine geometry*, circles and ellipses are equivalent. There is a still further generalization to *projective* transformations and *projective geometry*, in which all the conic sections are equivalent. In other words, circles, ellipses, parabolas, and hyperbolas can be transformed into one another by means of projective transformations.

As we have already stated, the Klein transformations are *one-to-one mappings of a set S onto itself*. Now in the previous chapter we indicated that a mapping of a *finite* set onto itself is necessarily one-to-one. But we shall now see that this is no longer true in the case of an infinite set. A mapping of such a set onto itself need *not* be one-to-one and, in reverse, a mapping can be one-to-one *into* a set and need *not* be onto the entire set. For an example of the former situation, consider the following mapping of the set of whole numbers, $S = \{0, 1, 2, 3, \ldots \}$, *onto* itself. Each even number is mapped onto its half and each odd number is mapped onto zero, as suggested below:

$$\begin{pmatrix} 0 & 1 & 2 & 3 & 4 & 5 & 6 & \ldots & 2n & 2n+1 & \ldots \\ 0 & 0 & 1 & 0 & 2 & 0 & 3 & \ldots & n & 0 & \ldots \end{pmatrix}$$

Here *every* whole number will appear in the lower row, so that the mapping is actually *onto S*. But since *many* numbers (zero and all odd numbers) are mapped onto zero, the mapping is *not* one-to-one. Again, let us consider the *one-to-one* mapping where each whole number is mapped onto its double, namely,

$$\begin{pmatrix} 0 & 1 & 2 & 3 & 4 & 5 & \ldots & n & \ldots \\ 0 & 2 & 4 & 6 & 8 & 10 & \ldots & 2n & \ldots \end{pmatrix}$$

Here the lower row contains only the nonnegative *even* integers, a proper *subset* of the whole numbers. Therefore we have a one-to-one mapping of the set *S into* but *not onto* itself.

We see therefore that if a Klein transformation is to permute an infinite set, it must be defined as a mapping that is both one-to-one and onto, since neither condition implies the other. From now on we shall use the term "transformation" only in this specialized sense. Then since we have shown in the previous chapter that "multiplication" or a succession of "one-to-one onto mappings" must be *associative*, we can say that this is true for the Klein transformations, a fact that will be useful when we wish to indicate that a certain set of transformations is a *group* under the operation of succession.

As an example of the sort of transformation which we shall be considering, there is a one-to-one mapping of the set of integers onto itself in which each integer is increased by 2. We can picture the mapping as follows:

$$\begin{pmatrix} \ldots & -4 & -3 & -2 & -1 & 0 & 1 & 2 & 3 & \ldots & n & \ldots \\ \ldots & -2 & -1 & 0 & 1 & 2 & 3 & 4 & 5 & \ldots & n+2 & \ldots \end{pmatrix}$$

One *invariant* under this transformation is the property of being an *even* integer, for the numbers -4, -2, 0, 2, 4, etc. are transformed into -2, 0, 2, 4, 6, etc., respectively. The characteristic of being an *odd* integer is also unchanged by the transformation. The *difference* between any two integers is invariant. Thus $10 - 3 = 7$; 10 is transformed into 12, 3 into 5, and $12 - 5 = 7$. In general, if m and n are any two integers, their transforms are $m + 2$ and $n + 2$, and

$$m + 2 - (n + 2) = m + 2 - n - 2 = m - n$$

As a further illustration, there is the transformation that doubles every real number (a one-to-one mapping of R onto itself). Then 0 is an *invariant number* under this transformation, since it is changed to $2 \cdot 0 = 0$. Also the ratio of any two numbers is invariant, for if x_1 is any real number, and x_2 is any real number except zero, the ratio of their transforms is $2x_1 : 2x_2$, which is equal to $x_1 : x_2$. Since Klein's classification was concerned with geometry, let us give our transformation a geometric interpretation. As explained earlier (Chapter 2), a Cartesian coordinate system can be set up on a straight line by establishing a one-to-one correspondence between its points and the set of real numbers. Then $x' = 2x$ represents the transformation we have been discussing, where x is a real number, namely, the Cartesian coordinate of point P in Figure 17.4, and x' is the co-

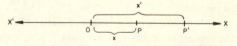

Figure 17.4

ordinate of P', the point into which P is transformed. The origin remains fixed and the distances of all other points from the origin are doubled by the transformation. (It is an example of a one-dimensional *affine* transformation.)

Because a one-to-one mapping has an *inverse* that is a mapping, every Klein transformation has an inverse which can be pictured as the transformation that carries the "points" or elements of a "space" back to their original positions. In the first example above, the *inverse* for an increase of 2 is a decrease of 2, and if the increase were followed by the decrease, the total effect would be nil. In the second illustration, if we transform all real numbers into their doubles and these

latter into their halves, each number will be back to its original value. The
net result of the succession of a transformation and its inverse is obviously the
identity transformation which maps each element of the fundamental set onto
itself.

We shall now show that the various sets of motions constitute transformation
groups. Let us first consider the set of all possible translations. It is evident that if
we walk, or translate ourselves two miles north, rest briefly, and then translate
ourselves another three miles north, we could also have reached our destination
by an uninterrupted walk of five miles north. Again, if we translate all points of
the plane (Figure 17.5) so as to bring A to A', and then perform a second trans-

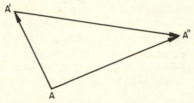

**Figure 17.5 The succession of translations
resembles vector addition.**

lation bringing A' to A'', the succession of the two translations is equivalent in
effect to a single translation bringing A to A''. Similarly, the succession of *any*
pair of translations is a translation, that is, a *transformation within the specified
set*. Hence succession is truly a binary operation on the set of all translations.
Moreover, each translation has an *inverse* that is a translation or member of the
fundamental set, since a trip two miles north can be followed by an about-face
through two miles south, and a translation from A to A' can be inverted by a
translation from A' to A, etc. with "no progress," the *identity* translation, as the
outcome. Since (1) succession is a binary associative operation on the set of all
translations and (2) that set includes an identity translation and (3) all translations
have inverses that are translations, therefore the set of all translations constitutes
a *group* under "succession."

It is readily seen that succession is a binary operation on the aggregate of all
rotations, since the succession of any two rotations is obviously equivalent to a
single rotation. For example, a counterclockwise rotation through 130° followed
by a clockwise rotation through 80° is equivalent to a single counterclockwise
rotation through 50°. Succession of transformations is always associative, as we
have said. Each rotation has an inverse that is also a rotation, namely, a rotation
of the same magnitude in the opposite direction. The identity is the 0° rotation.
Thus the set of all rotations is a *group* under "succession."

Now we shall indicate that the set of *all* rigid motions or displacements
constitutes a group. That this aggregate is closed can be deduced from the following
form of Euclid's definition of congruence: If there is a displacement bringing one
figure into coincidence with another, then the two figures are congruent; conversely, if two figures are congruent, there exists a displacement bringing one into

coincidence with the other. Let us imagine a displacement carrying a particular figure F_1 to coincidence with F_2, followed by a second displacement carrying F_2 (with F_1 superposed on it) to coincidence with F_3. Then F_1 is congruent to F_3, since it has been superposed on it. Since this is the case, the second part of Euclid's definition requires the existence of a *single* displacement that will superpose F_1 on F_3, that is, the succession of the two displacements F_1 to F_2, F_2 to F_3 is equivalent to one displacement, and the set of displacements is closed under succession, which is therefore a binary operation on the set. In addition, displacements can always be reversed, and figures can be brought back to the identical original position. For example, a displacement which consists of a three-mile north translation followed by a clockwise rotation through $20°$ and then a reflection in a line has as its inverse a reflection in the same line followed by a counterclockwise rotation through $20°$ and a three-mile south translation. Since (1) succession is a binary associative operation on the set of all displacements and (2) the set contains an identity displacement and (3) each motion has an inverse that is a member of the set, the system is a *group* under the operation of "succession." Note that our argument will still hold if we consider only *proper* rigid motions (where no reflections are involved). Hence those motions form a proper subgroup of the group of *all* rigid motions.

But the above discussion can be made more exact and more lucid by *algebraic* arguments. Klein made use of analytic geometry, and since we shall do likewise, we must digress briefly to develop the Cartesian representation for the transformations to be considered. Once again, then, we use a rectangular network of "streets" and "avenues." Suppose that the layout and numbering of streets in Figure 17.6 is merely tentative, pending the sale of lots and the types of buildings constructed on them. Suppose further that as things develop, property around $Q(-1, -2)$ in Figure 17.6 is reserved for building the town hall and the corner at $P(3, 1)$ is to be the site of a major office building. The planners decide on a modification of the layout of Figure 17.6 so that the town hall will be at the center of

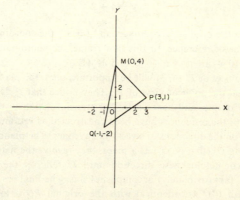

Figure 17.6

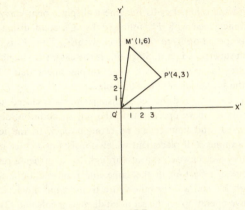

Figure 17.7

things. Preserving the directions of the streets and avenues in the figure, they *translate* the entire network diagonally downward to the left in the direction OQ until O occupies the position that was formerly Q. The result is pictured in Figure 17.7. To facilitate visualization of the translation of the meshwork, the triangle PQM has been outlined in Figure 17.6. This triangle is at rest as the rectangular framework slides diagonally beneath it but has been renamed $P'Q'M'$ (in Figure 17.7) in order to indicate that the vertices now have different coordinates. In the case of Q, the desired change has been effected. Now labeled as Q', its new coordinates are $(0, 0)$, that is, it is now at the origin of coordinates, or the center of town. Its coordinates have been increased by 1 and 2, respectively, and the translation of the network would have the same effect on the coordinates of any other point, so that

$$x' = x + 1$$
$$y' = y + 2$$

where (x, y) are the original coordinates and (x', y') the coordinates with respect to the new frame of reference. In this new frame, the coordinates of the original $P(3, 1)$ and $M(0, 4)$ are now $P'(4, 3)$ and $M'(1, 6)$.

Since the offices at $P'(4, 3)$ will be important ones, the planners consider the new layout from the point of view of P'. They notice that P' is not on the main street or main avenue, and hence a further transformation in the layout is considered. None of the residences have been constructed as yet, and thus there is no real restriction on the directions of streets and avenues. The planners keep $P'Q'M'$ at rest and using Q' (that is, O') as a pivot, they revolve the network beneath the triangle in a counterclockwise direction until P' falls on the main avenue. A rotation of $37°$ (approximately) accomplishes this. The final result is indicated in Figure 17.8 with $P''Q''M''$ identical with the original PQM except for the coordinates of vertices.

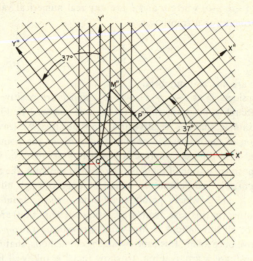

Figure 17.8

The city planners accomplished the desired transformations by displacement of the street-avenue mesh system, while the building sites P, Q, M stayed according to plans. If one wishes to picture this as the displacement of a *rigid body*, one must see the Cartesian network as the rigid object—a sort of iron grille that one can slide and rotate. Then the displacement will be a mapping of the plane onto itself, a transformation in the spirit of Klein. It is possible to effect the same transformation of the coordinates of P, Q, and M by keeping the framework at rest and displacing the triangle PQM. The mathematician Dirk Struik refers to the alternative points of view as "alias" and "alibi" because the first method merely gives different coordinates (names) to the same points, whereas the other technique shifts the points and geometric figures to other places. The "alibi" procedure is closer to Euclid's handling of congruence, except that in the Erlanger Program, in relativity, and in modern geometries in general, the displacement is not limited

to triangle PQM, but extends to *all* points of the space. All points are shifted diagonally upward to the right in the "alibi" picture, and the Cartesian network that remains at rest is pictured as a sort of fine web through which one observes the motion and sees that its effect is the one described by the algebraic formulas above, namely, to increase the coordinates of *each* point of the plane by 1 and 2, respectively.

A particular situation might require that the special point Q $(-1, -2)$ be translated to the origin (either by shifting the framework or else by sliding all points of the plane). In a different problem one might desire to have some other point as the new origin. There is infinite variety in the particular point one might want at the center of town, and therefore the original coordinates of Q might be symbolized as $(-a, -b)$ where a and b are *any* real numerical values. Therefore the general formula for a translation is

$$x' = x + a$$
$$y' = y + b$$

where a and b are any real numbers.

If one considers Q as $(-a, -b)$, where a and b are positive real numbers, then Q will occupy a position similar to the one pictured in Figure 17.6. But suppose, for example, that a is negative and $-a$ is positive. We leave to the reader the visualization of the translation of the framework or the points of the plane in such a case, or in the case where a is positive, b negative, or where both coordinates are positive, or one of them is zero, etc. At any rate, the above formulas represent *all* possible translations, as a and b range through all real numbers. *Constants* like a and b which, paradoxically, are *variable* from one particular instance to another (in this case, from translation to translation) are called *parameters*. The set of all translations is thus said to constitute a *two-parameter group*.

Let us prove once again, this time by algebraic methods, that the translations actually do constitute a group. First, to show that "is followed by" is a binary operation on the set of translations, we must show that the succession of any two translations is a translation. Consider then

$$T_1 = \begin{cases} x' = x + a_1 \\ y' = y + b_1 \end{cases} \quad \text{followed by} \quad T_2 = \begin{cases} x'' = x' + a_2 \\ y'' = y' + b_2 \end{cases}$$

The net result, $T_1 T_2$, is obtained by substituting the first pair of equations in the second to obtain

$$T_1 T_2 = \begin{cases} x'' = x + (a_1 + a_2) \\ y'' = y + (b_1 + b_2) \end{cases}$$

which is a *translation*, since it fits the translation formulas with values of the parameters $a_1 + a_2$, $b_1 + b_2$, respectively. If the reader wishes to analyze the composition of translations as carried out above, he can see it as strictly analogous to the method used for combining permutations in the previous chapter. Since it is impossible to list all points, because they form the continuum R^2 (Chapter 7), one follows the changes in position (or coordinates) of a typical point. Thus

$$T_1 = \begin{pmatrix} \cdots & (x, y) & \cdots \\ \cdots (x + a_1, y + b_1) & \cdots \end{pmatrix}$$

$$T_2 = \begin{pmatrix} \cdots & (x + a_1, y + b_1) & \cdots \\ \cdots (x + a_1 + a_2, y + b_1 + b_2) & \cdots \end{pmatrix}$$

Then, omitting the intermediate position of the typical point, we obtain

$$T_1 T_2 = \begin{pmatrix} \cdots & (x, y) & \cdots \\ \cdots (x + a_1 + a_2, y + b_1 + b_2) & \cdots \end{pmatrix}$$

Using the same method, let us now evaluate $T_2 T_1$, that is, T_2 followed by T_1. We have

$$T_2 = \begin{pmatrix} \cdots & (x, y) & \cdots \\ \cdots (x + a_2, y + b_2) \cdots \end{pmatrix}$$

$$T_1 = \begin{pmatrix} \cdots & (x + a_2, y + b_2) & \cdots \\ \cdots (x + a_2 + a_1, y + b_2 + b_1) \cdots \end{pmatrix}$$

so that

$$T_2 T_1 = \begin{pmatrix} \cdots & (x, y) & \cdots \\ \cdots (x + a_2 + a_1, y + b_2 + b_1) \cdots \end{pmatrix}$$

Since addition of real numbers is commutative, $a_2 + a_1 = a_1 + a_2$, $b_2 + b_1 = b_1 + b_2$, and $T_2 T_1 = T_1 T_2$. Thus, succession of translations is *commutative*. Let us note that (x, y), (x', y'), (x'', y''), (x''', y'''), etc. are usually used to represent successive coordinate pairs for a typical point under a succession of transformations of any kind. Then we could have computed $T_2 T_1$ as follows:

$$T_2 = \begin{cases} x' = x + a_2 \\ y' = y + b_2 \end{cases} \qquad T_1 = \begin{cases} x'' = x' + a_1 \\ y'' = y' + b_1 \end{cases}$$

Substituting the first set of equations in the second, we obtain

$$T_2 T_1 = \begin{cases} x'' = x + a_2 + a_1 \\ y'' = y + b_2 + b_1 \end{cases}$$

In summary, "is followed by" is a *binary commutative operation* on the set of all translations. That this operation is *associative* follows from the fact that succession of transformations is always associative.

Continuing with our proof that the translations form a group, we indicate that there is an *identity* translation, I, namely,

$$I = \begin{cases} x' = x \\ y' = y \end{cases}$$

Finally, every translation has an *inverse* within the set of translations; that is, if T is any translation, there exists a translation T^{-1} such that $T T^{-1} = T^{-1} T = I$. Thus if

$$T = \begin{cases} x' = x + a \\ y' = y + b \end{cases} \qquad T^{-1} = \begin{cases} x'' = x' - a \\ y'' = y' - b \end{cases}$$

then

$$T T^{-1} = \begin{cases} x'' = x + a - a = x \\ y'' = y + b - b = y \end{cases} = I$$

Since succession is commutative for translations, $T^{-1} T$ is also equal to I. In conclusion, the translations constitute an *infinite abelian two-parameter group* with respect to succession.

Sets of transformations do *not*, in general, qualify as groups, and if they are groups, they are not necessarily abelian (commutative) groups. For example, let us consider the one-parameter set of *reflections* in the lines $x = \frac{1}{2}a$ (parallels to the Y-axis), namely,

$$\begin{array}{l} x' = a - x \\ y' = y \end{array} \quad \text{with inverse} \quad \begin{array}{l} x'' = a - x' \\ y'' = y' \end{array}$$

If one examines the succession of the two transformations of the set, namely,

$$\begin{array}{l} x' = a_1 - x \\ y' = y \end{array} \quad \text{and} \quad \begin{array}{l} x'' = a_2 - x' \\ y'' = y' \end{array}$$

the result is

$$\begin{array}{l} x'' = a_2 - (a_1 - x) \\ y'' = y \end{array} \quad \text{or} \quad \begin{array}{l} x'' = (a_2 - a_1) + x \\ y'' = y \end{array}$$

But even if the value $a_2 - a_1$ is assigned to the parameter in the representative formula of the set, the "product" of the successive motions is not a reflection in any line parallel to the Y-axis. The result does *not* fit the required formula (except for points of the Y-axis for which $x = 0$). The transformation formulas call for *subtracting* x from a constant, whereas the combination of two successive motions *adds* x to a constant. The resultant of two motions of the set is, in general, a position unattainable by any *single* motion in the set. (Note that the position can be attained by *translation* through $a_2 - a_1$, parallel to the X-axis.) Hence succession is *not* a binary operation on the set and it is *not* a group.

In a modern treatment of Cartesian geometry, the Pythagorean formula is usually taken as an axiom, and then the invariance of distances and angles for all displacements is proved. Euclid's procedure was the reverse. He assumed the invariance and ultimately proved the Pythagorean theorem. If we accept the Pythagorean theorem, then, as we have seen in an earlier chapter, Figure 7.9 yields the distance formula

$$s = \sqrt{(\triangle x)^2 + (\triangle y)^2}$$

After a translation of the two points in the diagram to the new positions (x_1', y_1') and (x_2', y_2'), the distance between them is

$$s' = \sqrt{(x_2' - x_1')^2 + (y_2' - y_1')^2}$$

Substituting the formula for a translation,

$$s' = \sqrt{(x_2 + a - x_1 - a)^2 + (y_2 + a - y_1 - a)^2} = \sqrt{(x_2 - x_1)^2 + (y_2 - y_1)^2} = s$$

and the distance after translation is equal to the original distance. It is also possible to prove by algebraic methods that *angles* are invariant under the group of all translations.

It can be shown that the formulas for all rotations about the origin as pivot are

$$\begin{array}{l} x' = x \cos \theta + y \sin \theta \\ y' = -x \sin \theta + y \cos \theta \end{array}$$

where θ, a parameter, is the angle through which the original frame of reference is rotated counterclockwise.

Again, it can be proved that the above set of rotations is a group that leaves distances and angles invariant. To give a partial demonstration, let us show that distance from the origin is invariant. Originally the distance of any point (x, y) from the origin $(0, 0)$ is

$$s = \sqrt{(x-0)^2 + (y-0)^2} = \sqrt{x^2 + y^2}$$

Since the origin is the pivot of the set of rotations, its coordinates remain unchanged. (This can be proved by substituting $x = 0$, $y = 0$ in the rotation formulas.) A rotation changes the coordinates (x, y) to (x', y') as given by the formulas above. Hence the distance after a rotation, in the new frame of reference, is

$$s' = \sqrt{(x \cos \theta + y \sin \theta - 0)^2 + (-x \sin \theta + y \cos \theta - 0)^2}$$
$$= \sqrt{x^2(\cos^2 \theta + \sin^2 \theta) + y^2(\sin^2 \theta + \cos^2 \theta)}$$

Elementary trigonometry teaches that $\sin^2 \theta + \cos^2 \theta = 1$ for all values of the angle θ. Therefore

$$s' = \sqrt{x^2(1) + y^2(1)} = \sqrt{x^2 + y^2} = s$$

that is, the distance after rotation is the same as the original distance.

The invariance of distance under rotation is not surprising, since it is in accordance with our everyday experience with rotating wheels. But the algebraic formulas for the rotation group can also be interpreted as expressing the group of Lorentz transformations of the special theory of relativity, providing x, y, and θ are given suitable physical interpretations as follows:

$x =$ the position on a straight line at which some event takes place

$y = it$, where t is the time at which the event takes place and i is the imaginary unit, $\sqrt{-1}$

$\cos \theta = \dfrac{1}{\sqrt{1 - u^2}}$, where u is a relative velocity (measured in suitable units)

$\sin \theta = \dfrac{iu}{\sqrt{1 - u^2}}$

Since "imaginary" numbers are involved in this interpretation, a Lorentz transformation can be conceived in a pure mathematical sense as a rotation of the frame of reference through an "imaginary" angle. An actual *physical* meaning for such a transformation will be revealed in the next chapter. Just as the coordinates (x, y) of a point vary and are *not* usually invariant if the frame of reference is rotated, (x, it), that is, position and time, are not invariant or "absolute" in physics for observers in the different frames of reference with which the Lorentz group is concerned. We have proved above that $s = \sqrt{x^2 + y^2}$ is invariant under rotation. If we perform the substitution that converts rotations into Lorentz transformations,

$$s = \sqrt{x^2 + y^2} = \sqrt{x^2 + i^2 t^2} = \sqrt{x^2 - t^2}$$

is the corresponding invariant. Physicists term $\sqrt{x^2 - t^2}$ an "absolute" entity and indicate that its measurement is the same for all observers in uniform motion with respect to one another (Chapter 18). This is the fundamental invariant of the special theory of relativity. In the discussion above, $x^2 + y^2 = x^2 - t^2$ was proved invariant, but its negative, $-x^2 - y^2 = -x^2 + t^2 = t^2 - x^2$, could equally well have been proved invariant. In relativity the latter invariant may be used, and, in that case, $\sqrt{t^2 - x^2}$ is termed the *space-time interval* or *separation* between two events. Observers in all frames of reference (in uniform relative motion) will obtain identical measurements of the space-time separation, even though disagreeing on distance and time coordinates.

To return to the Program, the proper rigid motions which combine rotations and translations are described analytically as the *three*-parameter group

$$x' = x \cos \theta + y \sin \theta + a$$
$$y' = -x \sin \theta + y \cos \theta + b$$

and the similarity group is obtained by multiplying the right members of these formulas by a real parameter k. *Improper* rigid motions combine the proper ones with a reflection, which has the effect of changing the signs of the first two terms in the right member of the above formula for y'.

Affine geometry is governed by the group of linear transformations

$$x' = ax + by + c$$
$$y' = dx + ey + f$$

Because these transformations might be considered as the succession of

$$x' = ax + by$$
$$y' = dx + ey$$

and the *translation*

$$x'' = x' + c$$
$$y'' = y' + f$$

anything that is *novel* can be obtained by studying the *homogeneous* or *centered* affine group

$$x' = ax + by$$
$$y' = dx + ey$$

where a, b, d, and e are real parameters and $ae - bd \neq 0$. (The reason for this last condition will be indicated later.) The adjective "centered" is applicable because the origin or "center" is invariant under all transformations of the group, as the reader may check by substituting $x = 0$, $y = 0$. Note that rotations are special affine transformations where $a = e = \cos \theta$, $b = \sin \theta$, $d = -\sin \theta$, $ae - bd = \cos^2 \theta + \sin^2 \theta = 1$.

To prove that the set of homogeneous affine transformations does constitute a group involves considerable algebraic manipulation, and this very difficulty is a motivation for the abbreviated methodology of modern *linear algebra*. Let us merely suggest what is involved. First, one must show that succession is truly a binary operation on the centered affine set, or that if T_1 and T_2 are homogeneous affine transformations, then T_1T_2 is such a transformation. If

$$T_1 = \begin{cases} x' = ax + by \\ y' = dx + ey \end{cases} \quad \text{and} \quad T_2 = \begin{cases} x'' = Ax' + By' \\ y'' = Dx' + Ey' \end{cases}$$

then T_1T_2 can be obtained by substituting the equations for T_1 in those for T_2. Thus

$$T_1T_2 = \begin{cases} x'' = A(ax + by) + B(dx + ey) \\ y'' = D(ax + by) + E(dx + ey) \end{cases}$$

and

$$T_1T_2 = \begin{cases} x'' = (Aa + Bd)x + (Ab + Be)y \\ y'' = (Da + Ed)x + (Db + Ee)y \end{cases}$$

Now T_1T_2, however complicated in appearance, has the *form* of a centered affine transformation, and hence the set of such transformations is closed under succession. In other words, "is followed by" is actually a binary operation on the set.

Before continuing with our proof that the set is a group under succession, let us indicate the important by-product in terms of modern *linear algebra*. Now our study of the theory of games has made us familiar with the notion of a *matrix* or rectangular array. Then instead of considering the homogeneous affine transformations above, we can focus on the matrices of their coefficients. Thus

$$M_1 = \begin{pmatrix} a & b \\ d & e \end{pmatrix} \qquad M_2 = \begin{pmatrix} A & B \\ D & E \end{pmatrix}$$

We shall see shortly why the transformations can be symbolized as

$$X' = M_1X \qquad \text{and} \qquad X'' = M_2X'$$

If this is correct, then substitution of the first formula in the second yields for the product T_1T_2 the formula

$$X'' = M_2M_1X$$

But if this formula is to agree with the one previously obtained, then M_2M_1 must be the matrix of coefficients in that previous formula. Thus

$$M_2M_1 = \begin{pmatrix} Aa + Bd & Ab + Be \\ Da + Ed & Db + Ee \end{pmatrix}$$

If we think of M_2M_1 as meaning the *product* of the matrix M_2 by the matrix M_1, the following definition of such multiplication is suggested:

$$\begin{pmatrix} A & B \\ D & E \end{pmatrix}\begin{pmatrix} a & b \\ d & e \end{pmatrix} = \begin{pmatrix} Aa + Bd & Ab + Be \\ Da + Ed & Db + Ee \end{pmatrix}$$

This is, in fact, the definition of multiplication used for such matrices in linear algebra. We observe a simple rule for forming the *product matrix*, M_2M_1. To obtain the entry, say, in the first row, second column of that matrix, one multiplies the elements of the *first row* of M_2 (namely A, B) by the elements of the *second column* of M_1 (namely, b, e), respectively, and forms the sum of these products to yield $Ab + Be$. The general rule is: To obtain the entry in row r, column s of M_2M_1, multiply the elements of the rth row of M_2 by those of the sth column of M_1, respectively, and add these products. To illustrate this further, let us apply it to obtaining the entry in the second row, first column of M_2M_1. The second row of M_2 is $(D\ E)$ and the first column of M_1 is $\binom{a}{d}$. Therefore $Da + Ed$ should be the proper entry in M_2M_1. To apply the rule further, let us see why we used the symbolism

$$X' = M_1 X \qquad X'' = M_2 X'$$

If

$$M_1 = \begin{pmatrix} a & b \\ d & e \end{pmatrix} \qquad \text{and} \qquad X = \begin{pmatrix} x \\ y \end{pmatrix}$$

(where X is a 2×1 matrix), then the rule we gave leads to the first row, first column entry of $M_1 X$ by multiplication of the elements of $(a\ b)$ by those of $\binom{x}{y}$ to yield the entry $ax + by$. Likewise second row, first column of $M_1 X$ should be $dx + ey$. Since X has only one column, there will be no entries in $M_1 X$ other than those we have computed. Hence

$$X' = \begin{pmatrix} ax + by \\ dx + ey \end{pmatrix}$$

or if

$$X' = \begin{pmatrix} x' \\ y' \end{pmatrix}$$

this means

$$x' = ax + by$$
$$y' = dx + ey$$

which is the meaning we desire. We leave to the reader the task of applying the rule for multiplication of matrices to the proper interpretation of the symbolism, $X'' = M_2 X'$.

The algebra of matrices and, in particular, the rule for matrix multiplication are due to the noted British algebraist Arthur Cayley, whose creation of a special type of hypercomplex number was noted in an earlier chapter (Chapter 4). The multiplication rule extends to matrices of higher orders than those considered. We observe that to carry it out there must be as many elements in a row and hence as many *columns* in the first matrix as there are *rows* (elements in a column) in the second. Cayley's rule will give the product of an $m \times n$ matrix by an $n \times k$ matrix. Since matrix multiplication mirrors "multiplication," that is, succession of homogeneous affine transformations, and that multiplication must be associative, the same is true of matrix multiplication. Let us indicate that matrix multiplication is not commutative. In other words, if we seek the product $T_2 T_1$ which is defined by the matrix product, $M_1 M_2$, the results, in general, will be different from those for $T_1 T_2$ which was given by $M_2 M_1$. Thus, applying the multiplication rule,

$$\begin{pmatrix} a & b \\ d & e \end{pmatrix} \begin{pmatrix} A & B \\ D & E \end{pmatrix} = \begin{pmatrix} aA + bD & aB + bE \\ dA + eD & dB + eE \end{pmatrix}$$

which is, for most values of the constants a, b, A, B, etc., a different result from $M_2 M_1$.

For practice in matrix multiplication the reader can compute

$$\begin{pmatrix} 2 & -1 \\ 1 & 3 \end{pmatrix} \begin{pmatrix} 1 & 4 \\ 1 & -2 \end{pmatrix}$$

and then find the product when the order of factors is reversed, thereby computing $T_1 T_2$ and $T_2 T_1$, respectively, where

$$T_1 = \begin{cases} x' = x + 4y \\ y' = x - 2y \end{cases} \qquad T_2 = \begin{cases} x'' = 2x' - y' \\ y'' = x' + 3y' \end{cases}$$

But we must now complete the proof that the centered affine transformations form a group. It is obvious that there is an *identity* transformation in the affine set, namely, $\{x' = x, y' = y\}$, with *identity matrix*

$$I = \begin{pmatrix} 1 & 0 \\ 0 & 1 \end{pmatrix}$$

The reader can show that this matrix is actually a multiplicative identity by verifying that

$$\begin{pmatrix} a & b \\ c & d \end{pmatrix} \begin{pmatrix} 1 & 0 \\ 0 & 1 \end{pmatrix} = \begin{pmatrix} 1 & 0 \\ 0 & 1 \end{pmatrix} \begin{pmatrix} a & b \\ c & d \end{pmatrix} = \begin{pmatrix} a & b \\ c & d \end{pmatrix}$$

As for inverses, let us illustrate how to find them in the case of T_1 above. If (x,y) goes into (x',y') under T_1, then, to "invert" the situation, $(x',y',)$ should go back into (x,y). To find how this should be done, we can solve the simultaneous equations for T_1 to obtain x and y in terms of (x',y'). This would yield

$$x = 1/3 \; x' + 2/3 \; y'$$
$$y = 1/6 \; x' - 1/6 \; y'$$

Then the reader can verify that if

$$T_1 = \begin{cases} x' = x + 4y \\ y' = x - 2y \end{cases} \quad \text{and} \quad T_1^{-1} = \begin{cases} x'' = 1/3 \; x' + 2/3 \; y' \\ y'' = 1/6 \; x' - 1/6 \; y' \end{cases}$$

then $T_1 T_1^{-1} = I$. At any rate, we can symbolize the matrix of T^{-1} by

$$M_1^{-1} = \begin{pmatrix} 1/3 & 2/3 \\ 1/6 & -1/6 \end{pmatrix}$$

This matrix is said to be the *inverse* of the matrix M_1. We can prove that $M_1 M_1^{-1}$ $= M_1^{-1} M_1 = I$, the identity matrix. Thus the reader can check that

$$\begin{pmatrix} 1 & 4 \\ 1 & -2 \end{pmatrix} \begin{pmatrix} 1/3 & 2/3 \\ 1/6 & -1/6 \end{pmatrix} = \begin{pmatrix} 1 & 0 \\ 0 & 1 \end{pmatrix}$$

and then reverse the order of factors to obtain I once again.

But to return to the centered affine set, every transformation

$$x' = ax + by$$
$$y' = dx + ey$$

has the inverse homogeneous affine transformation

$$x'' = \frac{e}{ae - bd} x' - \frac{b}{ae - bd} y'$$
$$y'' = \frac{-d}{ae - bd} x' + \frac{a}{ae - bd} y'$$

providing $ae - bd \neq 0$, which is the condition stated on page 416. This completes the proof that the system of homogeneous affine transformations is a group.

Since the multiplicative system of all 2×2 matrices,

$$\begin{pmatrix} a & b \\ d & e \end{pmatrix}$$

in which $ae - bd \neq 0$, (the *invertible* 2×2 matrices) is *isomorphic* to the homogeneous affine group, the system of matrices is also a group.

The multiplicative system of $n \times n$ (*square*) invertible matrices is also a group when $n = 3$ or 4 or 5, etc. and that group is *isomorphic* to the homogeneous affine group in the Euclidean 3-space or 4-space or 5-space, etc.

Since a 2×2 or $n \times n$ matrix may not have a multiplicative inverse, some sets of such matrices are *not* groups with respect to multiplication. Also, although matrix multiplication is always associative, it is *not* necessarily commutative on a set of matrices. These facts make multiplicative systems of matrices available as concrete interpretations of some of the modern abstract algebraic structures whose multiplication is noncommutative. Also since square matrices are not all invertible a multiplicative system of such matrices may not admit division as a binary operation.

Cayley's algebra of matrices contains a definition of "scalar multiplication," a *unary operation* on matrices, and also a definition of *addition*, a binary operation on matrices. For an example of scalar multiplication one can treble all payoffs in the game whose matrix is given on page 251. Thus

$$3 \begin{pmatrix} 3 & 1 & 2 \\ 6 & 0 & -3 \\ -5 & -1 & 4 \end{pmatrix} = \begin{pmatrix} 9 & 3 & 6 \\ 18 & 0 & -9 \\ -15 & -3 & 12 \end{pmatrix}$$

Here the original matrix has been multiplied by the "scalar" (ordinary number) 3. This example illustrates that multiplication of a matrix by a scalar (number) is carried out by multiplying each matrix entry by that scalar.

For an illustration of addition of matrices, let us suppose that two children, who have been encouraged to keep small savings accounts, use the matrix form to tabulate interest received in a particular year. They add two matrices to total interest receipts for two years. Thus, if the matrices for the two years are

	Bank A	Bank B	Bank C				
Helen	2	3	1	and	2	4	2
James	4	1	1		5	2	3

then

$$\begin{pmatrix} 2 & 3 & 1 \\ 4 & 1 & 1 \end{pmatrix} + \begin{pmatrix} 2 & 4 & 2 \\ 5 & 2 & 3 \end{pmatrix} = \begin{pmatrix} 4 & 7 & 3 \\ 9 & 3 & 4 \end{pmatrix}$$

This example illustrates that two matrices of the same species, that is, having the same number of rows and the same number of columns, can be added by adding corresponding entries.

After the properties of matrix addition and scalar multiplication have been analyzed, one has available many examples of algebraic systems whose elements are *not* merely the numbers of ordinary algebra. Thus, let S represent the set of all 2×2 matrices (or all 3×3 matrices or all 4×4 matrices or all square matrices of any finite order) with entries that are real numbers or complex numbers or elements of some field. Further, if ● represents scalar multiplication on the matrices of S (that is, the multiplication of these matrices by a number in some field), and $+$, $\times$ represent addition and multiplication on S, then the system $\{S, +\}$ is an *abelian group*, the system $\{S, +, ●\}$ is a *vector space* (Chapter 4), the system $\{S, +, \times\}$ is a *noncommutative ring* which is special because it has a multiplicative unit or identity, and $\{S, +, \times, ●\}$ is an "algebra," or a *linear associative algebra.* We have already examined the "algebras" of ordinary complex numbers, of quaternions, of Cayley numbers (Chapter 4).

To return to the homogeneous affine transformations which motivated our discussion of matrix algebra, their geometric effect, if one takes the "alias" point of view, is to change from a rectangular Cartesian coordinate system to an *oblique system* (Chapter 7). Later on, we shall see the importance of this fact for differential geometry, general relativity, and post-relativity geometries, in which the "local" geometry of some surface or space may be affine. In relativity, the geometry of the homogeneous affine group is called *Euclidean* because a transformation to an oblique coordinate system does not change essential facts. If one takes the "alibi" instead of the "alias" viewpoint, the effect can be considered to be a stretching or distortion of the Euclidean plane to fit a "neighborhood" in one of the curved surfaces of differential geometry.

Plane projective geometry is the study of properties invariant under the linear fractional transformations,

$$x' = \frac{ax + by + c}{gx + hy + k}$$

$$y' = \frac{dx + ey + f}{gx + hy + k}$$

where

$$aek - afh + bfg - bdk + cdh - ceg \neq 0$$

The apparently (but not actually) complicated restriction on the coefficients is analogous to the condition $ae - bd \neq 0$ for the affine transformation, and is part of the "linear algebra" of the situation. Incidentally, if one substitutes $k = 1$ and $c = f = g = h = 0$, the new projective transformations reduce to the homogeneous affine transformations and the above limitation becomes $ae - bd \neq 0$. But the original inequality on the coefficients insures that every projective transformation is a one-to-one mapping of the plane onto itself with an *inverse* that is a projective transformation. Then it is a simple matter to show that the projective (linear fractional) transformations form a group.

The rigid motions constitute a subgroup of the similarity group which is a subgroup of the affine group which is a subgroup of the projective group. As the generality and the number of parameters increase, the number of invariants decreases, but there are still significant invariants under projective transformation. The *degree* of the equation of any curve is preserved. Thus first-degree equations, which represent straight lines, remain first-degree equations; that is, the property of being a straight line is invariant. To prove this, suppose that the equation of a straight line is

$$mx + ny + p = 0$$

Each point (x,y) on this line will have new coordinates (x', y') after a projective transformation, that is, after the point moves to the new position prescribed by such a transformation. Can one provide a Cartesian equation for the set of new positions? If the parameters are restricted in the way we have indicated, the formulas for a projective transformation can be solved for x and y in terms of x' and y', to yield the *inverse*,

$$x = \frac{Ax' + By' + C}{Gx' + Hy' + K}$$

$$y = \frac{Dx' + Ey' + F}{Gx' + Hy' + K}$$

where the letters A, B, C, etc. are, in general, different from a, b, c, etc. Substituting the inverse formulas in the equation of the line will express a relation between x' and y', the new coordinates, namely,

$$m\left(\frac{Ax' + By' + C}{Gx' + Hy' + K}\right) + n\left(\frac{Dx' + Ey' + F}{Gx' + Hy' + K}\right) + p = 0$$

or

$$(mA + nD + pG)x' + (mB + nE + pH)y' + (mC + nF + pK) = 0$$

This is a linear equation in x' and y', and hence indicates a straight line to be the configuration of points in the new positions.

In the same way, a second-degree equation in x and y remains a second-degree equation, that is, the property of being a conic section is preserved under projective transformation. A circle might be transformed into an ellipse or a parabola or a hyperbola, however, and any conic section may be changed into a conic of a different type.

If one considers any three points P_1, P_2, P_3 of a straight line, they will be converted by any projective transformation into three collinear points P_1', P_2', P_3' but, in general, both distances and ratios of distances will be changed. However, if the fate of *four* collinear points is examined, it can be proved that for every projective transformation the "ratio of ratios" or "double ratio" of distances is an invariant, that is,

$$\frac{P_1 P_3}{P_2 P_3} : \frac{P_1 P_4}{P_2 P_4} = \frac{P_1' P_3'}{P_2' P_3'} : \frac{P_1' P_4'}{P_2' P_4'}$$

We shall not establish the invariance in the general case of the *double ratio* or *cross ratio* or *anharmonic ratio*, as it is variously called, but we shall illustrate it for special cases. Thus the invariance is evident (by virtue of similarity properties) in the case of the special "central projection" of Figure 17.9.

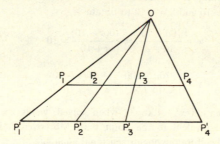

Figure 17.9   Central projection

The points P_1, P_2, P_3, P_4 with Cartesian coordinates 0, 1, 3, 9, respectively, in a straight-line "space" (Figure 17.10) have the double or cross ratio $3/2 : 9/8 = 3/2 \cdot 8/9 = 4/3$. Now consider a one-dimensional projective transformation in the straight-line "space," namely,

$$x' = \frac{2x - 2}{x + 1}$$

Substituting 0 in this formula,

$$x' = \frac{0 - 2}{0 + 1} = -2$$

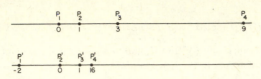

Figure 17.10

so that P_1' has the coordinate -2 in the new frame of reference. Substitutions of 1, 3, 9 give the coordinates of P_2', P_3', P_4', and the four new coordinates are -2, 0, 1, 1.6, respectively, with cross ratio $3/1 : 3.6/1.6 = 3/1 \cdot 16/36 = 4/3$, so that in this particular case the cross ratio has been shown to be invariant.

Klein and Cayley made use of the cross ratio in giving analytic interpretation to the non-Euclidean geometries of Lobachevsky and Riemann. They defined Lobachevskian geometry on a line by means of a special subgroup of the projective group $x' = (ax + b)/(cx + d)$, the subgroup leaving invariant two specified points of the X-axis. They defined a *plane* Lobachevskian geometry by means of a subgroup of the plane projective group, the subgroup leaving invariant a specified conic section—an ellipse or a circle, say.

In a straight-line space, the projective group has a one-parameter subgroup,

$$x' = \frac{ax + 100}{x + a}$$

which leaves the points $x = 10$ and $x = -10$ invariant. The reader can verify this by substituting $x = 10$ and $x = -10$ to obtain

$$x' = \frac{10a + 100}{10 + a} = \frac{10(a + 10)}{a + 10} = 10$$

and

$$x' = \frac{-10a + 100}{-10 + a} = \frac{-10(a - 10)}{a - 10} = -10$$

Since the equation $x^2 - 100 = 0$, when solved, yields $x = 10$ and $x = -10$, this equation is equivalent to the point pair. Moreover, it is a *second-degree* equation, and is therefore the one-dimensional analogue of a conic section, which is the figure to be left invariant in the two-dimensional Lobachevskian geometry. Cayley termed the invariant point pair or conic the *absolute*.

To define *distance* in their non-Euclidean geometries, Cayley and Klein proceeded by analogy with a discovery of Laguerre (1834–1866), who had shown that the distances and angles of ordinary *Euclidean* geometry can be expressed in terms of cross ratios, in other words, that Euclidean *metric* geometry is clearly a specialization of *projective* geometry. The concept of the "absolute" and the definition of distance unified Euclidean and non-Euclidean geometries into a single all-embracing theory.

In the Cayley-Klein non-Euclidean geometries, the distance from Q to P in a straight-line "space" (Figure 17.11) was defined in terms of the cross ratio of

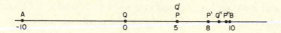

Figure 17.11

A, B, P, Q, where A and B are *invariant points* forming the Cayley "absolute." The distance was taken as

$$k \text{ logarithm } (ABPQ)$$

where k is some constant and $(ABPQ)$ symbolizes the cross ratio. The reason for using the logarithm was the customary one, namely, that a product is thereby converted into a sum, and this makes it true that

$$\text{distance } PQ + \text{distance } QR = \text{distance } PR$$

which is a property that one expects of anything described as a "distance." Thus, in accordance with this definition, if the invariant points A and B are $x = -10$ and $x = 10$, P is $x = 5$, Q is $x = 0$, the distance QP (Figure 17.11) is

$$k \text{ logarithm } \left[\frac{5+10}{5-10} \cdot \frac{0-10}{0+10} \right] = k \text{ logarithm } 3$$

(The constant k can be selected to suit special purposes, and logarithms to any base can be used.) If we select $k = 10$ and use common logarithms, the distance is $(10)(0.48) = 4.8$, which is not very different from our Euclidean measurement of the distance between 0 and 5.

Now let us carry out some transformation of the subgroup defining our Lobachevskian geometry. For example, in the formula for the subgroup, let $a = 20$. Then

$$x' = \frac{20x + 100}{x + 20}$$

Substitution of $x = 5$ and $x = 0$ shows that P and Q will be shifted to

$$x' = \frac{20(5) + 100}{5 + 20} = 8 \quad \text{and} \quad x' = \frac{20(0) + 100}{0 + 20} = 5$$

The Cayley-Klein distance between these new points will be

$$10 \text{ logarithm } \left[\frac{8+10}{8-10} \cdot \frac{5-10}{5+10} \right] = 10 \text{ logarithm } 3 = 4.8$$

which is the same distance as that between the original points, thus checking the invariance of distance. By *Euclidean* standards, the distance between the points 5 and 8 would be only 3 units, thus indicating the non-Euclideanism of our special geometry. To emphasize it still more, consider the transformation where $a = 11$, namely,

$$x' = \frac{11x + 100}{x + 11}$$

Then $x = 5$ and $x = 0$ are transformed into $x = 9\ 11/16$ and $x = 9\ 1/11$, respectively. In the Lobachevskian geometry these points are still 4.8 units apart whereas, by Euclidean standards, the distance is only 0.6 unit. If one marks off successive Lobachevskian lengths of 4.8 units, the corresponding Euclidean lengths get smaller and smaller as one approaches either $x = -10$ or $x = +10$. The Lobachevskian distance of either of these points (forming the "absolute") from any other point is "infinite." Thus, the Lobachevskian distance from 0 to 10 is

$$10 \text{ logarithm } \frac{10 + 10}{10 - 10} \cdot \frac{0 - 10}{0 + 10} = 10 \text{ logarithm } \frac{-200}{0}$$

It is impossible to divide -200 by 0, but if we divide -200 by a negative number very close to 0, the quotient will be very large and will have a large logarithm. By taking the divisor sufficiently close to 0, both the quotient and its logarithm can be made larger than any given number, however huge that number may be. That is what is meant when $-200 \div 0$ or its logarithm is said to be "infinite."

If a two-dimensional Lobachevskian geometry is set up, one chooses a conic like the ellipse in Figure 17.12 as the *absolute*, or figure to be left invariant. The

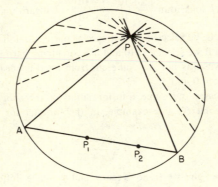

Figure 17.12   A Lobachevskian plane geometry

results are analogous to those in one dimension. All points within the absolute are at an infinite distance from that curve. Then if one defines parallel lines as lines meeting at infinity, Figure 17.12 shows how the world within the ellipse is a realization of Lobachevsky's geometry (Chapter 3). Through point P there are two "parallels" to P_1P_2, that is, two lines meeting it on the ellipse, and an infinite number of "nonintersectors."

Klein showed that the Riemannian species of non-Euclidean geometry can be developed in a fashion completely analogous to the Lobachevskian type by choosing an "imaginary" absolute, that is, an "imaginary" point pair or conic, and an imaginary value of the constant k. Euclidean geometry can also be treated in the same way by choosing a "degenerate" point pair or conic. In one dimension the point pair "degenerates" by having the points come closer and closer together until they are coincident. Thus $x^2 - 1 = 0$ represents a real point pair suitable for the absolute of a Lobachevskian geometry. If $x^2 - 0.01 = 0$ is selected, the absolute is still real, but the points $x = +0.1, x = -0.1$ are close together. If $x^2 = 0$, there are two coincident points, $x = 0, x = 0$, and this is the sort of absolute corresponding to Euclidean one-dimensional geometry. In the variation of absolute just described, the constant in the equation increased from -1 to -0.01 to 0. If a further increase to $+0.01$ and then to $+1$ is envisioned, the equations $x^2 + 0.01 = 0$ and $x^2 + 1 = 0$ will describe *imaginary* absolutes, namely, the point pairs $x = \pm 0.1$ i and $x = \pm i$, respectively, and these will correspond to one-dimensional Riemannian geometries. Thus Euclidean geometry is intermediate between the two types of non-Euclidean geometry. Similarly, Euclidean plane geometry is intermediate between the two-dimensional non-Euclidean geometries. A *real* absolute, like the circle $x^2 + y^2 = 1$, can be changed to the circle $x^2 + y^2 = 0.01$ with the tiny radius 0.1, and then can degenerate into $x^2 + y^2 = 0$, a circle of zero radius, and this can change further into the imaginary circles $x^2 + y^2 = -0.01$, $x^2 + y^2 = -1$, etc. The transitions we have indicated emphasize a point implied in the previous discussion, namely, that although there is essentially *only one* Euclidean geometry in each dimension (since one "degenerate" absolute is congruent to any other), there are many Lobachevskian and Riemannian geometries since the absolutes—the point pairs, the ellipses, etc.—can be chosen in an infinite number of different ways.

Klein developed projective geometries of 3, 4, 5, ..., n dimensions and associated with them the various higher-dimensional non-Euclidean geometries. But he also considered groups and geometries not directly related to the projective type. For example, around 1840, A. F. Moebius (1790–1868), whose name will appear frequently in the chapter on topology, the most modern of all "geometries," began a study of *inversive* geometry. To Klein this was merely the geometry associated with the one-parameter transformation group

$$x' = \frac{a^2 x}{x^2 + y^2}$$
$$y' = \frac{a^2 y}{x^2 + y^2}$$

The invariants (and hence the geometry) of this group differ from those in the projective case. Straight lines are not invariant but are converted into circles. The property of being a circle is invariant however—that is, inversion transforms every circle into a circle—and angles between lines and curves are also invariant. The latter property is described by saying that inversive geometry is *conformal*. This particular geometry is of great importance in the branch of analysis known as "functions of a complex variable."

The Erlanger Program, however significant from the point of view of bringing economy of thought into the treatment of geometry, does not cover everything that is called geometry today. It does include special relativity, but the *general* theory introduced by Einstein around 1916 cannot be described by Klein's 1872 definition of geometry, and the "post-relativity" geometries stimulated by the search for a "unified field theory" do not fit into the Erlanger Program. In all these geometries, the idea of *invariance* is still important, but the associated sets of transformations are not, in general, groups.

The discussion of these more general geometries will be postponed to a later chapter, and now a further excerpt from Young's biography of Klein will be offered. From the present point of view, possibly the most important facts in the extract are those concerned with the Göttingen mathematics institute and Klein's association with the Norwegian mathematician (Marius) Sophus Lie (1842–1899). Klein started a precedent at Göttingen that was to make it the mathematical center of the world until Hitler's day, when the mathematical division of our own Institute for Advanced Study carried on and improved the best features of Klein's plan. It is said that Klein and Sophus Lie, as young men, decided to divide the world of groups between them so as not to encroach on one another's territory. During most of his lifetime Klein received more recognition than Lie, whose greatest fame was probably posthumous. Today some of Klein's notions have been modified or replaced, and a few are considered passé. But one need merely examine current research periodicals in order to read about Lie groups, Lie algebras, etc. It is Lie's ideas that have stood up well under the critical test of time. But for a general reader who wants all the traditional geometries—Euclidean, non-Euclidean, conformal, etc.—in a nutshell, along with an elementary approach to the group concept, the Erlanger Program is still an eye-opener.

From the small Bavarian university of Erlangen, Klein passed in 1875 to the Technische Hochschule at the Bavarian capital, Munich. In Germany, where the different levels of society were kept rigidly distinct by what was termed the "Kastengeist," the level of professors at the technical schools was regarded as lower than that of university professors; but Klein looked at things from a broader standpoint. He says himself that there hovered before his mind's eye the vision of a polytechnic school like those of Paris and Zurich, and, in accepting this "call," he felt that he was taking a great step in advance.

The years at Erlangen had been only a preparation; it was at Munich that Klein began to feel his feet; here, to use his own words, he "worked himself through to a real mathematical individuality." But, in doing so, he undermined his health, and, at Leipzig, to which university he was called in 1880 as Professor of Geometry, he became seriously ill. His breakdown was probably accelerated by the antagonism he experienced at Leipzig. He was much younger than his colleagues, and they resented his innovating tendencies. In particular, there was opposition to his determination to avail himself of the vaunted German "Lehrfreiheit," and to interpret the word "Geometry" in its widest sense, beginning his lectures with a course on the Geometric Theory of Functions.

Klein was now at the acme of his mathematical powers. The theorem, which Klein himself prized highest among his mathematical discoveries, known as the "Grenzkreistheorem" in the theory of automorphic functions, flashed upon his mind suddenly at the seaside, in the small hours of the morning, as he sat, propped up on a sofa, because of his asthma. This was at Easter, 1882. The stimulus to this important extension of his earlier work on the icosahedron and the modular functions had been given by Poincaré's three notes in the "Comptes Rendus" of February and March, 1881.

It is interesting to note that the turning point in Klein's recovery from his breakdown at Leipzig was marked by his being invited to Baltimore, to take the chair vacated by

Sylvester in 1884. Indeed, he attributed to this invitation, and to the attractive vistas which it opened to his fancy, the tonification of his mind enabling him to throw off his malady. He eventually declined the invitation, which was not the last he received from an American university. But the days of his great productivity were over, although he had only reached what Dante calls "the middle of our walk in life"; from this time, for many years, he devoted himself to the development of his mathematical school.

In 1886 he was called to Göttingen. The great city was indeed no home for the true German, and it was with enthusiasm that the Kleins migrated to "the little garden-town," which at the time was still a medieval poem. It was here that the happiest home-days of the family were passed, in the comfortable villa, all their own, built by themselves, in its cool shady garden. On one side it was close to the Auditorium, on the other, near by, was the magnificent forest, the Göttingerwald, stretching to the Harz Mountains. Every day, after midday dinner and a short siesta, Klein used to walk up into the forest in mathematical converse with his colleagues and friends, after which he received his special students upstairs in his book-lined study, and discussed their work, or he prepared his own lectures.

Göttingen did not disappoint Klein and he refused every call elsewhere. He saw the university raised once more to a world-celebrity which even in the days of Gauss had never been surpassed, and to a material magnificence which had never been even imagined. The turning-point came when, as he himself humorously pointed out at a Christmas gathering of his students in 1893, the centre of the universe veered round to Göttingen, following as it always must, the advent of Eve's daughters. In fact, Prussian university life had been gently invaded. Two American women, inspired by Klein's visit to Chicago that year, had followed him to Europe, and a Girton girl (Grace Chisholm) had appeared at the same time independently; it was she who became eighteen months later the first female Doctor of Philosophy at a Prussian University.

At the Auditorium, on the top floor, was Klein's lecture room; in the days of his best lectures it rarely contained more than a dozen hearers, and these nearly all foreigners, but at the beginning of the present century there were near on a hundred students, mostly Germans, in his audience. Hard by was the wonderful mathematical reading-room, with the adjacent room lined with glass-cases full of models; here the professor held, after his lecture, his "Sprechstunde," when any of his pupils could come to him with requests or questions. For a very small fee the student received a pass-key enabling him to enter the reading-soom at any hour and avail himself of the books, which, unlike those at the University Library, might not be taken out, and of the manuscript and lithographed notes of lectures; in particular he could consult the written "Ausarbeitung" of Klein's lecture of the day before, made by Klein's assistant and previously submitted to himself; it happened, not infrequently, that rash statements of the professor were in this way controlled and corrected, these assistants, particularly the too early lost Ernst Ritter, being for the most part mathematical collaborators of a high order. All this, which had been organized by Klein, was later developed into a Mathematical Institute in a separate building.

Klein was the principal promoter of the Enzyklopädie der Mathematischen Wissenschaften. The Enzyklopädie was, from Klein's point of view, an effort to render accessible the bulk of existing mathematics. The progress of mathematics, he said, using a favourite metaphor, was like the erection of a great tower; sometimes the growth in height is evident, sometimes it remains apparently stationary; those are the periods of general revision, when the advance, though invisible from the outside, is still real, consisting in underpinning and strengthening. And he suggested that such was the then period. What we want, he concluded, is a general view of the state of the edifice as it exists at present. Klein undertook, for his own part, in accordance with his pronounced taste for Applied mathematics, the editorship of the volume on Mechanics.

Klein's appetite for work was such that, long before there was any prospect of the Enzyklopädie being completed, finding he was obliged to leave much of the organization in other hands, he had turned to the international question of mathematical teaching in the schools. He turned perhaps to something more concrete, which brought him into contact with minds other than those of the mathematical expert, something more comprehensible in fact than the huge colossus which was threatening to remain nothing but a torso. In the matter of the International Mathematical Teaching Commission Klein, as President, formed one of a triumvirate, of which another, the pioneer of the movement, was a world-travelled American, David Eugene Smith, and the third, as General Secretary, a native Swiss professor, Henri Fehr. This international work was checked and rendered abortive by World War I.

It is impossible even to touch on all Klein's schemes, yet we cannot but allude to a vast undertaking, still uncompleted, in which he was the prime mover, the editing and issuing of Gauss's Works. For Klein, Gauss represented the universal in Mathematics, and in this connexion he held Gauss before himself as the master to be copied.* But, if Gauss was here the prototype, it was buried in his individual person; the apostle of universality in Mathematics was Klein, and, without hyperbole, he may be termed the founder of what the 20th Century means by Mathematical Science, raised equally above departmental specialism and narrow nationalism. Indeed, during a period when the forces of material progress were, at first secretly, but afterwards openly, undermining the basis of the unity of civilised nations, generating jealousy, and inciting to war, Klein embodied the contrary principle, and, by his great power in his own country, and his wide-spread influence in the world of intellect, constituted one of the greatest forces tending to international amity and peace. In particular, it was he who revived the practice—prevalent in the Middle Ages, when Latin was the universal tongue of Learning—of inviting distinguished men from other countries to come and give lectures.†

In looking through Klein's Autobiography, one is struck by his constant references to his friendships. It was characteristic of him that they all bore directly, or indirectly, on his mathematical development and on his power of organisation. He was indeed a man without a hobby; in particular, and this is curious and interesting from a psychological point of view, although a German, and although endowed with an excessive acuteness of hearing,‡ he could not distinguish one tune from another.

Social intercourse for Klein meant the interchange of ideas, and for that he was as eager as an ancient Athenian. It was in such give-and-take that his own conceptions took form. There is, perhaps, no contradiction in saying that Klein was never the originator of his own ideas. He had not the generating force of a Cauchy or a Georg Cantor, but he had a phenomenal power of grasping the import of a suggestion, and working it out on a grander scale than any before him had imagined.

We have already mentioned incidentally several of those close friendships that directly influenced and even inspired Klein as a mathematician. The most striking of all, his friendship with Lie, merits a place by itself. It takes us back to his stay at Berlin in 1869. Klein writes: "The most important event of my time in Berlin was certainly that, towards

---

*There was hardly a new mathematical idea formulated, but Klein would cite a passage from Gauss in which it had been foreshadowed.

† In particular Poincaré came to Göttingen in 1894 or 1895, and gave a lecture in French to the mathematical professors and students.

‡ This enabled him to catch the faintest whisper during his lectures. Also, commonly, when away from home, he could not go to sleep unless he stopped up his ears with cotton-wool; this was the case, for instance, during his short stay at Trinity College in 1897, on the occasion of his receiving his honorary degree at Cambridge.

the end of October, at a meeting of the Berlin Mathematical Society, I made the acquaintance of the Norwegian, Sophus Lie. We had, in our work, been led from different points of view finally to the same questions, or, at least, to kindred ones. Thus it came about that we met every day and kept up an animated exchange of ideas. Our intimacy was all the closer, because at first we found very little interest for our geometrical interests in our immediate neighbourhood."

In the summer of 1870 Klein and Lie went together to Paris, where they had rooms side by side, and lived in the closest bonds of friendship. Their intention was to go on to England in the winter, but the outbreak of the Franco-German War prevented it. In this connection Klein writes: "This striving after the greatest possible breadth of scientific outlook, for which an acquaintance with what was being done in foreign countries seemed important, was very little understood at that time in Germany. Thus, for instance, when, urged by my father, I tried to obtain letters of recommendation from the Minister of Education in Berlin, I received the official answer: We have no use for French or English mathematics."

In Paris the two friends did not attend lectures, but worked together. It was, however, undoubtedly by the personal contact with the younger French mathematicians, particularly Camille Jordan, that the attention of the pair was directed to Galois' theory of groups. Jordan's treatise on the "Theory of Substitutions" had just appeared, and was to the two young men "a book with seven seals." We continue in Klein's own words: "It was, however, with Darboux that we were most intimate. At that time the French had been occupied with new investigations in metrical geometry (inversion, constant use of the circle at infinity), and this was quite unknown in Germany. Now it appeared that these were most closely related to our own work in line geometry. . . . For Lie, what was perhaps most interesting was the researches of the French into the geometrical theory of differential equations. It was by the coalescence of the two circles of ideas that Lie came to discover the connection between lines of curvature and minimal lines on a surface."

And here Klein gives a vivid sketch from this period of his life: "I had risen early one morning at the beginning of July, 1870, and I was just going out, when Lie, who was still in bed, called me into his room, and began explaining to me what he had found out during the night, the connexion between lines of curvature and minimal lines. I did not understand one word. Anyway, he assured me that it followed that the minimal lines on Kummer's surface must be algebraic curves of the 16th order. In the course of the morning it flashed upon me, while I was seeing over the Conservatoire des Arts et Métiers, that these must be the same curves of the 16th order which had already appeared in my theory of the Line-complexes of the first and second degree, and I succeeded at once in carrying out the geometrical proof, independently of Lie's transformation. When I got back in the afternoon, at 4 o'clock, Lie had gone out, and I left a letter for him containing my proof."

The outbreak of the war, Klein's consequent hasty departure, leaving his Norwegian friend behind, then the continuation of their mathematical conversations by letter, have an interest beyond that of the immediate consequences to Lie. While Lie remained at Paris, all went well; but when, in the heat of August, he started for Italy on foot, he was arrested at Fontainebleau as a spy, and kept in durance for four weeks, till, in fact, their friend Darboux had succeeded in explaining the inoffensive mathematical nature of the mysterious letters, in German, and in Klein's notoriously illegible handwriting, found in Lie's possession.

This episode was one of the manifestations of the unbroken continuance during the Franco-German war of friendly scientific relations between himself and his newly-won French friends, which made such an indelible impression on Klein's mind. We may say that those were still the days of scientific chivalry. When World War I broke out in 1914, Klein was incapable of conceiving any other state of affairs, but he was, as he had in fact been in 1870, carried away by the wave of patriotic fervour of the moment. When he therefore began to perceive that there was serious danger of what he regarded as a purely political cataclysm affecting international scientific relations, he lost, for one unfortunate moment, that dignified poise of judgment which was one of his most striking characteristics. Invited by telephone to sign a document, which, according to the impression left on his mind was to make an appeal to the scientists of Europe to maintain an objective attitude during the struggle, he impetuously acceded to the request. He did not see the text of the document until it appeared in the daily papers with his signature, as one of 93, printed below. It was very damaging to Klein's international position that he permitted his name to appear without protest at the foot of a document whose tenor was so entirely antagonistic to the whole tone of his life. And he was too proud, even after the Armistice, to withdraw publicly his signature. He desired his friends to know, and to tell, how it had been obtained, but he felt that, for himself, he stood upon his whole career, and not upon an isolated and misinterpreted action.

Lie was seven years older than Klein, but their standing in mathematical productivity was about the same. Lie had only begun mathematics seriously at 26, whereas Klein, at the age of 20, when they first met, had been for three years recognised as a mathematician and a writer. Lie was excessively tenacious of his own rights to his own ideas, and resented, as is well-known, any suggestion that they were in any sense second-hand. He quarrelled with Klein, his early friend, on this very point, and in the introduction to the third volume of the "Transformationsgruppen" he repudiates the notion that he was a pupil of Klein's; "rather," he says, "the contrary was the case." This irritated Klein, and justly. His generous nature was incapable of making the former of these suggestions, and his frank and healthy self-confidence prevented his accepting the latter of them. It is so rare in this world of ours that misunderstandings are ultimately cleared up, that it is a real pleasure to know that this was the case with Klein and Lie. In the vivid picture which Klein gives in his Autobiography and in the Collected Mathematical Papers of the interweaving of his own work with that of Lie, there is no bitterness.

Mrs. Klein wrote: "The relation to Lie was an intimate friendship in mathematical and in personal respects. That remained the same also later, when Lie visited us in Leipzig. And I, too, was fond of him, the powerful Northman with the frank open glance and the merry laugh, a hero in whose presence the common and the mean could not venture to show themselves.

"Then he became my husband's successor at Leipzig, and there he was seized with home-sickness. How well I understand that! He, the free, accustomed to his rough but beautiful northern home, how could he stand the great smoky city, the high houses, the narrow streets, the puny Saxon people. He suffered from melancholia, he was taken to a sanatorium, but there things only became worse, for they robbed him of his work and of his freedom. There, in his embittered state of mind, it must have been that he wrote the malicious things about my husband which were both painful and incomprehensible to him.

"But soon he understood that his best friend was ill and that he could not be held responsible for his actions. It was not only magnanimous and good, but also wise of my husband, not to enter into any polemic, but to let the thing simply rest. And he was not mistaken in his friend.

"One summer evening, as we came home from an excursion, there, in front of our door, sat a sick man. 'Lie!' we cried, in joyful surprise. The two friends shook hands, looked into one another's eyes, all that had passed since their last meeting was forgotten. Lie stayed with us one day, the dear old friend, and yet changed. I cannot think of him and of his tragic fate without emotion. Soon after, he died, but not before the great mathematician had been received in Norway like a king."

As for Klein himself, gradually, but irresistibly, the nervous malady of which from time to time he had had serious warnings, mastered and prostrated him. It was, he thought, partly due to heredity on his mother's side, partly to his own unbridled expenditure of mental energy—an energy which remained so indomitable that, even during the last two years of his life, when he lay helpless, becoming daily weaker and weaker in body, he never complained, and remained clear to the end, working and even correcting proof-sheets.

18

A Special Group and Its Application

A pure mathematician might state that the special or restricted theory of relativity, as presented by Einstein in 1905, was nothing more than one of Klein's "geometries," that is, the study of the invariants of a particular transformation group. As already explained for the case of observers in a linear cosmos, the group in question is the set of Lorentz transformations, or alternatively the group of *rotations*, providing the time coordinate is taken with an "imaginary" factor and the angle of rotation is also "imaginary." Although such imaginary entities are entirely acceptable from an abstract point of view, they do not completely satisfy the applied mathematician, the physicist, or the general reader. If relativity is to be associated with the world of reality, a different treatment is required. Therefore the present chapter will examine the classic concepts which required revision at the hands of Einstein, and will point out how he applied the group of Lorentz transformations to the resolution of paradoxes that plagued the nineteenth-century physicists.

Those men adhered to the postulates of Newton and to the theorems derived from his assumptions—the immense structure of classical physics. By 1880 this edifice began to topple. Just as the parallel postulate was the bone of contention which ultimately required a revolution in mathematical ideas, Newton's postulation of an *absolute space* and an *absolute time* was the source of the trouble that called for a new theory of the universe. The term "absolute" is akin to the concept of "invariant," but is somewhat more general. An examination of technical works on relativity would reveal the use of the word *covariant* to provide the greater generality. To illustrate an *absolute* quantity, one need only point to a result of counting. All observers would agree that a certain pamphlet contains 45 pages or that 8 people are seated at a table. The adjective *absolute* is applied to a physical phenomenon on which all observers are in agreement, in contrast to a *relative* result whose estimation may vary from observer to observer. Chicago is *west* for observers in New York City but *east* for those in Omaha. The purchasing power of a dollar is relative, because it varies with place and time.

Since Newtonian dynamics deals with motion, one must ask whether any judgments related to motion can be absolute. If you are a passenger in a subway train and another train is on the adjacent track, you may not be able to tell, momentarily, whether your train is at rest, moving backwards, or moving forward slowly. Only after the other train has passed and you see the platform of the station, can you realize that your train is at rest. But if two ships should pass in mid-ocean, an observer on one of them could not decide whether one or both ships are in motion; both might be moving in the same direction, one more rapidly than the other; they might be moving in opposite directions; one might be at rest and the other in motion. The same thought can be extended to celestial bodies. Since there are no landmarks or signposts in the far regions of space, there must be inevitable difficulty in deciding which bodies are at rest and which in motion. An astronomer may talk about the velocity of the sun relative to the "fixed stars," but are there any stars that are really stationary, or is there anything in the universe we can be certain is fixed?

It would appear that the answer is negative and that all motion is relative. Velocity could only be absolute if there should exist one frame of reference for which the laws of nature would appear simpler to observers stationary in it than for those using any other background. Then this would constitute a *unique* or fixed frame of reference that could be singled out by any observer in the universe, and motion referred to it could be considered absolute. Newton and his successors knew that the laws of mechanics could furnish no such frame, since the laws appear the same, for example, to observers on earth and to those moving at uniform speeds with respect to the earth. Scientists thought it might be otherwise with electromagnetic phenomena, in particular those involving light.

The "wave motion" of light was assumed to take place in a medium called the *ether*, which was pictured as filling all of space. The ether theory was gratifying because the ether could be thought of as carrying electromagnetic waves and because it seemed to offer some fixed frame of reference with respect to which there might be absolute motion. In 1887, however, the American physicists Michelson and Morley performed a crucial experiment whose outcome refuted this notion.

To understand the basis of the classic experiment they performed, consider the mundane matter of rowing a boat on a stream that flows at the rate of 3 ft. per sec. between parallel banks 120 ft. apart. Two oarsmen, who both row at the rate of 5 ft. per sec. in still water, set out from a point on one of the banks. The first man rows 120 ft. downstream and back. The current helps him on his way down and hinders him on his return. His rates of rowing are 8 ft. per sec. and 2 ft. per sec., respectively, in going and returning, and the total time is

$$\frac{120}{8} + \frac{120}{2} = 75 \text{ sec.}$$

The second oarsman sets out simultaneously with the first, proceeding straight across to a point on the opposite bank, and then rows back. To row straight, he heads diagonally backward (Figure 18.1) in order to allow for the effect of the

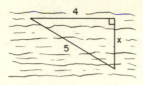

Figure 18.1

current. While he pushes the boat 5 ft., the current carries it 3 ft. (as in Figure 18.1), and his advance across the stream can be found by using the Pythagorean theorem:

$$x^2 + 3^2 = 5^2$$
$$x = 4$$

His rate of progress across the stream will be 4 ft. per sec. and the time required will be $120/4 = 30$ sec. The story will be the same for the return trip; therefore his total time will be 60 sec. and he will return ahead of the other oarsman.

In classical physics it was assumed that as the earth moves there is an "ether current" somewhat like the breeze when a man is driving a car. To measure the speed of the ether wind, Michelson and Morley devised an experiment in which the "oarsmen" were to be two light rays, one of which was to go down the ether stream and back, while the other was to travel the same distance across it and back. The rays were sent on the journey back by reflection from suitably placed mirrors. Instead of resulting in the earlier return of the transverse light ray, both rays arrived back *at the same time!* The experiment was repeated again and again to take account of any possible errors. Refinements were added so that an ether current as small as 1/5 kilometer per second would have been detected. The results were still negative! To all appearances, the earth stood permanently at rest in the ether, which, of course, contradicted the known fact that it speeds around the sun at nearly 20 miles a second.

Fitzgerald and Lorentz (Chapter 17) arrived independently at the same explanation of this phenomenon. They suggested that distances must *contract* in the direction of motion by an amount just sufficient to compensate for the handicap of the longitudinal light ray. If the rival rays have a normal speed of 5 units per second (each unit equal to about 37,200 miles) in some stream with a speed of 3 units per second, a contraction of longitudinal lengths to 4/5 of their original size would explain the simultaneity of return. The transverse ray would require 60 seconds for a 120-unit crossing and return. If the 120 units downstream contracted to $4/5 \times 120 = 96$, then the total trip down and back would take $96/8 + 96/2 = 60$ seconds, which checks with the outcome of the Michelson-Morley experiment.

Nineteenth-century physics indicated that the difficulties revealed by the Michelson-Morley experiment were not isolated failures of classic physical theories but actually pervaded electricity and optics. Although Lorentz and Fitzgerald pointed out a partial solution for the apparent contradictions, it was left to Einstein to proclaim in his revolutionary paper of 1905 that paradoxes could best be resolved by also scrapping the traditional concepts of space and time. The axioms of the special, or *restricted theory of relativity,* as enunciated by Einstein are:

I. *The principle of relativity*: The laws of physics are the same whether stated in one frame of reference or any other moving with uniform rectilinear velocity relative to the first.

II. The *velocity of light* is *constant,* that is, the same when measured by all observers in the universe, and is independent of its source.

The reason the theory is *special* is that the velocities considered are not general in nature since the relative speeds are assumed to be uniform or *constant* and motions must take place in a straight line.

The first of the above assumptions is equivalent to the impossibility of defining absolute motion, for to do so one would have to single out a special frame of reference and give evidence of laws holding only in this frame and no other; this would of course be contrary to postulate I, which calls for every natural law to hold in all systems in uniform motion relative to one another. As for the second axiom, if one considers it a *physical law* that light should have a certain velocity, its constancy would follow from postulate I and no second assumption would be needed. That the speed of light is independent of the source is confirmed by some empirical evidence. To understand this experimental basis, we recall that classic theory considered light a wave phenomenon in the ether. *Sound* is just such a phenomenon, since it is transmitted by waves propagated in air or some other material medium, and numerous examples can be adduced to show that its speed is neither increased nor decreased by motion of the source. In supersonic flight, planes arrive before they are heard. The velocity of the plane does not increase the speed of the sound it emits. This is in contrast to what happens to a bomb or some other material object dropped from a plane. Such an object acquires the same forward speed as the plane and, if there is little air resistance, hits the ground just below it. If the object is thrown forward, its speed of propulsion is added to that of the plane and it hits the ground in front of the plane.

Although the velocity of sound is uninfluenced, the plane's motion has some effect on sound issuing from it. It seems not to hurry the "spheres of disturbance" caused by the vibration but to push them closer together, that is, shorten the wavelength or *raise the pitch* of the sound advancing in the direction of the plane. For sound in the opposite direction, the spheres are pulled away so that pitch is lowered. A trained ear can actually discern whether a train is approaching or receding by the sound of its whistle. Light waves do not move in a material medium, but Einstein assumed the same independence of the source that exists for sound waves. That the "pitch" of light waves is affected by motion is backed up by the experimental *Doppler effect,* the shift in the spectrum of rapidly moving stars or nebulae. The same sort of change in spectrum can be observed for all types of electromagnetic waves, and this phenomenon is being used today for the *Doppler tracking* of man-made satellites and, in a sort of reverse procedure, for the new navigation by satellite. There is some empirical evidence, too, of the lack of effect of the source on the speed of light. *Double stars* are stars of about the same mass, close together and revolving about one another. If the speed of the source helped or hindered, then the time taken for light to reach the earth from the member of the pair moving toward us would be shorter than that for the receding partner. Actual observations show that there is no such effect.

By logical deduction from his two postulates Einstein arrived at the Lorentz transformations of Chapter 17. These are the formulas

$$x' = \frac{x - ut}{\sqrt{1 - u^2}} \qquad t' = \frac{t - ux}{\sqrt{1 - u^2}}$$

where the meaning of the letters is the same as in the previous explanation. The distance x and the time t are measurements made by an observer within a linear "cosmos." The consequences of the special theory of relativity indicate that distance and time are relative, and hence measurements of them by different observers need not agree. Now suppose that a second observer resides in a straight-line cosmos which is moving parallel to the first linear universe, with constant speed u. Then the distance and time measurements made by this second observer are x' and t', which are related to x and t by the Lorentz formulas. To simplify the form of these basic transformations, the *unit* of velocity has been taken to be the speed of light. In other words, a value $u = 1/10$ would signify a velocity one-tenth the speed of light, or in everyday units, a speed of about 18,600 miles per second.

The denominator, $\sqrt{1 - u^2}$, in the Lorentz formulas, indicates that for physical *reality*, u must be numerically less than 1. If $u = 1$, the denominators are zero and the formulas become meaningless. If u is numerically greater than 1, the denominators are imaginary. The exclusion of values for u numerically equal to or greater than 1 signifies that *no material body can have a velocity equal to or greater than the speed of light*. This is an important theorem of the special theory of relativity.

Consider a case where the speed of another observer, relative to us, has the permissible value $u = 3/5$. Consider the following two events:

$$\text{event I} \quad x = 0 \qquad t = 0$$
$$\text{event II} \quad x = 4 \qquad t = 2$$

This means that the first event occurred at our origin at "zero hour" and the second occurred 2 seconds later 4 units away ($186{,}000 \times 4 = 744{,}000$ miles away). Then, using the Lorentz transformation, we can find out what the other observer has to say. For event I,

$$x' = \frac{0 - \frac{3}{5} \times 0}{\sqrt{1 - \left(\frac{3}{5}\right)^2}} = 0 \qquad t' = \frac{0 - \frac{3}{5} \times 0}{\sqrt{1 - \left(\frac{3}{5}\right)^2}} = 0$$

We agree on this event, but for event II,

$$x' = \frac{4 - \frac{3}{5} \times 2}{\sqrt{1 - (\frac{3}{5})^2}} = 3\frac{1}{2} \qquad t' = \frac{2 - \frac{3}{5} \times 4}{\sqrt{1 - (\frac{3}{5})^2}} = -\frac{1}{2}$$

According to the other observer's space standards, the second event is not as far away. Even more startling is the estimate $t' = -\frac{1}{2}$, which signifies that in his opinion not only is the time interval between the two events much shorter, but event II occurred before event I, as indicated by the minus sign, since a positive t signifies time *after* the zero hour, and a negative t, time *before*. Thus we see that two observers in the universe may be in complete disagreement on *space* and *time* measurements as well as on the order of events.

When Einstein first pronounced these facts, a Vienna daily carried a headline "Minute in Danger!" To combat such a fear, we must point out that $u = \frac{3}{5}$ is not an ordinary speed like that encountered in everyday motions like walking, driving, flying, etc. because it represents $\frac{3}{5} \times 186,000 = 111,600$ mi. per sec. Even the earth as it speeds through the heavens covers a mere 20 mi. per sec. relative to the sun! And so the startling relativity results apply not to observers in trains, planes, even rockets, but to beta particles, for example, which are electrons shot out of the nuclei of radioactive atoms at speeds almost as huge as that of light. The point of view of an individual imagined to inhabit a beta particle would differ so radically from that of the scientist observing him that social life between them could be nothing but a series of violent disagreements!

To see if these observers can agree on anything or, in other words, to try to find some absolute or *invariant* entities, let us consider a set of observers moving with the respective speeds

$$u = \frac{4}{5}, \quad -\frac{4}{5}, \quad \frac{3}{5}, \quad \frac{12}{13}, \quad -\frac{7}{25}$$

relative to us, positive and negative velocities signifying motion to our right and left, respectively. Let our observations give

$$\begin{array}{lll}
\text{event I} & x = 0 & t = 0 \\
\text{event II} & x = 8 & t = 10
\end{array}$$

As before, all observers will agree on event I, Substitution in the Lorentz transformation will yield the following different estimates of the where and when of event II:

speed	u	0	$\frac{4}{5}$	$-\frac{4}{5}$	$\frac{3}{5}$	$\frac{12}{13}$	$-\frac{7}{25}$
distance	x	8	0	$\frac{80}{3}$	$\frac{5}{2}$	$-\frac{16}{5}$	$\frac{135}{12}$
time	t	10	6	$\frac{82}{3}$	$\frac{13}{2}$	$\frac{34}{5}$	$\frac{153}{12}$

No invariant is obvious from the table, but if we compute t^2, x^2, and $t^2 - x^2$ in each case, we obtain

$$t^2 \qquad 100 \quad 36 \quad \frac{6724}{9} \quad \frac{169}{4} \quad \frac{1156}{25} \quad \frac{23409}{144}$$

$$x^2 \qquad 64 \quad 0 \quad \frac{6400}{9} \quad \frac{25}{4} \quad \frac{256}{25} \quad \frac{18225}{144}$$

$$t^2 - x^2 \qquad 36 \quad 36 \quad 36 \quad 36 \quad 36 \quad 36$$

Thus, although time and space measurements vary from observer to observer, all will agree that

$$t^2 - x^2 = 36$$

$$\sqrt{t^2 - x^2} = 6$$

It can be shown that this is a general truth, that no matter what the speed u of an observer is, and no matter what event (x,t) we measure, his measurement (x',t') will be such that

$$(t')^2 - (x')^2 = t^2 - x^2$$

This can be proved by performing a little algebraic manipulation on the Lorentz transformation.

This means that although time and space are relative, a certain *mixture* of them is absolute. It is an error to think that because Einstein's theory is called Relativity, all measurements are relative. The mathematician Whitehead chose the name "separation" for the absolute invariant $\sqrt{t^2 - x^2}$. We can see why and can gain more understanding of this new space-time mixture if we chart events on a Cartesian graph. Figure 18.2 shows that when the events $(0, 0)$ and $(8, 10)$ are points

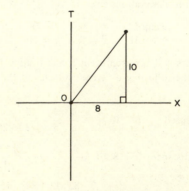

Figure 18.2

on such a graph, the actual *distance* between them as given by the Pythagorean theorem is

$$\sqrt{10^2 + 8^2} = \sqrt{164}$$

The separation is obtained by changing the sign in the Pythagorean theorem

$$\sqrt{10^2 - 8^2} = 6$$

and can be thought of as a sort of "pseudodistance."

Plotting events on a Cartesian graph and recognizing separation as algebraically similar to distance give one justification for the name of this invariant. The picture introduces another new element, for we see that to graph events occurring at points of a one-dimensional world gives rise to a two-dimensional picture. If we extend the world of events to all those taking place in a single *plane*, the Cartesian picture will be three-dimensional. If we picture the floor of a room as part of the plane of events (Figure 18.3), then the values of x and y will be the distances from the side (left) and back walls, respectively. The graph of an event (x, y, t) will be a point in the room, the value of the time t being represented by the distance above the floor. The graph of an event will be somewhere in the room on a vertical line through the point in the floor where it occurs.

If the two events are $(0, 0, 0)$—the corner at "zero hour"—and (x, y, t)—anywhere, anytime—then

$$\text{Pythagorean distance} = \sqrt{t^2 + x^2 + y^2}$$

$$\text{separation} = \sqrt{t^2 - x^2 - y^2}$$

Again, algebra will prove that Pythagorean distance varies with the observer, but that separation is invariant.

To picture events occurring at points of a three-dimensional space, a four-

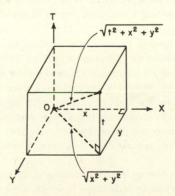

Figure 18.3

dimensional scheme would be necessary. We are not compelled to resort to visualization in this case, but can nevertheless use the associated invariant,

$$\text{separation} = \sqrt{t^2 - x^2 - y^2 - z^2}$$

numerically and algebraically, thus accepting the idea advanced in 1908 by the mathematician Minkowski. He was the first to state that it is logically simpler to regard the universe as a four-dimensional *space-time continuum* rather than to divide our experiences artificially according to unrelated spatial and temporal aspects, especially since different observers will hold divergent opinions on how the division should be effected.

To return to our one-dimensional world and two-dimensional space-time picture, let us compute the separation of $(0, 0)$ and $(4, 2)$. It is equal to

$$\sqrt{4 - 16} = \sqrt{-12}$$

This is an *imaginary* separation! Whenever x is greater in numerical value than t, the separation will be imaginary. Since the unit is equal to the speed of light, $x = 4$ is the distance light will travel in 4 seconds. Then, with the pair of events $(0, 0)$ and $(4, 2)$, if a light signal were sent from the place where the first event occurs (the origin), describing that event, it could not arrive at the place where the second event occurs *prior to* this second occurrence. Whatever happens at the first locality cannot be a "warning" to the second, influence it, or exert a "force" on it in any way. In other words, there can be no *causal* connection between the two events.

A little earlier we saw that a different observer would measure the event $(4, 2)$ as $(3\frac{1}{2}, -\frac{1}{2})$. Still he would obtain the separation as

$$\sqrt{\frac{1}{4} - \frac{49}{4}} = \sqrt{-12}$$

and so would all observers in uniform relative motion. Now we see why the fact that the second observer does not agree on the order of the two events does not matter in this case. The points where the events occurred will, in any frame of reference, be too far apart for causal connection or exertion of any "force," since the distance will have to be numerically greater than the time to ensure a negative value of $t^2 - x^2$ and an imaginary separation. This is just another way of saying that *no force can be propagated with a speed greater than that of light*.

In general, then, an imaginary separation signifies no causal connection. By contrast, a real separation would permit signaling. It can be shown algebraically, by using the Lorentz transformation, that in this particular case, although observers may disagree on the length of time intervals, they will never disagree on the *order* of events. When separation is real, then, a causal connection between two events is possible.

We have been considering events occurring at *different* places. If we were following the life history of an individual, however, different events might occur at the *same* place as far as his measurements are concerned. (If the individual were a human being with reference frame on earth, his personal motion in space would be so tiny in comparison with our unit of 186,000 miles that for practical purposes, he could consider himself permanently at rest.) Then for him event $I = (0, 0)$ and event $II = (0, t)$, and

$$\text{separation} = \sqrt{t^2 - 0^2} = t$$

The separation of the events would be identical with his measure of the time between them. Hence, for self-observation, *separation is a synonym for time*, and the choice of name for this invariant has additional justification. We remark, in passing, that separation could never be imaginary for events happening to the same individual, since $x = 0$ in such a case, and hence only t^2, which has to be positive, contributes to the value under the square root sign.

For an observer whose world is moving past with speed u, if $x = 0$,

$$t' = \frac{t - u \cdot 0}{\sqrt{1 - u^2}} = \frac{t}{\sqrt{1 - u^2}}$$

or the first observer's time t is given by

$$t = t' \sqrt{1 - u^2}$$

and hence is *less* than the time as recorded by any other observer because the factor $\sqrt{1 - u^2}$ is less than 1 in value.

The measurement by an individual of the time interval between two events occurring at the same place is called the *proper time* for the individual and, as just indicated, is less than the estimate made by any other observer. It is the time as measured by his own clock and, as it gives much smaller estimates of intervals than the clocks of observers moving at tremendous speeds relative to him, the latter observers will say his clock "runs slow."

In the early days of relativity, spinners of popular science yarns had as a favorite plot the tale of two friends parted in youth, one of the pair traveling into space at a velocity close to that of light, while the other remained quietly at home on earth. When the wanderer eventually returned from his lengthy journey he was still to all appearances a boy in his teens, although by earthly reckoning he had to be counted an octogenarian. His stay-at-home friend looked old enough to be his great-grandfather.

The explanation of this apparent paradox was that the traveler took his clock with him. It was at rest relative to his surroundings and measured his proper time. If we think of his speed as being 0.99 (that is 99% of the velocity of light), then he will think of himself as at rest and consider the earth to be receding from him with this speed. If the boys part at the age of eight, and 70 years elapse on earth, then the proper time for the wanderer as measured by his own clock will be given by

$$t = t' \sqrt{1 - u^2}$$
$$= 70 \sqrt{1 - (0.99)^2}$$
$$= 70 \times 0.14$$
$$= 9.8 \text{ years}$$

and he will be only eighteen years of age by his own standards, whereas his friend will be seventy-eight.

This is naturally a fantasy, but has some scientific justification in the fact that the human heart vibrates and so, like a pendulum, is a clock. Hence its beating should also be affected by rapid motion if we are to be consistent about the relativity of time. If an individual is in rapid motion relative to the earth, then by his own standards his heart may beat 72 times per minute. Since the results of *counting* are absolute, we will count 72 heart vibrations, but in a longer interval as measured by our own clocks, for the traveler was considering one minute of his proper time. In other words, we consider all his clocks to run slow, his heart being one of them. Since the same retardation affects all the metabolic processes in the body, it can be said that the wanderer "ages" less than the person remaining at home.

To consider other consequences of the special relativity theory, let us return for a moment to Minkowski's space-time picture, limiting ourselves once more to events on a one-dimensional straight line as graphed in a two-dimensional chart. Let the creatures existing in the one-dimensional space be insects crawling along OX (they cannot jump or fly). The events occurring in the life of any insect will form a continuous curve, the graph of each event lying on a line perpendicular to OX at the point x, where the event occurs. In Figure 18.4, P is the graph of the event at p, Q the representation of that at q, etc. Different insects will have different curves of events, or *world-lines*, as Minkowski called them. If this picture is generalized, individuals existing in 3-dimensional space will have world-lines in a 4-dimensional space-time continuum, which is a convenient mathematical abstraction, not something to be visualized.

In the 2-dimensional space-time continuum, an insect that remains permanently at rest at the origin will have the T-axis (OT) as its world-line, and if it is at rest at any other point, the world-line will be parallel to the T-axis. There are no world-lines parallel to the X-axis because this would signify that time stands still while the insect moves! Such lines are not world-lines, but can be considered "cross-sections" of the insect's space-time. They give all possible insect locations for a specific time, since t is constant along these lines, while x varies.

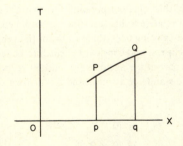

Figure 18.4

If the insect starts at the origin and moves to the right with a constant speed of $\frac{1}{2}$ unit per second, its progress chart will be

$$t \quad 0 \quad 1 \quad 2 \quad 4 \quad \text{etc.}$$
$$x \quad 0 \quad \tfrac{1}{2} \quad 1 \quad 2$$

This is no ordinary insect, to be sure, but some nuclear particle moving at $\frac{1}{2}$ unit = 93,000 miles per second. Then the formula connecting distance and time for this "insect" is

$$x = \tfrac{1}{2}t$$

This is the equation of its world-line, which is straight. If the insect moves with variable speed, its world-line is broken or curved. If its speed is any constant k, the equation of its world-line is

$$x = kt$$

Since material things can never achieve the speed of light, k must be numerically less than 1. The equations

$$x = t \quad \text{and} \quad x = -t$$

represent the world-lines of light rays moving right or left from the origin. All other world-lines, straight or curved, are thus restricted to the unshaded sector of the space-time continuum pictured in Figure 18.5. Only a portion of space-time is

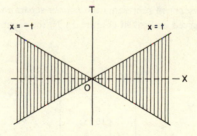

Figure 18.5

allotted to material objects. If we use a more *human* scale in our graph, this restriction will appear slight. If the unit on the X-axis is taken as 1 mile instead of 186,000, the shaded or "forbidden" area will look like that in Figure 18.6, except that it will be a much thinner sector, and therefore exclude only a small portion of the space-time continuum.

If we study a particular world-line like

$$x = \frac{5}{13}t$$

the chart for insect progress is

$$t \quad 0 \quad 13 \quad 26 \quad 39 \quad 52 \ldots 104$$
$$x \quad 0 \quad 5 \quad 10 \quad 15 \quad 20 \ldots 40$$

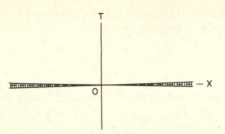

Figure 18.6

For the first pair of events, *separation* is

$$\sqrt{13^2 - 5^2} = 12$$

and since the intervals are equal, it will be 12 for every consecutive pair. Computing the separation of the first and last events

$$\sqrt{104^2 - 40^2} = 96$$

which is what common sense leads us to expect, since there are 8 intervals of 12 each.

In Euclidean geometry a straight line segment is the *shortest* distance between two points. In contrast to this it can be proved that separation obeys an opposite law. A straight line segment represents the *longest* separation between two points. If a world-line is curved or broken (Figure 18.7), the separation of the last event

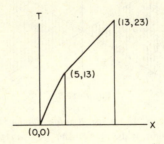

Figure 18.7

from the first is *greater* than the sum of the separations between successive events measured along the world-line. We shall not give a general proof but, for the sake of simplicity, we shall limit ourselves to an illustration. Take the events (0,0), (5, 13), and (13, 23). The separations are

$$
\begin{array}{lll}
\text{I to II} & \sqrt{13^2 - 5^2} = 12 \\
\text{II to III} & \sqrt{10^2 - 8^2} = 6 \\
\text{I to III} & \sqrt{23^2 - 13^2} = \sqrt{360} = 18.96
\end{array}
$$

The separation of I and III is almost 19, whereas the sum of the separations is only 18. To realize the significance of this fact, notice that the world-line from I to II represents a motion covering 5 units of distance in 13 seconds of time, that is, a motion with speed 5/13. Similarly, the world-line from II to III pictures a motion with speed 4/5. Then the broken world-line from I to II to III shows the progress of an insect moving *nonuniformly*. It would start with one velocity and later change abruptly to a more rapid one. By contrast, the straight world-line from I to III represents a *uniform* motion with speed 13/23.

So it is that insects whose world-lines are straight, signifying that they are either *at rest* or in *uniform* motion, will have greater separations than if these lines were broken or curved. For an object at rest, separation has been shown to signify personal or proper time. If a body is in uniform motion relative to an observer, relativity permits an individual situated on this body to judge himself at rest and the observer in motion. Therefore, once again separation becomes that individual's personal time. In the light of this equivalence of separation and personal time, the fact that separation is greatest along a straight line can be stated thus: *Proper time between two events will be a maximum for bodies at rest or in uniform motion.* This rule of behavior has sometimes been called the "law of cosmic laziness," and in relativity theory takes the place of the "path of least resistance."

One consequence of the Einstein theory is that, in addition to the fact that distance and time measurements depend on the observer making them, there is relativity of *mass* as well. Newton described the mass of a body as the quantity of matter it contains and defined it by means of his second law of motion in terms of "force" and change in *velocity*. Since velocity is no longer absolute in relativity, and "gravitational force" is ruled out by the general relativity theory, which we shall shortly discuss, we should expect a modification of the concept of mass.

Long before Einstein, physicists J. J. Thomson and W. Kaufmann had found that the movement of electrons with a very high velocity u causes an apparent increase of mass equal to approximately $\frac{1}{2}mu^2$. The mathematics of relativity agrees with this, for it indicates that if m is the mass of a body at rest in a particular frame of reference, which has speed u relative to an observer, this observer will obtain

$$\frac{m}{\sqrt{1-u^2}}$$

as his estimate of the mass. Since $u = 1$ for the speed of light, u is a fairly small fraction even for high electron speeds. For such small values of u, algebra indicates that

$$\frac{m}{\sqrt{1-u^2}} = m + \frac{1}{2}\,mu^2 \text{ (approximately)}$$

This means that the original mass m is increased by $1/2\,mu^2$, the same fact obtained in the experiments of Thomson and Kaufmann.

For example, if $m = 1$ and $u = 7/25$,

$$\frac{m}{\sqrt{1 - u^2}} = \frac{25}{24} = 1.04$$

$$m + \frac{1}{2} mu^2 = 1 + \frac{49}{1250} = 1.04 \text{ (a 4\% increase in mass)}$$

The expression $1/2\,mu^2$ is called the *kinetic energy* of a body, that is, the amount of work it is capable of doing by virtue of its motion. The term *potential energy* is used for the work it can do by virtue of its position, for example, the work possibility inherent in a compressed spring or a rock at the edge of a cliff. In the expression $m + 1/2\,mu^2$, if the second term is *energy* and the terms are to be combined, the first term must also represent some sort of energy. Since m is the *rest* mass, it is like a rock or spring at rest, and hence m can be thought of as potential energy. Then, in the expression above, the observer's measure of the mass of a body is partly potential and partly kinetic energy, but it is energy of one kind or another. This is the principle of the *equivalence of mass and energy* used in nuclear theory.

The term $1/2\,mu^2$ is a good approximation if u is small, but even if u is as small as $1/1000$, for example, its value in ordinary units like centimeters per second would be huge—3×10^7—and when squared, tremendously greater—9×10^{14}. If this figure is multiplied by $1/2\,m$ even for m as tiny as electron mass, the value of $1/2\,mu^2$, increase in mass, will be enormous in terms of ordinary energy units.

Although the Einstein formula

$$E = mc^2$$

is used in a somewhat different connotation, where matter is "annihilated" and converted into energy, there is a kinship in form to

$$\tfrac{1}{2}mu^2$$

in that we square a speed to obtain the energy associated with a mass, the speed c signifying that of light, which is 1 unit in relativity reasoning but 3×10^{10} measured in centimeters per second. Hence $c^2 = 9 \times 10^{20}$, so that even if m is the tiny mass of a nuclear particle, the product of its size by 9×10^{20} will be enormous, a fact basic in nuclear bombs as well as in more peaceful uses of nuclear energy.

The general theory of relativity, launched by Einstein in 1917, takes into account frames of reference with *variable* velocity relative to one another. The mathematics involved is not, as originally claimed, so difficult that it is comprehensible to only a dozen mortals, and in the next chapter, we shall give details of its fundamental concepts.

19

Geometry for Universe-Builders

The purpose of the general theory of relativity, like that of the special theory, is to distinguish the subjective impressions of individual observers from the reality of the thing being observed. By the addition of only one new postulate called the principle of equivalence, Einstein developed a physical concept more adequate than Newton's law of gravitation and one which is able to explain experimental facts that had defied classic theory and puzzled physicists and astronomers for a long time.

To see why "gravitational force" is not absolute but a matter of relative opinion depending on who is making the observation, consider the example of Newton and the famous falling apple. Suppose that you, instead of Newton, are an observer and that you are located in an elevator with opaque, shut doors. Unknown to you, the supports break and the controls fail. Emulating Newton, you let go of an apple. It appears to remain at rest, poised in mid-air, and hence *you* observe *no force* on it. But we say that this happens because you, the elevator, and the apple are all falling in the same way without your knowledge of it.

In your frame of reference, then, things do not fall, and no special force or influence makes its presence known. In relation to your frame of reference, there is no gravitation. Just as $+32$ will balance -32 in addition, giving zero as sum, so gravitational force, as we conceive it, is apparently neutralized by the elevator's motion, and hence must be numerically equivalent to it, but opposite in effect.

This is the basis of Einstein's *principle of equivalence*, which states: If attention is confined to a small region of space, *a gravitational field at rest is equivalent to a frame of reference moving with uniform acceleration in a field free of gravitation*, and it is impossible to devise any experiment that will distinguish between the two.

This means that any observer, like the man in the elevator, can select a frame of reference moving in such a way that in his *immediate neighborhood* all gravitational effects are neutralized, and consequently, in this small vicinity, the *restricted theory of relativity holds true*. By a small neighborhood we mean a tiny *space-time* region surrounding a point or event in his world-line, so that to apply the special theory in the presence of matter and its accompanying "gravitational field," the observer can pick a neutralizing frame of reference, but that will serve for a short time only. He must change that frame as time passes, for he will be in a new space-time region even if he considers himself at rest. For example, in the 2-dimensional space-time previously pictured, an observer at rest had a world-line parallel to the T-axis. As time passes, the events would be pictured on different portions of this line and would occupy different neighborhoods of the 2-dimensional space-time world.

In the observer's opinion, if his world-line is straight, signifying rest or uniform motion, his separation will be a maximum, as previously explained. Since separation is invariant, other observers will also agree that it is a maximum, but may not consider the world-line straight because they may not be in the same space-time neighborhood. Thus observers who exist far apart in 3-dimensional space or are widely separated in time—in other words, those who are remote when the customary mixture of space-time measures their separation—will agree on maximum properties but not on "straightness" versus *curvature*.

As a result, one fact of the restricted theory must be modified, namely, the "law of cosmic laziness." The route for maximum separation need not be straight. In the general theory, one says that the presence of matter *distorts* space-time in its neighborhood, and for that reason the world-line of maximum separation (the customary mathematical name for such a line is *geodesic*) is distorted into a *curve*. Let us picture a crumpling of the flat 2-dimensional space-time in which the world-lines of Chapter 18 were drawn. Then the straight lines in the diagrams would be wrinkled, possibly distorted into curves.

According to Newton, a planet describes an elliptic path around the sun because there is an attractive "gravitational force" between sun and planet, but according to the general theory of relativity, the matter in the sun causes distortion of the space-time in its neighborhood, and the planet's world-line of maximum separation has to be curved in order to conform to the curved continuum on which it is drawn. Thus the notion of gravitational force is relinquished. Curved paths or other effects (attributed to gravitation in Newtonian theory) occur only because following particular routes leads to maximum separation in space-time or, stated otherwise, because bodies follow the "law of cosmic laziness." Our education has conditioned us to believe to such a degree in a force tugging at things and making them fall that it is hard to get away from the idea. Surely it is easier to believe that our earth or the other planets are just lazy and fall "into the groove," than to believe that the sun exercises some remote control, holding invisible reins 93,000,000 miles or more in length.

Perhaps we can help ourselves out by an analogy. Before 550 B.C. the belief that our earth was flat was almost universal. Even later, the idea of sphericity was slow in gaining acceptance. Early maps made in accordance with the theory of a flat earth served very well until the new countries discovered and added to the flat map began to introduce inconsistencies. A *Mercator's projection* is a flat map whose distortions suggest what can happen. If you examine such a map, you will notice the apparent hugeness of Greenland in comparison to its size as mapped on a globe. The earliest cartographers did not use a sphere to map the earth, and hence in the days before Ptolemy, who understood the problem, people would have supposed that the flat maps gave the true size of faraway countries like Greenland. Inevitably, explorers of such a region would report that journeys there seemed much shorter than the maps indicated. Believing the map to be a correct representation of geographic facts, they might invent a theory that there was a *force* in this part of the world causing distances to shrink whenever explorers appeared on the scene. Our modern knowledge of the spherical shape of the earth would eliminate this theory, show that no force of the sort exists, and indicate that the effect is caused by the curvature of the earth. This is a 2-dimensional analogy indicating how a phenomenon produced by curvature might appear, at first conjecture, to be caused by some "force."

Space-time is curved here and there, wherever matter is present. The 2-dimensional analogy would be the surface of the skin with blisters on certain spots, or a golf course with hummocks drawing a ball off its straight course. The combined distortion of space-time by all the matter in the universe causes it to bend back on itself and close. As a crude 2-dimensional analogy, if we take a flat sector of a circle (Figure 19.1) and curve it sufficiently, we can close it so as to get a conical

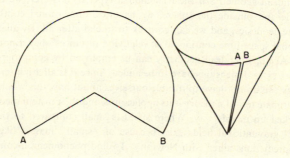

Figure 19.1

drinking cup. A cross-section of this cup would be circular or oval, a geometric curve that might be described as *finite, but unbounded*, since we can go round and round it in spite of its limited length. This is a one-dimensional cross-section of a closed 2-dimensional continuum. In the Einstein theory, *space* is considered a *finite but unbounded* 3-dimensional cross-section of a closed 4-dimensional space-time continuum. (It is not closed in the *time* direction, much as the conical drinking cup is not, for some of its cross-sections would be two straight lines in V-formation meeting at one end and open at the other.)

Still another item of the general relativity theory becomes clearer through a 2-dimensional analogy. If we consider geometry on the surface of a sphere, straight lines would be replaced by great-circle arcs, for these would be shortest distances, or geodesics (term used for minima *or* maxima, that is, for extrema). A triangle would be bounded by three great-circle arcs, and the sum of its angles would be greater than 180°. We can visualize, for example (Figure 19.2), the spherical triangle formed by the half of the prime meridian extending from North Pole to Equator,

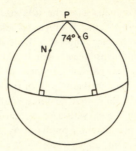

Figure 19.2

the same part of the meridian through New York City, and the slice of the Equator between the feet of these meridians. The angles at the feet would each be 90°, and the angle at the Pole would equal 74°, the longitude of New York, so that the total would be 254°. On the sphere we have a 2-dimensional continuum with a non-Euclidean geometry of the Riemannian type. If, however, we think of the space inside and outside the sphere, we have a 3-dimensional continuum with straight-line geodesics, and we can go back to Euclid after all. By analogy, in the 4-dimensional space-time continuum of relativity, distances obey a non-Euclidean geometry. Again, Euclidean geometry can be employed by going outside, but in this case, six extra dimensions have to be added. Since it is all a matter of abstract terminology, we do not have to trouble ourselves about how this would look.

The ultimate test of a theory is its application, and the Einstein theory has been amply backed up in this way. Where speed is small compared to that of light, or "weak" gravitational fields exist because of "small" masses like the earth, Einstein's predictions agree with Newton's. To find phenomena inconsistent with Newtonian mechanics, Einstein had to go beyond mundane space-time neighborhoods. His first case was that of the planet Mercury, close to the sun, a relatively large mass with a considerable gravitational field to which Mercury is strongly exposed. This planet does not move exactly as predicted by Newton's theory, for then it would describe virtually the same elliptical orbit over and over again in relation to the fixed stars. Newton allowed a slight "perturbation" or shift in the orbit because of the influence of other planets, but the *observed* amount of shift in the orbit was greater than the classic theory permitted, and the discrepancy had never been satisfactorily explained. The calculations of the orbit based on the Einstein theory agree more closely with observation.

The second test of the Einstein theory was more impressive because it was a matter of *complete prediction* from the theory, of facts never before observed. It concerned the effect of the curvature of space on the path of light rays. If the corpuscular theory of light is accepted, then as light from a distant star passes near the sun, it must, according to Newton, be bent toward the sun, because that body exerts gravitational attraction on the corpuscles. The amount of this deflection as computed by Einstein's general theory was double that figured from Newtonian theory. This cannot be checked ordinarily, since the sun's brillliance makes it impossible to see the star rays passing near, but total eclipses furnish an opportunity. British expeditions to observe such eclipses in Brazil and West Africa in 1919, as well as observations of later eclipses, confirmed Einstein's prediction beyond a doubt.

The third of the tests was connected with a fact mentioned in the previous chapter, namely, that the vibration of an atom is like the oscillation of a clock pendulum. The mathematics of the general theory indicates that the atom clock is slowed up near the sun, in comparison to a clock of the same atom on earth. The effect should be a shift in its spectrum toward the slow or red end. For the sun, this shift is too slight to lend itself readily to experimental verification, but there is a companion star of Sirius, known as a "white dwarf," for which it is believed that the density is as much as a ton per cubic inch. Consequently it sets up a gravitational field of so great an intensity in its neighborhood as to make it ideal for testing the hypothesis of spectral shift. Observations made by the astronomer Adams at Mt. Wilson agreed with Einstein's prediction. Very fine measurements made by Evershed in 1927 gave evidence of a shift toward the red even in the vicinity of the sun, where the shift is small.

Having sketched, in outline, the major features of the general theory of relativity, we shall now present some of the actual details of the mathematics involved. As in the case of restricted relativity, the subject matter can once more be described as *geometry*, in this case, the kind of geometry developed by Gauss and Riemann. Klein's ideas are still pertinent, because transformations and invariants are still important. However, the sets of transformations involved will not, in general, be groups.

Just as the surveying procedures of antiquity provided the physical basis for the principle of relativity of Euclidean metric geometry, modern geodetic problems stimulated the theoretic geometry that is the first step toward general relativity. Let us then consider what would be done by surveyors required to make measurements and subsequently to map a hilly region overgrown with forests. The transit would be of little use to them because the density of growth would make it difficult to make proper sights. It would become a matter of chaining—stretching the steel tape along the ground, setting up and measuring triangles. One triangle is contiguous with the next one, and a step-by-step advance through the woods is made. A triangle is determined by its three sides; that is, when surveyors have measurements for the lengths of sides, it is merely a matter of trigonometric computation to find the angles and the area.

Euclidean plane geometry would not be valid in an extensive geodetic survey, because the earth is a spheroid and not a flat surface. Even when the area under study is not huge, it may be impossible to consider the whole of it as the physical analogue of some figure in Euclid's plane. But regions that are relatively tiny portions of *any* curved surface may be considered as approximately flat. Applying Euclidean metric geometry to very small successive portions of such a curved surface, and then linking each region to the next, is the surveying tactic corresponding to the abstract mathematical subject called *infinitesimal* or *differential geometry*.

Some minor, very elementary aspects of this branch of mathematics were developed soon after Newton and Leibniz, but the tool that would ultimately survey the hills of the universe was developed in 1828 by the greatest "surveyor" of them all, Carl Friedrich Gauss. Mathematicians agree that Gauss should rank first among creative mathematical thinkers of the nineteenth century, and many scientists believe that Archimedes, Newton, and Gauss are the three all-time greatest. Gauss's creativity extended to many mathematical areas and, in particular, his *Disquisitiones generales circa superficies curvas* (1827) initiated a triumphal era for differential geometry. Concerning theoretical geometry, Gauss wrote to the astronomer Bessel: "All the measurements in the world are not the equivalent of a single theorem that produces a significant advance in our greatest of sciences."

Nevertheless, geodetic counterparts of concepts and propositions are clarifying. Hence one may envision surveyors who cover part of the mountainside with a network of curved lines (Figure 19.3) and mark intersections by stakes or fortuitously situated trees. The curved streets and avenues are numbered $u = 0, 1, 2, 3$, etc. and $v = 0, 1, 2, 3$, etc. Intermediate streets and avenues are thought of as having appropriate fractional or decimal values. Thus a one-to-one correspondence is set up between points of the region and u-v coordinate pairs. This seems very much akin to Descartes' scheme. The difference is that *Gaussian* coordinates give location but tell nothing about distance. Neither the u-curves nor the v-curves are necessarily equidistant and, in general, u-curves do not intersect v-curves at right angles. The values of u and v are just numbers that assign an order to the network curves. But when we state that a point P (Figure 19.3) has Gaussian coordinates (3, 4), we give no information about its distance from the origin or any other point. If these were Cartesian coordinates, we would know at once that the distance from the origin is 5 units.

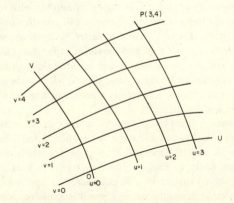

Figure 19.3   Gaussian coordinate system

That Gaussian coordinates serve to *locate* but *not to measure* should not seem strange to city dwellers, for if we state that the Metropolitan Museum of Art is located at 83rd Street and 5th Avenue in New York City or has coordinates (83, 5), we cannot automatically find its distance from (42, 8), since New York City blocks are not all the same length.

Long before Gauss's day, curvilinear coordinates were employed to locate points on the earth's surface. The meridians and the latitude circles form the traditional network; the longitude and latitude of a point are its coordinates, but do not represent distances from prime meridian or equator. Saying that a ship has traveled east through 6° of longitude does not tell us the length of the trip. Near the equator, such a journey would cover 400 miles, whereas at latitude 60°N. it would mean 200 miles. In both cases, the arc along which the voyage takes place is 6/360 = 1/60 of the entire circle of latitude—the equator or the sixtieth parallel, respectively. When a network is made up of straight lines or circles, each line of the network is alike in all its parts, or *homogeneous*, and that is why distances along such a line change in proportion to Gaussian coordinates. In previous chapters, changes in variables were symbolized by Δx, Δy, etc. We now make use of the symbols dx, dy, etc. whenever the changes are very small. Without going too deeply into mathematical meanings, we can say that the symbol d has *approximately* the same meaning as Δ for small changes. If the changes in distance and longitude of a ship traveling due east or west are small, they can be symbolized by ds and du (d for difference), and according to what we have just said $ds/C = du/360$ or $ds = k\, du$, where C is the length of the circumference of the circle of latitude and $k = C/360$.

In the same way, if a city block is short, and houses along it are fairly uniform in size, then (the front door of) number 810 will be approximately at the center of a block whose corner houses are numbered 800 and 820. If the block is 90 yards in length, 810 will be about 45 yards from either corner. In this case,

$$ds = 4.5\, du$$

and for other small blocks of the same sort,

$$ds = k\, du$$

where in each particular block k is a constant depending on the length of the block and the number of houses on it.

In general, the curves of a Gaussian network are not homogeneous. Since we are studying *infinitesimal* geometry, however, each mesh is relatively tiny (compared to the entire surface) and is considered, for purposes of approximation, to be a Euclidean parallelogram. Then distances along two sides of a single mesh— for example, $OABC$ in Figure 19.4—follow the same rule as for short city streets, and

$$ds = k_1 du \qquad \text{(along } OA \text{ or parallel to it)}$$
$$ds = k_2 dv \qquad \text{(along } OC \text{ or parallel to it)}$$

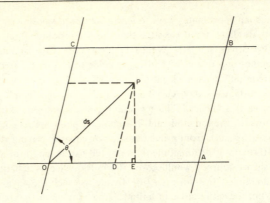

Figure 19.4

The surveyors now measure OA and OC and find their lengths to be 16.2 and 11.4 meters, respectively. Suppose that the rulings of the meshwork are numbered by consecutive integers. Then

$$du = 1 \quad \text{and} \quad dv = 1$$

Therefore, in parallelogram $OABC$,

$$16.2 = k_1\,(1) \quad \text{and} \quad 11.4 = k_2\,(1)$$

Hence

$$ds = 16.2\,du \text{ along } OA \text{ or parallel to it}$$

and

$$ds = 11.4\,dv \text{ along } OC \text{ or parallel to it.}$$

To obtain the distance from O of *any point* P within the small parallelogram, one can apply the Pythagorean formula (Figure 19.4) to right triangles OPE and DPE in order to derive what is called the *law of cosines* in trigonometry. From this law (which, we wish to emphasize, is merely a corollary of the Pythagorean theorem) it follows that

$$(ds)^2 = (OP)^2 = (OD)^2 + (DP)^2 + 2\,(OD)\,(DP)\cos\theta$$

where θ represents the size of angle AOC of the parallelogram. But OD and DP are distances along OA and parallel to OC, respectively. Hence

$$OD = 16.2\,du \text{ (meters)}$$

and

$$DP = 11.4\,dv \text{ (meters)}$$

Our surveyors may be able to measure angle AOC by using a theodolite. If this is difficult, they will measure OB by chaining. Then the three sides of triangle OAB will be known, and it will be possible to compute the size of angle A by a trigonometric formula. Suppose angle A is found to measure $108°$. Then angle

AOC is 72°. Now cosine 72° = 0.309 (approximately) and the above formula for ds^2 becomes

$$ds^2 = 262.44 \, du^2 + 114.13 \, du \, dv + 129.96 \, dv^2$$

This formula applies to every point of the small parallelogram; for example, if Q is the point for which $du = 0.2$, $dv = 0.1$, substitution in the distance formula gives

$$ds^2 = 14.08$$

and

$$ds = 3.75 \text{ meters}$$

The surveyors will now proceed from mesh to mesh, and by measuring two adjacent sides and an angle, or two sides and a diagonal, obtain a distance formula for each infinitesimal parallelogram. If it should happen that in some region the surveyors construct meshes that are *square*, then $k_1 = k_2$ and $\theta = 90°$, so that $\cos \theta = 0$. If $k_1 = k_2 = 1$, then in this region,

$$ds^2 = du^2 + dv^2$$

which is just the ordinary Pythagorean formula. This would occur for *all* meshes only in very special surveys, as we shall see. Again, in general, the proportionality factors k_1 and k_2 will vary in different meshes (as they do on different city blocks). Hence the *coefficients* of du^2, $du \, dv$, dv^2 will differ from mesh to mesh. These coefficients depend on the position of the infinitesimal parallelogram, or, in technical language, they are *functions* of the position, that is, of u and v.

The reader will be correct if he judges field work to be a long drawn-out process. But Gauss's *theory* of differential geometry is pure mathematics. Abstract surfaces are idealizations of the reality, and can be described by simple algebraic formulas. Then calculus, instead of the surveyor's tape, is used to obtain the co-efficients of the local distance formula, which is usually symbolized by

$$ds^2 = g_{11} du^2 + 2g_{12} du \, dv + g_{22} dv^2$$

where one or more of the coefficients g_{11}, g_{12}, g_{22} may be constants for some surfaces, but generally are expressions involving u and v, for, as we have just remarked, the coefficients usually vary with position on the surface. In other words, in Gaussian differential geometry, a *single* formula for ds^2, called the *metric*, in which g_{11}, g_{12}, and g_{22} are functions of u and v, serves for measurement in *all* small meshes of the surface.

For example, it can be shown that, if u and v represent the longitude and latitude of a point on a spherical surface whose radius is one unit of length, then

$$ds^2 = \cos^2 v \, du^2 + dv^2$$

Here $g_{11} = \cos^2 v$ and hence depends on v, the latitude; $g_{12} = 0$ and $g_{22} = 1$.

Again, on a surface called a *hyperbolic paraboloid* (whose appearance resembles the saddle of Figure 10.1) it is possible to select the curves of a network to be two sets of parabolas such that the metric for the surface (with the mesh system chosen) is

$$ds^2 = (1 + 4u^2)du^2 - 8uv\,du\,dv + (1 + 4v^2)dv^2$$

and

$$g_{11} = 1 + 4u^2$$
$$g_{12} = -4uv$$
$$g_{22} = 1 + 4v^2$$

so that all three coefficients vary with position on the surface, as is generally the case for the metric of a surface.

Since the values of g_{11}, g_{12}, g_{22} at each point of a surface determine the measurements in a small region surrounding the point, we may say that the expressions for the g's represent the entire metric geometry of the surface. As an illustration, consider on the above hyperbolic paraboloid the point for which $u = 1$, $v = 2$. The generalized Pythagorean theorem for a small region of the paraboloid close to this point is found by substituting these values of u and v in the above expressions for g_{11}, g_{12}, g_{22} to obtain 5, -8, 17, and

$$ds^2 = 5du^2 - 16\,du\,dv + 17\,dv^2$$

This specific formula will give any distances required in surveying or mapping the small area in question. If, however, one is interested in measuring distances on the part of the surface near $u = 2$, $v = 3$, then

$$ds^2 = 17\,du^2 - 48\,du\,dv + 37\,dv^2$$

that is, a different distance formula obtains. From the metric, all local distance formulas can be obtained by substituting particular values of u and v. Therefore the metric or general expression for ds^2 contains in a nutshell the entire geometry of a surface.

Since the two systems of curves forming a network are selected in a rather *arbitrary* fashion, it is remarkable that a single formula for the metric ds^2—that is, a set of coefficients g_{11}, g_{12}, g_{22}, resulting from a scheme chosen almost *at random* —should fix the measurement of an *entire* surface and thus determine, as it were, the nature of a 2-dimensional "space." That this is the case, however, has been indicated above and will now be emphasized in several different ways.

Each type of geometry discussed in a previous chapter required a special principle of relativity, with a group of permissible transformations of the network. The geometric properties of figures were the *invariants* for all transformations of the group, that is, for all permissible mesh systems. Except that transformation sets need *not* be groups, a similar relativistic point of view holds good for differential geometry. Different networks of curves may be chosen on a mountainside being surveyed. Thus meridians and latitude circles are not a unique framework for locating points of a sphere. By transforming a meshwork to another permissible frame of reference, one obtains a different set of coefficients, g_{11}', g_{12}', g_{22}', for the metric, but since the same distance is involved, ds^2 must be *invariant*. In differential geometry, there are transformations which express the new coefficients in terms of the original g_{11}, g_{12}, g_{22}. The curvilinear systems permitted by Gauss were of infinite variety and quite arbitrary except for two requirements—first, that they be continuously curved and, second, that they cover the surface singly without gaps. Similarly the transformations to other mesh works $u' = \phi(u, v)$, $v' = \psi(u, v)$ have great freedom, being restricted only by the need to have certain desirable analytic properties. Thus the functions $\phi(u, v)$, $\psi(u, v)$ must have continuous partial derivatives (Chapter 9) of the first and second orders.

While we are discussing Gaussian mesh systems, let us indicate that we may have *straight* lines in the framework even when a surface is curved. For example, the "meridians" on a cylinder or cone are straight. From another point of view, we can employ non-rectangular networks in the *plane*, like the *oblique* and *polar* systems in Figures 7.3 and 7.18 which have the respective metrics

$$ds^2 = du^2 + 2g_{12}\, du\, dv + dv^2$$

and

$$ds^2 = du^2 + u^2\, dv^2$$

In summary, then, the expression for the metric of any surface is dependent on two entirely different factors, the first being a "relative" matter—the particular network or frame of reference—and the second an "absolute" issue—the nature of the surface itself, that is, whether it is a plane or a sphere or an ellipsoid or a paraboloid, etc.

In discussing the methods of differential geometry, emphasis has been placed on the advantages of curvilinear networks for location and measurement on curved surfaces. But the feature of Gaussian geometry that has become most important in relation to relativity is its *intrinsic* nature, the fact that the surface leaves its stamp on the metric. The set of possible frameworks and the g's are *internal* aspects of the surface and have nothing to do with surrounding space. Longitude and latitude, for example, and all the geographical facts connected with them are defined in terms of the earth's surface. If one were to use, instead of a curvilinear network of longitude and latitude lines on the surface, a Cartesian system with origin at the earth's center and axes along three mutually perpendicular diameters, one would have to refer to *extrinsic* or external points, lines, and planes. This would be very inconvenient since it would necessitate the taking of measurements below the earth's surface, near its hot center, etc., and thus things are simplified by having reference frames right on the surface. When it is a matter of applying geometry to relativity theory, understanding is enhanced if one can avoid the conception of points external to physical 3-dimensional space or of points outside Minkowski's 4-dimensional space-time continuum. In the latter case, *intrinsic* observations of x, y, z are made by means of measuring rods, and clocks furnish the values of t; there is no notion of looking at things from a vantage point in some mystic, nonphysical 5-dimensional continuum.

A special advantage of intrinsic geometry is its ready generalization. Since it requires *two* coordinates, (u, v), to locate a point on an ordinary surface, we say that a surface is a two-dimensional "space" or *manifold*. If Cartesian coordinates are used, a relation among *three* of them, (x, y, z), is needed in order to describe such a manifold. For example, the equation $x^2 + y^2 + z^2 = r^2$ describes the surface of a sphere. Similarly, a 3-dimensional continuum would require four Cartesian coordinates for its expression but only three Gaussian coordinates (u_1, u_2, u_3). Again, four Gaussian coordinates (u_1, u_2, u_3, u_4) give location in abstract 4-dimensional spaces or manifolds. In a pure mathematical sense, we can have manifolds with any number, n, of dimensions. It was Riemann who extended Gauss's theory of 2-dimensional surfaces to cases where $n > 2$.

Location in an n-dimensional manifold is accomplished by a one-to-one correspondence between its "points" and the Riemannian n-tuples $(u_1, u_2, \ldots, u_n)$. But such location is mere charting, and there is need to know the "structure" of the manifold if it is to possess a geometry. For a *metric* geometry, the requisite is a distance formula, that is, a set of g's for the small n-dimensional region around each point. Just what algebraic form and mathematical properties a metric should have, that is, the generalization to n dimensions of ordinary 2-dimensional metric differential geometry, was worked out by the twenty-eight-year-old Riemann in 1854.

Riemann's explanation was given in the essay "On the Hypotheses at the Basis of Geometry," which is considered one of the greatest masterpieces in mathematical literature. This was his *Habilitationsschrift* (probationary essay) for the position of *Privatdozent* (lecturer) at the University of Göttingen. In compliance with academic tradition, the young mathematician had submitted three possible titles to the Göttingen faculty. He hoped one of the first two would be chosen, because he was well prepared to write and talk on these. In youthful fashion, he took his chances by listing as the third possibility a profound issue to which he had given very little thought, namely, the foundations of geometry. Fate, in the form of Gauss, designated this third topic for Riemann's probationary ordeal. The selection of the most difficult question was probably not a sadistic impulse on the part of the great master—but rather the natural desire, with his death imminent, to witness the torch he had kindled being carried forth brilliantly into the future.

The aging mathematician, never generous with praise, nevertheless found Riemann's doctoral dissertation of 1854 a work of gloriously fertile originality. Gauss fought weakness and pain to hear the *Habilitationsschrift* delivered, and pronounced it a surprise beyond his utmost expectations to see how such a difficult subject could be handled by such a young man. The words of the Titan were apparently unheard by mathematicians and physicists. Riemann achieved but modest recognition in his brief lifetime. His great essay on the foundations of geometry was published posthumously, and its applicability to physics was not fully realized until the day of general relativity.

Because men are 3-dimensional beings, they were able to realize, early in history, that the earth's surface is curved. This was revealed, for example, by the gradual appearance or disappearance of a ship in relation to the horizon. But could such a conclusion be reached by a 2-dimensional being confined to a spherical surface? If such a surface-dweller were to talk of "curvature" in any physical sense, he would be referring to the difference between straight lines and curves. He might provide analytic formulas to measure deviation from straightness. Then, *generalizing* these formulas so as to include more algebraic variables, he might, for the sake of clarification, describe the generalizations in quasi-geometric terminology. He might state how his generalizations would indicate whether or not his world or some other 2-dimensional cosmos is bent or "curved" in some enveloping 3-dimensional space which surface-dwellers cannot visualize. Going further he would derive formulas which would indicate whether 3-dimensional spaces are bent in some 4-dimensional continuum, which in turn may be curved in some external space, whose possible deviation from "flatness" might again be of interest, and so on *ad infinitum*.

Instead of speaking of a "curvature" which one cannot observe, one can express statements in alternative fashion by describing how a surface-dweller would be able to discover by suitable measurements whether or not the geometry of the triangles, polygons, circles, etc. in his cosmos is the Euclidean geometry of a "flat" surface or plane. Two-dimensional beings might plan a "geodetic" survey to settle the issue, in much the same spirit as Maupertuis, Clairaut, and La Condamine led expeditions into Lapland and Peru, to measure meridional arcs near the North Pole and the equator with the express purpose of choosing between oblate or prolate spheroid as the earth's shape. These great eighteenth-century scientists had wished to decide between Newton and Cassini. The former had indicated on purely *theoretic* grounds that the earth must be flattened at the poles. Cassini, the leading French astronomer of the day, based the alternative hypothesis of the prolate form on his geodetic measurement of the meridian arc between Dunkerque and Perpignan.

Let us imagine that a number of expeditions are organized by surface-dwellers and sent to survey widely separated regions of their space. One group, working in a desert, finds few obstacles to its operations and attempts to cover the ground with a Cartesian network of straight lines. To see whether the desert region is Euclidean (flat), the surveyors make distance measurements for points in the small meshes and substitute in the Pythagorean formula $ds^2 = du^2 + dv^2$. If the substitutions fail to check even approximately, this is no reason for rejecting the hypothesis of Euclideanism. The 2-dimensional scientists know that even in Euclidean geometry there are infinitely many different Gaussian networks with varying distance formulas and hence they proceed to a reexamination of the framework. Perhaps the surveying operations have not actually marked out rectangular meshes. The field workers start anew, measuring adjacent sides and diagonals in a sequence of infinitesimal parallelograms, in order to determine for each mesh the proper set of values for g_{11}, g_{12}, g_{22} and the appropriate distance formula

$$ds^2 = g_{11}\, du^2 + 2g_{12}\, du\, dv + g_{22}\, dv^2$$

After forty or fifty repetitions of this procedure for different infinitesimal parallelograms, the set of values obtained for g_{11} is examined. Suppose that the figures run somewhat as follows: -0.97, 1.02, 1.06, 1.00, 0.96, 0.94, 1.01, 0.98, 1.04, 1.03, 0.95, ... and that when all fifty are averaged and rounded off to two significant figures, $g_{11} = 1.0$. Let us say that g_{12} and g_{22}, like g_{11}, fluctuate very little from mesh to mesh and that, on the average,

$$g_{12} = 0.12 \qquad g_{22} = 1.0$$

Then a statistical metric for the entire region is

$$ds^2 = du^2 + 0.24\ du\ dv + dv^2$$

But this fits the standard form previously listed for an oblique network, and is merely an expression of the law of cosines. Therefore in each parallelogram of the mesh, the cosine of one of the angles has the value 0.12, and this angle is $83°$ in size. Then in every mesh two of the angles measure $83°$, and the other two $97°$, so that the lines of the mesh system do *not* meet at right angles. But the desert region can be pronounced Euclidean (flat) and the surveyors realize that this fact was obscured at first because they had erroneously believed the frame of reference to be rectangular.

While the first expedition has carried on as described, a second group is at work in totally different territory. Obstacles have made it impossible to set up markers for a Cartesian reticulation, but with a polar scheme of radiating lines and concentric circles it has been possible to skirt impassable areas. Measurements substituted in the plane polar distance formula

$$ds^2 = du^2 + u^2\ dv$$

check and thus confirm that the framework used is actually a polar one.

The net result of the two geodetic surveys is that the two regions measured are *Euclidean* or, from the viewpoint of 3-dimensional observers, flat. If very numerous surveys of a similar sort were conducted in areas all over the world of the 2-dimensional folk, and in every case led to a metric formula possible for some permissible network in a plane, then the infinitesimal geometry of the 2-dimensional cosmos would be pronounced Euclidean by its inhabitants.

It is possible to imagine a Gaussian "survey" of the Minkowski 2-dimensional space-time of the previous chapter. Then the derivation of the metric would start with the "Pythagorean formula" of special relativity,

$$s^2 = t^2 - x^2$$

and with the simplest mesh system one would have

$$ds^2 = dt^2 - dx^2$$

or, using the transformation

$$u = ix$$
$$v = t$$

one would obtain the usual Pythagorean formula and the corresponding metric,

$$s^2 = v^2 + u^2 = u^2 + v^2$$

or

$$ds^2 = dv^2 + du^2 = du^2 + dv^2$$

so that the space-time continuum of *special relativity* has a Euclidean or flat geometry. Because imaginary numbers are used in the above transformation to a Euclidean meshwork, just as in converting the Lorentz transformations to Euclidean rotations of axes (Chapter 17), some mathematicians refer to the restricted relativity space-time continuum as "pseudo-Euclidean" or "semi-Euclidean."

To us, in 3-dimensional space, a 2-dimensional surface that is pronounced flat or Euclidean by surveys like those just considered would *not* necessarily look like a plane. If the 2-dimensional beings lived on the surface of a cylinder or a cone, millions of surveys of different regions would all lead to the verdict "Euclidean." But to 3-dimensional observers, cylinders and cones are *not* planes. However, the local or differential geometry of such a 2-dimensional cosmos is Euclidean, because very small regions are for all practical purposes flat, exactly like parts of a plane. Hence 2-dimensional surveys of many small regions would yield the same sort of results for worlds that are cylindric or conic as for those that are plane, and would not lead to any distinction among them or between a plane and any other surfaces that are locally Euclidean. To find any failures in Euclideanism, the 2-dimensional folk would need to attempt surveys of *large* areas, as we shall see.

To see why the infinitesimal geometry of a cylinder or cone is the same as that of the plane, we can think of rolling a plane rectangular sheet of paper into a cylinder, or a plane sector into a cone. In fact, if a surface can be constructed from any plane paper pattern without stretching or tearing this pattern, its local geometry will be Euclidean. If any network is drawn on the plane pattern, it will remain after the pattern is rolled into a surface. Since there is no stretching or tearing in the process, the distances and the corresponding metric formula are also unaltered. All surfaces that can be formed from plane patterns in this way are termed *developable*, because from an opposite point of view such surfaces can be unfolded into plane forms, or *developed* on a plane (see Figure 19.5).

Although many surveys of small regions will not enable creatures bound to the surface of a cylinder to distinguish it from a plane, "global" operations would bring awareness. We imagine the radius of a cylindric world to be relatively large, so that in the early stages of the 2-dimensional civilization there are many unexplored and unmeasured regions. But a time arrives when there are means of transportation that make long journeys possible. Then it is observed that in certain cases a lengthy trip in one direction brings the traveler back to his starting point. In Figures 19.1 and 19.2, traveling "east" or "west" will ultimately bring an individual home. This sort of thing *cannot* happen in a plane, since trips "east" or "west" carry the traveler to ever-new territory. In a cylindric or conic universe, the 2-dimensional creatures would say that although the local geometry of their

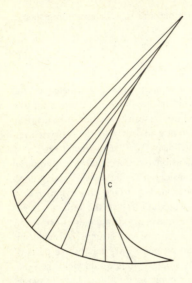

Figure 19.5 Developable surface generated by tangents to a twisted curve *C*. (A *plane* curve with its tangents can be twisted or "fluted" to form the surface pictured and, conversely, the surface can be flattened out.)

world is flat or Euclidean, its "global" or macroscopic geometry is not that of the Euclidean plane.

One point at issue is how the small parts of a 2-dimensional universe are connected with one another. The nature of a surface *in toto* depends on its *connectivity*, that is, the effect created by connecting any two points of the surface by a continuous path, or of drawing some closed, continuous path like the circumference of a circle or ellipse. If one were to draw such a closed path on the surface of a sphere, for example, the Arctic circle on the earth, then the circumference would act as a sort of fence dividing the surface into two parts, so that one could not pass continuously from one part to the other because the fence would interfere. To make a journey from a point in the North Frigid Zone to some point in the Torrid Zone (or vice versa) and remain on the surface of the earth, it would be necessary to cross the Arctic circle. On the other hand, on the surface of a doughnut (Figure 19.6) certain closed curves would *not* fence in the portions of the surface and would not interfere with the possibility of connecting points with one another. The doughnut or *torus* is said to possess different connectivity from the sphere. All such "global" issues, however, are outside the province of classic differential geometry. The discussion of *topology* in a later chapter will provide a fuller treatment of geometry "in the large" whereas the present chapter will be limited to geometry "in the small."

If the task of cartographers were the mapping of *developable* surfaces, their job would be an easy one. As we have mentioned, there is no stretching or tearing in the unrolling of such surfaces. Not only are distances invariant under such "development," but relative directions are also unchanged. The countries of the cosmos would not necessarily have the correct shape, but they would have the correct area in the development on a plane. But, alas for human mapmakers, the earth's surface is not developable! Spheres, spheroids, paraboloids and, in fact, most surfaces are *not* developable, and therefore not even locally flat. Because the methods of Gaussian differential geometry are internal or *intrinsic*, their practical counterpart—surveying in a 2-dimensional manifold—would enable beings confined to a nondevelopable surface to discover that the local geometry of their world is non-Euclidean, in other words that no one of the many different forms of ds^2 that apply to different Gaussian networks in a plane can fit the local distance measurements of their cosmos.

Let us return to the previous picture, where members of a geodetic survey attempted to cover a region with a polar network. If things had turned out differently, and distance measurements had *not* checked in the polar Pythagorean formula, it would have been necessary, as it was in the desert survey, to use statistical methods to arrive at the correct metric. Now imagine that, in this new situation, a sequence of forty or fifty determinations of g_{22} for different meshes yields measurements all close to one another in value and that they average 1.0, whereas measurements of g_{12} are either zero or close to zero. This would mean that the cosine of the angle between curves of the network is zero, so that this angle measures 90° and the curves are orthogonal (perpendicular). Now suppose that the values of g_{11} do not exhibit the same sort of statistical concentration as g_{12} and g_{22}, but gradually decrease as the v-coordinate of meshes increases through positive values $v = 1, 2, 3, 4, 5, \ldots$. This would indicate that the value of g_{11} depends on the value of v or that g_{11} is a function of v. Advanced mathematical methods can be used to find the function that is the "best fit," in the statistical sense, of the set of figures for g_{11}. Now suppose that the function that fits is $g_{11} = \cos^2 v$. Then the metric would be

$$ds^2 = \cos^2 v \, du^2 + dv^2$$

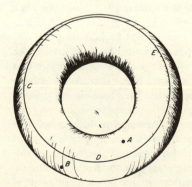

Figure 19.6 Connectivity on a torus. Curve *CDE* does not fence off an area since *A* can be connected with *B* by a path proceeding around the inside of the torus.

This happens to be the distance for a longitude-latitude network on an ordinary spherical surface, and hence the region being surveyed would be declared "spherical." The surface-dwellers might not have the word "sphere" in their vocabulary, but their knowledge of differential geometry would tell them that, even with the infinite variety of plane Euclidean mesh systems, there is not a single one to which the above distance formula corresponds. In other words, they would be able to show that no permissible transformation of coordinates would convert the above metric into the simple Pythagorean formula

$$ds^2 = du^2 + dv^2$$

They would therefore say that the region being surveyed is *non-Euclidean*.

Gauss provided an exact definition of the property that he termed *curvature*, the one we have described as the "non-Euclideanism" of a surface in a region surrounding some point. Gauss was able to prove that this attribute can be expressed in terms of g_{11}, g_{12}, g_{22} and their partial derivatives of the first and second orders, a fact that he called a *theorema egregium* (a most remarkable theorem). An alternative way of stating Gauss's *theorema egregium* is to say that curvature or non-Euclideanism is an *intrinsic* or *absolute* characteristic of a surface. And, in general, any property of a surface involving the g's, their derivatives, and arbitrary functions of u, v, the Gaussian coordinates, is considered *absolute*. Still a third way of expressing all these facts is to say that ds^2 and "non-Euclideanism" are *bending invariants*, characteristics unchanged by deformations of small regions of a surface.

In all the hypothetical experiments we have assigned to the surface inhabitants, we have had them base their tests on infinitesimal operations and the definition of the fundamental metric. But to find out whether a surface geometry is Euclidean or non-Euclidean, or from a 3-dimensional viewpoint, flat or curved, we can in theory perform a *crucial test* based on any metric theorem that is valid for Euclidean geometry only, for example, the proposition asserting that the sum of the angles of a triangle is equal to 180°. As mentioned in Chapter 3, Gauss himself measured the triangle formed by the peaks of three mountains, Brocken, Höher Hagen, and Inselberg, in order to test whether our space is Euclidean (flat). Since the angle sum came out close to 180°, he could not tell whether the deviation from the Euclidean sum was an error of observation or the result of non-Euclideanism.

We think of a Euclidean triangle as a figure bounded by three straight-line segments. But what would a triangle or a "straight line" be on the surface of a sphere? From the modern abstract point of view the term "straight line" can be left completely undefined and can be given appropriate interpretations in particular applications of an abstract geometry. Hence, 2-dimensional surveyors may use their own special idea of what a straight line is. We imagine them to hold the tape taut in laying out a "straight line." But if they are working on a spherical surface their straight line would thus appear to 3-dimensional observers as a great-circle arc. In any case, whether on a Euclidean plane or on some curved surface, a surveyor's "straight-line" segment is the shortest distance between two points of a small area. In differential geometry such a line is called a *geodesic*, and a triangle is a figure bounded by three geodesics. On a sphere, the sides of a triangle will be three great-circle arcs. Now suppose once more that on a sphere (Figure 19.2), a triangle is formed by joining the North Pole to two points on the equator so that the sum of the angles of this triangle is 254°. Hence, measurements of angles in such a triangle or, at any rate, in some fairly large triangle, made by creatures limited to a spherical surface, would indicate that the "global" geometry of their world cannot be that of a Euclidean plane or a developable surface.

But beings on a spherical surface would be able to conduct other experiments to reveal that their universe is not even locally Euclidean, providing always that they did not limit themselves to a very small region of their world. They need not involve themselves in "global" considerations but must go beyond the small neighborhood of each individual point. Starting from such a point, let us say that they measure off numerous "straight-line" segments of equal length, and join the end-points of these segments to form a circle. They are aware that in Euclidean plane geometry the length of the circumference should conform to the formula $C = \pi d$, or that for every circle the ratio of the circumference to the diameter is the same number, the constant π. But they find that, in their universe, the ratio of the circumference to the diameter is *not* constant but varies with the diameter of the circle, and in every case is less than π. If the circle drawn is very tiny, they find a ratio almost equal to π. If they draw a somewhat larger circle, say, that one which 3-dimensional observers call the 60th parallel of latitude (Figure 19.7a), they find the ratio to be 3. If they were to draw a still larger circle, the 45th parallel (Figure 19.7b), for example, the ratio would be about 2.83, and if they drew the equator, the ratio would be 2. The reason for these results is that in relativity all observers must agree on measurements, and in the diagrams the lengths of various "straight lines" like the "diameters" have been deduced from the realization that these are extrinsically great-circle arcs, and extrinsic length must equal intrinsic length.

Riemann dealt with a 3-dimensional spherical space which is analogous to the 2-dimensional spherical world just considered. Then by analogy with the procedures above, we, the creatures who live in a 3-dimensional cosmos and cannot view it from some external point in a 4-dimensional space, could nevertheless discover whether the geometry of our world is Euclidean or not, and in the latter case whether it is spherical. Thus, starting at a point, we might mark off very numerous "straight-line" segments of equal length so that their end-points would form a 2-dimensional spherical surface. If tiny squares are applied to this surface and enumerated as a device for measuring its area, it will turn out in cases where the radius is very small that the area $A = \pi d^2$, where d is the "straight-line" diameter. If the diameter is then increased in size, the formula may always remain the same, and this would prove that the portion of space being surveyed is Euclidean. If, however, our entire space should be a Riemannian 3-dimensional spherical surface, the area A would turn out to be less than πd^2 for any diameter that is not tiny; that is, the ratio of A to d^2 would be less than π, and would diminish as the diameter of the sphere is increased.

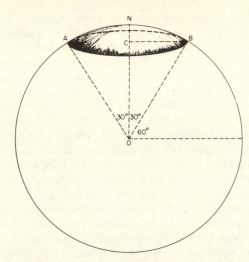

Figure 19.7(*a*)   The *intrinsic* diameter of the 60th parallel is ANB, a 60° arc of a great circle. If the radius of the
sphere is taken as the unit of measure, $ANB = 1/6 \cdot 2\pi =
\pi/3$, $CB = \frac{1}{2}$, and the circumference of the 60th parallel
measures $2\pi \cdot \frac{1}{2} = \pi$. Thus the ratio of the circumference to
the intrinsic diameter is $3 : 1$.

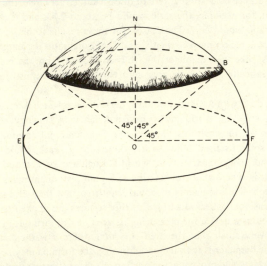

Figure 19.7(*b*) ANB, the intrinsic diameter of the 45th
parallel, measures $\frac{1}{4} \cdot 2\pi = \pi/2$; $CD = \sqrt{2}/2$ and the circumference of the parallel $= 2\pi(\sqrt{2}/2) = \pi\sqrt{2}$. The ratio of the
circumference to the intrinsic diameter is $2\sqrt{2} : 1 = 2.83$
(approximately).

In all the illustrations we have given, spherical spaces were chosen for the sake of simplicity. The uniformity of a spherical space causes the experiments to yield the same results for every region of such a universe. This uniformity would not occur for other types of non-Euclidean space, but detection of non-Euclideanism would nevertheless be possible. The theorems, however, would vary with the locality. If the universe were an elongated ellipsoid, for example, the value of the ratio C/d would depend not only on the size of d but on the location of the circle on the ellipsoidal surface. Or 2-dimensional creatures might live on a surface somewhat irregularly curved, but not too different from a plane, a cosmos that to 3-dimensional observers would look like the rippled surface of a lake, or like a flat surface with a few ridges, blisters, and puckers. The intrinsic measurements in the former case would show the surface to be approximately Euclidean, whereas in the latter case, Euclideanism would be established except for specific regions which would manifest various degrees of non-Euclideanism.

At long last, what does all the preceding discussion of the intrinsic geometry of spaces have to do with relativity? The answer is that if we substitute Riemann for Gauss, a Minkowski 4-dimensional space-time manifold for a 2-dimensional surface, *ourselves* for the surface creatures, certain observations from nineteenth- and twentieth-century experimental physics for surveying procedures, then we shall have Einstein's general relativity of 1916. The theoretical and empirical steps by which the great universe-builder arrived at his theory of gravitation are, *mutatis mutandis*, the 4-dimensional analogues of the operations carried out by creatures confined to an ordinary surface.

For differential geometry on a surface, Gauss provided the metric or fundamental form

$$ds^2 = g_{11}du_1du_1 + g_{12}du_1du_2 + g_{21}du_2du_1 + g_{22}du_2du_2$$

with the understanding that $g_{12} = g_{21}$, and the order of all multiplications is immaterial. Since it is the *coefficients*, the g's, that determine the formula, it is customary to omit the du_1 and du_2 and write the coefficients in a square arrangement, or square *matrix*:

$$\begin{pmatrix} g_{11} & g_{12} \\ g_{21} & g_{22} \end{pmatrix}$$

When this matrix is accompanied by the law of transformation to g_{11}', g_{12}', etc., that is, the set of transformations to all permissible mesh systems or frames of reference, then it is termed a *tensor*, and $g_{11}, g_{12}, g_{21}, g_{22}$, are called *components* of the tensor. Thus we see that the whole issue of Euclidean versus non-Euclidean geometry, flat versus curved space for surface-dwellers, depends on the tensor components, and it is the hypothetical surveyor's task to determine these components as functions of u_1 and u_2.

Analogous considerations apply to the 4-dimensional spaces of Riemann. In relativity, which is a physical geometry, it seems meaningless to speak of 3-dimensional or 4-dimensional "surfaces" because this would seem to imply that they are "surfaces" of something, in other words that there is some sort of external 4- or 5-dimensional space in which they are embedded. Hence it is preferable to refer to a *continuum* (of any number of dimensions) with an *intrinsic* geometry.

The physicists surveying the 4-dimensional space-time continuum have, corresponding to the *points* where 2-dimensional surveyors place markers, *events* (x, y, z, t); the x, y, z of an event are determined by measuring rods, and the t by a clock. In the arbitrary curvilinear Gaussian or Riemannian networks used in general relativity, one may specify an event by (u_1, u_2, u_3, u_4). But the mere specification of points or events does not make a geometry. In other words, the space-time continuum is amorphous, structureless, until ds^2 is specified. In the case of a general Riemannian 4-dimensional continuum, this is a matter of furnishing the *matrix* (or the *tensor*, when the all-important transformation law is included) which is

$$\begin{pmatrix} g_{11} & g_{12} & g_{13} & g_{14} \\ g_{21} & g_{22} & g_{23} & g_{24} \\ g_{31} & g_{32} & g_{33} & g_{34} \\ g_{41} & g_{42} & g_{43} & g_{44} \end{pmatrix}$$

where $g_{12} = g_{21}$, $g_{13} = g_{31}$, $g_{14} = g_{41}$, $g_{23} = g_{32}$, $g_{24} = g_{42}$, $g_{34} = g_{43}$. When the first g in each of these equations is substituted for the equivalent g in the matrix above, the number of g's is reduced from 16 to 10. The corresponding metric is

$$ds^2 = g_{11} du_1{}^2 + 2g_{12} du_1 du_2 + 2g_{13} du_1 du_3 + 2g_{14} du_1 du_4 + g_{22} du_2{}^2$$
$$+ 2g_{23} du_2 du_3 + \cdots + g_{44} du_4{}^2$$

There are now *ten* g's to be fixed instead of the three of ordinary surface theory, and this is one reason why Einstein's task was so difficult. In our previous discussion, after the g_{11}, g_{12}, g_{22} functions for a surface had been determined, its intrinsic geometry was complete; all the rest would have been inference from the fundamental formula. Einstein's restriction on his ten g's, and their formulation as functions of space and time provide the complete story of our physical universe as far as "gravitational" phenomena are concerned.

In the earlier discussion of the special theory of relativity, there was the metric (separation formula), $s^2 = t^2 - x^2$. This metric was used in describing a 1-dimensional motion as recorded in a 2-dimensional space-time continuum. Generalizing this metric so that it can be used to describe *local* motion in 3-dimensional space, one obtains the formula in the 4-dimensional space-time continuum, namely,

$$ds^2 = dt^2 - dx^2 - dy^2 - dz^2$$

or, in Riemannian form, if we take time as the *fourth* coordinate,

$$ds^2 = -du_1{}^2 - du_2{}^2 - du_3{}^2 + du_4{}^2$$

Comparing this with the general form above, we see that all elements of the matrix except the diagonal ones are equal to zero, that is, $g_{11} = g_{22} = g_{33} = -1$, $g_{44} = +1$, and all the other g's are zero. This is a specially simple Riemannian tensor, and the corresponding metric is Euclidean or pseudo-Euclidean (page 463) because a permissible transformation of coordinates (involving imaginaries) will transform it to the Pythagorean distance formula

$$ds^2 = du_1{}^2 + du_2{}^2 + du_3{}^2 + du_4{}^2$$

In 2-dimensional Gaussian differential geometry, the metric of any surface can be derived, as explained earlier in this chapter, by applying the Pythagorean formula (or its corollary, the law of cosines) to measurements made on the infinitesimal meshes of a Gaussian network. Einstein's *principle of equivalence* furnishes the analogous starting point for deriving the metric of the 4-dimensional space-time continuum of general relativity. It will be recalled that the principle of equivalence, reduced to geometric form, states that in infinitesimal regions of space-time the restricted theory of relativity holds; that is, in such regions, if a suitable local frame of reference is chosen, it is true that

$$ds^2 = dt^2 - dx^2 - dy^2 - dz^2$$

One may apply this rule to a small "curvilinear" Riemannian mesh of space-time, much as was done with Gaussian meshes on a surface where $k_1 du$ and $k_2 dv$ represented distances "parallel" to the sides of the mesh, the proportionality factors k_1 and k_2 being determined by surveying. By analogy, general relativity must make use of $k_1 du_1$, $k_2 du_2$, $k_3 du_3$, $k_4 du_4$, where the k's are determined by "surveying" a particular mesh in space-time. Application of the Pythagorean formula gave the *local* values of g_{11}, g_{12}, g_{22} in

$$ds^2 = g_{11} du^2 + 2g_{12} du\, dv + g_{22} dv^2$$

Similarly, in general relativity, application of the formula $ds^2 = dt^2 - dx^2 - dy^2 - dz^2$ will give the *local* values of the *ten* coefficients or ten tensor components, $g_{11}, g_{12}, g_{13}, \ldots, g_{44}$.

We conceived of the Gaussian surveyors as proceeding from mesh to mesh obtaining different coefficients, or g's for each locality, and finally expressing the g's as *functions* of u and v to obtain a formulation of ds^2, the metric in which the entire geometry of the surface is implicit. Similarly, to obtain the metric of the space-time continuum of general relativity is a matter of proceeding from mesh to mesh of space-time, "surveying" to obtain the k's or proportionality factors, and applying again and again the restricted relativity formula for ds^2, the space-time separation formula of the previous chapter. After we have discussed the g's abstractly, this "surveying" problem still remains.

Einstein postulated the principle of equivalence in order to make the metric of special relativity applicable to the surveying of small regions of space-time. But, in order to determine the g's, the tensor components, for all points of the space-time continuum and all possible Riemannian frames of reference, he found it necessary to make additional assumptions based on theoretical or empirical facts of physics and astronomy. The *invariance* of ds^2, the metric tensor, under arbitrary transformations of the Riemannian coordinates is one axiom governing the selection of suitable functions of u_1, u_2, u_3, u_4 for the various g's. Another restriction arises from the distribution of material bodies in space, since "gravitational" phenomena are to be explained as the effect of such bodies on space-time structure, that is, on the g's. Einstein made a third assumption, based on the views of the great mathematical physicists who had preceded him, namely, that the laws of physics, in particular the "field" laws, are partial differential equations of the second order, and hence that the g's must satisfy such equations, this in effect being the expression of the new law of gravitation. The "surveying" of those regions of space-time where the presence of matter produces "gravitational" phenomena, or, alternatively, non-Euclideanism in the geometry, becomes a matter of *solving* the associated differential equations for the g's. To find *particular* or specific solutions calls for boundary and/or initial conditions (Chapter 9).

The physicist M. Schwarzschild succeeded in solving the partial differential equations for radially symmetric "gravitational fields" like those around the sun. In the boundary conditions, he introduced m, the "gravitational mass" of the sun (or other particular material body involved), and also specified that for events at a great distance from the body there would be no "gravitational" effects, that is, that the space-time in such regions would conform to the metric of special relativity so that $g_{44} = 1, g_{11} = g_{22} = g_{33} = -1$ and all the other g's would be zero for the values of u_1, u_2, u_3, u_4 in these regions. Using polar coordinates with the origin at the center of mass of the body, he obtained the following solution of the differential equations:

$$g_{44} = 1 - \frac{2m}{r}$$

$$g_{11} = -\frac{1}{1 - 2m/r}$$

$$g_{22} = -r^2$$

$$g_{33} = -r^2\sin^2\theta$$

The corresponding metric is

$$ds^2 = \left(1 - \frac{2m}{r}\right) dt^2 - \frac{dr^2}{1 - 2m/r} - r^2 d\theta^2 - r^2 \sin^2\theta \, d\phi^2$$

or, if one wishes to use the symbols u_1, u_2, u_3, u_4,

$$ds^2 = \left(1 - \frac{2m}{u_1}\right) du_4{}^2 - \frac{du_1{}^2}{1 - 2m/u_1} - u_1{}^2 \, du_2{}^2 - u_1{}^2 \sin^2 u_2 \, du_3{}^2$$

If u_1, that is, r, the radial distance, is very large, then $2m/u_1$ will be negligible and the metric will reduce to that of special relativity (in polar coordinates) as required by the boundary condition. This will also occur for space-time regions where $m = 0$, that is, where no matter is present. Hence the Schwartzschild metric is very general, applying to events occurring near masses like the sun as well as to occurrences remote from matter.

One might have expected that with the acclaim of the general theory, Einstein could have rested from his labors, but instead he continued along the lines of still further generalization in order to include electromagnetic phenomena, since he believed that electric and magnetic field strengths are also *relative* and not absolute. Students of elementary classical physics learn that there is an analogy between certain mechanical and electromagnetic laws. To take a specific instance, consider Newton's formula for gravitational attraction,

$$F = G\,\frac{Mm}{d^2}$$

where M, m are the masses of the two attracting bodies, d is their distance apart, F is the force of attraction, and G is a constant depending on the particular units of mass, distance, force that are used.

This becomes Coulomb's law of *magnetic* attraction between the poles of two magnets if we replace M and m by the magnitudes of the poles and G by a suitable magnetic constant. It becomes *electric* attraction if we replace M and m by electric charges and G by another suitable constant. Moreover, if we take Newton's law and think of it as applied to constant masses like those of the earth and sun, it reduces to the inverse-square law, the attraction between the earth and sun varying inversely as the square of the distance between them. But we have seen that the inverse-square law also arises in the theory of light. Thus, phenomena in gravitation, magnetism, electricity, and light seem to be manifestations of one general principle. Matter and electromagnetic entities exert influences or create a *field* in the space around them.

Most of the present chapter was devoted to the mathematics which enabled Einstein to handle the gravitational field, that is, the influence of matter on the space-time continuum. For thirty years or so, Einstein, Hermann Weyl, Sir Arthur Eddington, and others tried to obtain a single theory that would include gravitational and electromagnetic fields as two special cases. Finally, in February 1950, Einstein released to the world a "unifield field theory," which still awaits the sort of experimental verification that Einstein's earlier work has had. There is no doubt, however, that the new theory, expressed in the language of *tensors*, is one of the greatest contributions to mathematical physics in modern times.

If we trace facts to their sources, then we must say that Einstein's signal contribution to modern physics was made possible by one of the many very great advances in modern mathematics made by Carl Friedrich Gauss, that towering figure whose diversified mathematical achievements are, of necessity, described at many points of our story. We have already discussed his fundamental discoveries in algebra, astronomy, non-Euclidean geometry, complex analysis, probability theory, and differential geometry. His contributions to number theory and topology will be discussed later. Thus there is actually no most appropriate subject with which to unify the story of his life. Although he himself pronounced number theory the "queen" of mathematics, general readers may be more impressed by how he paved the path to relativity. Hence, having reached the end of that road, we are presenting his biography at its terminus.

Gauss was born in Braunschweig (Brunswick) in 1777, where his father, a bricklayer and gardener, would have been quite willing to have his son follow the same pursuits. The boy's exceptional numerical talent manifested itself very early—in fact, before he was three years of age, as legend has it. At that early age he found an error committed by his father in making out the payroll for the bricklayers whom he supervised.

The child prodigy's school experiences were fortunate in one respect, for the assistant schoolmaster was none other than Johann Martin Bartels (1769–1836), whose later influence on Lobachevsky has already been mentioned (Chapter 3). When Gauss was ten and Bartels was eighteen, they became acquainted, and from then on studied elementary and advanced mathematics together. Bartels soon recognized that Carl was no ordinary child but a full-fledged genius who could improve on the analytic techniques of Newton, Euler, and Lagrange. The young teacher publicized his student's mathematical feats by recounting them to prominent citizens of Braunschweig. Eventually the stories reached (Carl Wilhelm) Ferdinand, Duke of Braunschweig, to whom Gauss was presented at the age of fourteen. At once the Duke became the kindest and most generous of patrons. He financed all of Gauss's education for the next eight years, right through the doctorate at the University of Helmstädt in 1799. The Duke paid for the publication of his protégé's doctoral dissertation, and granted him a small regular allowance thereafter in order to enable him to carry on mathematical research without the need to teach or seek some other means of earning a living. When Gauss married in 1805, his patron increased the amount of the annual stipend. This support ceased only in 1807, after Duke Ferdinand died from wounds received in the battles of Auerstädt and Jena.

In 1807, then, Gauss was faced with the practical problem of finding a job. This was, in fact, no problem at all, since his renown was widespread by that time and he received excellent offers from leading academic institutions all over Europe. Alexander von Humboldt (1769–1859), the famous traveler and scientist, sought to have Gauss remain in Germany, and was influential in obtaining his appointment as director of the Göttingen Observatory. Laplace was Gauss's "reference" for the position. "Who is the leading mathematician in Germany?" von Humboldt asked Laplace. "Johann Friedrich Pfaff," was the answer. Von Humboldt was crestfallen. "What is your opinion of Gauss?" he inquired. "Oh, he is, of course, the greatest mathematician in the world," Laplace explained.

What had Gauss done to earn such a reputation? He had already made a multitude of discoveries, among them the proof of the fundamental theorem of algebra, published in his Ph.D. dissertation of 1799. Then there was the 1801 *Disquisitiones Arithmeticae*, whose contents will be discussed later (Chapter 21). Gauss was contemplating the preparation of a second volume for this work when there occurred an event which caused him to swerve from pure to applied mathematics.

Although astronomic theory had indicated the probable existence of tiny planets or "planetoids," the very first observation of such a heavenly body was made at Palermo on January 1, 1801, by the Italian astronomer Giuseppe Piazzi

(1746–1826). The planetoid he discovered was named Ceres. It was possible to obtain only a very few exact observations of Ceres' positions in the course of its motion around the sun, and therefore the computation of its orbit from such limited empirical data presented a problem which aroused Gauss's interest. In formulating the issue, he had to solve (approximately) an algebraic equation of the eighth degree. Thus he provided a "practical" application of the very equations which the Babylonians of Hammurabi's day had solved merely because they enjoyed algebra.

When in 1802 Pallas, another planetoid, was discovered, Gauss embarked on the formulation and solution of the *general* problem of computing the orbit of *any* such tiny member of our solar system. The ultimate result was his 1809 *Theoria motus corporum coelestium*, in which he showed just how to chart the complete orbits of planets and comets from limited observational data; he dealt with the difficult matter of "perturbations," and provided the techniques that were to govern computational astronomy until the recent era. Observations of the diminutive planets, Ceres, Pallas, Vesta, Juno, etc. confirmed the accuracy of Gauss's methods, which were enormous time-savers as well.

Only in 1820 did Gauss develop a new interest by embarking on that theoretical research in geodesy which we have pictured as the applied background for his differential geometry. Incidental to the computational aspects of geodesy was his use of the statistical method of "least squares" which, in the *Theoria motus*, he had claimed as one of his early discoveries. However, from the point of view of publication, Adrien-Marie Legendre (1752–1833) held priority, since his exposition of the method had appeared in 1806, three years earlier.

Next Gauss made discoveries in optics, in electromagnetic theory, and in the theory of Newtonian attraction. In connection with gravitational attraction, he initiated the mathematical subject which came to be known as "potential theory." In 1833 he and his young associate, Wilhelm Weber (1804–1891), invented the electric telegraph, and used it to transmit messages. Weber, in fact, foretold the eventual importance of this method of communication.

But it was only long after Gauss's death that the mathematical world became aware of the full extent of his mathematical creativeness. In 1898 the Royal Society of Göttingen borrowed Gauss's *Notizenjournal* (scientific diary) from one of his grandsons. An analysis of the 146 brief summaries which this record contained indicated that during the period from 1796 to 1814 Gauss anticipated practically all the major discoveries of his contemporaries and immediate successors and that he held a priority he never claimed in discovering non-Euclidean geometry, complex variable theory, elliptic functions, quaternions, etc. In delaying publication, Gauss was like Newton. But whereas Newton was tardy, Gauss *never* published some of his most important findings. Why he failed to announce his discoveries has not been adequately explained. There are those who claim that Gauss did not publish because he was a perfectionist and never found his work good enough; they point to his motto, *Pauca sed matura* (Few but ripe). Others suggest that the feeling that none of his research was ever *druckreif* was due to a trauma produced by the rejection which the French Academy of Sciences accorded to his first and possibly greatest masterpiece, the *Disquisitiones Arithmeticae.*

As has been said repeatedly at other points, the purely personal aspects of the life of a man like Gauss are less significant than his role in the history of science. Nevertheless, for the sake of human interest, some facts on record should be recounted. At age twenty-eight Gauss married Johanne Osthof of Braunschweig, who bore him two sons and a daughter. She died in 1809, shortly after the birth of the second of her sons. Her husband's letters and the statements of those who knew him well reveal that the loss of his young wife was a source of permanent sorrow to him. Museum researchers have performed chemical analyses of certain markings that appear in letters where Gauss referred to the death of Johanne. These spots have been proved to be tear stains, and it has been assumed that the tears were those of Gauss. It is true that he married Johanne's friend, Minna Waldeck, less than a year after his loss, but this appears to have been a marriage of convenience, for the sake of his three motherless children. There were two sons and a daughter by the second marriage. Two of Gauss's sons emigrated to the United States, and their numerous descendants are citizens of our country today.

Gauss was "all mathematician," as he himself said. His only nonmathematical activities were a few intellectual hobbies like studying foreign languages. He knew English and Russian, for example, and enjoyed reading literature in both languages. He liked to correspond with scientists in their native languages. Thus he carried on a correspondence with a certain Monsieur Leblanc, whose discoveries in the higher arithmetic Gauss lauded and encouraged. In 1807 the French mathematical correspondent intervened in Gauss's behalf with the French General Pernety, whose troops were occupying Hanover. This resulted in Gauss's discovery that Monsieur Leblanc was in reality a woman, Sophie Germain (1776–1831). Hers is one of the few important feminine names in mathematical history. Prior to her day there were only Hypatia, Maria Gaetana Agnesi (1718–1799), and the Marquise du Châtelet (1706–1749), the last named being more of an expositor than a creative mathematician. Sophie Germain made notable discoveries not only in number theory but also in acoustics and the theory of elasticity.

There is a record of the letter which Gauss wrote to thank Mademoiselle Germain for her intercession with the French military authorities, and to express his surprise concerning her true identity. "How can I describe my admiration and amazement in learning that the esteemed Monsieur Leblanc is now transformed into someone so important?" wrote the foremost world mathematician. ". . . A talent for abstract thought in general, and for mathematics in particular, is exceedingly rare. . . . But when a member of the sex which, according to our prejudices, must find enormous difficulties in mathematical research, actually succeeds in penetrating the most problematic issues, then one must acknowledge her extraordinary courage and her great genius. . . ." Finally, the letter closes with a date bearing a sentimental allusion, "My birthday—April 30, 1807."

Several months after this birthday, Gauss wrote to another friend: "Lagrange is deeply interested in the higher arithmetic; he considers the two special theorems

I formulated 'most beautiful and most difficult to prove.' But nevertheless Sophie Germain has succeeded in proving them."

Gauss and the French woman mathematician continued to carry on their mathematical correspondence. In 1831 he influenced the faculty of Göttingen in their decision to grant an honorary doctorate to Sophie Germain, but she died just a month before the degree was to be conferred. Her illustrious mentor, who was just a year her junior, was to survive her by almost a quarter of a century. He was active in significant research almost to the day of his death, February 23, 1855.

20

Post-Relativity Geometry

Riemann extended Gauss's notion of geometry in a way that was vital for the general relativity of 1916. But in the quest for a unified field theory, further generalizations of geometry arose. *Non-Riemannian geometry* was essentially the result of concepts shaped in 1917 by Tullio Levi-Civita (1873–1941), whose biography and contributions to mathematics will be considered in the present chapter.

That certain geometrical or physical entities are affected by motion should be no surprise to those who have read all the preceding chapters. But now come the added revelations that, in non-Euclidean or "curved" spaces, parallelism is a relative concept and it is impossible to maintain directions in the course of motion. Thus, suppose that at some point P of a curved surface like a sphere, two small rods are initially coincident so that they have the same direction, and then while one rod remains at rest the other moves by infinitesimal steps so that in each tiny displacement it maintains its direction. If the bottom of the moving rod is attached to a closed curve starting at P and if the moving rod makes the complete circuit, one might expect that on its return to P it would have the same direction as the rod that remained at rest, marking the original direction of the moving rod. But it can be proved that this is *not* the case and, if the tour of the moving rod has enclosed a considerable area, the discrepancy in direction on its return will be very large. Of course this would not happen if the rod were to move around a closed curve in a Euclidean plane (Figure 20.1) for then it would return to its exact original direction. Nontransference of direction is just another manifestation of the non-Euclideanism of a space. Levi-Civita attacked this problem in 1917 by generalizing the ordinary concept of parallelism. His notion was to be invaluable in progress toward a "unified field theory" and a powerful stimulus to the development of post-relativity geometries.

If, prior to the study of plane geometry, a child is questioned on the meaning of parallel lines, he will usually say that they are lines that have the same direction, an idea more accurately formulated by the Euclidean theorems which prove that lines forming equal corresponding angles with a transversal are parallel, and conversely (Figure 20.2). We shall now, in *three* steps of increasing generality, extend the elementary Euclidean concept to the parallelism of Levi-Civita.

In Chapter 3 we indicated that the assumptions about parallelism constitute the essential difference between Euclidean and non-Euclidean geometries "in the

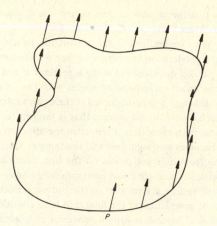

Figure 20.1

large," that is, over an entire surface. Hence for geometry "in the small," that is, differential geometry, it seems logical to consider the question of parallel motion in small regions of a curved surface and to relate parallelism in one "neighborhood" to that in an adjacent small area. Therefore we imagine, once again, creatures confined to a two-dimensional curved surface, and we observe their geometric activities from our vantage point in "outer space." If, in a small region of their universe, they attempt to construct Figure 20.2, that is, if they attach one end of a small "straight" rod to a "straight" line and try to move it parallel to itself, we may see both the line and the rod as *curved* geodesics. But then we may decide that, for a small portion of the surface, there is little difference between the small curved rod and a straight line vector *tangent* to it (and the surface). Hence we agree to reason about parallelism in terms of such small tangent vectors.

With the point of view adopted, our first generalization can deal with parallelism on a *developable* surface like a cylinder, a cone, or the surface of Figure 19.5. When such surfaces are rolled out on a plane, all their geodesics become straight lines, and conversely. If our fictitious creatures live on a developable surface, and if the "flattening" process causes the successive positions of the small rod, or tangent vector, to form a diagram like Figure 20.2 where corresponding angles

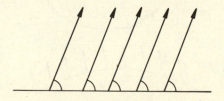

Figure 20.2

are equal, one says that the *original* tangent vector was displaced *parallel* to itself *in the sense of Levi-Civita.*

The next step is to generalize the above situation to a *definition* of parallel displacement along a geodesic on *any curved surface* whatsoever: A tangent vector is said to undergo parallel displacement along a geodesic if and only if the moving tangent vector forms equal corresponding angles with the geodesic. The thought in this case is that, although a general curved surface cannot be flattened out on a plane, one can attach a developable surface that is tangent to the general surface all along the geodesic path of motion. Then when the attached surface is unrolled, the geodesic path becomes a straight line and the tangent vectors are transformed into vectors issuing from different points on the line. Since angles are invariant in the flattening process, there are equal corresponding angles in the plane figure, and the transformed vectors are parallel in the ordinary Euclidean sense.

In the process of generalization the final step is to consider parallel displacement along *any curve C* (not necessarily a geodesic) on a general curved surface. Once again the formulation uses parallelism on the developable surface attached to the general surface and tangent to it all along the curve *C*. If that curve is twisted, that is, does not lie in a single plane, the attached surface might look something like the one pictured in Figure 19.5. When the developable surface is rolled out on a plane, the tangent vector *v* issuing from *P* of curve *C* is mapped onto *v'* issuing from point *P'* of curve *C'* in the plane (Figure 20.3). In the plane, *P'* traces curve *C'*, and the vector *v'* is required to remain actually parallel to itself in the elementary Euclidean sense. Finally the plane is turned back into the developable surface so that all points go back to their original positions. The various positions of *v'* now become positions of the vectors *v* tangent to the surface along the curve *C*, and these vectors are called *parallel in the sense of Levi-Civita.* It must be emphasized that the vectors *v* will *not*, in general, be parallel in the usual sense. We have given a geometric description, but the *exact* expression of the Levi-Civita notion of parallel displacement is furnished in terms of differential equations.

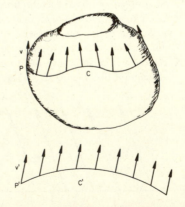

Figure 20.3 Levi-Civita parallel displacement

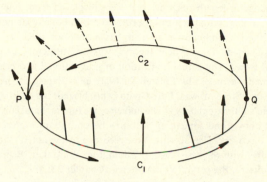

Figure 20.4

Even the parallelism of Levi-Civita is *not* an absolute property, but is *relative* to the curve of transport. Thus, if points P and Q on a general curved surface are joined by a geodesic and a nongeodesic, then, as we have indicated, parallel displacement of a vector along the former is defined somewhat differently from that on the latter, and therefore it seems logical that the total effect on the direction of the vector might possibly be different in the two cases. Again, suppose that it did not matter along which curve the vector moves in the Levi-Civita parallel displacement from P to Q (Figure 20.4) of a curved surface. Then it would reach the same direction at Q by moving along C_1 as along C_2. If this were true, it could be carried "parallel" to itself along C_1 and then *back* along C_2, so that it would return to its original direction, which, as stated at the beginning of this chapter, is *not*, in general, possible. Therefore, Levi-Civita parallel displacement along C_1 leads to one direction at Q, whereas parallel displacement along C_2 leads to a *different* direction at Q.

The *non-Riemannian* geometry that grew from this idea is like the Riemannian in that it deals with n-dimensional continua with different meshworks ($u_1, u_2, \ldots, u_n$). Only there is no metric ds^2, since the new type of geometry has nothing to do with distance, but is based either on the idea of *displacement*, a generalization of Levi-Civita's parallelism, or on the concept of *paths*, a generalization of the notion of *geodesics*. The latter kind of non-Riemannian geometry was founded about 1922 by Oswald Veblen and Luther P. Eisenhart. The displacement generalization was begun about 1918 by Hermann Weyl in his attempt to create a unified field theory. He developed a logical, beautiful type of pure geometry, but it did not fit the facts of our physical universe.

Weyl worked on the effects of non-Euclideanism on nontransference of *length*. In other words, instead of concentrating on the fact that vectors will have different directions when transported from P to Q by parallel displacement along different paths, he assumed that two unit measuring rods would have different *lengths* if they were displaced from P to Q along different paths. Later, Eddington created an elegant generalization of Weyl's idea by assuming that the measuring rod would change in length not only if it were displaced from one point of the continuum to another, but also if it were rotated around one of its points as pivot. And again, although Eddington's geometry is a superior piece of pure mathematics, the associated "unified field theory" was unsatisfactory as a description of the physical world. The actual unification of gravitational and electromagnetic theories was finally achieved by Einstein in 1950, as explained at the close of the previous chapter. Since it was Levi-Civita who blazed the mathematical trail toward Einstein's ultimate theory of the universe, the fundamental concept involved was explained above, and something will now be said about the creative thinker who made the discovery. The facts are those recorded by the noted British geometer W. V. D. Hodge, with the assistance of Eddington, Whittaker, Beniamino Segre, and Enrico Volterra, the son of Vito Volterra (Chapter 23).*

Tullio Levi-Civita and Vito Volterra were both mathematicians who in the course of active lives contributed greatly to the high reputation enjoyed by Italian mathematics in general, and the school of mathematics in Rome in particular; both had made many contributions which have found a permanent place in mathematical literature, and both ended their days as victims of a political system which destroyed institutions and liberties in which they were firm believers.

Levi-Civita was born in Padua on 29 March 1873, the son of Giacomo Levi-Civita and his wife, Bice Lattis. The family was a wealthy one, well known for its strong liberal traditions. Giacomo Levi-Civita was a barrister, jurist and politician, and was for many years mayor of Padua, and a Senator of the Kingdom of Italy. As a young man he had served as a volunteer and fought with Garibaldi in the campaign of 1866, and he had played an important part in the Risorgimento.

Giacomo Levi-Civita was anxious that his son should follow in his footsteps as a barrister, but Tullio's interest in the physical and mathematical sciences was apparent even in early childhood, and when he expressed a wish to follow his own inclinations his father never opposed him; and in later years the son's eminence in the scientific world was a source of great pride to the father. Consequently, when he completed his classical studies at the Ginnasio-Liceo Tito Livio in his native city at the age of seventeen, Tullio Levi-Civita entered the faculty of science at the university of Padua as a student of mathematics, and four years later he took his degree.

Amongst his teachers at the university of Padua were D'Arcais, Padova, Veronese, and Ricci-Curbastro (known to the scientific world simply as Ricci). The two last-named were the most distinguished, and both had considerable influence on the future career of their brilliant pupil. The influence of Ricci is the more obvious, since it developed into active collaboration, but probably Veronese's influence was quite as important, since it is largely to him that Levi-Civita owed the remarkable spacial intuition and familiarity with multidimensional space which characterizes the younger man's contributions to the Ricci-Levi-Civita partnership in the absolute differential calculus.

Levi-Civita's undergraduate days were not over before he began to write mathematical papers, and his ability was quickly recognized. At the age of twenty-five he was appointed professor of mechanics at Padua. For twenty years he held that post, and these were among the most productive of his life. In 1918 he was called to the chair of mechanics at Rome, a post which he held for another twenty years, until racial

* Royal Society of London, *Obituary Notices of Fellows*, Vol. 4, No. 11 (November 1942), pp. 151 ff.

discrimination, introduced into Italy in 1938, brought about his removal from office. Until then his life was uneventful, spent in the happy pursuit of his mathematical interests.

The work by which Levi-Civita is certainly best known is that on the absolute differential calculus, with its applications to relativity theory. The study of the particular class of invariants known as tensors goes back to the work of Riemann and Christoffel on quadratic differential forms (though the name tensor was only introduced by Voigt in 1898). In 1887 Ricci published his famous paper in which he developed the calculus of tensors including the important operation of covariant differentiation. For a considerable number of years following the publication of this paper he was engaged in working out his "absolute differential calculus" aided by a number of able pupils, foremost among them being Levi-Civita. The results of the work of Ricci, Levi-Civita and others were finally published in a joint memoir by Ricci and Levi-Civita which appeared in 1900 and which presented the theory of tensors essentially in the form used by Einstein and others fifteen years later.

In 1917 Levi-Civita made an advance in the absolute differential calculus of fundamental importance, with the introduction of the concept of parallel displacement. Few mathematical ideas have found such diverse applications so quickly as that of parallel displacement. It is the basis of the unified representation of gravitational and electromagnetic fields in relativity theory, and there are still more far-reaching consequences which are not yet fully recognized in physics. Levi-Civita's direct contributions to relativity theory are substantial, but they are of a less conspicuous nature. In addition to the large number of papers which Levi-Civita published on the absolute differential calculus and relativity he published two books, *Questioni di meccanica classica e relativistica* (1924) and *Lezioni di differentiale assoluto* (1925). Both of these have become standard works, and the latter was translated into English and has been widely read in the translation as well as in the original.

Although he is most famous for his work in relativity theory, Levi-Civita has had an important influence on many other branches of mathematics. It is particularly necessary to mention his work in analytical dynamics, to advance which subject he did as much as any one during the earlier years of the twentieth century. The largest individual group of these papers deals with the problem of three bodies [Chapter 25]. In 1923 Levi-Civita, in association with Professor Amaldi, published a three-volume work on rational mechanics, which is now one of the accepted classics on the subject, and which has been translated into various other languages, including Russian.

Hydrodynamics is another subject which attracted Levi-Civita's attention, and to which he made a considerable number of contributions. His work on hydrodynamics is to a certain extent bound up with his work on the general theory of systems of partial differential equations. His work on this subject forms an important addition to the well-known Cauchy-Kovalevsky theory.

It was natural that the efforts made to find a common framework to contain both quantum mechanics and the general theory of relativity should prove of the greatest interest to Levi-Civita. In 1933 he published a paper in which he proposed to replace Dirac's first order equations by a set of second order equations which took into account the gravitational field. When the two sets of equations are compared in the case in which there is no gravitational field it is found that Levi-Civita's equations are reconcilable with Dirac's when the electromagnetic field is either purely electric or purely magnetic, but not in the general case.

Though reference has been made to some of Levi-Civita's more outstanding researches there are very many which must be passed over. It ought to be mentioned, however, that in addition to his work in the realm of pure science he was frequently consulted by technicians and engineering firms on problems of practical engineering. In this way he was brought to do work of considerable value to the outside world, notably in connexion with the construction of submarine cables and the vibration of bridges.

Again, there is a great deal of evidence of Levi-Civita's interest in branches of mathematics in which he never published any researches. Old pupils frequently consulted him on their problems, and always found him acquainted with their work, even though it differed widely from anything that Levi-Civita had taught them, and he was always ready both with encouragement and with useful advice.

Levi-Civita was not at first appearances a very striking figure. The first impression received was of an exceedingly small man (he was only about five feet tall) who was very short-sighted. His bearing was quite unpretentious; but, after having talked to him for a while, one was particularly struck by the vivaciousness and precision of his discourse and his very wide knowledge extending over pure and applied mathematics, astronomy and physics, and also by his remarkable acquaintance with the literature on these subjects, both old and new. One noticed in particular his quick grip on a problem, and his passionate interest in all sorts of scientific questions.

In spite of his small frame Levi-Civita was very robust, and enjoyed excellent health until he was well over sixty. What energy he had to spare from his work was devoted to his three great hobbies—mountaineering, cycling, and foreign travel. As a young man he devoted most of his vacations to mountaineering in the Dolomites, and in spite of his physical handicaps he was a good climber. As he grew older his worsening eye-sight curtailed his climbing activities, but he kept them up as far as he could for many years. He enjoyed cycling, and was often to be seen cycling round the countryside near Padua while he was professor there, and subsequently during his frequent visits to his parents; and when he was no longer able to climb he continued to visit his beloved mountains on his bicycle.

He was singularly well placed for indulging his third passion, foreign travel. His private fortune and freedom from domestic worries removed many obstacles from his way, and opportunities were regularly given him by invitations to visit countries in all parts of the world in order to deliver lectures to scientific gatherings. These visits he enjoyed intensely; he could see new places, meet new people, and, thanks to his own personal charm, make a host of new friends. Indeed, Levi-Civita was one of the personally best known and best liked mathematicians of his time.

Levi-Civita was also fortunate in his home life. There was a strong bond of affection between his father, who died in 1922, and himself. While the father was intensely proud of his son's scientific achievements, the son was equally proud of the father's record in the Italian wars of liberation. Though not himself an active politician, Levi-Civita was extremely interested in politics, and remained true to the noble traditions of his family. A visitor to his study could be in no doubt as to his beliefs; while three of the walls were lined by bookcases, the fourth remained empty save for two solitary portraits, one of his father and the other of Garibaldi. Levi-Civita always felt a great tenderness for his mother, and visited her regularly, either at the family house in Padua, or at her villa in Vigodarzere, a nearby village where she lived for several years before her death in 1927. After her death the villa was kept on by her daughter, Ida Senigaglia, and Levi-Civita continued to spend a part of each year there.

In 1914 Levi-Civita married Libera Trevisani. She had been his pupil at the university of Padua, and had taken her doctorate in mathematics. She proved herself a clever and affectionate companion to him, and she was a very gracious hostess, not only to the many eminent scientists who came to visit them from afar, but to the more humble students who regularly visited their home. She accompanied her husband on his many travels and shared with him the many friendships which he made on these journeys. There were no children of the marriage.

For Levi-Civita, research and teaching went hand in hand, and he guided a great number of pupils in fields in which he was the pioneer. His teaching was not circumscribed by any curriculum, as was usual in an Italian university, but was freely given to all who came to consult him. During his vacations, either in the Dolomites or at Vigodarzere, his former pupils would come to be near him, and he would follow their researches with the utmost interest. With infinite patience and unselfishness he would enter into the problems which they brought to him; nothing gave him more pleasure than to have an opportunity of helping them, and it was with the greatest pride that he would present their works for publication by one or other of the many learned societies of which he was a member. Indeed, as one of his pupils once remarked, no one ever merited more than he did the title of Maestro.

He was a born teacher. His scientific papers are models of lucidity, and his books are amongst the easiest reading on their specialized topics. In conversation he could give in simple terms a very simple account of an abstract theory. In acquiring his command over such vast fields of science he was greatly aided by being the possessor of an unusually good memory.

Many honours came to Levi-Civita. He received honorary degrees from many universities throughout the world, including Toulouse, Aachen, Amsterdam and Paris. Academies in all countries of the world honoured him by election to honorary membership. The list is too long to repeat, but included the Institut de Paris and the Berlin academy, and societies in Leningrad, Madrid and South America. In his own country he was a member of the Reale Accademia dei Lincei, the Reale Accademia d'Italia and the Pontificia Accademia delle Scienze dei Nuovi Lincei. When, in 1936, Pope Pius XI dissolved the last named and replaced it by the Pontificia Accademia delle Scienze, an international body, Levi-Civita became an original member of the new academy. The Sylvester medal of the Royal Society was conferred on him in 1922, and in 1930 he was elected a foreign member. He was elected an honorary fellow of the Royal Society of Edinburgh in 1923, and of the London Mathematical Society in 1924. He was also an honorary member of the Edinburgh Mathematical Society, and attended one of its colloquia in St. Andrews.

Although he did not take any active part in politics, he could not remain indifferent to the rise of fascism in Italy, and in 1925, after the "Matteotti affair" he was, with Volterra and many other Italian scientists, a signatory of the "Manifesto Croce," which stigmatized fascism and deprecated its growing power in Italy. For some time, however, his scientific renown protected him from persecution. But in September 1938 the government issued decrees removing from office all professors of Italian universities who were of Jewish origin, and dismissing them from Italian academies. Levi-Civita, who came under the ban, found himself cut off from all that made life interesting to him. It is recorded that he learned of the decrees while staying with his sister at Vigodarzere, when someone happened to switch on the radio as they were being announced. His expression did not change, and he went out for his afternoon walk as usual. The blow soon told on him, however. His health began to fail. After he returned to Rome severe heart trouble developed, and as he was forbidden by his doctors to take any long journey he could not accept any of the offers of asylum which came to him from foreign universities. The remaining years of his life were very sad; he was confined to his room, and unable to continue his work. He died on 29 December 1941 of a stroke. At first the Roman newspapers, except the *Osservatore Romano*, the organ of the Vatican, ignored his death, and it was only after the Pontifical Academy had used its influence that the family were able to announce in the newspapers the fact of his death and the arrangements for his funeral.

If one is interested in *pure* geometry, then the ultimate generalization is to be found in the work of Élie Joseph Cartan (1869–1951), one of the foremost mathematicians of the recent period. Cartan's concept includes those of Levi-Civita,

Weyl, and Eddington as special cases. For a description of the transition to post-relativity geometry, and, later, for an account of some phases of Cartan's research, we shall quote statements made by the late J. H. C. Whitehead (1904–1960), whose specialties were related to Cartan's. John Henry Constantine Whitehead who was himself a leading mathematician of our day, specialized in topology (Chapter 25) and, like Cartan, applied that subject to the study of generalized spaces. Whitehead was the author of many research papers, some written in collaboration with American mathematicians. With Oswald Veblen of the Institute for Advanced Study he wrote *The Foundations of Differential Geometry* (1932). He was born in Madras, India, and was the son of the Bishop of Madras and the nephew of the mathematical philosopher Alfred North Whitehead. J. H. C. Whitehead taught at Oxford University from 1932 to 1946, and then from 1947 to 1960. In between times he was at Princeton and the Institute for Advanced Study. He received his Ph.D. from Princeton in 1930, where he was a research fellow from 1929 to 1932, and then visiting lecturer from 1946 to 1947. He was a visiting member of the Institute in the spring of 1960, when he died suddenly on May 8 of that year. His description of the transition from pre-relativity to post-relativity geometry will now be presented.*

Let us briefly recall the state of geometry just before and just after the discovery, in 1916, of general relativity. During the period between this time and 1870 ideas concerning the foundations of geometry had been dominated by Felix Klein's Erlanger Program. Riemann's philosophical concept of geometry had been ignored by most geometers though the analytical side of his theory had been extensively developed, notably by Ricci and Levi-Civita. After the discovery of general relativity, which was based on Riemannian geometry, it was realized that the Erlanger Program was no longer adequate as a general description of geometry. The first person to understand the mathematical implications of this, in anything like their full generality, seems to have been Weyl, when he introduced generalized affine, projective, and conformal geometries, whose relation to their classical counterparts was analogous to that of Riemannian to Euclidean geometry.

After the publication, in 1918, of Weyl's first paper in this field, and of one by J. A. Schouten, the development of the subject was guided by two rather different sets of ideas. One of these, due to Cartan, is described below. According to the other, which was first formulated explicitly by Veblen in 1928, at the Bologna International Congress, a geometry is the study of a "geometric object," such as a tensor or projective connexion, from which others can be derived by certain formal processes. One of the main objectives is to define a complete set of *invariants*, such as the curvature tensor and its successive covariant derivatives in generalized affine geometry.

Relativity may have been the "practical" motivation for Cartan's creation of generalized spaces, but his methodology in geometry and, in fact, in all his specialties, drew its strongest inspiration from the geometric and group-theory creations of Sophus Lie. Therefore, if Levi-Civita's life and mathematical

---

* Royal Society of London, *Obituary Notices of Fellows*, Vol. 8, No. 21 (November 1952), pp. 71 ff.

contributions are one preliminary to a discussion of Cartan's work, the ideas and activities of Lie should logically form another.

Part of Lie's story was related in connection with Klein. We recall how the two young mathematicians met in 1869 and then went to Paris to confer with leading French mathematicians, in particular with Darboux. Then, in 1870, when the Franco-Prussian war broke out, Klein returned to Germany and Lie, an inveterate hiker and mountain climber, set out for Italy *on foot*. Like Archimedes, he had no interest in the war being waged around him, and was unaware that military precautions might limit one's freedom to practice mathematics. Hence, at age twenty-eight, Lie *almost* duplicated the last hour of Archimedes' life. The Syracusan's mathematical notes and geometric diagrams (drawn on the sand along the Sicilian shore line) made him suspect to a Roman soldier, who stamped on the figures and slew Archimedes when he protested. Two millennia later, the Norwegian geometer paused in the first lap of his walking tour, rested in the shade of a tree, and entered in a notebook some of his thoughts on how a geometry of lines could be transformed into a geometry of spheres. He was seized by a French soldier who thought the towheaded giant was a German spy, and mistook his diagrams for copies of French military plans, his symbols for a secret code, the mathematicians named in his notes for a list of guilty parties. Lie, imprisoned at Fontainebleau, appealed for help to Darboux, Chasles, Bertrand, and other French mathematicians who knew him. It took almost a month before they could obtain his release.

According to Darboux, the Fontainebleau prison was a comfortable place, and Lie was able to make use of his enforced leisure to enlarge on the ideas in the notes which had caused his arrest. In this way he made his first important mathematical discovery, which he presented in a short paper to the French Academy a month after he left prison. Of this brief note, Darboux said, "Nothing resembles a sphere less than a straight line and yet . . . Lie found a transformation which makes spheres correspond to straight lines and hence makes it possible to derive theorems about spheres from those about lines and vice versa."

The 1870 paper was Lie's mathematical debut, a relatively late one as mathematicians' careers go. (Marius) Sophus Lie was born in 1842 at Nordfjordeid (near Flöro), Norway, where his father was a minister. He showed no special interest in mathematics at school, and as late as 1865, when he had completed his courses at the University of Christiania (in preparation for teaching in the secondary schools), he was undecided between philology and mathematics. A reading of the works of the geometer Plücker first aroused his interest and made him aware of his own mathematical talent. After travels in Germany, Italy, and France from 1869 to 1871, he returned to Christiania, took his doctorate in 1872, and became "extraordinary" (associate) professor of mathematics, a position he held until 1886. In that year he was invited to Leipzig, where he was offered a full professorship and the directorship of the mathematical institute. He held this position until 1898. In an earlier chapter, we have had Frau Klein's account of Lie's 1898 visit to Göttingen. Shortly thereafter, the Norwegian parliament created a special chair for him at Christiania, with an exceptionally high salary. But he lived only a half year after receiving this honor, dying in February 1899, at age fifty-six.

Other recognition had come to him before that, however. He had been invited to give a special address in Paris in 1895 for the centenary of the founding of the great École Normale, which along with the École Polytechnique, has always supplied France and the world with leading mathematicians and scientists. He spoke on the importance of the work of Galois, his idol. Then, in 1898, Lie was the first recipient of the Lobachevsky Prize, an award of 500 rubles given by the Physico-Mathematical Society of the University of Kazan for the best original research work submitted in the field of geometry, preferably non-Euclidean.

The book which led to this prize was indeed a remarkable one containing in one section the solution of the so-called Riemann-Helmholtz space problem, a result important not only in group theory and differential geometry, but also in applications to physiology. What Lie did was *correct* and *rigorize* Hermann von Helmholtz's 1868 and 1884 attempts at solution. Helmholtz had sought some justification of Riemann's use of the *quadratic* metric;

$$ds^2 = g_{11}du_1{}^2 + 2g_{12}du_1du_2 + \ldots + g_{nn}du_n{}^2$$

Why, he asked, is it not equally logical to postulate ds^2 as the sum of *fourth* powers, or to express it by some entirely different type of formula? His answer to this question was a kinematic theory of space, that is, an explanation characterizing space according to the movements possible within it. (The idea is much the same as that of Klein's transformations.) Helmholtz drew the foundation of his theory from experimental sources and set down four basic postulates from which he was able to deduce Riemann's metric. Lie's 1898 treatment of the same problem was vastly superior, at least from the point of view of pure mathematics. His assumptions referred to transformations and invariants; they enabled him to establish rigorously and exhaustively the general results of Riemann and Helmholtz.

Lie's period of mathematical creativity (1870–1899) was devoted entirely to the study of *continuous* transformation groups and their invariants. He showed how these two related ideas can furnish principles for simplifying and unifying diverse mathematical subjects—geometry, the theory of *algebraic* invariants (largely the work of Cayley and Sylvester), mechanics, ordinary and partial differential equations. The "contact transformations" he conceived in 1870 during his sojourn in prison were the means of bringing all of Hamiltonian dynamics under the wing of group theory.

Application of these same contact transformations also shed light on some of the most difficult theoretical questions associated with partial differential equations. Like Poincaré, who felt that groups are "all of mathematics," and indicated that "physical laws are differential equations," Lie devoted himself to groups and asserted that "differential equations constitute the most important branch of modern mathematics."

The Norwegian was a prodigious worker, and with the assistance of his disciples, Friedrich Engel (1861–1941) and Georg Scheffers (1866–1945), he set down his ideas in full in colossal tomes. His *Theorie der Transformationsgruppen* appeared in three large volumes between 1888 and 1893. Then there were *Differentialgleichingen* (Differential Equations) in 1891, *Kontinuierliche Gruppen* (1893), and *Berührungstransformationen* (Contact Transformations) in 1896. In the era between 1900 and 1930 these works, with all their excessive detail, were studied by graduate mathematics students all over the world. But since that time Lie's group-theory and geometric notions have been distilled, and their abstract essence extracted. The "pure" part of his work is thus seen in all its profundity, and is being extended in the vigorous research in today's abstract algebra. The study of "Lie groups" and "Lie algebras" is receiving the attention of the best mathematical minds.

In personality, Lie conformed to the conventional idea of the genius. He was shy and withdrawn, but nevertheless outspoken on occasion and given to temper tantrums. In relation to Klein, we have described Lie's tendency toward depression and notions of persecution. To "relax his nerves," he went on hikes or read the novels of Walter Scott and Frederick Marryat. George Abram Miller (1863–1951), an American researcher in group theory in the early years of its development, attended Lie's classes in Leipzig, and later reminisced about them. He recounted that Lie was careless in dress, and sometimes appeared in class without collar or necktie, or even with his hat on. Miller described one of Lie's verbal outbursts, when the latter stated, "Mathematics may be the royal road to learning, but unfortunately we mathematicians do not receive royal pay." Lie's students respected him, but gossiped about his eccentricities. Nevertheless the doctoral candidates at Leipzig flocked to Lie, who assigned easy topics for research and provided a tremendous amount of assistance. But pupils found him personally crude and were wont to contrast him with the suave Klein. They pronounced the latter a "cultured gentleman" but said, "The best you can say for Lie is that he is a good fellow." The consensus today is that he was a very good fellow indeed.

Now, to link the geometric notions of Lie, Levi-Civita, and Cartan, let us recall from the previous chapter that in the small "neighborhood" of a point in a Riemannian space the geometry is approximately Euclidean, so that for an ordinary 2-dimensional surface, the geometry of the surface and that of the tangent plane are the same in the locality of the point of contact. In the case of an *n*-dimensional Riemannian manifold, this space can be considered as the aggregate of its tangent spaces or, in the words of Cartan, as a multitude of tiny facets each of which is a Euclidean space. But all these facets or tangent spaces must be fused into a whole, the *Riemannian manifold*. There must be a "connection" between one tangent space and a "neighboring" or "infinitesimally close" tangent space, that is, an analytic formulation of the "displacement" of the former into the latter. Cartan's generalized spaces consist of (1) an underlying manifold with (2) a tangent space at each point of this manifold in which (3) a Kleinian geometry obtains (that is, an affine or projective or conformal geometry related to the affine, projective, or conformal transformations of the tangent space, respectively), and (4) a connection or displacement formula defining the displacement of any tangent space along any curve from the point of contact to a neighboring point.

When these four points are treated less intuitively and more mathematically, the influences of Levi-Civita and Lie become more apparent. The Levi-Civita tangent spaces are Euclidean, and his parallel displacement is a special case of (4) above. As for Sophus Lie, the notions he contributed are numerous. In the first place, he emphasized the importance of the principle of *linearity* in analytic approximations. The idea is similar to the numerical integration techniques, in which one uses a chord or straight-line segment as an approximation to a small curved arc. Lie indicated that in arbitrary analytic transformations, however complicated the functional formulations, the differentials dx, dy, etc., representing the differences in the coordinates of neighboring points, undergo only a homogeneous affine transformation. In Chapter 17 the plane group of such transformations was expressed by

$$x' = a_1 x + b_1 y$$
$$y' = a_2 x + b_2 y$$

Then if u, v are Gaussian coordinates on a surface, and these coordinates undergo arbitrary analytic transformations, the corresponding Lie "infinitesimal transformations" are

$$d(u') = a_1\, du + b_1\, dv$$
$$d(v') = a_2\, du + b_2\, dv$$

where, for each analytic transformation of the entire surface, the values of a_1, b_1, a_2, b_2 depend only on the particular point of the surface, that is, are functions of u and v. The different points of the tangent plane are obtained by varying du and dv. Similarly, in a Riemannian space of n dimensions, $d(u_1')$, $d(u_2')$, ... $d(u_n')$ would be expressible as linear functions of du_1, du_2, ..., du_n, and the points of the tangent space would be obtained by varying du_1, du_2, ..., du_n. Then to each arbitrary analytic transformation of the underlying manifold there corresponds a homogeneous affine transformation in each tangent space, and corresponding to the set of analytic transformations of the manifold there is the affine group and affine geometry in each tangent space. If the manifold is Riemannian, the metric

$$ds^2 = g_{11}dx_1^2 + 2g_{12}dx_1 dx_2 + \ldots + g_{nn}dx_n^2$$

fuses the affine geometries of all the tangent spaces. The non-Riemannian geometries are not metric, and in them the generalizations of Levi-Civita's parallel displacement integrate the geometries of the tangent spaces in a different way. The Cartan generalized spaces have nonaffine geometries (projective or conformal, etc.) in the tangent spaces, and different connections or displacements from those in the Riemannian or Levi-Civita manifolds.

Concerning Élie Cartan and his mathematical work Professors Shing-Shen Chern and Claude Chevalley state*:

---

* Shing-Shen Chern, and C. Chevalley, "Élie Cartan and His Mathematical Work," *Bull. of the Amer. Mathematical Society*, Vol. 58, No. 3 (March 1952).

Undoubtedly one of the greatest mathematicians of this century, Élie Cartan's career was nevertheless characterized by a rare harmony of genius and modesty. He was born on April 9, 1869, in Dolomieu (Isère), a village in the south of France. His father was a blacksmith. Cartan's elementary education was made possible by one of the state stipends for gifted children. In 1888 he entered the École Normale Supérieure, where he learned higher mathematics from such masters as Tannery, Picard, Darboux, and Hermite. His research work started with his famous thesis on continuous groups, a subject suggested to him by his fellow student Tresse, recently returned from studying with Sophus Lie in Leipzig. Cartan's first teaching position was at Montpellier, where he was *maître de conférences;* he then went successively to Lyons, to Nancy, and finally in 1909 to Paris. He was made a professor at the Sorbonne in 1912. The report on his work which was the basis for this promotion was written by Poincaré; this was one of the circumstances in his career of which he seemed to have been genuinely proud. He remained at the Sorbonne until his retirement in 1940.

Cartan was an excellent teacher; his lectures were gratifying intellectual experiences, which left the student with a generally mistaken idea that he had grasped all there was on the subject. It is therefore the more surprising that for a long time his ideas did not exert the influence they so richly deserved to have on young mathematicians. This was partly due to Cartan's extreme modesty. But in 1939, the celebration of Cartan's scientific jubilee, J. Dieudonné could rightly say to him: "... *vous êtes un 'jeune,' et vous comprenez les jeunes"*—it was then beginning to be true that the young understood Cartan.

In foreign countries, particularly in Germany, his recognition as a great mathematician came earlier. It was perhaps H. Weyl's fundamental papers on group representations published around 1925 that established Cartan's reputation among mathematicians not in his own field. Meanwhile, the development of abstract algebra naturally helped to attract attention to his work on Lie algebra.

Cartan was elected to the French Academy in 1931. In his later years he received several other honors. Thus he was a foreign member of the National Academy of Sciences, U.S.A., and a foreign Fellow of the (British) Royal Society. In 1936 he was awarded an honorary degree by Harvard University.

Closely interwoven with Cartan's life as a scientist and teacher had been his family life, which was filled with an atmosphere of happiness and serenity. He had four children, three sons, Henri, Jean, and Louis, and a daughter, Hélène. Jean Cartan oriented himself towards music, and already appeared to be one of the most gifted composers of his generation, when he was cruelly taken by death. Louis Cartan was a physicist; arrested by the Germans at the beginning of the Resistance, he was murdered by them after a long period of detention. Henri Cartan followed in the footsteps of his father to become a leading mathematician.

Cartan's mathematical work can be roughly classified under three main headings: group theory, systems of differential equations, and geometry. These themes are, however, constantly interwoven with each other in his work. Almost everything Cartan did is more or less connected with the theory of Lie groups.

Sophus Lie introduced the groups of transformations which were named after him. The idea of considering the *abstract group* which underlies a given group of transformations came only later; it appears quite explicity in the first paper by Cartan. Whereas, for Lie, the problem of classification consisted in finding all possible transformation groups on a given number of variables (a far too difficult problem in the present stage of mathematics as soon as the number of variables is not very small), for Cartan the problem was to find all possible abstract structures of continuous groups. He solved the problem completely for "simple" groups (those having no proper normal subgroups). Once the structures of all simple groups were known, it became possible to look for all possible realizations of any one of these structures by transformations of a specific nature, and, in particular, for their realizations as groups of linear transformations. This is the problem of the determination of the representations of a given group; it was solved completely by Cartan for simple groups. The solution led in particular to the discovery, as early as 1913, of the *spinors*, which were to be rediscovered later in a special case by the physicists.

Cartan also investigated the infinite Lie groups, i.e., the groups of transformations whose operations depend *not* on a finite number of continuous parameters, but on arbitrary functions. In that case, one does not have the notion of the abstract underlying group. Cartan and Vessiot found, at about the same time and independently of each other, a substitute notion which consists in defining when two infinite Lie groups are to be considered as isomorphic. Cartan then proceeded to classify all possible types of non-isomorphic infinite Lie groups.

Cartan also paid much attention to the study of topological properties of groups considered in the large. He showed how many of these topological problems may be reduced to purely algebraic questions; by so doing, he discovered the very remarkable fact that many properties of the group in the large may be read from the infinitesimal structure of the group, i.e., are already determined when some arbitrarily small piece of the group is given. His work along these lines resembles that of the paleontologist reconstructing the shape of a prehistoric animal from the peculiarities of some small bone.

The idea of studying the abstract structure of mathematical objects which hides itself beneath the analytical clothing was also the mainspring of Cartan's theory of differential systems. He insisted on having a theory of differential equations which is invariant under arbitrary changes in variables. Only in this way can the theory uncover the specific properties of the objects one studies by means of the differential equations they satisfy, in contradistinction to what depends only on the particular representation of these objects by numbers or sets of numbers. In order to achieve such an invariant theory, Cartan made a systematic use of the notion of the *exterior differential* of a differential form, a notion which he helped to create and which has the required property , of being invariant with respect to any change of variables.

Raised in the French geometrical tradition, Cartan had a constant interest in differential geometry. He had the unusual combination of a vast knowledge of Lie groups, a theory of differential systems whose invariant character was particularly suited for geometrical investigations, and, most important of all, a remarkable geometrical intuition. As a result, he was able to see the geometrical content of very complicated calculations, and even to substitute geometrical arguments for some of the computations.

In the 1920's the general theory of relativity gave a new impulse to differential geometry. This gave rise to a feverish search for spaces with a suitable local structure. The most notable example of such a local structure is a Riemann metric. It can be generalized in various ways, by modifying the form of the integral which defines the arc length in Riemannian geometry (Finsler geometry), by studying only those properties pertaining to the geodesics or paths (geometry of paths of Eisenhart, Veblen, and T. Y. Thomas), by studying the properties of a family of Riemann metrics whose fundamental forms differ from each other by a common factor (conformal geometry), etc. While in all these directions the definition of a parallel displacement is considered to be the major concern, the approach of Cartan to these problems is most original and satisfactory. Again the notion of group plays the central role. Roughly speaking, a generalized space in the sense of Cartan is a space of tangent spaces such that two infinitely near tangent spaces are related by an infinitesmal transformation of a given Lie group. Such a structure is known as a connection.

Besides several books, Cartan published about 200 mathematical papers. His major specialties, in addition to geometry, were group theory and differential equations. Cartan's papers on group theory fall into two categories, distinguished from each other both by the nature of the question treated and by the time at which they were written.

The papers of the first cycle are purely algebraic in character; they are more concerned with what are now called Lie algebras than with group theory proper. The work of Cartan's second group-theoretic period is concerned with the groups themselves, and not with their Lie algebras, and in general with the global aspect of the group.

For an account of his algebraic discoveries we return to J. H. C. Whitehead once more.*

In the years 1897, 1898 Cartan turned his attention from Lie algebras to linear associative algebras. In 1898 he proved the Wedderburn structure theorem [Chapter 28] for algebras over the real and complex fields. The methods which Wedderburn (1908) used in proving his theorem are more suitable than Cartan's to the problems of linear associative algebra. Indeed this paper of Wedderburn's is one of the outstanding contributions to the subject and it is reasonable to associate the theorem with his name. But the fundamental importance of Cartan's paper, which Wedderburn duly acknowledged, should not be forgotten.

During the years 1904 to 1909, there are his papers on infinite transformation groups, as defined by Lie. Such a group is "infinite" in the sense that its general transformation cannot be expressed in terms of a finite set of parameters. In general it is only defined as a local group, or pseudo-group, whose transformations operate on different open subsets of Cartesian space. Finally a (local) group, G, of this kind is defined as the totality of transformations which leave invariant a given set of differential equations. The transformations in G are themselves given by a set D, of differential equations. Since G is infinite the general solution of D is not expressed in terms of a finite set of parameters.

There have been very few, if any, new contributions to the general theory of infinite groups since these papers of Cartan. This is doubtless due to the difficulty of the subject and also to the appearance of temporary finality in Cartan's work. That is to say, there does not seem to have been much hope of greatly extending his theory with the methods which have been available during the last forty years. At the present time the obvious questions are those concerning global infinite groups, acting on n-dimensional manifolds. It may be that a theory of such groups will be constructed on the basis of Cartan's local theory. In this case it would not be surprising if the latter were eventually considered to be his greatest work.

From 1916 onwards Cartan's papers, with one or two exceptions, were on differential geometry, including the theory of generalized spaces and differential geometry in the large. This work on differential geometry would, by itself, have been sufficient to establish Cartan among the leading mathematicians of this half-century. It is remarkable that he embarked on it when he was nearly fifty years old and maintained a steady output of first-class work throughout the subsequent thirty years.

As for Cartan's work on differential equations, probably the best authority is his own 1945 book, *Les systèmes différentiels extérieurs et leurs applications géométriques*. But we now return to the Chern-Chevalley biography for further details of Cartan's geometric work, which is the major theme of the present chapter.†

Einstein's theory of general relativity gave a new impetus to differential geometry. In their efforts to find an appropriate model of the universe, geometers have broadened

---

* Royal Society of London, *Obituary Notices of Fellows, loc. cit.*

† Shing-Shen Chern and C. Chevalley, *loc. cit.*

their horizon from the study of submanifolds in classical spaces (Euclidean, non-Euclidean, projective, conformal, etc.) to that of more general spaces intrinsically defined. The result is an extension of the work of Gauss and Riemann on Riemannian geometry to spaces with a "connection," which may be an affine connection, a Weyl connection, a projective connection, or a conformal connection. In these generalizations, sometimes called non-Riemannian geometry, an important tool is the absolute differential calculus of Ricci and Levi-Civita. The results achieved are of considerable geometric interest. For instance, in the theory of projective connections, developed independently by Cartan, Veblen, Eisenhart, and Thomas, it is shown that when the space has a system of paths defined by a system of differential equations of the second order, a generalized projective geometry can be defined in the space which reduces to ordinary projective geometry when the differential system is that of the straight lines. Numerous other examples can be cited. The problem at this stage is twofold: (1) to give a definition of "geometry" which will include most of the existing spaces of interest; (2) to develop analytic methods for the treatment of the new geometries, it being increasingly clear that the absolute differential calculus is inadequate.

For this purpose Cartan developed what seems to be the most comprehensive and satisfactory program and demonstrated its advantages in a decisive way. This contribution clearly illustrates his geometric insight and we consider it to be the most important among his works on differential geometry. It can be best explained by means of the modern notion of a "fiber bundle."

A *fiber bundle* is merely the generalization of a simpler concept considered earlier in this book (Chapter 7), namely, the idea of a *Cartesian product*. Let us recall that if there are two sets of numbers, $A = \{1, 2, 3\}$ and $B = \{7, 8\}$, then the Cartesian product symbolized by $A \times B$ and read as "*A cross B*" consists of all possible ordered number pairs formed from selecting the first number from A and the second from B. Thus, "*A cross B*," or $A \times B$, is the set of *six* number pairs, $\{(1, 7), (1, 8), (2, 7), (2, 8), (3, 7), (3, 8)\}$. Again, if A is the set of all real numbers between 0 and 2—that is, the interval [0, 2] on the X-axis in Figure 20.5— and B is the set of all real numbers in the interval [0, 1] on the Y-axis, then $A \times B$ is the shaded rectangle in Figure 20.5.

If A consists of all points of a circle and B of all points of a line segment, then $A \times B$ is the surface of a cylinder. To clarify this, suppose that the equation of the circle A is $x^2 + y^2 = 25$ (Figure 20.6), and that B is the interval [0, 1] on

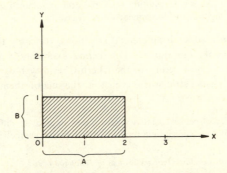

Figure 20.5 $A \times B$ for $A = [0, 2]$ and $B = [0, 1]$

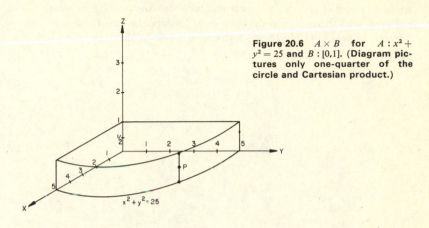

Figure 20.6 $A \times B$ for $A : x^2 + y^2 = 25$ and $B : [0,1]$. **(Diagram pictures only one-quarter of the circle and Cartesian product.)**

the Z-axis of the diagram. One point of this circle has $x = 3$, $y = 4$ as its coordinates. If this is combined with the point $\frac{1}{2}$ of the interval $[0, 1]$, the result is the number triple $(3, 4, \frac{1}{2})$, or point P of Figure 20.6. Combining $(3, 4)$ with *any* number z on $[0, 1]$ leads to $(3, 4, z)$, a point on that *vertical* element of the cylinder which goes through $(3, 4, 0)$ $(3, 4, \frac{1}{2})$, $(3, 4, 1)$ etc. in Figure 20.6. Now to form $A \times B$ one must combine *every* point on the circle with *every* point of the line segment in the way just indicated, and this will lead to all points on all elements of the cylinder.

In the language of fiber bundles, the circle would be called the *fiber*, the line segment would be termed the *base space*, and the cylindric surface would be the *total* or *bundle space*, which can be transformed into the *fiber* by *projection*. But the cylindric surface is a trivial example of a fiber space because it is too special, much in the way that a plane would be a trivial instance of a surface. The *general* fiber bundle is only *locally* a Cartesian product, that is, only the small "neighborhoods" of points of the bundle are required to be Cartesian products, and the bundle as a whole need *not* be such a product. This permits the presence of "twists" or "torsion" in the bundle or total space. In fact, a fiber bundle is called a *skew product* by one of the Soviet mathematical leaders of our day, L. Pontrjagin, whose specialty is topology, a "geometric" subject in which his blindness has apparently not handicapped him in making discoveries.

In addition to having small subsets that are Cartesian products, a fiber bundle must meet certain continuity and group-theory requirements concerning the way the "neighborhoods" or tiny subsets are linked with one another to make up the total space. This part of the definition will become clearer after the concepts of "analytic topology" have been developed in later chapters. The idea of a fiber bundle has been treated here not only because it is important in Cartan's ultimate answer to "What is geometry?" but also because it is fundamental in other branches of mathematics today. Its connection with elementary analytic geometry has been specified, and it is interesting to note that, although the term "fiber bundle" is not used, the idea of a Cartesian product is developed in detail and given considerable emphasis in United States high school courses in the "new mathematics" today.

21

East Meets West in the Higher Arithmetic

The story of geometry which reached its final abstract climax in the work of Élie Cartan started with the ideas of Gauss and Riemann. Those two nineteenth-century Titans exerted an equally powerful influence in other mathematical areas, and, as in the case of geometry, the Gauss and Riemann beginnings evolved into some of the most significant tools of twentieth-century pure mathematical research. As an instance, one can consider the effect of Gauss and Riemann on the theory of numbers (the higher arithmetic). In fact, of all the great contributions of Gauss, his number-theory discoveries are often considered the greatest of all. To him we owe the saying, now classic: Mathematics is the queen of the sciences, and arithmetic the queen of mathematics.

The "queen of mathematics," that is, the higher arithmetic, is often described as the theory of the natural numbers. In the theory one must define certain binary arithmetic operations. If subtraction is to be one such operation (or, what is the same thing, if additive inverses are to exist), then zero and the negative integers must be adjoined to the natural numbers. Hence the theory of numbers is actually the study of properties of the set of integers, $\{\ldots, -3, -2, -1, 0, 1, 2, 3, \ldots\}$. In the present chapter we shall discuss important aspects of that theory.

The theory of the integers is a worthwhile study for many reasons. For one thing, there is perhaps no other branch of mathematics whose issues form an unbroken chain of thought linking early man with his descendants in the nuclear age. Again, the higher arithmetic holds a strong appeal for scholar and amateur alike. The proverbial man in the street can readily understand the meaning (if not the proofs) of all theorems. There is also the intriguing and mystifying challenge of the numerous unsolved problems, issues whose "truth" seems "self-evident" to the uninitiated just because so much numerical evidence can be found to lend credence to the hypotheses involved. Another reason for devoting a detailed chapter to the theory of numbers is that the subject is the purest of *pure* mathematics and thus offers a contrast with geometry, differential equations, mathematical statistics, and all the other mathematical fields whose applications to physics and practical affairs have been stressed in previous chapters. In spite of its purity, however, number theory does have some applicability to the physical world. Finally, as the present chapter will show, the higher arithmetic offers some exceedingly simple illustrations of the advanced concepts of modern abstract algebra.

The introduction to the "queen of mathematics" involves the very elementary notion of the divisibility of one natural number by another. At school, children frequently consider that question in working with fractions. In cancellation, it is helpful to break integers down into divisors or *factors*. Such experience shows that there are two cases, the first illustrated by $5 = 5 \cdot 1, 7 = 7 \cdot 1, 11 = 11 \cdot 1, 29 = 29 \cdot 1$, etc., the second by $4 = 2 \cdot 2, 6 = 2 \cdot 3, 12 = 2 \cdot 2 \cdot 3, 30 = 2 \cdot 3 \cdot 5$, etc. Natural numbers like 5, 7, 11, 29, in which the factoring appears trivial since the number is expressed as the product of itself and unity (the number *one*), are called *primes*. More exactly a prime number is an integer greater than 1 whose only positive integral divisors are itself and unity. Natural numbers like 4, 6, 12, 30, which have positive integral divisors other than themselves and unity, are called *composite*. The factoring process expresses a composite natural number as the product of its prime divisors.

Let us see what an illustrious Alexandrian, Eratosthenes (*ca.* 230 B.C.), had to say about primes. His famous *sieve* was designed to remove the composite numbers from the list of positive integers and leave the set of all primes. He is probably best known for his measurement of the earth. But his talents were not limited to number theory and geography. He was also a poet, philosopher, and historian, who was summoned to Alexandria to direct the great Library. There is much lore associated with his name. For example, his colleagues at the Library and Museum appear to have endowed him with a number of nicknames. One of these was *Pentathlos*, a tribute to his "five-sided" nature, that is, to the universality of his interests; another labeled him a second Plato. At any rate, he must have had a likable personality, since a sobriquet is usually the product of friendly, informal relationships.

In constructing his sieve, Eratosthenes saw immediately that 4, 6, 8, 10, 12, ..., that is, all *even* natural numbers greater than 2, are composite. Therefore his formula for sifting was as follows: List the integers starting with 2, and then strike out every *second* number after 2. Then from the list of integers remaining, strike out every *third* number after 3. Then return and strike out every *fifth* integer after 5, every *seventh* integer after 7, every *eleventh* integer after 11, etc. In the process, certain integers may be crossed out several times, for example, 15 will be stricken out in both the second and third siftings, 35 in the third and fourth, etc. We indicate several steps in Eratosthenes' method for straining out composite numbers.

Step 1: 2 3 ~~4~~ 5 ~~6~~ 7 ~~8~~ 9 ~~10~~ 11 ~~12~~ 13 ~~14~~ 15 ~~16~~ 17 ~~18~~ 19...

Step 2: 2 3 5 7 ~~9~~ 11 13 ~~15~~ 17 19 ~~21~~ 23 25 ~~27~~ 29
31 ~~33~~ 35 37 ~~39~~ 41 43 ~~45~~ 47 49...

Step 3: 2 3 5 7 ~~9~~ 11 13 ~~15~~ 17 19 ~~21~~ 23 ~~25~~ ~~27~~ 29
31 ~~33~~ ~~35~~ 37 ~~39~~ 41 43 ~~45~~ 47 49...

Step 4: 2 3 5 7 ~~9~~ 11 13 ~~15~~ 17 19 ~~21~~ 23 ~~25~~ ~~27~~ 29
31 ~~33~~ ~~35~~ 37 ~~39~~ 41 43 ~~45~~ 47 ~~49~~...

When the sifting is done, one is left with the succession of primes

2	3	5	7	11	13	17	19	23	29	31	37	41	43	47
					53	59	61	67	71	73	79	83	89	97...

An observation of this list indicates that among the first six integers after 1, namely, 2, 3, 4, 5, 6, 7, two-thirds are primes; among the first twelve after 1, one-half are primes; among the first 96 after 1, only one-quarter are primes. The incidence of primes becomes less and less frequent the farther we go, for the obvious reason that more integers precede any particular number and are available as possible divisors. If, for example, one continues the list of natural numbers as far as 1,000,000,000, only about 5 per cent of the integers up to that point will be primes. With such continued thinning out of the primes, it seems possible that after a certain position in the list of integers there may be no more primes and all integers will be composite. Euclid *proved* that such a conjecture is false. His theorem is: *The number of primes is infinite.*

His reasoning, although much more rigorous, was somewhat as follows. In the above list of primes, multiply the first two prime numbers, and add 1 to this product; then multiply the first three primes and add 1, etc. This yields

$$2 \cdot 3 + 1$$
$$2 \cdot 3 \cdot 5 + 1$$
$$2 \cdot 3 \cdot 5 \cdot 7 + 1$$
$$2 \cdot 3 \cdot 5 \cdot 7 \cdot 11 + 1$$

We could compute the results of these operations and find the values 7, 31, 211, 2311, etc., but it is better merely to *reason* about the nature of the results so that we can come to conclusions about

$$2 \cdot 3 \cdot 5 \cdot 7 \cdot 11 \cdot 13 \cdot 17 \cdots 89 \cdot 97 + 1$$

without resorting to tedious multiplication.

At any rate, Euclid considered each expression above with a view to finding possible prime divisors (that are exact). Thus, in the first expression, one might perform division by 2 or by 3, with the results

$$2)\overline{2 \cdot 3 + 1} \qquad\qquad 3)\overline{2 \cdot 3 + 1}$$
$$\quad 3 + \tfrac{1}{2} = 3\tfrac{1}{2} \qquad\qquad\quad 2 + \tfrac{1}{3} = 2\tfrac{1}{3}$$

Since the quotients are fractional, 2 and 3 are *not* exact divisors. In the same way it can be shown that the second expression above, $2 \cdot 3 \cdot 5 + 1$, does *not* have the primes 2, 3, or 5 as exact divisors. Finally,

$$2 \cdot 3 \cdot 5 \cdot 7 \cdot 11 \cdot 13 \cdot 17 \cdots 89 \cdot 97 + 1$$

does not have the primes 2, 3, 5, 7, 11, 13, 17, ..., 89, or 97 as exact divisors. Hence if one were to "multiply out" to obtain the value of the above expression, the result might be composite, but in this case its prime factors would not be any of the prime numbers in the list from 2 to 97, and could only be primes greater than 97.

Now, Euclid argued, if at any stage in listing the primes in order of size (as Eratosthenes did), one believes he has arrived at the "last prime," let him form an expression like those above, that is, the product of all the primes up to the point in question, increased by 1, thus:

$$N = 2 \cdot 3 \cdot 5 \cdot 7 \cdot 11 \cdot 13 \cdot 17 \cdots 89 \cdot 97 \cdots l + 1$$

where l is the "last prime." Consider the number N, Euclid said. It may be prime or it may be composite. In the latter case it can be broken down into prime factors. But then these factors could *not* be any of the primes 2, 3, 5, 7, ..., 89, 97, ..., l, as the previous argument has indicated. Hence, since the prime factors or divisors of N are not in the list from 2 to l, they can only be prime numbers beyond l, so that l cannot be the last prime. But N may not be composite. It may be a prime number. But N is obviously larger than l, as its expression above indicates. Then, if N is a prime, it is a prime greater than l. In this way, Euclid provided a *reductio ad absurdum* for the hypothesis that some prime number is the last on the list. Since there is no last prime, the number of primes must be infinite, and Euclid had established his theorem.

Let us continue with another of Euclid's famous number-theory propositions, the one called the *fundamental theorem of arithmetic*. It states that *every integer greater than 1 can be factored into a product of primes in only one way.*

It may seem obvious that if one factors 10,500 into $2 \cdot 2 \cdot 3 \cdot 5 \cdot 5 \cdot 5 \cdot 7$, there is no other way to effect a breakdown into primes. But in modern algebra there are systems (integral domains—see Chapter 4) whose elements are like the integers because they obey the same fundamental arithmetic laws. Nevertheless, such systems may lack the unique factorization property. Euclid revealed the profundity of his thought by realizing that the characteristic in question is not an obvious one but a very special attribute that must be demonstrated.

We shall not immediately consider an integral domain but, for a simple example, shall take the semigroup $\{S, \bullet\}$ where S is the set of positive even integers, $\{2, 4, 6, 8, \ldots\}$, and $\bullet$ is ordinary multiplication (hence *associative*). Multiplication is truly a binary operation on S because the product of any two elements in S is an element of S. Now consider the factorization of the numbers of S into smaller numbers of S. We exclude statements like $6 = 2 \cdot 3$, $14 = 2 \cdot 7$, etc. since 3 and 7 are *not* elements of the set S. Let us follow the example of Professor Ivan Niven* and call a number in the set S a *pseudoprime* if it cannot be factored into two or more smaller numbers of the aggregate. Then we can obtain a set of pseudoprimes, $\{2, 6, 10, 14, 18, 22, 26, 30, 34, 38, 42, 46, \ldots\}$. All other numbers of the original aggregate S can be factored into pseudoprimes, but there are cases where this factorization is *not* unique, for example, $60 = 2 \cdot 30$ or $60 = 6 \cdot 10$, and $132 = 2 \cdot 66$ or $132 = 6 \cdot 22$.

Euclid's interest in prime numbers was emulated by mathematicians and amateurs alike throughout the ages. What has been sought and never found is a mathematical *formula* for *all* prime numbers. Thus $f(n) = n^2 - n + 41$ will lead to a prime

---

* I. Niven, "The Concept of Number," in *Insights into Modern Mathematics*, Twenty-Third Yearbook, National Council of Teachers of Mathematics, Washington, D.C. (1957) pp. 7–35.

if any of the integers $1, 2, 3, \ldots, 39, 40$ is substituted for n. But if $n = 41$, a composite number results, since $41^2 - 41 + 41 = 41^2$. Another formula, $f(n) = n^2 - 79n + 1601$, yields primes when any natural number less than 80 is substituted for n.

It was inevitable that Fermat, considered by many the greatest arithmetician in history, would seek a formula to yield prime numbers. He conjectured that

$$F_n = 2^{2^n} + 1$$

would produce *only* primes for $n = 0, 1, 2, 3, \ldots$. It is true that the first five of these *Fermat numbers* are primes. Thus there are the five *Fermat primes*,

$$F_0 = 2^{2^0} + 1 = 2^1 + 1 = 3$$
$$F_1 = 2^{2^1} + 1 = 2^2 + 1 = 5$$
$$F_2 = 2^{2^2} + 1 = 2^4 + 1 = 17$$
$$F_3 = 2^{2^3} + 1 = 2^8 + 1 = 257$$
$$F_4 = 2^{2^4} + 1 = 2^{16} + 1 = 65,537$$

But Fermat's hunch was ultimately proved to be erroneous. In 1732 Euler showed that $F_5 = 2^{32} + 1$ is *composite*, with 641 as a divisor. As time passed, F_n was shown to be composite for fifteen other values of n and, to date, no Fermat primes other than the classic five have been revealed. Even with electronic computers, the task of investigating new Fermat numbers is formidable because of their size. Among the most recent to be proved composite, there are F_{36}, which contains more than $20,000,000,000,000$ digits, and F_{73}, which is so enormous that, if its digits were typed on a tape, that tape could be wound around the equator some $60,000,000,000$ times!

Fermat primes intrigued Gauss, who was able to apply them, surprisingly, to an age-old problem of Euclidean *geometry*—the question of just which regular (equilateral and equiangular) polygons are constructible with straight-edge and compasses. Euclid's Book IV discusses constructions for regular polygons of 3, 4, 5, and 15 sides. The reader doubtless carried out some of those constructions at school by the method of dividing a circle into a number of equal parts and then joining the points of division. Having constructed 4 equal arcs, it is possible to bisect each of these to obtain 8 equal arcs whose chords form a regular octagon. By bisecting arcs again and again, a regular polygon of 16 or 32 or 64, etc. sides can be constructed. Thus Euclid's Book IV implies the constructibility, with straight-edge and compasses, of regular polygons of $2^k \cdot 3$, $2^k \cdot 4$, $2^k \cdot 5$, $2^k \cdot 15$ sides where $k = 0, 1, 2, 3, \ldots$. Not until the day of Gauss, however, was it known that there are other regular polygons which are constructible by means of those limited tools which Plato had prescribed as a proper discipline for geometry.

A month before his nineteenth birthday Gauss established the fact that a regular polygon of 17 sides can be constructed with straight-edge and compasses alone. This discovery is said to have swayed young Gauss in favor of a career in mathematics rather than in languages and philology. He extended his theorem by proving that Platonic constructibility is possible whenever n, the number of sides of a regular polygon, is a *Fermat prime* or, more generally, when $n = 2^k p_1 p_2 \cdots p_r$, where $p_1, p_2, \ldots, p_r$ are *different* Fermat primes. This general theorem implies, for example, the constructibility of regular polygons of 257, 65537, $2^k \cdot 3 \cdot 17 \cdot 257$, $2^k \cdot 3 \cdot 5 \cdot 17 \cdot 65537$, etc. sides.

Although there is no single formula for all primes, the use of Eratosthenes' sieve leads to a rather involved rule for the *number of primes* up to a certain point in the natural number scale. Thus we know that there are 50,847,478 primes below 1,000,000,000 and the use of the newest electronic computers will probably provide more data of the same sort.

There are factor tables available for researchers and amateurs. Some of these were ground out "by hand" before the era of mechanical calculators, and there are many stories about those who were willing to undertake the taxing arithmetic computations involved. One of these was the German calculating prodigy Zacharias Dase (1824–1861), whose feats of mental computation surpassed all those of other arithmetical prodigies known up to his time. Dase enjoyed the calculation of logarithmic, trigonometric, and other tables which he prepared in his leisure time. The compilation of factors of numbers starting with 7,000,000 was a volunteer activity to which he gave the last few years of his short life.

After the noted French mathematician Émile Borel had advanced certain statistical hypotheses concerning what the proportion of primes should be when exceedingly high integers are reached, a young Bulgarian student named Sougarev prepared enormous statistical tabulations of facts about primes between 2,000,000 and 2,100,000 in order to test the Borelian thesis. Sougarev did this during the years 1937–1939 to pass the time while he was an invalid in various hospitals in France.

Neither Dase nor Sougarev can be called mathematicians, but the best available factor tables are those prepared by a leading American number-theorist, Derrick Norman Lehmer (1867–1938), whose son, Derrick Henry Lehmer, has followed in his father's footsteps by making discoveries in higher arithmetic and applying them to the facilitation of computations and tabulations connected with primes and other number-theoretic issues.

History tells us that Dase was dull and unintelligent except for his computational ability. Sougarev is not at all known in the world of mathematics. The elder Lehmer, however, who was as fascinated by the factorization of numbers as those two amateurs, was an entirely different type of computer, more closely akin to Euclid and Eratosthenes. Like the latter, Lehmer was versatile, interested in poetry and music as well as mathematics. He edited and contributed to a literary periodical. He composed two operas and a number of songs based on legends and musical themes of Indians of the western United States.

In our present scientific era, the various tabulations of both Lehmers are considered invaluable for researchers throughout the world, and here again history offers a contrast. In 1776, that all-important date for Americans, the Austrian government published a factor table that had been prepared by a Viennese schoolteacher, Anton Felkel. This was a remarkable compilation of factors of numbers up to 408,000, but it is not surprising that it was not a best seller. Austrian government authorities, however, appear to have expected wide sales, and when only a small number of subscriptions were received, all the remaining copies of Felkel's tables were declared to be "scrap paper" to be used in preparing cartridges for the war against Turkey. Even more ill-starred, from the point of view of usefulness, is the factor table of J. P. Kulick (1793–1863) which was once in the hands of the Vienna Academy, but has *never* been published. Kulick, a Prague professor, devoted twenty years to preparing the factors of numbers up to 100,000,000. (The Lehmer tables extend to 10,000,000.)

To return to Euclid's theory of primes, no real progress beyond the Alexandrian's ideas was made until the 1850 research of the noted Russian mathematician Pafnuti Lvovich Tchebycheff (1821–1894), whom the British algebraist James Joseph Sylvester (1814–1897), called "the greatest mathematician of this (nineteenth) century or any other age." Sylvester, who obtained a refinement of one of Tchebycheff's number-theoretic approximations, exerted a tremendous influence on American mathematics through his activities from 1877 to 1883. During that period he was professor of mathematics at Johns Hopkins University, and founded (1878) the *American Journal of Mathematics*, the first scholarly mathematical publication in the United States. He also wrote poetry, which he erroneously judged to be of the same high quality as his mathematical discoveries, and was generally given to a poetic style of expression, *vide* his description of Tchebycheff's contribution to the theory of prime numbers: "He was the only man ever able to cope with the refractory character and erratic flow of prime numbers and to confine the stream of their progression within algebraic limits, building up, if I may so say, banks on either side which that stream, devious and irregular as are its windings, can never overflow." The precise nature of Tchebycheff's discovery will be explained below.

The Russian mathematician had been elected to membership in the Royal Society of London and, whenever he visited England to attend Society meetings, he also conferred with Sylvester on research problems, in particular those associated with *applied mathematics*. The two mathematicians would meet at Lincoln's Inn Fields to talk about "linkages" (mechanical devices for drawing straight lines). "His special object," wrote Sylvester, "was to obtain sight of some of the old machines of Watt. . . . He found that the intuitive mechanical sagacity of Watt had anticipated his own mathematical deductions. Tchebycheff has found the best form of 3-bar motion, of which Watt's parallel motion is only a special case, and has arrived at more accurate results than Watt's, with a more compact arrangement, and, in some cases, more convenient of application."

But Tchebycheff is doubtless better known to students of mathematics today as the founder of the highly creative Russian school of probability theory. There is the fundamental "Tchebycheff inequality," for example, which generalizes "Bernoulli's theorem" and leads directly to the important "law of large numbers."

In spite of the important research of Tchebycheff and his followers, a precise formula for all primes was not found. Then Borel and other mathematicians decided to use a statistical approach. The most remarkable result of this type had preceded Borel's efforts. The *prime number theorem* was proved in 1896 by the French mathematician Jacques Hadamard, (1865–1963) and the Belgian Charles J.

de la Vallée Poussin (1866–1962). There are many tales of the genius of Hadamard. Among these is the fact that he achieved the highest score ever obtained in the entrance examination to the *École Polytechnique*, France's greatest school of science and, in Hadamard's youth, the foremost world institution of its type. Simplifications and modifications of the proof of the famous theorem were provided by the German Edmund Landau (1877–1938) and others. Then, in 1932, Norbert Wiener deduced a much simpler proof from certain of his own discoveries which G. H. Hardy named "Tauberian theorems" (after the German analyst Tauber). Perhaps the most elementary demonstration to date is the 1949 proof of Atle Selberg, professor at the Institute for Advanced Study. These proofs are called "elementary" in a special technical sense; their content is exceedingly difficult.

The prime number theorem derives from a conjecture of Gauss. He evidently arrived at it empirically by making an actual count in a table of primes. At any rate, logarithmic tables which he used as a lad of fourteen have been found, with a complete statement of the theorem, written in his hand, as a marginal note. The theorem provides a *formula* giving an approximation to the proportion or density of primes among the integers, and indicates that the approximation must become better and better, the greater the number of integers considered. For example, it is known that there are 50,847,478 primes among the first 1,000,000,000 integers. The prime number theorem gives the estimate 48,254,942, an error of only 5 per cent. We do *not* know how many primes there are in still larger intervals but the Hadamard-De la Vallée Poussin theorem proves that for all practical purposes there will be almost no error in the formula after a while. This remarkable prime number theorem states that the proportion or density of primes among the first n integers when n is large, is approximately equal to $1/\ln n$ where $\ln n$ signifies the natural logarithm of n (Chapter 22).

To return to Tchebycheff once more, the nature of his contribution is more readily understood if the prime number theorem is stated in the alternative form: If $P(n)$ is the number of primes equal to or less than the integer n, then the ratio of $P(n)$ to $n/\ln n$ approaches the number 1 as a *limit*, as n gets larger and larger. The Russian was unable to establish the existence of this limit, but he did show that, for sufficiently large values of n, the ratio in the theorem lies between two positive constants, a and A. These constants are Sylvester's "banks on either side which that stream . . . can never overflow." He himself subsequently brought these "banks" closer together, thus improving Tchebycheff's approximation. But ultimately it was seen that such techniques would not establish the prime number theorem. Hence Hadamard and De la Vallée Poussin resorted to other methods, namely, those of modern *analysis*.

If Gauss initiated the drive for a proof of the great theorem, one must again go back to antiquity for origins, since it was Euclid who first sought a formulation of the primes. His *Elements*, moreover, contained the germ of most of the still unresolved problems of number theory. Although answers are not yet available, these challenging puzzles have served an important purpose in stimulating the creation of an enormous amount of new mathematics through the centuries. One of these problems grew out of the last theorem of the ninth book of the *Elements*, which states: If $2^n - 1$ is a prime, then $2^{n-1}(2^n - 1)$ is equal to the sum of its proper divisors. (A proper divisor is any factor other than the number itself.)

Euclid was able to give a simple proof of this theorem by making it dependent on a fact known to the Pythagoreans, and to the Babylonians before them, namely, that

$$1 + 2 + 2^2 + 2^3 + \cdots + 2^{k-1} = 2^k - 1$$

The reader can derive this rule by applying the formula for the sum of a geometric progression, or he can use "mathematical induction" on the particular observations

$$1 + 2 = 3 = 2^2 - 1$$
$$1 + 2 + 2^2 = 7 = 2^3 - 1$$
$$1 + 2 + 2^2 + 2^3 = 15 = 2^4 - 1$$

where the sum of *two* terms is 1 less than the *second* power of 2, the sum of *three* terms is 1 less than the *third* power of 2, etc.

In Euclid's theorem, one is given that $2^n - 1 = p$, a *prime*, and must prove that $2^{n-1}p$ is equal to the sum of its proper divisors. Now the sum of the proper divisors is

$$1 + 2 + 2^2 + \cdots + 2^{n-2} + 2^{n-1}$$
$$+ p(1 + 2 + 2^2 + \cdots + 2^{n-2})$$

If we use the formula of the previous paragraph, the sum becomes

$$2^n - 1 + p(2^{n-1} - 1)$$

or, since $2^n - 1 = p$,

$$p + p(2^{n-1} - 1) = p + 2^{n-1}p - p = 2^{n-1}p$$

where $2^{n-1}p$ is the desired result.

If $n = 2$, then $2^n - 1 = 2^2 - 1 = 3$, which is a *prime*. Therefore $2^{n-1}(2^n - 1) = 2^1(2^2 - 1) = 2 \cdot 3 = 6$. The fact that $6 = 1 + 2 + 3$ is in agreement with Euclid's theorem.

The Pythagoreans or other early Greek mathematicians had provided the name *perfect* for numbers like 6 that are equal to the sum of all proper divisors. The brotherhood immediately invested perfect numbers with mystic qualities, and other religious groups followed suit. It came to be said that God created the universe in six days because 6 is a perfect number. Alcuin of York (735–804), who established a school at Charlemagne's court, taught that the human race is imperfect because it is descended from the *eight* souls of Noah's ark, and 8 is *imperfect* since it is not equal to the sum of its proper factors, $1 + 2 + 4$.

To obtain examples of perfection, the reader may substitute $n = 3$ and $n = 5$ in Euclid's formula. This will yield the perfect numbers $28 = 1 + 2 + 4 + 7 + 14$ and $496 = 1 + 2 + 4 + 8 + 16 + 31 + 62 + 124 + 248$. (We note that for $n = 4$,

$2^n - 1 = 15$, which is *composite*. Hence Euclid's theorem does not apply to this case.) Perfection is very hard to find since, even with the aid of today's digital computers, Euclid's formula has not yielded too many examples of perfect numbers. The twenty-second perfect number in order of size was given computer verification not long ago. That number, corresponding to $n = 9941$, is the perfect number $2^{9940}(2^{9941} - 1)$. It contains 5985 digits! Still more recently (1963) the twenty-third perfect number was found.

Now all the perfect numbers which computers have checked are *even*. Euler proved in fact that Euclid's formula will include *all* even perfect numbers. These facts suggest that, possibly, an odd number cannot be perfect. But no rigorous proof has been found for this conjecture. Hence, although it is known that there is no odd perfect number less than 10,000,000,000, there is the famous unsolved problem of number theory: Can a perfect number be odd or must *all* perfect numbers be even?

The condition for Euclid's formula to yield a perfect number is that $2^n - 1$ be *prime*. Hence the search for *primes* of the form $2^n - 1$ is a problem in itself. The first person to devote himself to that question was Father Marin Mersenne (1588–1648), a close friend of Descartes. Mersenne wrote on a variety of other mathematical subjects as well, and carried on a scientific correspondence with Fermat, Pascal, and all the leading mathematicians of his day. He transmitted the letters of each one to all the others, thus performing single-handedly the sort of service rendered by scholarly journals today.

It is easy to show (Mersenne did so) that if $2^n - 1$ is to be a prime, then n must be a prime. The converse theorem is false, for if, for example, $n = 11$ (a prime), $2^{11} - 1 = 2047 = 23 \cdot 89$ is composite. Knowing this fact, Mersenne proceeded to examine successive numbers, $M_p = 2^p - 1$ where p is a prime, that is, $p = 2, 3, 5, 7, 11, 13, 17, \ldots$. He worked steadfastly until he reached the prime $p = 257$. Then in 1644 he published a list of *eleven* values of p, which he claimed to be the *only* primes up to $p = 257$ for which M_p is a prime.

But Mersenne erred in two respects. He omitted three values of p (less than 257) for which $M_p = 2^p - 1$ is actually prime. On the other hand, he listed M_{67} and M_{257} as primes when, in fact, they are composite. The following story will reveal why one must forgive Mersenne and all those *human* computers who operated before the era of electronic brains. When, in 1903, M_{67} was first proved composite by Frank Nelson Cole (1861–1927), a leading American algebraist, he confided to a friend that it had required "three years of Sundays" to discover that

$$2^{67} - 1 = (193,707,721)(761,838,257,287)$$

In 1931 D. H. Lehmer proved, by theoretical methods, that M_{257} must be composite, but his proof, unlike Cole's, did not produce a pair of numerical factors.

Today we can assert correctly that, among the 55 prime numbers between 2 and 257 inclusive, there are 12 primes, namely,

$$p = 2, 3, 5, 7, 13, 17, 19, 31, 61, 89, 107, 127$$

for which $M_p = 2^p - 1$ is a prime. Such numbers are called *Mersenne primes* and when multiplied by 2^{p-1}, yield perfect numbers. The numbers not on Mersenne's, original list are M_{61}, M_{89}, M_{107}. They were not revealed as primes until the years 1883, 1911, and 1914, respectively. The first Mersenne prime for which $p > 257$ is M_{521}, discovered in 1952, more than three centuries after the good friar published his numerical results.

Most number-theoretic issues had their origins in antiquity, as we have seen, with Euclid, the Hellenic mathematician who provided two important sources of inspiration—namely, proofs for certain crucial theorems, and concepts leading to problems that are still unsolved. The story of Euclid's influence must be repeated in relation to the problem of finding all "Pythagorean triples," that is, all sets of natural numbers that satisfy

$$x^2 + y^2 = z^2$$

and hence can be sides of a right triangle. There are tabulations of such triples in Babylonian cuneiform tablets predating Pythagoras by a thousand years. Pythagoras himself did not solve the problem, in the sense of obtaining a formula for all triples, but, in Neugebauer's opinion, the cuneiform documents give indirect evidence that the Babylonians did find a general formula. By Euclid's time the problem was completely solved, and in Book X of the *Elements* there is a geometric wording of the formula

$$x = 2\,uv$$
$$y = u^2 - v^2$$
$$z = u^2 + v^2$$

where u is a natural number greater than v, u and v have no common factor and one of them is odd, the other even. Thus if $u = 3$, $v = 2$, then $x = 12$, $y = 5$, $z = 13$. If $u = 4$, $v = 1$, then $x = 8$, $y = 15$, $z = 17$. If $u = 5$, $v = 2$, the triple is $\{20, 21, 29\}$.

Strictly speaking, the above formula provides only *primitive* Pythagorean triples, those having no common factor. Every primitive triple leads to an infinite number of other triples that are multiples of the primitive set. Thus from 5, 12, 13 one obtains 10, 24, 26, and 50, 120, 130, etc. Substituting the various triples in $x^2 + y^2 = z^2$ leads to an infinite number of different squares which are equal to the sum of other squares, that is, an infinite number of right triangles with integral sides.

Reflecting on the plethora of integral solutions of $x^2 + y^2 = z^2$, Fermat went on to investigate the possibility of such solutions for $x^3 + y^3 = z^3$, $x^4 + y^4 = z^4$, ..., $x^n + y^n = z^n$, where n is any integer greater than 2. Then, in a marginal note opposite a discussion of Pythagorean triples in his copy of Bachet's *Diophantus*, Fermat made an entry that is as famous in the history of mathematics as, say, the French Revolution in the history of modern man. Fermat wrote: "By contrast it is impossible to separate a cube into two cubes, a fourth power into two fourth powers, or in general any power above the second into two powers of the same degree. I have found a truly marvelous proof of this theorem but this margin is too narrow to contain it."

If only Fermat had published that marvelous proof! But from his day up to the present time, the greatest mathematical minds have attempted and have failed to produce a demonstration of what has come to be known as *Fermat's Last Theorem*: There is no solution in natural numbers of the equation $x^n + y^n = z^n$ if n is an integer greater than 2.

It would be more accurate to describe Fermat's theorem as a *conjecture* since it has never been proved, that is, demonstrated for *all* natural numbers in the aggregate, $\{3, 4, 5, 6, 7, \ldots\}$. Proofs have been given, however, for many special values of n in that set. Thus it is fairly easy to show, for example, that $x^4 + y^4 = z^4$ has no positive integral solutions (Fermat himself did so). This implies that $x^n + y^n = z^n$ has no such solutions if n is any higher power of 2, that is, if $n = 8$ or 16 or 32 or 64, etc. To see why, consider, for example, $x^8 + y^8 = z^8$, which is the same as $(x^2)^4 + (y^2)^4 = (z^2)^4$ or $a^4 + b^4 = c^4$, if one substitutes $a = x^2$, $b = y^2$, $c = z^2$. Then if there were a triple of natural numbers, $\{x, y, z\}$ satisfying $x^8 + y^8 = z^8$, the squares of these integers could satisfy $a^4 + b^4 = c^4$, which is impossible because Fermat's theorem is true for $n = 4$.

To prove Fermat's theorem for all cases where $n > 4$ it would suffice to demonstrate it where n is any odd prime, that is, any prime greater than 2. For then if n were composite, say, $n = 3000$, one could say that $x^{3000} + y^{3000} = z^{3000}$ or $(x^{1000})^3 + (y^{1000})^3 = (z^{1000})^3$ cannot possibly have a solution in natural numbers because then $a^3 + b^3 = c^3$ (where $a = x^{1000}$, $b = y^{1000}$, $c = z^{1000}$) would have such a solution, which cannot be, once Fermat's theorem is established for odd primes, including $n = 3$.

Although no solution has ever been given for all odd primes, Euler did prove Fermat's theorem for the special prime, $n = 3$. About 1825, proofs for $n = 5$ were given by Legendre and P. G. Lejeune-Dirichlet (1805–1859), working independently. Some fifteen years later, Gabriel Lamé (1795–1870), who was as distinguished in the higher arithmetic as he was in analysis, geometry, and mathematical physics, gave a proof for $n = 7$. Sophie Germain, whom we have already named as one of the few women mathematicians of all time, proved that if the integral solutions x, y, z are restricted to those which are prime to one another and to the exponent n, then it is impossible to solve $x^n + y^n = z^n$ if n is any prime less than 100. In 1908, the American algebraist Leonard Eugene Dickson (1874–1954) extended Germain's theorem to all primes less than 7000, and in more recent times Barkley Rosser extended the upper limit to 41,000,000. Then Emma Lehmer and Derrick H. Lehmer gave a further extension to primes less than 253,747,889. If, in addition to the above restriction on the integers x, y, z, the prime n must be of a certain special type, recent research establishes Fermat's theorem for all such primes less than $2^{3617} - 1$, a number of 1089 digits.

There would perhaps be no point in continuing the list of great names associated with proofs of special cases of Fermat's Last Theorem if one did not at the same time indicate how attempts to prove the elusive theorem gave rise to some of the most potent concepts of modern mathematics. Thus Ernst Kummer (1810–1893)—like many amateurs and some leading mathematicians, such as Cauchy—thought, at first, that he had arrived at a demonstration. But he discovered he had made the error of assuming that Euclid's fundamental theorem of arithmetic holds for entities that generalize the integers, in other words, that those entities have a unique factorization into "pseudoprimes" (page 500). Hence he next created *ideal numbers* such that composite numbers of the new species do satisfy unique factorization into "prime" ideal numbers. Kummer's ideal numbers led to Dedekind's concept of an *ideal*, a notion we have mentioned previously (Chapter 4). An ideal is *not* a number but a set of numbers qualifying as a special kind of subring (page 81) of a ring. The choice of "ideal" as the name of such a system was an intuitive anticipation of things to come, for, in the hands of later algebraists, in particular, Emmy Noether, the notion was to be revealed as one of the most beautiful and most fruitful in modern algebra.

But the magnificent conception which Kummer launched did *not* enable him to arrive at a proof of Fermat's Last Theorem. However, the theory of ideal numbers made it possible for him to establish certain general conditions under which $x^n + y^n = z^n$ would be unsolvable by means of integers. Kummer's conditions were basic in all subsequent investigations of the problem up to those of the American number theorist H. S. Vandiver, the foremost living authority on Fermat's Last Theorem. His extensions of the Kummer criteria and the creation of further conditions of his own are the basis of recent numerical results that extend the upper boundary of primes for which Fermat's theorem holds. Nevertheless all Vandiver's evidence does *not* constitute proof, and the theorem remains an unsolved problem, the most puzzling challenge in the higher arithmetic.

Mathematicians are still hopeful about obtaining a proof, and logicians have come forth with "metamathematical" theorems offering the possibility that Fermat's Last Theorem is an *undecidable* proposition, one which cannot be proved to be either true or false. Such a state of affairs is likely to destroy a layman's faith in the utter power of mathematical reasoning! At any rate, it could not have been considered by the German professor Paul Wolfskehl, who, in 1908, left 100,000 marks to be awarded to the first person who would give a complete proof of Fermat's Last Theorem. The bequest encouraged countless amateurs to devise ingenious but erroneous proofs. The years have been rolling by since 1908 without any discovery of a proof, but what a handsome prize the Wolfskehl Fund might have become, with all the accumulated compound interest, if World War I had not put an end to the value of the mark!

It was Euclid's formula for Pythagorean triples, we recall, that was the point of departure for Fermat's problem. But now we must go on to details of the mathematical contributions of later number theorists. After Euclid and Fermat, the next major contributors were Euler and Legendre. Then there were Gauss and Riemann, whose influence on number theory was as important as that on differential geometry (and hence on relativity). Each number-theorist in our sequence of great names inspired all his successors. In fact, there is a story which illustrates both

Legendre's effect on Riemann and the latter's genius. As a student at the *gymnasium*, the boy Riemann borrowed from the school library Legendre's *Théorie des Nombres*, a scholarly work of 859 pages. When he returned it in less than a week, his mathematics teacher inquired, "How did you like it?" Riemann responded, "That was a marvelous book. I mastered it completely." This was no idle boast, for when, some months later, examiners quizzed the youth in some detail on the contents of the "marvelous book," he answered perfectly.

Let us now illustrate some of Gauss's number-theoretic ideas. We have already spoken of "clock arithmetic" (Chapter 16) which will now be dignified by Gauss's concept of *congruence*. By his definition, two integers are said to be *congruent* with respect to a positive integral *modulus m* if they leave the same remainder when divided by *m* or, equivalently, if the difference of the integers is exactly divisible by *m*. Thus, if 7 is selected as the modulus, 37 is congruent to 16, and 5 is congruent to 19. If 3 is selected as the modulus, 85 is congruent to 4. The symbol $\equiv$ is used for "is congruent to," and the above statements are written as

$$37 \equiv 16 \ (\text{mod } 7)$$
$$5 \equiv 19 \ (\text{mod } 7)$$
$$85 \equiv 4 \ (\text{mod } 3)$$

Once more, the fundamental question of arithmetic divisibility is at issue. One advantage of Gauss's concept and symbolism is that congruence statements can be manipulated in a manner somewhat analogous to the technique used in solving algebraic equations. Thus congruence statements, like equations, can be added and multiplied, providing the modulus remains the same throughout such manipulations. It is easy to prove algebraically that this is so. Thus, let us show that if

$$a \equiv b \ (\text{mod } m)$$

and

$$c \equiv d \ (\text{mod } m)$$

then

$$a + c \equiv b + d \ (\text{mod } m)$$

and

$$ac \equiv bd \ (\text{mod } m)$$

By Gauss's definition, the given facts are

$$a - b = km$$
$$c - d = lm$$

Since *equations* may be added,

$$(a + c) - (b + d) = (k + l)m$$

and hence

$$a + c \equiv b + d \ (\text{mod } m)$$

The given equations may also be written in the form

$$a = b + km$$
$$c = d + lm$$

Since these equations may be multiplied,

$$ac = (b + km)(d + lm)$$
$$= bd + blm + dkm + klm$$
$$= bd + (bl + dk + kl)m$$

Hence $ac - bd = (bl + dk + kl)m$ and therefore

$$ac \equiv bd \pmod{m}$$

To give a few examples of how Gauss's congruences can be used to reveal divisibility properties, one can provide proofs for rules that are sometimes given in elementary arithmetic without any indication as to *why* they are valid. Thus there is the theorem that a number is divisible by 3 or 9 if and only if the sum of its digits is divisible by 3 or 9. For example 7,104,288,015 is seen to be divisible by 9 (and *a fortiori* by 3) because the sum of its digits is 36. The proof of the theorem will now be given. In Chapter 1 we saw that any positive integer N expressed in the decimal system has the form

$$N = a_0 + a_1 \cdot 10 + a_2 \cdot 10^2 + \cdots + a_n \cdot 10^n$$

Now, multiplying the congruence $10 \equiv 1$ (mod 3 or mod 9) by itself repeatedly, one obtains

$$10^2 \equiv 1 \text{ (mod 3 or mod 9)}$$
$$10^3 \equiv 1 \text{ (mod 3 or mod 9)}$$
$$\cdot \quad \cdot \quad \cdot$$
$$10^n \equiv 1 \text{ (mod 3 or mod 9)}$$

Therefore

$$.a_1 \cdot 10 \equiv a_1$$
$$a_2 \cdot 10^2 \equiv a_2$$
$$a_3 \cdot 10^3 \equiv a_3$$
$$\cdots$$
$$a_n \cdot 10^n \equiv a_n$$

and then $N = a_0 + a_1 \cdot 10 + a_2 \cdot 10^2 + \cdots + a_n \cdot 10^n \equiv a_0 + a_1 + a_2 + \cdots + a_n$

or

$$N \equiv a_0 + a_1 + a_2 + \cdots + a_n \text{ (mod 3 or mod 9)}$$

which means that $N = a_0 + a_1 + a_2 + \cdots + a_n + 3k$ (or $9l$)

Thus the right member (and hence N) is divisible by 3 (or 9) if and only if $a_0 + a_1 + \cdots + a_n$ is so divisible.

In a similar way, a rule for testing divisibility by 11 can be derived. Thus

$$10 \equiv -1 \pmod{11}$$

because

$$10 - (-1) = 10 + 1 = 11 \text{ (which is divisible by 11)}$$

Multiplying

$$10 \equiv -1 \pmod{11}$$

by itself repeatedly yields

$$10^2 \equiv +1 \ (\mathrm{mod} \ 11)$$
$$10^3 \equiv -1 \ (\mathrm{mod} \ 11)$$
$$10^4 \equiv +1 \ (\mathrm{mod} \ 11)$$
$$\text{etc.}$$

Therefore, multiplying these congruence statements by $a_1, a_2, \ldots, a_n$, and adding the results to $a_0 \equiv a_0$ yields

$$N \equiv a_0 - a_1 + a_2 - a_3 + a_4 - a_5 + \cdots (\mathrm{mod} \ 11)$$

that is,

$$N = a_0 - a_1 + a_2 - a_3 + a_4 - a_5 + \cdots + 11k$$

Hence an integer is divisible by 11 if and only if the sum of its digits with signs alternately plus and minus is divisible by 11. For example, 91,720,453,803 is seen to be divisible by 11 in accordance with this rule.

Although congruences for a fixed modulus may be added, subtracted, and multiplied, they do not obey all the laws which hold for equations. Thus there is the cancellation law for equations: If $ca = cb$, where $c \neq 0$, then $a = b$. This law does *not* transfer to all congruences, $ca \equiv cb$. Thus although $26 \equiv 8 \ (\mathrm{mod} \ 6)$, that is, $2 \cdot 13 \equiv 2 \cdot 4 \ (\mathrm{mod} \ 6)$, it is *not* true that $13 \equiv 4 \ (\mathrm{mod} \ 6)$, since $13 - 4 = 9$, which is *not* a multiple of 6. There is, however, a modified cancellation law for congruences: If c is prime to the modulus m, the $ca \equiv cb$ implies $a \equiv b$.

Again, although a *linear equation* with integral coefficients always has a unique rational solution, a linear congruence may have no integral solution whatsoever or, at the other extreme, may have two or more noncongruent solutions. For example, let us show that $2x \equiv 1 \ (\mathrm{mod} \ 4)$ has no integral roots. Now any integer will be congruent modulo 4 to a remainder or "residue" in the set $\{0, 1, 2, 3\}$, and hence we need only consider these four numbers as possible roots. Substituting them in the term $2x$, we have $2 \cdot 0 = 0$, $2 \cdot 1 = 2$, $2 \cdot 2 = 4$, which is congruent to 0 (mod 4), $2 \cdot 3 = 6$, which is congruent to 2 (mod 4). No one of these results is congruent to 1, as required. Therefore the given congruence has *no* integral solutions.

But let us next consider the linear congruence $3x \equiv 6 \ (\mathrm{mod} \ 12)$, and seek integral solutions among the residues $\{0, 1, 2, 3, \ldots, 11\}$. We find the solutions $x = 2$, $x = 6$, $x = 10$, which check because $3 \cdot 2 = 6$, $3 \cdot 6 = 18 \equiv 6 \ (\mathrm{mod} \ 12)$, $3 \cdot 10 = 30 \equiv 6 \ (\mathrm{mod} \ 12)$.

Is there any type of linear congruence that has one and only one solution (modulo m) and is therefore truly analogous to a linear equation? The answer is affirmative for a linear congruence where the modulus and the coefficient of x have no common divisor except 1, that is, are relatively prime. There is a theorem indicating that $ax \equiv b \ (\mathrm{mod} \ m)$, where a and m are relatively prime, has a unique solution (mod m). For example, in $2x \equiv 1 \ (\mathrm{mod} \ 9)$, 2 and 9 are relatively prime. This congruence has the unique solution $x = 5 \ (\mathrm{mod} \ 9)$, as the reader can check by substituting $x = 0, 1, 2, \ldots, 8$ in the formula for the congruence.

An important special case of the above theorem occurs when the modulus m is a prime. In that case, all integers not divisible by m (not congruent to zero) are relatively prime to m and hence, if used as values of a in $ax \equiv b \pmod{m}$, will lead to a unique solution of that congruence. Hence there is the theorem: If p is a prime, and if $a \not\equiv 0 \pmod{p}$, then $ax \equiv b \pmod{p}$ has a solution that is unique, modulo p. Thus changing to a *prime* modulus in each of the original examples above, we find that $2x \equiv 1 \pmod{5}$ and $3x \equiv 6 \pmod{11}$ have the unique solutions $x = 3 \pmod{5}$ and $x = 2 \pmod{11}$, respectively.

If we consider polynomial congruences of higher degree, we surely cannot expect theorems about algebraic equations to carry over, since roots of congruences must be integers. But if, guided by the results with linear equations, we insist on a *prime modulus*, then there is an analogue for the *corollary* to Gauss's *fundamental theorem of algebra* (Chapter 4). That corollary states: A polynomial equation of the nth degree (with coefficients in the complex domain) has at most n roots in the complex domain. The analogous theorem for congruences is due to Lagrange and can be stated as follows: If the modulus is a *prime* p, then a polynomial congruence of the nth degree has at most n roots among the integers, $\{0, 1, 2, \ldots, p-1\}$.

The fundamental theorem for congruences indicates that a congruence of higher degree may have *no* integral roots whatsoever even though the modulus is prime. Also, if $p < n$, the congruence has, of necessity, fewer than n integral roots in the set $\{0, 1, 2, \ldots, p-1\}$. Thus, as an example of a quadratic congruence, there is

$$x^2 \equiv 1 \pmod{3}$$

which has two roots, the maximum number possible, in the set $\{0, 1, 2\}$, namely, $x = 1$ and $x = 2$. Again, there is the cubic congruence

$$x^3 + x^2 \equiv 1 \pmod{5}$$

which has the unique solution $x = 3$ in the set $\{0, 1, 2, 3, 4\}$. And finally there is

$$x^4 \equiv 2 \pmod{5}$$

with *no* solutions in the set $\{0, 1, 2, \ldots, 4\}$.

Returning to the congruence $x^2 \equiv 1 \pmod{3}$, which had the maximum number of solutions possible, we might try to generalize that happy result in a number of ways. For example, if we observe that the degree of the congruence is 1 less than the prime modulus, we might next examine $x^4 \equiv 1 \pmod{5}$, $x^6 \equiv 1 \pmod{7}$, $x^{10} \equiv 1 \pmod{11}$, etc. In the first of these congruences, $x = 1, 2, 3, 4$ are *all* roots because $1^4 = 1$, $2^4 = 16 \equiv 1 \pmod{5}$, $3^4 = 81 \equiv 1 \pmod{5}$, $4^4 = 256 \equiv 1 \pmod{5}$. In the case of $x^6 \equiv 1 \pmod{7}$, the solutions (mod 7) are the residues in the set $\{1, 2, 3, 4, 5, 6\}$ because $1^6 = 1$, $2^6 = 64 \equiv 1 \pmod{7}$, $3^6 = 729 \equiv 1 \pmod{7}$, $4^6 = 4096 \equiv 1 \pmod{7}$, $5^6 = 15625 \equiv 1 \pmod{7}$, $6^6 = 46656 \equiv 1 \pmod{7}$. We leave to the reader the task of proving that the first ten natural numbers are solutions of $x^{10} \equiv 1 \pmod{11}$.

The above numerical examples all suggest Fermat's "little theorem": If p is a prime, then the first $p-1$ natural numbers are the solutions, modulo p, of the congruence $x^{p-1} \equiv 1 \pmod{p}$.

The theorem in question may be described as "little" in comparison with Fermat's more famous theorems, but his "small" result is truly remarkable because there is nothing analogous to it in the classic theory of polynomial equations. What the "little theorem" tells us is that in *every* system $\{S, \otimes\}$ where $\otimes$ is multiplication modulo p, a prime, and $S = \{1, \ldots, p-1\}$, the non-zero residues modulo p, *every* number of the system (which is readily shown to be an abelian group) is a root of the congruence $x^{p-1} \equiv 1 \pmod{p}$. This signifies that *all* integers except those divisible by p are roots of the congruence.

Alternative statements of Fermat's theorem point up its potential applications to problems of divisibility. Thus one can subtract $1 \equiv 1 \pmod{p}$ from the congruence $x^{p-1} \equiv 1 \pmod{p}$ to obtain $x^{p-1} - 1 \equiv 0 \pmod{p}$ and one can then multiply by $x \equiv x \pmod{p}$, where $x \not\equiv 0$, to yield the result $x^p - x \equiv 0 \pmod{p}$. Then Fermat's theorem implies: If p is a prime and x is any integer such that $x \not\equiv 0 \pmod{p}$, then $x^{p-1} - 1$ and $x^p - x$ are exactly divisible by p.

As early as 500 B.C. the Chinese were aware of one divisibility fact included in Fermat's theorem, for their manuscripts asserted that $2^p - 2$ is divisible by p when p is a prime, Thus $2^{11} - 2 = 2046$ is divisible by 11, which can readily be checked, and $2^{9941} - 2$ is divisible by the prime 9941, a fact which no one would care to verify "by hand."

But Fermat's theorem implies an infinite number of other divisibility statements. For example, $3^{9941} - 3$, $4^{9941} - 4$, $5^{9941} - 5$, $\ldots$, $9940^{9941} - 9940$ must all be divisible by 9941, and $2^{65537} - 2$, $3^{65537} - 3$, $\ldots$, $65536^{65537} - 65536$ are all divisible by the Fermat prime 65537.

Although $2^n - 2$ must be divisible by n if n is a prime number, the early Chinese (and even, much later, the great Leibniz) erred in conjecturing that the converse statement would be true. They believed that if $2^n - 2$ is divisible by n, then n would, of necessity, be prime, so that the divisibility property could then be used as a test of primality. The conjecture was discovered to be false only in 1819, when it was shown that $2^{341} - 2$ is exactly divisible by $341 = 11 \cdot 31$, a *composite* number. Subsequently it was found that $2^n - 2$ is divisible by n for an infinite number of other composite values of n.

We have discussed the general *linear* congruence and one special type of *binomial* congruence, $x^{p-1} \equiv 1 \pmod{p}$ where p is a prime. Theory shows that the question of solving any *quadratic congruence*, $Ax^2 + Bx + C \equiv 0 \pmod{m}$, can be reduced to a matter of solving *linear* congruences and *binomial quadratic* congruences. Therefore it is not surprising that in the initial sections of the great *Disquisitiones Arithmeticae* Gauss gave a complete and elegant treatment of the binomial quadratic congruence $x^2 \equiv a \pmod{m}$. Its theory is much more difficult than that for the binomial equation $x^2 = a$, which has, immediately, the roots $x = +\sqrt{a}$ and $x = -\sqrt{a}$. In connection with binomial quadratic congruences, Gauss proved (when he was only nineteen years of age) a fact that Euler and Legendre had observed but were unable to demonstrate rigorously. This was the law of *quadratic reciprocity*, one of the most famous theorems in all of number theory. It states that if p and q are different prime numbers other than 2, then

$$x^2 \equiv q \pmod{p} \qquad \text{and} \qquad x^2 \equiv p \pmod{q}$$

are *both* solvable, or *both* unsolvable, unless *both* p and q are congruent to 3 (mod 4), that is, leave the remainder 3 when divided by 4, in which case one of the congruences is solvable and the other is not.

Some numerical examples will clarify the meaning of this law. Observe first that if any integer is divided by 4, the possible remainders are $\{0, 1, 2, 3,\}$. But since the moduli in the reciprocity law must be *odd* primes, the remainders from division by 4 cannot be 0 or 2 since the former remainder would imply that the modulus is exactly divisible by 4 and the latter would make it divisible by 2. Hence 1 and 3 are the only possible remainders in connection with the reciprocity theorem. Let us then take $p = 5$, $q = 11$. These moduli leave the remainders 1 and 3, respectively, when divided by 4. Hence, by the reciprocity law, *both* or *neither* of the following must be solvable:

$$x^2 \equiv 11 \ (\text{mod } 5) \qquad \text{and} \qquad x^2 \equiv 5 \ (\text{mod } 11)$$

Since $11 \equiv 1 \ (\text{mod } 5)$, the first congruence can also be expressed as $x^2 \equiv 1 \ (\text{mod } 5)$. This congruence and $x^2 \equiv 5 \ (\text{mod } 11)$ are *both* solvable, the former being satisfied by $x = 1$ and $x = 4$, the latter by $x = 4$ and $x = 7$. Again, take $p = 5$, $q = 13$, both of which leave the remainder 1 when divided by 4. Once more, both or neither of the following should be solvable:

$$x^2 \equiv 13 \ (\text{mod } 5), \text{ that is, } x^2 \equiv 3 \ (\text{mod } 5) \qquad \text{and} \qquad x^2 \equiv 5 \ (\text{mod } 13)$$

Here *neither* is solvable, as the reader can check by substituting $x = 0, 1, 2, 3, 4$ successively in the first, and $x = 0, 1, 2, 3, \ldots, 12$ in the second.

The second case of the quadratic reciprocity law is illustrated when $p = 7$ and $q = 3$, since *both* these moduli leave a remainder of 3 when divided by 4. We can see by substituting $x = 0, 1, 2, \ldots, 6$ that

$$x^2 \equiv 3 \ (\text{mod } 7)$$

is not solvable, whereas

$$x^2 \equiv 7 \ (\text{mod } 3), \text{ that is, } x^2 \equiv 1 \ (\text{mod } 3)$$

is solvable, having the roots $x = 1$ and $x = 2$.

The theory of numbers is a vast domain, and hence we can consider only certain selected aspects of the subject. Thus we have discussed, in the main, the elementary classic treatment of the divisibility and primality of integers. But we have mentioned that Tchebycheff, Hadamard, De la Vallée Poussin, Landau, Wiener, Hardy, and all the latter-day number theorists used *analysis*, the mathematics of the *continuum*, in order to shed light on the *discrete* set of integers. This is just one more of the paradoxes of the higher arithmetic! We cannot go deeply into the subject of *analytic* number theory, but even a superficial discussion will indicate that the concepts and tools are of a type that would appear (except to a mathematical genius) exceedingly unlikely to reveal properties of the numbers of everyday arithmetic.

To begin with, we shall go back to the ideas of the mathematician who represented "analysis incarnate," namely, Leonhard Euler. Among other infinite series he considered, there was the series

$$1 + \frac{1}{2} + \frac{1}{3} + \frac{1}{4} + \cdots + \frac{1}{n} + \cdots$$

He proved that this series is *divergent*. The reader can show, as follows, that the sum in question cannot be defined because its value would have to exceed that of any set of half-dollars, however large their number. Thus we shall group the terms of the series as follows in order to indicate that the sum of each subset is equal to or greater than 1/2 (a half-dollar). Thus we have

$$\frac{1}{2} + \frac{1}{2} + \frac{1}{2} + \frac{1}{3} + \frac{1}{4} + \frac{1}{5} + \frac{1}{6} + \frac{1}{7} + \frac{1}{8} + \frac{1}{9} + \frac{1}{10} + \cdots + \frac{1}{16}$$
$$+ \frac{1}{17} + \frac{1}{18} + \cdots + \frac{1}{32} + \cdots$$

To see why this is so, let us examine the set $\{1/3 + 1/4\}$. There $1/3 > 1/4$ and hence $1/3 + 1/4 > 1/4 + 1/4$, that is, $1/3 + 1/4 > 1/2$. In the same way, each of the first three fractions in the sum, $1/5 + 1/6 + 1/7 + 1/8$, is greater than 1/8, and hence the total is greater than 4/8. Similarly, the next collection of numbers in the above series has a sum exceeding 8/16, etc., and the group after that has a total greater than 16/32, etc. The next group (not printed) would contain 32 fractions ending with 1/64, and its sum would exceed 32/64, etc. Thus, if one continued the series to the point where there were 2,000,000,000 such groups, each one of which (except for the first three) totals more than 1/2, then the grand total would be greater than 1,000,000,000, and far, far greater sums could be obtained by continuing the series. Therefore it is meaningless to speak of a definite, finite sum for the entire infinite series.

But Euler went farther. He showed that if the series is modified slightly by raising the denominators of the fractions to a power greater than one, then the series "converges," that is, has a sum (Chapter 22). Thus

$$1 + \frac{1}{2^{1.1}} + \frac{1}{3^{1.1}} + \cdots + \frac{1}{n^{1.1}} + \cdots$$

$$1 + \frac{1}{2^2} + \frac{1}{3^2} + \cdots + \frac{1}{n^2} + \cdots$$

$$1 + \frac{1}{2^3} + \frac{1}{3^3} + \cdots + \frac{1}{n^3} + \cdots$$

all have finite sums, and hence all are *convergent* series. Euler incorporated all convergent series of this type into a single function, namely,

$$\zeta(s) = 1 + \frac{1}{2^s} + \frac{1}{3^s} + \cdots + \frac{1}{n^s} + \cdots$$

where s is any real number such that $s > 1$, and ζ is the Greek letter zeta, so that $\zeta(s)$ is read as "zeta of s."

Euler developed properties of the function $\zeta(s)$ and related them to the theory of primes. But the importance of the function in analytic number theory occurs through a generalization which Riemann made in an epochal memoir published in 1859. In that paper, Riemann considered the meaning of

$$\zeta(s) = 1 + \frac{1}{2^s} + \frac{1}{3^s} + \cdots + \frac{1}{n^s} + \cdots$$

when the exponent s is a *complex variable*, $x + iy$, where x and y are real numbers. To a general reader who is not a specialist in mathematics, formal symbols like 2^{5-4i} or 3^{-1+i} will surely seem puzzling. But a complex "power" of a natural number is meaningful in the branch of analysis called "theory of functions of a complex variable," a subject founded by Riemann and analysts who were his contemporaries. The facts of the matter are that, for convergence of the above series when the variable s is complex, that is, when $s = x + iy$, it is necessary to impose the restriction $x > 1$. In other words, if the complex numbers $x + iy$ are represented as number pairs (x, y) and plotted as points in the XY-plane, only points to the right of the line $x = 1$ are permissible if $\zeta(s)$ is to converge. But Riemann extended the domain of the complex function $\zeta(s)$. Using a method that is termed "analytic continuation," he assigned values to $\zeta(s)$ for $s = x + iy$ and $x \le 1$, that is, for points of the XY-plane on or to the left of the line $x = 1$. [Only one point had to be excluded from the domain, namely, $(1, 0)$.]

The complex function $\zeta(s)$, defined as we have just explained, is known as the *Riemann zeta-function*. Its properties were applied by Riemann, Hadamard, and most modern researchers on the distribution of primes. In his memoir of 1859, Riemann asked: For what values of s will $\zeta(s) = 0$? He *conjectured* that all values of $s = x + iy$ for which $\zeta(s) = 0$ and x lies between 0 and 1 would be of the form $\frac{1}{2} + iy$. This is the "Riemann hypothesis," one of the famous unsolved problems of mathematics. In 1914 Godfrey Harold Hardy made considerable progress with this difficult problem, but neither he nor anyone else has obtained a proof up to the present time.

Some of the great analytic number-theorists after Riemann treated his conjecture as a postulate and, on assuming its truth, were able to deduce some remarkable theorems of "additive number theory," a subject which treats the decomposition of natural numbers into sums of primes or the expression of integers in other interesting forms. All such theorems are conjectures, since they are based on a conjecture.

But the Russian I. M. Vinogradov and his disciples initiated a new type of analytic methodology. They applied it to providing substantial backing for another famous conjecture about primes, the hypothesis which the Russian C. Goldbach (1690–1764) confided to Euler in 1742. Goldbach had observed for many special cases that an even number greater than 2 is the sum of two primes. Thus $4 = 2 + 2$, $6 = 3 + 3$, $8 = 3 + 5$, $10 = 5 + 5$, $12 = 5 + 7, \ldots, 100 = 3 + 97$, $102 = 5 + 97, \ldots$.

Goldbach conjectured that *every* even number is the sum of two primes but was unable to *prove* it. However, about 1937, Vinogradov was able to establish that every "sufficiently large" even number is the sum of *at most four* primes. This result, called the Vinogradov-Goldbach theorem, seems to "approximate" the truth of Goldbach's conjecture.

Since the theory of the zeta-function, the analytic maneuverings of Hardy, and the "trigonometric sums" of Vinogradov would all take us far beyond the scope of the present book, we cannot discuss them. But because analytic number theory is related to the theory of functions of a complex variable, we now offer a capsule introduction to the elements of that subject. In truth it is *not* a difficult branch of analysis and has a special beauty of its own, as well as important physical applications. Today an engineer will study its elementary aspects and a Cambridge don will develop its purest, most advanced issues. Its origins must be credited to Gauss, Cauchy, Riemann, and Weierstrass. The subject matter of "complex variable theory" consists of the analytic geometry, the differential and integral calculus, and the general infinite processes associated with functions where independent and dependent variables are complex.

Let us begin with the Cartesian geometry of such functions. We know that $y = x^2$, where x and y are *real* variables, has a parabola as its graph. But what is the picture of $w = z^2$ where w and z are complex variables, that is, where $z = x + iy$ and $w = u + iv$, x, y, u, v being real variables?

In ordinary Cartesian graphs, there is an X-axis of *real* numbers and a Y-axis of *real* numbers. Then the domain of $w = z^2$, that is, the set of values which z may assume, will be located in the XY-plane. But if z may have any complex values whatsoever, its domain will be the *entire* XY-plane. In order to provide a graphic picture of the range, a second plane is needed, in which there are *real* U and V axes, and the range is the set of points in this plane representing the values assumed by $w = u + iv$. The function $w = z^2$ may also be expressed as

$$u + iv = (x + iy)^2$$

or

$$u + iv = x^2 + 2ixy - y^2$$

Since the real and imaginary parts of each member of this equation must be respectively equal,

$$u = x^2 - y^2$$

and

$$v = 2xy$$

These equations represent a *mapping* of the *points* of the XY-plane into those of the UV-plane since, corresponding to each point (x, y), there is defined a unique point (u, v).

In the UV-plane (Figure 21.1*a*) the equations of the "streets," or lines parallel to the V-axis, are $u = 1$, $u = 2$, $u = 3$, etc., and the equations of the "avenues," or lines parallel to the U-axis, are $v = 1$, $v = 2$, $v = 3$, etc. Substituting these values of u and v in the pair of equations above, one obtains the *hyperbolas* $x^2 - y^2 = 1$,

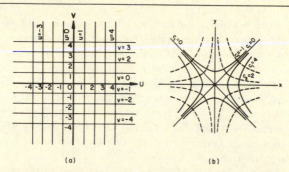

(a)   (b)

Figure 21.1   **The complex function** $w = z^2$ **as a conformal mapping. The lines** $u = c_1$, $v = c_2$ **correspond to the hyperbolas** $x^2 - y^2 = c_1$, $2xy = c_2$, **respectively.**

$x^2 - y^2 = 2$, $x^2 - y^2 = 3$, etc. and $2xy = 1$, $2xy = 2$, $2xy = 3$, so that as another *geometric* interpretation of the complex function $w = z^2$ there is the *correspondence* between the meshwork of hyperbolas (Figure 21.1b) and the rectangular network of straight lines. It can be proved that the *mapping* of the hyperbolas onto the straight lines is *conformal*, that is, *angles* are *invariant* under the mapping (providing the origin $x = 0$, $y = 0$ is excluded from the domain of the function). The perpendicularity of the straight lines corresponds to the orthogonality of the two systems of hyperbolas.

After the above example of the analytic geometry of complex functions, let us illustrate the calculus of such functions. Thus from experience with real functions, one might conjecture that for $w = z^2$ the derivative $dw/dz = 2z$. Such an assertion would be a mere formalism, since there is no meaning involved until an exact *definition* of the derivative of a complex function has been furnished. This definition is strictly analogous to that in the case where y is a real function of x and

$$\frac{dy}{dx} = \text{limit } \frac{\Delta y}{\Delta x}$$

as Δx approaches zero, *providing* such a limit exists (Chapter 22). Then if w and z are complex variables, and w is a function of z,

$$\frac{dw}{dz} = \text{limit } \frac{\Delta w}{\Delta z}$$

as Δz approaches zero, providing such a limit exists.

It looks as if, after all, the differential calculus of complex functions must be much the same as the elementary calculus of real functions. But a remarkable new feature is concealed in the clause "as Δz approaches zero." The *geometric* interpretation of $\Delta y / \Delta x$ for a real function is the slope of a secant to the curve representing the function, and "Δx approaches zero" is associated with the fact that a variable point of the curve approaches coincidence with the fixed point as the secant approaches the tangent (page 187) to the curve at point P. Now "Δz approaches zero" means that a variable point in the XY-plane approaches coincidence with a fixed point, and if the complex function is continuous, a similar motion toward coincidence occurs in the UV-plane (Δw approaches zero). However, there is *not* a unique motion toward coincidence analogous to the real case where a variable point moves along a real curve toward a fixed point P. There are an infinite number of paths along which Δz can approach zero, that is, which will produce the desired coincidence. The question arises: Will the limit of $\Delta w/\Delta z$ have the *same* value regardless of how $\Delta z \to 0$? It seems natural that the path of approach might influence the final result, and this is actually the case. It is only when the limit exists, and is the *same* for every path of approach, that the complex function is said to have a derivative at the point in question.

If the path of approach is along the X-axis, then $\Delta z = \Delta x$ and

$$\frac{\Delta w}{\Delta z} = \frac{\Delta u + i\Delta v}{\Delta x} = \frac{\Delta u}{\Delta x} + i\frac{\Delta v}{\Delta x}$$

If, on the other hand, the path of approach is along the Y-axis where $z = x + iy$ becomes $z = 0 + iy$, then $\Delta z = i\Delta y$ and

$$\frac{\Delta w}{\Delta z} = \frac{\Delta u}{i\Delta y} + \frac{\Delta v}{\Delta y} = -i\frac{\Delta u}{\Delta y} + \frac{\Delta v}{\Delta y}$$

If limits exist for both paths of approach, and if these limits have the same value, then

$$\frac{\partial u}{\partial x} + i\frac{\partial v}{\partial x} = -i\frac{\partial u}{\partial y} + \frac{\partial v}{\partial y}$$

Equating real and imaginary parts of the left and right members, one arrives at the *Cauchy-Riemann* differential equations, which pervade the theory of functions of a complex variable, namely,

$$\frac{\partial u}{\partial x} = \frac{\partial v}{\partial y} \quad \text{and} \quad \frac{\partial v}{\partial x} = -\frac{\partial u}{\partial y}$$

If a function has a derivative at a point, then these equations must be satisfied at that point. Conversely, if these equations are satisfied and if the partial derivatives are continuous, then $f(z)$ has a derivative. If the Cauchy-Riemann equations are satisfied throughout some region of the XY-plane, then $f(z)$ is said to be *analytic* in that region. What happens when the fundamental equations are *not* satisfied? There may be *singular points* at which this occurs, and this requires special study. But there may be failure of the Cauchy-Riemann conditions in general. Then the function is *nonanalytic*. This may occur because the derivative is not uniquely defined for all the different paths along which Δz may approach 0. Then in contrast to being *monogenic*, the function is *polygenic*.

If, in addition to the conditions satisfied above, the functions $u(x, y)$ and $v(x, y)$ have continuous partial derivatives of the second order, then (Chapter 9)

$$\frac{\partial^2 u}{\partial x \partial y} = \frac{\partial^2 u}{\partial y \partial x} \quad \text{and} \quad \frac{\partial^2 v}{\partial x \partial y} = \frac{\partial^2 v}{\partial y \partial x}$$

Differentiating the first Cauchy-Riemann equation with respect to x and the second with respect to y so that

$$\frac{\partial^2 u}{\partial x^2} = +\frac{\partial^2 v}{\partial x \partial y}$$

$$\frac{\partial^2 v}{\partial y \partial x} = -\frac{\partial^2 u}{\partial y^2}$$

one obtains

$$\frac{\partial^2 u}{\partial x^2} = -\frac{\partial^2 u}{\partial y^2}$$

or

$$\frac{\partial^2 u}{\partial x^2} + \frac{\partial^2 u}{\partial y^2} = 0$$

This means that $u(x, y)$ is a solution of Laplace's equation (Chapter 9) and, in a similar fashion, it can be shown that $v(x, y)$ satisfies the same equation. Thus the theory of functions of a complex variable is another field in which partial differential equations, the Cauchy-Riemann set and the equation of Laplace, play a basic role.

To understand the integral calculus of complex functions, the reader would need a more profound understanding of certain ways in which the Cauchy-Mengoli integral (Chapter 27) can be generalized. The more important aspects of that question will be treated in later chapters. Here we might remark that whereas the Cauchy-Mengoli integral

$$\int_a^b f(x)\, dx$$

represents the limit of a sum where a typical term can be taken as $f(x_i)\, \Delta x$, where x_i is a point on an interval or segment of the X-axis from a to b, it is possible to form a definite integral where $f(x)$ varies along a curve. Such an interpretation is possible for the complex integral

$$\int_C f(z)\, dz$$

Then the fundamental proposition about complex integrals is the *Cauchy integral theorem,* which establishes the fact that if $f(z)$ is differentiable at all points in a certain region of the plane and C is a closed curve (like a circle or ellipse or some oval) in that region, then

$$\int_C f(z)\, dz = 0$$

From this theorem various Cauchy integral formulas are derived, and these are associated with the development of complex functions into infinite series in z. It turns out that if $f(z)$ is differentiable at all points on a circle C with center at $(0, 0)$, then

$$f(z) = a_0 + a_1 z + a_2 z^2 + \cdots + a_n z^n + \cdots$$

(where the coefficients are complex numbers determined by the Cauchy integral formulas) is a series that converges for all values of $z = x + iy$, corresponding to points (x, y) within the circle C.

If, as in the case of a real variable, a function $f(z)$ is defined to be *analytic* (in a region) provided that it can be expressed as a power series convergent in the region, then by virtue of the Cauchy formulas, one can say that for *complex functions*, a *differentiable* function is an *analytic* function. One cannot make a similar assertion for real functions.

Although the discussion of the last few pages takes us past the kindergarten stage of the theory of functions of a complex variable, it cannot transport us to such summits as the Riemann zeta-function. Since all the material was motivated by the *analytic* theory of numbers, it seems fitting to end with the human story of the mathematician who, probably more than any other during the twentieth century, helped to foster analytic methods for the resolution of the discrete problems of the higher arithemtic. That man was Godfrey Harold Hardy (1877–1947). For some facts about his life, personality, and general contributions to mathematics, we shall now quote one of his great successors, Professor E. C. Titchmarsh of Cambridge University.* Then we shall follow with the story of the Hindu number-theory genius, Srinivasa Ramanujan (1887–1920), who, through Hardy's efforts, became known to the mathematical world and was enabled to leave a remarkable arithmetical legacy to posterity.

Godfrey Harold Hardy was born on 7 February 1877, at Cranleigh, Surrey. He was the only son of Isaac Hardy, Art Master, Bursar and House Master of the preparatory branch of Cranleigh School. His mother, Sophia Hardy, had been Senior Mistress at the Lincoln Training College. Both parents were extremely able people and mathematically minded, but want of funds had prevented them from having a university training.

The future professor's interest in numbers showed itself early. By the time he was two years old he had persuaded his parents to show him how to write down numbers up to millions. When he was taken to church he occupied the time in factorizing the numbers of the hymns, and all through his life he amused himself by playing about with the numbers of railway carriages, taxi-cabs and the like.

As soon as he was old enough G. H. went to Cranleigh School, and by the time he was twelve he had passed his first public examination with distinctions in mathematics, Latin and drawing. His mind was turned in the direction of Cambridge by a curious incident, which he has related in *A Mathematician's Apology*. He happened to read a highly coloured novel of Cambridge life called *A Fellow of Trinity*, by "Alan St Aubin" (Mrs Frances Marshall), and was fired with the ambition to become, like its hero, a fellow of Trinity. He went up to Trinity College, Cambridge, as an entrance scholar in 1896.

Hardy was fourth wrangler in 1898. He took Part II of the Tripos in 1900, being placed in the first division of the first class, Jeans being then below him in the second division of the first class. In the same year he was elected to a Prize Fellowship at Trinity, and his early ambition was thus fulfilled. Hardy and Jeans, in that order, were awarded Smith's Prizes in 1901.

* Royal Society of London, *Obituary Notices of Fellows*, Vol. 6, No. 18 (November 1949), pp. 446 ff.

His life's work of research had now begun, his first paper apparently being that in the *Messenger of Mathematics*, Vol. 29, 1900. It is about the evaluation of some definite integrals, a subject which turned out to be one of his permanent minor interests, and on which he was still writing in the last year of his life.

In 1906, when his Prize Fellowship was due to expire, he was put on the Trinity staff as lecturer in mathematics, a position he continued to hold until 1919. This meant that he had to give six lectures a week. He usually gave two courses, one on elementary analysis and the other on the theory of functions. The former included such topics as the implicit function theorem, the theory of unicursal curves and the integration of functions of one variable. This was doubtless the origin of his first Cambridge tract, *The Integration of Functions of a Single Variable*. This work is so well known now that it is often forgotten that its systematization was due to Hardy. He also sometimes took small informal classes on elementary subjects, but he was never a "tutor" in the Oxford sense.

In 1908 Hardy made a contribution to genetics which seems to be little known by mathematicians, but which has found its way into textbooks as "Hardy's Law." There had been some debate about the proportions in which dominant and recessive Mendelian characters would be transmitted in a large mixed population. The point was settled by Hardy in a letter to *Science*. It involves only some simple algebra, and no doubt he attached little weight to it. As it happens, the law is of central importance in the study of Rh-blood-groups and the treatment of haemolytic disease of the newborn. In the *Apology* Hardy wrote, "I have never done anything 'useful.' No discovery of mine has made, or is likely to make, directly or indirectly, for good or ill, the least difference to the amenity of the world." It seems that there was at least one exception to this statement.

He was elected a Fellow of the Royal Society in 1910, and in 1914 the University of Cambridge recognized his reputation for research, already worldwide, by giving him the honorary title of Cayley Lecturer.

To this period belongs his well-known book *A Course of Pure Mathematics*, first published in 1908, which has since gone through numerous editions and been translated into several languages. The standard of mathematical rigour in England at that time was not high, and Hardy set himself to give the ordinary student a course in which elementary analysis was for the first time done properly. *A Course of Pure Mathematics* is hardly a *Cours d'Analyse* in the sense of the great French treatises, but so far as it goes it serves a similar purpose. It is to Hardy and his book that the outlook of present-day English analysts is very largely due.

Another turning point in Hardy's career was reached about 1912, when he began his long collaboration with J. E. Littlewood. There have been other pairs of mathematicians, such as Phragmén and Lindelöf, or Whittaker and Watson, who have joined forces for a particular object, but there is no other case of such a long and fruitful partnership. They wrote nearly a hundred papers together, besides (with G. Pólya) the book *Inequalities*.

Soon afterwards came his equally successful collaboration with the Indian mathematician Ramanujan, though this was cut short six years later by Ramanujan's early death. An account of this association is given by Hardy in the introductions to Ramanujan's collected works and to the book *Ramanujan*. In a letter to Hardy in 1913, Ramanujan sent specimens of his work, which showed that he was a mathematician of the first rank. He came to England in 1914 and remained until 1919. He was largely self-taught, with no knowledge of modern rigour, but his "profound and invincible originality" called out Hardy's equal but quite different powers. Hardy said, "I owe more to him than to any one else in the world with one exception, and my association with him is the one romantic incident in my life."

Hardy was a disciple of Bertrand Russell, not only in his interest in mathematical philosophy, but in his political views. He sympathized with Russell's anti-war attitude, though he did not go to the lengths which brought Russell into collision with the authorities. In a little book *Bertrand Russell and Trinity*, which he had printed for private circulation in 1942, Hardy has described the Russell case and the storms that raged over it in Trinity. It was an unhappy time for those concerned, and one may think that it all would have been better forgotten. It must have been with some relief that, in 1919, he heard of his election to the Savilian Chair of Geometry at Oxford, and migrated to New College.

In the informality and friendliness of New College Hardy always felt completely at home. He was an entertaining talker on a great variety of subjects, and one sometimes noticed every one in common room waiting to see what he was going to talk about. Conversation was one of the games which he loved to play. He played several games well, particularly real tennis, but his great passion was for cricket. A vivid account of Hardy's affection for cricket and his life in his later Cambridge years is given by C. P. Snow, in an article entitled "A mathematician and cricket," in *The Saturday Book, 8th Year*.

In *A Mathematician's Apology* Hardy is at some pains to show that [pure] mathematics is useless, or at any rate harmless. He says, "It is true that there are branches of applied mathematics, such as ballistics and aerodynamics, which have been developed deliberately for war . . . but none of them has any claim to rank as 'real'. They are indeed repulsively ugly and intolerably dull; even Littlewood could not make ballistics respectable, and if he could not, who can?" His views on this subject were obviously coloured by his hatred of war, but in any case his whole instinct was for the purest of mathematics.

Nevertheless, he was a Fellow of the Royal Astronomical Society, which he joined in 1918 in order that he might attend the meetings at which the theory of relativity was debated by Eddington and Jeans. He even once, in 1930, took part in a debate on stellar structure, which involved R. H. Fowler's work on Emden's and allied differential equations. On this he made the characteristic remark that Fowler's work, being pure mathematics, would still be of interest long after all the physical theories, which had been discussed had become obsolete. This prophecy has since been very largely fulfilled.

His likes and dislikes, or rather enthusiasms and hates, can be listed as follows:

Enthusiasms
   (i) Cricket and all forms of ball games.
   (ii) America, though perhaps he only came into contact with the pleasanter side of it.
   (iii) Scandinavia, its people and its food.
   (iv) Detective stories.
   (v) Good literature, English and French, especially history and biography.
   (vi) Walking and mild climbing, especially in Scotland and Switzerland.
   (vii) Conversation.
   (viii) Odd little paper games, such as making teams of famous people whose names began with certain combinations of letters or who were connected with certain countries, towns or colleges. These were played for hours in hotels or on walks.
   (ix) Female emancipation and the higher education of women (though he opposed the granting of full membership of the university to Oxford women).
   (x) *The Times* cross-word puzzles.
   (xi) The sun.
   (xii) Meticulous orderliness, in everything but dress. He had a large library and there were piles of papers all about his rooms, but he knew where everything was and the exact position of each book in the shelves.
   (xiii) Cats of all ages and types.

Hates
   (i) Blood sports of all kinds, war, cruelty of all kinds, concentration camps and other emanations of totalitarian governments.

(ii) Mechanical gadgets; he would never use a watch or a fountain pen, and the telephone only under compulsion. He corresponded chiefly by prepaid telegrams and post cards.

(iii) Looking-glasses; he had none in his rooms, and in hotels the first thing he did in his room was to cover them over with bath-towels.

(iv) Orthodox religion, though he had several clerical friends.

(v) The English climate, except during a hot summer.

(vi) Dogs.

(vii) Mutton—a relic of his Winchester days, when they had by statute to eat it five days a week.

(viii) Politicians as a class.

(ix) Any kind of sham, especially mental sham.

In 1928–1929 he was Visiting Professor at Princeton and at the California Institute of Technology, O. Veblen coming to Oxford in his place. In 1931 Hardy returned to Cambridge to accept the Sadleirian chair of Pure Mathematics.

Perhaps the most memorable feature of this period was the Littlewood-Hardy seminar or "conversation class." This was a model of what such a thing should be. Mathematicians of all nationalities and ages were encouraged to hold forth on their own work, and the whole thing was conducted with a delightful informality that gave ample scope for free discussion after each paper. The topics dealt with were very varied, and the audience was always amazed by the sure instinct with which Hardy put his finger on the central point and started the discussion with some illuminating comment, even when the subject seemed remote from his own interests.

He also lectured on the calculus of variations, a subject to which he had been drawn by his work on inequalities.

After his return to Cambridge he was elected to an honorary fellowship at New College. He held honorary degrees from Athens, Harvard, Manchester, Sofia, Birmingham, Edinburgh, Marburg and Oslo. He was awarded a Royal Medal of the Royal Society in 1920, its Sylvester Medal in 1940 and the Copley Medal in 1947. He was President of Section A of the British Association at its Hull meeting in 1922, and of the National Union of Scientific Workers in 1924–1926. He was an honorary member of many of the leading foreign scientific academies.

Some months before his death he was elected "associé étranger" of the Paris Academy of Sciences, a particular honour, since there are only ten of these from all nations and scientific subjects. He retired from the Sadleirian chair in 1942, and died on 1 December 1947, the day on which the Copley Medal was due to be presented to him.

For the biography of the mathematician who furnished the "one romantic incident" in Hardy's life, there is the statement of Professor Robert D. Carmichael, one of whose numerous mathematical specialities is number theory.*

During our generation no more romantic personality than that of Srinivasa Ramanujan has moved across the field of mathematical interest. Indeed it is true that there have been few individuals in human history and in all fields of intellectual endeavor who draw our interest more surely than Ramanujan or who have excited more fully a certain peculiar admiration for their genius and their achievements under adverse conditions. There is nothing particularly noteworthy about Ramanujan's ancestry to account for his great gifts. His father and paternal grandfather were petty accountants in Kumbakonam, an important town in the Tanjore district in India. His mother was a woman of strong

---

* Robert D. Carmichael, "Some Recent Researches in the Theory of Numbers," *American Mathematical Monthly*, Vol. 39, No. 3 (March 1932), pp. 139 ff.

common sense. For some time after marriage she was without children. Her father prayed to a famous local goddess to bless his daughter with children; and shortly afterwards Ramanujan, her eldest child, was born.

His school and college education were meager and somewhat irregular. On January 16, 1913, at 25 years of age, he wrote to G. H. Hardy a letter containing the following words:

"I have had no University education but I have undergone the ordinary school course. After leaving school I have been employing the spare time at my disposal to work at Mathematics. I have not trodden through the conventional regular course which is followed in a University course, but I am striking out a new path for myself."

On February 27 of the same year he wrote to Hardy as follows:

"I have found a friend in you who views my labours sympathetically. This is already some encouragement to me to proceed. . . . I find in many a place in your letter rigorous proofs are required and you ask me to communicate the methods of proof. . . . The sum of an infinite number of terms of the series $1 + 2 + 3 + 4 + \cdots = -1/12$ under my theory. If I tell you this you will at once point out to me the lunatic asylum as my goal. . . . What I tell you is this. Verify the results I give and if they agree with your results . . . you should at least grant that there may be some truths in my fundamental basis. . . . To preserve my brains I want food and this is now my first consideration. Any sympathetic letter from you will be helpful to me here to get a scholarship either from the University or from Government. . . ."

These passages from Ramanujan's letters will give a faint indication of the human side of a correspondence which stands as a remarkable one in the history of our science. Along with the letters numerous astonishing mathematical results were communicated by Ramanujan to Hardy. Many of these were right; not a few were wrong; even the errors themselves were sometimes brilliant. The association thus opened up with Hardy became one of the prime factors in the life of Ramanujan. He went to England in 1914.

Hardy has put on record the fact that his appearance set a great puzzle for solution. What was to be done in teaching such an intellect the spirit and methods of modern mathematics? In some directions his knowledge was profound. In others his limitations were quite as startling. He could work out modular equations and theorems of complex multiplication, to orders unheard of. His mastery of continued fractions was remarkable. He had found for himself properties of the Zeta-function and applied them to many famous problems in the analytic theory of numbers. Yet he had never heard of a doubly periodic function or of Cauchy's theorem the (most fundamental fact of complex variable theory). Indeed, he had but the vaguest idea of a function of a complex variable. His conception of mathematical proof was quite inadequate. His results, new or old, right or wrong, had been obtained by argument, intuition, induction, mingled in a way which he himself could not coherently describe. Many of his proofs were invalid. Not a few of his results were false. His approximations were not as close as he supposed. But, notwithstanding the fact that he had never seen a French or German book and that his command of English was meager, he had conceived for himself and had treated in an astonishing way problems to which for a hundred years some of the finest intellects in Europe had given their attention without having reached a complete solution. That such an untrained mind made mistakes in dealing with such questions is not remarkable. What is astonishing is that it ever occurred to him to treat these problems at all.

In a few years after reaching England Ramanujan had a fair knowledge of the theory of functions and the analytic theory of numbers. He had learned to know when he had proved a theorem and when he had not. And his flow of original ideas kept up without abatement during this period of acquisition.

In the spring of 1917 he began to show evidence of being unwell. He went to a Nursing Home at Cambridge early in the summer of 1917 and was never afterwards out of bed for long at a time. He died April 26, 1920, a little more than seven years after writing his initial letter to Hardy, the event which brought him for the first time into touch with modern mathematical ideas. During this brief interval, about one-third of which was spent in bed on account of illness, he added to the achievements made unaided during the first years of his life a body of contributions which is sure to exert a marked influence on the development of certain chapters of mathematics, indeed has already exerted such an influence.

Under the title *Collected Papers of Srinivasa Ramanujan* there has been brought together in a single volume everything published by Ramanujan with the exception of a few solutions of questions proposed by other mathematicians and answered by him in the *Journal of the Indian Mathematical Society*. He left behind him a large mass of unpublished material in note-books now famous, and these have been edited and published.

Although complex variable theory led Hardy and Ramanujan to discoveries which the former liked to describe as useless, another part of modern analysis was designed to cope with one of the most constant and pressing of human issues, the need to be thrifty. It appeared to scientists that nature also observed frugality, and the physical principles involved were expressed in analytic language. Because the analysis of the late nineteenth century did not suffice for this purpose, new analytic concepts had to be created and, as we shall see in the next chapter, the results were notions as "pure" and abstract as those employed by Hardy and all the analytic number-theorists.

22

The Reformation of Analysis

One reason for the present primary position of pure mathematical theories is that time and again in mathematical history logical crises have arisen from failure to purify fundamental notions completely, that is, to divorce ideas from their origins in the world of reality. In this respect, calculus was no exception. In the long period of its development from Archimedes to Newton and Leibniz, basic conceptions were related to physical situations and procedures were derived by intuitive arguments. Kepler measured an area swept out by the *motion* of a (celestial) point on a physical "continuous curve." Newton's derivative was a "fluxion of a fluent," usually illustrated by an instantaneous velocity or acceleration. Again, the derivative was visualized as the limit of the slope of a variable secant as one point of a *curve* moves to coincidence with another. Thus, definitions were dependent on the meaning of "motion" and "continuous curve." That our intuitions about motion will produce logical paradoxes was indicated by Zeno of Elea about 450 B.C. (Chapter 24). But even earlier, the Pythagoreans had become aware of the great danger of putting one's faith in intuition.

The "common sense" view of a line segment will never suggest its possible incommensurability, and by the end of the nineteenth century there were a number of examples of how our perception of motion along ordinary paths may lead to faulty mathematical judgments. Surely no one (except a modern mathematician) would ever expect a continuous line or curve to lack a definite direction (that is, a tangent) at each and every one of its points, and thereby make the geometric definition of the derivative as the slope of a tangent fall apart. Nor would one think it possible for a curve to fill an *area* completely. Yet Giuseppe Peano (1858–1932) showed that both these things can happen. To obtain a *Peano curve* (Figure 22.1), let a square be divided into four equal squares numbered 0, 1, 2, 3, counterclockwise, starting at the bottom left-hand corner. Now divide each of these squares into four equal squares, and run around the 0 square clockwise, ending opposite a square of 1; pass into the square of 1, and round 1 counterclockwise into 2, and so on. Carry the division further, stage by stage *forever*, running continuously round the squares, starting always at the bottom left-hand corner, running counterclockwise round the first newly divided square after an odd number of divisions and clockwise after an even number, and alternating passages round the squares indicated in the diagram. At each stage of the process there is ambiguity of direction at each "corner" of the path because there the curve has two perpendicular directions. As the process

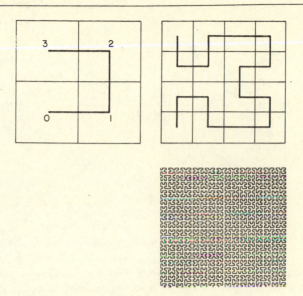

Figure 22.1   Construction of the Peano curve

continues there are more and more corners, closer together, so that ultimately the points of ambiguity are all points of the path.

In an address* to the American Mathematical Society at the turn of the century, James Pierpont (1866–1938), professor at Yale and a leading analyst of his day named *eight* properties which we automatically assign to a curve on the basis of geometric intuition. He demolished each of these in turn by the production of monstrosities indicating that intuition is erroneous. Two of Pierpont's eight points have been illustrated by the Peano curve. He stated further that our most firmly rooted intuitive notion of a curve, the one often used as a definition, is that it can be generated by the *motion of a point*. Here, he said, two ideas are essential, namely that motion is continuous and that at each instant it has a definite direction and speed. Now, as we have seen, direction and speed are obtained from the *derivatives* dy/dx and ds/dt. But suppose that the curve "generated by the motion of a point" is like the four-leaved rose (Figure 7.21). What happens when the moving point reaches the origin? There is no unique answer since there are apparently eight different paths possible. Suppose the path of a moving point to be the continuous curve of Figure 22.2 whose equation is

$$y = 0 \quad \text{for} \quad x = 0$$
$$y = x \sin \frac{1}{x} \quad \text{for} \quad x \neq 0$$

* See James Pierpont, "On the Arithmetization of Mathematics," *Bulletin of the American Mathematical Society*, Vol. 5 (May 1899), p. 398.

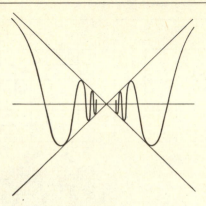

Figure 22.2

That curve lies between two straight lines each making an angle of 45° with the X-axis, and oscillates with indefinitely increasing frequency (and *change of direction*) as it approaches the origin. The result is that at the origin *no* direction (tangent) can be specified. We ask: How does the point move as it passes through the origin? Or if it is to start at the origin and is to proceed to right or left, in what direction should it move?

After indicating other fallacious notions, Pierpont emphasized that any definition that is to serve as a basis for *rigorous* deduction must be an idealization of physical experience and can only correspond approximately to hazy intuitive notions. He prescribed that the definitions of analysis be framed by using the "epsilon, delta" (ε, δ) criterion of Augustin-Louis Cauchy (1789–1857) and Karl Wilhelm Theodor Weierstrass (1815–1897).

It is the purpose of this chapter to explain the nature of this criterion, and to reveal how the two mathematicians named by Pierpont succeeded in banishing "monsters" by making analysis more rigorous. Pierpont stated that the "ε, δ" definitions make it possible to reason with absolute precision, and nevertheless arrive at deductions in accord with the evidence of our intuition. It is the custom to say that Weierstrass "arithmetized" analysis, by which is meant that he replaced definitions and proofs based on geometric pictures or sensory observations by concepts expressed in terms of *real numbers*. To obtain a rigorous formulation of limits, continuity, and the like, Weierstrass was forced to create his own special theory of irrational numbers.

In analysis, the fundamental concept is that of *limit*, and it is to Cauchy that we owe the first rigorous definition. We have seen that integral calculus defines lengths, areas, volumes as the *limiting* values (when such exist) of certain approximating sums, and that the *derivative*, the basic notion of differential calculus, is likewise a limit. Up to this point we have handled the fundamental concepts of calculus in an informal, intuitive fashion, in much the same spirit as the founders of the subject. But even a layman today should be made aware of the extreme lack of rigor in the reasoning of Newton and Leibniz. We must then reexamine our previous arguments and present a more tenable treatment of limits, one that will not produce "monsters."

We have associated calculus concepts with approximative processes. In such estimation, if a first result seems too crude, there is an attempt to improve it, and if a second approximation does not satisfy standards of precision, there is a third, and so on. We shall now suppose that we are dealing with techniques of estimation that are not hit-or-miss, but possess a definite formula or algorithm that will always provide a *better* approximation than some approximation previously made, so that successive estimates will provide a sequence of better and better approximations to the "ideal" measurement we are seeking. Thus, for our present purposes, let us consider a hypothetical sequence in which successive estimates are

$$\frac{1}{2}, \frac{2}{3}, \frac{3}{4}, \frac{4}{5}, \frac{5}{6}, \frac{6}{7}, \frac{7}{8}, \frac{8}{9}, \cdots$$

Suppose that some new light on the situation indicates that all further estimates will be governed by the law $n/(n + 1)$, where n is the integer giving the position of the estimate in the above sequence. Thus the next estimate would be the *ninth*, and since $n = 9$, this approximation would equal 9/10, and if we continued approximating, the millionth result would be 1,000,000/1,000,001. This millionth estimate may satisfy our standard of precision, and we can then use it as an approximation to the measurement we are seeking, or we may assert that for all practical purposes, 1,000,000/1,000,001 is so close to 1 in value that we may as well assert that the entity in question measures 1 unit. Our conviction that this is a good thing to do is strengthened if we reexamine the sequence. We see that the fractions constantly increase in size, so that approximations after 1,000,000/1,000,001 will be larger, but never quite 1, since the denominator exceeds the numerator. Yet the difference from 1 will diminish if we continue the sequence of approximations. Algebraically, this difference for the *n*th term is

$$1 - \frac{n}{n+1} = \frac{n+1-n}{n+1} = \frac{1}{n+1}$$

If $n = 1,000,000,000,000$, the difference will be about 0.000,000,000,001, which is exceedingly slight. The sum total of all these numerical facts is implied in the statement that 1 is the *limit* of the sequence whose law is $n/(n + 1)$.

About standards of precision, we shall say that any set of numbers approximates a particular number c within a *standard* ε (where ε is a positive number), if the numerical difference between c and every number of the set is less than ε. Thus the set {2, 9, 3, 8, 4, 7, 5} approximates the number 6 within the *standard* 5. But this is not the smallest standard, since the set also approximates 6 within the standard 4.1 or 4.01 or 4.001. Again, the sequence whose law is $n/(n + 1)$ approximates 1 within the standard 0.6 because all terms of the sequence differ from 1 by less than 0.6. But here we could also say the sequence approximates 1 within the standard 0.51 and also within the even finer standard 0.50001. The reader will readily see that the set of terms $n/(n + 1)$ for which n is greater than a million approximates 1 within the standard 0.000,001 or an even smaller standard, since, for example, if $n = 1,000,000$, then $1,000,000/1,000,001$ differs from 1 by $1/1,000,001$, which is less than our standard, and successive terms will differ from 1 by even less.

Now we are ready for the *definition*: The sequence of real numbers $\{a_1, a_2, a_3, \ldots, a_n, \ldots\}$ is said to have the *limit L* if, corresponding to each (positive) standard of approximation ε, however small, a term a_m of the series can be found so that all succeeding terms approximate L within the standard ε.

To clarify this definition, consider the sequence

$$1\frac{2}{3}, 1\frac{8}{9}, 1\frac{26}{27}, 1\frac{80}{81}, \ldots$$

whose law of formation is

$$a_n = 2 - \frac{1}{3^n}$$

This means that the *first* term is found by subtracting $1/3^1$ from 2, the second by subtracting $1/3^2$ from 2, the third by subtracting $1/3^3$ from 2, etc. Then the eighth term will be

$$a_8 = 2 - \frac{1}{3^8} = 1\frac{6560}{6561}$$

By inspection, we might say that 2 is the *limit* of the sequence. But taking the formal requirement of the definition, suppose that we assign the standard $\varepsilon = 0.01$, what selection must we make for m so that all terms subsequent to a_m should differ from 2 by less than 0.01? If we examine the first four terms of the sequence, no one of them will come within the standard. But the fifth term is $2 - 1/3^5$, which differs from 2 by $1/3^5$ or about 0.004. Therefore we select $m = 4$. Then a_5 differs from 2 by less than 0.01, and subsequent terms differ from 2 by $1/3^6, 1/3^7, 1/3^8, \ldots$ which are all smaller than $1/3^5$. Then *a fortiori* they will differ from 2 by less than 0.01.

Now suppose that someone assigns the standard $\varepsilon = 0.0001$. Then we find that we can choose $m = 8$, since $a_9 = 2 - 1/3^9$ differs from 2 by $1/3^9$, or by 0.00005, approximately, and subsequent terms of the sequence differ from 2 by still less.

No matter how small the value assigned to the positive number ε, a value of m can be selected so that $1/3^{m+1}$ will be smaller than ε, and therefore $a_{m+1} = 2 - 1/3^{m+1}$ will differ from 2 by less than ε, and for terms after a_{m+1} the difference will be even less.

This seems like a lengthy argument to establish a fact which is "intuitively" obvious, but we have indicated at some length that modern mathematics demands the arithmetization of analysis lest various "monsters" develop from intuitive reasoning. In each of the following sequences, arithmetic proof can be given to indicate the existence or nonexistence of a limit:

Sequence	Law for a_n
(1) $1, \dfrac{1}{2}, \dfrac{1}{3}, \dfrac{1}{4}, \dfrac{1}{5}, \dfrac{1}{6}, \cdots$	$\dfrac{1}{n}$
(2) $1, \dfrac{1}{4}, \dfrac{1}{3}, \dfrac{1}{8}, \dfrac{1}{5}, \dfrac{1}{12}, \cdots$	$\dfrac{1}{n}$ (if n is odd)
	$\dfrac{1}{2n}$ (if n is even)
(3) $1, -\dfrac{1}{2}, \dfrac{1}{3}, -\dfrac{1}{4}, \dfrac{1}{5}, -\dfrac{1}{6}, \cdots$	$\dfrac{1}{n}$ (if n is odd)
	$-\dfrac{1}{n}$ (if n is even)
(4) $4, 4, 4, 4, 4, 4, \ldots$	4
(5) $2, 4, 8, 16, 32, 64, \ldots$	2^n
(6) $-2, 2, -2, 2, -2, 2, \ldots$	-2 (if n is odd)
	2 (if n is even)

We remark that it is essential to know the law of formation of a sequence if one wishes to draw conclusions about terms beyond those specifically listed. On intelligence tests a type of question frequently asked calls for the next term of a sequence like

$$1, \frac{1}{4}, \frac{1}{9}, \cdots$$

The answer given in the key is 1/16. But we claim that 1/22 would also be a correct response. No law of formation is stated in the test, and therefore there is no reason to assume that $a_n = 1/n^2$. Why not, then, use

$$a_n = \frac{1}{n^3 - 5n^2 + 11n - 6}$$

since it fits the three terms given? To show this, let

$$n = 1 \qquad a_1 = \frac{1}{1 - 5 + 11 - 6} = 1$$

$$n = 2 \qquad a_2 = \frac{1}{8 - 20 + 22 - 6} = \frac{1}{4}$$

$$n = 3 \qquad a_3 = \frac{1}{27 - 45 + 33 - 6} = \frac{1}{9}$$

Then if our alternative law of formation is used, for

$$n = 4 \qquad a_4 = \frac{1}{64 - 80 + 44 - 6} = \frac{1}{22}$$

Why is it not *more* intelligent to select a less obvious formula that is possible on the evidence of a_1, a_2, a_3? We could furnish many other alternatives in addition to the above.

To return to the six sequences whose law of formation we have listed definitely, it seems evident that a limit exists in the first three instances and that this limit is *zero* in each case. The sequences differ, however, in appearance and in formula for a_n. If we wish to provide this fact with empirical meaning, we may think that three different approximative procedures are used for the measurement of the *same* entity. As for the proofs that zero is the limit in the sequences (1), (2), (3), above, they rest on the fact that if *any* positive value is assigned to ε, then m can be selected in each case so that $1/m$, $1/2m$, $-1/m$ will each differ from zero by less than the value of ε. Then a_m and *a fortiori* all terms after a_m will approximate zero within the standard ε.

The limit of the fourth of the illustrative sequences is 4 since, however small a positive value of ε is assigned, we can select $m = 1$, $a_m = 4$ and all subsequent terms are also 4, and therefore the difference from 4 is equal to zero, which is definitely less than ε. The sequence in question illustrates the theorem: The *limit* of a constant sequence is the constant itself.

Lest we think that every sequence has a limit, there are (5) and (6) above. In sequence (5) the terms just grow larger and larger without any upper bound. If we have the feeling that the terms will never exceed 1,000,000,000,000, say, all we need do is compute a_{40}. To approximate its value we can readily compute $2^{10} = 1024$, which exceeds 1000. Then $2^{40} = (2^{10})^4$ and so will exceed $(1000)^4$, which is the figure named above. Then the fortieth term and every term thereafter will exceed this huge number. If we name any figure, however large, 2^m will exceed it for some value of m. Then it is hopeless to seek any number as a possible limit, that is, a number to which all terms of the sequence after a certain one will be very close. It might seem that there could be no empirical counterpart for such a sequence, but we can imagine that cells in some malignant growth double in number repeatedly at regular short intervals of time. In mathematics, if not in life, this process could continue indefinitely, and the number of cells would exceed all bounds.

Sequence (6) fails to have a limit by virtue of oscillation. The terms are alternately very close to -2 and 2—in fact, they coincide with these numbers—but the terms after a particular one are not *all* close to -2, or *all* close to 2, or close to any other number. Then if we think of -2 as a possible limit we *cannot* choose m so that a_m and all subsequent terms differ from -2 by an amount $\varepsilon = 0.01$, say, and the same difficulty exists if we think of 2 as a possible limit. For a physical suggestion of this sequence, we can picture a clock pendulum (that never slows down). The numbers describing the succession of its extreme left and extreme right positions would form an oscillatory sequence like (6) above.

As one more illustration we adduce one of the most important sequences in mathematics:

$$2^1, \left(1\frac{1}{2}\right)^2, \left(1\frac{1}{3}\right)^3, \left(1\frac{1}{4}\right)^4, \left(1\frac{1}{5}\right)^5, \left(1\frac{1}{6}\right)^6, \ldots$$

with the law of formation

$$a_n = \left(1 + \frac{1}{n}\right)^n$$

To reveal that this law applies we rewrite the above terms of the sequence as

$$\left(1 + \frac{1}{1}\right)^1, \left(1 + \frac{1}{2}\right)^2, \left(1 + \frac{1}{3}\right)^3, \left(1 + \frac{1}{4}\right)^4, \left(1 + \frac{1}{5}\right)^5, \left(1 + \frac{1}{6}\right)^6, \ldots$$

The sequence has a limit, but it is difficult to establish this fact by using the definition of a limit, and it becomes advisable instead to use a special theorem on limits.

As a preliminary, we recall that sequence (5) above failed to approach a limit because a_n increased *without bound*. But let us picture a different type of increasing sequence. Let us say that a child has just learned to walk and the maximum distance he can proceed without falling is 9 feet. The proud parents take movies of Baby's first steps and project them. Then one slow-motion sequence might show the child at the following distances (in feet) from the starting point

$$1, 1\frac{1}{2}, 2, 2\frac{1}{2}, 3, 3\frac{1}{2}, 4, 4\frac{1}{2}, 4\frac{3}{4}, 4\frac{7}{8}, 4\frac{15}{16}, \ldots$$

Here the terms increase steadily, but the situation is not the same as in (5) above, because there is an upper bound. Baby cannot advance beyond 9 feet. The sequence is empirical and we cannot give a definite rule of formation, but if we examine the last few terms listed, it appears that Baby is beginning to totter, and will soon reach his "limit" somewhere in the neighborhood of 5 feet. The picture suggests the following theorem for sequences: If the terms of a sequence increase steadily, but are less than some real number A, then the sequence has a limit, and this limit is either A or a number smaller than A.

Now, returning to the important sequence $a_n = (1 + 1/n)^n$, the approximate values of the first few terms are 2.00, 2.25, 2.37, 2.44, 2.49, ... Use of a table or computing machine or logarithms will yield the approximations

$$a_{100} = (1.01)^{100} = 2.705$$
$$a_{1000} = (1.001)^{1000} = 2.717$$

We assert, and it can actually be proved, that the terms of the sequence increase steadily, but that after a while the steps, like Baby's, become very tiny. We observe that between the hundredth and thousandth term, the "distance" progressed is only about 0.012. If we go beyond the thousandth term, the steps become even smaller, and the progress above 2.717 is infinitesimal. For example, the ten-thousandth term is 2.718, which makes the progress after 9000 steps only about 0.001. It would appear safe to name 3 as an upper bound beyond which the terms cannot go. Then, according to the theorem, the sequence has a limit, and this limit is either 3 or a number smaller than 3. Since the ten-thousandth term (to four decimal places) is 2.7183 (approximately), the limit lies between this number and 3.

It can be proved that this limiting number is irrational, but the proof is not as simple as that for $\sqrt{2}$. Moreover, the limit is not a run-of-the-mill irrational, expressible as a square root, cube root, or root of any order. Nor is it a finite combination of such roots. But before we say more about this particular limiting number, let us call it by its traditional name. The symbol e is customarily used for the *limit* of the sequence whose law is $(1 + 1/n)^n$, and since it is a difficult irrational but, like π, a very important one, we emphasize that $e = 2.7183$ is merely an approximation to its value. To continue the story of e from the *pure* mathematician's point of view, it is not only an irrational without a *finite* expression in terms of roots, but it is not even possible to write an algebraic equation (with integer coefficients) of any degree—millionth, trillionth, etc.—that has e as one solution. For this reason e is called a *transcendental* number, in the sense that it *transcends* classic algebra. This remarkable fact was proved in 1873 by the French mathematician Charles Hermite (1822–1905). Although the Babylonians and other peoples of antiquity had good rational approximations to π, its "utter irrationality" was not established until 1882, when the German C. L. F. Lindemann (1852–1939) proved that π, like e, is transcendental. But ignorance of the ultimate character of π did not interfere with its usefulness for some 3500 years prior to the demonstration of its "nonalgebraic" nature. The transcendental character of both e and π is a *theoretical* issue whose primary interest lies in the domains of philosophy and abstract mathematics.

The transcendental e is much younger than π; it made its first appearance in the seventeenth century not long after the Scotsman John Napier (1550–1617) formulated his concept of logarithms. A modification of the Napierian system led to *natural logarithms*. They are exponents of the base e, that is,

$$x = \text{natural logarithm of } y$$

usually symbolized by

$$x = \ln y \quad \text{or} \quad y = e^x$$

Natural logarithms are the simplest for calculus and modern analysis in general. That the term "natural" is appropriate can be illustrated by their appearance in various laws of growth and decay. For example, bacteria in a certain culture may grow (approximately) according to the law

$$A = A_0 e^{0.8t} \quad \text{or} \quad t = 1.25 \ln \frac{A}{A_0}$$

where A_0 and A are the initial number of bacteria and the number at time t, respectively. Again, an *approximate* formula for the decay of a certain radioactive substance might be

$$A = A_0 e^{-0.0004t} \quad \text{or} \quad t = 2500 \ln \frac{A_0}{A}$$

where A_0 and A are the initial amount and the amount left after t years, respectively. Suppose that we wish to find the half-life of the substance, that is, the time in which it will decay to half the original amount. At that time $A_0/A = 2$. Then from the formula, $t = 2500 \ln 2$; if we reveal that the approximate value of the natural logarithm of 2 is equal to 0.69, then $t = 1725$ years *approximately*.

Let us now indicate one application of limits which has been anticipated in earlier chapters—namely, the matter of finding the sum of an infinite series. Thus we used an algebraic trick (Chapter 2) to show how a repeating decimal can be converted into a common fraction. Such a procedure will indicate, for example, that $0.3333\ldots = 1/3$. But instead, we can consider $0.3333\ldots$ as equivalent to the infinite series,

$$0.3 + 0.03 + 0.003 + 0.0003 + \cdots$$

and can inquire whether or not that series has a finite sum.

In connection with the present discussion we shall associate an appropriate *sequence* with a series, and if that sequence has a *limit*, we shall say that the series is *convergent* and the limit is its *sum*. Thus, in any infinite series, we shall choose the sequence $S_1, S_2, S_3, S_4, \ldots, S_n, \ldots$ where the S's are the sums corresponding to $1, 2, 3, 4, \ldots, n$ terms, respectively.

Before studying the sequence corresponding to the above series, we observe that it belongs to a type described as a *geometric progression*, where the ratio of each term (after the first) to its predecessor is a constant. Thus, in the series under consideration, $0.03/0.3 = 0.003/0.03 = 0.0003/0.003 = 0.1$, and the reader will observe that the "common ratio" in the following geometric progression is 2:

$$1 + 2 + 4 + 8 + \cdots + 2^{n-1} + \cdots$$

Now in elementary algebra it is proved that S_n, the sum of the first n terms of a geometric progression, that is, the nth term of its *associated sequence* is,

$$S_n = \frac{a_1 - a_1 r^n}{1 - r}$$

where a_1 is the first term of the progression and r is the common ratio. Hence, in the sequence corresponding to the repeating decimal,

$$S_n = \frac{0.3 - 0.3(0.1)^n}{1 - 0.1} = \frac{0.3 - 0.3(0.1)^n}{0.9}$$

In other words,

$$S_n = \frac{1}{3} - \frac{1}{3}(0.1)^n$$

Thus S_n approximates $1/3$ closely for large values of n; in fact, S_n differs from $1/3$ by the amount, $1/3(0.1)^n$, which can be made arbitrarily small. For example, if we choose the standard of approximation, $\varepsilon = 0.000,001$, it would suffice to choose $n = 6$, for then

$$\frac{1}{3}(0.1)^6 = \frac{0.000,001}{3} = \frac{\varepsilon}{3}$$

which is surely less than the standard ε. However small ε is chosen, it will be possible to find an m such that S_m approximates $1/3$ within the specified standard. Therefore the sequence has the *limit* $1/3$, or the series has the *sum* $1/3$, and this is the value of $0.3333\ldots$.

Applying the geometric progression formula once again, we find that for the sequence associated with the series

$$1 + 2 + 4 + 8 + \cdots + 2^n + \cdots$$

$$S_n = \frac{1 - 1(2)^n}{1 - 2} = \frac{2^n - 1}{2 - 1} = 2^n - 1$$

As n grows greater and greater, 2^n increases beyond all bounds. For example, if $n = 20$, 2^n is greater than $1,000,000$ and $S_n = 2^n - 1$ is almost as huge. Thus S_n has no limit, and in that case the series is said to *diverge*. The present example resembles that of the harmonic series illustrated in the previous chapter.

Infinite series need not be geometric progressions. Consider the infinite series:

$$1 - \frac{1}{2} + \frac{1}{2} - \frac{1}{3} + \frac{1}{3} - \frac{1}{4} + \frac{1}{4} - \frac{1}{5} + \cdots \text{ forever}$$

where the general term is $1/n$ for n odd, $-1/n$ for n even. The sum of the terms listed is

$$1 - \frac{1}{5} = \frac{4}{5}$$

since all the terms balance except the first and last. If we take an odd number of terms of this series, even as many as $1,000,001$, S_n, the sum of n terms, will obviously be 1. If we go out to $-1/1,000,000$, S_n will be

$$1 - \frac{1}{1,000,000} = 0.999999$$

We see that no matter how far out we go, the sums S_n are alternately 1 and numbers closer and closer to 1. Therefore, this infinite series converges and its sum is said to be 1.

Consider the infinite series

$$1 + 1 - 1 - 1 + 1 + 1 - 1 - 1 + \cdots$$

Notice that the sums will fluctuate taking the values 1, 2, 1, 0 no matter how far out in the series we go. Therefore, the sums do not get close and remain close to any single number, that is, S_n does not have a limit. The series is *divergent*, and there is no meaning to the term sum as applied to it.

To come back to the concept of a *limit*, we have treated only the case of the limit of a sequence. To allude to terminology we have used earlier, this is the *discrete* case of a limit, because a sequence is a *function* whose *domain* is the set of natural numbers ordered according to size, $\{1, 2, 3, \ldots, n, \ldots\}$. The *range*, that is, the set of corresponding functional values, listed in the corresponding order, is $\{a_1, a_2, a_3, \ldots, a_n, \ldots\}$, the ordered set which is a *sequence*. The limit of a sequence is then the limit of a *special* type of *function*, one with a discrete domain.

But what about a function with a "continuous" domain like the real number continuum or an interval on the real number line? If we reexamine our previous discussion of the calculus, we find that, in the main, this was the type of function occurring in the limiting processes we used. We shall now proceed to an elementary example, but before doing so, we state once and for all that every function mentioned in an illustration or definition will be assumed to have the real number continuum or one of its subsets (made up of one or more "continuous" intervals) as its domain.

Let us then consider the function defined by $y = f(x)$ where

$$(x) = \frac{2x^2 - 2x}{x - 1} \qquad \text{for } x \neq 1$$

(We have excluded $x = 1$ from the domain because it would lead to a zero denominator, and division by zero is impossible.) If then we draw the Cartesian graph of $y = f(x)$, we obtain a straight line with a puncture at the point $(1, 2)$ (Figure 22.3).

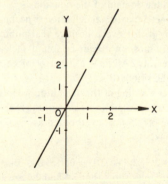

Figure 22.3

The graph shows that for values of x slightly smaller or slightly larger than 1, the corresponding values of y or $f(x)$ are close to 2, and $f(x)$ approximates 2 within smaller and smaller standards as x is taken closer to the value 1. This would seem to justify the statement that "the limit of f at 1 is equal to 2," symbolized by

$$\lim_1 f = 2$$

Perhaps the following alternative wording and symbolism will hold more appeal: The limit of $f(x)$ as x "approaches" 1 is equal to 2, or

$$\lim_{x \to 1} f(x) = 2$$

The reason this statement is considered less rigorous than the other is that the term "approaches" has an intuitive connotation associating it with motion, and if there is anything rigor must prevent, it is one of the monstrous paradoxes brought on by our ideas of physical motion.

To return to our example, we can justify the same limiting value of the function by means of an algebraic argument. Now

$$f(x) = \frac{2x^2 - 2x}{x - 1} = \frac{2x(x - 1)}{(x - 1)} = 2x \qquad \text{for } x \neq 1$$

Just as long as we do not take $x = 1$, we may substitute any other values of x in

$$f(x) = 2x$$

Taking values close to 1, say $x = 0.9$ and $x = 1.1$, we obtain $f(0.9) = 1.8$ and $f(1.1) = 2.2$, which are values close to 2. If x approximates 1 more closely, $f(x)$ will approximate 2 more closely. Thus for $x = 0.99$ and $x = 1.01$, we have $f(0.99) = 1.98$ and $f(1.01) = 2.02$. We could continue to select better and better approximations of 1 as values of x and then the corresponding values of $f(x)$ would be better and better approximations of 2, and this justifies our conclusion that $\lim_1 f = 2$. We observe that although $\lim_1 f$ is defined and equal to 2, the symbol $f(1)$ is meaningless because $x = 1$ does *not* belong to the domain of the function.

Our illustration suggests the formal definition: The function f is said to have the limit L at a if and only if, for every positive real number ε (however small), a positive number δ (depending on ε) can be found such that for all values of x except $x = a$ in the interval $(a - \delta, a + \delta)$, the set of corresponding values, $f(x)$, approximate L within the standard ε.

For a more compact wording of this definition we introduce the term *absolute value* of a real number, which signifies its magnitude regardless of sign (direction). Thus, the absolute value of -5 is 5, symbolized as

$$|-5| = 5$$

Similarly we see that $|+2| = 2$, $|-0.01| = 0.01$, $|3.4 - 3| = 0.4$, $|2.6 - 3| = 0.4$, etc. The last two of these examples show that 3.4 and 2.6 are equally good approximations of 3.

Now we can abbreviate our definition as follows:

$$\lim_a f = L$$

if and only if, for every positive ε (however small), a positive δ (depending on ε) can be found such that

$$|f(x) - L| < \varepsilon$$

for all $x \neq a$ satisfying the inequality

$$|x - a| < \delta$$

Let us apply the formal definition to the problem of finding $\lim_0 f$ where f is defined by

$$f(x) = \frac{x^3 + 5x}{x} \qquad \text{for } x \neq 0$$

We can guess the value of $\lim_0 f$ and then give formal proof. We see that

$$\text{for } x \neq 0, \qquad f(x) = \frac{x(x^2 + 5)}{x} = x^2 + 5$$

Hence, for values of x close to 0, $f(x)$ will be close to 5. Therefore we guess that $\lim_0 f = 5$. For *proof* we must show that if a positive ε is given, a positive δ (depending on ε) can be found such that

$$|f(x) - 5| < \varepsilon$$

for all $x \neq 0$ satisfying the inequality

$$|x - 0| < \delta$$

or

$$|x| < \delta$$

To set about finding δ in terms of ε, we observe that if $x \neq 0$, we can substitute $x^2 + 5$ for $f(x)$ in the first inequality above, and that requirement becomes

$$|x^2 + 5 - 5| < \varepsilon$$

or

$$|x^2| < \varepsilon$$

which is satisfied if $|x| < \sqrt{\varepsilon}$, an inequality of the same form as $|x| < \delta$. Therefore, if we choose $\delta = \sqrt{\varepsilon}$, the conditions of the definition are satisfied. To clarify this point, suppose that the assigned standard of approximation is $\varepsilon = 0.25$. Then $\sqrt{\varepsilon} = 0.5$, and

$$|f(x) - 5| < 0.25$$

that is,

$$|x^2 + 5 - 5| < 0.25 \text{ or } |x^2| < 0.25$$

for all $x \neq 0$ satisfying the inequality $|x| < 0.5$. Perhaps the original wording of the definition will lend further clarification. Hence we say, alternatively, that $\lim_0 f = 5$ because for all x except $x = 0$ in the interval $(0 - 0.5, 0 + 0.5)$, that is, $(-0.5, 0.5)$, the set of corresponding values, $f(x)$, approximates 5 within the standard 0.25.

Let us choose our next illustration from the calculus. If $s = 16t^2$ is given as the distance formula for a falling body, and we require the *derivative* or *instantaneous velocity* at the time $t = 2$, we must find the *limit* of the average velocity as t "approaches" 2. To find this limit, we first formulate the average velocity during the interval between $t = 2$ and $t = 2 + h$ where $|h|$ is a small positive number. If h itself is positive, we have an interval just after $t = 2$ seconds. If h is negative, we have an interval just before $t = 2$. From the tabulation,

$$\Delta t \begin{bmatrix} & t & & s & \\ 2 & & & 64 & \\ 2 + h & & & 16(2 + h)^2 & \end{bmatrix} \Delta s$$

we find that $\Delta s = 16(4 + 4h + h^2) - 64 = 64h + 16h^2$, and $\Delta t = h$ so that the average velocity is

$$\frac{\Delta s}{\Delta t} = \frac{64h + 16h^2}{h} = 64 + 16h \qquad \text{if } h \neq 0$$

Thus, in common usage, the average velocity would be described as f, a "function of h" whose domain is a *continuum*. By definition, the *derivative* or *instantaneous velocity* we are seeking is $\lim_{0} f$, or $\lim_{h \to 0} f(h)$, which is read as "the *limit* of $f(h)$ as h approaches 0." Since $f(h) = 64 + 16h$, we see that as $|h|$ gets smaller and smaller or closer to 0, $f(h)$ gets closer to 64. Hence our intuition tells us that the limit is equal to 64. To establish it rigorously, we must show that given any standard ε, we can determine δ so that

$$|f(h) - 64| < \varepsilon$$

for all $h \neq 0$ satisfying the inequality

$$|h - 0| < \delta, \qquad \text{that is,} \quad |h| < \delta$$

Substituting $64 + 16h$ for $f(h)$ in the first inequality above expresses that conditon as

$$|64 + 16h - 64| < \varepsilon \qquad \text{or} \qquad |16h| < \varepsilon$$

which will be satisfied if

$$|h| < \frac{\varepsilon}{16}$$

This inequality is of the same form as $|h| < \delta$ above. Hence we can meet the requirements of the definition by selecting $\delta = \varepsilon/16$.

To indicate the "common sense" of the rigorous proof, suppose that the standard assigned is $\varepsilon = 0.16$. We choose $\delta = 0.01$ and must show that

$$|f(h) - 64| < 0.16$$

for all $h \neq 0$ satisfying the inequality

$$|h| < 0.01$$

But, as indicated above, $|f(h) - 64| = |16h|$. Now if $|h| < 0.01$, $|16h| < 0.16$ and we have proved our point.

Since some of our examples of the hazards of intuition were associated with the meaning of a continuous curve or a continuous motion, let us now identify both these ideas with the continuity of a function, and provide an "ε, δ" definition for that concept. Perhaps counterexamples can be of some help in suggesting a definition of continuity. For instance, examination of the punctured line of Figure 22.3 would reveal a discontinuity of the line or the corresponding function at $x = 1$. In fact, with $f(x)$ as defined for the graph, $f(1)$ was a meaningless symbol. This suggests that if a function f defined by $y = f(x)$ is to be continuous at $x = a$, then $f(a)$ must be defined, that is, a must belong to the domain of the function. But that is not enough. If we continue our study of Figure 22.3, we see that the discontinuity at $x = 1$ would not be removed merely by defining $f(1)$. For if we define $f(1) = 4$, the graph would then consist of the punctured line and the single lonely point $(1,4)$ "vertically" above the puncture. To fill the gap we would have not only to give meaning to $f(1)$ but also to choose $f(1) = 2$, the constant which we proved to be $\lim_{1} f$.

Our illustration shows that for continuity of a function at $x = a$, $f(a)$ must be defined and also it must be true that $f(a) = \lim_{a} f$. But there is one final issue: Is it possible that $\lim_{a} f$ does not exist? That is the difficulty at the "jumps" in the "step function" pictured in Figure 14.10. There, if a jump occurs at $x = a$, the functional value $f(a)$ *is* defined. (It corresponds to the height of the upper step.) But $\lim_{a} f$ does *not* exist because, while $f(x) \rightarrow f(a)$ as x approaches *a from the right*, $f(x)$ approaches a lower value than $f(a)$ as x approaches *a from the left*.

In the hyperbola of Figure 7.17 (which Omar Khayyám used in his solution of a cubic equation), there is a discontinuity at $x = 0$. We can consider it to be produced by two types of failure. In the first place, $f(0)$ is not defined. The reader can examine the effect of choosing $f(0) = 20$ or $f(0) = -10$, etc. and plotting in the graph of Figure 7.17 the corresponding points, $(0, 20)$ or $(0, -10)$, etc. No matter what real number he might assign to $f(0)$, the point $(0, f(0))$ would be an isolated point and would not serve the purpose of making the left branch of the hyperbola continuous with the right. In the present case, if x approaches 0 from the right, $f(x)$ gets larger and larger and never remains close to any constant value. As x approaches 0 from the left, $f(x)$ takes values like -100, -1000, $-1,000,000$, etc. These numbers represent greater and greater "debts," as it were, so that $f(x)$ gets smaller and smaller as x approaches 0 from the left, but $f(x)$ never remains close to any constant value. Hence there are neither right nor left limits for f and $\lim_{0} f$ does *not* exist.

Our illustrations suggest the following definition: The function f is *continuous* at a if and only if (1) a belongs to the domain of f, that is, $f(a)$ is defined, (2) $\lim_{a} f$ exists, and (3) $f(a) = \lim_{a} f$. Or we can express this definition in ε, δ form by stating: A function f is continuous at a if and only if $f(a)$ is defined and if, corresponding to

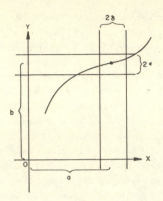

Figure 22.4

every positive number ε (however small), a positive number δ (depending on ε) can be found such that

$$|f(x) - f(a)| < \varepsilon$$

for all x satisfying the inequality

$$|x - a| < \delta$$

Figure 22.4 shows that the continuity of a graph or function at a or at the point $(a, b = f(a))$ signifies that, for each pair of horizontal lines, one at a distance ε above, the other at the same distance below (a, b), there is a pair of vertical lines, one at a distance δ to the right, the other at the same distance to the left of (a, b) such that every point of the graph between the vertical lines is also between the horizontal lines. In Figure 22.5 there is a jump discontinuity at (a, b), and no pair of vertical lines can be found so that the property just described is fulfilled.

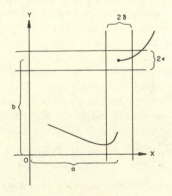

Figure 22.5

The rigorization of analysis was begun in 1821 when Cauchy gave the "ε, δ" definition of limit and then defined continuity, derivative, and integral in terms of his limit concept. The further arithmetization of analysis was carried out, in the main, by Weierstrass, and was inspired by the various analytic "freaks" he and other analysts were able to derive from intuitive notions. With Weierstrass the "ε, δ" technique became the basic language of modern analysis.

The nonspecialist, however, rarely acquires any fondness for the Cauchy-Weierstrass formal procedures, chiefly because determining δ in terms of ε does, in instances less elementary than those we have illustrated, require considerable ingenuity and algebraic skill. The layman has a leaning toward intuitive formulation, however faulty, and an "ε, δ" proof may seem an esoteric technique introduced to handle the calculus of exceptional, abnormal functions, produced *ad hoc* by pure mathematicians. But a brontosaurus is no anomaly in a Paleozoic age, and epsilons and deltas, far from being exceptional, populate modern analysis and actually prevent the multiplication of logical "monsters." Some readers may feel that treating them in general works or in elementary texts is part of the "new mathematics" movement, but this is not the case. The author's high school mathematics teacher and lifelong inspiration, John A. Swenson, presented the Cauchy conception of a limit to high school students, and dealt with it in textbooks written years ago.*

Leaving for later consideration the details of how analysis progressed after its arithmetization, we conclude our chapter with biographies, personal and mathematical, of the two mathematicians who rigorized analysis.

Augustin-Louis Cauchy can be cited both for the quantity and for the quality of his mathematical output. He produced almost eight hundred papers, which filled twenty-four large quarto volumes. These memoirs presented his rigorization of analysis, his pioneer work in the theory of finite groups, his contributions to differential equations and functions of a complex variable, and his applied mathematical discoveries—in mechanics and astronomy. Cauchy accomplished all of this in spite of such handicaps as physical frailty and the hazards of the period of political unrest in which he lived.

His place and date of birth are equally famous—Paris, 1789. During his childhood, French schools were closed, and Augustin, the eldest of six children, was educated by his father, as were all the other children. This had definite advantages. In the first place, Cauchy senior removed his family from Paris in order to escape the Terror, and they settled in the village of Arcueil. The great Laplace, who was a friendly neighbor at Arcueil, soon discovered Augustin's remarkable mathematical talent, and encouraged him, in fact, to follow a career in pure mathematics rather than in engineering, which, for practical reasons, was the boy's first choice.

A second advantage that accrued from schooling with Cauchy *père* was that Augustin-Louis' intellectual development extended to literary as well as scientific areas. In later life he thought nothing of shaping his addresses before learned

---

* See, for example, John A. Swenson, *Integrated Mathematics*, Book V, Edwards Bros., Ann Arbor, 1937.

academies in the form of (lengthy) poems. He was a remarkable linguist, mastering not only the classical, but also the Semitic and modern languages. When he taught in Italy and Austria, he had no difficulty in lecturing in Italian and German. Also, unlike many great mathematicians, who affect weakness in computation, Cauchy was a lightning calculator, able to defeat the arithmetic prodigy Mandeux.

After the early education at home, Cauchy attended two of France's great schools, the École Polytechnique and the École des Ponts et Chaussées. Having acquired an engineering degree from the latter, he served as a civil engineer from 1808 to 1813, first assisting with the construction of the Ourcq canal and then with the Cherbourg dike. Then, at age twenty-four, he decided to follow Laplace's advice, and renounced engineering in favor of pure mathematical research. Within a couple of years he was appointed to a chair in analysis and mechanics at the École Polytechnique, where he taught until 1830. It was in the lectures at the Polytechnique that he began the rigorization of analysis.

By the time he was twenty-seven, in 1816, Cauchy had made many brilliant contributions to various branches of mathematics and was recognized as one of the leading mathematicians in Europe. He was therefore worthy of admission to membership in the great Académie des Sciences. There were no vacancies, however, and here some ugly politics came into play. Gaspard Monge (1746–1818), the great geometer whose name is associated with the creation of that *descriptive geometry* which is so useful in mechanical drawing, had been a close friend to Napoleon. Lazare Carnot (1753–1823), disciple of Monge and father of Sadi Carnot (1796–1832) who became truly renowned in the history of science, had opposed Napoleon. In 1816 both Monge and Carnot were purged from the Académie and Cauchy was assigned by royal decree although he had not been elected by the members of the Académie. This created an enormous scandal in scientific circles and Cauchy became very unpopular.

In 1830 the tables were turned; Louis-Philippe came to power and Cauchy, as a loyal Bourbon, refused to swear allegiance to the new government. He went into voluntary exile and accepted a professorship in mathematical physics at the University of Turin. In 1833 the deposed King Charles X, living in exile in Prague, summoned Cauchy to tutor his grandson, the thirteen-year-old Duke of Bordeaux. For five years the great analyst served as a sort of baby-sitter to the pampered youth, and Charles made him a baron for this martyrdom. Finally, in desperation, Cauchy escaped to Paris by saying he had to attend the celebration of his parents' golden wedding anniversary. Once back in France, he was permitted to resume his post at the Académie, but when he was unanimously elected to the Collège de France, he refused the proffered chair because it involved the taking of an oath of allegiance.

The Bureau des Longitudes saw fit to place him "temporarily" without making an issue of loyalty. For four years Cauchy was able to carry on important work at the Bureau and make his own major contributions to mathematical astronomy. In 1843 the government stepped in and decided to replace Cauchy, since his position was only "temporary." Cauchy then addressed a brilliant literary open letter to the people, stating his case and the ethical principles he felt were involved. From this time on, French scholars, who on the whole disagreed with Cauchy's political views, came to see him in a different light—as an eccentric, perhaps, but also as a man of principle, willing to suffer to uphold what he believed to be right.

In 1848 Louis-Philippe was ousted, and oaths of allegiance were abolished. Napoleon III restored them in 1852, but he took a liberal attitude which exempted Cauchy, who was able therefore to live the remaining five years of his life in security, lecturing at the Sorbonne and continuing his endless researches in every area of mathematics and mathematical physics.

On the German front, the battle for rigorization was waged by Weierstrass. He was born at Ostenfelde in the district of Münster in 1815 just about the time of Napoleon's Waterloo and Cauchy's appointment to a professorship at the École Polytechnique. Whereas Cauchy brought sounder logical methods into analysis almost casually, that is, without any formal declaration or manifesto, Weierstrass, with German thoroughness, carried on an intellectual campaign to indicate the mathematical disasters inherent in failing to rigorize analysis.

It would be hard to conceive of a confirmed bachelor and an ardent advocate of extreme mathematical rigor as a romantic figure. Nevertheless one cannot give Weierstrass' biography without describing his role as teacher and sponsor of a great and beautiful woman mathematician, Sonya Kovalevsky (1850–1891). The Russian girl was Weierstrass' favorite pupil, and in one of her discoveries she was associated with that other advocate of rigor, Cauchy. The first basic proposition in the theory of partial differential equations is the Cauchy-Kovalevsky theorem.

Whereas Cauchy had been favored by an understanding and loving father who was able to direct his education properly, Weierstrass was hindered by a domineering parent, a customs officer who failed to recognize his son's mathematical genius and tried to force him into the rigid mold of the civil service. Karl had been an excellent student at the *gymnasium* (secondary school) and at the end of each term had received all the prizes, six or seven, except the award for good penmanship. Ironically, teaching calligraphy was to be one phase of his earning his bread and butter later on. While he was at the *gymnasium* it was ham and butter that were a source of income, for, at age fifteen he started to do the accounting for a female merchant who dealt in these commodities.

He attended the University of Bonn, where his father sent him for the purpose of studying accounting, commerce, and law. But Karl spent all his time there in fencing and enjoying the *gemütlichkeit* of the beer halls. When he returned home in 1839 without a diploma, one can imagine what his father's reaction was. Next, young Weierstrass decided to prepare himself for a career in teaching and entered the Academy of Münster, which was near his home. This was a fortunate step, for there he encountered an able and inspiring mathematics professor, Christoph Gudermann (1798–1852), who aroused his interest in the newly created *elliptic functions* of Abel and Jacobi. Gudermann confided his own ideas to his pupil and told him that he had striven fruitlessly for years to use *power series* (infinite series involving variables) to represent elliptic functions. Where his teacher failed Weierstrass succeeded, and such series are the keynote to all his subsequent analytic discoveries.

From 1841 to 1854, or almost until the age of forty, Weierstrass taught school and simultaneously carried on remarkable mathematical research without ever offering any of it for publication or discussing it with another living soul. The one exception was his *Abelian Integrals*, a memoir that appeared in the catalogue of studies for 1848–1849 of the Royal Catholic Gymnasium at Braunsberg. This article was completely unintelligible to the students, parents, and teachers who read the catalogue, but their only reaction was "That's math. What do you expect?" What no one expected or suspected was that Weierstrass' article contained epochal discoveries.

In 1854 Weierstrass decided to submit a memoir on *Abelian Functions* to the *Journal für die reine und angewandte Mathematik* (usually called Crelle's Journal, after the name of the founder, A. L. Crelle [1780–1855]). This periodical was one of the two (the other was the Frenchman Liouville's Journal) scholarly mathematical publications on the continent. The research paper of the obscure secondary school teacher created a furor in scientific circles. The University of Koenigsberg conferred an honorary doctorate on Weierstrass, and the Ministry of Education gave him a year's leave with pay so that he could continue his research. Then he became a professor at the Royal Polytechnic School in Berlin, and eventually at the University of Berlin, where, from 1864 to 1897, he carried on those investigations for which he was recognized as the leading world figure in the field of analysis.

When he first met Sonya Kovalevsky he was in his middle fifties and she was twenty. She had enjoyed the very advantages he had lacked in his own youth, in particular, prosperous and kindly parents. She was the daughter of a Russian general, Ivan Sergyevitch Krukovsky. The family circle and friends, among whom the writer Dostoevsky was included, are pictured charmingly in Sonya Kovalevsky's own literary account of her childhood, *The Sisters Rajevsky*. Sonya and her sister Anyuta were part of a young people's movement to promote the emancipation of women in Russia. It was customary for one of the girls to contract a marriage of convenience to enable her to leave the country and study at a foreign university. Then her whole retinue of friends could also escape—on the pretext of visiting the "married" friend in France or Germany. At eighteen Sonya contracted just such a nominal marriage with Vladimir Kovalevsky.

Vladimir escorted her to Heidelberg, where she studied mathematics and he geology. He soon left for Jena and she for Berlin, where she was befriended by Weierstrass. Since Sonya, as a woman, could not be admitted to university lectures, Weierstrass gave her sets of notes on his lectures, directed her reading, held conferences with her, and provided critical appraisals of her original mathematical efforts. Because his academic schedule was a very crowded one, the supervision of Sonya's mathematical education was an added burden. To make this task easier, Weierstrass' sisters invited Sonya to the family home on Sundays, and following a lengthy afternoon mathematical discussion between teacher and pupil, the entire family would meet for Sunday night supper. It is recorded that Weierstrass' sisters, who were little interested in mathematics, nevertheless derived great pleasure from Sonya's company and the regular Sunday evening get-togethers.

Sonya worked diligently at mathematical studies for a full four years and finally was awarded a doctorate in 1874 at the University of Göttingen (*in absentia*, because she was a woman). In spite of this degree, and Weierstrass' letters of recommendation, she was unable to obtain an academic position anywhere in Europe. For this reason she returned to her family in Russia. Very soon she became involved in the gay social whirl at St. Petersburg. When Tchebycheff, the leading Russian mathematician of the day, paid a visit to Berlin, Weierstrass had news of Sonya's activities, and was keenly disappointed to hear of her apparent frivolity. He wrote a letter in which he reproached her gently, but in no uncertain terms. His scolding had little effect for some time, but after a lapse of several years, she returned to mathematics. She wrote to Weierstrass to ask for assistance and information on technical problems, and once again he supervised her research, this time from afar, by means of regular correspondence.

In the interim, Sonya had met Vladimir Kovalevsky again; this time he wooed and won her in earnest, and they lived happily together. Their first and only child, a daughter, Foufie, was born in 1878. Vladimir became professor of paleontology at the University of Moscow, and all went well until he became involved in shady speculations that eventually resulted in his disgrace and suicide.

Sonya found herself once more in a foreign country, this time on no intellectual lark, but faced with a serious problem—that of earning a living for herself and her little daughter. Once again Weierstrass came to her assistance. Through Mittag-Leffler, another of Weierstrass' distinguished students, she secured a position as lecturer at the University of Stockholm. The advanced attitude of the Swedish toward women worked to her advantage. In 1889 Mittag-Leffler succeeded in obtaining a life professorship for her.

During her years in Stockholm Sonya Kovalevsky did her best work. Weierstrass, always her friend, continued his inspirational mathematical correspondence with her; the great teacher once again advised from afar when he could not be present in person. His distinguished pupil wrote a paper on the refraction of light in a crystalline medium, and followed this with the memoir that won the Prix Bordin of the French Academy, namely, *On the Rotation of a Solid Body about a Fixed Point*. The judges considered this paper so exceptional that they raised the prize from 3000 to 5000 francs, a large sum of money in those days.

Two years later (1891) there was an epidemic of influenza in Stockholm and Sonya succumbed to it. She died at the age of forty-one. Weierstrass survived her by half a dozen years. His life span was double hers, for he died peacefully in Berlin in his eighty-second year. Although he is renowned in mathematical history mainly for his role as a very great analyst, nevertheless, in the opinion of some, he made a comparable contribution in helping Sonya Kovalevsky to become one of the foremost women mathematicians of all time.

23

Royal Roads to Functional Analysis

It should have become increasingly clear that the function concept is a thread which runs through all of mathematics. At different points of our story we have in fact discussed propositional functions, logical functions, real functions, complex functions, analytic functions, set functions, payoff functions, decision functions, frequency (density) functions, distribution functions, the Riemann zeta-function, etc. All these and still other kinds of functions which have played a role in our development show the advantage of *pure* (abstract) theories of functions (and relations) where the domain and range need not be number sets but may be aggregates of arbitrary nature. Such generality would permit the domain, in special instances, to be a *set of functions*, and the range to be a set of numbers. In the present chapter we shall see how the ideas of the great Italian mathematician Vito Volterra (1860–1940) led him to a study of such "functions of other functions," as he called them. Jacques Hadamard (Chapter 21) named them *functionals*, and Paul Lévy gave the name *functional analysis* to the study of analytical properties of functionals.

But the Volterra-Hadamard-Lévy "royal road" is a modern, abstract, advanced one. Therefore let us begin by indicating how, early in mathematical history, issues in the physical world motivated the construction of paths which converged to larger routes that ultimately flowed into the broad highway of functional analysis.

According to legend, the first "royal road" in the direction of functional analysis was blazed by Queen Dido about 800 B.C. In Vergil's *Aeneid* there is a reference to the story in question. Dido, fleeing from the Phoenician city of Tyre ruled by King Pygmalion, her tyrannical brother, and arriving at the site that was to become Carthage, sought to purchase land from the natives. They asserted that they would sell only as much ground as she could surround with a bull's hide. She accepted the terms and made the most of them by cutting a bull's hide into narrow strips which she pieced together to form a single, very long strip. Then, by sheer intuition, she reasoned that the *maximum* area could be encompassed by shaping the strip into the circumference of a circle. A rigorous mathematical proof that Dido made the optimum choice was not achieved until the nineteenth century. But today we speak of the general *problem of Dido*, namely, the question of finding what closed curve of specified length will enclose a maximum area.

A slight modification of an example already considered (Chapter 8) will solve Dido's problem for a special case. Suppose then that we ask: What are the dimensions of the *rectangle* of maximum area that has a perimeter of 40 ft. (or, more generally, of $4a$ linear units, where $a > 0$)? If we represent the unknown width by x, then two sides of the rectangle will have length x, and $40 - 2x$ (more generally, $4a - 2x$) will be left for the two remaining sides. Hence each will measure $20 - x$ (or, in the general case, $2a - x$). Hence the area will be

$$y = x(20 - x) \qquad \text{or} \qquad y = x(2a - x)$$

that is,

$$y = 20x - x^2 \qquad \text{or} \qquad y = 2ax - x^2$$

As we have indicated earlier, the graph of such a formula is a quadratic parabola with a highest point corresponding to the maximum value of y, the area. Instead of reading the coordinates of the maximum point from a graph, it is more convenient to apply the fact that the slope of the tangent is zero at the maximum point. Thus

$$\frac{dy}{dx} = 20 - 2x \qquad \text{or} \qquad \frac{dy}{dx} = 2a - 2x$$

$$\begin{aligned} 20 - 2x &= 0 & & \text{or} & 2a - 2x &= 0 \\ x &= 10 & & & x &= a \\ 20 - x &= 10 & & & 2a - x &= a \end{aligned}$$

The dimensions would be 10 ft. by 10 ft. (a by a), and the maximum area would be 100 sq. ft. (a^2). Thus the *rectangle of maximum area for a given perimeter is a square*.

Dido's problem is described as *isoperimetric*. The story of her ingenuity may be fictional, but it is *true* that Pappus of Alexandria (*ca.* 300 A.D.) considered isoperimetric problems in Book V of his greatest work, the *Mathematical Collection*. Not only did he treat the areas of figures with the same perimeter, but also the volumes of solids bounded by the same surface area. At one point Pappus proved that the honeybee's choice of a *regular hexagonal* cell makes possible the storage of a *maximum* amount of honey with a *minimum* use of wax.

Today mathematicians tend to apply the adjective "isoperimetric" to any problem in which an extremum, that is, a maximum or a minimum, is to be determined subject to one or more *constraints*. As an example of such constraints, we have just required (1) that a figure be a rectangle and (2) that it have a perimeter of 40 ft. Loosely speaking, even a game-theory issue where a player seeks to maximize or minimize a payoff $E(x_1, x_2, \ldots, x_n)$ is "isoperimetric," with constraints requiring $x_1 \geq 0$, $x_2 \geq 0$, ..., $x_n \geq 0$, and $x_1 + x_2 + \cdots + x_n = 1$.

But our ultimate objective is to discuss *functionals*. Hence let us illustrate such functions in connection with Dido's problem. In the special case with the constraints (1) and (2) of the previous paragraph, the *area* of the rectangle is a function of its boundary, or if we use another of Volterra's descriptive phrases, the area is a "function of a line." One can picture (Figure 23.1) a few of the infinite variety of rectangles possible under restriction (2). Three of the rectangles in the diagram

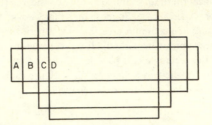

Figure 23.1

enclose a smaller area than that bounded by the square of the same perimeter. But our emphasis is on the fact that a functional, one of Volterra's "functions of lines," is involved. Some of the ordered pairs belonging to the functional appear in the diagram and can be tabulated as follows.

Boundary Line	A	B	C	D (square)
Area Enclosed	51	75	96	100

Strictly analogous to Dido's problem is the matter of finding the form which will give the greatest volume within a fixed surface area. The solution to this problem is a *sphere*, and again a proof would involve consideration of a functional. If one selects an area of 150 sq. in., one can picture it as belonging to a cube whose edge is 5 in., or to a box with dimensions $7\frac{1}{2}$, 5, 3 (measured in inches), or as belonging to some polyhedron in the form of a crystal or to some oval surface, or, in fact, to an infinite variety of different surfaces. The volumes of the above cube and box are 125 and 112.5 cubic inches, respectively, so that there is more volume for the *same* surface area in the former case. Using the formulas $S = 4\pi r^2$ and $V = 4/3 \pi r^3$, the reader can verify that a sphere with radius 3.46 in. (approximately) will have a surface area of about 150 sq. in. and a volume of about 173 cu. in., so that the content enclosed by the same area is far greater than in the two instances above. This does *not* prove, however, that 173 cu. in. is the greatest volume which 150 sq. in. can enclose. Cartesian geometry will furnish equations or functions for the different closed surfaces having an area of 150 sq. in. As such functions vary, so will the volume enclosed, and to each function there will correspond a unique volume; or the volume is a function of a function, that is, it is a *functional*.

In Hellenic antiquity, Dido (according to legend) and Pappus, in actual fact, were not alone in applying functionals to geometric and physical problems involving maxima and minima. Heron of Alexandria postulated a minimum principle for optics more than two centuries prior to Pappus' isoperimetric theorems. To give a popular version of a major question in Heron's *Catoptrica* (*ca.* 60 A.D.), let us suppose that A and B in Figure 23.2a are two houses and that line CD is a straight river bank. A boy living in house A is to proceed to the river to fill a pail of water.

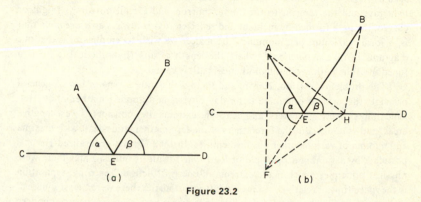

(a)   (b)

Figure 23.2

Then he is to bring the full pail to his neighbor in house *B*. What point on the bank will enable the boy to walk the minimum distance? The reader may try to solve this problem by trial, but it is not difficult to prove, by elementary geometry, that if the boy walks to a point *E* such that angle α = angle β (Figure 23.2*a*), the path *AEB* is the shortest possible. The proof is suggested by Figure 23.2*b*, where the allegedly optimal path *AEB* is equal in length to the straight line segment *FEB*. In the diagram, the dotted line *AHB* represents any other path the boy might choose. Now the path *AHB* has the same length as *FHB*, and *FHB* is longer than *FEB* because the straight line segment is the shortest distance between two points.

Suppose that in Figure 23.2*a*, *CD* represents a *mirror*. Then, according to Heron's assumption, a ray of light starting at *A* and reaching point *B* after reflection in the mirror must follow the shortest route, namely *AEB* in Figure 23.2*a*, where angle α = angle β. Hence Heron asserted as the law of reflection of light: *The angle of incidence is equal to the angle of reflection.*

Heron's assumption of a *minimal* path is correct for a straight mirror, but if the reflecting surface should be curved as in Figure 23.3, it is still true that the angle of incidence equals the angle of reflection. However, the path followed is *maximal*. Figure 23.3 suggests that the actual path *AEB* is longer than any neighboring path

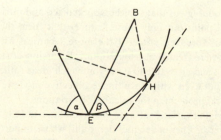

Figure 23.3   Reflection from a concave mirror

such as *AHB*. In the general case, the equal angles α and β are those between the light rays and the tangent to the curved mirror, and for all mirrors, straight or curved, the path formed by incident and reflected rays is always an *extremum*, that is, a local maximum or minimum. The matter we wish to emphasize is that our diagrams involve a *functional*, where the length of any (hypothetical) path is a function of the path, as that hypothetical path is varied.

The problems of Dido, Heron, and Pappus illustrate a type of law or general principle that was to dominate latter-day physics and applied mathematics. Hellenic thinkers were geometers, and hence the extrema they considered were lengths, areas, and volumes. But the forefathers of modern mathematics dealt with maxima and minima of various other physical entities. In 1662 Fermat generalized Heron's postulate by assuming a principle of *least time* for all of optics. Later there were physical principles which made extrema of energy, "action," entropy, separation in the space-time of relativity. In probability and statistics there were "least squares" and maximum likelihood. *Minimax* principles, as we have seen, are basic in modern game theory, statistical decision, and mathematical economics. In fact, scientific models through the centuries appear to have been formulated as one long "game against nature" (Chapter 12). In actuality, it is not a case of nature's behavior, but rather of man's tendency to project his own characteristics and thus to see principles of thrift in nature's organization of the universe. In some instances it has become apparent that nature will not be disciplined into conformity with the parsimony prescribed by man. Nevertheless the numerous precepts of economy have been a means of unifying whole branches of science, linking past developments to further progress, furnishing orientation to students of science who might go completely astray in reading learned treatises or consulting modern "introductions" to new branches of science.

We have illustrated that differential calculus will handle some types of extrema. The more advanced species gave rise to the *calculus of variations*, a highly technical subject which can nevertheless be described as a generalization of the treatment of maxima and minima in the elementary calculus. In his *Methodus inveniendi lineas curvas minimive proprietate gaudentes* (1744), Euler explained and extended maximinimal notions of Newton, the Bernoullis, Maupertuis. His 1753 *Dissertatio de principio minimae actionis* associates him with Lagrange as co-inventor of the calculus of variations.

Since the methodology of that subject is specialized and intricate, we shall not develop it but shall point out only those features that form the background for functional analysis. Moreover, some of the more technical aspects of applying minimax principles were treated at length in our discussion of statistical decision, and hence it would seem advisable to go beyond such ideas rather than to present more of the same. But even with our ultimate objective in view, we must indicate the nature of a typical problem in the calculus of variations.

The first challenging variational issue was formulated in 1696 by Jean Bernoulli (1667–1748), younger brother of Jacques Bernoulli, whose important contributions to probability have already been noted. The younger Bernoulli posed the problem of determining the shape of the most thrilling slide, that is, the path of fastest descent or *brachistochrone*. (In Greek *brachys* means short, *chronos* signifies time.) To state the problem exactly, suppose that one is given two points not in the same vertical line. How can one select from all the many curves joining the points that one along which the time taken by a particle to slide without friction under the action of gravity from the higher to the lower point will be a *minimum*? In Figure 23.4 we have indicated only three of the infinite variety of possible paths, namely, a

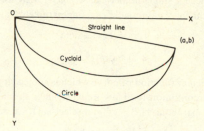

Figure 23.4   Brachistochrone problem

straight line, an arc of a *circle*, and an arc of a *cycloid* (Chapter 8). A race among three steel ball bearings, say, rolling down *smooth* inclines with the three shapes of Figure 23.4 would (in theory) always result in victory for the object on the cycloidal path. This would be the case even if there were hundreds or thousands of other paths (elliptic, parabolic, cubic, sinusoidal, etc.) in the competition. We must now give some idea of what Bernoulli had to do in order to *prove* that the cycloid is the brachistochrone.

In Figure 23.4 we have selected the origin as the higher of the two points to be joined. Hence its coordinates are $(0, 0)$. Let us call the coordinates of the lower point (a, b), and consider an arbitrary curve through the two points, namely $y = f(x)$. Since the two points are on the curve, we must have $f(0) = 0$, $f(a) = b$. Elementary mechanics and the facts of the calculus indicate that, when a definite function $y = f(x)$ is specified, the time of descent is equal to the definite integral (page 203).

$$T = \int_0^a \sqrt{\frac{1 + (y')^2}{2gy}} \, dx$$

where y' represents the derivative of the function $y = f(x)$, and $g = 32$ (approximately) is the acceleration due to gravity.

As the function $y = f(x)$ is varied, the time T varies and the left member of the above formula becomes $T(y)$. *Time* is a function of the function y, that is, time is a *functional* dependent on the different expressions for $y = f(x)$ [and hence for $y' = f(x)$] substituted in the definite integral according to whether the two points are joined by a straight line segment, a parabolic arc, an elliptic arc, a cycloidal arc, etc. But to each function or curve, $y = f(x)$, there will correspond a unique value of the time T, and hence time is a function of the curve selected. Bernoulli's problem was typical, since the general problem of the calculus of variation takes the form of solving for the unknown function $y = f(x)$ which is to minimize or maximize a functional that is expressed as a definite integral of the form

$$I(y) = \int_0^a F(x, y, y')dx$$

But Bernoulli, Euler, and Lagrange did *not* handle such integrals as functionals. Instead they introduced a single numerical *parameter* α by using for $y(x)$ variable curves (functions) which are expressible as a one-parameter family

$$y(x) = Y(x) + \alpha \eta(x)$$

with slope

$$y'(x) = Y'(x) + \alpha \eta'(x)$$

where $Y(x)$ is a particular function or curve which is subjected to continuous variation by changing α continuously. Thus imagine, in a special problem, that the original curve is the parabola $Y(x) = x^2$ and select $\eta(x) = x$, say. Then

$$Y(x) = x^2 + \alpha x$$

a one-parameter family of parabolas with slope formula

$$Y'(x) = 2x + \alpha$$

If these formulas are substituted in the integral above, it will become a function of the numerical variable α. Thus

$$I(\alpha) = \int_0^a F(x, x^2 + \alpha x, 2x + \alpha)\, dx$$

This example suggests the calculus of variation procedure. At any rate, if $I(\alpha)$ is just an ordinary function of a real variable, elementary calculus applies and $I'(\alpha) = 0$ is a necessary condition for a maximum or a minimum. The corresponding values of α can then be substituted to see whether, in a particular problem, the above condition suffices and the geometric or physical entity—length, area, volume, separation, energy, "action"—is actually an extremum for the value (or values) of α obtained.

But our own purpose is to get away from the classic procedure with its reduction to functions of a numerical variable. Instead of forcing problems into the artificial mold of the calculus of elementary functions, there is the alternative of generalizing the meaning of functionality, and this objective leads us to Volterra's work. Although he was not primarily concerned with extrema, nevertheless just such integrals as we have been considering are featured in the questions he raised, and his problems once again call for finding an unknown function which appears under the integral sign.

Volterra's pure mathematical researches were almost always motivated by applications—from mechanics, mathematical physics in general, biology, or economics. Some of the problems drawn from these special fields called for the solution of *integral equations*, and such equations involved functionals. Crudely put, an integral equation is one where the unknown is a function which appears under the integral sign. When we carried out the calculus process of differentiation, we were solving a very elementary integral equation. Instead of being asked to find the derivative of x^4, the question might be put in the form: What function when integrated over the interval $(0, x)$ will give x^4 as answer? In the Leibnizian symbolism, this problem would be written thus:

$$x^4 = \int_0^x f(t)\, dt$$

and would be considered as an *integral equation* of the general type

$$g(x) = \int_0^x f(t)\, dt$$

where $g(x)$ is a given, specified function. This equation, when interpreted, expresses *all* of *differential calculus*, since finding $f(t)$ for any $g(x)$ is identical with finding the derivative of $g(x)$. Thus an entire branch of analysis becomes a mere study of one type of functional, or one kind of integral equation.

But the solution of more general integral equations involves much more than differential calculus. Before proceeding to such issues, we might indicate that the life of the creator of the concept of functional was almost as unusual as his mathematical ideas. The following are biographical details as recorded by one of Volterra's admirers, the mathematical physicist Sir Edmund Whittaker (1873–1956).*

Vito Volterra was born at Ancona on May 3, 1860, the only child of Abramo Volterra and Angelica Almagià. When he was three months old the town was besieged by the Italian army and the infant had a narrow escape from death, his cradle being actually destroyed by a bomb which fell near it.

When he was barely two years old his father died, leaving the mother, now almost penniless, to the care of her brother Alfonso Almagià, an employee of the Banca Nazionale, who took his sister into his house and was like a father to her child. They lived for some time in Terni, then in Turin, and after that in Florence, where Vito passed the greater part of his youth and came to regard himself as a Florentine.

At the age of eleven he began to study Bertrand's *Arithmetic* and Legendre's *Geometry*.† and from this time on his inclination to mathematics and physics became very pronounced. At thirteen, after reading Jules Verne's scientific novel *Around the Moon*, he tried to solve the problem of determining the trajectory of a projectile in the combined gravitational field of the earth and moon: this is essentially the "restricted Problem of Three Bodies," and has been the subject of extensive memoirs by eminent mathematicians both before and after the youthful Volterra's effort: his method was to partition the time into short intervals, in each of which the force could be regarded as constant, so that the trajectory was obtained as a succession of small parabolic arcs. Forty years later, in 1912, he demonstrated this solution in a course of lectures given at the Sorbonne.

When fourteen he plunged alone, without a teacher, into Joseph Bertrand's *Calcul différentiel*; he does not seem to have had access to any work on the integral calculus at this time, and when he attacked various special problems relating to centres of gravity, he

---

* Royal Society of London, *Obituary Notices of Fellows*, Vol. 3, No. 10 (December 1941), pp. 691 ff.

† The standard *liceo* or secondary school mathematics texts in the late nineteenth and early twentieth centuries.

discovered for himself that they could be solved by means of an operation (integration) which was the inverse of differentiation.

His family, whose means were slender, wished him to take up a commercial career, while Vito insisted on his desire to become a man of science. The struggle between vocation and practical necessity became very acute, and the family applied to a distant cousin, who had succeeded in the world, to persuade the boy to accept their views. This man, Edoardo Almagià, who died at the age of eighty in 1921, was one of the most celebrated civil engineers and financiers in Italy in the latter part of the nineteenth century; as a contractor of public works he constructed many important railways and harbours at home and abroad, including the harbours of Alexandria and Port Said in Egypt; the proprietor of vast estates in Tuscany and the Marches, he was renowned for his charities; and it was in the course of excavations at his palace in the Corso Umberto, once the Palazzo Fiano-Ottoboni, that the discovery was made of the sculptures of the Ara Pacis of the Emperor Augustus, which are now among the treasures of the Museo delle Terme in Rome.

His interview with his young relative turned out differently from what the family had expected. Impressed by the boy's sincerity, determination and ability, the older man threw his influence on the side of science, and turned the scale. Professor Roiti offered a nomination as assistant in the Physical Laboratory of the University of Florence, though Vito had not yet begun his studies there; it was accepted, and now the die was cast. The young aspirant entered the Faculty of Natural Sciences at Florence, following the courses in Mineralogy and Geology as well as in Mathematics and Physics. In 1878 he proceeded to the University of Pisa, where he attended the lectures of Dini, Betti and Padova; in 1880 he was admitted to the Scuola Normale Superiore, where he remained for three years; and here, while still a student, he wrote his first original papers. Under the influence of Dini he had become interested in the theory of aggregates and the functions of a real variable, and he gave some examples which showed the inadequacy, under certain circumstances, of Riemann's theory of integration, and adumbrated the developments made long afterwards by Lebesgue.

In 1882 he graduated Doctor of Physics at Pisa, offering a thesis on hydrodynamics in which certain results, actually found earlier by Stokes, were rediscovered independently. Betti at once nominated him as his assistant. In 1883, when only twenty-three years of age, he was promoted to a full professorship of Mechanics in the University of Pisa, which after the death of Betti was exchanged for the Chair of Mathematical Physics. He now set up house in Pisa with his mother, who up to that time had continued to live with her brother. In 1888 he was elected a non-resident member of the Accademia dei Lincei; in 1892 he became professor of mechanics in the University of Turin, and in 1900 he was called to the Chair of Mathematical Physics in Rome, as the successor of Eugenio Beltrami. In July of that year he married Virginia Almagià, one of the daughters of the distinguished relative who had first made it possible for him to follow a scientific career. She had inherited intellectual brilliance from her father, and great beauty from her mother, and as the wife of Vito Volterra took upon herself all the cares which might have distracted her husband from his scientific work, undertaking the education of their children and the administration of all their possessions. Six children were born of the union, of whom four now survive. His mother still lived with them, and died at the age of eighty at the Palazzo Almagià in March 1916.

As a comparatively young man he was appointed by the Government as Chairman of the Polytechnic School at Turin, and Royal Commissioner. The way was open for him to become a great figure in political and administrative life; but he preferred the career of a pure scientist, and took an active part in public affairs on only two occasions—the World War of 1914–1918, and the struggle with Fascism.

In July 1914 he was, according to his custom at that time of year, at his country house at Ariccia, when the war broke out. Almost at once his mind was made up that Italy ought to join the Allies; and in concert with D'Annunzio, Bissolati, Barzilai and others, he organized meetings and propaganda which were crowned with success on the 24th of May in the following year, when Italy entered the war. As a Lieutenant in the Corps of Engineers he enlisted in the army, and, although now over fifty-five years of age, joined the Air Force. For more than two years he lived with youthful enthusiasm in the Italian skies, perfecting a new type of airship and studying the possibility of mounting guns on it. At last he inaugurated the system of firing from an airship, in spite of the general opinion that the airship would be set on fire or explode at the first shot. He also published some mathematical works relating to aerial warfare, and experimented with aeroplanes. At the end of these dangerous enterprises he was mentioned in dispatches, and decorated with the War Cross.

Some days after the capitulation of Gorizia he went to this town while it was still under the fire of Austrian guns, in order to test the Italian instruments for the location of enemy batteries by sound. At the beginning of 1917 he established in Italy the Office for War Inventions, and became its Chairman, making many journeys to France and England in order to promote scientific and technical collaboration among the Allies. He went to Toulon and Harwich in order to study the submarine war, and in May and October 1917 took part in the London discussions regarding the International Research Committee, to the executive of which he was appointed. He was the first to propose the use of helium as a substitute for hydrogen, and organized its manufacture.

When in 1917 some political parties—especially the Socialists—wanted a separate peace for Italy, he strenuously opposed their proposals; after the disaster of Caporetto, he with Sonnino helped to create the parliamentary *bloc* which was resolved to carry on the war to ultimate victory.

On the conclusion of the Armistice in 1918, Volterra returned to his purely scientific studies and to his teaching work in the University. The most important discoveries of his life after the war were in the field of mathematical biology, and of these we must now give an account.

His discourse at the opening of the academic year following his election to the Roman Chair, shows that already in 1901 he was interested in the biological applications of mathematics. His own researches in this field, stimulated by conversations with the biologist Umberto D'Ancona of the University of Siena, began at the end of 1925; his first and fundamental memoir on the subject appeared soon thereafter, and was followed by several further papers. In the winter of 1928–1929 it was made the subject of a course of lectures delivered by Volterra at the Institut Henri Poincaré in Paris. These lectures, together with a historical and bibliographical chapter compiled by D'Ancona, were published in 1931.

The entities studied in these investigations were *biological associations*, i.e. systems of animal (or plant) populations of different species, living together in competition or alliance in a common environment; and the theory is concerned with the effects of interaction of these populations with one another and the environment as expressed in their numerical variations.

Volterra's scientific activity overflowed in many domains quite outside his more usual fields of research. For instance, the student of topology (Chapter 25) who reads Professor Lefschetz's admirable monograph on *Analysis Situs* (Paris, 1924), finds therein the photographs of a number of ingenious models constructed by Volterra in order to show how two manifolds, defined in very different ways, may nevertheless be homeomorphic (Chapter 25) to each other.

There remains to be told the melancholy story of his later years. In 1922 Fascism seized the reins in Italy. Volterra was one of the very few who recognized from the beginning the danger to freedom of thought, and immediately opposed certain changes in the educational system, which deprived the Italian Middle Schools of their liberty. When the opponents of the Fascist Government in the House of Deputies withdrew altogether from the debates, a small group of Senators, headed by Volterra, Benedetto Croce and

Francesco Ruffini, appeared, at great personal risk, at all the Senate meetings and voted steadily in opposition. At that time he was President of the Accademia dei Lincei and generally recognized as the most eminent man of science in Italy.

By 1930 the parliamentary system created by Cavour in the nineteenth century had been completely abolished. Volterra never again entered the Senate House. In 1931, having refused to take the oath of allegiance imposed by the Fascist Government, he was forced to leave the University of Rome, where he had taught for thirty years; and in 1932 he was compelled to resign from all Italian Scientific Academies. From this time forth he lived chiefly abroad, returning occasionally to his country-house in Ariccia. Much of his time was spent in Paris, where he lectured every year at the Institut Henri Poincaré; he also gave lectures in Spain, in Roumania, and in Czechoslovakia. On all these journeys he was accompanied by his wife, who never left him, and learned typewriting in order to copy his papers for him; he was accustomed to say that the signature "V. Volterra" in his later works represented not Vito but Virginia Volterra.

In the autumn of 1938, under German influence, the Italian Government promulgated racial laws, and his two sons were deprived of their University positions and their civil rights; at their father's suggestion, they left their native country to begin a new life abroad. (One of them, Enrico Giovanni, is Professor of Mathematics at the University of Texas.)

Vito Volterra had a remarkable power of inspiring affection. When in the last months of his life the new laws forbade him to have Italian servants, all of them refused to leave; a maid who had been with him for more than twenty years, and who was forced to leave, died of sorrow a week afterwards.

In December 1938 he was affected by phlebitis: the use of his limbs was never recovered, but his intellectual energy was unaffected, and it was after this that his two last papers were published by the Edinburgh Mathematical Society and the Pontifical Academy of Sciences respectively. On the morning of 11 October 1940 he died at his house in Rome. In accordance with his wishes, he was buried in the small cemetery of Ariccia, on a little hill, near the country-house which he loved so much and where he had passed the serenest hours of his noble and active life.

To return to the research in integral equations that Volterra had begun half a century before the last sad period of his life, we illustrate one type that he considered, namely,

$$g(x) = \int_0^x K(x, t) f(t) \, dt$$

where $g(x)$ and $K(x, t)$ are *known* functions, and $f(t)$ is the unknown function that is to be found. If, for example, the given functions are

$$g(x) = x^4$$
$$K(x, t) = xt$$

then the equation becomes

$$x^4 = \int_0^x xt \, f(t) \, dt$$

By trial, we can obtain the solution

$$f(t) = 3t^2$$

We can see that this will check, since it is true that

$$x^4 = \int_0^x 3xt^2 \, dt$$

The reader can verify, by using antidifferentiation in place of integration in the right member, that the result is equal to x^4. The antiderivative of $3t^2$ is $t^3 + c$, and since x is treated as a constant in the right member of the equation, the antiderivative of $3xt^2$ is $xt^3 + c$. Following the pattern of Chapter 8, we might imagine the latter expression to be the formula for a growing area, volume, or quantity of work, etc. Let us picture then that an area is involved and that

$$A = xt^3 + c$$

The Leibnizian symbol $\int_0^x$ would signify that growth starts with $t = 0$ and stops when $t = x$. Substituting the first condition,

$$0 = 0 + c \qquad \text{and} \qquad c = 0$$

Then substituting $t = x$, the value of t when growth terminates,

$$A = x(x^3) = x^4$$

which is the desired result.

But a general theory of such integral equations would require consideration of whether there are solutions for all selections of $g(x)$ and $K(x, t)$, and if the answer is in the negative, what restrictions must be placed on these two functions in order that solutions may exist. Also, there is the matter of a method other than trial and error for finding a solution when it is known to exist.

A still more difficult type of Volterra equation is

$$g(x) - f(x) = \int_0^x K(x, t) f(t) \, dt$$

where once again $g(x)$ and $K(x, t)$ are given, and $f(t)$ (or $f(x)$), which is essentially the same function) is to be found. If, for example, the given functions are

$$g(x) = x^7 + 5x^3$$
$$K(x, t) = x^2 t$$

then

$$f(t) = 5t^3 \qquad \text{and hence} \qquad f(x) = 5x^3$$

will check in the above integral equation, since it is true that

$$x^7 + 5x^3 - 5x^3 = \int_0^x x^2 t(5t^3) dt = \int_0^x 5x^2 t^4 dt$$

Again, using antidifferentiation for the expression under the integral sign, $A = x^2 t^5 + c$. Now the condition, $A = 0$ when $x = 0$, gives $c = 0$. The value of A is required for $t = x$, so that $A = x^2(x^5) = x^7$, the value of the left member.

Instead of studying the general theory, we shall examine integral equations that arise naturally from problems in mathematical physics. Here we wish to point out the connection of integral equations with the concept of *functional*. In both types of integral equation above, the *kernel*, $K(x, t)$, is given and then the value of the integral

$$\int_0^x K(x, t) f(t) \, dt$$

depends on the selection of the unknown function $f(t)$. For each permissible selection of $f(t)$ there will be a unique value of the integral, that is, the integral is a function of the function $f(t)$, hence a functional. The second type of Volterra equation illustrated above can also be expressed in the form

$$g(x) = f(x) + \int_0^x K(x, t) f(t) \, dt$$

and therefore in this case as well the right member is a functional dependent on f. In both types of equation the problem is to select the independent variable f so that the value of the functional in the right member will be equal to the given function $g(x)$. Therefore the concept involved is just a generalization of algebra with the most elementary type of function. In school algebra pupils are asked: If $y = 8 - 3x$ and $y = 2$, what is the value of x? The integral equation poses a similar problem, namely: Given the value of the dependent variable, find the corresponding value of the independent variable.

Then the reader may not be surprised to learn that Volterra and Hilbert were able to reduce integral equations to algebra of a type that is familiar, namely, the solution of a set of linear equations. In school days one learned how to find x and y when, for example,

$$3x + 2y = 7$$
$$2x + 5y = 12$$

and also solved systems of three or more linear equations in three or more unknowns. But the systems derived from integral equations involve an *infinite* number of equations in an infinite number of unknowns. Nevertheless the methodology is not more difficult than in the finite case, but naturally has to be followed up with some consideration of limits, convergence, etc., as is usual in infinite processes.

The general technique of solving such infinite systems was something new in Volterra's day, for although some historians claim that Fourier was the first to solve such an infinite system, it appears that George William Hill (1838–1914) provided the first rigorous solution to receive any publicity. In 1877 he solved an infinite number of linear equations involving an infinite number of unknowns by using the method of "determinants." He did this in connection with research analyzing the motion of the moon's perihelion. Subsequently Hill's method was made rigorous by Poincaré and by the Scandinavian mathematician von Koch (1870–1924). Hill rated high among American scientists of his day; the *Biographical Directory* of such men of science paired him with Simon Newcomb (1835–1909) for first rank among U.S. astronomers, and it classified him as second only to E. H. Moore as a pure mathematician. The ability of all three men was recognized by their mathematical contemporaries, who elected Hill, Newcomb, and Moore as third, fourth and sixth presidents of the American Mathematical Society in 1895, 1897, 1901, respectively.

If integral equations can be converted into Hill sets of linear equations, a converse fact is true. Turning to a physical example, as Volterra always did, one can consider a finite system at first, a system derived from the study of a mechanical configuration in stable equilibrium where very tiny oscillations about the equilibrium state can occur. If the configuration can be determined by $1, 2, 3, 4, \ldots, n$ *generalized coordinates* (angles, distances, areas, etc.), it has $1, 2, 3, 4, \ldots, n$ "degrees of freedom," as physicists say. It can be proved that in this case any tiny oscillation is the resultant of $1, 2, 3, 4, \ldots, n$ simple harmonic vibrations, a situation reminiscent of Fourier's technique (Chapters 7 and 9). To specify a simple harmonic vibration, that is, a sine or cosine function, we must know its amplitude and period. Dynamic theory shows that the necessary facts can be obtained by solving $1, 2, 3, 4, \ldots, n$ linear equations, respectively. To see how this is done, consider the system of two equations in two unknowns illustrated above. If the set of real numbers is the domain of variables x and y, then either trial and error or elementary algebraic techniques lead to the unique solution

$$x = 1 \qquad y = 2$$

Next, try to solve

$$3x + 2y = 7$$
$$6x + 4y = 11$$

and you will readily see that there is no answer because the second equation contradicts the first. An equation consistent with the first equation would be

$$6x + 4y = 14$$

and not

$$6x + 4y = 11$$

But then if we had the system

$$3x + 2y = 7$$
$$6x + 4y = 14$$

where x and y represent real numbers, either equation would be equivalent to the other, so that it is actually a case of one equation in two unknowns, an *indeterminate* problem. Then

$$x = a$$
$$y = \tfrac{1}{2}(7 - 3a)$$

where a is any real number.

Thus a system of two linear equations in two unknowns may have no solutions, a unique solution, or an infinite number of solutions. One equation of such a system might be

$$2x + 3y = 0$$

and the other

$$\lambda x - 4y = 0$$

The reader can readily obtain the solution $x = 0$, $y = 0$. The value of the coefficient λ does not matter. But if other solutions for x and y are sought, they can only exist when the two equations reduce to one, that is, when the corresponding coefficients in the two equations are proportional, and

$$\frac{\lambda}{2} = \frac{-4}{3}$$

$$3\lambda = -8$$

$$\lambda = -\frac{8}{3}$$

The second equation now becomes

$$-\frac{8}{3}x - 4y = 0$$

or, multiplying by $-3/4$,

$$2x + 3y = 0$$

which is identical with the first.

Returning now to the tiny oscillations of a mechanical system with two "degrees of freedom," the associated linear equations would look something like

$$(5 - \lambda)x + 2y = 0$$
$$2x + (5 - \lambda)y = 0$$

(where the figures have been selected somewhat unrealistically in order to make the arithmetic easy).

It would be interesting to consider the physical origin of these equations. The mechanical configuration in this case is two-dimensional, and the small oscillations are represented by a plane linear transformation of homogeneous or centered affine type (Chapter 17). Thus if x' and y' are the generalized coordinates after a slight displacement from the state of equilibrium,

$$x' = 5x + 2y$$
$$y' = 2x + 5y$$

To be up to date, such a transformation is now called a *linear operator*.

From a geometric point of view, as Klein described it, one should seek the points, lines, etc. that are unchanged by the affine transformation or operator. If a point is unchanged, its new coordinates are the same as the old and this signifies $x' = x$, $y' = y$ in the above equations. Now, substituting x' for x and y' for y, the solution of the simultaneous equations thus obtained is

$$x = 0$$
$$y = 0$$

This means that the only point unchanged by the operator is the origin.

If any *lines* are to be invariant or unchanged by the operator, this does *not* require that their individual points be unchanged; they may merely be shifted along these lines. For example, if a line like $y = 2x$ (Figure 23.5) were invariant under homogeneous affine transformation, the point P (1, 2) might shift to $P'(\frac{1}{2}, 1)$

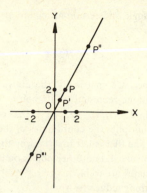

Figure 23.5

or to $P''(3, 6)$ or to $P'''(-2, -4)$. Thus if a point lies on a line through the origin and is to remain on this same line, its coordinates after transformation are equimultiples of its original coordinates, that is, $\lambda x = x'$, $\lambda y = y'$, where λ is some real number. But the linear operator above states that $x' = 5x + 2y$ and $y' = 2x + 5y$. Hence

$$\lambda x = 5x + 2y$$
$$\lambda y = 2x + 5y$$

or

$$(5 - \lambda)x + 2y = 0$$
$$2x + (5 - \lambda)y = 0$$

This is the pair of equations in the illustration with which we started. The origin, that is, $x = 0$, $y = 0$, will satisfy this pair of equations, whatever the value of λ. But if the equations are to be satisfied for other values of x, this can occur, as previously explained, only if the coefficients in the equations are proportional. Therefore

$$\frac{5 - \lambda}{2} = \frac{2}{5 - \lambda}$$

or

$$25 - 10\lambda + \lambda^2 = 4$$
$$\lambda^2 - 10\lambda + 21 = 0$$

and

$$\lambda = 3 \quad \text{or} \quad \lambda = 7$$

If $\lambda = 3$ is substituted in the two linear equations above, they reduce to the same equation,

$$2x + 2y = 0$$

or

$$x + y = 0$$

and the net result is a single indeterminate equation with an infinite number
of real solutions,

$$x = a$$
$$y = -a$$

where a is any real number. Geometrically, the above equation, $x + y = 0$, is a
straight line, and the solutions represent all its points, some of which are $(0, 0)$,
$(1, -1), (2, -2), (-1, 1), (5/3, -5/3), (-2, 2)$, etc. In accordance with our assumption, the linear operator multiplies the coordinates of these points by $\lambda = 3$, Thus,
$x + y = 0$, or $y = -x$, is a line that is left invariant by the linear operator, because
$(0, 0)$, a point on it, remains fixed, and other points on it are merely shifted along
the line when their coordinates are trebled.

For $\lambda = 7$, both equations reduce to

$$x - y = 0$$

and the solutions are

$$x = b$$
$$y = b$$

the set of points of the invariant line $x - y = 0$ or $y = x$. The effect of the linear
operator on the individual points of $y = x$ is to leave the origin unchanged and
shift the other points by multiplying their coordinates by $\lambda = 7$.

In problems like the above, the two values $\lambda = 3$ and $\lambda = 7$ are usually referred to as *eigenvalues*, from the German *Eigenwerte*. The hybrid term has maintained its own in spite of attempts to use names like *characteristic values* or *proper
values*. In the theory of vector spaces the invariant lines are called *eigenvectors*.

In the physical interpretation of the present problem (the tiny oscillations of a
mechanical system), an eigenvalue is equal to the square of the period of a harmonic
vibration, and there are two component harmonic vibrations with periods $\sqrt{3}$ and
$\sqrt{7}$, or frequencies $1/\sqrt{3}$ and $1/\sqrt{7}$. This means that in the first component about
0.6 of a vibration is completed in one second, and in the other component, about
0.4 of a vibration. The sine formulas for the component vibrations are

$$A \sin \frac{2\pi}{\sqrt{3}} t \quad \text{and} \quad B \sin \frac{2\pi}{\sqrt{7}} t$$

The invariant lines (eigenvectors), namely, $y = x$ and $y = -x$, are perpendicular,
and their inclinations, 45° and 135°, are called the principal directions associated
with this mechanical problem.

For the sake of practice, the reader can show that $\lambda = 2$ and $\lambda = 7$ are the eigenvalues of the linear operator

$$x' = 3x - 2y$$
$$y' = -2x + 6y$$

and that the principal directions are given by the (perpendicular) invariant lines,
$y = \frac{1}{2}x$ and $y = -2x$.

There are many algebraic issues that may arise in such problems—for example, the matter of *equal* values of λ, or the question of what numerical restrictions the linear operator must obey so that the values of λ should be real and positive (so that the periods, $\sqrt{\lambda}$, may be real), etc. There are also the associations of the algebra with elementary analytic geometry, for example, in the transformation of the equations of a conic section to standard or normal form. Thus, the *matrix* of coefficients in the original linear operator, namely,

$$\begin{pmatrix} 5 & 2 \\ 2 & 5 \end{pmatrix}$$

can be interpreted to apply to the equation of the conic section, $5x^2 + 4xy + 5y^2 = 1$, if we write the left member as the rectangular array

$$\begin{pmatrix} 5x^2 & 2xy \\ 2xy & 5y^2 \end{pmatrix}$$

Here the matrix coefficients make their appearance. Theory shows that a suitable *rotation* (Chapter 17) will transform this equation into

$$3(x')^2 + 7(y')^2 = 1$$

where the *coefficients* are the *eigenvalues*, 3 and 7. This is the equation of an ellipse, with semiaxes (found by substituting $y' = 0$ or $x' = 0$) equal to the "frequencies" $1/\sqrt{3}$ and $1/\sqrt{7}$. Thus the analytic geometry of an ellipse is associated with the above problem in mechanics. Similarly, in assocation with the practice example suggested for the reader, the conic section

$$3x^2 - 4xy + 6y^2 = 1$$

can be transformed into the ellipse

$$2(x')^2 + 7(y')^2 = 1$$

If a mechanical system has $4, 5, \ldots, n$ degrees of freedom, then a set of $4, 5, \ldots, n$ linear equations will give the solution for its small oscillations. In general there will be $4, 5, \ldots, n$ real, distinct eigenvalues for λ and an equal number of principal directions in an abstract hyperspace of $4, 5, \ldots, n$ dimensions. But other vibrations —those of continuous media like a musical string, a membrane, an elastic solid, the "ether" of classical physics—will have a *countably infinite* (Chapter 24) number of degrees of freedom, and if the procedure for mechanical systems is generalized, there will be an infinite set of linear equations with an infinite number of unknowns. Volterra and Hilbert showed that such an infinite set could be replaced by a single integral equation.

To make this fact seem reasonable, one need only recall the applications of integral calculus where an area is approximated by a sum of rectangles or squares or sectors, and the limit of such a sum, as the number of terms became infinite, defines the area. In this limiting process the sum of an infinite series

$$\Delta t(f_1 + f_2 + f_3 + \cdots + f_n + \cdots)$$

led to the definite integral

$$\int_a^b f(t)\, dt$$

where the domain of $f(t)$ is a *continuum*, the interval of real numbers from a to b. (In what follows we shall adjust linear units so that $a = 0$, $b = 1$.) This suggests that if we had a series, $K_1 f_1 + K_2 f_2 + K_3 f_3 + \cdots + K_n f_n + \cdots$, where the underlying limiting process converges, a replacement by

$$\int_0^1 K(t) f(t)\, dt$$

might prove to be valid.

If the issue of vibrations of continuous media is just a matter of extending the linear operator from a finite number n to an infinite number of variables and equations, then the eigenvalues and principal directions (eigenvectors) will be infinite in number and will be obtained by solving

$$\lambda f_1 = K_{11} f_1 + K_{12} f_2 + K_{13} f_3 + \cdots + K_{1n} f_n + \cdots$$
$$\lambda f_2 = K_{21} f_1 + K_{22} f_2 + K_{23} f_3 + \cdots + K_{2n} f_n + \cdots$$
$$\lambda f_3 = K_{31} f_1 + K_{32} f_2 + K_{33} f_3 + \cdots + K_{3n} f_n + \cdots$$

$$\lambda f_m = K_{m1} f_1 + K_{m2} f_2 + K_{m3} f_3 + \cdots + K_{mn} f_n + \cdots$$

where the unknowns are $\lambda, f_1, f_2, f_3, \ldots, f_n, \ldots$ and the K's are numerical constants making the infinite series in the above equations *converge* (Chapter 22). Also, because the operator is *symmetric*, the values of the K's must be such that $K_{12} = K_{21}$, $K_{13} = K_{31}$, $K_{23} = K_{32}$, etc. and, in general, $K_{ij} = K_{ji}$.

If we assume the validity of replacement of series by definite integrals, we can substitute

$$\lambda f_1 = \int_0^1 K_1(t) f(t)\, dt$$

for the first equation above, and similarly for the other equations—

$$\lambda f_2 = \int_0^1 K_2(t) f(t)\, dt$$

$$\lambda f_3 = \int_0^1 K_3(t) f(t)\, dt$$

etc. The subscripts 1, 2, 3, in $K_1(t)$, $K_2(t)$, $K_3(t)$, etc. are meant to indicate that these are different functions. The reason for this is that the K coefficients in the right member of each linear equation are, in general, different.

If we wish to symbolize the entire set of linear equations by a *single* integral equation, we can do so by introducing another variable in addition to t. The continuous variation in t as it proceeds from 0 to 1 corresponds to the progress from term to term in each of the above series, for example, from $K_{11} f_1$ to $K_{12} f_2$ to

$K_{13}f_3$ to $K_{14}f_4$, etc. in the first series. We now introduce an additional continuous variable s (whose domain is also the interval from 0 to 1) to replace the progress from row to row in the set of integral equations above, where K_1 changes to K_2 to K_3, etc., and simultaneously in the *left* member f_1 changes to f_2 to f_3, etc. Then the integral equation corresponding to the entire system is

$$\lambda f(s) = \int_0^1 K(s,t)f(t)\,dt$$

where λ is a real numerical parameter, and $f(t)$ is an unknown function but $K(s,t)$, the kernel, is a known function. The integral in the right member is called a *linear operator* by analogy with the algebraic situation. The integral equation is classified as linear in type as might be expected from the fact that it is a composite of an infinite set of linear algebraic equations.

Such an infinite set, or else a single linear integral equation, describes the vibrations of a musical string of unit length when a suitable choice is made for the kernel $K(s,t)$. Then the solution is possible only for proper values of λ, that is, for the eigenvalues. It turns out that the eigenvalues are $\lambda = (1/kc)^2$, and hence the periods $\sqrt{\lambda} = 1/kc$, where c is a constant determined by the physical condition of the string and k is any one of the positive integers, 1, 2, 3, ... The corresponding frequencies are kc, and this result indicates the basic *law* of musical *harmony*, namely, that a musical tone is composed of a fundamental note and overtones whose *frequencies* (kc) are all integral multiples of the fundamental frequency (c). As in the case of the elementary linear operators, the eigenvalues will now be substituted either in the set of linear equations or in the corresponding integral equation. Then the solutions obtained are

$$f(t) = \sin k\pi t \quad \text{or its equivalent,} \quad f(s) = \sin k\pi s$$

where k is one of the positive integers, 1, 2, 3, ...

After later developments in the theory and application of integral equations, due mainly to the Swedish mathematician Ivar Fredholm (1866–1927), optical rather than acoustical vocabulary came into vogue, and the set of values for λ was referred to as the *spectrum* of eigenvalues. (A strict analogy with a *frequency* spectrum would use $1/\sqrt{\lambda}$.) It was Hilbert who established the general *spectral theory of symmetric linear operators*. Prior to this particular research it required mathematics of the greatest difficulty to attack even the simplest problems in the field, and therefore Hilbert's methods were an enormous contribution to mathematics and theoretical physics.

The term "linear operator" has been applied to the right member of

$$\lambda f(s) = \int_0^1 K(s,t)f(t)\,dt$$

by analogy with the corresponding set of linear algebraic equations. But such a set of equations can also be considered as representing a *transformation* (in the special case considered, a homogeneous affine transformation). Then the right member of the integral equation is described as an *integral transform*, and $\lambda f(s)$ is called the transform of $f(t)$ by the kernel $K(s,t)$. In the above integral equation the transform is equal to a numerical multiple of the function $f(t)$ (or $f(s)$, which is the same function).

Integral transforms take a more general form in modern applied mathematics. If $f(t)$ and the kernel $K(s, t)$ are known functions, then the transform of $f(t)$ in some interval (a, b) may be found, namely,

$$\int_a^b K(s, t) f(t)\, dt$$

In general, the transform is *not* a mere numerical multiple of $f(t)$, but is some entirely different function, $g(s)$, that is,

$$g(s) = \int_a^b K(s, t) f(t)\, dt$$

Often, in applied mathematics, $K(s, t)$ and $g(s)$ are known functions, and $f(t)$ is unknown. In this case one must solve an integral equation to find $f(t)$.

There are a number of different types of integral transform which are of enormous importance in facilitating solution of differential equations and boundary value problems (mathematical physics), as well as in other areas of applied mathematics, for example, mathematical statistics. In particular, there is the *Laplace transform*

$$g(s) = \int_0^\infty e^{-st} f(t)\, dt$$

where the kernel is always e^{-st}, and $f(t)$ and $g(s)$ vary with the problem to be solved. The symbol ∞ makes the integral "improper." What that means is that $\int_0^A$ must be found and then one must see whether the result has a limit as A gets larger and larger. If a limit exists, it is the value of $g(s)$.

There is a *Fourier transform* which generalizes that of Laplace, and there are still other integral transforms. Laplace was aware of the need to "invert" his transform, that is, to find $f(t)$, given $g(s)$, and Fourier actually did invert his more general transform, although he did not conceive of his procedure in terms of an integral equation. But he did accomplish what was equivalent to such a solution and thus, in a sense, was the first mathematician in history to accomplish the solution of an integral equation.

Now to return to the general question of *functional analysis*, one must point to a very significant concept that developed in relation to Hilbert's contributions to that subject. This was the idea of *Hilbert space*, the infinite analogue of a *Euclidean space* of 1 or 2 or 3 or ... n dimensions.

We shall approach the concept of Hilbert space through some very elementary considerations. Thus the function expressed by the equation $y = x^2$ can be represented by a plane Cartesian graph in which the *domain* or set of values of the independent variable x is the real number continuum, that is, the X-axis, a *Euclidean space* of one dimension. A function of two numerical variables like $z = x - y^2$ has a surface as its Cartesian picture if a *Euclidean plane* is its domain, that is, if its domain is the set of real number pairs (x, y). Next we can adduce a function like

$$w = x^2 + y^2 - 4z^2$$

which can be named a *hypersurface* because geometric language is a convenience. The domain of this last function is an ordinary *Euclidean space* (of three dimensions). We can go on to functions whose domains are Euclidean 4-space, 5-space, etc. so that

$$y = g(x_1, x_2, x_3, \ldots, x_n)$$

represents a function g whose domain is a Euclidean space with n dimensions, where n is a *finite* positive integer.

The next step would be to consider

$$y = g(x_1, x_2, x_3, x_4, \ldots, x_n, \ldots)$$

where the domain is a *space* with a *countable* infinity of dimensions (Chapter 24). Hilbert endowed this sort of abstract space with the type of structure that seemed desirable for *functional* problems he encountered in his study of the spectra of linear operators. By a *structure* we mean the fundamental set of *definitions and postulates* that governs the "points," $(x_1, x_2, x_3, x_4, \ldots, x_n, \ldots)$ of the infinite-dimensional space and from which the theorems of its "geometry" can be deduced.

Hilbert's constant use of Fourier series in spectral analysis emphasized in his mind the fact that a function which can be represented by a Fourier series determines the coefficients of the terms in such an expansion and conversely, given a sequence of permissible coefficients, a function is determined. Acoustically, this signifies that a musical tone can be broken down into a fundamental note and its overtones, and conversely a synthesis of simple harmonic vibrations with specific amplitudes will constitute a musical tone. Thus, Hilbert observed, there is no intrinsic difference between a function capable of Fourier series representation and the infinite sequence of its Fourier coefficients. If we symbolize the sequence by $(x_1, x_2, x_3, x_4, \ldots, x_n, \ldots)$ we can consider it as a "point" of an infinite-dimensional space, named a *Hilbert space*, after its creator. Since a sequence is equivalent to the function it represents, a point in Hilbert space corresponds to both sequence and function, and the Hilbert space can be described as a *function space*, that is, a space whose points are in one-to-one correspondence with the set of all functions of a certain type.

In functional analysis such a space is analogous to the X-axis of plane Cartesian geometry. If one is studying functions of a single numerical variable like $y = x^2$, $y = 3x - 7$, etc., the X-axis contains the domain, the set of values which x may take. But the calculus of variations, integral equations, general spectral analysis, etc. deal with functions of a function f, and if f can be expressed by a Fourier series, then it corresponds to a point $(x_1, x_2, x_3, x_4, \ldots)$ of *Hilbert space*, which is thus the *domain of the functional*. Here there is subtlety similar to the replacement of an infinite set of linear algebraic equations by an integral equation, the former being a discrete aggregate, and the latter containing a function defined on a continuum. In the same way the sequence of coordinates (Fourier coefficients) of a function or point of Hilbert space is discrete, whereas the function it represents has a continuous domain. But it is not such philosophical issues that make Hilbert space mathematically useful. Its applicability derives from the axioms that govern the algebra and the analysis of this space, the basic rules for performing arithmetic operations with the different elements symbolized by $(x_1, x_2, x_3, \ldots)$, the assignment of a *metric*, that is, a distance formula, etc. In all considerations, matters of convergence—that is, the existence of limits—are involved since the points correspond to *infinite* sequences. We have seen that in Euclidean plane and solid geometry the square of the distance of a point from the origin is represented by $x^2 + y^2$ and $x^2 + y^2 + z^2$, respectively, by virtue of the Pythagorean theorem. If metric properties of Hilbert space are to be analogous to those in Euclidean spaces of finite dimension, then the square of the "distance" of a point from the origin should be

$$x_1{}^2 + x_2{}^2 + x_3{}^2 + x_4{}^2 + \cdots x_n{}^2 + \cdots$$

and this infinite series will have meaning only if it *converges*. This fact must be incorporated in the definition of Hilbert space, which is therefore *not* the set of all sequences $(x_1, x_2, x_3, x_4, \ldots)$ but only the set of those sequences for which $x_1{}^2 + x_2{}^2 + x_3{}^2 + x_4{}^2 + \cdots x_n{}^2 + \cdots$ *converges*. Thus the sequence $(1, 1, 1, 1, \ldots, 1, \ldots)$ is not a point in Hilbert space since

$$1^2 + 1^2 + 1^2 + 1^2 + \cdots + 1^2 \cdots = 1 + 1 + 1 + 1 + \cdots + 1 + \cdots$$

does not converge, because the sum of the first n terms is n, which does not approach a limit but merely gets larger and larger with increasing values of n.

Hilbert space is only one type of generalized space. The great Polish mathematician Stefan Banach (1892–1945) created and investigated a whole class of function spaces more general than the particular space of Hilbert. Moreover, in the first years of this century, the American E. H. Moore (1862–1932) and the French mathematician Maurice Fréchet both sought to generalize still further Volterra's and Hilbert's work with functionals, in order to establish a genuine *general analysis*, a calculus of relations between two aggregates whose elements are completely abstract or arbitrary in nature. General analysis is a study of functions in which *neither* independent nor dependent variable need be a number. For that matter, the generality is so great that the independent variable need not even be a function, so that one is no longer dealing with a function of a curve or a function of a surface or a function of any function, but merely with a function of some completely abstract entity. Of course the generalized functions and their analysis include the other types as special cases.

The first task in setting up a general analysis is similar to that of Hilbert in imparting a structure to his function space. Moore and Fréchet had to provide definitions and postulates that would apply to functions where the nature of the variables is completely unspecified. Fréchet attempted to formulate a structure that would resemble as closely as possible the corresponding definitions and postulates of classic analysis. This is analogous to Hilbert's efforts to create for his function space a structure similar to that of Euclidean spaces. But let us first see how Moore proceeded. He used a principle of generalization that was practiced by mathematicians long before he asserted it explicitly. He stated that *the existence of analogies between central features of various theories implies the existence of a general abstract theory which includes the particular theories and unifies them with respect to those central features.*

Applying his principle, Moore developed a theory of integral equations and eigenvalues which is a generalization of the analogous theory of Hilbert. Fréchet's form of general analysis, described by its author in a series of memoirs starting in 1904, is better known and more readily applicable than the Moore theory. This is due to several factors, one of which is that Moore did *not* generalize *continuity* or *differentiability* of ordinary functions. But these attributes of functions are the essence of any form of analysis and are the very ones which Fréchet took great pains to specify.

Fréchet emphasized the infinitesimal (analytic) properties of a functional, and concentrated on the functional relation itself rather than on the nature of the independent variable. When that variable is a curve, surface, or function of some kind, one can emulate Hilbert and conceive of it as a point of some function space. Fréchet indicated that if such a space is to have any resemblance to ordinary spaces, however, one needs a way of recognizing when two points (or the two functions they represent) are neighboring. This can be done if one can define a *distance* between the two points (two functions). But what meaning can be given to "nearness" if variables are completely abstract or unspecified?

Fréchet's answer to this last question had best be considered later, in our discussion of *analytic topology* (Chapter 25). He does, however, give a simple example of a suitable definition of the distance between two functions that are defined and continuous in some interval. Figure 23.6 indicates that in the interval [0, 1] the maximum difference in functional values for the pair of continuous functions graphed is 2 units. This, Fréchet indicates, would be a suitable measure of the distance between the two functions or their corresponding points in function space.

Like Moore, Fréchet listed a number of basic principles and salient features of his brand of general analysis, for example: "Those properties which, in the theory of ordinary numerical functions, arise only in the handling of the most complicated, most difficult, least practical functions, are the characteristics that are the

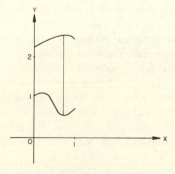

Figure 23.6

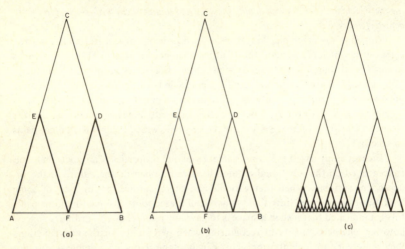

Figure 23.7

simplest, most practical, in fact indispensable attributes of functionals or functions of generalized variables."*

To illustrate this aspect of his theory, Fréchet alludes to the property of *semicontinuity*. This concept is due to René Baire (1874–1932), one of the most brilliant contributors to real analysis (*not* general analysis). If one presents this "advanced" semicontinuity of *functionals*, there are available illustrations which are both simple and profound. For example, in the isosceles triangle ABC (Figure 23.7a) let $AB = 5$ in., $AC = BC = 10$ in. Joining the midpoints as indicated gives $AE = EF = FD = DB = 5$ in. so that the zigzag line $AEFDB$ measures 20 in. Joining midpoints of the triangles AEF and FDB as indicated (Figure 23.7b) will produce 8 segments, each 2.5 in. long, and a 20 in. zigzag line. Joining midpoints in the four triangles formed will again give a zigzag line measuring 20 in., etc. As the process continues, the zigzag line will get closer and closer to the 5 in. base of the triangle (Figure 23.7c). We can see this intuitively or we can use the above definition of Fréchet to fix the distance between the two functions—the zigzag line and the straight line base. The maximum vertical height of the zigzag line will measure this distance, and this will get smaller and smaller. In books on mathematical recreations, this example is often used to produce the fallacy 20 = 5. But from our present point of view, interest lies in the fact that the *length* of the zigzag line is a function of this line, that is, the length is a *functional*. To each zigzag line there corresponds a specific length. If we wish to add some variety to our picture, we can carry out the above process with a taller isosceles triangle, one with legs 40 in. or 400 in. each. To every zigzag line thus obtained, there would then correspond a length of 80 in.

---

*Maurice Fréchet, *L'analyse générale et les ensembles abstraits*, *Revue de métaphysique et de morale*, Vol. 32 (1925).

(800 in.), and we could obtain 80 in. (800 in.) lines as close as we please to the 5 in. base. We can repeat all of this by using shorter or longer legs in the isosceles triangle, for example, legs measuring 2.6 in. with a zigzag line of 5.2 in. The *functional* (the length of any such zigzag line) is *not continuous* at the base of the triangle, since the length of the zigzag line does *not* necessarily approximate closely the length of the base as the position of the line gets close to the position of the base. However, the functional has the property that Baire called *lower semicontinuity*. Put very crudely, this means that all the different zigzag lines (with lengths 5.2, 20, 80, 1000, 1,000,000, 1,000,000,000, etc.) we can possibly draw, as close to the base as we please, have a *lower bound*, namely 5 in., but *no* upper bound. (By analogy, for *upper semicontinuity* there would be an *upper* but *not* a lower bound.)

The definition of continuity of an ordinary real function, $y = f(x)$, at the point (a, b) (Chapter 22) requires that y approximate b within an arbitrarily small standard ε, providing x is in the interval $(a - \delta,\ a + \delta)$, This definition and Figure 22.4 indicate that the inequalities,

$$y > b - \varepsilon \qquad y < b + \varepsilon$$

must both be satisfied when x is in the interval selected. It was Baire who reached a generalization by making the requirement less stringent. If we require that only the first inequality, $y > b - \varepsilon$, be satisfied (when x is in a sufficiently small interval containing a), we have the definition of *lower semicontinuity* at (a, b). If, on the other hand, we insist only on the second inequality, *upper semicontinuity* is defined. In the case of the functional in the above illustration, y is the length of the zigzag line, and $b = 5$. It is evident that $y > 5 - \varepsilon$ for every choice of ε (with corresponding value of δ yielding zigzag lines close to the base of the triangle) but that $y < 5 + \varepsilon$ cannot be satisfied for all such (ε, δ) pairs.

Now we must reveal that the concept of lower semicontinuity enabled the Italian analyst, Leonida Tonelli (1885–1946) to effect a revolution (1921) in the methodology of the calculus of variations. We note, in passing, the marked contribution of the Italian school to functional analysis. After Volterra there was Salvatore Pincherle (1853–1936), who considered the "qualitative" aspects of Volterra's functions of lines. Tonelli's ideas embedded the calculus of variations in the more general functional analysis, and they were carried further by Silvio Cinquini, a leading contemporary Italian analyst. In the United States, the direct methods of Tonelli were extended by Lawrence M. Graves and led to very significant results in the work of Edward J. McShane. For a lucid explanation of the new methods and their superiority in the theoretical aspects of the subject there is no better exposition than that of Cinquini,* or for a complete history of their development, on an advanced level, an account by McShane.†

With Volterra's "functions of lines" and his integral equations, and with the

---

* S. Cinquini, *Il Calcolo delle Variazioni*, Zanichelli, Bologna, 1940.

† E. J. McShane, "Recent Developments in the Calculus of Variations," *Semicentennial Addresses of the American Mathematical Society*, New York, 1938, pp. 69 ff.

general analysis of Fréchet and Moore, we have carried to recent times one of those two main streams that started with Pythagoras, namely the channel of the "infinite processes," that is, the mathematics of the continuum. But, in a sense, we have got ahead of our story, for it was Georg Cantor, as pointed out in one of the first chapters in this book, who created the modern philosophy of the infinite that is the foundation for any type of general analysis. Once more we have arrived at a point where, in accordance with the program explained in the Introduction, we must go round the spiral of analysis again, but at a more advanced level than that of the previous full circle.

24

Infinite Hierarchy

There are mysteries within mystery, gods above gods . . .
That's what is called infinity.

Jean Cocteau

Before discussing those notions which, as Einstein once said, precipitated the fiercest "frog-mouse" battle in mathematical history, let us see what manner of man created them. Georg Cantor (1845–1918) came by his imaginative traits naturally, for his heritage was an artistic one. On his mother's side there had been a long line of painters, pianists, violinists, and conductors. Cantor's father tried to force him to be practical and insisted at first that he train for the profession of engineering, but consented finally to his pursuing his bent for pure mathematics. Just before he was thirty, Cantor published his first revolutionary paper on the theory of the *infinite*. The facts as well as the method were fundamentally new in mathematics and marked the man of genius.

Cantor's personal life, like that of so many gifted men, was unhappy. He earned his living by teaching at second-rate colleges. The most tragic fact of all was the mathematical and personal animosity of his teacher, Leopold Kronecker (1823–1891). The latter disagreed violently with Cantor's point of view, and to this day pitched battles are being fought regularly between Kronecker and Cantor adherents. No decisive victory has been achieved by either side as yet. Poor Cantor lacked the stamina to be a revolutionary; he took things so much to heart that he went out of his mind and spent his last days in a mental hospital.

Whether or not Cantor's notions will stand the test of time, they are presented here for their beauty and because they have probably been one of the greatest stimuli in recent years toward placing mathematics on a firm logical foundation. No discussion of modern mathematics would be complete without what Hilbert called "the most admirable fruit of the mathematical mind, and one of the highest achievements of man's intellectual processes."

The trouble in the household of the infinite actually started some 2400 years before Cantor's day when, as we have seen, the Pythagoreans attempted to measure the diagonal of a unit square in a finite number of steps. During the same early period the paradoxes of the sophist Zeno of Elea added considerable fuel to the flames. Zeno was a follower of the Greek philosopher and mystic, Parmenides. Teacher and pupil alike stressed the doctrine that the world of sense is nothing but an illusion, and therefore Zeno concentrated on the destruction of a single but ubiquitous concept associated with the external world, namely *motion*. "If motion, which pervades everything," thought Zeno, "can be shown to be self-contradictory, and hence unreal, then everything else must assume the same unreal quality. By convincing people of the unreality of motion, I can best teach the doctrine of Parmenides and successfully discredit the world of the senses."

Thus it was that Zeno propounded four famous paradoxes on motion, which philosophers and mathematicians are still discussing. Zeno, in pushing Parmenides' notions, may or may not have been guilty of "sophistry." Nevertheless it takes a pretty clever man to get himself talked about for over two thousand years. Among those who have given serious thought to these puzzling questions are Thomas Aquinas, Descartes, Leibniz, Spinoza, and Bergson. Zeno's paradoxes in some form have been used as arguments for all the theories of space, time, and infinity that have been propounded from his day to ours.

The point at issue in the first two of Zeno's arguments is essentially the same. He presented these puzzles as:

I. The Dichotomy. There is no motion because that which is moved must arrive at the middle (of its course) before it arrives at the end. (And of course it must traverse the half of the half before it reaches the middle, and so on *ad infinitum*.)

In other words, to cover a distance of 1 yard, one must first reach the $\frac{1}{2}$ yard point, before that the $\frac{1}{4}$ yard, before that the $\frac{1}{8}$, and so on, *ad infinitum*. How is it possible to reach an infinite number of positions in a finite time?

II. The Achilles. The slower when running will never be overtaken by the quicker; for that which is pursuing must first reach the point from which that which is fleeing started, so that the slower must necessarily always be some distance ahead.

A modern version of this states that Achilles can run 1000 yd. a minute while a turtle can run 100 yd. a minute. The turtle is placed 1000 yd. ahead of Achilles. Zeno's argument states that Achilles can never overtake the turtle, for when Achilles has advanced 1000 yd,. the turtle is still 100 yd. ahead of him. By the time Achilles has covered these 100 yd., the turtle is still ahead of him, and so on, *ad infinitum*, as the following table shows.

Position	Achilles	Tortoise
1	0	1 0 0 0
2	1 0 0 0	1 1 0 0
3	1 1 0 0	1 1 1 0
4	1 1 1 0	1 1 1 1
5	1 1 1 1	1 1 1 1. 1
6	1 1 1 1. 1	1 1 1 1. 1 1
7	1 1 1 1. 1 1	1 1 1 1. 1 1 1
	etc.	etc.

The trouble here is essentially the same as in the first paradox. If Achilles must occupy an *infinite* number of positions in order to overtake the tortoise, will he ever be able to do this in a *finite* time? Thus the solution of both these puzzles will be a matter of clarifying notions about "infinity."

The first real solution of such problems connected with infinity was propounded by Cantor in 1882. He attacked the core of all the trouble by furnishing a new number concept to replace the usual idea of *counting*. Unconsciously man assumes that, in order to have complete numerical knowledge concerning a collection, we should be able to pass its terms in review, one by one. In the case of the Dichotomy, if we wish to count the positions that must be reached—the halfway mark, the quarter, the eighth, the sixteenth, etc.—we shall never finish. Or, in the case of Achilles and the tortoise, if we imagine an umpire to say "Stop—Go!" when Achilles reaches the 1000 yd. mark (and the turtle the 1100 yd. mark), to repeat this when he reaches the 1100 yd. mark (and the turtle the 1110 yd. mark), to repeat it again at the third stage, the fourth stage, and so on, then Zeno's contention would be true in practice and Achilles would never overtake the tortoise.

But, as we have seen in previous chapters, it is actually possible or even necessary to reason about a collection without going through such an enumeration as that described. This may be illustrated even in the case of finite collections. We can speak of "Americans" without having a personal acquaintance with each citizen of the United States. We can do this because we know the essential characteristic that each individual has if he belongs to the group, and that he lacks, if he does not. So it is with infinite aggregates. They may be known by their characteristics even though we cannot complete the task of counting them, one by one.

To appreciate the substitute Cantor offered for ordinary counting, let us review the process of matching, or one-to-one correspondence, which was used even by primitive man. Cantor was the first to formalize that notion as a criterion for determining the cardinal number of a set. To illustrate a matching, Cantorian style, let us pair each natural number with its double, as follows:

1	2	3	4	5	6	7	8	$\cdots$	n	$\cdots$
2	4	6	8	10	12	14	16	$\cdots$	$2n$	$\cdots$

Then, since we have agreed that two sets of things are equal in number when they can be paired in one-to-one fashion, there are as many numbers in our first set as in the second; that is, the cardinal number of the entire set of natural numbers is the same as the cardinal number of the *proper subset* of *even* natural numbers. Evidently we have illustrated a collection that is *numerically equal to a part of itself!*

Again, let us take a part $A'B'$ of line segment AB, and compare the number of points in the two. We move $A'B'$ to the position indicated, draw AA' and BB' meeting in P (Figure 24.1). Then we pair the points of $A'B'$ with those of AB as follows.

To find the partner of any point on $A'B'$, such as C', draw PC' and extend it to meet AB in C. Then C and C' are partners. On the other hand, to find the partner of any point on AB, as D, draw PD and let it meet $A'B'$ in D'. Then D and D' are partners. For every point on $A'B'$ we have a point on AB, and for every point on

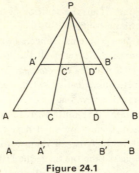

Figure 24.1

AB we have one on *A′B′*. In other words, there is a *one-to-one correspondence between the points of AB and those of A′B′*; that is, there are as many points in part of *AB* as in all of *AB*. Once again, *the whole of something is numerically equal to one of its parts.*

The Cantorian point of view was actually anticipated by Galileo, who showed that there is a one-to-one correspondence between the integers and their squares, although the latter set of numbers is only part of the first. The correspondence is

1	2	3	4	5	$\cdots$	n	$\cdots$
1	4	9	16	25	$\cdots$	n^2	$\cdots$

Galileo merely saw in this discovery a puzzling fact and did not develop the matter further. Infinity did not appear in mathematics as a mature concept until it was discovered by Bernhard Bolzano (1781–1848), who paved the way for Dedekind and Cantor.

A set of things, whether they are numbers or curves, points or instants of time, speeds or temperatures, is said to be *infinite* if it can be put into *one-to-one correspondence with one of its parts*. In Zeno's paradox, the Dichotomy, we see that the argument hinges on the fact that a moving point will occupy an infinite number of positions. The positions to which allusion is made in the paradox are

$$1/2, \ 1/4, \ 1/8, \ 1/16 \cdots$$

In the fact that these can be put into one-to-one correspondence with the infinite set of positive integers lies the proof that the number of positions to be occupied is infinite. Thus we have the correspondence

1/2	1/4	1/8	1/16	$\cdots$	$1/2^n$	$\cdots$
1	2	3	4	$\cdots$	n	$\cdots$

Then if one-to-one correspondence is a criterion of equality, there is nothing remarkable about having a finite length contain an infinite number of positions. Again, in the Achilles paradox, we have an infinite set of positions. The tortoise's positions can be put into one-to-one correspondence with Achilles', as we see in the table on page 578, and each can be put into one-to-one correspondence with the positive integers. We have

Achilles	0	1000	1100	1110	1111	1111.1	1111.11	etc.
Tortoise	1000	1100	1110	1111	1111.1	1111.11	1111.111	etc.
Integers	1	2	3	4	5	6	7	etc.

Therefore both the turtle and Achilles occupy an infinite number of positions. The fact that puzzled Cantor's predecessors is this: If Achilles were to catch up with the tortoise, the places where the tortoise had been would, on the one hand, be only part of the places where Achilles had been, and on the other, would be equal in number to Achilles' positions (on account of one-to-one correspondence). Thus, by stating as an axiom that the whole of something *can* equal a part, Cantor was able to deny the argument of this paradox.

We have just proved that the natural numbers form an infinite set. They are said to constitute a *countable* or *denumerable infinity*, and any set that can be put into one-to-one correspondence with them is also countable or denumerable. Since the even natural numbers (or the odd) can be put into one-to-one correspondence with all the natural numbers, there is a countable infinity of even natural numbers and a countable infinity of odd natural numbers. Likewise, since the positions

$$1/2, \ 1/4, \ 1/8, \ 1/16, \ \text{etc.}$$

of Zeno's Dichotomy can be put into one-to-one correspondence with the integers, they constitute another illustration of a countable set.

Just as the mathematician has numerals, 1, 2, 3, etc., that name the finite numbers, he has symbols for the *transfinite* ones as well. He represents a countable infinity of objects by the first letter of the Hebrew alphabet $\aleph_0$ (read *aleph null*). The null or zero subscript was used because Cantor hoped to be able to find other examples of infinite sets and then be able to arrange them in order of magnitude: $\aleph_0, \ \aleph_1, \ \aleph_2$, etc.

Is the set of all positive rational numbers countable? If we consider the vast number of rational fractions *between* 0 and 1, those between 1 and 2, those between 2 and 3, etc. we begin to suspect that we are on the verge of something new. Consider the fractions between 4 and 5 and select any two, such as 4 1/4 and 4 3/4. Note that between these two there are others, such as 4 3/8 and 4 5/8. Between the last two are 4 7/16 and 4 9/16 and so on, *forever*. Between each pair of whole numbers we have an amazing assemblage of fractions! Surely, since there is an infinity of whole numbers and between each pair there appears to be an infinity of rational fractions—in other words, an *infinity of infinities*—surely this aggregate should tower over the *countable infinity*! But such is not the case. The positive rational fractions are merely a one-to-one match for the positive natural numbers, and therefore merely constitute a countable infinity! There are only $\aleph_0$ positive rational fractions.

The argument follows. Take a common fraction at random, say 3/5. Notice that the sum of numerator and denominator is 8. Notice, too, that there are many fractions having this same sum of numerator and denominator. Arranged in the order of increasing numerators, they are

$$1/7, \ 2/6, \ 3/5, \ 4/4, \ 5/3, \ 6/2, \ 7/1$$

Any other sum will similarly give rise to a set of fractions. Thus one can conceive of a tabulation of the various finite sets corresponding to different sums:

Sum	Rational Numbers
2	1/1
3	1/2, 2/1
4	1/3, 2/2, 3/1
5	1/4, 2/3, 3/2, 4/1
6	1/5, 2/4, 3/3, 4/2, 5/1
7	1/6, 2/5, 3/4, 4/3, 5/2, 6/1
etc.	

By this procedure we are bound to arrive sooner or later at any positive rational number the reader may designate. If he mentions 99/100, we shall reach that with the sum 199. If he mentions 157/211, we shall arrive at that with the sum 368. Whole numbers will appear among the fractions, such as 4/2 or 4/1, for example. Also the same integer may be repeated, as in 2/1 and 4/2. The same fraction may appear repeatedly as 1/2, 2/4, 3/6, etc. We shall agree, however, to take each but once in the matching process that follows, in which each fraction is matched with an integer and each integer with a fraction.

1/1	1/2	2/1	1/3	3/1	1/4	2/3	3/2	4/1	1/5	$\cdots$
1	2	3	4	5	6	7	8	9	10	$\cdots$

In this way a one-to-one correspondence is set up, and we see that the set of positive rational fractions can be matched with the set of natural numbers. Thus the set of positive rational fractions, even though an infinity of infinities, is after all merely *countable* or *denumerable*.

Let us now carry out some transfinite arithmetic and see whether in the process we can obtain further illustrations of countable sets. If we were to add one element to a countable set, what would the effect be? For example, let us add the element 0 to the countable set of all even natural numbers, to form the set of even whole numbers (non-negative even integers). Then we can set up the one-to-one correspondence

0	2	4	6	8	$\cdots$	$2n-2$	$\cdots$
1	2	3	4	5	$\cdots$	n	$\cdots$

The bottom row contains $\aleph_0$ (aleph null) elements. Then, since there are $\aleph_0$ even natural numbers, as a result of the matching on page 579, the top row contains $1 + \aleph_0$ elements. Therefore

$$1 + \aleph_0 = \aleph_0$$

If we were to add two, three, four, or any finite number of elements to the aggregate of even integers, we could once more match the enlarged set with the natural number aggregate. For example, the latter set can be matched with the collection, $\{-6, -4, -2, 0, 2, 4, 6, \ldots\}$. Therefore, for any natural number k,

$$k + \aleph_0 = \aleph_0$$

An analogous equation can never arise in dealing with *finite* cardinals. In fact, $k + a \neq a$ if k and a are finite cardinals greater than zero.

Again, since the odd and even subsets of the natural numbers are both countable, there are $\aleph_0$ odd natural numbers as well as $\aleph_0$ even natural numbers. The union of those two sets is the aggregate of *all* natural numbers. Hence

$$\aleph_0 + \aleph_0 = \aleph_0$$

or

$$2\aleph_0 = \aleph_0$$

Again, for a *finite*, nonzero cardinal a, $a + a \neq a$ and $2a \neq a$.

By uniting the sets in the three rows below, one obtains the set of natural numbers:

$$1, 4, 7, 10, \ldots, 3n - 2, \ldots$$
$$2, 5, 8, 11, \ldots, 3n - 1, \ldots$$
$$3, 6, 9, 12, \ldots, \quad 3n \quad , \ldots$$

Therefore

$$\aleph_0 + \aleph_0 + \aleph_0 = \aleph_0$$

or

$$3\aleph_0 = \aleph_0$$

and similarly, if m is any positive integer, $m\,\aleph_0 = \aleph_0$

If we examine the tabulation of positive rational numbers on page 582, we observe that the first column contains all the rationals with numerator 1. These rationals can be matched with the set of natural numbers thus:

$$\frac{1}{1}, \frac{1}{2}, \frac{1}{3}, \frac{1}{4}, \ldots, \frac{1}{n}, \ldots$$

$$1, \quad 2, \quad 3, \quad 4, \ldots, n, \ldots$$

Therefore there are $\aleph_0$ rationals in the first column. In the second column there are all the rationals with numerator 2, and they can also be matched with the set of natural numbers so that there are $\aleph_0$ of these, etc. for *each* column. Hence

$$\aleph_0 + \aleph_0 + \aleph_0 + \cdots = \aleph_0$$

There are as many columns as there are different numerators. The possible numerators are

$$1, 2, 3, 4, \ldots, n, \ldots$$

or there are $\aleph_0$ numerators and hence $\aleph_0$ columns. Therefore the above equation says that a countable infinity of countable infinities is still countable, or

$$\aleph_0 \times \aleph_0 = \aleph_0$$

that is,

$$(\aleph_0)^2 = \aleph_0$$

Multiplying both sides of this equation by $\aleph_0$ yields

$$(\aleph_0)^3 = (\aleph_0)^2 = \aleph_0$$

Continuing in the same way,

$$(\aleph_0)^m = \aleph_0$$

where m is any positive integer.

Are there *noncountable* sets? We shall now show that the set of all real numbers between 0 and 1 is such a set. This set includes all the rational fractions between 0 and 1 as well as all the irrationals within the same boundaries. In Chapter 2 *decimal* expansions were used for these real numbers. It was shown that a rational fraction may have a terminating expansion like $1/4 = 0.25$ or a *repeating* pattern like $3/11 = 0.272727 \ldots$ but that for irrational numbers the expansion is always an infinite, nonrepeating one. Now in the case of terminating fractions like $1/4 = 0.25$, we shall always use the nonterminating decimal $0.25000 \ldots$ and avoid the alternative expansion $0.24999 \ldots$.

But this is all preliminary to proving that the set of real numbers between 0 and 1 is *not* countable. Suppose that someone claims that he has found a scheme for putting this set into one-to-one correspondence with the aggregate of natural numbers, and suppose further that the first few real numbers matched with natural numbers are as follows:

1	0 . 4 1 5 2 3 7 . . .
2	0 . 2 8 9 9 9 9 . . .
3	0 . 7 3 2 0 6 5 . . .
4	0 . 1 8 3 7 3 7 . . .
5	0 . 3 5 8 0 9 1 . . .

. . . .

. . . .

No matter what sort of plan anyone uses, there will always be at least one real number missing from his list. It is possible to construct such a decimal fraction as follows: For the tenths place, select a digit different from the tenths place of the first number in the list and different from 0 and 9*; for the hundredths place, select a digit different from 9 and different from the hundredths place of the second number on the list; for the thousandths place, select a digit different from 9 and from the thousandths place of the third number on the list; etc. Thus the first digits selected might initiate a decimal like 0.15326 . . . . When (conceptually) the so-called *diagonal process* (see diagonal lines in above list) is continued on and on, a real number is obtained that differs from each number in the list in at least one decimal digit (the one between the diagonal lines) and hence is not equal to any number on that list. Thus we have seen how to construct a real number between 0 and 1 which cannot be in any list like the one suggested above. Therefore the cardinal number of the set of reals between 0 and 1 must be greater than $\aleph_0$. The symbol for this new transfinite number is C, chosen because the interval in question is a *continuum* (page 38).

We shall see that there are transfinite cardinals greater than C, in fact, infinitely many of them. It might be natural, then, to ask whether there are any transfinite cardinal numbers between $\aleph_0$ and C. Only recently have mathematicians been able to provide a satisfactory answer to this question. Cantor was unable to discover any set of the desired intermediate cardinality, and hence he guessed that there was no such aggregate. His conjecture came to be known as the *continuun hypothesis*, namely, that there is no cardinal number greater than $\aleph_0$ and smaller than C.

After Cantor, different mathematicians "purified" set theory, that is, based it on an abstract postulate system. One such axiomatization, for example, is due to Ernst Zermelo (1871–1956) and Abraham Fraenkel. Then, in 1938, using the Zermelo-Fraenkel system, Kurt Gödel demonstrated that one can safely *assume* Cantor's continuum hypothesis as an additional postulate in set theory. In other words, he proved that the continuum hypothesis is consistent with the Zermelo-Fraenkel axioms. Gödel had broken substantial ground, but he had not said the last word, since he had neither proved the continuum hypothesis nor shown that it is undemonstrable, in other words, *independent* of the other set theory postulates. That the latter fact holds was definitively established by Paul J. Cohen in 1963 in the same fashion that had been used to demonstrate that the parallel postulate is independent of the other axioms of Euclidean geometry (Chapter 3). Thus Cohen exhibited a model, a very subtle one, of a *non*-Cantorian set theory in which the continuum hypothesis is negated (while the other postulates of set theory hold true). Later we shall see how Cohen's non-Cantorian theory handles other logically troublesome issues of Cantor's theory. At this point we can summarize the situation by saying that the continuum hypothesis can be assumed or denied, depending on the applications one has in mind.

---

* The reason we avoid 9 in each case is that we wish to produce a real number *not* in our list. If 9's are permitted, our selection of digits might merely lead to an alternative representation of some rational number actually in our list. We exclude 0 as a selection for the first place lest the selection of digits lead to 0.000 . . . = 0, which is a boundary of the interval (0, 1) but not a number within that interval.

But let us return to the development of *Cantorian set theory*. The transfinite number C is the cardinal number corresponding to the interval $I = (0, 1)$. Using that interval as our universal set, let us now consider some of its subsets, for example, R, the set of rational numbers between 0 and 1, and R', the complement of the rational set, that is, the set of irrationals between 0 and 1. Then, using the symbolism of Chapter 6,

$$I = R \oplus R' = R \cup R'$$

Now select any (irrational) number in R', for example $\sqrt{5}$, and form the countable set,

$$E = \left\{ \sqrt{5}, \frac{1}{2}\sqrt{5}, \frac{1}{3}\sqrt{5}, \ldots, \frac{1}{n}\sqrt{5}, \ldots \right\}$$

Let $\bar{E}$ represent the complement of E in R', that is, the residual set of irrationals between 0 and 1 after the numbers in E are removed. Now the set $R \cup E$ is countable because it is the union of two countable sets, the rationals and the set E described above. Let $F = R \cup E$. Then $I = F \cup \bar{E}$. Also we know that $R' = E \cup \bar{E}$. Now we see that I and R' can be matched because their component subsets can be matched, $\bar{E}$ as a subset of I with $\bar{E}$ as a subset of R', the countable subset F with the countable subset E. Hence I and R' have the *same cardinal number* C. In other words, "whole equals part" once more. There are C irrationals in (0, 1) even though the irrationals form a proper subset of all the C real numbers in (0, 1).

The statement $R \cup R' = R' \cup R = I$ now has the corollary

$$\aleph_0 + C = C + \aleph_0 = C$$

Put otherwise, $C = C - \aleph_0$, which means that removing a countable infinity from a continuum does *not* diminish the latter's magnitude. Consider

$$C - \aleph_0 - \aleph_0$$

The number C can be substituted for the difference represented by the first two terms, and thus we arrive at the result

$$C - \aleph_0$$

which, as we have shown, is equal to C. By the same sort of reasoning we can see that

$$C - \aleph_0 - \aleph_0 - \aleph_0 - \cdots = C$$

What this means is that if from *any* aggregate whose cardinal number is C (the set of reals or the set of irrationals in the interval (0, 1), for example) we *remove* a countable infinitude of elements and then another countable infinity and then another—forever, thus taking away a countable infinity of countable infinitudes—the power of the original set is undiminished and its cardinal is still that of the continuum. When a Pythagorean or a school student in any age first encountered $\sqrt{2}$, $\sqrt[3]{5}$, or some other irrational number, the experience seemed strange. But now we see that, from a statistical point of view, an irrational should be considered the *usual* type of number and a rational as exceptional, since there are incredibly more incommensurables than commensurables in the world of real numbers.

Since we speak of a "universe" of real numbers, we had better extend our story to show that irrationals will predominate in a larger interval of real numbers as well as in the entire set of real numbers. As a minor preliminary, we remark that if instead of taking reals *between* 0 and 1, we include the number 0 or the number 1 or both, the cardinal of the set is still C, that is, $1 + C = C + 1 = 2 + C = C + 2 = C$. In fact, it can be shown that $m + C = C + m = C$ where m is any natural number.

By pairing every real number between 0 and 1 with its double, and every number in $(0, 2)$ with its half, we find that C is the cardinal number for all reals in the interval $(0, 2)$; by pairing the reals of $(0, 1)$ with their trebles and those of $(0, 3)$ with their thirds, C is shown to be the cardinal for reals in the interval $(0, 3)$, etc. so that C is the cardinal for reals in $(0, n)$ where n is any positive integer. Put otherwise,

$$C + C + C + C + \cdots (n \text{ times}) = C$$

or

$$n\,C = C$$

By pairing 0.3 with 1.3, 0.72 with 1.72, $\sqrt{0.5}$ with $1 + \sqrt{0.5}$, x with $1 + x$, where x is any real between 0 and 1, we establish a one-to-one correspondence between the reals of $(0, 1)$ and those of $(1, 2)$. By pairing x with $\frac{1}{2}x - \frac{1}{3}$, we establish a one-to-one correspondence between the reals of $(0, 1)$ and those of $(-\frac{1}{3}, \frac{1}{6})$. In general, we can match the reals of $(0, 1)$ with the reals in *any* finite interval. Therefore C is the cardinal for any such interval. We shall use the symbol $[0, 1]$ to signify reals between 0 and 1 inclusive, and the symbol $[0, 1)$ if 0 is included but not 1, and $(0, 1]$ if 1 is included but not 0. Then

$$[0, 1] \cup (1, 2] \cup (2, 3] \cup \ldots$$

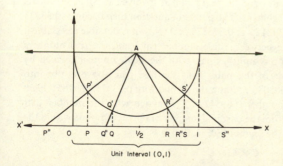

Unit Interval (0,1)

Figure 24.2   One-to-one correspondence between the point set $(0, 1)$ and the real line. A typical point P of the interval $(0, 1)$ is matched with a point P' of the semicircle, and the latter point is then matched with P'' on the real line (a procedure which can be carried out in reverse).

will signify *all* non-negative reals and

$$(0, -1] \cup (-1, -2] \cup (-2, -3] \cup \cdots$$

all negative reals. In both these cases the corresponding cardinal number will be

$$C + C + C + C + \cdots \text{ (forever)} = \aleph_0 \times C$$

and hence the cardinal for *all* reals is $2\,\aleph_0 \times C$, which reduces to $\aleph_0 \times C$. But the diagram of Figure 24.2 indicates that the cardinal number of the set of *all* reals is C, because this set is in one-to-one correspondence with the reals between 0 and 1. Therefore

$$\aleph_0 \times C = C$$

Now let us show that

$$C \times C = C^2 = C$$

If we consider the unit square of Figure 24.3 there is a unit segment correspond-

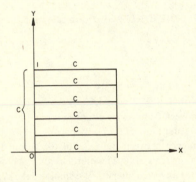

Figure 24.3

ing to each point between 0 and 1 on the *Y*-axis. Each segment has C points and there are C segments, therefore $C \times C = C^2$ points. The Cartesian coordinates of any point within the square will be a *pair* of real numbers between 0 and 1, such as (1/3, 1/2) or ($\pi/4$, 3/8). We shall write the real numbers involved as *infinite* decimals just as we did in a previous argument. The points in question then become (0.33333 . . . , 0.50000 . . .) and (0.7853981 . . . , 0.375375 . . .). Now it is possible to establish a one-to-one correspondence between these number pairs and the real numbers between 0 and 1. We match any number pair with the *single* number formed by alternating the decimal digits of the pair. Thus the first pair above is put into correspondence with 0.3530303030 . . . and the second with 0.7387553397851 . . . . Conversely, the single real number 0.695427 . . . is mated with the number pair (0.652 . . . , 0.947 . . .). Therefore the cardinal number of the set of points within the square is the same as that of the segment (0, 1), that is

$$C \times C = C^2 = C$$

A similar argument will show that the number of points within a unit cube is C, or

$$C \times C \times C = C^3 = C$$

and, in general, $C^n = C$, where n is any positive integer.

The fact that there are no more points in a square or cube than in a line segment is rather a shock to our intuitive notion of what facts ought to be. Let us point out then that the sort of matching used to show that two sets of things have the same *cardinal* number is not at all concerned with the *order* of the elements in the sets. Points that are close together on the segment (0, 1) may be mated with points far apart in the unit square. When, in Chapter 2, it was stated that (0, 1) constitutes a *linear continuum*, this fact depended in an essential way on the arrangement of the numbers in the *order* of increasing magnitude.

Having examined addition and multiplication combinations involving $\aleph_0$ and C, we turn to the issue of still greater "infinities," cardinals larger than C. There are infinitely many of these larger cardinals, as a consequence of the fact that given any set, finite or infinite, a set with a higher cardinal number can always be obtained by taking all possible subsets of the given set. For example, the set of elements $\{\alpha, \beta, \gamma\}$ has the cardinal number 3. By proceeding in the fashion of Chapter 13, it can be shown that there are 8 subsets (proper or improper) of the given set. Again, a set with cardinal number 4 has 16 subsets. In general, if the cardinal number of a finite set is n, there will be 2^n subsets. One can prove this by applying the fundamental principle of permutations (Chapter 13) and reasoning that in forming a subset there are two possibilities for each member of the given set, namely, to include the member in a subset or to exclude it. Therefore there will be $2 \cdot 2 \cdot 2 \cdots$ (n times) possibilities in all. There is also a special theorem which proves that $2^n > n$ for every cardinal number n, *finite* or *transfinite*.

As a consequence of this theorem, $2^{\aleph_0} > \aleph_0$, and $2^C > C$. There is a *special theorem* establishing that $2^{\aleph_0} = C$. Then by the first of the two inequalities just listed, $C > \aleph_0$, and we have three distinct transfinite cardinals in increasing order of size, namely, $\aleph_0$, C, and 2^C. Let us use S to symbolize 2^C. Then by the special inequality theorem to which we have alluded, $2^S > S$, and there is a fourth and greater transfinite cardinal. If this fourth cardinal is used as an exponent for 2, a fifth transfinite cardinal results, and so on *ad infinitum*, producing an infinite sequence of transfinite cardinals arranged in increasing order of magnitude.

Our many illustrations of the arithmetic of cardinals, finite and transfinite, might be summarized by saying that their addition and multiplication satisfy closure, commutative, associative, and distributive laws. But the "peculiarities" of the transfinite cardinals make it impossible to carry out "normal" subtractions and divisions with them. These results occur because, in reasoning about cardinal numbers, one-to-one matching is the sole criterion. But another very important factor, that of *order*, is essential to a handling of questions related to Cantor's theory and to other mathematical subjects as well. In the case of infinite sets, we have just seen how vital order can be in matters of geometric dimension, where a mere matching of elements without consideration of their order can make a line segment the equivalent of a square or a cube or an n-dimensional cube with $n > 3$. This destruction of dimensionality was accomplished by changing the natural order of the real numbers in the $(0, 1)$ interval.

It must be noted that one cannot literally arrange or rearrange the huge infinitude of numbers of a continuum like $(0, 1)$; it is rather that one conceives of some particular arrangement or order, like that of increasing magnitude, or that of the rationals on page 582. One focuses on a *binary relation* containing pairs of elements in a particular class, a relation like "less than" or "earlier than," and then this relation gives rise to a certain "arrangement." But one might select the relation "greater than" instead of "less than" and thus generate a different "arrangement" or ordering of the same set. In fact, as some examples cited earlier have indicated, changing the ordering of an infinite set may sometimes produce startling results. The point to be emphasized is that a definition of order will not be found in the nature of the class to be "arranged," since the ordering need not be unique. Instead, the issue is one of defining some sort of *relation* that provides the protocol by which certain elements can be conceived as lower or earlier, and others as later or higher.

Particular types of relation were discussed earlier (Chapter 7). But now we must define a kind of relation suitable to our next objective, which is an exposition of Cantor's theory of transfinite ordinal numbers. Therefore we define a *full preference relation*, sometimes called a *simple ordering* or a *linear ordering*. Given a particular aggregate, a *full preference relation* is one with the three attributes listed below. By describing its effect on the elements of the given class, these characteristics tell how the order rule works.

(1) *Asymmetry*: If x and y are elements of the given class such that x precedes y, then y cannot precede x.

(2) *Transitivity*: If x, y, z are three distinct elements of the class such that x precedes y and y precedes z, then x precedes z.

(3) *Connectedness*: (*Comparability* of *all* pairs of distinct elements in the given set, thus making the rule "full.") If (x, y) is *any* pair of distinct elements of the class, then x must precede y, or y must precede x.

It is obvious that these properties are satisfied by the relation $<$ (less than) within various sets of real numbers, and hence $<$ is a full preference relation for such sets. But one can adduce specific instances where one or more of the required characteristics is lacking. For example, property (1) is lacking if the relation "is brother or sister of" is applied to the children of a family. If x is the brother or sister of y, then y is the brother or sister of x. The relation is *symmetric*, not asymmetric. There is a failure in transitivity when the relation "is the father of" is applied to an aggregate of people. If x is the father of y, and y is the father of z, then it is not true that x is the father of z. Hence the particular relations cannot establish simple orderings of the set of people.

An ordering of a set by a given relation may fail to be "full," that is, not all pairs of distinct elements may be comparable. Thus, connectedness is lacking in the following example: The members of a certain organization are to be seated in "order" at a banquet table. The rule of protocol states that members of Committee I are to be seated ahead of those of Committee II, and these are to be seated ahead of members of Committee III, etc. Now if x and y are members of the same committee, the rule will *not* tell whether x is to precede y or y is to precede x. Hence the rule does *not* establish a *full* preference relation for seating.

Cantor required that order be added to the structure of infinite sets prior to assigning an *ordinal* number to these aggregates. In defining the ordinal number of such an ordered class, he first considered its cardinal number. Thus the finite cardinal number "five" or the symbol 5 is not limited to particular desciptions like 5 pictures or 5 neutrons, etc. but is a way of characterizing *all* possible sets of a certain type, namely those with "fiveness" as a common attribute. The elements of any one of these sets are in *one-to-one correspondence* with those in another. Now Cantor required a particular *ordinal number* to characterize all *ordered* aggregates of a certain type, namely, those having the *same cardinal* number and "arranged" in *similar fashion*. Another way of stating this is to say that two ordered sets have the same ordinal number if they are *isomorphic* with respect to order. This means that in matching the sets, *order is preserved.* Thus order is preserved in the one-to-one matching if when element x precedes element y in the first ordered aggregate, then x', in the second ordered set, the mate of x, precedes y', the mate of y. Thus the ordered sets $\{A, B, C\}$, $\{C, A, B\}$, $\{A, C, B\}$, etc. are isomorphic with respect to order. To prove this, consider any pair of such sets, for example, $\{A, B, C\}$ and $\{C, A, B\}$. We note that, in the second ordered set, C, A, B are mates of A, B, C, respectively, in the first ordered set. In the first sequence we can make the three statements: A precedes B, B precedes C, A precedes C. The test of preservation of order requires that we substitute the mates of A, B, and C, respectively, in these statements and find that the results are true in the second sequence. Doing so, we have C precedes A, A precedes B, C precedes B. This is a correct description of the second arrangement.

In the same way, one can establish order isomorphism between *any* pair of the six different ordered sets formed by permuting the elements of $\{A, B, C\}$. Hence those six sets all have the same order type, so that the *ordinal* 3 corresponds to any one of the six arrangements. In the same way the $n!$ arrangements of n elements are all of the same Cantor ordinal type, namely, n. Then if we consider the non-negative integers or whole numbers arranged in order of increasing magnitude, $\{0, 1, 2, 3, 4, 5, \ldots, n, \ldots\}$. each number after the first is the *ordinal* number for the set of numbers preceding it. Thus 5 is the order type of $\{0, 1, 2, 3, 4\}$ and n is the ordinal for $\{0, 1, 2, \ldots, n-1\}$. In other words, the set of natural numbers $\{1, 2, 3, \ldots, n, \ldots\}$ can be considered to represent *either* all nonzero finite cardinals or all finite ordinals. When arranged in order of increasing magnitude, this most basic set can be described not only as ordered, but also as *well-ordered*, by which is meant that *every* nonempty subset has a first element. Examples of well-ordered sets are $\{\alpha, \beta, \gamma, \delta\}$, $\{6, 5, 4, 3, 2\}$, or any *finite* order.

But there are ordered sets that are not well-ordered. For example, the rational numbers in order of size is not a well-ordered set because it contains subsets like the set of rationals greater than 4. That subset (arranged in order of size) has no first number, since thare is no next rational that comes immediately after 4. Again,

$$\{\ldots, -4, -3, -2, -1, 0, 1, 2, 3, 4, \ldots\}$$

is ordered but not well-ordered, for such subsets as the integers less than 2, or integers less than 3, or integers less than -4, etc. have no first elements.

An *ordinal number* is often defined as the order type of a well-ordered set. Well-ordered sets have an ordinal number and other special properties that make them very important in mathematical reasoning. Among these is the fact that they have a definite beginning and a next term after any term except the last (if there is a last). One of the chief uses of well-ordered sets is in the method of *mathematical induction*, which readers may have studied at school. That method makes use of the well-ordering of the natural numbers. Cantor planned to generalize the technique to a process of *transfinite induction*.

In creating *transfinite* ordinal numbers, Cantor sought to enlarge the set of finite ordinals so that the extended set would be well-ordered. Since the set of ordinals *after* all the finite ordinals was to be nonempty, the well-ordering property required that there must be a *first* ordinal immediately after the entire sequence of finite ordinals (natural numbers). The successor to the natural number (finite ordinal) sequence is a *transfinite* ordinal describing the order type of that sequence. Cantor symbolized that first transfinite ordinal by ω. Then if the enlarged set of ordinals is to be well-ordered and if there are to be ordinals after ω, there must be a *first* ordinal immediately after ω. Cantor labeled this new transfinite ordinal $\omega + 1$. It is the order type of $\{1, 2, 3, \ldots; \omega\}$. After $\omega + 1$, there must be the ordinals $\omega + 2, \omega + 3$, etc. Then after the entire sequence

$$\{\omega, \omega + 1, \omega + 2, \ldots, \omega + n, \ldots\}$$

there is a first or next ordinal which Cantor labeled 2ω. It is followed by $2\omega + 1$, $2\omega + 2, \ldots$. After the entire sequence thus formed there are the further transfinite ordinals 3ω, $3\omega + 1$, $3\omega + 2, \ldots$. Thus Cantor conceived of his ordinals as the following well-ordered set, where each ordinal gives the type of the ordered set preceding it:

$$
\begin{array}{llll}
1, & 2, \ldots, & n, \ldots; \\
\omega, \ \omega + 1, \ \omega + 2, \ldots, \ \omega + n, \ldots; \\
2\omega, 2\omega + 1, 2\omega + 2, \ldots, 2\omega + n, \ldots; \\
3\omega, 3\omega + 1, 3\omega + 2, \ldots, 3\omega + n, \ldots;
\end{array}
$$

$$
\begin{array}{llll}
\omega^2 & , \omega^2 + 1 & , \omega^2 + 2 & , \ldots, \omega^2 + n & , \ldots; \\
\omega^2 + \omega, \ \omega^2 + \omega + 1, \ \omega^2 + \omega + 2, \ldots, \omega^2 + \omega + n, \ldots; \\
\omega^2 + 2\omega, \ \omega^2 + 2\omega + 1, \ \omega^2 + 2\omega + 2, \ldots, \omega^2 + 2\omega + n, \ldots;
\end{array}
$$

As we have said, Cantor's ordinal type ω describes the order type of the natural numbers arranged in order of magnitude, that is, the order type of each of the following *sequences* (Chapter 22) and many others as well:

$$1, 2, 3, 4, \ldots, n, \ldots$$
$$0, 1, 2, 3, \ldots, n - 1, \ldots$$
$$2, 1, 3, 4, \ldots, n, \ldots$$
$$3, 2, 1, 4, \ldots, n, \ldots$$

Observe that a countable set need *not* be a sequence, that is, an ordered set with cardinal $\aleph_0$ need not be of order type ω, for if we consider the first two rows of our lengthy tabulation of ordinals, we have a single ordered set containing $\aleph_0 + \aleph_0 = \aleph_0$ elements, and yet its order type is 2ω, the ordinal which follows the first two rows. We can also consider the first three rows, with a total of $\aleph_0$ elements and order type 3ω, etc. Thus a countable set has a single cardinal, $\aleph_0$, but an infinite number of ordinals, depending on the way its elements are arranged.

Not all sums of finite and transfinite ordinals are to be found in our lengthy list of such ordinals. Thus, whereas $\omega + 1$ is listed, $1 + \omega$ does not appear. The former is the order type of the set formed by placing a single symbol ω *after* all the natural numbers, and $1 + \omega$ is now defined to be the order type of the sequence formed by placing a single element *ahead* of any sequence. Thus one can place the symbol 0 ahead of the sequence of natural numbers to obtain the sequence of whole numbers

$$\{0, 1, 2, 3, \ldots, n - 1, \ldots\}$$

But that is just one of the sequences already cited as having order type ω. This signifies that $1 + \omega$ is *not* a new ordinal, since it is identical with ω. As a further consequence, $1 + \omega \neq \omega + 1$. Our example illustrates what is true in general: Addition involving sums of finite and *transfinite* ordinals is *not commutative*.

We have seen that a finite set has only one order type but that an infinite set may have many orders. Thus since well-ordering is desirable for some important purposes, perhaps when a set is *not* well-ordered, the desired property can be achieved by rearrangement of terms. The ordered set

$$\{\ldots, -5, -4, -3, -2, -1, 0, 1, 2\}$$

(reading from left to right) is not well-ordered, but if we reverse it to form

$$\{2, 1, 0, -1, -2, -3, -4, -5, \ldots\}$$

this result (reading from left to right) is well-ordered. The positive rationals arranged in order of size do not form a well-ordered set, but the rearrangement on page 582 is one way of accomplishing well-ordering. Since any *countable* set can be put into one-to-one correspondence with the natural number sequence and that sequence is well-ordered, every countable set, if not already well-ordered, can be arranged in such a way as to be well-ordered. Finally, there is a famous theorem generalizing this fact to all infinite sets. The proof would carry us beyond the scope of this work, but the statement is: *Every set can be well-ordered.*

Cantor's theory of the infinite was his answer to Zeno's paradoxes and other antinomies, "monsters," scandals, that revealed the need of a good foundation for calculus and analysis in general. But his *Mengenlehre* (theory of sets) generated its own paradoxes and that was why Zermelo and Fraenkel, as well as P. Bernays, von Neumann, and others, saw the need to place the subject on a firmer foundation. Since we have just been talking about transfinite ordinals, let us examine a contradiction which plagued Cantor in 1895–1896 and which Burali-Forti (1861–1931) revealed in 1897. Cantor believed the infinite aggregate of *all* ordinals (ordered according to increasing magnitude) to be well-ordered. Then, using his own principle of formation, he realized it would be possible to create a new ordinal, the next *after* those in the set under consideration and symbolizing the type of order in that class. Herein lies the contradiction: If we are considering the class of *all* ordinals, how can there be a new ordinal outside this class, *not* part of the "all"?

To take an antinomy arising out of the sort of reasoning we used with cardinals, we recall from previous chapters that a given set may be a class of classes. In other words, the elements of a set may themselves be sets. As such a set Bertrand Russell offered for consideration the class of *all* those classes that are *not* members of themselves, and posed this question: Is the set of sets so described a member of itself? If the answer is negative, then it is a class that does *not* contain itself, hence must belong to the class of classes in question, thus does contain itself, and we have a contradiction. If the answer is affirmative, the class is a member of the "all" and is therefore a set that does *not* contain itself, which contradicts the fact that we have just affirmed that it is a member of itself.

Since the above argument may be difficult to follow, we offer another version due to Russell. In a certain village the village barber shaves *all* those men and only those who do not shave themselves. Who shaves the barber? If he is one of those men who do not shave themselves, he must be shaved by the barber (himself) and hence he does shave himself, a paradoxical situation. If, on the other hand, it is claimed that he is someone who shaves himself, then he cannot be shaved by the village barber, which contradicts the fact just asserted namely, that he, the barber, is someone who does shave himself.

These "all" paradoxes of *Mengenlehre* arose around the beginning of the twentieth century, but in the sixth century B.C. Epimenides the Cretan said, "All Cretans are liars," and in the fourth century B.C. Eubulides gave a better formulation of the same "liar paradox" by asserting, "What I am now saying is a lie." This paradox occurs in each case because the quoted sentence, if true, must by what it says be false, and if false, must by what it says be true.

The paradoxes were the motivation for Zermelo, Fraenkel, and all the other mathematicians who provided pure set theory with a foundation that would exclude dilemmas of the kind we have illustrated. The general idea is to *restrict* the wide latitude in Cantor's conception of a set as any collection whatsoever, an idea we have used repeatedly (in informal discussions). As early as 1908, Russell proposed to overcome contradictions by saying that one may *not* gather all sorts of entities into a single aggregate, but must limit a set to things of the *same type*. He presented a *theory of types* by which "individuals" are considered of type 0, classes of individuals are considered of type 1, classes of classes are of type 2, sets of classes of classes are of type 3, etc. Then individuals may be elements of a class; classes may be elements of a class of classes; a class of classes may be an element of a set of classes of classes, etc. Anything of type n may be an element of a class of type $n + 1$, so that the latter class will contain as members only entities of the same type.

Russell's theory of types explains paradoxes by stating that they involve a confusion of types; that is, they disobey the membership rules just stated. Thus when, in the first version of the barber paradox, we speak of a "class that is (or is not) a member of itself, "we are committing the error of considering a class (type 1) to be (or not to be) a member of itself (type 1). This is wrong because an entity of type 1 can only be an element of a set of type 2. In other words, a class *cannot* be a *member* of itself. It can only be an element of a class of classes. (It can, of course, be a subset of itself, but that is not what the barber paradox asserts.)

There is also a hierarchy of types of language. At level 0 there are words referring to individuals; at level 1 there are statements about individuals; at level 2 there are assertions about statements, etc. If one keeps the levels distinct, no statement can refer to itself, since it must refer to an entity at the level just beneath it. Thus Eubulides' claim, "*This* statement is false," is ruled out of consideration, and the "liar paradox" is resolved. One may, for example, permissibly state, at level 1, "Epimenides was a Cretan," then at level 2, "The statement just made is true," then at level 3, "Our illustrative level 2 statement has a permissible logical format."

In 1904 Zermelo had dealt with an issue involving a type 2 aggregate. He had given a proof of the famous *well-ordering theorem*, in which he made an assumption about a class of classes. This postulate has come to be known as Zermelo's *axiom of choice* and, like the continuum hypothesis, was a logical challenge to logicians and pure mathematicians from Zermelo's day until just recently, when discoveries of Gödel and Paul Cohen shed new light on the issue. Gödel and Cohen accorded the selection axiom the same treatment that they had given the continuum hypothesis. In other words, Gödel proved Zermelo's axiom to be *consistent* with the other postulates of abstract set theory, and Cohen showed that the axiom of choice is *independent* of the other postulates. Therefore one is free to include the moot axiom in a *Cantorian* set theory, or to substitute for it an assumption *negating* it in a *non-Cantorian* set theory. Cohen offered different substitutes for the axiom of choice, just as the parallel postulate is negated in different ways in Lobachevskian and Riemannian non-Euclidean geometries.

But let us now examine the actual content of the axiom of choice. In that assumption Zermelo extended to infinite classes a familiar procedure with finite sets. Thus we regularly elect *representatives* to city councils. We may consider each district as a set of people and then our set of districts becomes a class of non-overlapping classes; one representative is *chosen* from each district, and the city council is the class of the representatives. Zermelo idealized this *finite* empirical situation and assumed that such a set of representatives could also be selected when the number of classes is infinite. His axiom states: For *any* class of non-overlapping classes there exists at least one class which contains one representative element from each class of the aggregate of classes.

Some of the foremost analysts of this century—Borel, Baire, Lebesgue, and others—objected to Zermelo's assumption on various technical grounds. Thus one can choose representatives for Congress simultaneously, or one after another. The critical analysts are willing to permit the latter procedure (in theory) even when the number of classes is denumerably infinite, but object to selecting all representatives simultaneously, by a *single* act of choice, as it were. A major bone of contention is therefore the lack of a *constructive* method for obtaining the representative class in certain situations. A suggestion as to the nature of the difficulty is found in the following story, which is part fact, part legend. In the early days of Cantorian theory, the son of an American multimillionaire was a graduate student in mathematics at Göttingen. So great was his talent that the university professors predicted he would be a leading mathematician of the next generation. Then suddenly he announced he would return to the United States to become a partner in his father's business and would renounce mathematics except as an amateur hobby. One of his fellow students decided to remonstrate with him and arrived on the scene as the American was directing his valet in the packing. The German youth was astounded to see that several trunks were filled with layers and layers of shoes, and his amazement increased when the valet opened a closet that revealed shelves completely filled with *more* shoes.

In spite of his innate courtesy, the advocate of pure science could not help exclaiming "So many pairs!"—whereupon he received the rejoinder, "Nevertheless a countable number!"

"Why, oh why are you doing this foolish thing? Why give up a career for which you are so well qualified?" the young man pleaded.

Legend has it that the departing American felt obliged to put up a strong defense and responded, "I am discouraged because mathematics is ultimately baffling. For example, my father has had a pair of shoes delivered to me almost every other day since I have been at the university, and my valet has been keeping track by numbering the pairs, writing the integer 1 inside the left shoe of the first pair to arrive, the integer 2 inside the left shoe of the second pair to arrive, and so on, inscribing successive integers for successive pairs in order of arrival. But with each pair of shoes comes a pair of socks; hence I know that I have as many pairs of socks as pairs of shoes. Still, I can't give my valet a constructive rule telling exactly how to choose the sock which is to bear the number 1 or 2 or 3 or 4, etc. Therefore, even if a mathematician like Cantor were to conceive of an infinity of pairs or sets of indistinguishable entities (like socks), mathematics does not give constructive laws of selection needed to prove facts related to such sets."

Actually, since even a multimillionaire could not provide an *infinity* of pairs of *socks*, the difficulties with Zermelo's axiom is only suggested by the story, but does not really exist for a *finite* number of pairs however large this finite number. But more valid objections to Zermelo's axiom arose from the following situation. He used his postulate in the proof of the well-ordering theorem, which states that *every set* can be well-ordered. From that theorem it follows that the aggregate of real numbers between 0 and 1 inclusive can be well-ordered. Nevertheless, no one to date has been able to exhibit such a well-ordering. This failure caused mathematicians to examine Zermelo's proof to see whether he had erred in some way. In the process they proved the well-ordering theorem equivalent to the axiom of choice. What this means is that not only can the well-ordering theorem be deduced from the axiom, but conversely, if the possibility of well-ordering every set is assumed to be true, Zermelo's axiom can be deduced as a theorem. Thus, doubts about the universal possibility of well-ordering entail doubts about Zermelo's axiom. Another feature in the situation is the use of an "existence proof." Some modern logicians refuse to accept a proof that something exists or something can be done (a set can be well-ordered, for example). Such critics say that a valid proof not only must show that a thing can be done but must contain an exact method for carrying out the process.

Cohen, however, settles the issue, as already stated, by showing how one can consistently *assume* either the axiom of choice or its negation. It all depends on whether one's attitudes or the applications one desires call for Cantorian or non-Cantorian set theory. Cohen has even presented different forms of the negation and, in each case, has specified how large an infinity of sets it is possible to "choose from" as well as which infinities are too large to choose from (for the particular negation assumed).

This chapter has emphasized the Cantorian theory of *infinite* aggregates, but sets whose cardinal number is finite are not to be considered of minor importance in mathematics. In fact, we shall see later that the "intuitionist" school of logic employs only the latter type of aggregate and would bar infinite classes except for those that can be *constructed* by adding one element at a time *ad infinitum*. Again, methods for determining cardinal number and ordinal type have been stressed, but properties of *operations* with sets, like addition or union, were more important when we treated the *algebra* of classes in Chapter 6. Now, as an immediate sequel to the present discussion, the theory of infinite sets will be related to the modern subject of *topology*.

25

Angelic Geometry

Hermann Weyl (1885–1955), a leading mathematician of the recent era, once remarked that the angel of topology and the demon of abstract algebra are engaged in a war where each battle is a struggle for the soul of another branch of mathematics. The two subjects in question do, in fact, lie at the root of almost all of modern pure mathematics. The "fiendish" member of the fundamental pair has been developed to some extent at various points in previous chapters, and now the "loftier" of the two opponents will be considered.

Topology might be described as the *general* study of *continuity*, that concept whose manifold aspects have challenged philosophers and mathematicians from Pythagoras' day to our own. In the nineteenth century Cantor and Dedekind considered many types of *geometric* continua which were point sets in a plane or in space or on a surface, on a hypersurface, etc. Since the associated problems were always couched in geometric language, topology was at first considered a sort of geometry, and the Cantorian study of general continua, where different point aggregates are specified and limiting points, neighborhoods, and the like are defined, came to be called *point-set topology* or, alternatively, *analytic topology* because the concepts and methods are characteristic of modern analysis. Today all such topics, in particular those aspects we have treated in Chapters 22, 23, and 24, are generalized in the subject which provides the conceptual background for modern analysis and which, appropriately, is called *general topology*.

A different technique for handling the properties of continua is found in the *combinatorial topology* whose modern phase was launched by Henri Poincaré in a series of memoirs from 1895–1904. At that time topology was still considered a branch of geometry and was called *analysis situs*, analysis of situation. Poincaré wrote:*

Analysis situs is a purely *qualitative* subject where quantity is banned. In it two figures are always equivalent if it is possible to pass from one to the other by a continuous deformation, whose mathematical law can be of any sort whatsoever as long as continuity is respected. Thus a circle is equivalent to any sort of closed curve but not to a straight-line segment, because the latter is not closed. Suppose that a model is copied by a clumsy craftsman so that the result is not a duplicate but a distortion. Then the model and the copy are not equivalent from the point of view of metric geometry or even from that of projective geometry. The two figures are, however, topologically equivalent. . . .

---

* H. Poincaré, *La valeur de la Science*, Flammarion, Paris, 1905, p. 44.

It has been said that geometry is the art of applying good reasoning to bad diagrams. This is not a joke but a truth worthy of serious thought. What do we mean by a poorly drawn figure? It is one where proportions are changed slightly or even markedly, where straight lines become zigzag, circles acquire incredible humps. But none of this matters. An inept artist, however, must *not* represent a closed curve as if it were open, three concurrent lines as if they intersected in pairs, nor must he draw an unbroken surface when the original contains holes.

The above quotation from Poincaré indicates that topology is not concerned with the issues of Euclidean geometry—the measurement of lengths, areas, volumes, angles, the making of scale drawings or enlargements. Again, topology is not limited to the rules of projective geometry, where straight lines may change in position but may never be distorted into curves, and circles may be transformed into ellipses and vice versa but may not acquire humps or altogether arbitrary closed contours. All of this would naturally suggest the question: Can there be any use in a "geometry" like topology that permits enormous freedom in transforming figures into others that are to be considered equivalent?

It is not our purpose to answer the above question fully at this point, but we shall indicate briefly that some important aspects of physical theories are rooted in topology. There is, for example, the "finite but unbounded" space of relativity. Einstein's *unbegrenzt* signifies "without limitation in the form of boundaries." This is the equivalent of Poincaré's term *closed*, which makes curves like the circle and ellipse, lines like the periphery of a polygon, surfaces like a sphere all closed (*unbegrenzt*). A line segment or arc is not closed, because its end-points are *boundaries*. Surfaces that are not closed are the circular disk, which has a boundary in its peripheral circle, and the circular annulus, which is bounded by two concentric circles. Einstein used the term "finite" in a metric sense, and therefore, in accordance with Poincaré's statement, the property involved is not a topological one. In relativity, a finite space is one that is not inconceivably large but where one can say, for example, in the case of a circle, that its length is less than 50 ft., or in the case of a sphere, that its area is less than 1000 sq. ft. Topologists handle finiteness of geometric figures somewhat differently.

Again, other features of physical reality are topological. Thus we see that in a straight-line universe (Figure 25.1) an insect journeying from *A* to *C* would have to pass through *B* because *B* is *between A* and *C*. Thus "between-ness" is a *quality* associated with geometric situation (*analysis situs*). In a circular universe (Figure 25.1) an insect would *not* have to pass through *B* to reach *C*, since he could travel around the circle by an alternative route. In fact, we can make *no* unique

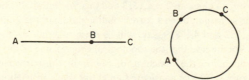

Figure 25.1

"between-ness" statement about three points on a circle, for instead of saying that *B* is between *A* and *C*, we could equally well claim that *A* is between *B* and *C*, or that *C* is between *B* and *A*. If we are thinking of a geometric *time* continuum, for example, and we picture it as a *directed* straight line, we can say for three instants of time— *A*, *B*, *C*,—that the time instant *B* comes *after A* but *before C* (or the reverse if the line has the opposite direction), but we cannot make such a statement if we conceive of the time continuum as a circle, an ellipse, or a polygonal periphery.

To leave Poincaré's ideas and return to Cantor's, we recall that one-to-one correspondence is the criterion for equivalence of sets as far as their cardinal number is concerned. One result of this mode of comparing aggregates is that the point sets corresponding to a line segment, a cube, a "hypercube," have the same cardinality. This affront to intuition is alleviated by the observation that, to establish the equivalence of the point sets named, the figures must be shattered, so to speak; that is, the matching is *not* continuous, because points originally close together may have as mates points far apart. The situation emphasizes that topology is not a mathematics of measuring and that the cardinal number of points in a figure is not an adequate description. Therefore one-to-one correspondence is supplemented by the requirement of *bicontinuity*, or continuity both ways, before two figures may be called topologically equivalent.

Figure 24.1 illustrated a one-to-one correspondence between the line segments *AB* and *A'B'*. In Chapter 22 we gave a rigorous definition of a continuous function of a real variable, but here we shall merely state informally that the one-to-one correspondence pictured in Figure 24.1 is a *continuous* mapping of *AB* onto *A'B'* because points close together on *AB* are mapped onto points close together on *A'B'*. The correspondence is also a continuous mapping of *A'B'* onto *AB* because points close together on *A'B'* are mapped onto points close together on *AB*. Just as the ε, δ technique rigorized the concept of continuity in Chapter 22, the present notion of continuity can also be given suitable mathematical rigorization. But we shall merely state informally that Figure 24.1 illustrates a *bi-unique*, *bicontinuous* correspondence between two point sets.

Two sets are said to be topologically equivalent if and only if a bi-unique (one-to-one), bicontinuous correspondence between the sets can be established. Such a correspondence is called a *homeomorphism*, and topologically equivalent sets are termed *homeomorphic*. Figure 25.2 illustrates some topologically equivalent figures and the homeomorphisms matching their points. Figure 25.3 shows figures no two of which are homeomorphic. The circle of Figure 25.4 is homeomorphic to either one of the knots, as we can see intuitively if we number points on the circle as if it were a clock, $1, 2, 3, \ldots, 12$ and imagine infinitely fine subdivisions. We can assign the same numbers to corresponding points in the knots. Then points with numbers 4.1782 and 4.17819, say, will be close together on the clock as well as on the knots. We observe that the topological equivalence of the three figures is in no way dependent on the matching of points in the surrounding space. In topology each figure would be considered an entire space in itself, as it were.

This fact leads to a warning about the traditional "continuous deformation" or "stretching and bending without tearing" or "rubber-sheet" descriptions of topological equivalence, where deformations may possibly involve an external space in which a configuration is embedded. If figures can be deformed into one another, they are homeomorphic, but Figure 25.4 indicates that the converse is false. Either knot is equivalent to the circle, but no amount of stretching, bending, etc. would deform it into a circle. Such a knot would have to be broken or else deformed in a 4-*dimensional space* to convert it into a circle by mere bending and stretching. Therefore it is better not to consider deformation as the definition

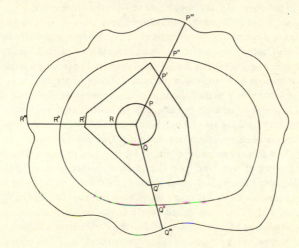

Figure 25.2

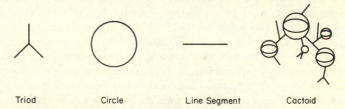

Triod   Circle   Line Segment   Cactoid

Figure 25.3

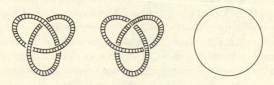

Figure 25.4

of homeomorphism and to rely instead on the original concept of bi-unique, bi-continuous correspondence between two configurations. For those who are fond of the "rubber-sheet" notion there is consolation, since this idea is embodied in a special, exceedingly important concept called *homotopy*. If one does wish to consider configurations as embedded in some space—for example, the curves on a surface—then deformation within the surface can be considered. A curve which (remaining within the surface) can be deformed continuously into another curve is said to be *homotopic* to the latter. Thus the knots and the circle (Figure 25.4) are *not* homotopic in the plane, but are homotopic in 4-dimensional space.

Difficulties arose in topology because continua were conceived as *geometric* and there are two essentially different ways of regarding a geometric figure. One can see it from Cantor's point of view, as an ensemble of its individual points, or one can consider it as a whole, as Euclid did. In the latter case it is a point or a line segment or a triangle or a circle or a plane or a sphere or a cylindric surface or a cube, etc.; or it may be a *finite* collection of such geometric elements. But if one replaces the segment or triangle or cube, etc. by its (infinite) set of points, one is back to Cantor. From a lay viewpoint it would certainly seem easier to think of the *finite* combination of elements in a tetrahedron or triangular pyramid—the 4 points or vertices, the 6 line segments or edges, the 4 triangles or faces. The number of elements or "cells" of each kind, plus the mode of *combining* them or fitting them together (the paper pattern, as it were) is often called the *combinatorial skeleton* of the particular continuum, in this case a tetrahedral surface. The study of properties of continua after this fashion is called *combinatorial* or *algebraic* topology, since algebraic symbols and formulas are used for the combinatorial schemes and the reasoning is carried out by means of algebraic concepts and operations. If issues of general (point-set) topology arise, one covers the set with a combinatorial mesh-work of cells, then subdivides these repeatedly into finer cells until ultimately the points are reached, much in the manner suggested by the nested intervals of Chapter 2 or by the way a foot-rule can be divided into inches, then half-inches, quarter-inches, etc. (*ad infinitum*, in theoretical mathematics). Sometimes it is not necessary to carry out the process of subdivision indefinitely in order to trap all points as the net grows finer and finer. If a continuum has a property called "compactness," it turns out to be possible to cover all its points by means of a *finite* number of cells. In some continua, one traps the points by a finite number of *overlapping* patches. In all that we have just said it is evident that the two kinds of topology, combinatorial and general, are not divorced from one another but that there is an interplay of concepts and methods. However, the matter of fusing the two types into a single logical whole has not yet been accomplished, and all attempts to do so have thus far been frustrated by the excessive difficulties that arise. Nevertheless one can say that either kind of topology is concerned with *topological spaces*.

What then is a *topological space*? It is just a set (space), finite or infinite, endowed with a "topology." Crudely speaking, the "topology" assigned can be considered to specify the subsets that are "neighborhoods," and such specification makes it possible to arrive at definitions of "limit" and "continuity." But the idea of a "neighborhood" is not purely qualitative (topological) since it has a *metric* connotation, for some measurement of distances would seem to be required to decide which points are close enough to a given point to be part of its "neighborhood." Hence the more general nonmetric conception of an "open set" is introduced. Specifically, one assigns a "topology" to a given universal set *I*, that is, one converts it into a *topological space*, by selecting a collection of subsets of *I* and postulating that these subsets are to be labeled *open sets*. The subsets of the collection can be chosen somewhat arbitrarily except that the collection must be closed with respect to certain operations, and the subsets pronounced to be "open" must also have properties resembling those of the metric "neighborhoods" which they generalize. Hence the following three conditions must be satisfied:

(1) The empty set $\varnothing$ and the universal set *I* are open sets.
(2) The *union* (logical sum) of any aggregate of open sets (that is, the set of elements belonging to at least one open set of the aggregate) is an open set.
(3) The *intersection* (logical product) of any two open sets is an open set.

To provide an illustration and also to show that a set *S* can be provided with *different* "topologies," in other words, can be structured to form *different* topological spaces, let us return to the universe $I = \{\alpha, \beta, \gamma\}$ of Chapter 6. In that chapter we saw that *I* contains eight subsets. We can endow *I* with a topology by specifying that only $\varnothing$ and *I* are to be considered open sets. (Observe that the three requirements listed above would be satisfied.) At the other extreme, we might topologize *I* by postulating that all eight subsets are "open." Or *I* might be converted into a topological space by defining the class of open sets as

$$\left\{ \varnothing, I, A = \{\alpha, \beta\}, C' = \{\alpha\} \right\}$$

where the symbols *A* and *C'* have the same meaning as in Chapter 6. The reader can readily show that this "topology" has the required properties and that the same is true of the "topologies" $\{\varnothing, I, A, A'\}$, $\{\varnothing, I, A, B'\}$.

A fundamental problem of either combinatorial or general topology is one of classification, of subdivision of topological spaces into different types. This is done in many scientific situations by means of a special kind of binary relation called an *equivalence relation* (illustrated in Chapter 21 by "equivalence modulo *m*"). If, for example, we have on cards the names of all subscribers to a magazine, and wish to arrange the cards in alphabetic order, we may consider the binary relation on the set of subscribers defined by $\rho = $ "has the same initial letter in his surname." Then as we sort the cards we place them in different packs according to this criterion, all cards in one pack being *equivalent* in the sense that the surnames start with A or with B, etc. To subdivide each pack further we could use other equivalence relations. In a sense, names like Bentley, Billings, Brown are equivalent, if, say, we are talking about a fictitious character, Mr. B_____, or guessing a name in a game or puzzle where the initial letter is the only hint.

To give a formal definition, we now state that a binary relation ρ on a set S is an equivalence relation if, and only if, ρ is

(1) *Reflexive*: $x \rho x$ where x is any member of S;

(2) *Symmetric*: If $x \rho y$, then $y \rho x$, where x and y are members of S;

(3) *Transitive*: If $x \rho y$ and $y \rho z$, then $x \rho z$, where x, y, z are members of S.

Parallelism is an equivalence relation dividing the lines of a plane into sets in each of which the lines are "equivalent" in the sense of having the same direction. Likewise, congruence, similarity, and projectivity effect a subdivision of geometric figures into sets of equivalent figures. In similarity geometry, for example, a scale drawing or an enlargement is equivalent to the original figure and may be substituted for it, and thus only a single category is involved. In topology, homeomorphism is an equivalence relation; a sphere is topologically equivalent to a polyhedron, an arc to a line segment, a knot to a circle. Just as we speak of the types {0, 1, 2, 3, 4} when congruence modulo 5 is the equivalence relation, we may speak of *the* line segment, *the* circle, *the* sphere as types. In number theory with modulus 5, we may substitute 6 or 21 for 1; 200 or 315 for 0; 14 or 1019 for 4; etc. In topology we may substitute a convex polygon or a star polygon for a circle, or a Euclidean plane for a punctured sphere. That the last of these substitutions is correct can be proved by the process of *stereographic projection* (Figure 25.5) (used in making maps that represent angles correctly and hence give a true representation of small neighborhoods of each point on the earth). The diagram indicates that when one point is removed from the spherical surface (in this case, the North Pole), there is a bi-unique, bicontinuous correspondence between its points and those of the tangent plane at the South Pole.

Thus we have already mentioned several types of configuration or topological

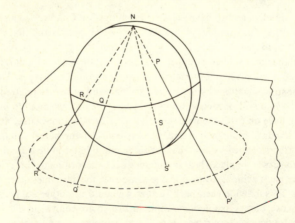

Figure 25.5   Stereographic projection

equivalence class—*the* arc, *the* circle, *the* sphere, *the* punctured sphere. It might seem natural to extend the classification so as to include *the* sphere in Euclidean 4-space, *the* sphere in Euclidean 5-space, etc., or since we cannot visualize these, to their analytic equivalents,

$$x^2 + y^2 + z^2 + w^2 = 1$$
$$\ldots x_1{}^2 + x_2{}^2 + x_3{}^2 + \ldots + x_n{}^2 = 1$$

To limit the classification of types so that it is not overpowering, it would seem sensible to follow a procedure similar to that used with alphabetization, where we first subdivide according to initial letters, then take each pack or equivalence class and subdivide it according to second letters, etc.

If we think of topology in its original sense as a sort of geometry, we might consider the first crude subdivision as that into plane geometry, solid geometry, 4-dimensional geometry, or more exactly into 1-dimensional, 2-dimensional, 3-dimensional, ..., n-dimensional configurations. The only trouble is that we are not sure what we mean by *dimension* in the topological sense. We have already explained what is meant by the term as applied in "Euclidean" spaces. But topological spaces are far more general in nature and in some instances dimension cannot be specified by the definitions used for special cases. Let us then defer temporarily the consideration of how dimension is defined in topology. A satisfactory definition can be given, and the dimension of a topological space can be proved to be a topological property of the space, that is, an invariant under homeomorphism, so that spaces can be classified according to dimension. Since we may imagine that all the topological spaces have now been sorted into packs labeled 0, 1, 2, 3, 4, ..., according to dimension, let us study the subclassification. Topology has advanced to the point where complete classification for dimensions 0, 1, 2, has been achieved (at least for spaces which are "compact manifolds"). For more complicated spaces or those of dimensionality 3 or higher, the classification problem has not yet been solved.

When a surface is called a *manifold*, the term signifies that the neighborhood of every point on the surface is homeomorphic to the interior of a circle. A Euclidean plane and a sphere are manifolds. In the former case a neighborhood of a point is identical with the interior of a circle. On a sphere, the neighborhood of a point can be deformed by being flattened out into the interior of a circle. A mathematical cone (Figure 7.11), we recall, looks like an hourglass. It is *not* a manifold because the neighborhood of the vertex, the point where the two "nappes" of the cone meet, is not homeomorphic to the interior of a circle. The reason for this failure is suggested if we try to deform the neighborhood by flattening it out. Except at the vertex, two points of the cone would go into one point of the flat area so that the correspondence would be two-to-one instead of one-to-one.

We have explained (page 602) the importance of "compactness" as a property making it possible to cover certain manifolds with a *finite* number of "cells." Hence, let us remark further that for the special topological spaces we shall consider, whenever we speak of a *compact* 2-dimensional manifold, the term *compact* will correspond to what we have called a *closed* surface confined to a finite portion of space. The topologist defines compactness more generally by avoiding the metric concept of "finite portion of space." He bases his definintion on *intrinsic* properties of a surface, that is, those that do not depend on the surrounding space. In what follows, then, whenever we refer to a *surface*, we shall mean a compact 2-dimensional manifold.

The classification of surfaces is almost as simple as that of modular congruence, where $\{0, 1, 2, 3, 4\}$ represents all types of integer modulo 5, or $\{0, 1, 2, 3, \ldots, m-1\}$ represents all types of integer modulo m. In Figure 25.6 we have indicated an

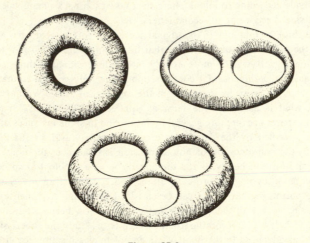

Figure 25.6

ordinary *torus* (shaped like a doughnut or inner tube), a torus with two holes (like a pretzel), and a torus with three holes. The fundamental theorem for surfaces states that the types are *the* sphere, *the* torus with one hole, *the* torus with two holes, . . ., *the* torus with m holes, etc., where m is any positive integer. This is seen to be strictly analogous to the classification of integers modulo m.

In a simple classification problem like alphabetization we can place any name in the proper category if we can read, and know the alphabet, and it is not difficult to place 1019 modulo 5. One merely divides by 5 and obtains the remainder 4, which makes 1019 equivalent to 4 (modulo 5). But to classify some strange abstract topological surface by proving it homeomorphic to a sphere or a torus with m holes might be exceedingly difficult for a novice, or even an expert in topology. Unless we have special information about the surface, or a "hunch," how can we tell whether to try to set up the correspondence with an ordinary torus or one with a thousand holes?

In any case, the direct resort to formulating homeomorphisms is not an easy one. However, the use of topological *invariants* is of great assistance in the problem of classification, numerical invariants in particular. What we mean by a numerical *invariant* can be illustrated in elementary mathematics. Every cubic polynomial has exactly 3 linear factors. There are an infinite number of different cubic polynomials but the *number* of linear factors, 3, is the same in every case. This is a numerical invariant for the set of all cubic polynomials. By analogy, what we seek is a *number* that is the same for sphere, ellipsoid, polyhedron, and all surfaces topologically equivalent to these. Again, 2 is the *numerical* invariant which specifies for quadratic polynomials the number of linear factors that *any* such polynomial must have. The analogous numerical invariant is different for quartic, quintic, etc. polynomials, so that all algebraic polynomials can be *classified* according to the number of linear factors. Thus anyone who asks for a quadratic polynomial with linear factors $(x + 1)$, $(x - 3)$, and $(2x + 5)$ is talking nonsense.

In topology, after we find a numerical topological invariant for *the* sphere, we must find the analogous number for *the* torus. If, as in the case of algebraic equations, we obtain a different number for each type of surface, these numerical invariants will furnish a criterion for classification.

Let us start with polyhedra (surfaces whose faces are polygons). In Figure 25.7 we have illustrated a few of those studied in school geometry. There are the five regular (Platonic) polyhedra, a prism, and a frustum. These are all simple (or "simply connected") polyhedra, which means intuitively that they contain no

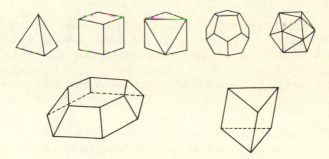

Figure 25.7 Simple polyhedra

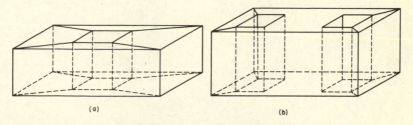

(a)
(b)

Figure 25.8 Nonsimple polyhedra

holes. Figure 25.8 illustrates some nonsimple polyhedra. We tabulate the number of vertices, edges, faces, for each simple polyhedron:

	Vertices (n_0)	Edges (n_1)	Faces (n_2)
Regular tetrahedron	4	6	4
Cube	8	12	6
Regular octahedron	6	12	8
Regular dodecahedron	20	30	12
Regular icosahedron	12	30	20
Triangular prism	6	9	5
Frustum of a hexagonal pyramid	12	18	8

These numbers vary from figure to figure, but it is easy to check that in every case

$$n_0 - n_1 + n_2 = 2$$

where n_0, n_1, n_2 represent the number of vertices, edges, faces, respectively. If a count is made for other simple polyhedra, the same result will be obtained, and it can be *proved* that, for all simple polyhedra,

$$n_0 - n_1 + n_2 = 2$$

This fact was first observed by Descartes, and rediscovered a century later by Euler. It is customary to refer to the proposition as *Euler's theorem*. If we deform a polyhedron or if we inflate it so that edges and faces become curved, if we continue the inflation until the surface becomes spherical (Figure 25.9), the numbers n_0, n_1, n_2 will remain the same, and no matter what type of polyhedron we deform or inflate,

$$n_0 - n_1 + n_2 = 2$$

for all the resulting surfaces, and finally for the sphere (subdivided into the curvilinear polygons into which the polyhedral faces have been deformed).

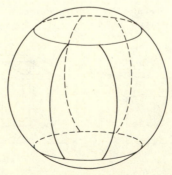

Figure 25.9

We have found our first numerical topological invariant. It is usually called the *Euler characteristic*, or just the *characteristic* of a surface. We can now state: The Euler characteristic of the sphere is 2.

Figure 25.8 illustrates nonsimple polyhedra with Euler characteristics, $16 - 32 + 16 = 0$ and $24 - 44 + 18 = -2$, respectively. When inflated sufficiently they can be deformed into an ordinary torus and a torus with two holes, respectively (where both surfaces are subdivided into curvilinear polygons). Then the characteristics of the torus and the torus with two holes are 0 and -2, respectively.

In topology it is proved that if the number of holes in a torus is p, its Euler characteristic is $2 - 2p$. The number p can be given a more precise and fundamental topological meaning than "number of holes" and is then appropriately named the *genus* of the surface. For a sphere, the genus is 0. When $p = 0$ is substituted in the formula $2 - 2p$, the correct Euler characteristic, 2, results. All ordinary surfaces can then be classified according to either genus or characteristic. In the former case, one of the numbers $\{0, 1, 2, 3, 4, 5, \ldots\}$ can be assigned to the surface, and in the latter, one of the numbers $\{2, 0, -2, -4, -6, -8, \ldots\}$.

One does not usually picture a surface as a composite of vertices, edges, polygonal faces, but merely as the set of its points. Still it is possible to cover a surface with curvilinear polygons (in many different ways). Then, in order to standardize procedures, one can break the polygons down into triangles so that there is a *triangulation* of the surface. The surface becomes a polyhedron, and one can obtain its Euler characteristic by simple counting. Then one can classify the surface as to type. It can be proved that no matter what subdivision into polygons (triangles) is used, the characteristic has the same value, as seems natural if it is really an intrinsic property of the surface.

This was the whole story of surfaces up to the day of A. F. Moebius (1790–1868). After his contribution to topology, it was revealed that the "ordinary" surfaces are just *half* the story. They came to be called *oriented* surfaces in contrast to the *nonoriented* surfaces of Moebius. We shall now explain the difference between these two types.

If we use the rectangle $ABCD$ of Figure 25.10 as the pattern for a surface and bring AB and CD together so that A coincides with D and B with C, we obtain a finite portion of a cylindric surface. Let us start again, and this time, giving the paper a slight twist, reverse the directon of DC so that now D coincides with B and C with A. The resulting surface is known as a *Moebius strip*, and the reversal of the direction of DC involved in its construction is a heuristic clue to its *nonorientability*. This surface and similar ones (made by giving two, three, four, etc. twists to the rectangle before identification of the edges) have been a happy source of recent parlor magic. If the reader should attempt the "tricks" we now describe, we advise that he draw the median line EF, that he make BC relatively long. AB narrow, in order to facilitate handling, that he use Scotch tape to make the seam binding AB and CD.

Now if one wishes to contrast cylinder and Moebius band, one can cut the

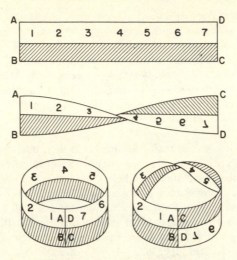

Figure 25.10   Cylinder and Moebius strip

cylindric surface along the median line. As expected, it falls into two parts. If the Moebius strip is cut along its median line, the unexpected results; it does not fall in two, but becomes a Moebius band with two twists instead of one. It can be proved that this two-twisted band is homeomorphic (topologically equivalent) to a cylindric surface. We can suggest why this is so by starting with the rectangular pattern of Figure 25.10, twisting it twice before seaming AB and DC together. This time we observe that we must identify D with A, and C with B, as in the case of the cylinder, so that the Moebius band with two twists and the cylindric surface have the same *combinatorial* pattern.

Now let us return to the original Moebius strip with a single twist, where the direction of CD is reversed before seaming. We imagine that a watch is placed with its center at the midpoint of the seam $AB(CD)$, and that the time indicated on its face is 11:45, so that the minute hand points to 9 and the hour hand to 12. We start sliding the watch to the left, that is, in the direction indicated by the minute hand. We keep going so that the center of the watch remains on the median line and its back is in contact with the surface. When the watch first returns to the seam, it will be on the side of the paper invisible to us, with its back towards us. If the paper and the watch case were transparent, we would see the minute hand still pointing to the left, but the hour hand would point down instead of up. From *our point of view* the direction of rotation has been reversed, since originally the direction along the minor arc from 9 to 12 was *clockwise* and now it is *counterclockwise*.

In all of this we are thinking *not* of the physical paper band and a real watch but of an abstract mathematical surface and watch, without any thickness and completely transparent. The physical experiment can be carried out with a transparent pliable plastic pattern for the strip and a watch crystal with the hands painted on it as pointing to 9 and 12.

The Moebius strip is called *nonorientable* because it is not possible to *fix* the direction of rotation around a point of this surface. A surface is called *orientable* if it is possible to draw a small directed circle at each point (either clockwise or counterclockwise) so that the little circles at points close together will have the same direction. We can see that this is desirable if we think of the analytic processes in mathematics. In trigonometry the positive sense of rotation is conventionally fixed as *counterclockwise*. An angle $+30°$ signifies one-twelfth of a revolution in the counterclockwise direction, and $-30°$ implies the same rotation in the opposite direction, and the trigonometric functions associated with the two rotations will, in general, be different. For example, sine $30° = 0.5$, sine $(-30°) = -0.5$. Calculus and other analytic subjects use the trigonometric facts and formulas in addition to other processes where orientation is important. None of this mathematics would be valid in physical applications if circles and hence angles might change their algebraic signs without the awareness of the surveyors, astronomers, and draftsmen, who shift their tools slightly or observe objects in some small region of the earth or heavens. In a cosmos like a Moebius strip some of our science and mathematics would have to be modified.

We emphasize that the Moebius surface is nonorientable because we cannot arrange to have circles at neighboring points agree on direction. The device of the moving watch seems to guarantee, in the beginning, that neighbors will agree, but in the end we see that some close to the seam may pronounce a certain rotation clockwise, whereas others nearby, but on the other side of the seam, will declare the same rotation to be counterclockwise.

Another way of indicating the nonorientability of the Moebius band is to compare it once again with the cylindric surface. We would say that the latter surface has two sides, in the sense that one could paint the inside of a cylindric tank white and the outside red. In drawing a circle or other continuous closed path starting on the red side, one remains on that side, and similarly for the white side. One cannot go from red to white or vice versa without crossing over the top. But the center of the watch in our illustration has described a continuous closed path on the Moebius strip and succeeded in getting to the other side without crossing "over the top," that is, over any boundary. Since a continuous closed path must always be confined to one side of a surface, mathematicians call the Moebius strip one-sided or unilateral. To show this in "parlor magic" style, one may decide to color a Moebius band by taking a piece of colored chalk whose length is equal to the width of the strip, then placing it along the seam and sliding it to the left, always perpendicular to the median line. The chalk will color the surface as it progresses, and if one continues on and on for a sufficient distance, he will find that he has colored the entire surface and that he might have saved effort by dipping the band into a pail of paint. Thus nonorientability and one-sidedness are just two different ways of describing a particular topological aspect of a Moebius band. Mathematicians prefer the notion of nonorientability because it is an *intrinsic* property and does not depend on the surrounding space. This is the same issue discussed in the differential geometry used in relativity. In the present case we would say that an insect confined to a Moebius strip could discover its nonorientability by making a closed tour around it. To see that it is one-sided requires observation from a vantage point in the surrounding space.

Since we have said that a strip with two twists is homeomorphic to a cylinder, the former must be two-sided or orientable. The reader may experiment with coloring such a band in order to see that the continuous motion of the chalk, no matter how prolonged, will not color the entire strip, and that it is possible to start with a different color on the reverse side of the seam and never run into the first. This suggests that for 2, 4, 6, 8, ... or any even number of twists a Moebius band is two-sided or orientable, and for 1, 3, 5, 7, ... or any odd number of twists, onesided or nonorientable.

As stated earlier, any compact orientable 2-dimensional manifold can be classified according to either its genus or its Euler characteristic; that is, it can be typed as a sphere or some sort of torus. There is, in theory, a way of closing up a Moebius strip so as to form a compact manifold. Then for the resulting surface and other nonorientable, one-sided types of surface, there is a complete classification by numerical invariants which is parallel to that explained earlier for the ordinary surfaces. Finally, the fundamental theorem of the topology of surfaces is: *Two surfaces are equivalent if and only if they are both orientable or both nonorientable and have the same characteristic number.* This, speaking crudely, makes the classification of surfaces as clear-cut as the alphabetization of cards. When some exceedingly abstract 2-dimensional point set is involved, one must first determine whether or not it is orientable. Then after covering its surface with polygons (triangles), one counts vertices, edges, faces, in order to obtain the characteristic. The two facts place it in a specific category.

Returning to the pattern (Figure 25.10) for the cylinder, we can follow the identification of edges *AB* and *CD* of the rectangle by the identification of the other

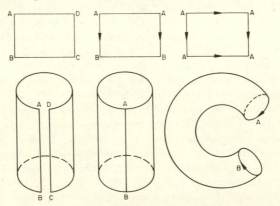

Figure 25.11   Torus formed from combinatorial pattern

pair of opposite sides as indicated in Figure 25.11. The cylindric surface is no longer bounded by two edges, but is closed by identification and becomes a torus. All the vertices of the rectangular pattern coalesce into a single one, so that $n_0 = 1$. The identification of opposite sides leaves only 2 edges, so that $n_1 = 2$. There is only one face, the rectangle itself, so that $n_2 = 1$. Then the characteristic is

$$n_0 - n_1 + n_2 = 1 - 2 + 1 = 0$$

which checks with the previous result for the torus.

If we consider removing boundaries, that is, closing the cylindric surface, we can carry out the first two steps in Figure 25.11 and then identify the free edges of the rectangle after giving them *opposite* directions (Figure 25.12). This construction

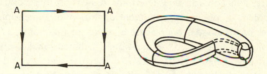

Figure 25.12   Klein bottle formed from combinatorial pattern

cannot be carried out in physical space without having the surface intersect itself as indicated in Figure 25.12, which pictures a *nonorientable torus* or *Klein bottle*, named after Felix Klein, who was the first to point out its pattern and properties. The nonorientability can be seen to arise from identifying the lengths of the rectangular pattern in the same way one seams the widths in constructing the Moebius strip. In 4-dimensional space it would be possible to carry out the construction of the twisted nonorientable torus without having it intersect itself. Lewis Carroll, the mathematical author of *Alice in Wonderland*, described the Klein bottle in his story "Sylvie and Bruno," where he called it the Purse of Fortunatus. We observe from its rectangular pattern that this Purse has the same Euler characteristic, 0, as the untwisted torus. That is why, in the fundamental theorem of the topology of surfaces, the characteristic is *not* sufficient for classification, and the orientability must also be known.

If we were 4-dimensional creatures, we could readily see how, for the Moebius strip, the identification indicated in Figure 25.13 must close it, producing a surface which can be proved to be homeomorphic to the *projective plane*, the plane in which the 2-dimensional figures of projective geometry can be located. The *pattern* shows that, after identification, $n_0 - n_1 + n_2 = 2 - 2 + 1 = 1$, so that the Euler characteristic of the *projective plane* is 1. We remark that the projective plane is an alternative choice to the closed Moebius strip for the fundamental compact nonorientable 2-dimensional manifold, playing the same role as the sphere does among orientable manifolds.

Not all point sets or topological spaces are 2-dimensional, however. Before determining the facts just mentioned, one must know the dimension of the set

Figure 25.13   Combinatorial pattern for the projective plane

involved. The Peano curve (Figure 22.1), which is a helpful illustration in many of the logical issues of topology, will assist us once more. It is described by the continuous motion of a moving point. Hence, intuition says it should be 1-dimensional. On the other hand, it fills a square, which would make it appear 2-dimensional. After all, we have just seen how one might convert a square into a cylinder or Moebius strip, and the Peano curve will then cover such a surface. What is its dimensionality, and how can one determine dimension in general?

The problem of obtaining a suitable definition of dimension for the very general spaces of topology is once again a matter of taxonomy, of classifying all possible topological spaces into 0-dimensional, 1-dimensional, 2-dimensional, ..., n-dimensional aggregates. Let us return to the line segment and square which Cantor proved to contain the same number of points and see if we can find a topological property of each that will help to define dimension. What we seek is some qualitative property of the segment that will hold for an arc, zigzag line, wavy curve, that is, for any topologically equivalent figure, but will not be true for the square and its equivalents. Earlier we mentioned the case of an insect traveling in a straight line cosmos. Let us say that it has traveled from A to C on the straight-line segment in Figure 25.1 and now wishes to make the return journey. In the interim, however, some catastrophe has destroyed B. There is now a hole where B was originally located, and the return trip is impossible. Suppose that an insect lives, instead, within a square (Figure 25.14a). Then if the original trip took place along the path ABC, the removal of B does not prevent a return journey from C to A, because a detour is possible. To make the trip impossible would require a disaster destroying all points along some entire path (Figure 25.14b) like DEF, so that no avenue of return in any direction is possible. If the insect is to be prevented from returning to A, it will take not merely a great many points but a certain arrangement of these points to block the way. In Figure 25.14b we notice that the simplest kind of point set or "space" that will cut point C off from point A is what we are tempted to term *1-dimensional*. This gives us the idea of saying that a

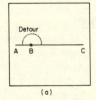

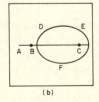

Figure 25.14

square or some other space is 2-dimensional, if when you wish to join two of its points, it requires a one-dimensional set to prevent you from doing so. We cannot call this statement a definition because we don't precisely know what it means for a set to be one-dimensional, and hence it would not be logical to use the idea of one-dimensionality to explain the notion of two-dimensionality. However, we have made very definite progress. For the situation which confronts us now is a very familiar one in mathematics. We are faced with an *induction*. To explain three-dimensionality we need the notion of two-dimensionality; to explain two-dimensionality it is sufficient to understand one-dimensionality. When we have done this, we are prepared at once, *by induction*, (that is, one step at a time) to define all higher dimensions.

You notice that there are two features to this projected dimension theory. First, it is in some sense a negative approach. We converted the idea of having enough "room" into an idea of "obstacles," of being unable to do something when certain point sets are removed from our space. Second, we have hitched our wagon to mathematical induction, a very powerful and beautiful mathematical tool. This approach to the problem is due to Poincaré, who gave it in a form considerably more precise than we have.

In mathematical inductions it is customary to start with the case $n = 1$, but in the dimension theory of topological spaces, one goes back to the completely empty set $\emptyset$ for which a *negative* dimensionality, -1, is postulated. Now let us consider a set consisting of two isolated points, A and B (Figure 25.15). If you wish to move

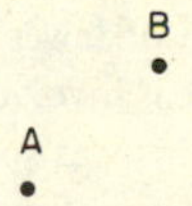

B
•

A
•

Figure 25.15   A 0-dimensional space

from A to B, you cannot, since there are no points in between. You are blocked from the very start by the empty set, $\emptyset$. Because motion is blocked by the empty set, which has dimension -1, the dimensionality of our space of two points is one higher. Hence it is 0-dimensional. A space consisting of one point is *a fortiori* 0-dimensional. Since it is not empty, it cannot have dimensionality -1, and if its dimensionality were higher than zero, we would have the contradiction that part of the above two-point space is of higher dimension than the whole. If we have a space of three points or any finite number of points, it is readily seen that the argument for the two-point space can be repeated. Hence any space containing a finite number of points is 0-dimensional. If the countably infinite set of natural numbers $\{1, 2, 3, \ldots\}$ is graphed on the real number line, there are gaps or empty spaces between the points. Therefore motion between any pair of points is impossible; that is, it is blocked by the empty set. Hence the set of natural numbers is 0-dimensional. Any countable set can be matched with it and given a matching topology so that all countable sets become *equivalent* topological spaces of the *same dimension*. Thus the countable set of all rational points, for example, would be 0-dimensional. Although the set of irrational points is not countable, it can be proved that it is 0-dimensional.

This brings us to the meaning of a 1-dimensional space. It is one where movement between any pair of points can be blocked by a 0-dimensional space. Then a straight line is 1-dimensional because, as we have seen, a single point blocks motion. What about a circle or any closed curve? On the circle of Figure 25.1 the removal of the single point B will not block motion between A and C. If, however, in Figure 25.1 we also remove a point from the longer arc from A to C, motion from A to C will be impossible. Since it requires the removal of two points, that is, a 0-dimensional set, to block motion between any pair of points on the circumference, the circle is 1-dimensional. Hence any closed curve is also 1-dimensional because such a curve is homeomorphic to a circle. Next we pronounce the square 2-dimensional because it requires the removal of a 1-dimensional space to prevent the motion between any pair of points of the square. In Figure 25.14b we see that removal of a circle or closed curve will accomplish the blockage, but how do we know that there is no 0-dimensional set that would be equally effective? When we think of a point set as dense as the irrationals, for example, we might imagine that this or some similar sort of very dense 0-dimensional set might do the trick. With our limited development of general topology we are unable to prove the non-existence of any such set, but it can actually be demonstrated that no 0-dimensional set will block the way. Similarly, when a space is to be proved 8-dimensional, one must show that motion between any pair of its points will be blocked by a 7-dimensional set but not by one of 6 or fewer dimensions.

We may feel that a lot of trouble is involved in proving a square 2-dimensional, when we were intuitively sure that this was so right along. A criterion like the one of our definition is important, however, for the varied types of sets that can be converted into topological spaces, many of which defy intuition as well as simpler methods of defining dimension. We saw that even the more or less elementary aggregate of irrational points is puzzling. Its points are as numerous as those in the entire real set and differ from it by the relatively negligible set of rationals; yet the modern topological theory of dimensionality assigns a lower dimension to the irrationals than to the reals. This is another indication that topological properties like dimensionality are qualitative and may appear, at times, to conflict with results obtained from counting or measurement. Again, we are reminded that *analysis situs* is concerned with situation, and it is the relative situation of the irrationals, rather than the length they cover, that determines a topological property.

Topology is too new a subject to have many practical applications. Perhaps its most important service, however, has been that of solving baffling problems in *other* branches of mathematics where the traditional quantitative methods have failed or are incredibly arduous. As Weyl humorously put it, the "soul" of each mathematical subject in turn is likely to be captured by topology. In applied mathematics there has been "capture" of parts of electrical network theory, industrial management science, linear programming and game theory, statistical mechanics, social psychology and other behavioral sciences by today's *graph theory*, a subject which applies the *combinatorial topology* of certain specialized 1-dimensional sets, called *graphs*.

The origin of graph theory, like that of some other combinatorial topological questions, can be attributed to Euler. In his day there were seven bridges crossing the Pregel River in Königsberg, Germany (Figure 25.16). Walking along the

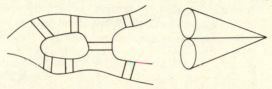

Figure 25.16   Königsberg bridges and their graph

bridges from one bank of the river to islands in the river and then to the other bank and back was a favorite Sunday pastime. The arrangement of the bridges is indicated in the *graph* (set of vertices and connecting edges) of Figure 25.16, in which there are 4 vertices and 7 edges. Euler asked: Is the graph *unicursal*, that is, can one traverse *all* seven bridges without crossing any bridge twice? By combinatorial reasoning not unlike that used in our earlier treatment of permutations and combinations, Euler arrived at a negative answer and the following theorems:

(1) In any graph the number of *odd* vertices (an odd vertex being one that lies on an odd number of edges) is even.

(2) A graph that has no odd vertices can be described unicursally by a circuit or re-entrant route that starts at any vertex and returns to that vertex.

(3) A graph that has exactly two odd vertices can be described unicursally by a route starting at one of the odd vertices and ending at the other.

(4) A graph that has more than two odd vertices *cannot* be described unicursally.

The Königsberg bridge graph exhibits 4 odd vertices and hence, by the last theorem quoted, cannot be traversed unicursally.

In any graph the unicursal routes which pass over each *edge* exactly once are called *Euler lines* (Euler *circuits* if they are re-entrant). A "dual" concept is that of a *Hamilton line* (Hamilton circuit) which passes through every *vertex* exactly once. Since the existence of Hamilton lines (circuits) is important in many practical applications of graph theory, it is unfortunate that, in contrast to the situation with Euler lines, there are no general theorems solving the existence problem for Hamilton lines. One must still rely, to considerable extent, on trial and error.

Hamilton lines arose in connection with the recreational game to which allusion was made in the footnote on page 80. In a paper presented before the British Mathematical Association meeting at Dublin in 1857, Hamilton proposed and (using combinatorial analysis and group theory) solved some special cases of the following problem: In a regular dodecahedron is there a continuous route (along the edges) that passes through each of the 20 vertices exactly once? Since our problem involves only *topological* aspects of a 1-dimensional point set, we need not consider the regular dodecahedron as pictured in Figure 25.7 or as constructed in one of the commercial "go round the world" versions of Hamilton's original puzzle. In such toys there is usually a wooden dodecahedron with a nail at each vertex (which bears the name of a city—Frankfort, Brussels, Delhi, etc.) and a string to wind from nail to nail along the edges. For our purposes we can imagine the network of dodecahedron edges to be deformed and flattened into the graph of Figure 25.17. In that graph the reader can verify that the following are Hamilton

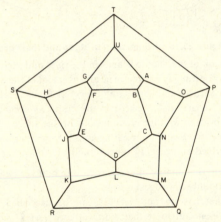

Figure 25.17   Graph of the 20 vertices and 30 edges of a regular dodecahedron

circuits:

$$ABCDEFGHJKLMNOPQRSTUA$$
$$FBAUTPONCDEJKLMQRSHGF$$
$$FBAUTSRKLMQPONCDEJHGF$$

He can also observe that, although the circuits pass through all vertices, they do *not* traverse all edges.

The matter of Euler and Hamilton lines is purely topological, but when graph theory assigns lengths to edges or "weights" to vertices, quantitative elements are added to the associated problem. For example, in Hamilton's problem one might give the distance (by air) between each pair of adjacent cities (that is, cities joined by an edge). Then there arise maximum and minimum problems, questions of optimization reminiscent of the game theory and linear programming questions of Chapter 11, and the variational issues of Chapter 23. Thus the methodology of graph theory has found the shortest route (by air) for a complete circuit (without duplication) of all the capital cities in the United States. This is a special case of the "traveling salesman problem," where the salesman must visit certain towns before returning to his home office and wishes to accomplish his task in the shortest time or as inexpensively as possible.

All of these facts indicate that even if topology has not yet reached the man in the street, it is at present a useful tool and an intriguing one to those who enjoy puzzles and mathematical recreations. Additional aspects of the subject can be revealed by considering the history of its development. We have already mentioned Euler repeatedly as a sort of founding father. Gauss, who was too modest to place his claim in the discovery of non-Euclidean geometry, appears to have also withheld most of his thoughts on analysis situs. In one of his posthumous papers, however, there is the prediction that the subject would become a matter of the greatest importance in the years ahead. In 1833 he had complained, "Leibniz initiated the geometry of situation, Euler and Vandermonde gave it momentary consideration, but after a century and a half we still know practically nothing about it." Gauss himself actually did very little to alleviate this situation. His topological work consisted of analyzing knots associated with a problem of electrostatics where linked and unlinked networks are involved. The physicist Kirchhoff (1824–1887) developed the matter fully, and his results are still of practical importance in electric network theory. Some of the more complicated problems lend themselves to elegant solutions provided, once more, by the new, fashionable graph theory.

It was Gauss's most outstanding student, Riemann, who made the first major contribution to *analysis situs*, namely, the formulation of certain abstractions which are still called *Riemann surfaces*. They solved the problem of converting general relations between complex variables into (one-valued or uniform) functions. Instead of representing the variable z in one Cartesian plane, Riemann conceived of 2 or 3 or 4 or more z-planes or sheets, according to whether the relation is a one-to-two or one-to-three or one-to-four correspondence, etc. (Figure 25.18). He pictured the sheets as attached at certain special points ("branch points") and also envisioned certain abstract "bridges" enabling passage from one sheet to another. In this way he established the homeomorphism or topological equivalence of the w-plane and the many-sheeted z-plane or Riemann surface. In short, Riemann *topologized analysis*, as mathematicians are wont to say.

As far as combinatorial *analysis situs* is concerned, Poincaré is the next mathematician in chronological order. In fact, researchers today usually consider pre-Poincaré topology as trivial and name the French leader as the creator of the whole subject. Since Poincaré was a universalist who applied his magic touch to so many mathematical domains, what motivated his decade of intensive concentration on the new "geometry"? It was his attempt to solve one of the most difficult problems in all mathematics, the so-called n-body problem. For the simplest

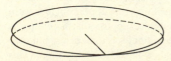

Figure 25.18 Riemann surface for the values of z when z and w are related by a one-to-two correspondence ($w^2 = z$, for example).

case, $n = 2$, that is, the 2-body problem, the solution was given by Newton. This is merely the story we told previously about his *proof* of Kepler's third law, that is, his solution of the differential equations obtained from his second law of motion and his inverse-square law of gravitational attraction. His solution gave the *approximate* orbits of two heavenly bodies, the sun and the earth, say, or the earth and the moon, etc. We emphasize the word *approximate* because the sun and a planet, or a planet and a satellite, are not two isolated bodies exerting gravitational tugs on one another. The pull of all the other bodies exerts an influence, and hence the n-body problem is that of determining what the *exact* course of all celestial bodies will be for all future time, or what it was in all the past, given exceedingly precise observational data *now*, or at any particular instant of time.

In 1887 King Oscar II of Sweden offered a prize for the solution of this problem. Poincaré did not solve the problem, but he was awarded the prize for his formidable attack on the special case $n = 3$, the 3-body problem (the determination of all-time orbits for sun, earth, moon, for example). It was the overpowering difficulty of the quantitative approach to the problem that motivated qualitative considerations, and this required Poincaré to formulate a rigorous geometry of situation. To explain the full meaning of the mathematics would involve a lengthy digression. Hence let us merely remark that the 3-body problem involves the solution of a system of nine differential equations.

Poincaré showed that if two of the bodies have masses that are small compared to the third, *periodic* solutions exist. Shortly before his death in 1912, he proved that certain orbits would be periodic, *provided* that a simple geometric theorem, topological in nature, is true. He was ailing and fatigued at the time and found himself unable to demonstrate the proposition on which the whole issue depended. The theorem *sounds* like an easy one, for it concerns the annular area between two concentric circles, the problem being to prove that when a certain topological transformation of this area is carried out, two of its points must remain fixed. It is hard to believe that a serious issue of dynamical astronomy can depend on such an apparently simple question. In 1913 George D. Birkhoff (whose ergodic theorem was discussed in Chapter 15) gained international fame in the mathematical world by *proving* the last theorem of Poincaré, a feat considered one of the greatest achievements in the American mathematician's long, distinguished career. Like many of the most difficult topological problems (some are still unsolved), the particular theorem sounds like an exercise in high school geometry.

In the same way, many of the famous theorems of topology sound obvious or trivial. For example, there is the *Jordan curve theorem*. Any curve homeomorphic to a circle is called a Jordan curve, and the theorem establishes that in the plane, or on the sphere, a Jordan curve divides the surface into two distinct regions which have no point in common (an "inside" and an "outside"). Figure 19.6 indicates that the theorem does not hold on the torus. Some things that appear obvious form the still unsolved problems of topology. For example, suppose that a country is subdivided into states or districts, and a map of this country is to be colored in such a way that no two states with a common boundary have the same color. (This common boundary must be a line segment and may not be a single point.) By diagramming the most intricate kinds of subdivision, the reader can convince himself, by trial, that *four* colors will always suffice. To date no proof of this "obvious fact" has been obtained even though the problem has been under investigation ever since Moebius proposed it more than a century ago.

It has been proved, however, that *seven* colors suffice for any map on the torus, however complicated the scheme of subdivision. The minimum number of colors required for a map on a specified surface is called the *chromatic number* of that surface, and furnishes another example of a topological invariant. Thus the chromatic number of the torus is 7, and topologists hope to be able to prove that the chromatic number of the plane is 4.

As emphasized earlier, the topological invariants connected with a surface or space help to characterize it, and therefore by establishing the inequality of Euler characteristics, chromatic numbers, or other numerical invariants, one can demonstrate that two abstract manifolds are not homeomorphic, and thus belong to different classifications. Such use of invariants is much easier than attempting to prove directly that no homeomorphism exists for two configurations.

The chromatic number is an invariant arising through the subdivision of a surface, and in general, topology arrives at characterizations of manifolds by studying the nature of the submanifolds which can be embedded in them. For example, invariants of a surface can be derived from the curves which can be drawn on these surfaces. Thus, on a sphere, an ordinary closed curve like a circle or an ellipse can be contracted continuously into a point without leaving the surface. On the other hand, a meridional or a latitudinal circle on a torus (Figure 25.19*a*)

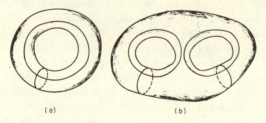

(a)   (b)

Figure 25.19   Variation in connectivity and Betti number on different surfaces

cannot be contracted continuously into a point without breaking away from the surface. This situation was described earlier (Chapter 19) as a matter of *connectivity*, a property associated with the possibility of *connecting* any point of a surface with any other by means of a continuous path. Thus, on a torus, even if a meridional and a latitudinal circle are drawn, it is still possible to connect any point of the surface with any other without crossing these circles; that is, they do not fence in any portion of the surface. By contrast, a single circle, or any simple closed curve, can serve as a dividing line on a spherical surface. Hence the sphere is said to be *simply connected*, whereas the torus is described as being *multiply connected* or having a higher order of connectivity. But another way the topologist handles this question is to apply the notion of *homotopy* or continuous deformation. On a simply connected surface any simple closed curve can be deformed continuously into any other, or all closed curves are equivalent, if homotopy is the criterion. This is not the case on nonsimply connected surfaces.

Instead of studying closed curves from the point of view of homotopy, one may relate them to the Jordan curve theorem, and examine their bounding properties. The fact that in the plane or on the sphere a closed curve bounds, that is, has an "inside" and an "outside," although this need not be the case on the torus or other surfaces of higher genus, suggests a way of establishing the nonhomeomorphism of surfaces. If a third closed curve were drawn on the torus of Figure 25.19*a*, it would divide the surface into two parts, or be a boundary. But in the torus with two holes (Figure 25.19*b*), one may draw meridional and latitudinal circles around each hole (4 closed curves in all) without obtaining a boundary, that is, without dividing the surface into disjoint areas. Therefore the torus with two holes is *not* homeomorphic to the ordinary torus.

These facts are associated with numerical invariants of each surface. The maximum number of closed curves which can be drawn on a surface without dividing it into disjoint areas is a topological invariant called the 1-dimensional *Betti number* of the surface, after the Italian Enrico Betti (1823–1892). From what was stated above, the Betti numbers for the sphere, the simple torus, and the torus with two holes must be 0, 2, 4, respectively. But Riemann held priority over Betti in indicating the topological invariance for each surface of the number of non-bounding closed curves. Since the German geometer actually wished to *divide* a surface by means of closed curves, he always drew one more closed curve than Betti or Poincaré and therefore the *Riemannian connectivity number* is the minimum number of closed curves that will divide a surface into disjoint regions. Thus this number is always one greater than the corresponding Betti number. Hence the sphere, torus, and two-holed torus possess connectivity 1, 3, 5, respectively.

It seems intuitive that the Riemann-Betti-Poincaré notion should be capable of generalization to invariants associated with the dividing or bounding properties of sets or closed surfaces in a 3-dimensional manifold, and then to analogous invariants for abstract spaces of higher dimensionality. Poincaré did, in fact, develop his *homology theory* on this basis, and since he believed in the omnipotence of the group concept, it is not surprising that he applied it with fruitful results in this instance. If a closed curve C or "cycle," as it is called in topology, is the *boundary* of a region on a surface, the Poincaré theory states that "C is homologous to zero," which is symbolized by $C \sim 0$. If cycles are oriented (clockwise and counterclockwise), then $-C$ has a meaning. Since one can go round and round a closed curve, $2C$, $3C$, $4C$, etc. also have meaning. Cycles may be added and subtracted. Thus $C + C'$ and $C - C'$ are indicated in Figures 25.20*a* and 25.20*b*. If one selects a convention concerning the region bounded and this assumption is intuitively equivalent to having the area to the *left* of the bounding curves, then $C - C'$ bounds the annulus in Figure 25.20*b*, or $C - C' \sim 0$. In this case, C is said to be homologous to C'. From

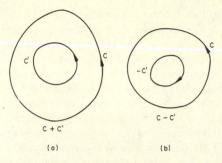

C + C'   C - C'

(a)   (b)

Figure 25.20

the point of view of homology, C and C' are equivalent. If all cycles homologous to a particular one are considered equivalent, then they can be classified as a single type. Thus homology is an equivalence relation for the classification of cycles, just as earlier in this chapter homeomorphism was an equivalence relation for spaces. The number of types of cycle, that is, the number of *homology classes*, on a manifold is an invariant of the manifold. From what has been stated above, it is easy to show that the cycle types on a manifold form a commutative or abelian group with respect to the operation of addition. Poincaré named this the *Betti group*. The concept of cycles and their homologies may be extended to 3-dimensional, 4-dimensional, . . . , n-dimensional manifolds. On higher dimensional manifolds one has not only closed curves but also closed surfaces, etc.; that is, the cycles have 1, 2, 3, 4, etc. dimensions.

Poincaré formulated the concepts of cycle and homology more exactly by developing an algebraic or combinatorial scheme of representation for a manifold. Then he showed that the fundamental problem of homology theory is to determine the structure of the Betti groups for the 1-, 2-, 3-, . . . , n-dimensional cycles, and in particular to determine the numerical invariants associated with each Betti group. These are called the n-dimensional *Betti numbers* and the *coefficients of torsion* (in nonorientable manifolds). There is a fundamental theorem: Two orientable manifolds which have equal Betti numbers of all dimensions are topologically equivalent. After Poincaré, other significant contributions to homology theory were made by members of the Princeton school, in particular by Oswald Veblen (1880–1960), J. W. Alexander, and Solomon Lefschetz. Alexander proved the so-called duality theorem, which is a broad generalization of the Jordan curve theorem. Then this theorem was extended still further by Lefschetz and by P. Alexandroff and Pontrjagin. Lefschetz established certain "intersection" theorems and applied these to developing very important formulas for the number of points which remain fixed when a manifold is subjected to a continuous mapping onto itself. This theory of fixed points was applied in the previously mentioned research of Birkhoff and Morse.

There are a number of great names in topology today. The Princeton school is still notable, but there are also distinguished topologists outside the United States.

Some of these contribute directly to the subject, whereas others apply its concepts to various branches of mathematics much as Cartan used topological concepts in differential geometry. In fact, as the twentieth century advanced, both analytic and combinatorial topology progressed far beyond their initial phases. Major developments in point-set topology are due to the modern Polish analysts and to the school of the American mathematician R. L. Moore. But the United States has contributed even more notably to the growth of combinatorial topology through the remarkable discoveries mentioned above, and those made more recently by mathematicians at Princeton and the Institute for Advanced Study. One can name specifically H. Marston Morse, A. W. Tucker, Hassler Whitney, N. E. Steenrod. Their research, exceedingly advanced in nature, contributes to the rigorization, generalization, and application of the elementary topological concepts considered in this book.

In the present and preceding chapters there has been some indication of how the "angel of topology" is a guiding spirit in modern geometry, analysis, physics, and mathematical astronomy, but the *bon mot* of Weyl quoted at the outset of this chapter implies that *analysis situs* has penetrated into other mathematical subjects and pervades current activity. Thus topology is likely to be a most potent factor in the immediate future of mathematics.

26

The Leonardos of
Modern Mathematics

Leonardo da Vinci is the classic example of the universal genius, an increasingly rare type in human history, primarily because the advance of civilization entails more and more specialization of interests. Carl Friedrich Gauss, the towering figure who provided a glorious ending to the eighteenth century of mathematical history and propelled the nineteenth century into vital activity, was not a universalist in the same sense as Leonardo. However, his creative thought was all-encompassing within the domain of mathematics and physical science. His biography was given in Chapter 19 and his name has been mentioned in almost every chapter. His contributions to number theory, differential geometry, non-Euclidean geometry, astronomy, probability theory, the theory of functions, and topology have been discussed. Now, a century after Gauss, there is probably no living mathematician who is able to contribute so notably to so many different mathematical specialties. Twentieth-century science has reached a point where our leading thinkers usually find it necessary to devote themselves to a subspecialty of a subspecialty of a field already highly specialized. This is markedly true in mathematics, where Henri Poincaré and David Hilbert are often considered to have been the last of the universalists, with the concession that more recent additions to the mathematical Pantheon, like Hermann Weyl and John von Neumann, were still able to place the stamp of their genius on many branches of the mother science.

What manner of man can embrace all or most of mathematics in his research activities? Perhaps some clue is to be found in the biographies of the modern mathematical "Leonardos." The personal stories of Gauss and von Neumann have already been told, but there remains the matter of examining the biographies of Poincaré, Hilbert, and Weyl. If the mathematical ideas and discoveries of all five men were presented in complete detail, they would offer a fairly complete survey of modern mathematics.

Henri Poincaré was born in Nancy in 1854, the son of Léon Poincaré, a physician whose native town was Rouen. The family was a distinguished one; Henri's first cousin, Raymond, was president of France during World War I; his sister, several years his junior, married the philosopher Émile Boutroux, and her son was a budding young mathematician when death cut his career short.

Poincaré was a delicate child and was affected badly by diphtheria and other serious illnesses of childhood. There was little romping and outdoor sport for him. Instead he received his first education at home under the kindly guidance and direction of his mother. At elementary school his greatest interest was nature study, and this was not replaced by mathematics until he was fifteen years of age. His literary aptitude manifested itself early, for a composition written by him at age nine was pronounced a "small masterpiece" by his teacher. It is not surprising then that his mature years brought membership in the *literary* section of the Institut de France, an honor never accorded to any other mathematician.

His scientific education began with a short period at the School of Forestry, where he was awarded first prize in mathematics. He left this school shortly in order to enter the École Polytechnique, and after graduating from this famous institution, he decided to become an engineer and hence attended the School of Mines. He served as an engineer at Vesoul, not far from Nancy, where on one occasion he was a member of a rescue crew after a mine explosion and fire had occurred. But his very first research papers—in the field of differential equations—attracted the attention of leading French mathematicians, and he was offered a teaching position at Caen. In 1881, when he was twenty-seven, he was called to the University of Paris, where his academic advancement was rapid. In 1886 he became professor of mathematical physics and probability theory, and later occupied the chair in celestial mechanics. From 1904 to 1908 he held a chair in astronomy at the École Polytechnique. From 1908 to 1912 his health was poor, and at age fifty-eight he died suddenly from an embolism following surgery.

In the field of research Poincaré's total creative output was so varied in subject matter and so colossal in volume that it is almost impossible to survey it. Hence it seems best to indicate three or four of his major specialties. He carried on research in differential equations as related to both physical science and pure mathematics. Earlier in this book we quoted his belief that mathematical astronomy is a major scientific subject. The fact that he was the primary figure in the creation of modern combinatorial topology has been emphasized and illustrated.

Poincaré was an exceedingly modest man. That, along with his belief that the world of mathematics should be an international one, made him often give credit to foreign (non-French) mathematicians for work that he had done himself. To honor the Italian mathematician Enrico Betti, Poincaré named certain topological entities Betti numbers (Chapter 25), although Poincaré, not Betti, had discovered them. The story is similar for the Fuchsian functions, where Poincaré wished to compliment the German analyst Lazarus Fuchs (1833–1902). There is a mathematical joke to the effect that Poincaré called certain functions Kleinian because Felix Klein had nothing to do with them. This is not quite accurate, however. About 1881 Poincaré had initiated his theory of *automorphic functions*, a generalization of elliptic functions (Chapter 7), which are themselves a generalization of ordinary trigonometric functions. He was unaware that Klein held priority. When the latter wrote a letter to apprise Poincaré of this fact, he immediately gave Klein's name to all subsequent automorphic functions, and in fact to much of their theory which he himself developed, single-handedly, after Klein had a nervous breakdown. One thing is certain:

the French leader extended himself in generosity to mathematicians of other nationalities.

During the years when Poincaré was working on automorphic functions he also carried on research in algebra and number theory, and during every period of his professional life he gave some of his best thought to mathematical astronomy. His results appeared in the three volumes of his *Méthodes nouvelles de la mécanique céleste*, published in 1892, 1893, and 1899. Then he published another three volumes in 1905, 1909, 1910, this work being entitled *Leçons de mécanique céleste*. This was more pedagogical and practical than his earlier three-volume treatise, which had been largely mathematical and theoretical, and so revolutionary in its point of view that it was considered comparable to Laplace's *Mécanique céleste*. As sequels to these astronomical works there were his *Leçons sur les figures d'équilibre d'une masse fluide* and his *Leçons sur les hypothèses cosmogoniques*. But it is impossible to make a suitable selection of Poincaré titles from some five hundred research papers, thirty books on mathematical physics and astronomy, numerous philosophic and popular works on science.

A remark on these last is, however, in order. Around the turn of the century Poincaré started to write a series of monographs to popularize important aspects of mathematics, physical science, and the philosophy of science. In these books he exhibited his literary talent and his ability as a lucid expositor. E. T. Bell claims that "workmen and shopgirls could be seen in the parks and cafés of Paris avidly reading one or other of Poincaré's popular masterpieces in its cheap print and shabby paper cover."* If this sounds like an exaggeration, let us say that these very monographs are very popular in France at the present time, and in their English translations are best sellers among paperbacks on science in the United States today.

It is evident that Poincaré's life was "all work and no play." One would like to think that he found his trips for the purpose of lecturing at mathematical or scientific congresses a sort of vacation. One of these brought him to the United States in 1904, when he was one of the guest speakers at the St. Louis Exposition. Like most mathematicians he enjoyed music, and he was very happy in the midst of his family—his wife, son, and three daughters.

David Hilbert was born just eight years after Poincaré, on January 23, 1862, in the city of Königsberg in East Prussia. We have the word of Hermann Weyl that Hilbert's career was comparable to that of the other mathematical "Leonardos" we have named: "The world looked upon David Hilbert as the greatest mathematician living in the first decades of the twentieth century." Weyl gave one important reason for this evaluation:

At the International Mathematical Congress in Paris in 1900 David Hilbert, convinced that problems are the life-blood of science, formulated twenty-three unsolved problems which he expected to play an important role in the development of mathematics during the next era. How much better he predicted the future of mathematics than any

---

* E. T. Bell, *Men of Mathematics*, Simon and Schuster, New York, 1937, p. 530.

politician foresaw the gifts of war and terror that the new century was about to lavish upon mankind! We mathematicians have often measured our progress by checking which of Hilbert's questions have been settled.*

Because Weyl was himself a universalist, it was recognized that he was the person best qualified to write a critical biography of Hilbert. Therefore, when the latter died in 1943, leading scientific societies throughout the world called on Weyl to wrote an account of the great leader's life and work. One of the best and fullest of such eulogies was the one Weyl wrote for the Royal Society of London.† We quote some excerpts from that biography.

Hilbert was descended from a family which had long been settled in Königsberg and had brought forth a series of physicians and judges. During his entire life he preserved uncorrupted the Baltic accent of his home. For a long time Hilbert remained faithfully attached to the town of his forbears, and well deserved its honorary citizenship which was bestowed upon him in his later years. It was at the University of Königsberg that he studied, where in 1884 he received his doctor's degree, and where in 1886 he was admitted as Privatdozent; there, moreover, he was appointed Ausserordentlicher Professor, in 1892, succeeding his teacher and friend Adolf Hurwitz, and in the following year advanced to a full professorship. The continuity of this Königsberg period was interrupted only by a semester's studies at Erlangen, and by a travelling scholarship during the year before his habilitation, which took him to Felix Klein at Leipzig and to Paris where he was attracted mainly to Ch. Hermite. It was on Klein's initiative that Hilbert was called to Göttingen in 1895; there he remained until the end of his life. He retired in the year 1930.

In 1928 he was elected to foreign membership of the Royal Society.

Beginning in his student years at Königsberg a close friendship existed between him and Hermann Minkowski, his junior by two years, and it was with deep satisfaction that in 1902 he succeeded in bringing Minkowski also to Göttingen. Only too soon did the close collaboration of the two friends end with Minkowski's death in 1909. Hilbert and Minkowski were the real heroes of the great and brilliant period, unforgettable to those who lived through it, which mathematics experienced during the first decade of this century in Göttingen. Klein ruled over it like a distant god, "divus Felix," from above the clouds; the peak of his mathematical productivity lay behind him. Among the authors of the great number of valuable dissertations which in these fruitful years were written under Hilbert's guidance we find many Anglo-Saxon names, names of men who subsequently have played a considerable role in the development of American mathematics. The physical set-up within which this free scientific life unfolded was quite modest. Not until many years after the first world war, after Felix Klein had gone and Richard Courant had succeeded him, towards the end of the brief period of the German Republic, did Klein's dream of the Mathematical Institute at Göttingen come true. But soon the Nazi storm broke and those who had laid the plans and who taught there with Hilbert were scattered over the earth, and the years after 1933 became for him tragic years of ever deepening loneliness.

Hilbert was of slight build. Above the small lower face with its goatee there rose the dome of a powerful, in later years bald, skull. He was physically agile, a tireless walker, a good skater, and a passionate gardener. Until 1925 he enjoyed good health. Then he fell ill

---

* Hermann Weyl, "A Half-Century of Mathematics," *American Mathematical Monthly*, Vol. 58, No. 8 (October 1951), p. 525.

† Royal Society of London, *Obituary Notices of Fellows*, Vol. 4, No. 13 (November 1944), pp. 547 ff.

of pernicious anaemia. Yet this illness only temporarily paralyzed his restless activity in teaching and research. He was among the first with whom the liver treatment, inaugurated by G. R. Minot at Harvard, proved successful; undoubtedly it saved Hilbert's life at that time.

Hilbert's research left an indelible imprint on practically all branches of mathematical science. Yet in distinct successive periods he devoted himself with impassioned exclusiveness to but a single subject at a time. Perhaps his deepest investigations are those on the theory of number fields. His monumental report on the "Theorie der algebraischen Zahlkörper", which he submitted to the Deutsche Mathematiker-Vereinigung, is dated as of the year 1897, and as far as I know Hilbert did not publish another paper in this field after 1899. The methodical unity of mathematics was for him a matter of belief and experience. It appeared to him essential that—in the face of the manifold interrelations and for the sake of the fertility of research—the productive mathematician should make himself at home in all fields. To quote his own words: "The question is forced upon us whether mathematics is once to face what other sciences have long ago experienced, namely to fall apart into subdivisions whose representatives are hardly able to understand each other and whose connexions for this reason will become ever looser. I neither believe nor wish this to happen; the science of mathematics as I see it is an indivisible whole, an organism whose ability to survive rests on the connexion between its parts." Theoretical physics also was drawn by Hilbert into the domain of his research; during a whole decade beginning in 1912 it stood at the centre of his interest. Great, fruitful problems appear to him as the life nerve of mathematics. "Just as every human enterprise prosecutes final aims," says he, "so mathematical research needs problems. Their solution steels the force of the investigator." In his famous lecture at the International Congress of Mathematicians at Paris in 1900 Hilbert tries to probe the immediate future of mathematics by posing twenty-three unsolved problems; they have indeed, as we can state to-day in retrospect, played an eminent role in the development of mathematics during the subsequent forty-three years. A characteristic feature of Hilbert's method is a peculiarly direct attack on problems, unfettered by any algorithms; he always goes back to the questions in their original simplicity. When it is a matter of transferring the theory of linear equations from a finite to an infinite number of unknowns he begins by getting rid of the calculatory tool of determinants. A truly great example of far reaching significance is his mastery of Dirichlet's principle which, originally springing from mathematical physics, provided Riemann with the foundation of his theory of algebraic functions and Abelian integrals, but which subsequently had fallen a victim of Weierstrass's pitiless criticism. Hilbert salvaged it in its entirety. The whole finely wrought apparatus of the Calculus of Variations was here consciously set aside. We only need to mention the names R. Courant and M. Morse to indicate what role this direct method of the Calculus of Variations was destined to play in recent times. It seems to me that with Hilbert the mastering of single concrete problems and the forming of general abstract concepts are balanced in a particularly fortunate manner. He came of a time in which the algorithm had played a more extensive part, and therefore he stressed strongly a conceptual procedure; but in the meantime our advance in this direction has been so uninhibited and with so little concern for a growth of the problematics in depth that many of us have begun to fear for the mathematical substance. In Hilbert simplicity and rigour go hand in hand. The growing demand for rigour, imposed by the critical reflections of the nineteenth century upon those parts of mathematics which operate in the continuum, was felt by most investigators as a heavy yoke that made their steps dragging and awkward. Full of longing and with uneasiness they looked back upon Euler's era of happy-go-lucky analysis. With Hilbert rigour figures no longer as enemy, but as promoter of simplicity. Yet the secret of Hilbert's creative force is not plumbed by any of these remarks. A further element of it, I feel, was his sensitivity in registering hints which revealed to him general relations while solving special problems. This is most magnificently exemplified by the way in which, during his theory of numbers period, he was led to the enunciation of his general theorems on class fields and the general law of reciprocity.

We shall now recall Hilbert's most important achievements. In the years 1888–1892 he proved the fundamental finiteness theorems of the *theory of invariants* for the full projective group. His method, though yielding a proof for the existence of a finite basis for the invariants, does not enable one actually to construct it in a concrete individual case. Hence the exclamation by the great algorithmician P. Gordan, at the appearance of Hilbert's paper: "This is not mathematics; this is theology!" It reveals an antithesis which reaches down to the very roots of mathematics. Hilbert, however, in further penetrating investigations, furnished the means for a finite execution of the construction.

His papers on the theory of invariants had the unexpected effect of withering, as it were overnight, a discipline which so far had stood in full bloom. Its central problems he had finished once and for all. Entirely different was his effect on the *theory of number fields*, which he took up in the years 1892–1898. It is a great pleasure to watch how, step by step, in a succession of papers ascending from the particular to the general, the adequate concepts and methods are evolved and the essential connexions come to light. These papers proved of extraordinary fertility for the future. On the pure theory of numbers side I mention the names of Furtwängler, Takagi, Artin, Hasse, Chevalley, and on the number-and-function theoretical one, those of Fueter and Hecke.

During the subsequent period, 1898–1902, the *foundations of geometry* are nearest to Hilbert's heart, and he is seized by the idea of axiomatics. The soil was well prepared, especially by the Italian school of geometers. Yet it was as if over a landscape, wherein but a few men with a superb sense of orientation had found their way in murky twilight, the sun had risen all at once. Clear and clean-cut we find stated the axiomatic concept according to which geometry is a hypothetical deductive system; it depends on the "implicit definitions" of the concepts of spatial objects and relations which the axioms contain, and not on a description of their intuitive content. A complete and natural system of geometric axioms is set up. They are required to satisfy the logical demands of consistency, independence, and completeness, and by means of quite a few peculiar geometries, constructed *ad hoc*, the proof of independence is furnished in detail. The general ideas appear to us to-day almost banal, but in these examples Hilbert unfolds his typical wealth of invention. While in this fashion the geometric concepts become formalized, the logical ones function as before in their intuitive significance. The further step where logic too succumbs to formalization, thus giving rise to a purely symbolic mathematics—a step upon which Hilbert already pondered at this epoch, as a paper read to the International Congress of 1904 proves, and which is inevitable for the ultimate justification of the role played by the infinite in mathematics—was systematically followed up by Hilbert during the final years of his mathematical productivity, from 1922 on. In contrast to L. E. J. Brouwer's intuititionism, which finds itself forced to abandon major parts of historical mathematics as untenable, Hilbert attempts to save the holdings of mathematics in their entirety by proving its formalism free of contradiction. Admittedly the question of truth is thus shifted into the question of consistency. To a limited extent the latter has been established by Hilbert himself in collaboration with P. Bernays, by J. von Neumann, and G. Gentzen. In recent times, however, the entire enterprise has become questionable on account of K. Gödel's surprising discoveries. While Brouwer has made clear to us to what extent the intuitively certain falls short of the mathematically provable, Gödel shows conversely to what extent the intuitively certain goes beyond what (in an arbitrary but fixed formalism) is capable of mathematical proof. The question of the ultimate foundations and the ultimate meaning of mathematics remains open; we do not know in which direction it will find its final solution nor even whether a final objective answer can be expected at all. "Mathematizing" may well be a creative activity of man, like language or music, of primary originality, whose historical decisions defy complete objective rationalization.

A chance occasion, a lecture in 1901 by the Swedish mathematician E. Holmgren in Hilbert's seminar dealing with the now classical paper of Fredholm's on *integral equations*, then but recently published, provided the impulse which started Hilbert on his investigations on this subject which absorbed his attention until 1912. Fredholm had limited himself to setting up the analogue of the theory of linear equations, while Hilbert recognized that the analogue of the transformation on to principal axes of quadratic forms yields the theory of the eigenvalues and eigenfunctions for the vibration problems of physics. He developed the parallel between integral equations and sum equations in infinitely many unknowns, and subsequently proceeded to push ahead from the spectral theory of "completely continuous" to the much more general one of "bounded" quadratic forms. To-day these things present themselves to us in the framework of a general theory of Hilbert space. Astonishing indeed is the variety of interesting applications which integral equations find in the most diverse branches of mathematics and physics.

It was also due to his influence that the theory of integral equations became a world-wide fad in mathematics for a considerable length of time, producing an enormous literature, for the most part of rather ephemeral value. It was not merit but a favour of fortune when, beginning in 1923 (Heisenberg, Schrödinger) the spectral theory in Hilbert space was discovered to be the adequate mathematical instrument of quantum physics. This later impulse led to a re-examination of the entire complex of problems with refined means (J. von Neumann, M. Stone, and others).

This period of integral equations is followed by Hilbert's physical period. Significant though it was for Hilbert's complete personality as a scientist, it produced a lesser harvest than the purely mathematical phases, and may here be passed over. I shall mention instead two single, somewhat isolated, accomplishments that were to have a great effect: his vindication of Dirichlet's principle; and his proof of a famous century-old conjecture of Waring's, carrying the statement that every integer can be written as a sum of four squares over from squares to arbitrary powers. The physical period is finally succeeded by the last one, already mentioned above, in the course of which Hilbert gives an entirely new turn to the question concerning the foundation and the truth content of mathematics itself. A fruit of Hilbert's pedagogic activity during this period is the charming book by him and Cohn-Vossen, *Anschauliche Geometrie*.

This summary, though far from being complete, may suffice to indicate the universality and depth of Hilbert's mathematical work. He impressed the seal of his spirit upon a whole era of mathematics. And yet I do not believe that his research work alone accounts for the brilliance that radiated from him, nor for his tremendous influence. Gauss and Riemann, to mention two other Göttingers, were greater mathematicians than Hilbert, and yet their immediate effect upon their contemporaries was undoubtedly smaller. Part of this is certainly due to the changing conditions of time, but the character of the men is probably more decisive. Hilbert's was a nature filled with the zest of living, seeking the intercourse of other people, and delighting in the exchange of scientific ideas. He had his own free manner of learning and teaching. His comprehensive mathematical knowledge he acquired not so much from lectures as in conversations with Minkowski and Hurwitz. "On innumerable walks, at times undertaken day by day," he tells in his obituary on Hurwitz, "we browsed in the course of eight years through every corner of mathematical science." And as he had learned from Hurwitz, so he taught in later years his own pupils—on far-flung walks through the woods surrounding Göttingen or, on rainy days, as peripatetics, in his covered garden walk. His optimism, his spiritual passion, and his unshakable faith in the value of science were irresistibly infectious. He says: "The conviction of the solvability of each and every mathematical problem spurs us on during our work; we hear within ourselves the steady call: there is the problem; search for the solution. You can find it by sheer thinking, for in mathematics there is no *ignorabimus*." His enthusiasm did get along with criticism, but not with scepticism. The snobbish attitude of pretended indifference, of "merely fooling around with things" or even of playful cynicism, did not exist in his circle. Hilbert was enormously industrious; he liked to quote Lichtenberg's saying: "Genius is industry." Yet for all this there was light and laughter around him. Under the influence of his dominating power of suggestion one readily considered important whatever he did; his vision and experience inspired confidence in the fruitfulness of the hints he dropped. It is moreover decisive that he was not merely a scientist but a scientific personality, and therefore capable not only of teaching the technique of his science but also of being a spiritual leader. Although not committing himself to one of the established epistemological or metaphysical doctrines, he was a philosopher in that he was concerned with the life of the idea as it realizes itself among men and as an indivisible whole; he had the force to evoke it, he felt responsible for it in his own sphere, and measured his individual scientific efforts against it. Last, not least, the environment also helped. A university such as Göttingen, in the halcyon days before 1914, was particularly favourable for the development of a living scientific school. Once a band of disciples had gathered around Hilbert, intent upon research and little worried by the toil of teaching, it was but natural that in joint competitive aspiration of related aims each should stimulate the other; there was no need that everything come from the master.

His homeland and America among all countries felt Hilbert's impact most thoroughly. His influence upon American mathematics was not restricted to his immediate pupils. Thus, for instance, the Hilbert of the foundations of geometry had a profound effect on E. H. Moore and O. Veblen; the Hilbert of integral equations on George D. Birkhoff.

A picture of Hilbert's personality should also touch upon his attitude regarding the great powers in the lives of men; social and political organization, art, religion, morals and manners, family, friendship, love. Suffice it to say here that he was singularly free from all national and racial prejudices, that in all questions, political, social, or spiritual, he stood forever on the side of freedom, frequently in isolated opposition to the compact majority of his environment. Unforgotten by all those present remains the unanimous and prolonged applause which greeted him in 1928 at Bologna, the first International Congress of Mathematicians at which, following a lengthy struggle, the Germans were once more admitted. It was a telling expression of veneration for the great mathematician whom every one knew to have risen from a severe illness, but at the same time an expression of respect for the independent attitude, "au dessus de la mêlée", from which he had not wavered during the world conflict. He died in Göttingen, Germany, in 1943, at the age of eighty-one, when he succumbed to a compound fracture of the thigh brought about by a domestic accident.

Weyl, Hilbert's biographer, is the last, in chronological order, of the modern Leonardos with whom the present chapter is concerned. Because Weyl was a universalist, it was a group of mathematicians, rather than a single specialist who wrote his biography, namely, Professors M. H. A. Newman, H. Davenport, P. Hall, G. E. H. Reuter, and L. Rosenfeld. Each one of these men is an expert in a particular one of Weyl's specialities and hence is able to explain Weyl's role in the area in question.*

Hermann Weyl was born on 9 November 1885, the son of Ludwig and Anna Weyl, in the small town of Elmshorn near Hamburg. When his schooldays in Altona ended in 1904 he entered Göttingen University as a country lad of eighteen, and there remained (except for a year at Munich), first as student and then as Privatdozent, until his call to Zurich in

---

* Royal Society of London, *Biographical Memoirs of Fellows*, Vol. 2 (1956), pp. 305 ff.

1913. Of these days he said (in the obituary of Hilbert for this Society, 1944), "Hilbert and Minkowski were the real heroes of the great and brilliant period which mathematics experienced during the first decade of the century in Göttingen, unforgettable to those who lived through it. Klein ruled over it like a distant god, 'divus Felix,' from above the clouds." Among those nearer to his own age whom he found there were Carathéodory and Harald Bohr, Courant, Zermelo, Erhard Schmidt.

While still a schoolboy he had picked up in his father's house an old copy of the *Critique of Pure Reason*, and absorbed with enthusiasm Kant's thesis of the *a priori* nature of Euclidean geometry. But in Göttingen Hilbert had just completed his classical work on the foundations of geometry, with its host of strange "counter"-geometries. Kantian philosophy could not survive this blow: Weyl transferred his allegiance to Hilbert. "I resolved to study whatever this man had written. At the end of my first year I went home with the 'Zahlbericht' under my arm, and during the summer vacation I worked my way through it—without any previous knowledge of elementary number theory or Galois theory. These were the happiest months of my life, whose shine, across years burdened with our common share of doubt and failure, still comforts my soul."

In spite of the great variety of mathematical stimulation of the Göttingen years, this was the only period of comparable length in which he devoted himself to a single branch of mathematics—analysis, and to a single theme, the problems that arose naturally out of his dissertation, on singular integral equations. Towards the end of this period two causes combined to turn his attention to wider fields. First, in the session 1911–1912 he lectured on the theory of Riemann surfaces, and was led by his sense of the inadequacy of existing treatments to plunge deep into the topological foundations. Secondly, in 1913 he accepted the offer of a chair at the Institute of Technology in Zurich, where his colleague for one year was Einstein, who was just then discovering the general theory of relativity. Weyl was soon launched on the series of papers on relativity and differential geometry which culminated in the book *Raum-Zeit-Materie*. Later still this work led on, through his analysis and generalization of the Lie-Helmholtz space-problem, to his third great theme, the representation theory of the classical groups, and its application to quantum theory.

In the decade 1917–1927, he was at the height of his powers. A stream of papers appeared, not only on his main themes, but on any mathematical topics that interested him—and that meant in almost all parts of mathematics. (A glance at the papers listed in the bibliography under the year 1921 will give some impression of the spread of his interests at this time.) He was ready to defend this universality, this refusal to put all his effort into making steady and systematic progress in one field. "My own mathematical works were always quite unsystematic, without method or connexion. Expression and shape are almost more to me than knowledge itself. But I believe that, leaving aside my own peculiar nature, there is in mathematics itself, in contrast to the experimental disciplines, a character which is nearer to that of free creative art. For this reason the modern scientific urge to found Institutes of Science is not so good for mathematics, where the relationship between teacher and pupil should be milder and looser. In the fine arts we do not normally seek to impose the systematic training of pupils upon creative artists."

The years at Zurich were happy ones, during which, he says, the worst that happened to disturb his peace was a series of offers of chairs by foreign universities; "for such decisions worried me." In the *Rückblick* he tells amusingly how, when he received his first invitation to Göttingen, to succeed Klein in 1923, he walked his wife Hella round and round a block in Zurich till nearly midnight, then jumped on the last tram to telegraph acceptance—and refused. The second invitation, to succeed Hilbert in 1930, he accepted, after still more painful hesitations. But his short stay as professor in Göttingen was clouded over by the threat of coming political events. In 1933 he decided that he could not stay in Germany after the dismissal of his colleagues by the Nazis, and he accepted an offer of permanent membership of the Institute for Advanced Study, then newly founded in

Princeton. There he worked as a member till his retirement in 1951, and he remained an emeritus member till his death in 1955, spending half his time there and half in Zurich. Of the Institute he said that it is the finest workshop for a mathematician that it is possible to imagine.

He married in 1913 Helene Joseph, the daughter of a doctor in Ribnitz in Mecklenburg, and there were two sons of the marriage. All who were visitors at the Weyls' house in Mercer Street will remember her charm and gaiety. She shared to the full his taste for philosophy and for imaginative and poetical literature, and was the translator of many Spanish works, including the writings of Ortega y Gassett, into German. She died in 1948.

In 1950 he married Ellen Bär, born Lohnstein, of Zurich, and from that time had the happiness of spending half of each year in Zurich. He died suddenly, of a heart attack, on 9 December 1955.

The last public event of his life was the seventieth birthday gathering, at which he was presented with a volume of "Selecta" from his own works. A wider circle of his friends had a last happy glimpse of him at the Amsterdam Congress in 1954, where he delivered the address on the work of the Fields Medallists (Kodaira and Serre), a *tour de force* which showed him, in his sixty-ninth year, well abreast of those new theories which have changed the face of mathematics in the last twenty years.

Few mathematicians have left so clear an impression of themselves in their work. His life-long interest in philosophical problems, and his conviction that they cannot be separated from the problems of science and mathematics, has left its mark everywhere in his work. In the last year of his life he wrote a brief philosophical autobiography, which he called "Erkenntnis und Besinnung," a title which he explained in these words. "In the intellectual life of man there can be clearly distinguished two domains: the domain of *action* (Handelns), of shaping and construction to which active artists, scientists, technicians and statesmen devote themselves; and a domain of *reflexion* (Besinnung) of which the fulfilment lies in insight, and which, since we struggle in it to find the *meaning* of our activity, is to be regarded as the proper domain of the philosopher." The essay itself traces with affectionate detachment the philosophical progress of the young Weyl, from Kant through Husserl's "Phenomenology" and Fichte's Idealism to his discovery in 1922 of the medieval mystic Eckehart, who gave him for a time "that access to the religious world which I had lacked ten years earlier.... But my metaphysical-religious speculations, aroused by Fichte and Eckehart, never came to a clear conclusion; that was in the nature of things." He turned, under the stimulus of writing his book on the philosophy of science (1927) to the astringent pages of Leibniz. "Auf den metaphysischen Hochflug folgte die Ernüchterung."

In mathematical logic, too, things seemed a little less sure at the end of his life than at the beginning, "the world becomes strange, the pattern more complicated," but he held steadily to his view that postulation cannot replace construction without the loss of significance and value. This belief he held so seriously that he deliberately kept away throughout his life from those mathematical theories which make essential and systematic use of the Axiom of Choice.

The literary graces with which he liked to adorn his work gave it an unmistakable flavour. Who else, in the austere pages of the London Mathematical Society, would sum up the outcome of Minkowski's "unexpected difficulties" in the geometry of numbers in the words: "But of him it might be said, as of Saul, that he went out to look after his father's asses, and found a kingdom"? And who else would have found in *Der Rosenkavalier*, of all places, an apt and serious quotation for his chapter on "Ars Combinatoria" (genetics)? It was a particularly cruel fate that imposed upon him "the yoke of a foreign tongue that was not sung at my cradle." Under this handicap he did not resign himself to writing in a drab and timid style, but used his adopted language as boldly and colloquially as his own—sometimes, it is true, with results which were surprising as well as pleasing.

The mathematical form of his presentations was even more characteristic than his literary style. His strong preference for arguments that stem from the central core of the problem, rather than verifications—even easy verifications—by computation, and his liking for pregnant verbal statements where others might use symbols, more easily seized by the mathematician's eye, sometimes made close demands on the reader's attention, but the reward was doubly great when the passage was understood. But his absorption in conceptual analyses and general theories never extinguished his zest for formal mathematical detail. He considered examples to be the life-blood of mathematics, and his books and papers are full of them. In the obituary of Hilbert he mentions with admiration the many examples by which Hilbert illustrated the fundamental theorems of his algebraic papers—"examples not constructed *ad hoc*, but genuine ones worth studying for their own sake!" The *dictum* of Hilbert on the subject, in *Mathematical Problems*: "The solution of problems steels the forces of the investigator: by them he discovers new methods and widens his horizon. One who searches for methods without a definite problem in view is likely to search in vain:"—this he took as a precept in his own mathematical life.

He was indeed not only a great mathematician but a great mathematical writer. His style was leisurely by modern standards, but it had a wonderful richness of ideas. His discoveries will surely not only long survive as mathematics, but will be read in his own incomparable accounts of them.

Throughout his life Weyl continued to write papers from time to time on topics in *analysis*, but the long series of papers (1908–1915) in which (following Hilbert's precept) he applied the new theory of integral equations to eigenvalue problems of differential equations establishes most clearly his stature as an analyst.

The papers written in this period fall into two groups. In the first (1909–1910) the theory of singular eigenvalue problems is developed for ordinary differential equations.

The second early group of papers in analysis are on the asymptotic distribution of natural frequencies of oscillating continua, *e.g.* membranes, elastic bodies and electromagnetic waves in a cavity with reflecting walls.

Weyl's outstanding quality as an analyst is already shown by these early papers. In all of them the argument moves by clearly visible steps, each involving difficult work.

In 1913 there appeared *Die Idee der Riemannschen Fläche*. This book, in which Weyl revealed his full powers for the first time, marked the beginning of the widening of his mathematical interests. By its declared subject it belonged to analysis, and indeed it contained a masterly exposition of the classical theory of algebraic and analytic functions on Riemann surfaces, culminating in a proof of the uniformization theorem. But it was the plan, revolutionary at that time, of placing "geometrical" function theory on a basis of rigorous definition and proof, hitherto enjoyed only by the Weierstrass theory, that gave the book its unique character, and forced Weyl to plunge deep into the *topology* of manifolds. In his Lectures on Algebraic Functions of 1891–1892 Felix Klein had shown that the notion of a Riemann surface need not be tied to the multiply-sheeted coverings of a sphere to which Riemann had confined himself, but could be extended to include any surface provided with *local uniformizing variables* (conformal maps of the members of an open covering onto a circular domain). When Klein delivered his lectures there were no means available of giving exact form to these ideas: the lack of topological notions made it impossible even to define a Riemann surface precisely. In 1910 and 1911 L. E. J. Brouwer published his papers on the topology of simplicial manifolds. Weyl saw at once that here was the basis for an exact treatment of Klein's ideas, with Hilbert's proof of the Dirichlet Principle as the instrument for establishing the existence of differentials on the surface. To these ingredients, which, as might be expected, he modified and simplified to suit his purpose, he added others of his own. In order to prove, as he wished, that the "analytisches Gebilde" can itself be regarded a surface, he needed a thoroughgoing axiomatic definition of a surface, which should make it clear that the "points" can be mathematical objects of any kind (in this case pairs of power series). The notion of a *neighbourhood-space*, as a set in which certain subsets are associated with each point as its neighbourhoods, had been introduced by Hilbert in 1902, but his definition remained unused and almost unnoticed. Weyl revived and clarified Hilbert's definition, and showed for the first time how it could be applied. The conditions which make the restriction to manifolds were not separated, as they would be today, from the general topological axioms, but the notion of a *topology* as a designated family of subsets, was clearly brought into view. He could now define a *surface* to be a (connected) triangulable 2-manifold, and a *Riemann surface* to be a surface on which, for each point p_0, certain complex-valued continuous functions are designated *regular at p_0*, again subject to suitable axioms.

Another of the new ideas which Weyl brought to his task had to wait more than twenty years to be independently rediscovered in more general form by topologists. This was the isolation of the topological part of the proof of the duality between the differentials and the 1-cycles on the surface.

Still another substantial contribution made in this book to the topology of the subject is the treatment of the covering surface. This notion had been used by Poincaré, but only Weyl's exact definitions and proofs made clear what precisely are the parts played by the topological and the function-theoretic properties.

It is natural that some of these pioneering topological chapters should appear somewhat rugged to modern readers. Weyl himself published a new edition in 1955 in which he got rid of the troublesome triangulation condition, and cast many of the definitions into the clearer forms which forty years of progress in topology had made possible.

It is convenient to mention here another contribution of Weyl to topology, though its connexion with continuous manifolds would have seemed remote when it was published. This was the short series of papers (1923–1924) written in Spanish, on "combinatorial analysis situs," that is the axiomatic theory of cell-complexes. A good deal of the material was of an expository character and Weyl himself seems to have attached little importance to the papers; but in fact this was the first appearance in the literature of a homology theory based on an axiomatic definition of abstract cell-complexes.

Weyl's interest in *general relativity*, and through it in *differential geometry*, began through his giving a course of lectures on the subject in Zurich after the departure of Einstein to Berlin—lectures which were the nucleus from which the book *Raum-Zeit-Materie* grew, through a series of revisions and expansions, to the great treatise of 1923 (5th edition). This book is too well-known to need lengthy description. It gave Weyl his first opportunity to combine discussion of the philosophical questions in which he was so deeply interested with technical mathematics. On the mathematical side, it is distinguished, as might be expected, for the precision of the results. Nowhere else, for example, is there to be found so thorough and exact a discussion of the central orbit, finishing with rigorous inequalities for the maximal and minimal distances—a useful piece of information for discussion of the motion over long periods of time.

Weyl's own principal contribution to the subject was his "unified field theory" of gravitation and electricity—the beginning of the quest on which Einstein spent so many fruitless years. The two papers of Weyl on the subject have been more influential in differential geometry than in relativity theory. Weyl took up Levi-Cività's idea (1917) of the "parallel displacement" of a vector, but made the decisive innovation of freeing it entirely from dependence on a Riemann metric. This was the starting point of the rapid development of projective differential geometry ("geometry of paths") which took place, particularly in the United States under the leadership of Oswald Veblen.

In his papers on gravitation and the electron Weyl took up the notion introduced by Einstein in one of his attempts at a "unified field theory" of attaching a local set of axes (*Vierbein*) to each point of space-time. By using this method to express the relation between a spinor-field and the metric of general relativity, Weyl found a natural interpretation of the Vierbein in terms of the spin, which has proved fruitful in the hands of later workers. He also pointed out in these papers the possibility of representing a fermion [Chapter 13] field by a pair of 2-component spinors distinguished by definite, and opposite, senses of rotation which on spatial reflexion go over into each other. When the mass is zero the 4-component wave equation degenerates into two uncoupled 2-component equations describing these two oriented spinors. This non-conservation of parity, thought at first to be little more than a mathematical curiosity, has turned out to be an essential property of the so-called weak interactions between fermions of small mass ("leptons"). It is now established that neutrinos are indeed describable by the 2-component formalism introduced by Weyl.

From relativity Weyl turned, by a natural transition, to the problem of finding the "inner reason" for the structure of general metric space, i.e., deducing the Riemann assumption of a metric based on a quadratic form from axioms about the group of "movements" in the space [Chapters 17, 19, 20]. For the classical constant-curvature spaces Helmholtz had characterized the group of movements as the smallest which allows free mobility, i.e., contains just one element which carries a point, a directed line through the point, and so on, into another arbitrary system of the same kind. He sketched a proof, which was made exact by Lie, that such a group coincides with the group of linear transformations leaving a quadratic form invariant. Weyl's problem was to formulate and prove a corresponding theorem for infinitesimal geometry. This he did by analyzing the meaning of the assumption that the metric, i.e., the group of movements, uniquely determines the affine connexion, and he showed that this assumption and the conservation of volume suffice to characterize the group of infinitesimal rotations at a point as the set of linear transformations that leave a non-degenerate quadratic differential form invariant.

In Hilbert's development of axiomatic mathematics two stages can be seen. The first is the setting up of branches of mathematics as theories of pure structure, that is to say, as theories about sets in which some subsets are distinguished, in accordance with certain conditions. The distinguished subsets may be given such picturesque names as "open set," "straight line," or "pair of elements and their product," but it is the conditions they satisfy, not their names, that make one subject differ from another. The second stage is the formalization of set theory itself, which in the first stage was treated "naively."

Axiomatic theories in the first sense were quite congenial to Weyl, who, as we have seen, was himself the first to give recognizable axiomatic characterizations of two central concepts of modern mathematics, topological neighbourhood-spaces and cell-complexes, and in the *Mathematical analysis of the space-problem* found axioms for Riemannian geometry. No logical problems need be posed by such theories, provided that illustrative examples or "models" can be made out of the real numbers, and provided, of course, that the real numbers are accepted as secure. Hilbert's treatment of the second stage, which amounted to reducing mathematics to the properties of the grammar of the sentences expressing its theorems, was highly repugnant to Weyl. He perhaps took too seriously the comparison sometimes made between making a proof according to Hilbert's "rules of procedure," and playing a game of chess. The object of the comparison is to illustrate the surprising and important fact that formal logical systems can be described objectively without reference to their intended meaning, but Weyl saw in this a degradation of mathematics. "Hilbert's mathematics may be a pretty game with formulae, more amusing even than chess; but what bearing does it have on cognition, since its formulae admittedly have no material meaning by virtue of which they could express intuitive truths?" He quotes with approval Brouwer's aphorism. "To the question, where shall mathematical rigour be found, the two parties give different answers. The intuitionist says: in the human intellect, the formalist: on paper."

His own first incursion into the field was in the monograph *Das Kontinuum,* and he never departed greatly from the position he there took up (though he published a more extended exposition in 1921). A characteristic opening paragraph declared his purpose. "In this little book I am not concerned to disguise the 'solid rock' on which the house of analysis is built with a wooden platform of formalism, in order to talk the reader into believing at the end that this platform is the true foundation. What will be propounded is rather the view that the house is largely built on sand. I believe I can replace this shaky part of the foundation by strong and reliable supports, but they will not carry everything that is nowadays generally believed to be secure. The rest I abandon: I can see no other possibility."

The "sandy part" was the part of mathematics which (he said) involves a "vicious circle," namely the kind of definition called "impredicative"* by Russell, who also saw here a danger to the stability of mathematics. As a measure of protection against the appearance of "semantic" paradoxes Russell had enunciated the principle: "No totality can contain members defined in terms of itself." It was to give formal expression to his principle that Russell introduced the "ramified" theory of types, of which Weyl's *Kategorien* and *Stufen*, in *Das Kontinuum*, are a version. The proliferation of real numbers of different types makes analysis quite intractable and led Russell to the desperate expedient of the Axiom of Reducibility, which simply postulates that for every sentence with a single free variable, x, of level a, there exists a "first-order" sentence defining the same class, that is a sentence of the lowest type possible for sentences with the free variable x. In 1949 Weyl rejected this way out of the difficulty. "Russell, in order to extricate himself from the affair, causes reason to commit hara-kiri, by postulating [the existence of an equivalent first order sentence] in spite of its lack of support by any evidence."

His own cure was the drastic one of allowing only "first order" definitions, and throwing away the parts of mathematics that failed to survive the purge. This means that bounded sets of real numbers need not have least upper bounds, and that we may not, in general, form the set, $P(E)$, of all subsets of a given set. He carried out in detail, in the book, the development of analysis as far as it would go on this basis. His set theory was of a genetic kind, only such sets being admitted as could be built up from "ground categories" by the use of allowed principles of construction. It thus resembled the Zermelo system (to which he refers) later developed by von Neumann; but in place of the more powerful Zermelo operations, he uses only the combination of Boolean operations and quantification of type-0 variables, by means of a recursive iteration scheme. The theory of real numbers to which this leads is of the sort that has been made familiar by the Intuitionists. The extent of the sacrifice involved is much more accurately known at the present day. The movement of logic is now towards a re-interpretation of classical proofs in a constructive sense, rather than a policy of voluntarily jettisoning certain of the most powerful instruments of proof, which is unlikely to recommend itself to mathematicians in general. Nevertheless a return to the "naive" acceptance of the axioms of classical set-theory as self-evident truths, on which we can confidently build up mathematics, is now out of the question; and in this change of opinion Weyl's writings certainly played an important part. Brouwer's analysis goes deeper, but his theories are to many impenetrable. It was Weyl's advocacy of the intuitionist views and his clear and attractive expositions of them, in his papers and books that first made them accessible to many mathematicians, and turned the revolutionary doctrine of his time into the orthodoxy of today.

---

* When a set M and an object m are so defined that m is a member of M but is defined only by reference to M, the definition of M or of m is called *impredicative*. An impredicative definition can be found in each of the paradoxes of logic and the theory of sets.

27

Twentieth-Century Vistas—Analysis

The "Leonardos" of modern mathematics contributed to many of its subdivisions, but the mother science also progressed notably through the research of mathematicians who devoted themselves to particular specialties. Therefore, in this and the next two chapters, we shall name some of the greatest recent mathematical specialists and describe the nature of their discoveries. The sample will be a representative one. "Scene I" will discuss modern analysts and phases of analysis. "Scene II" will feature the lives and ideas of some "demons of abstract algebra." "Scene III" will indicate aspects of the crisis in the subject which is, according to one's point of view, the basis of all mathematics or else not mathematics at all, but *metamathematics*.

Our first topic, the recent *theory of measure*, has its roots in metric geometry, where a number is assigned to a length or an area or a volume. We have seen that, in antiquity, measurement was at first considered just a case of comparison with a standard unit. Then the problem of incommensurables revealed that the question was not as simple as intuition suggests, and that it requires a consideration of infinite processes. Then when calculus was fully developed, there came the more sophisticated point of view that, for most figures, measures do *not* exist a priori but are contingent on the existence of associated limits. The evaluation of such limits became the task of integral calculus, a tool that also gives measures for many physical entities like mass, work, pressure. As scientific history advanced, both physical and abstract geometrical concepts became more complicated, and there was greater need for precise mathematical formulation.

In Cantor's theory of the infinite, one-to-one correspondence is the criterion for determining whether two sets have the same cardinal number or whether one aggregate has more elements than the other. But this does not give the "length" (or area or volume) of a point set. In fact, the interval [0, 3] has the same cardinal number of points, C, as [0 ,1], although Euclidean geometry says the former is three times the latter in length. (We have used the bracket [ ] symbolism to indicate that the intervals in question include the end-points, 0 and 3 in the first case, 0 and 1 in the second.)

What sort of foot rule can we apply to infinite sets to obtain a suitable measure for such abstractions so that there may be application to their counterparts in the physical world? In dealing with the set of all real numbers, the Cartesian picture is the number axis. Here one uses a standard length for the unit interval [0, 1] and marks it off repeatedly to obtain the intervals [1, 2], [2, 3], etc. If *measure* is to be just a generalization of length, it would seem a good idea to say that the *measure* of the set of real numbers in each of these intervals is also one unit. Thus one can readily measure point sets that are intervals or finite aggregates of nonoverlapping intervals. But certain other questions present themselves naturally: If from the set of all real numbers between 0 and 1 we remove the end-points 0 and 1, what is the length or measure of the remaining set, that is, the open interval (0, 1)? Or suppose that we remove all the rational fractions that have 1 as numerator: 1/1, 1/2, 1/3, 1/4, 1/5, . . . ; what is the measure of the residue? Or suppose that we go further and remove all the rational points in [0, 1]; what length remains? In the last years of the nineteenth century, Émile Borel gave much creative thought to these and many more difficult questions of the same kind. Then, in 1902, with Borel's ideas as background, Henri Lebesgue (1875–1941) established a *general* theory of measure. He abstracted its structure from all the particular "measure theories" of the past—empirical, abstract geometrical, Borelian, etc.

In measurement in geometry, perimeters of polygons are obtained by totaling the lengths of individual sides, and area is sometimes found by subdividing a polygon into triangles, measuring these, then adding up. Such procedures assume: *The whole is equal to the sum of its parts.* In Lebesgue's *generalization* this becomes: *The measure of the logical sum or union of a finite or countably infinite number of non-overlapping sets is equal to the sum of their measures.* (Where the number of sets is countably infinite, the existence of a measure will involve the question of convergence of a series.)

On the other hand, generalizations of measures do *not* illustrate the classic postulate which asserts that the whole is greater than one of its parts. That axiom applies to finite sets only. The essence of infinity is that the whole of an infinite collection does equal a part. Then, in general measure theory, a statement which combines finite and infinite attributes is: *The measure of a set is either equal to or greater than the measure of a proper subset.*

Again, abstraction from particular measures leads to the assumption: *The measure of a set is zero or some positive real number.* Also, it is postulated that the measure of the empty set Ø is zero. That "nothing" and zero are *not* identical is indicated by the fact that the converse of the axiom just stated is false. If the Lebesgue measure of a class is zero, this aggregate is *not* necessarily empty. In general measure theory, a set consisting of a single point like the origin or the point $x = 1$ on the X-axis or the point $x = 3$, etc. measures zero. (The reason for this is that the interval is the basis of linear measure, and a single point can be covered by an arbitrarily small interval.) If we accept a postulate stating that for finite and countably infinite sets the whole is equal to the sum of its parts, then the measure of a set containing two isolated points is zero, as is the measure of any finite or countably infinite collection of isolated points. Now, referring to a question raised earlier, the aggregate of all fractions with numerator 1 can be shown to measure zero, and the same is true of the class of all rational numbers between 0 and 1. For this reason, when such sets are removed from the interval [0, 1], the residual set still measures 1 unit. Thus the aggregate of irrational numbers in [0, 1] measures 1 unit. More formally, if we symbolize the class removed by E and the residue by E', the logical sum or union of E and E' is the unit interval. Therefore

$$\text{(measure of } E) + \text{(measure of } E') = \text{measure of } [0, 1]$$
$$0 + \text{measure of } E' = 1$$
$$\text{measure of } E' = 1$$

Here the measure of the whole interval is equal to the measure of the part E', but is greater than the measure of the part E.

In our discussion of the calculus it was indicated that elementary measures (like areas, volumes, pressure, etc.) are expressible as definite integrals. Lebesgue observed that this was a common feature of earlier measures and generalized it for inclusion in his theory. For example, a new type of definite integral had been created by the Dutch analyst T. J. Stieltjes (1856–1894), and Lebesgue used this to define an important type of measure, now named Lebesgue-Stieltjes measure in honor of both mathematicians. The Lebesgue-Stieltjes measure is in fact the backbone of probability theory today.

In the history of classic analysis, elementary measure and integral calculus preceded differential calculus partly because practical problems called for measurement and partly because the concepts involved are essentially easier in the case of integration. History repeated itself: measure and integration of functions with arbitrary variables were developed first because they presented fewer difficulties. To return now to the idea of integration in elementary calculus, the Mengoli-Cauchy definite integral (Chapter 8) is defined in a fashion suggested by Figure 27.1a, where the sum of a set of rectangles is used to approximate the area bounded by the segment of the X-axis between a and b, the curve $y = f(x)$, and the ordinates at $x = a$ and $x = b$. Then the area or definite integral can be defined as the *limit* of

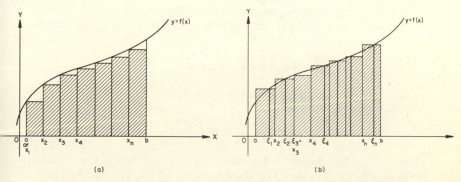

Figure 27.1   Approximating (Darboux) sums for the Mengoli-Cauchy integral

the sum of rectangular areas, as the base of each rectangle gets smaller and smaller (and the number of rectangles gets larger and larger). More formally,

$$\int_a^b f(x)\, dx = \lim\, [f(x_1)\Delta x + f(x_2)\Delta x + \cdots + f(x_n)\Delta x]$$

as Δx approaches zero. Here we have taken the bases of the approximating rectangles as equal in length, for each base $= \Delta x = (b - a)/n$. This is not necessary, however. The bases may vary in length as long as each base approaches zero as a limit. Also, the approximating rectangles may be drawn with altitudes at any point ξ_i of an interval so that in Figure 27.1b the approximating sum is

$$(x_2 - x_1)\, f(\xi_1) + (x_3 - x_2)\, f(\xi_2) + \cdots + (x_{n+1} - x_n) f(\xi_n)$$

and it can be proved that as the lengths of the intervals approach 0, the same limit is reached as in the original definition, that is,

$$\int_a^b f(x)\, dx$$

In the Mengoli-Cauchy definition, $y = f(x)$ was assumed to be a function continuous at each point of the interval $[a, b]$. Also, if one wishes to use condensed symbolism, the sum of the rectangles is indicated by

$$\sum_{i=1}^n f(x_i)\Delta x$$

or, alternatively, by

$$\sum_{i=1}^n (x_{i+1} - x_i)\, f(\xi_i)$$

where $\sum$, the Greek sigma, symbolizes the sum of a finite number of elements. Then

$$\int_a^b f(x)\, dx = \lim_{\Delta x \to 0}\, \sum_{i=1}^n f(x_i)\, \Delta x$$

In 1854 Riemann generalized this definition so that continuity of $f(x)$ in the interval $[a, b]$ was no longer required; that is, certain types of discontinuity were permitted at a finite number of points in this interval. Riemann merely required that $f(x)$ be *bounded* (finite in value) in the interval $[a, b]$. Then he formed the *approximating sum* (later called a *Darboux sum*)

$$L_1\Delta x + L_2\Delta x + \cdots + L_n\Delta x = \sum_{i=1}^n L_i\Delta x$$

where $L_1, L_2, \ldots, L_n$ are the (greatest) *lower bounds* of $f(x)$ in each interval. He also formed the approximating sum

$$U_1\Delta x + U_2\Delta x + \cdots + U_n\Delta x = \sum_{i=1}^n U_i\Delta x$$

where $U_1, U_2, \ldots, U_n$ are the (least) *upper bounds* of $f(x)$ in each interval. Then if the above approximating sums have the *same* limit as $\Delta x \to 0$, this common limit is the *Riemann integral* of $f(x)$ in the interval (a, b).

The great French geometer Gaston Darboux (1842–1917) showed that a function $f(x)$, which is bounded in (a, b), will have a Riemann integral in (a, b) if, and only if, the discontinuities of $f(x)$ in (a, b) constitute a set of *measure zero*.

The next generalization of the definite integral was made in 1902 by Lebesgue, whose theory of integration created a veritable revolution in analysis. Before explaining Lebesgue's idea, we must remark that the creation of this new type of integration was one more example of simultaneous discovery. Hardy pointed out that William Henry Young, Lebesgue's British contemporary,

working independently, arrived at a definition of the integral different in form from, but essentially equivalent to Lebesgue's. He had not made Lebesgue's applications; the great theorems about integrals and derivatives are Lebesgue's and his alone. But naturally Young's integral, being equivalent to Lebesgue's, "has them in it." If Lebesgue had never lived, but the mathematical world had been presented with Young's definition, it would have found Lebesgue's theorems before long. In the definition itself Young was anticipated by about two years, and it must have been a heavy blow to a man who was just beginning to find himself a mathematician; but he recognized the anticipation magnanimously, and set himself wholeheartedly to work at the further development of the theory. The phrase "the Lebesgue integral" is Young's.*

To understand the nature of the Lebesgue integral, one should compare it with the Riemann integral which it generalizes. To obtain a pertinent example of a Riemann integral, consider that in a diagram like Figure 27.1a the X-axis represents *time* and the interval [a, b] measures 1 month. If the interval [a, b] should be subdivided into 30 equal parts, then $\Delta x = 1/30$, that is, represents one day. Now if we think of $f(x)$ as a thermographic record of temperatures for a particular month, then $f(x_1), f(x_2), \ldots, f(x_n)$ would represent temperatures at the same hour each day. In that case, the Darboux sum

$$f(x_1)\, \Delta x + f(x_2)\, \Delta x + \cdots + f(x_n)\, \Delta x$$

would be formed by multiplying each daily temperature by 1/30 and adding the products so obtained. This leads to the same result as adding the 30 temperatures for different days and dividing by 30. The latter procedure indicates that in the present instance the Darboux sum would give an "average" or mean temperature. A more representative average of temperatures during the month would be obtained from a diagram like Figure 27.1b, that is, from

$$f(\xi_1)\, \Delta x + f(\xi_2)\, \Delta x + \cdots + f(\xi_n)\, \Delta x$$

where $\xi_1, \xi_2, \ldots, \xi_n$ represent any instants of time whatsoever on different days, so that $f(\xi_1). f(\xi_2), \ldots, f(\xi_n)$ would be temperatures on different days but *not* necessarily at the same hour each day.

A better statistical summary would be obtained by finding the average when $\Delta x = 1$ hour or 1/720 of a month, and an even better one by taking $\Delta x = 1$ minute or 1/43,200 of a month, $\Delta x = 1$ second or 1/2,592,000 of a month, etc. Thus one can consider Δx as getting smaller and smaller, that is, $\Delta x \to 0$. Then the limit of

---

* Royal Society of London, *Obituary Notices of Fellows*, Vol. 4, No. 12 (November 1943).

the resulting averages is the Riemann integral (or the Mengoli-Cauchy integral, since we have assumed $f(x)$ to be continuous), symbolized by

$$\int_0^1 f(x)\, dx$$

Now, although the average sought can be adequately represented by the simplest type of integral, let us see how emulators of Lebesgue might proceed. (A Mengoli-Cauchy integral or a Riemann integral is a Lebesgue integral, but not conversely.) Suppose that the extreme temperatures during the month in question are 5° C and 25° C. We have indicated these two readings so as to form an interval on the Y-axis, the axis of temperatures (Figure 27.2). Then one procedure leading

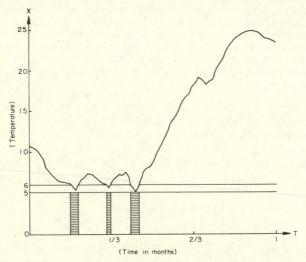

Figure 27.2   Hypothetical thermograph with rectangles whose areas total $5m_1$ in the sum approximating a Lebesgue integral

to the Lebesgue integral calls for subdivision of the interval [5, 25] on the Y-axis. Let us first divide it into 20 equal parts. Then the first small interval represents temperatures from 5° to 6°, the next one temperatures from 6° to 7°, etc. Corresponding to the first interval, one now forms the product $5m_1$, where m_1 is the total length of time during the month when temperatures were somewhere between 5° and 6°, that is, equal to or greater than 5° but less than 6°. Perhaps such temperatures occurred on a few days (Figure 27.2), so that the total time is about 1/12 of a month. Then the product required is (5) (1/12) = 5/12. Next the product $6m_2$ is formed, where m_2 is the total time when temperatures were equal to or greater than 6° but less than 7°. If $m_2 = 1/20$, the product required is (6) (1/20) = 6/20. Next $7m_3$ is formed, etc., and the Lebesgue approximating sum or mean temperature would then be

$$5m_1 + 6m_2 + 7m_3 + \cdots + 24m_{20}$$

(Note that this mean is computed by using relative frequencies in the fashion explained on page 266.) A more representative temperature would be obtained by taking

$$\eta_1 m_1 + \eta_2 m_2 + \eta_3 m_3 + \cdots + \eta_n m_n = \sum_{i=1}^{n} \eta_i m_i$$

where η_1 is *any* temperature between 5° and 6°, η_2 is *any* temperature between 6° and 7°, etc.

Next, the interval from 5° to 25° can be divided into smaller parts, say 100 equal intervals, the first ranging from 5° to 5.2°, the second from 5.2° to 5.4°, etc., and a still better average, that is, a more representative monthly temperature, will then be

$$\sum_{i=1}^{100} \eta_i m_i$$

If finer and finer subdivisions are considered so that $\Delta y \to 0$, and if the approximating sums (averages) approach a limit, then this limit is called the *Lebesgue integral* of the temperature function over the interval [0, 1].

In the above case, the limiting average temperature will be the same whether the approximating sums (made up of rectangular areas) are figured by the Mengoli-Cauchy (Riemann) method or by that of Lebesgue, as one can see by carrying out the arithmetic approximations with some actual thermograph, using fine subdivisions of the X-axis with the classic method, and then using fine subdivisions on the Y-axis with the Lebesgue method. But the reason the Lebesgue integral is a powerful generalization is that, for many advanced functions, with amazingly numerous discontinuities distributed in a fashion not encountered in elementary mathematics, there is a limit for the Lebesgue sums, that is, the peculiar functions are integrable in the Lebesgue sense, but *not* in the sense of Riemann and *a fortiori* not in the sense of Mengoli-Cauchy. For example, there is the function which consists of all ordered pairs (x, y) such that x is any real number in the interval [0, 1], $y = 1$ when x is rational, $y = 0$ when x is irrational. Some of the pairs belonging to the function are $(0, 1), (0.2, 1), (1, 1), (\frac{1}{4}\sqrt{2}, 0), (\pi/5, 0), (\frac{1}{2}\sqrt{3}, 0)$. Figure 27.3 suggests the function,

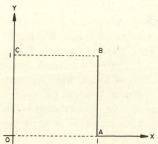

Figure 27.3 Cartesian representation of the function,
$$\begin{cases} y = 1 \text{ for } x \text{ a rational number in } [0, 1] \\ y = 0 \text{ for } x \text{ an irrational number in } [0, 1] \end{cases}$$

but in truth it is not possible to provide a satisfactory physical representation. The best one can do is to picture two parallel rows of infinitesimal, "densely" packed beads. The domain of the function is the whole interval [0, 1], and the range contains only the two numbers 0 and 1. Because the strange function we have described is discontinuous at *every point* of [0, 1] since the values of y jump back and forth between 0 and 1, the set of discontinuities has the same measure as the length of the interval, namely 1. Hence, by Darboux's theorem, our freakish function cannot have a Riemann integral over the interval [0, 1]. But let us prove directly that no Riemann integral exists. If the Darboux sums

$$L_1\Delta x + L_2\Delta x + L_3\Delta x + \cdots + L_n\Delta x$$
$$U_1\Delta x + U_2\Delta x + U_3\Delta x + \cdots + U_n\Delta x$$

are formed, they will have the values

$$0\cdot\Delta x + 0\cdot\Delta x + 0\cdot\Delta x + \cdots + 0\cdot\Delta x$$
$$1\cdot\Delta x + 1\cdot\Delta x + 1\cdot\Delta x + \cdots + 1\cdot\Delta x$$

respectively. The first of these sums has the constant value 0. Therefore its *limit* will be 0, as Δx gets smaller and smaller. The second sum has the value $n\Delta x$. If $\Delta x = 1/n$, this sum has the value $(n)(1/n) = 1$. Then if we allow n to become larger and larger as Δx gets smaller and smaller, the corresponding Darboux sum is always equal to 1. Hence its *limit* is 1. Thus the limits of the two sums, being 0 and 1, are *not* the same, and hence there is no Riemann integral.

Now consider the Lebesgue method. On the Y-axis there are just two distinct points, 0 and 1, which form the complete range of the function. The Lebesgue approximating sum is then

$$0\cdot m_1 + 1\cdot m_2$$

where m_1 is the measure of the set of points for which $f(x) = 0$ and m_2 is the measure of the set for which $f(x) = 1$. The first aggregate is the set of irrational points, the second is the set of rationals, and the corresponding measures are 1 and 0, respectively (page 641). Therefore the Lebesgue approximating sum is

$$0\cdot 1 + 1\cdot 0 = 0$$

Because the range of the function under consideration consists of the two isolated values 0 and 1, no finer subdivision of that range is possible, and the Lebesgue approximating sum does not vary but instead is equal to the constant 0. Thus there exists a Lebesgue integral of the function over the interval [0, 1] and the value of that integral is 0. Then the "area" bounded by "beads" and ordinates in Figure 27.3 measures 0, whereas intuition might mislead us into seeing a one-inch square with area 1. If, however, we think of the Lebesgue integral as an "average," then the 0 figure is justified because 0° "temperatures" occur at C points (of time), a much more highly infinite number than the countable infinity of rational points (of time) where "temperatures" of 1° occur. Thus 0° is the *mode* or most frequently occurring "temperature."

In the above illustration, the Lebesgue measures of certain point sets were essential. In the previous example of a temperature average, the numbers m_1, m_2, m_3, etc., the total times for which temperatures were within certain small ranges, were also measures. Specifically, m_1 is the *measure* of the set of points on the X-axis (time axis), corresponding to temperatures between 5° and 6°, say, or between 5° and 5.2°, etc. (depending on the fineness of subdivision on the Y-axis), and similarly for m_2, m_3, etc. In the *general case*, one is *not* necessarily dealing with functions of time (or measuring the set of rationals or the set of irrationals as in the second illustration above). Then one must either assume or prove that there actually is a *measure* for each point set involved in forming the Lebesgue approximating sums. Therefore, existence of a *Lebesgue integral* is dependent on the existence of a *Lebesgue measure*.

What exactly is Lebesgue measure? Before giving a formal answer, it would be well to summarize and add to the informal explanation given earlier in this chapter. In the first place, then, we must emphasize that the motivation for the research of Borel and Lebesgue was twofold, namely, to generalize Cantor's theory of sets, and also to investigate "difficult" point sets like those defined by *discontinuous functions*. Such aggregates began to appear toward the end of the nineteenth century—in Fourier series, in function theory, in dynamics, in probability theory, etc. The problem was to assign to each "difficult" point set an appropriate measure, that is, a *real nonnegative* (positive or zero) *number*, which is analogous to a length or an area or a volume.

Thus a measure is a *function* whose domain is some class of sets and whose range is an aggregate of nonnegative real numbers. It would then seem logical to divide an explanation of measure into two parts, first, a discussion of the *domain*, the class of sets to be measured, and second, the rules which are to govern the range.

Borel proceeded in just this way, as will now be indicated. First he formulated a class of sets, and then he presented his theory of measure for this class. He started with the elementary figures, namely, line segments or intervals, rectangles, rectangular solids (box forms), etc. on a line, in a plane, in space, etc., respectively. Restricting ourselves momentarily to a linear universe, an obvious generalization should be that to point sets made up of unions or logical sums of intervals, finite or countably infinite in number. Such intervals might be "closed" like [0, 1] or "open" like (0, 1) or "degenerate," that is, consisting of a single point like the origin or the point $x = 2$, etc. In particular, the entire X-axis or the set of real numbers would be a Borel set because it can be considered as the union of the countable infinity of intervals, [0, 1], [0, −1], [1, 2], [−1, −2], etc. Having conceived of all such logical sums of intervals, Borel next thought of "subtractions." For example, the interval [0, 1] can be removed from the real axis, leaving the set of real numbers less than 0 or greater than 1. Using the terminology of Chapter 6, one can call this set the *complement* of [0, 1]. Again, the complement of the degenerate interval consisting of the origin alone is a punctured X-axis. Also the null or empty set $\emptyset$ is the complement of the entire X-axis (set of real numbers). Borel considered that all complements of intervals and unions of intervals should be included in his special class of sets. Then with all those sets formed originally by means of union and those formed later by complementation, finite or countably infinite unions would be formed. Now if new sets should result, complements of these would be taken, and then unions formed again, etc., again and again indefinitely. The aggregate of *all* sets formed in this way is known as the *class of linear Borel sets*. Borel considered analogous procedures with rectangles in the plane, rectangular solids in space, and *n*-dimensional "rectangular solids" in Euclidean *n*-space. The resulting aggregate for all dimensions is known as the *class of Borel sets*.

What Borel did was to give a *constructive* method of assigning measures (nonnegative real numbers) to these sets such that (1) the measure of an interval or rectangle, etc. is the same as its length or area, etc., respectively; (2) sets congruent in the sense of elementary Euclidean metric geometry have equal measures; and (3) the measure of a finite or countably infinite union of nonoverlapping sets is equal to the sum of the measures of the individual sets.

But Lebesgue was able to enlarge the class of Borel sets, that is, the domain of the Borel measure function. In particular, Lebesgue was able to show that the domain of the Borel measure function did *not* include all subsets of sets whose Borel measure is zero. Just as Borel's measure reduces to a length or an area or a volume for an "elementary" figure, Lebesgue's measure agrees with Borel's when the set being measured is a Borel set. But even Lebesgue's class of sets, that is, the domain of his measure function, does *not* include all sets, as G. Vitali showed in 1905. The general problem of determining whether, in a given class of sets, a measure can be defined so as to satisfy postulates analogous to those for Borel and Lebesgue measure is a very difficult question. Nevertheless, by furnishing an abstract postulational foundation for the measure of point sets, Lebesgue provided a method more readily applicable than Borel's constructive definition of 1894. Arnaud Denjoy, a leading analyst of our day, has this to say about the matter:*

Lebesgue believed he had surpassed Borel in his own conception of the measure of sets. But the fact is that any sets for which Lebesgue provided a means of measurement were already susceptible of measurement by Borel's method. But what Lebesgue added to the picture, what Borel never suspected, is that there are *nonmeasurable* sets. . . . Also, after Borel had discovered how to measure sets, he did not realize that he could apply his discovery to the measurement of functions. Lebesgue's greatest claim to fame arises from the fact that he was able to use the concept of measure in this way and to deduce from it a definition of his integral.

The way Lebesgue brought the measure of sets into his formulation of the integral has already been indicated, and it has been emphasized that the existence of the Lebesgue integral is contingent on the existence of the measures m_1, m_2, m_3, etc. occurring in the Lebesgue approximating sums. A crude definition of a *measurable function* (as Denjoy and other modern analysts use the term) describes it as one for which all the measures m_1, m_2, m_3, . . . etc. in the Lebesgue approximating sums

---

* A. Denjoy, *La théorie des fonctions de variables réelles*, Conférences du Palais de la Découverte, Université de Paris, p. 29. (Translated by the present author.)

actually exist. Borel, as Denjoy says, never had any doubts about the existence of such measures, and therefore it was necessary for Lebesgue to furnish a counterexample.

To see that there are nonmeasurable sets even within an ordinary interval, consider the linear interval $[0, \pi]$ and fit it around a circle of unit diameter (and circumference π). Now subdivide the points of the circumference into subsets with a special property as follows. If point P is a member of a subset, then so are the points whose distances from P (measured counterclockwise along the circumference) are $0, 1, 2, 3, 4, \ldots$. Observe that if $P = 0$ the point 4 will belong to the subset and will fall between points 0 and 1 on the circumference since a point at distance $\pi = 3.14$ (approximately) will coincide with 0, and therefore a point at distance 4 will coincide with a point at distance 0.86 (approximately). Similarly, a point at distance 2π will coincide with 0, and therefore a point at distance 7 will fall on top of one at distance 0.72 (approximately), etc. Again, one must assign to the same subset the points whose distances from P are 1, 2, 3, 4, ... measured *clockwise* along the circumference. Thus starting at *any* point P on the circumference, a subset of the circumference could be formed by taking the points at distances 0, 1, 2, 3, 4, ..., measured in either direction along the circumference. An infinite number of disjoint subsets could be formed by the method described. Then, if we accept the axiom of Zermelo (Chapter 24), one representative of each of the subsets could be selected to form an aggregate E. If E is rotated through a distance of one unit counterclockwise around the circumference, a set E_1 is formed which has no points in common with E (because each representative element in E is rotated into a *different* point of the set it represents). If E is rotated through a distance of -1, one obtains a set E_{-1} which has no points in common with E or E_1. Rotating E through a distance of 2 in either direction leads to E_2 or E_{-2}, two sets with no points in common with E, E_1, E_{-1}, or each other. Continuing the process, indefinitely, one obtains $\{E, E_1, E_{-1}, E_2, E_{-2}, \ldots, E_n, E_{-n}, \ldots\}$, a countable collection of nonoverlapping sets which fills out the whole circumference. Moreover, since each set is formed from one of the preceding sets by a rotation, which is a rigid motion (Chapter 17), the sets are all *congruent* and should have the same measure m, where m must be zero or some positive real number. If the measure m were zero, then, by virtue of property (3) on page 648, the measure of the union of all the sets would be

$$0 + 0 + 0 + 0 + \cdots = 0$$

which contradicts the fact that the union is the whole circumference, whose measure is in truth π. Hence m cannot be equal to zero. If on the other hand, m were positive, the measure of the union would be

$$m + m + m + m + \cdots$$

This sum exceeds all numerical bounds, is "infinite," so to speak, a result which once again contradicts that fact that the circumference has the finite length π. Since m can be neither zero nor positive, no numerical measure whatsoever can be assigned to E (or to the congruent sets $E_1, E_{-1}, E_2, E_{-2}, \ldots$) and E is *unmeasurable* (as are $E_1, E_{-1}, E_2, E_{-2}, \ldots$).

What manner of man first constructed such nonmeasurable sets, and then, in a more positive spirit, created a revolutionary concept of integration? The story of Lebesgue's life is in fact mainly an account of his studies and mathematical achievements. We shall follow his biography with that of the co-inventor of the Lebesgue integral, William Henry Young, who appears to have led a more colorful existence than Lebesgue's. Perhaps it is merely that he was fortunate to have a biographer like Geoffrey Harold Hardy, who was aware that life has a personal as well as a mathematical side, and possessed the literary ability to write a "human interest story" about a professional colleague.

Henri Léon Lebesgue was born in Beauvais in 1875. His father was a printer in modest circumstances, but a bibliophile nevertheless, who accumulated and studied from a small library of serious reading. This cultured father died while Henri was still at school, and the boy's further education was only made possible when he was granted a scholarship to the lycée of his native town, and later to a lycée in Paris. At the age of nineteen he was admitted to the École Normale Supérieure, and his first post appears to have been that of *maître des conférences* at Rennes, which he held until 1906. He then moved to Poitiers, where he was described first as *chargé de cours à la Faculté des Sciences* and later as professor. About 1912 he was called to Paris as *maître des conférences*, and he afterwards became professor at the Collège de France. He was elected to the Académie des Sciences in 1922. He was made honorary member of the London Mathematical Society in 1924, and a foreign member of the Royal Society in 1934.

His greatest contribution, as we have seen, was the replacement of some classic ideas of analysis by more general and powerful concepts. But his critical spirit first manifested itself in connection with an issue of geometry connected with developable surfaces (Chapter 19). It was explained earlier that cylinders, cones, and the set of tangents to a skew curve constitute such surfaces because they can be rolled out or "developed" on a plane, or because they are locally Euclidean. Classic differential geometry had required that developable surfaces satisfy certain regularity conditions, like possessing a continuously turning tangent plane. Lebesgue demonstrated that such conditions are not necessary. To justify this on intuitive grounds and to convince his fellow students at the École Normale, he crumpled a piece of paper in his hand in order to obtain an extremely irregular surface that fails to meet the classic requirements, but nevertheless was originally a plane surface, and hence is "developable" even when deformed. Lebesgue's formal proof, published in 1899, is said to have "*scandalisé*" Darboux. Just as Darboux was shocked by one of Lebesgue's geometric ideas, other leading mathematicians at the turn of the century reacted unfavorably to his revolutionary notions in analysis, and the French "big wigs" came close to flunking him in the defense-of-thesis examination.

In 1904 Lebesgue's *Leçons sur l'Intégration* was published. Its 1928 revision is still the standard work on the Lebesgue integral. In 1905 a paper on *Fonctions Représentables Analytiquement* contained some of the most beautiful of Lebesgue's analytic findings. The problem considered had already been solved in one way by the great René Baire. It involved the expression of discontinuous functions in terms of easier functions and, if possible, in terms of continuous functions.

Baire had classified functions as follows: In the lowest category there are the continuous funtions, which Baire defined as constituting C_0, the "class zero." The limits of sequences of continuous functions are, in some instances, continuous functions and hence members of the class zero. But in other cases, the limits are discontinuous. Baire defined C_1, the "class one," as consisting of those discontinuous functions which are limits of sequences of continuous functions. Again, limits of sequences of functions in this new class are, in some cases, functions of this class. When they are not, they are functions in C_2, the Baire "class two," etc. Lebesgue gave a number of new proofs of Baire's basic theorem and proved the existence of functions in a Baire class of any assigned order. He also showed that it was possible to designate (*nommer*) a function not belonging to any Baire class. Thus Baire's classification was incomplete, and did not, from Lebesgue's point of view, include all possible real functions.

Lebesgue's integral did not generalize completely the original calculus idea of integration as *both* summation and antidifferentiation. For this reason Arnaud Denjoy, in 1912, generalized integration further; the German Perron created still another type of integral in 1914, and later there were generalizations by Stieltjes (Dutch), A. Khintchine (Russian), and P. J. Daniell (British). Lebesgue worked with the Denjoy and Stieltjes integrals as well as with his own.

Young never emphasized his own practically simultaneous discovery of the famous Lebesgue integral, Hardy recounts, among other biographical details of the English analyst's life.*

William Henry Young (1863–1942) was one of the most profound and original of the English mathematicians of his era.

Young was born in London on 20 October 1863. His ancestors were Ipswich people, but had been bankers in the City for some generations. His early education was at the City of London School: the headmaster, Edwin A. Abbott, had been a schoolfellow of his father. Abbott was the author of the entertaining mathematical fantasy *Flatland*, and was enough of a mathematician to recognize Young's exceptional talents, which seem indeed to have been understood much better at school than at home.

He came up to Cambridge, as a scholar of Peterhouse, in 1881. He came with a reputation to sustain and, if we are to judge him as an undergraduate and by the standards of the time, he hardly lived up to it. It is easy now to see why. The whole system of mathematical education in Cambridge was deplorable. The college teaching was negligible, the professors were inaccessible, and an undergraduate's only chance of learning some mathematics was from a private coach. Young, like nearly all the best mathematicians of his time, coached with Routh, from whom he could learn a lot. But he had many other interests, and no doubt he wasted much of his time. He was a good, though unsystematic, chess player, and an enthusiastic swimmer and rower; and his greatest disappointment as an undergraduate seems to have been his failure to get a place in the college

* Royal Society of London, *Obituary Notices of Fellows*, Vol. 4, No. 12 (November 1943), pp. 307 ff.

boat. He had always immense physical as well as mental energy, and remained an ardent oarsman all his life.

Whatever disappointments he may have had, Young seems to have been happy as an undergraduate: he said afterwards that " he never *began* to live until he went to Cambridge"; and he formed one friendship which was important throughout his career. This was with George and Foss Wescott, the sons of Professor Westcott, afterwards Bishop of Durham. Both of the Westcotts became bishops themselves later, and their influence led Young to turn his attention to theology. His family were Baptists, but he was baptized into the Church of England, became for a time superintendent of a Sunday school, and won a college theological award, the "Butler prize."

Young was a Fellow of Peterhouse from 1886 to 1892, but was never given any permanent position either by the college or the university. It was not until he was an old man that his college, in 1939, elected him an Honorary Fellow.

The next thirteen years of Young's life were spent, almost exclusively, in teaching and examining. It was common enough then for Cambridge mathematicians to earn quite large sums by private coaching, and Young set himself resolutely to do so. Here the Westcott connexion was a help, bringing him a good many pupils. He also went twice to Charterhouse as a temporary assistant master. It is difficult for any one who knew Young only later to imagine him as a schoolmaster, but he seems to have enjoyed the experience. He did much examining, at Eton and at other big schools; but primarily, through all these years, he was a coach. His position became a little more official in 1888, when Girton made him a lecturer in mathematics. "Lecturing" at Girton meant, in effect, more coaching, and after this coaching absorbed practically all his time and energy. He was working from early morning till late at night, sometimes taking two classes simultaneously in adjacent rooms, and often going without lunch.

His preferences in mathematics seem very surprising to any one familiar with his later work. He had read widely, and could teach anything in reason, but astronomy was his pet subject. "Astronomy" was the mathematical astronomy required for the tripos of those days, and a man who could make that stimulating must have been a teacher indeed. This interest lasted, and his first suggestion for "independent work," in the early days of his marriage, was one of a textbook of astronomy to be written in collaboration with his wife.

And all this time we hear not one word of research. Young was the most original of the younger Cambridge mathematicians; twenty years later he was the most prolific. Yet no one suggested to him that he might have it in him to be a great mathematician; that the years between twenty-five and forty should be the best of a mathematician's life; that he should set to work and see what he could do. The Cambridge of those days would seem a strange place to a research student transplanted into it from to-day.

I still find it difficult to visualize Young's own attitude during these early years of unproductivity. The productivity, when it did come, was so astonishing; it seems at first as if it must have been the sequel to years of preparation, by a man who had succeeded at last in finding his subject and himself. One would have supposed that any one so original, however he might be occupied, must surely have found something significant to say, but actually, the idea of research seems hardly to have occurred to Young. Mr. Cowell says that Young once told him that he "deliberately accepted ten years of drudgery," that he "fancied his knowledge of the Stock Exchange," and that he thought that he could "win his leisure" by thirty-five; but "leisure" meant freedom, comfort, reading, and travel, not a life of mathematical research. The truth seems to be that Young had really no time to think of much but his teaching; that the atmosphere of Cambridge was mathematically stifling; that no one was particularly anxious to look out for or encourage originality; and that he was too much absorbed in his routine, in his pupils and their performance, to dream of higher ambitions.

However that may be, the dreams were to come and the "drudgery" to end, and the end came quickly after Young's marriage. In 1896 he married Grace Chisholm, the second of his wrangler pupils. Mrs. Young's father was H. W. Chisholm, for many years Warden of the Standards, and her brother, Hugh Chisholm, was editor of the *Encyclopaedia Britannica*. The family carries on this tradition of distinction, and two of Young's six children are well known to us as mathematicians. The eldest son, Frank, was killed as an airman in France in 1917.

The great break in Young's life came, quite suddenly, in 1897; and here perhaps I had better quote Mrs. Young's own words. "At the end of our first year together he proposed, and I eagerly agreed, to throw up lucre, go abroad, and devote ourselves to research." It implies a revolution in Young's whole attitude to mathematics. But Mrs. Young had studied in Göttingen before her marriage, and knew what the air of a centre of research was like, so that possibly the revolution was a little less abrupt than it appears. At any rate, the Youngs left Cambridge for Göttingen in September. "Of course all our relations were horrified, but we succeeded in living without help, and indeed got the reputation of being well off." Young's "banker's instincts" had served him well.

Young's permanent home was abroad for the rest of his life, in Göttingen until 1908 and then in Switzerland, first in Geneva and afterwards in Lausanne; but the continuity of home life was much broken by his many activities. In 1901 he came back to Cambridge, and had rooms in Peterhouse during term time for some years, returning home for vacations. During 1902–1905 he was Chief Examiner to the Central Welsh Board, and seems to have thrown himself into the work with all his usual enthusiasm. His reputation was now rising, and he became a Fellow of the Royal Society in 1907, but it was not until 1913 that he obtained any definitely academic position. He was still not properly appreciated, and I can remember that, when he was a candidate for the Sadleirian chair in 1910, no one in Cambridge seemed to take his candidature very seriously.

The next few years were his years of greatest activity. He wrote a great deal—there are forty papers of his in volumes 6–18 of the *Proceedings of the London Mathematical Society* only; and in 1913 he at last became a professor. His first posts were of an "occasional," though honourable, kind. He was the first Hardinge Professor of Mathematics in Calcutta; this involved residence in India for the three winter months of the next three years. He also became Professor of the Philosophy and History of Mathematics in Liverpool, and lectured there during the summer. This was a special post created for him, and he held it until 1919, when he was appointed Professor of Pure Mathematics at Aberystwyth. Here he was as energetic as ever, but his residence abroad was sometimes a source of trouble, and disagreements about this, and about appointments to the staff, led to his resignation in 1923.

By this time Young had almost ceased his activity as an original mathematician. It had slackened for a time about 1915, but the death of his son in 1917 caused him great distress and drove him back to mathematics "as a drug." It was in this year that the London Mathematical Society awarded him the De Morgan medal. He did good work after, but none quite equal to his best, and after 1923 he wrote little. He was President of the London Mathematical Society during 1922–1924, and his last papers (including his Presidential address) were printed in its *Proceedings* in 1925. It was a little later, in 1928, that the Royal Society gave him the Sylvester medal in recognition of a life of invincible mathematical activity. Young was well over sixty now and regarded his career as a constructive mathematician as finished, but there were other openings for his activities. He had always been keenly interested in the international organization of mathematics. In 1929 he became President of the International Union of Mathematicians.

In 1932 he turned back to his other old interests: law (he had been a member of Lincoln's Inn from early days), finance, and above all languages. These included Jugoslav and Polish, two of the most intricate of Central Europe.

The end of Young's life was rather tragic, since he was cut off from his family completely by the war. Mrs. Young had left him, as they meant for a few days only, and the collapse of France prevented her return; and his children were settled in London, South Africa and Paris. He had always been the centre of a family, and found himself imprisoned in Switzerland and practically alone. Little had been heard from him, but it is known that he died quite suddenly, that he "just went out," and that the University of Geneva, of which he was an honorary doctor, did him every honour.

It is not particularly difficult to estimate Young's rank as a mathematician. His work consists of three books and over 200 papers. Two features of it stand out on almost every page, intense energy and a profusion of original ideas. Indeed it is obvious to any reader that Young has a superabundance of ideas, far too many for any one man to work out exhaustively. One feels that he should have been a professor at Göttingen or Princeton, surrounded by research pupils eager to explore every bypath to the end. It may be that he had hardly the temperament or the patience to lead a school in this way, but he never had a chance to try.

It is not surprising that a good many of Young's theorems should have been missed and rediscovered. At his best, he can be as sharp and concise as any reader could desire; and he (or he and his wife together) could write an excellent historical and critical résumé, with just the right spice of originality. There is one particular compliment which I find it easy to pay to Young. His work stands up stoutly to critical examination. There are men who seem to me admirable mathematicians so long as they write about geometry or physics —it is easy to be impressed by what one does not understand very well. Young's best work seems to me to be his work on the subjects which I myself know best, on the theory of Fourier and other orthogonal series, on the differential calculus, and on certain parts of the theory of integration.

I will say something first about the last of these subjects, not because his work here (except that on the "Stieltjes integral") seems to me his very best, but because it is the most widely known, and because it was the occasion of a disappointment which, coming as it did right at the beginning of his active career, might easily have broken the spirit of a weaker man.

The theory of functions of a real variable has been written afresh during the last forty or fifty years. In particular, the foundations of the integral calculus have been entirely remodelled; and it is acknowledged by every one that, among those who have reconstructed them, Lebesgue stands first. The "Lebesgue integral" opens the blocked passages and smooths the jagged edges which disfigured the older theories, and gives the integral calculus the aesthetic outlines of the best "classical" mathematics. In particular it brings integration and differentiation into harmony with one another. It is Lebesgue's theorems about integrals and derivatives, the core of any modern treatment of the subject, which are his greatest achievement. Young, working independently, arrived at a definition of the integral different in form from, but essentially equivalent to, Lebesgue's.

It may seem a paradox, but it is possible that Young's work on integration, fine as it was, actually impeded his recognition. These subjects were not popular, even in France, with conservatively minded mathematicians. In England they were regarded almost as a morbid growth in mathematics, and it was convenient for men out of sympathy with Young's interests, and perhaps a little jealous of his growing reputation, to dismiss him as "the man who was anticipated by Lebesgue." It is easy enough now to recognize the absurdity of such a view: if Young had never given his definition of an integral, his reputation would not be very materially affected.

Most of this work is set out in four special memoirs. In the last of these the "Young integral" is actually defined. All these papers were written in ignorance of Lebesgue's work, and recast when Young discovered it. This spoils their continuity a little, and it is perhaps a pity that he did not leave them as they stood and add the acknowledgments necessary in appendices; the genesis and progress of his own ideas would then have stood out more clearly. In later papers, Young developed the whole theory differently, by the "method of monotone sequences." This procedure is particularly well adapted to *proofs* of theorems in the integral calculus. But I mention it here less for this reason than because it led Young to one of his admitted triumphs.

There is another important generalization of the Riemann integral, the integral first defined by Stieltjes in 1894. The Stieltjes integral covers sums as well as ordinary integrals, and has come rapidly into vogue since about 1909, primarily because of its outstanding importance in the theory of "linear functionals." In Stieltjes integration we integrate one function f with respect to another function g; the classical case is that in which f is continuous and g monotone. It was inevitable, after Lebesgue's work, that mathematicians should try to combine the two generalizations, and define the integral of any Lebesgue-integrable f with respect to any monotone g. In particular, Lebesgue had tried, but his results were not altogether satisfactory. Young solved the problem with complete success: he showed that his method of monotone sequences could be applied to this more general problem with little more than verbal changes.

The best tribute to Young's work that I can quote is that of Lebesgue himself. Referring to his own attempt, he says "En réalité, je n'avais que très imparfaitement compris ce rôle [that of monotone sequences], sans quoi je n'aurais pas écrit . . . qu'il serait très difficile d'étendre la notion d'intégrale de Stieltjes par un procédé différent de celui que j'employais. Peu de temps après que j'eus commis cette imprudence, M. W. H. Young montrait que mon procédé était loin d'être indispensable, et que l'intégrale de Stieltjes se définit exactement comme l'intégrale ordinaire par le procédé des suites monotones. . . . Ce travail de M. Young est le premier de ceux qui ont finalement bien fait *comprendre* ce que c'est qu'une intégrale de Stieltjes. . . ."

Young's work on integration, which reaches its peak in the paper on the Stieltjes integral, was preceded and accompanied by a whole flood of papers on the theory of sets of points and its application to the general theory of functions. A considerable part of the contents of these papers is incorporated in the Youngs' book published in 1906 and, unfortunately, never revised and reprinted. But there are two other fields in which Young shows his powers at their highest. These are the theory of Fourier series (and other special orthogonal series), and the elementary differential calculus of functions of several variables.

28

Twentieth-Century Vistas— Algebra

Let us now proceed to the next sample from the modern repertory. Since "Scene II" will present certain concepts that are important in twentieth-century *abstract algebra*, it seems appropriate to precede the discussion with a biography of one of the chief contributors to that subject, namely, Emmy Noether (1882–1935)—a biography written by one of the mathematical "Leonardos," Hermann Weyl. "Emmy Noether was a great mathematician, the greatest . . . that her sex has ever produced," he wrote.*

She was born the 23rd of March, 1882, in the small South German university town of Erlangen. Her father was Max Noether, himself a great mathematician who played an important role in the development of the theory of algebraic functions as the chief representative of the algebraic-geometric school. He had come to the University of Erlangen as a professor of mathematics in 1875, and stayed there until his death in 1921. Besides Emmy there grew up in the house her brother Fritz, younger by two and a half years. He turned to applied mathematics in later years, then became professor at the Technische Hochschule in Breslau, and by the same fate that ended Emmy's career in Göttingen was driven off to the Research Institute for Mathematics and Mechanics in Tomsk, Siberia. The Noether family is a striking example of the hereditary nature of mathematical talent, the most shining illustration of which is the Basle Huguenot dynasty of the Bernoullis.†

Side by side with Noether at Erlangen was another mathematician, Gordan, under whom Emmy wrote her doctor's thesis in 1907: "On complete systems of invariants for ternary biquadratic forms." Besides her father, Gordan must have been well-nigh one of the most familiar figures in Emmy's early life, first as a friend of the house, later as a mathematician also; she kept a profound reverence for him though her own mathematical taste soon developed in quite a different direction. I remember that his picture decorated the wall of her study in Göttingen. These two men, the father and Gordan, determined the atmosphere in which she grew up.

The scientific kinship of father and daughter—who became in a certain sense his successor in algebra, but stands beside him independent in her fundamental attitude and in her problems—is something beautiful and gratifying. The father was a very intelligent, warm-hearted harmonious man of many-sided interests and sterling education.

Emmy Noether took part in the housework as a young girl, dusted and cooked, and went to dances, and it seems her life would have been that of an ordinary woman had it

---

*Hermann Weyl, "Emmy Noether," *Scripta Mathematica*, Vol. 3, No. 3 (July 1935), pp. 201 ff.

† For a recent instance of this phenomenon, see Chapter 30, where the mathematical activities of today's Neumann family are described.

not happened that just about that time it became possible in Germany for a girl to enter on a scientific career without meeting any too marked resistance. There was nothing rebellious in her nature; she was willing to accept conditions as they were. Her dependence on Gordan did not last long; he was important as a starting point, but was not of lasting scientific influence upon her. Nevertheless the Erlangen mathematical air may have been responsible for making her into an algebraist. Gordan retired in 1910; he was followed first by Erhard Schmidt, and the next year by Ernst Fischer, whose field was algebra. He exerted upon Emmy Noether, I believe, a more penetrating influence than Gordan did. Under his direction the transition from Gordan's formal standpoint to the Hilbert method of approach was accomplished. She refers in her papers at this time again and again to conversations with Fischer. This epoch extends until about 1919.

Already in Erlangen about 1913 Emmy lectured occasionally, substituting for her father when he was taken ill. She must have been to Göttingen about that time, too, but I suppose only on a visit with her brother Fritz.

During the war, in 1916, Emmy came to Göttingen for good; it was due to Hilbert's and Klein's direct influence that she stayed. Hilbert at that time was over head and ears in the general theory of relativity, and for Klein, too, the theory of relativity and its connection with his old ideas of the Erlangen program brought the last flareup of his mathematical interests and mathematical production. To both Hilbert and Klein, Emmy was welcome as she was able to help them with her invariant-theoretic knowledge. For two of the most significant aspects of the general relativity theory she gave at that time the genuine and universal mathematical formulation.

During World War I, Hilbert tried to push through Emmy Noether's "Habilitation" in the Philosophical Faculty in Göttingen. He failed due to the resistance of the philologists and historians. It is a well-known anecdote that Hilbert supported her application by declaring at the faculty meeting, "I do not see that the sex of the candidate is an argument against her admission as Privatdozent. After all we are a university and not a bathing establishment." Probably he provoked the adversaries even more by that remark. Nevertheless, she was able to give lectures in Göttingen, that were announced under Hilbert's name. But in 1919, after the end of the War and the proclamation of the German Republic had changed conditions, her Habilitation became possible. In 1922 there followed her nomination as a "nichtbeamteter ausserordentlicher Professor"; this was a mere title carrying no obligations and no salary. She was, however, entrusted with a "Lehrauftrag" for algebra, which carried a modest remuneration.

During the wild times after the Revolution of 1918, she did not keep aloof from the political excitment, she sided more or less with the Social Democrats: without being actually in party life she participated intensely in the discussion of the political and social problems of the day.

She always remained a convinced pacifist, a stand which she held very important and serious.

In the modest position of a "nichtbeamteter ausserordentlicher Professor" she worked in Göttingen until 1933, during the last years in the beautiful new Mathematical Institute that had risen in Göttingen chiefly by Courant's energy and the generous financial help of the Rockefeller Foundation. I have a vivid recollection of her when I was in Göttingen as visiting professor in the winter semester of 1926–1927, and lectured on representations of continuous groups. She was in the audience; for just at that time the hypercomplex number systems and their representations had caught her interest and I remember many discussions when I walked home after the lectures, with her and von Neumann, who was in Göttingen as a Rockefeller Fellow, through the cold, dirty, rain-wet streets of Göttingen. When I was called permanently to Göttingen in 1930, I earnestly tried to obtain from the Ministerium a better position for her, because I was ashamed to occupy such a preferred position beside her whom I knew to be my superior as a mathematician in many respects. I did not suceed, nor did an attempt to push through her election as a member of the Göttinger Gesellschaft der Wissenschaften. Tradition, prejudice, external considerations, weighted the balance against her scientific merits and scientific greatness, by that time denied by no one. In my Göttingen years, 1930–1933, she was without doubt the strongest center of mathematical activity there, considering both the fertility of her scientific research program and her influence upon a large circle of pupils.

Her development into that great independent master whom we admire today was relatively slow. Such a late maturing is a rare phenomenon in mathematics; in most cases the great creative impulses lie in early youth. Sophus Lie, like Emmy Noether, is one of the few great exceptions. Not until 1920, thirteen years after her promotion, there appeared in the Mathematische Zeitschrift that paper of hers written with Schmeidler, "Über Moduln in nicht-kommutativen Bereichen, insbesondere aus Differential- und Differenzen-Ausdrucken," which seems to mark the decisive turning point. It is here for the first time that the Emmy Noether appears whom we all know, and who changed the face of algebra by her work. Above all, her conceptual axiomatic way of thinking in algebra becomes first noticeable in this paper dealing with differential operators as they are quite common nowadays in quantum mechanics. In performing them, one after the other, their composition, which may be interpreted as a kind of multiplication, is not commutative. But instead of operating with the formal expressions, the simple properties of the operations of addition and multiplication to which they lend themselves are formulated as axioms at the beginning of the investigation, and these axioms then form the basis of all further reasoning. A similar procedure remained typical for Emmy Noether from then on.

Emmy Noether had many pupils, and one of the chief methods of her research was to expound her ideas in a still unfinished state in lectures, and then discuss them with her pupils. Sometimes she lectured on the same subject one semester after another, the whole subject taking on a better ordered and more unified shape every time, and gaining of course in the substance of results. It is obvious that this method sometimes put enormous demands upon her audience. In general, her lecturing was certainly not good in technical respects. For that she was too erratic and she cared too little for a nice and well arranged form. And yet she was an inspired teacher; he who was capable of adjusting himself entirely to her, could learn very much from her. Her significance for algebra cannot be read entirely from her own papers; she had great stimulating power and many of her suggestions took final shape only in the works of her pupils or co-workers. A large part of what is contained in the second volume of van der Waerden's "Modern Algebra" must be considered her property. The same is true of parts of Deuring's book on algebras in which she collaborated intensively. Hasse acknowledges that he owed the suggestion for his beautiful papers on the connection between hypercomplex quantities and the theory of class fields to casual remarks by Emmy Noether. She could utter far-seeing remarks in her prophetic lapidary manner, out of her mighty imagination that hit the mark most of the time and gained in strength in the course of years; and such a remark could then become a signpost to point the way for difficult future work. And one cannot read the scope of her accomplishments from the individual results of her papers alone: she originated above all a new and epoch-making style of thinking in algebra.

She lived in close communion with her pupils; she loved them, and took interest in their personal affairs. They formed a somewhat noisy and stormy family, "the Noether boys" as we called them in Göttingen. Artin and Hasse stand beside her as two independent minds whose field of production touches on hers closely, though both have a stronger arithmetical texture. With Hasse above all she collaborated very closely during her last years. Richard Brauer and she dealt with the profounder structural problems of algebras, she in a more abstract spirit, Brauer, educated in the school of the great algebraist I. Schur, more concretely operating with matrices and representations of groups; this, too, led to an extremely fertile cooperation. She held a rather close friendship with Alexandroff in Moscow, who came frequently as a guest to Göttingen. I believe that her mode of thinking has not been without influence upon Alexandroff's topological investigations. About 1930 she spent a semester in Moscow and there got into close touch with Pontrjagin also. Before that, in 1928–1929, she had lectured for one semester in Frankfurt while Siegel delivered a course of lectures as a visitor in Göttingen.

In the spring of 1933 the storm of the National Revolution broke over Germany. The Göttinger Mathematisch-Naturwissenschaftliche Fakultät, for the building up and consolidation of which Klein and Hilbert had worked for decades, was struck at its roots. After an interregnum of one day by Neugebauer, I had to take over the direction of the Mathematical Institute. But Emmy Noether, as well as many others, was prohibited from participation in all academic activities, and finally her *venia legendi*, as well as her "Lehrauftrag" and the salary going with it, were withdrawn. A stormy time of struggle like this one we spent in Göttingen in the summer of 1933 draws people closer together; thus I have a particularly vivid recollection of these months. Emmy Noether, her courage, her frankness, her unconcern about her own fate, her conciliatory spirit, were, in the midst of all the hatred and meanness, despair and sorrow surrounding us, a moral solace. It was attempted, of course, to influence the Ministerium and other responsible and irresponsible but powerful bodies so that her positon might be saved. I suppose there could hardly have been in any other case such a pile of enthusiastic testimonials filed with the Ministerium as was sent in on her behalf. At that time we really fought; there was still hope left that the worst could be warded off. It was in vain. Franck, Born, Courant, Landau, Emmy Noether, Neugebauer, Bernays and others—scholars the university had before been proud of—had to go because the possibility of working was taken away from them. Göttingen scattered into the four winds! This fate brought Emmy Noether to Bryn Mawr; she taught there for a short time and was also a guest at the Institute for Advanced Study in Princeton. She harbored no grudge against Göttingen and her fatherland for what they had done to her. She broke no friendship on account of political dissension. Even in the summer of 1934 she returned to Göttingen, and lived and worked there as though all things were as before. She was sincerely glad that Hasse was endeavoring with success to rebuild the old, honorable and proud mathematical tradition of Göttingen even in the changed political circumstances. But she had adjusted herself with perfect ease to her new American surroundings, and her girl students there were as near to her heart as the Noether boys had been in Göttingen. She was happy at Bryn Mawr; and indeed perhaps never before in her life had she received so many signs of respect, sympathy, friendship, as were bestowed upon her during her last one and a half years at Bryn Mawr.

I shall close with a short general estimate of Emmy Noether as a mathematician and as a personality. Her strength lay in her ability to operate abstractly with concepts. It was not necessary for her to allow herself to be led to new results on the leading strings of known concrete examples. This had the disadvantage, however, that she was sometimes but incompletely cognizant of the specific details of the more interesting applications of her general theories. She possessed a most vivid imagination, with the aid of which she could visualize remote connections: she constantly strove toward unification—she possessed a strong drive toward axiomatic purity.

Of her predecessors in algebra and number theory, Dedekind was most closely related to her. For him she felt a deep veneration. She expected her students to read Dedekind's appendices to Dirichlet's "Zahlentheorie" not only in one, but all editions. She took a most active part in the editing of Dedekind's works; here, the attempt was made to indicate, after each of Dedekind's papers, the modern development built upon his investigations. Her affinity with Dedekind, who was perhaps the most typical Lower Saxon among German mathematicians, proves by a glaring example how illusory it is to associate in a schematic way race with the style of mathematical thought. In addition to

Dedekind's work, that of Steinitz on the theory of abstract fields was naturally of great importance for her own work. She lived through a great flowering of algebra in Germany, toward which she contributed much. Her methods need not, however, be considered the only means of salvation. In addition to Artin and Hasse, who in some respects are akin to her, there were algebraists of a still more different stamp, such as I. Schur in Germany, Dickson and Wedderburn in America, whose achievements are certainly not behind hers in depth and significance.

Two traits above all determined Emmy Noether's nature. There was first, the native productive power of her mathematical genius. She was not clay, pressed by the artistic hands of God into a harmonious form, but rather a chunk of human primary rock into which He had blown his creative breath of life. Second, her heart knew no malice; she did not believe in evil—indeed it never entered her mind that it could play a role among men. This was never more forcefully apparent to me than in the last stormy summer, that of 1933, which we spent together in Göttingen. The memory of her work in science and of her personality will not soon pass away.

As Weyl's biography states, Emmy Noether's greatness is found not only in her own research but also in her stimulating effect on other mathematicians and, above all, in the fact that "she originated a new and epoch-making style of thinking in algebra." As a tribute to that new mode of thought, we have already given much space to the discussion of modern abstract algebraic systems. We have explained the nature of groups, rings, integral domains, fields, vector spaces, linear algebras (hypercomplex number systems), and Boolean lattices. Our spiral scheme of organization permits us to proceed further in the treatment of any or all of these structures. But we shall backtrack and advance only in relation to two of Emmy Noether's specialties, namely, the theory of ideals and the theory of hypercomplex number systems.

Weyl tells us that she was a disciple of Dedekind and compares some of her research with that of Wedderburn and Artin. Let us therefore focus on some of the more elementary concepts which appear in her work, as well as in that of the other three mathematicians named. Since we have already discussed some hypercomplex systems in connection with Hamilton, we shall postpone enlargement of the subject to our second illustration and shall begin the present discussion by talking about *ideals*. We must show how an ideal in a ring is analogous to a normal subgroup in a group, a fact making Galois the inspiration for both Dedekind and Noether. We must also indicate how ideals enabled Dedekind to realize the objective of providing unique factorization into "primes" (actually *prime ideals*) for certain integral domains where unique factorization into "primes" is lacking.

Since an ideal is a special type of subring of a ring, we must recall the technical meaning of the term "ring." For a classic example, we return to the set of integers

$$S = \{\ldots, -3, -2, -1, 0, 1, 2, 3, \ldots\}$$

Then the system $\{S, +, \times\}$ is a *ring*, that is, one can perform addition and multiplication within the system in such a way that traditional rules are satisfied. More exactly, the system is an *abelian group* with respect to addition, a *semigroup* with respect to multiplication, and multiplication is distributive with respect to addition.

If S is the set of all $n \times n$ matrices (Chapter 17), then the system $\{S, \oplus, \otimes\}$ where $\oplus$ and $\otimes$ are matrix addition and matrix multiplication, provides another example of a ring. This second instance is a more general type of ring since its multiplication is *not* commutative. Hereafter we shall refer to a ring with commutative multiplication as a *commutative ring*.

The ring of integers is specialized not only by the commutativity of its multiplication but also by the fact that one can carry out cancellation within the system, that is, if $ca = cb$, where a, b, and c are integers and $c \neq 0$, then $a = b$. Any ring having the two special properties described is called an *integral domain*. An additional example of an integral domain is the system $\{S, +, \times\}$ where S is the set of all numbers of the form $a + b\sqrt{2}$, where a and b are ordinary integers. By applying school algebra, the reader can demonstrate that all the required postulates are satisfied. To provide instances that are simpler for present purposes, let us return to "clock arithmetic," that is, to addition and multiplication of integers modulo m, where m is some fixed integer. Let us choose $m = 3$ and consider the following tables for the arithmetic of "residues" or "residue classes" modulo 3. The first tabulation repeats that on page 378. In the second table, the only fact that may call for review is $2 \otimes 2 = 1$. The reader must recall that $2 \otimes 2$ signifies the remainder or "residue" when the ordinary product, 2×2, is divided by the modulus 3.

Addition Modulo 3					*Multiplication Modulo* 3		
$\oplus$	0	1	2		$\otimes$	0	1	2
0	0	1	2		0	0	0	0
1	1	2	0		1	0	1	2
2	2	0	1		2	0	2	1

If $S = \{0, 1, 2\}$ is the set of residues modulo 3, then we have already shown (page 378) that those residues constitute an abelian group with respect to $\oplus$, addition modulo 3. The above multiplication table indicates that $\otimes$ is closed on the set S. The reader can readily verify that $\otimes$ is associative and commutative, also that it is distributive with respect to $\oplus$. But we shall presently demonstrate those special properties in a somewhat different way by showing that they must hold of necessity because they must "mirror" the corresponding properties of $\times$ and $+$, ordinary multiplication and addition of integers. At any rate, $\{S, \oplus, \otimes\}$, as defined by the tables above, is a *commutative ring*.

The *finite* modulo 3 ring is derived from the *infinite* ring of integers and has a similar structure. The two rings are *not* isomorphic however, since their elements are not in one-to-one correspondence. Then just how are the rings related? To explain this, one picture will be worth a thousand words. As a start, we reintroduce the number line (Figure 28.1) with the integers represented as equidistant points on that line. Next, we wind the line around a circular "clock" of circumference 3 (Figure 28.1) so that the points 0, 1, 2, of the number line are superimposed on the

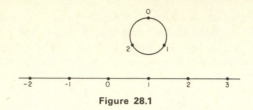

Figure 28.1

same numbers on the "clock" and the rest of the number line is wound round and round in both directions forever. Then 3, 4, 5 will fall on the hours 0, 1, 2, respectively. Winding in the other direction will place -1, -2, -3, on 2, 1, 0, respectively, and members of the three residue classes

$$S_0 = \{\ldots, -6, -3, 0, 3, 6, \ldots\}$$
$$S_1 = \{\ldots, -5, -2, 1, 4, 7, \ldots\}$$
$$S_2 = \{\ldots, -4, -1, 2, 5, 8, \ldots\}$$

will fall on 0, 1, 2, respectively.

We have just mapped the whole world of integers onto the three points $\{0, 1, 2\}$ on the clock of Figure 28.1. The associated mapping or function is a many-to-one correspondence, particularly impressive because it is an infinity-to-one correspondence that telescopes an infinite structure into a finite one. Although the mapping considered is not one-to-one as in the Klein geometric transformations or in the homeomorphisms of topology, a pertinent question would once again ask whether any features of the original picture remain *invariant*. The whole ring structure of the integers is, in fact, invariant when the integers are projected onto the modulo 3 system. This occurs because the *addition and multiplication operations are preserved*. By this we mean that $+$ corresponds to $\oplus$, $\times$ to $\otimes$, that the modulo 3 image of a sum of integers is the (modulo 3) sum of their images, the modulo 3 image of a product of integers is the (modulo 3) product of their images. For example, there are the following features of the mapping:

$$5 \to 2$$
$$7 \to 1$$
$$(5 + 7) \to (2 \oplus 1) \text{ because } 12 \to 0$$
$$(5 \times 7) \to (2 \otimes 1) \text{ because } 35 \to 2$$

Since addition and multiplication of integers are preserved, so are the ring properties of those operations. For example, if a, b, c are any three integers, and a', b', c' are their images under the mapping, then

$$[(a + b) + c] \to [(a' \oplus b') \oplus c']$$
$$[a + (b + c)] \to [a' \oplus (b' \oplus c')]$$

and hence

$$(a' \oplus b') \oplus c' = a' \oplus (b' \oplus c')$$

In other words, the associativity of addition of integers is preserved in the modulo 3 map. The preservation of the other properties of + and × can be demonstrated in similar fashion.

In more technical language, the type of mapping illustrated is called a *homomorphism* (not to be confused with the homeomorphism of topology). A *homomorphism* or *homomorphic mapping* of a ring is a mapping that preserves sums and products. In general, a ring homomorphism is a many-to-one correspondence, but includes, as a special case, a one-to-one correspondence preserving sums and products, namely, a ring *isomorphism*. Isomorphic rings, like other isomorphic structures, are considered abstractly identical. But in many cases one of two isomorphic models may be easier to handle than the other. For example, we have the case of analytic Euclidean geometry versus the isomorphic classic synthetic model.

The fact that a homomorphism may effect a substantial condensation without changing structure makes that type of mapping an extremely important tool, the more so since the ring is not the only algebraic system susceptible to such telescoping. A homomorphism can be applied to structures with fewer than two operations—groupoids, semigroups, groups—and to systems with two or more operations—vector spaces, linear algebras, etc. Since a homomorphic mapping preserves *all* operations, it maps groups onto groups, vector spaces onto vector spaces, etc.

In passing, let us point out that "preserving operations" has a meaning more subtle than that phrase indicates. Thus consider the mapping of the set of integers onto the set of integral powers of 2, as indicated below:

$$\begin{pmatrix} \ldots, & -3, & -2, & -1, & 0, & 1, & 2, & \ldots \\ \ldots, & 2^{-3}, & 2^{-2}, & 2^{-1}, & 2^0, & 2^1, & 2^2, & \ldots \end{pmatrix}$$

The first row is a group with respect to *addition*; the second row is a group with respect to *multiplication*. *Sums* in the first row are "preserved" as *products* in the second. For example, $(-3) + 2 = -1$ is mapped onto $2^{-3} \times 2^2 = 2^{-1}$. Hence if S is the set of integers and T is the set of integral powers of 2, then the *group* $\{S, +\}$ is homomorphic, in fact isomorphic, to the group $\{T, \times\}$. The addition of the first group is preserved as multiplication in the second. In a homomorphism, all that is required is that there be matching of operations and results of operations in a system and its image.

For present purposes, let us concentrate on the group homomorphism where the additive group of integers is mapped onto the additive group of residues modulo 3. Then, by examining the classes S_0, S_1, S_2 (page 662), which are mapped onto 0, 1, 2, respectively, it will be seen that S_0 possesses special properties that are lacking in S_1 and S_2. It is easy to observe that $\{S_0, +\}$ is an abelian group, in fact a *normal subgroup* of the additive group of integers. But +, ordinary addition, is *not* even closed on S_1 (or S_2). Thus, for example, the sum of 1 and 4 (two elements of S_1) is 5, a member of S_2, and in S_2, $2 + 5 = 7$ (a member of S_1). Hence S_1 and S_2 are not even semigroups under ordinary addition.

The special property possessed by S_0 illustrates a general theorem which states that, in *every* homomorphic mapping of one group onto another group, the set of elements mapped onto the identity of the second group constitutes a *normal subgroup* of the first group. There is also a converse theorem indicating that every normal subgroup of any group will lead to a homomorphism where the normal group and its cosets are the "residue classes," and the normal subgroup is mapped onto the identity of the image group. Since a group homomorphism is thus always associated with a *normal subgroup*, the latter is called the *kernel* of the homomorphism. In the case of finite groups, the index of the normal subgroup in the whole group (Chapter 16) gives the order of the image group and indicates the degree of condensation effected by the homomorphism. Thus the *maximal* proper normal subgroup of a group would offer the greatest opportunity for telescoping the group into a smaller group. Of course the whole group is an "improper" normal subgroup of itself, and if it is the kernel of a homomorphism, the whole group is mapped onto a single image element. That would represent condensation to the state of complete collapse, where a trivial picture results—a singleton group. At the other extreme, the group identity constitutes a normal subgroup, and if that is the kernel of a homomorphism, there are only single elements in the cosets or residue classes, and the homomorphic image is an isomorphic or abstractly identical group. This result is not a condensation but a duplication.

Let us now consider S_0 in relation to the *ring* of integers, where two operations, $+$ and $\times$, are involved. We have seen that S_0 is closed under $+$, ordinary addition. But now we observe that S_0 consists of all integral multiples of 3, and thus S_0 is also closed under $\times$, ordinary multiplication, since the product of any two multiples of 3 is a multiple of 3. Then, because S_0 is an abelian group under $+$ and closed under the associative multiplication operation $\times$, which is distributive with respect to $+$, the system $\{S_0, +, \times\}$ is a *ring*, in fact, a proper *subring* of the ring of integers.

In the homomorphic mapping of the ring of integers onto the modulo 3 ring, the class S_0 is still the kernel, that is, S_0 is mapped onto 0, the additive identity, and the role of $\{S_0, +\}$ as a special type of subgroup of the additive group of integers, namely, a normal subgroup, suggests that $\{S_0, +, \times\}$ may be a special type of subring of the ring of integers. Hence, let us search for special multiplicative properties of S_0. We readily see that if a member of S_0 is multiplied by *any integer whatsoever*, whether or not the integer is in S_0, the product is a member of S_0. Thus it is not surprising that $3(-6) = -18$, a member of S_0, since S_0 is closed under $\times$. But we also have $3 \times 4 = 12$, a member of S_0, $3 \times 5 = 15$, a member of S_0, although 4 and 5 are not members of S_0.

A subring possessing the property just described is called an *ideal*. In other words, a subset of a ring with a commutative multiplication is called an ideal if it is a subgroup of the additive group of the ring and if it contains all products of its elements by arbitrary elements of the ring. (In rings with a noncommutative multiplication, there are *left ideals* and *right ideals*.)

It can be proved that in any ring homomorphism, the set of elements mapped onto the zero (additive identity) of the image is an ideal of the original ring, and conversely, an ideal of a ring can be made the *kernel* of a homomorphic mapping. Then, as in the case of normal subgroups of a group, the ideals of a ring determine the ring homomorphisms and the possibilities for condensation. Again, if a ring has no ideals other than itself and the subset consisting of the additive identity, it has only "improper" homomorphisms.

What has just been said can be applied to showing that a *field*, the fundamental structure of arithmetic and algebra, cannot be properly condensed, that is, has no proper homomorphisms. For proof, one can consider a field as a special kind of ring and then show that it has no ideals except itself and the additive identity (zero). Suppose that a field has an ideal other than zero. Then that ideal contains at least one nonzero element, which we shall symbolize as a. Because $a \neq 0$, it has a multiplicative inverse, a^{-1}, in the field (Chapter 4). Now the product of a member of an ideal by any ring element whatsoever must belong to the ideal. Hence $aa^{-1} = 1$ is an element of the ideal and, by virtue of the property just applied, $b \cdot 1 = b$ belongs to the ideal, where b is any element whatsoever in the field. Thus the ideal is identical with the whole field. In summary, in any field the field itself and the zero ideal are the only ideals, and the only homomorphisms are therefore either an isomorphism or a map onto a single element.

In the ring of integers, the set of all multiples of 3 constitutes an ideal, a fact which suggests that all multiples of any integer m will form an ideal and therefore be the kernel of a homomorphism onto the ring of residues modulo m. The reader can check that this is actually the case by studying homomorphisms of the ring of integers onto modulo 2, modulo 4, modulo 5, etc. rings and showing that the sets

$$(2) = \{\ldots, -2, 0, 2, 4, \ldots\}$$
$$(4) = \{\ldots, -4, 0, 4, 8, \ldots\}$$
$$(5) = \{\ldots, -5, 0, 5, 10, \ldots\}$$
$$(6) = \{\ldots, -6, 0, 6, 12, \ldots\}$$

constitute ideals in the ring of integers. The ideals we have symbolized by (2), (4), (5), and (6) are specialized in type because the membership in each case is made up of multiples of a singe element—2 or 4 or 5, etc. Such ideals are described as *principal*.

A *principal ideal* in a ring is an ideal whose elements are all the multiples of a single element a which is said to *generate* the ideal. In other words, the principal ideal (a) is the set of all elements xa, where a is a fixed member of the ring and x is any other member. It turns out that in the ring of integers *all* ideals are principal. But the same property need not obtain in more general rings. There may, for example, be an ideal "generated" by two elements. Such an ideal would be symbolized by (a, b) and, in a commutative ring with unit element (multiplicative identity), would consist of all elements $xa + yb$, where a and b are fixed members of the ring and x and y are *any* elements whatsoever in the ring.

The principal ideals (2), (3), (4), etc. are all subrings of the ring of integers, but one can go further by showing that some of these ideals are *subideals* of others. Thus, since multiples of 4, for example, 8, 12, 16, 20, etc., are obviously also multiples of 2, the ideal (4) is a subideal of the ideal (2). This fact can be symbolized as (4) $\subseteq$ (2), where $\subseteq$ is the standard symbol for "is contained in." Or using $\supseteq$ the symbol for "contains," one has (2) $\supseteq$ (4); that is, the principal ideal (2) contains the principal ideal (4) as a subideal. The reader can verify that (3) $\supseteq$ (6) $\supseteq$ (12), (2) $\supseteq$ (10), (5) $\supseteq$ (10), etc. The illustrative examples suggest that if a and b are given integers, then $(a) \supseteq (b)$ if and only if a divides b, symbolized as $a|b$. Thus, in the case of principal ideals, the relation of class inclusion, $\supseteq$, and the relation of ordinary arithmetic divisibility, $|$, are apparently interchangeable.

Let us see how far the analogy between $\supseteq$ and $|$ can be pushed. We recall that an integer m is a prime if its only divisors are itself and the unit 1, that is, the only possible divisibility statements are $m|m$ and $1|m$. By analogy, the corresponding relations involving inclusion of ideals would be $(m) \supseteq (m)$ and $(1) \supseteq (m)$. (Note that (1) signifies the aggregate of multiples of the unit 1, in other words, the entire set of integers.) The inclusion statements define a *maximal ideal*, for they assert that (m) is a subideal only of itself and the entire ring, (1). Or one can say, by analogy with arithmetic, that (m) is "divisorless," that is, lacks proper divisors. At this point the analogy with elementary arithmetic begins to weaken since, in the very general rings of Dedekind's theory, being maximal (divisorless) is a sufficient but *not* a necessary conditon for an ideal to be prime. According to Dedekind's definition, of which we shall give a sophisticated formulation later, if an ideal is maximal (divisorless), it is surely prime, but if it is prime, it need not be maximal.

If we continue with everyday questions of arithmetic divisibility and the associated decomposition of an integer into a product of other integers, for example, $6 = 2 \times 3$, we can inquire about the corresponding factorization of ideals. Thus, if (x) and (y) are principal ideals, their product, $(x)(y)$, is *defined by*

$$(x)(y) = (xy)$$

that is, the product of two principal ideals is generated by the product of the generating elements of the two ideals. Hence

$$(2)(3) = (2 \cdot 3) = (6)$$

which is the analogue of the arithmetic factorization cited above. As another example, the reader can verify that

$$1155 = 3 \cdot 5 \cdot 7 \cdot 11$$
$$(1155) = (3)(5)(7)(11)$$

Our illustrations suggest a complete parallel between factorization of integers and factorization of ideals, so that for the ring of integers (or for any ring all of whose ideals are *principal*) the analogy is perfect. But the trouble starts when more general integral domains are examined. For example, consider S, the set of all numbers, $a + b\sqrt{-5}$, where a and b are ordinary integers. Then $\{S, +, \times\}$ can be shown to be an integral domain, and in that domain, 6 is not only the product of 2 and 3 but also of $1 + \sqrt{-5}$ and $1 - \sqrt{-5}$, where $1 + \sqrt{-5}$, $1 - \sqrt{-5}$ can be proved to be "primes," that is, irreducibles of the integral domain. Euclid's fundamental theorem asserting unique factorization fails for this integral domain and for many others. But when Dedekind arrived at a definition of a *prime ideal* suitable for his purposes, he was able to show that (2), (3), $(1 + \sqrt{-5})$, $(1 - \sqrt{-5})$ are *not* prime ideals and can be broken down into prime ideal factors as follows:

$$(2) = (2, 1 + \sqrt{-5})(2, 1 - \sqrt{-5})$$
$$(3) = (3, 1 + \sqrt{-5})(3, 1 - \sqrt{-5})$$

Then, by Dedekind's definition, the *ideal* (6) does have a *unique* factorization into prime ideals

$$(6) = P_1 P_2 P_3 P_4$$

where $P_1 = (2, 1 + \sqrt{-5})$ and P_2, P_3, P_4, are the other ideals in the right members of the above equations for (2) and (3). Let us remind the reader of the exact meaning of an ideal with two generators, for example,

$$P_2 = (2, 1 - \sqrt{-5})$$

As stated on page 665, the ideal P_2 must consist of all the numbers $2x + (1 - \sqrt{-5})y$ where x and y are any numbers whatsoever in the domain, that is, any numbers of the form $a + b\sqrt{-5}$, with a and b ordinary integers. Thus, for example, $2(3 - 2\sqrt{-5}) + (1 - \sqrt{-5})(2 + \sqrt{-5}) = 13 - 5\sqrt{-5}$ would be a member of the prime ideal P_2. Also we must explain that the product of ideals with more than one generator can be defined in a manner illustrated by

$$(a, b)(c, d) = (ac, ad, bc, bd)$$

We shall continue discussion of the generalization of number-theory concepts through the theory of ideals after we have probed a little more deeply into some side-effects of ring homomorphisms. When a homomorphic mapping is not an isomorphism, it is a many-to-one correspondence, and as a result some algebraic properties that depend on the individuality of the ring elements may be lost. This may produce a representation that is more general or freer than the original system because it is bound by fewer laws. On the other hand, the fusion of elements into a single image by means of a many-to-one map may produce a type of blurring where the image has certain properties not present in the original. In that case the map is more specialized or less general than the system it represents. A homomorphic projection is like a shadow. Some features of the original system appear clearly, but some may be lost and others exaggerated. Both types of shadow effect are found in different modulo m maps of the ring of integers, as we shall now see.

In order to study a property that may be preserved in some homomorphic maps of the ring of integers and lost in others, let us start with the elementary problem of finding integral solutions of a quadratic equation like

$$x^2 + x - 6 = 0$$

Using Harriot's method (Chapter 5), we factor the left member and obtain

$$(x - 2)(x + 3) = 0$$

Harriot reasoned that each parenthesis represents a number that *must* be zero for the first parenthesis or for the second or for both parentheses because *the product of two numbers is zero only if at least one of the numbers is zero.* This fact seems obvious to us, but it is actually one of the laws governing the world of integers and *not* a characteristic of rings in general.

The application of this special law enables one to say that

$$x - 2 = 0 \quad \text{or} \quad x + 3 = 0$$

and

$$x = 2 \quad \text{or} \quad x = -3$$

so that there are two integers meeting the conditions expressed by the equation.

Let us now prove that the law assumed by Harriot in solving polynomial equations must hold in every ring that is an *integral domain* and can in fact be deduced from the *cancellation law* (Chapter 4), the property distinguishing an integral domain from more general commutative rings. We must show then that if a and b are elements of an integral domain such that

$$ab = 0$$

then

$$a = 0 \quad \text{and/or} \quad b = 0$$

Now either a and b are both equal to zero, or at least one of them is not equal to zero. Suppose then that $a \neq 0$. We would have

$$ab = 0 \quad \text{and} \quad a \cdot 0 = 0$$

so that

$$ab = a \cdot 0$$

In the last equation a may be cancelled to yield

$$b = 0$$

(Similarly, if we started by assuming $b \neq 0$, we could prove that $a = 0$. In any case, at least one factor must vanish.)

Conversely, if Harriot's law is assumed, the cancellation law follows. Thus, given

$$ab = ac, \quad a \neq 0$$

then

$$ab - ac = 0$$

or

$$a(b - c) = 0$$

By Harriot's law, either one factor or both factors must vanish. Since $a \neq 0$,

$$b - c = 0$$

and

$$b = c$$

so that the cancellation law has been deduced.

As stated in our chapter on symbolic logic, $p \Leftrightarrow q$ signifies that propositions p and q are logically equivalent. In the present instance, the cancellation law and Harriot's law are logically equivalent. For that reason, an integral domain is sometimes *defined* as a commutative ring in which Harriot's law is satisfied.

Let us now see whether the above equivalent properties of the ring of integers are preserved when that ring is mapped homomorphically on $\{0, 1, 2, 3\}$, the ring of residues (or residue classes) modulo 4, whose addition and multiplication tables are as follows:

Addition Modulo 4						*Multiplication Modulo* 4				
$\oplus$	0	1	2	3		$\otimes$	0	1	2	3
0	0	1	2	3		0	0	0	0	0
1	1	2	3	0		1	0	1	2	3
2	2	3	0	1		2	0	2	0	2
3	3	0	1	2		3	0	3	2	1

To furnish ammunition, we now seek solutions of the first-degree equation

$$2 \otimes x = 0$$

in the modulo 4 ring, that is, when the domain of x is the set of residues $\{0, 1, 2, 3\}$. Our problem is equivalent to solving the congruence

$$2x \equiv 0 \ (\mathrm{mod} \ 4)$$

(See Chapter 21). In any case,

$$x = 0 \quad \text{or} \quad x = 2$$

will solve either the equation or the congruence, since

$$2 \otimes 0 = 0 \quad \text{and} \quad 2 \otimes 2 = 0$$

or alternatively

$$2 \cdot 0 \equiv 0 \ (\mathrm{mod} \ 4) \quad \text{and} \quad 2 \cdot 2 \equiv 0 \ (\mathrm{mod} \ 4)$$

Here the corollary to the fundamental theorem of algebra fails because a first-degree equation has two roots. This will not surprise us, because Chapter 21 indicated that this sort of thing may occur for congruences whose modulus is *not* prime. But we are more interested in showing that the failure of the fundamental theorem results from the fact that a particular homomorphic image of the ring of integers—namely, the modulo 4 ring—is *not* an integral domain. The modulo 4 multiplication table indicates that $2 \otimes 2 = 0$. Here a product vanishes although neither factor is equal to zero, and hence Harriot's rule is *not* a universal law for the modulo 4 ring. Then that ring will also lack a cancellation law. The modulo 4 multiplication table shows that $2 \otimes 3 = 2$ and $2 \otimes 1 = 2$. Therefore

$$2 \otimes 3 = 2 \otimes 1$$

Someone not aware that the cancellation law fails in the modulo 4 ring might cancel 2 from the last equation and thus arrive at the paradox, $3 = 1$.

In the statement $2 \otimes 2 = 0$, 2 is said to be a *divisor of zero*. Thus an integral domain is often described as a commutative ring in which there are no divisors of zero. For further examples of rings with divisors of zero, the reader can verify that in arithmetic on an ordinary clock, that is, in the modulo 12 ring, $2 \otimes 6 = 0$ and $3 \otimes 4 = 0$ so that 2, 3, 4, and 6 are all divisors of zero. It is readily seen that there are always divisors of zero in modulo m rings when m is *not* prime. Therefore, such homomorphic maps of the integers are rings but *not* integral domains.

It is in fact a theorem of modern algebra that a modulo m ring is an integral domain if, and only if, m is a *prime* number. Thus we may wish to avoid modulo 6, modulo 8, modulo 9, and modulo 10 systems, for example, since 6, 8, 9, and 10 are composite numbers. Suppose that we use as homomorphic images of the integers only the modulo m systems with $m = 3$ or 5 or 7 or 11 or some other prime number. Some parts of the theory of equations are safe, but will other aspects of the algebra of integers be blurred or distorted? The axioms defining a ring permit addition, subtraction (by means of additive inverses), and multiplication, but say nothing about division, that is, about multiplicative inverses. Every division question can be converted into one about multiplication. For example, $6 \div 2 = ?$ becomes $2 \times ? = 6$ or a matter of solving $2x = 6$. In the domain of integers the cancellation law permits us to cancel 2 if we write the last equation as

$$2x = 2 \times 3$$

and thus

$$x = 3$$

But suppose that we ask for an answer to $2 \div 5$, that is, seek a solution to $5x = 2$. Our world is that of the integers, and hence we cannot solve this problem in that domain. No integer can properly fill the blank in (5) ( ) = 2, nor can we carry out an integral factorization of the right member of $5x = 2$ that will permit cancellation of 5 from both members. Infinitely many other equations of the form $ax = b$, where a and b are integers, are impossible to solve within the integral domain.

Since 3 is a prime number, the modulo 3 ring is an integral domain and does not present special difficulties with respect to cancellation or divisors of zero. But if the integers are mapped on the modulo 3 system, does that map reflect the impossibility of performing division with certain pairs of integers? Suppose that $1 \oplus 2$ is called for, that is, the solution in the modulo 3 system of $2 \otimes x = 1$. The answer is $x = 2$ because $2 \otimes 2 = 1$ in the modulo 3 ring. In other words, $1 \oplus 2 = 2$ in that ring, and this fact does not reflect the impossibility of carrying out in the original aggregate of integers the following combinations (and an infinite number of others) of which $1 \oplus 2$ is the image: $1 \div 2$, $4 \div 11$, $13 \div 5$, $22 \div 17$. What has happened is that these division combinations, impossible in the domain of integers, have become fused with the possible combinations projected onto $1 \oplus 2$. Some of the latter are $4 \div 2$, $25 \div 5$, $16 \div 8$, $88 \div 11$. In the modulo 3 ring, except for division by zero, which is always excluded in arithmetic, all division combinations can be carried out, because the modulo 3 ring is in fact a *field*.

To prove that a commutative ring is a field, one must show that it has a unit element (multiplicative identity) and that every element except zero has a multiplicative inverse (Chapter 4). If we examine the modulo 3 multiplication table (page 661), we can see that 1 is a multiplicative identity because its product with any element of the ring leaves that element unchanged. The table also shows that 1 and 2 have themselves as multiplicative inverses since $1 \otimes 1 = 1$, $2 \otimes 2 = 1$. Hence the modulo 3 ring is a field.

It can be proved that every modulo p ring where p is a prime number is a field and thus that the homomorphism of the integers onto such modulo p fields leads to a representation with additional properties not present in the original system. Here the shadow distorts the original structure so that it assumes a more specialized form. This mapping effect is opposite in type to that in the case of the modulo 4, modulo 6, modulo 8, etc. images, which remove original properties and present a more general picture than the original domain of integers.

In homomorphisms of the ring of integers, the images are either fields or else commutative rings that are not even integral domains, and there is no middle ground where the image is an integral domain but not a field. But in homomorphic mappings of integral domains more general than the set of integers, the intermediate result may occur, namely, preservation of the property of being an integral domain without distortion into a field. That fact is used in one modern definition of Dedekind's conception of a *prime ideal*: An ideal in an integral domain is said to be prime if, when the ideal is the kernel of a homomorphic mapping, the image is an integral domain. If the image satisfies the stronger condition of being a field, the ideal is not only prime but divisorless (maximal).

Our exposition of the nature of rings and ideals, homomorphisms of rings and other algebraic structures, would make it possible for a reader to examine one of Emmy Noether's most important research papers, namely, her 1926 *Abstrakter Aufbau der Idealtheorie*. A year after that memoir was published she was collaborating with Richard Brauer (today one of the leading American algebraists) in a different but not entirely unrelated area of abstract algebra. Then, in 1929, she began her advanced research on hypercomplex number systems. Although we shall not discuss her particular contributions to that subject, we shall honor her once again by explaining the background material that motivated her research, and by quoting the ideas and opinions of those whose fields of investigation were closest to her own.

For example, there is Emil Artin (1898–1962), about whom Hermann Weyl wrote, "He is in some respects akin to Emmy Noether." In a biographical eulogy of Artin in the January 1967 *Bulletin of the American Mathematical Society*, Richard Brauer says, "Whatever standards we use, he was a great mathematician." Brauer, describing Artin's manifold contributions to number theory, field theory, topology, and other mathematical subjects, offers the characterization, "For Artin to be a mathematician meant to participate in a great common effort, to continue work begun thousands of years ago, to shed new light on old discoveries, to seek new ways to prepare the developments of the future."

Let us see how Artin shed new light on the algebraic discoveries which preceded Emmy Noether's work and his own. The ideas he elucidated were those of the Scottish-American Joseph Henry Maclagen Wedderburn (1882–1948). Artin wrote:

> To give a really exhaustive account of the influence of Wedderburn on the development of modern algebra would require years of preparation. In order to present at least a modest account of this influence it is necessary to restrict oneself rather severely. If we were to mention all the consequences and applications of his theorems we could easily fill a whole volume. . . . For the understanding of the significance that Wedderburn's paper *On hypercomplex numbers* (1907) had for the development of modern algebra it is imperative to look at the ideas his predecessors had on the subject.*

It seems advisable to obey Artin's injunction, and consider the notions that paved the way for the 1907 memoir, *On hypercomplex numbers*. The date of the earliest hypercomplex numbers was 1843. These were the quaternions (Chapter 4), created by Hamilton when he was seeking to formulate an algebra of vectors in three-dimensional space that would be analogous to the algebra of 2-dimensional vectors, or their pure mathematical representation, the ordinary complex numbers $a + bi$, where a and b are real numbers and $i = \sqrt{-1}$ is the imaginary unit or the "normal" unit vector. From the point of view of modern mathematics, the most significant fact about the ordinary complex numbers is that they illustrate what Artin and other algebraists today term a *linear associative algebra* or, more briefly, an *algebra*, a system $\{S, \oplus, \otimes, \circ\}$, which can be described as a *ring* with an additional operation $\circ$, called an "external" or "scalar" multiplication in contrast to the ring multiplication, $\otimes$. Alternatively, as indicated at the close of Chapter 4, and also on page 421, an algebra can be considered to be a special type of *vector space*, one for which a binary vector multiplication $\otimes$ is defined, in addition to the unary scalar multiplication $\circ$ which every vector space must possess.

One can think of an ordinary complex number like $5 + 4i$ as a vector with components (5, 4). Hence the algebra of such complex numbers or vectors is a set of rules for combining or operating on such number pairs. But to explain the adjective *linear* one can use the alternative form $a + bi$ which indicates any complex number as a linear combination

$$a \cdot 1 + b \cdot i$$

of the two basic elements or unit vectors 1 and i, where the coefficients a and b are numbers in the real number field. As for the term *associative*, it refers to the

---

* Emil Artin, "The Influence of J. H. M. Wedderburn on the Development of Modern Algebra," *Bulletin of the American Mathematical Society*, Vol. 56, No. 1, Part 1 (January 1950).

postulate that the ring multiplication obeys an associative law. Again, we have seen (Chapter 4) that any real quaternion is a linear combination of the "units" $1, i, j, k$.

A *hypercomplex* number of *order* or *dimension n* can be thought of either as an ordered n-tuple $(a_1, a_2, a_3, \ldots, a_n)$, or else as a linear combination

$$a_1e_1 + a_2e_2 + a_3e_3 + \cdots + a_ne_n$$

of n "independent" elements of unit length, namely, the *unit vectors* $e_1, e_2, e_3, \ldots, e_n$, where the coefficients $a_1, a_2, a_3, \ldots, a_n$ are numbers in some *field*, called the scalar or external field. Note that in our description, we have, for the sake of simplicity, used $+$ instead of $\oplus$ (to signify vector addition) and have omitted the symbol $\circ$ for scalar multiplication, merely employing a juxtaposition like a_1e_1 or a_2e_2, etc. instead of $a_1 \circ e_1$ or $a_2 \circ e_2$. We shall continue such simplified notation in the pages ahead.

This concept of hypercomplex numbers and their algebras was initiated by Hermann Grassmann (1809–1877), a German geometer, who formulated the notion in his *Ausdehnungslehre* (Theory of Extension) published in 1844, just a year after Hamilton presented his quaternions. As we shall see, Hamilton's algebra is just a special instance of the broader Grassmann concept, which embraces even the tensor algebra of general relativity. The German's ideas were overlooked in the main during his lifetime, and their greatness was not recognized until the twentieth century.

In any Grassmann linear algebra the addition operation and the external or scalar multiplication are defined much as one might expect by analogy with elementary algebra. But the distinguishing feature of the different algebras, in addition to dimension, is the definition of the ring (or vector) multiplication. Thus the sum of

$$a_1e_1 + a_2e_2 + \cdots + a_ne_n$$

and

$$b_1e_1 + b_2e_2 + \cdots + b_ne_n$$

is taken to be

$$(a_1 + b_1)e_1 + (a_2 + b_2)e_2 + \cdots + (a_n + b_n)e_n$$

which is just a generalization of such facts of elementary algebra as $(4 - 5i) + (-3 + 2i) = 1 - 3i$. Again, *scalar* multiplications like $6(-7 + 4i) = -42 + 24i$ are generalized to

$$k(a_1e_1 + a_2e_2 + \cdots + a_ne_n) = ka_1e_1 + ka_2e_2 + \cdots + ka_ne_n$$

where k is any number of the external or scalar field.

Suppose that the ring (or vector) multiplication is to be defined in an algebra of order 2. Then

$$(a_1e_1 + a_2e_2)(b_1e_1 + b_2e_2) = a_1b_1e_1{}^2 + a_1b_2e_1e_2 + a_2b_1e_2e_1 + a_2b_2e_2{}^2$$

must be a hypercomplex number of the ring and therefore a linear combination of e_1 and e_2. To make this possible, $e_1{}^2, e_1e_2, e_2e_1, e_2{}^2$ must each be such a combination, and it suffices to postulate a multiplication table for these basic vectors or units. (Observe that we have been using juxtaposition to abbreviate *vector multiplication*. Thus $e_1{}^2$ signifies $e_1 \otimes e_1$, and $e_1 e_2$ signifies $e_1 \otimes e_2$, etc.)

In the algebra of ordinary complex numbers where $e_1 = 1$ and $e_2 = \sqrt{-1}$, the table is

$\otimes$	e_1	e_2
e_1	e_1	e_2
e_2	e_2	$-e_1$

which can be read $e_1 e_1 = e_1{}^2 = e_1$, $e_1 e_2 = e_2$, $e_2 e_1 = e_2$, $e_2 e_2 = e_2{}^2 = -e_1$. By substituting 1 for e_1 and $\sqrt{-1}$ for e_2, the reader can verify that this table agrees with the facts of algebra as learned at school.

But the above is not the only algebra of second order. If one attaches no concrete meanings to e_1 and e_2, but treats them as pure abstractions, it is permissible to postulate many different multiplication tables. For example, let us assume the table below:

$\otimes$	e_1	e_2
e_1	0	e_1
e_2	e_1	e_2

From the table one obtains $e_1 e_1 = e_1{}^2 = 0$, $e_1 e_2 = e_1$, $e_2 e_1 = e_1$, $e_2 e_2 = e_2{}^2 = e_2$. The multiplication is commutative since $e_1 e_2 = e_2 e_1$.

It can be observed that the ring of complex numbers so defined is *not* an integral domain, because $e_1 e_1 = 0$, that is, e_1 is a divisor of zero because the product in question is zero, although neither factor e_1 is zero. In fact, e_1 is a *unit* vector. To introduce some of the vocabulary associated with linear algebras, the element e_1 would be called *nilpotent* in the present case because, although it is not itself zero, one of its powers (in this case, the second) is zero. Such an element will be called *properly nilpotent* if its product with every hypercomplex number of the algebra is nilpotent. Thus, in the algebra under consideration it is possible to prove that the product e_1 with *any* hypercomplex number $a_1 e_1 + a_2 e_2$ is

$$e_1(a_1 e_1 + a_2 e_2) = a_1 e_1{}^2 + a_2 e_1 e_2 = 0 + a_2 e_1 = a_2 e_1$$

and the second power of $a_2 e_1 = a_2{}^2 e_1{}^2 = a_2{}^2(0) = 0$ so that the product is nilpotent, and hence e_1 is *properly nilpotent*.

In the Wedderburn theorems, and in other major results on linear associative algebras, the *properly nilpotent* elements play an important part. If *all* the elements of an algebra are properly nilpotent, it is called a *nilpotent algebra*. If not, some of its elements may be properly nilpotent and the aggregate of these can be shown to be a *subalgebra*. All of this is the starting point of Wedderburn's theories. If one follows the further consequences of nilpotency, it is revealed that the study of linear associative algebras reduces itself to the consideration of three types called *nilpotent* algebras, total *matric* algebras, and *division* algebras.

Our brief discussion of nilpotency has given some idea of how elements in a nilpotent algebra may behave. A total matric algebra was in fact illustrated in Chapter 17, where definitions of matrix operations were given. When matrix addition, multiplication, and scalar multiplication are limited to the set of all 2×2 or 3×3 or ... $n \times n$ matrices with entries from a specified field, the resulting system is a linear associative algebra often described as a *total matric algebra* of order 2^2 or 3^2 or ... n^2. There remains the problem of illustrating the third type of algebra, the *division algebra*.

Whether one conceives of an algebra as a ring with a supplementary scalar multiplication or as a vector space with a supplementary vector multiplication, the fact is that one *cannot* invert vector multiplication, that is, carry out division, in a general linear associative algebra. Hence there is specialization in a *division algebra*, one in which it is possible to carry out division by any element except zero. The algebra of the ordinary plane complex numbers is such an algebra. The other 2-dimensional algebra cited above has a commutative multiplication, but it is *not* a division algebra, as one can realize from the fact that it is not even an integral domain where cancellations can be performed. We shall emphasize the fact by showing that it would not be possible, for example, to carry out the division

$$\frac{6e_1 + 2e_2}{2e_1}$$

If some complex number $xe_1 + ye_2$ of the algebra were an answer, then

$$xe_1 + ye_2 = \frac{6e_1 + 2e_2}{2e_1}$$

and

$$2e_1(xe_1 + ye_2) = 6e_1 + 2e_2$$

$$2xe_1{}^2 + 2ye_1e_2 = 6e_1 + 2e_2$$

$$0 + 2ye_1 = 6e_1 + 2e_2$$

which is impossible. Even if one selects $y = 3$, the last equation above leads to $2e_2 = 0$, which is false, since e_2 is a unit vector and $2e_2$ signifies a vector of magnitude 2 and the same direction as e_2.

For still other second-order algebras one might assign multiplication tables like

$\otimes$	e_1	e_2
e_1	0	0
e_2	0	0

and

$\otimes$	e_1	e_2
e_1	e_1	e_2
e_2	0	0

The first of these might be said to describe a *zero algebra* of order 2 since the product of any two complex numbers would be zero, because

$$(a_1e_1 + a_2e_2)(b_1e_1 + b_2e_2) = a_1b_1e_1{}^2 + a_1b_2e_1e_2 + a_2b_1e_2e_1 + a_2b_2e^2$$

and each real coefficient in this result is thus to be multiplied by zero, in accordance with the multiplication table.

The reader can verify that the second multiplication table above indicates that the unit e_2 is nilpotent and that the ring is *not* an integral domain, since $e_2e_1 = 0$ and $e_2e_2 = 0$ where none of the factors is zero. Since it is not an integral domain, the algebra is *not* a division algebra, a fact which can be further illustrated by attempting to find a meaning, within the algebra, for the "quotient," $(3e_1 - 4e_2)/e_2$. What we wish to stress above all is that this second algebra is *noncommutative*, because its multiplication table indicates that e_1e_2 is *not* equal to e_2e_1. Therefore, to show the impossibility of carrying out any kind of division above, the reader should consider *two* cases, namely, $e_2(xe_1 + ye_2) = 3e_1 - 4e_2$ and $(ze_1 + we_2)e_2 = 3e_1 - 4e_2$.

One might continue to postulate arbitrary multiplication tables for algebras of order 2—with the purpose of seeing what variety is possible, what the effect of changing the scalar field may be, whether any 2-dimensional algebra other than that of the ordinary complex numbers is a division algebra, and whether the abstract complex numbers $a_1e_1 + a_2e_2$ of the particular algebras of order 2 can be provided with geometrical or physical interpretations. Instead of mere enumeration or trial and error, perhaps some *theorems* on all such algebras should be sought. The last idea would have been typical of Wedderburn's thought, for he always sought general "structure theorems" to put an end to the enumeration of more and more particular instances of algebras with or without specified properties.

In the present case, some of the questions just raised can be answered by an earlier structure theorem, proved both by the German G. Frobenius (1849–1917) and by Charles Saunders Peirce (1839–1914), who was the son of the better known Harvard mathematician Benjamin Peirce (1809–1880). The younger Peirce was granted recognition posthumously, but some of his contributions to pure mathematics are still overlooked, possibly because his influence on logic and philosophy was so much greater (Chapter 6). The Peirce-Frobenius structure theorem reveals that if the scalar or coefficient field is the system of real numbers, then the ordinary complex numbers constitute the *only* division algebra of order 2, and that if one proceeds to higher orders, there will be one more division algebra, of order 4, that of the quaternions with real coefficients. Morever, there are no other division algebras with real coefficients except that of the real numbers themselves (an algebra of the first order with the single basic unit 1). The structure theorem in question indicates what a time-saver such a proposition can be. It makes unnecessary the examination of the tremendous variety of multiplication tables one could postulate for algebras of each order, for example, the different rules for

$$e_1{}^2, e_1e_2, e_1e_3, \ldots, e_1e_5, e_2e_1, e_2{}^2, e_2e_3, \ldots, e_5e_1, e_5e_2, \ldots, e_5{}^2$$

in an algebra of the fifth order. The Peirce-Frobenius theorem tells us that, with *all* such tabulation, *forever*, no division algebras would be found, after the particular one of the fourth order discovered by Hamilton.

In 1961, the algebraic topologists J. F. Adams, R. H. Bott, M. A. Kervaire, and J. W. Milnor arrived at a more general theorem than that of Peirce and Frobenius by proving that the real *nonassociative* case provides only one more division algebra, namely, that of the Cayley 8-tuples discussed in Chapter 4.

The multiplication table for the quaternions of Hamilton appears in Chapter 4. We shall apply that table in showing that it is always possible to divide by any quaternion except zero. We shall employ a method similar to that learned at school for carrying out divisions with ordinary complex numbers where divisors are multiplied by conjugates, thus:

$$\frac{3-i}{4+2i} = (3-i)\left(\frac{1}{4+2i} \cdot \frac{4-2i}{4-2i}\right) = (3-i)\left(\frac{4-2i}{16-4i^2}\right) = (3-i)\left(\frac{4-2i}{20}\right)$$

$$= \frac{12-10i+2i^2}{20} = \frac{10-10i}{20} = \frac{1}{2} - \frac{1}{2}i$$

If the quaternion divisor (denominator) is $a_1 + a_2i + a_3j + a_4k$, we shall thus convert the division into multiplication by

$$\frac{1}{a_1 + a_2i + a_3j + a_4k}$$

Then, either left or right multiplication of the numerator and denominator of this fraction by the *conjugate* $a_1 - a_2i - a_3j - a_4k$ yields

$$\frac{a_1 - a_2i - a_3j - a_4k}{a_1{}^2 + a_2{}^2 + a_3{}^2 + a_4{}^2}$$

where the new denominator will be a positive real number, a scalar which we shall symbolize by r. Hence instead of dividing by the above quaternion, one can multiply by the quaternion

$$\frac{a_1}{r} - \frac{a_2}{r}i - \frac{a_3}{r}j - \frac{a_4}{r}k$$

The algebra of quaternions can be used to illustrate some of the questions raised about algebras in general. In the first place, if the field of the coefficients or scalars is changed, an algebra may cease to be a division algebra. In the second place, quaternions do have geometric and physical significance. If the field of the quaternion coefficients is that of the ordinary *complex numbers*, the method used above will indicate that division may not always be possible, for as we shall now show, r may turn out to be zero in many cases, and then

$$\frac{a_1}{r} - \frac{a_2}{r}i - \frac{a_3}{r}j - \frac{a_4}{r}k$$

would be meaningless. We recall that $r^2 = a_1{}^2 + a_2{}^2 + a_3{}^2 + a_4{}^2$. If we seek to divide by the quaternion $(\sqrt{-4} + 3i - 2j + \sqrt{-9}k)$, then $a_1 = \sqrt{-4}$ and $a_1{}^2 = -4$, $a_2 = 3$ and $a_2{}^2 = 9$, $a_3 = -2$ and $a_3{}^2 = 4$, $a_4 = \sqrt{-9}$ and $a_4{}^2 = -9$, and

$$-4 + 9 + 4 - 9 = 0$$

Hence division by the given quaternion is impossible. The reader can provide other examples where division fails, if he will choose figures similar to the above—say, division by $1 + \sqrt{-25}i + 5j + \sqrt{-1}k$.

The application of quaternions to physics arises from Hamilton's efforts to design them for such a purpose. We may recall (Chapter 4) that he was seeking to extend a 2-dimensional vector algebra to three dimensions. A 2-dimensional vector may be symbolized by (a_1, a_2) or by the Gaussian $a_1 + a_2\sqrt{-1}$ where a_1 and a_2 are the vector components in the directions of the X- and Y-axes. Then a 3-dimensional vector, (a_1, a_2, a_3), was considered by Hamilton to be the quaternion $0 \circ 1 + a_1 i + a_2 j + a_3 k$. The product of two 3-dimensional vectors (a_1, a_2, a_3) and (b_1, b_2, b_3) was defined by Hamilton to be the product of the corresponding quaternions, that is,

$$(0 \circ 1 + a_1 i + a_2 j + a_3 k)(0 \circ 1 + b_1 i + b_2 j + b_3 k)$$

If the reader will find this product in accordance with the quaternion multiplication table, he will obtain the result

$$-(a_1 b_1 + a_2 b_2 + a_3 b_3) + i(a_2 b_3 - b_2 a_3) + j(a_3 b_1 - b_3 a_1) + k(a_1 b_2 - b_1 a_2)$$

The trinomial $a_1 b_1 + a_2 b_2 + a_3 b_3$ is a *scalar* and may be given many physical interpretations, for example, as *work*, *electromotive force*, *circulation* in a fluid. For one geometric interpretation of this scalar, take the case where a quaternion is multiplied by itself or squared. Then $a_1 = b_1$, $a_2 = b_2$, etc. and the above trinomial becomes $a_1{}^2 + a_2{}^2 + a_3{}^2$, whose square root represents the magnitude (length) of the vector, just as in two dimensions $\sqrt{a_1{}^2 + a_2{}^2}$ is the length or magnitude (Chapter 4) of a vector with components a_1 and a_2. As for the coefficients of i, j, and k in the product above, it can be shown that they are the components of a vector whose magnitude is equal to the *area of the parallelogram* whose adjacent sides are the vectors (a_1, a_2, a_3) and (b_1, b_2, b_3).

Broad as Grassmann's concept of hypercomplex numbers was, his algebras did not include all possible multiplication tables, but only those satisfying the conditions

$$e_1{}^2 = 0, e_2{}^2 = 0, e_3{}^2 = 0, \ldots, e_n{}^2 = 0$$

$$e_1 e_2 = -e_2 e_1, e_1 e_3 = -e_3 e_1, \ldots, e_r e_s = -e_s e_r$$

It is the second set of conditions which makes Grassmann's algebras noncommutative, and the requirements $e_1^2 = e_2^2 = e_3^2 = \cdots = e_n^2 = 0$ show that each of the basic units is *nilpotent*.

In fact, every element of a Grassmann algebra is nilpotent. That this is so can be inferred indirectly by realizing that Grassmann's work was "space analysis," where products had geometric interpretations. Thus, in 3-dimensional space, $a_1e_1 + a_2e_2 + a_3e_3$ is a *vector*, $Ae_1e_2 + Be_2e_3 + Ce_3e_1$ is an *area*, $Ke_1e_2e_3$ is a *volume*, and the product of any *four* units or basic vectors, for example, $e_1e_2e_3e_1$ must be equal to zero, since a 4-dimensional entity is meaningless within a 3-dimensional space. One can prove directly, of course, that

$$e_1e_2e_3e_1 = e_1e_2(-e_1e_3) = e_1(-e_2e_1)e_3 = e_1(e_1e_2)e_3 = e_1^2e_2e_3 = 0(e_2e_3) = 0$$

and similarly that any other product of four units, for example, $e_3e_2e_1e_3$, must also be equal to zero. Therefore if the second or third power of an element of a 3-dimensional Grassmann algebra is not equal to zero, its fourth power must surely vanish, and hence, as stated above, every element is *nilpotent*.

In passing, a word about Grassmann himself. He came from a family of Protestant ministers, and it was therefore natural for him to specialize in theology at the University of Berlin. He was also greatly interested in philology, but his mathematical study was more of a sideline or hobby, and his first published work on a mathematical subject (the theory of tides) appeared in 1839, when he was thirty. In the interim he had done some teaching, and in 1842 he settled down to a career as a mathematics teacher at the *gymnasium* of the Baltic port of Stettin, his native city. He continued in this position all his life, and never attained the rank of a university professor. As stated previously, his mathematical stature has been recognized only in recent times. Before that he was considered something of a dilettante, who dabbled in politics, church affairs, music and art, and who placed his Sanskrit dictionary for the *Rig Veda* (one of the Hindu sacred writings) in the same category as his creation of hypercomplex numbers.

But to return to the ultimate development of Wedderburn's algebraic ideas, they are incorporated in the Wedderburn structure theorems:

Every algebra without properly nilpotent elements is uniquely expressible as the "direct sum" of algebras without proper ideals, and conversely.

Every algebra without proper ideals is the "direct product" of a division algebra and a total matric algebra.

The Wedderburn theory reveals the abstract structure of any linear associative algebra whatsoever as follows. First the properly nilpotent elements are removed. Then to the "difference" algebra that remains, the first of the above structure theorems applies, and that difference algebra is expressible in terms of algebras without proper ideals. These in turn are expressible, according to the second theorem above, in terms of division algebras and total matric algebras.

Thus the structure of *any* algebra is revealed by studying (1) the set of properly nilpotent elements initially removed (it can be proved that they form an algebra, hence a *nilpotent* subalgebra of the given one) and (2) the division and total matric algebras in terms of which the "difference algebra" is expressible. In this way the *infinite* variety of linear associative algebras or hypercomplex number systems was reduced by Wedderburn to a *finite* study of only three types of algebra, the nilpotent, division, and total matric types.

Within the mathematical world, Wedderburn's worth is fully recognized, but unfortunately hardly a single man in the street has ever heard of him. Therefore the reader should find the following biography of particular interest. It was written by Professor Hugh S. Taylor, Chairman of the Chemistry Department and Dean of the Graduate Division of Princeton University, and many of the facts included were provided by other associates and friends of Wedderburn, both at Princeton and at the Institute for Advanced Study.*

Wedderburn was born in 1882, in Forfar, Scotland, the tenth child in a family of fourteen children, which included seven brothers and six sisters. His father was Alexander Stormonth Maclagen Wedderburn, M.D., of Pearsie. His mother was Anne Ogilvie. On his father's side his grandfather was Parish Minister of Kinfauns and Professor of Exegesis in the Free Church College of Aberdeen. His paternal great-grandfather was Parish Minister of Blair Atholl and a Chaplain in the Black Watch. On his maternal side his grandfather was a lawyer in Dundee, as had been true of the family for several preceding generations. The detailed history of the Wedderburn family has been recorded in the *Wedderburn Book*. Of his many brothers and sisters only two survived him, a brother, Ernest Wedderburn of Edinburgh, Scotland, and a sister, Miss Elizabeth Wedderburn, of Paris, France.

Wedderburn's early school years were spent at the Forfar Academy from 1887 to 1895. In the latter year he transferred for the last three years of his school education to George Watson's College in Edinburgh, 1895–1898. From this college at the age of sixteen and a half years he obtained the leaving scholarship and entered Edinburgh University in the autumn of 1898. After five years as a student in this university he obtained, in 1903, the M.A. degree with First-Class Honours in Mathematics. His first research "On the isoclinal lines of a differential equation of the first order" was published in the *Proceedings of the Royal Society of Edinburgh* in the same year, when the author had just reached his twenty-first year. Two other communications to the Royal Society of Edinburgh date from the same year. During the year 1902–1903 Wedderburn had served as Nichol Assistant in the Physical Laboratory of the University.

In the following year Wedderburn pursued postgraduate studies in Germany. In the winter semester of 1903–1904 he attended the University of Leipzig; in the summer semester of 1904 he was at work in the University of Berlin. Appointed Carnegie Fellow for the year 1904–1905 he proceeded to the United States where the American phase of his scientific career began in postgraduate study at the University of Chicago. Two papers stem from this period, one "On the structure of hypercomplex number systems" and a second entitled "A theorem on finite algebras."

Following the year in the United States was a period of four years, 1905–1909, once more in Edinburgh as Lecturer in Mathematics and assistant to Professor George Chrystal. Communications to the Royal Society of Edinburgh, to the Edinburgh and London Mathematical Societies, to the *Annals of Mathematics* and, in collaboration with Oswald Veblen, to the *Transactions of the American Mathematical Society* characterize these years as a creative scientist. His joint authorship with Veblen stemmed from the

* Royal Society of London, *Obituary Notices of Fellows*, Vol. 6, No. 18 (November 1949), pp. 618 ff.

year in Chicago and was an advance notice of a future association of both in Princeton University in the years ahead. Meanwhile the researches already accomplished earned for Wedderburn the degree of Doctor of Science in Edinburgh University in 1908. From 1906 to 1909 he served as editor of the *Proceedings of the Edinburgh Mathematical Society*.

Wedderburn entered on the Princeton phase of his life's work in the autumn of 1909. He was called to Princeton as Preceptor in Mathematics, joining a goodly company of distinguished members of the Princeton faculty. The "preceptor guys" as several generations of Princeton students have labelled them, not without deep affection and esteem, were the creation of Woodrow Wilson in his reorganization of the Princeton Faculty. That revolution "required a large scale infusion of new blood, of scholars who would assume an intimate personal relation with small groups of undergraduates and impart to them something of their own enthusiasm for things of the mind." To Wilson, the preceptorial system seemed "the only effectual means of making university instruction the helpful and efficient thing it should be." With a body of fifty preceptors which he brought to Princeton in 1905 he planned to "transform the place where there are youngsters doing tasks to a place where men are doing thinking."

It was thus that L. P. Eisenhart and Oswald Veblen, Gilbert A. Bliss, George D. Birkhoff and Joseph H. M. Wedderburn joined the preceptors of mathematics in Princeton in the years 1905 to 1909. Bliss left after three years to pursue his distinguished career in Chicago. Birkhoff stayed three years and proceeded to Harvard to become there eventually leader in mathematics and Dean of the Faculty of Arts and Sciences. Eisenhart, Veblen and Wedderburn stayed on at Princeton, building up its standards and traditions in mathematical research. Veblen left Princeton University in 1932 to establish, by a process of fission from the Department of Mathematics in the University, a Department of Mathematics in the newly organized Institute for Advanced Study also located in Princeton. Veblen, Alexander and von Neumann made the transition. Eisenhart and Wedderburn stayed behind.

Five happy years as preceptor at Princeton were succeeded by the five grim years of World War I. Wedderburn responded immediately to the call to arms. In view of his already great scientific achievements it would not have been difficult at all for him to have obtained a commission in the British Army at the outset of hostilities. He had already seen service in the Territorial Forces before the war. It is altogether characteristic of the man and of his exceptional modesty that he volunteered at the beginning of hostilities as a private. The Princeton records show that he was the first resident of Princeton to leave for the war on the outbreak of hostilities, and had the longest war service of any member of the University or resident of the town. By November 1914, he had obtained his commission as Lieutenant in the 10th Battalion of the Seaforth Highlanders, receiving his Captaincy in January 1915. From 1915 to February 1918 he continued as Captain in the Seaforth Highlanders but transferred then to the Royal Engineers as Captain in the 4th Field Survey Battalion. In this capacity he saw service in France from January 1918 to March 1919. To this service he brought his exceptional gifts as mathematician and physicist taking an active part in the development of sound-ranging techniques for the location of enemy batteries and the practical employment of those techniques on the field of battle. His services were mentioned in despatches from the British Army in France.

Wedderburn returned to Princeton after the conclusion of the war still in the capacity of preceptor. In 1920–1921 he was Assistant Professor of Mathematics, and was promoted in the spring of 1921 to an Associate Professorship on permanent tenure. Seven years later he was promoted to Professor of Mathematics, a position which he held until his retirement in 1945, with a continuous faculty record of thirty-seven years of service.

His colleagues among the Princeton preceptors recall with enthusiasm the rich human qualities of Wedderburn. They recall his passion for play as well as for work, his desire for companionship and association with men. He loved the out-of-doors, found deep satisfaction in the wilderness, in the woods, canoeing along rivers and streams in the company of thoughtful men. As in his scientific work, he brought to the construction of the camp-site, the erection of the tent, the paddling of the canoe up- and down-stream the qualities of a complete perfectionist. In the wilds of Northern Canada, with congenial men, he found complete happiness. In the tasks of leisure hours he utilized the same resources of knowledge in mathematics, mechanics and physics that marked him out among his scientific colleagues in the class-room and the mathematics seminar. His taste in literature ran to books of travel and he accumulated a large library of travel. He admired John Buchan both for his native Scottish background and his appreciation of the open spaces of the earth.

While Wedderburn had a broad and comprehensive knowledge of pure and applied mathematics, his research work was done exclusively in algebra. Indeed, algebra was his all-embracing scientific passion and it is in algebra that he made his eminent contributions, distinguished no less by their originality than by their depth. It was these characteristics which earned for him, from the Royal Society of Edinburgh, his first major distinction, the award to him in 1921 of the MacDougall-Brisbane Gold Medal and Prize for the period 1918–1920, in recognition of his investigations on hypercomplex numbers. The Royal Society of London, in 1933, elected him Fellow, also by reason of his distinction in the field of algebra.

His mathematical discoveries include two famous theorems which bear his name, frequently quoted by mathematicians. One of his Princeton colleagues in mathematics has said:

"Before Wedderburn began his investigations, the classification of the semi-simple algebras was done only if the ground field was the field of real or complex numbers. Interesting as this investigation was, it did not lead to a deeper insight into the nature of the hypercomplex numbers. Wedderburn attacked the problem in full generality and introduced new methods. This led to a complete understanding of the structure of semi-simple algebras over any given ground field. He showed that they are a direct sum of simple algebras and finally proved the celebrated theorem that a simple algebra consists of all matrices of a given degree with elements taken from a division algebra. These results were found in 1907 and stated in his famous paper entitled 'On hypercomplex numbers.' This paper contains many other results and ideas and was the beginning of a new era in the theory.

"The second important contribution concerns the investigation of skew fields with a finite number of elements. The commutative case had been investigated before by E. H. Moore and had led to a complete classification of all commutative fields with a finite number of elements. Since a non-commutative finite field had never been found one could suspect that they do not exist. It was Wedderburn who showed in 1905 that this is indeed the case.

"This second theorem gives at once the complete classification of all semi-simple algebras with a finite number of elements. But it also had numerous other applications in number theory and in projective geometry. It gave at once the complete structure of all projective geometries with a finite number of points and showed the interesting fact that in all these geometries the theorem of Pascal is a consequence of the theorem of Desargues."

It is of the article "On hypercomplex numbers" that Hermann Weyl, in his book *The classical groups*, remarks "of paramount importance for the modern development is Wedderburn's paper of the year 1908, where he investigates associative algebras in an arbitrary number field k."

After the war period and during the 1920's came a group of papers dealing with other aspects of modern algebra. Wedderburn's greatest contributions were made prior to 1930, and his most important publication thereafter was undoubtedly his book *Lectures on Matrices* (1934). The work is a synthesis that is Wedderburn's own. It contains a number of original contributions to the subject. Though he did not follow the abstract point of view that had just become dominant, neither did he commit the error made by others of treating matrix theory as an art of juggling elements in an array. The important ideas of linear transformations, vector spaces, bilinear forms, appear in Wedderburn's book. Also, as in his best work, one finds here some neat and suggestive algebraic devices that make the book a very valuable reference book, even at the present time.

Of the papers that Wedderburn published after 1930, the best one was *Non-commutative domains of integrity* (Crelle 1931). In this paper the author gave an improvement and extension of known results on reducing a matrix with elements in a Euclidean domain to diagonal form. This stimulated further work on this problem and led to a complete solution a few years later. Also Wedderburn considered in this paper the important problem of imbedding a non-commutative domain in a division ring. He showed that this can be done if common multiples exist. This result was found independently by Oystein Ore.

Towards the close of the 1920's Wedderburn suffered what some of his friends considered a mild nervous breakdown. More and more he withdrew from normal companionships, turned away from such leisurely pursuits and distractions as he had hitherto enjoyed. He followed an increasingly solitary life, dining alone at his club, making no effort to seek out his friends. Even on holidays in the Berkshire Mountains he would spend long intervals in solitude and silence. Those who deliberately penetrated beyond this self-imposed barrier found the same friendly human person that his earlier preceptor colleagues had known, with the same intellectual vigour, originality, depth of understanding and quality of effort that characterized all his scientific work. To one such visitor he occasionally remarked that he "lived too much alone." To reverse the trend seems, however, to have been beyond his capacities if not his desires. In the last years his solitariness increased so that a new generation of undergraduates and research scholars saw less and less of him. He was made Professor Emeritus in 1945 and spent the remaining three years of his life, seeing little of colleagues and friends. It was possible occasionally to break down the barriers of his reserve as, for example, on the occasion of the Princeton Bicentennial Celebrations; at the conference on the physical sciences, when, as it happened, a dozen or so of the Fellows of the Royal Society were at one time in town he accepted an invitation to meet with them at tea. But these were rare and exceptional occasions, less and less frequent as the years passed, and in October 1948, Wedderburn was found dead in his residence at Princeton, where for many years he had lived alone.

He had suffered a heart attack, and no one was present to cheer his last moments.

Wedderburn's work has been presented in order to indicate the nature of the ultimate generalization of elementary arithmetic. In addition, the discussion of linear algebras supplements the general treatment of abstract algebra by providing specific details of the subject matter in one very important area.

29

Twentieth-Century Vistas— Logic and Foundations

Having probed some of the heights and depths of modern algebra, we now proceed to our third sample of recent mathematical activity, namely, the attempt to place all mathematical reasoning on a firmer foundation. Various foundational "crises" arose in the course of mathematical history but were, as we have seen, successfully resolved. In the earliest instance, Eudoxus showed how subtle assumptions and definitions would handle the problem of the incommensurable, the issue which confronted the Pythagoreans when they encountered $\sqrt{2}$. Much later, Lobachevsky settled the matter of the parallel postulate and his solution led to the definite formulation of the modern axiomatic method (Chapter 3). By substituting rigorous arithmetic procedures for geometric intuition, Cauchy and Weierstrass liquidated the classical analytical monstrosities (Chapter 22). More recently Kurt Gödel and Paul Cohen tangled with set-theory foundations and, among a number of brilliant results, provided an elegant answer to the challenges raised by the continuum hypothesis and the axiom of choice (Chapter 24). Some of the paradoxes associated with Cantor's theory of infinite sets had, in fact, been accorded a partial resolution in the Whitehead-Russell theory of types (Chapter 24) and in the gradual building of a restrictive foundation for pure (abstract) set theory by Zermelo, Fraenkel, von Neumann, Bernays, and others.

All such solutions to critical problems can be considered triumphs of the axiomatic method. Even the answers to lesser questions have been related to special properties of the postulate systems forming the bases of mathematical sciences. Thus, *completeness* was seen to be desirable in Euclidean geometry (Chapter 3) if one wished to be able to decide whether or not every triangle is isosceles or, more generally, to be able to prove or refute any statement involving permissible (defined or undefined) terms of Euclidean geometry. Completeness, then, might be desirable in other deductive systems as well, or even in a system of logical reasoning whose purpose would be the deduction of *all* of mathematics.

From what we have said, it might seem that if one were careful enough in the selection of the basic assumptions serving as the foundation of mathematics, then all paradoxes could be avoided and all problems could be solved. That this is *not* the case was indicated in 1931 by a devastating discovery, namely, Gödel's "metamathematical" *incompleteness theorem*, whose background we shall now present prior to considering the content of the theorem itself. It can come as no surprise that Gödel is the author of the most revolutionary result in modern mathematical philosophy, since his name has occurred previously in our story whenever some recondite issue of mathematical foundations was discussed (Chapters 3 and 24).

After Lobachevsky's day, the axiomatization of various parts of mathematics and logic was eminently successful and reached a peak of purification (abstractness) in Hilbert's *formalist* school of thought. Hilbert held that "mathematics is a game played according to certain simple rules with meaningless marks on paper." In a Hilbertian *formal system* one has a collection of symbols, along with rules for combining these symbols into "well-formed" formulas. Some of these formulas are the initial postulates and some are "derived" or "deduced" from the postulates by rules of proof. Since the symbols are devoid of meaning, the manipulations with them might turn out to be utterly senseless, that is, might ultimately lead to *contradictions* unless *consistency* of the whole system of rules can be established.

In relation to consistency proofs, Hilbert introduced the idea of *metamathematics*, a theory which studies the properties of formal axiomatic systems. Metamathematics is, in a sense, *external* to all formal systems, that is, outside or beyond Hilbert's "game with meaningless marks." Then metamathematical statements are *not* among the formulas of the "game" but, instead, are *meaningful* assertions about the properties and relations of symbols, axioms, theorems, etc. in a formalized system. Thus the following four formulas might occur *within* some formalization of arithmetic:

$$1 + 1 = 2$$
$$1 + 1 = 3$$
$$y > 2$$
$$3 > 2$$

On the other hand, the following five examples would be *metamathematical* statements about the system:

"y" is a variable

"2" is a numeral

The formula "$3 > 2$" is obtained from the formula "$y > 2$" by substituting the numeral "3" for the variable "y."

"$1 + 1 = 3$" is a formula but *not* a theorem of the system.

The formal arithmetic system is consistent.

How can one prove or refute the last metamathematical statement illustrated above? In general, as we have already stated, consistency is the essential problem of Hilbert's formalist philosophy. Now, in Chapter 3, we indicated that the method of *models*, that is, interpretations, is the usual technique for establishing the consistency of a set of postulates. But, if one uses that method, one furnishes only a *relative* proof of consistency. Thus one may establish the consistency of Lobachevskian or Riemannian geometry by using parts of Euclidean geometry as models, and the consistency of Euclidean geometry by using Cartesian geometry as a model. But then how is Cartesian geometry to be proved consistent? One might interpret that in terms of algebra (arithmetic). But then all such models merely shift the responsibility for consistency from one branch of mathematics to another and thus would require that, ultimately, the consistency of some system (usually arithmetic) be *assumed*. This idea seemed unsatisfactory to Hilbert, and hence he sought "absolute" (*not* relative) consistency proofs, above all, for the formal system on which the consistency of other parts of mathematics rests, namely the arithmetic of the whole numbers. An *absolute* consistency proof is one that does *not* assume the consistency of some other system. To indicate what such a proof requires, let us form a crude and oversimplified analogy. Thus one knows *directly* from the rules of poker that a hand can contain four aces but cannot possibly contain five aces. On a higher level, Hilbert hoped to be able to formalize mathematical systems so that direct examination of the rules of those "games" and the inevitable consequences of the rules would indicate the impossibility of deducing as theorems *both* a well-formed formula and its formal negation.

Hilbert, Bernays, and their followers were able to provide absolute proofs of consistency for certain limited formal systems, for example, a formalized additive (but not multiplicative) arithmetic of the whole numbers. Again, when the truth-table logic of Chapter 6 is properly formalized, it is not difficult to give an absolute proof of its consistency (as well as its completeness). It was hoped, therefore, that an absolute proof of the consistency of the arithmetic (additive and multiplicative) of the whole numbers would soon be found, but all efforts in that direction failed. Hilbert's most serious troubles arose in attempting to construct an absolutely consistent, complete formal system for the derivation of *all* of classical mathematics including arithmetic). Then, in 1931, the twenty-five year-old Kurt Gödel delivered the body blow to the formalist program by proving that Hilbert's formalization or, in fact, *any* formal system equivalent to the Russell-Whitehead *Principia Mathematica* must, of necessity, be incomplete. What Gödel did was show that there must be "undecidable" statements in *any* such system, that is, propositions which can neither be proved nor refuted using the rules of the system, and that *consistency* is one of those *undecidable* propositions. In other words, the consistency of an all-embracing formal system can neither be proved nor disproved within the formal system itself. Thus Gödel's "incompleteness theorem" put an end to Hilbert's dream of providing for all of mathematics a formal axiomatization that is both complete and absolutely consistent.

Gödel's proof that one cannot establish the consistency of any formalization of elementary number theory is a consequence of his and Hilbert's use of "finitistic" constructive methods. Naturally, if one is "playing a game" part of which is to make an exhaustive examination of relations among the formulas of that "game" in order to search for possible contradictions producing inconsistency, he will have to limit himself to formulas with a *finite* number of symbols and to *finite* sets of properties or relations (unless he makes a "permissible" induction by "constructive" procedures). Also, "finitistic" techniques avoid the type of crisis which, as we have seen, may occur in reasoning with infinite sets. But if one is fearless, he may proceed in the fashion of the German logician, Gerhard Gentzen, who, in 1936, gave a proof for the consistency of arithmetic by using transfinite induction up to a sufficiently large transfinite ordinal (Chapter 24). Thus his demonstration was *not* a "finitistic" absolute proof within formal arithmetic itself.

Gödel's 1931 discovery was revolutionary in the same negative fashion as Galois' proof of the impossibility of finite solvability for polynomials of degree higher than 4 (Chapter 16) and Hamilton's relinquishment of commutative multiplication (Chapter 4). But Gödel's incompleteness theorem and its corollary concerning consistency are even more destructive, for they attack the very *foundation* of *all* pure mathematics. Although Gödel's later work has also been revolutionary (Chapter 24), no other one of his contributions is so *final* for mathematical methodology as the 1931 theorem. As recognition of that achievement, he was made a permanent member of the Institute for Advanced Study at Princeton in 1938, and was awarded an honorary degree at Harvard University in 1952.

Apropos of one of Hilbert's major discoveries, the algebraist P. Gordan (1837–1912), one of Emmy Noether's teachers, proclaimed, "This is not mathematics; it is theology!" Perhaps Gordan's statement would be more applicable to Gödel's theorem for, if one insists on *absolute* consistency proofs, and if these are unattainable, then consistency becomes a matter of *faith* rather than a question of logical reasoning.

Gödel's proof is highly technical and employs a novel methodology. Nevertheless, since the work is a modern classic, we shall outline his demonstration and feature its high points. But first, let us remark that his famous argument is patterned, in part, on a logical paradox formulated by the Frenchman Jules Richard in 1903. The Richard paradox arises as follows. Suppose that one wishes to list a sequence of properties of the whole numbers, characteristics with descriptions like "prime" or "even" or "perfect square" or "multiple of 4." Having formulated definitions, one will require some criterion for ordering them. Thus, observing that each definition contains a finite number of words each containing a finite number of letters, we can pick out the definition containing the smallest number of letters and assign to it position (1) in the serial order. The next shortest definition can then be given position (2), and so on. If two definitions contain the same number of letters, we can place them on the basis of the alphabetical order of the letters in each. Thus each definition will have its own order number.

Now we note a certain possibility. Perhaps a particular order number may possess the characteristic described in the definition to which it corresponds. Thus it might possibly occur that 19 has been assigned to the property of "being prime," whose definition has been formulated, as "*x*, a whole number, is prime if and only if it is not exactly divisible by any natural number except itself and 1." Observe that 19 actually has the property described. By contrast, the order number 12 may have been assigned to the property of "being odd," whose definition is "*x*, a whole number, is odd if and only if it leaves a remainder 1 when divided by 2." Note that 12 does *not* have the characteristic defined. An order number like 12 that does *not* possess the property described in the definition to which it corresponds is called *Richardian*. A number like 19, then, is *not* Richardian.

But consider the property of being Richardian. When that is carefully defined, its definition will contain a certain number of letters of the alphabet and hence can be assigned a position in our sequence of definitions. Let us symbolize the number assigned by the letter *n*. Our challenge now arises: *Is the number n Richardian?* If the answer should in fact be affirmative, then *n* cannot possess the property to which it corresponds (thus, the property of being Richardian). In other words, if we assume that *n* is Richardian, then we must conclude that it is not Richardian. Let us start anew and this time assume that *n* is not Richardian. Then *n* must possess the property to which it corresponds (thus, the property of being Richardian). In the present case, the hypothesis that *n* is not Richardian leads to the conclusion that it is Richardian. In summary, we have a *paradox* because there are only two possible answers to the question raised and either one leads to the conjunction of a statement *R* and its negation $\sim R$. The situation violates the law of contradiction, $\sim(p \land \sim p)$ (page 118), which asserts that a statement and its negation cannot both be true.

Now one refutation* of the Richard paradox argues that "being Richardian" is a notational property depending on *metamathematical* concepts like the number of letters in a sentence, etc. and hence that its definition does *not* belong on a list of purely arithmetic properties. If the said definition is deleted from our list (or, at any rate, transferred to some metamathematical theory about that list), then there can be no argument like the one presented above and hence the paradox is resolved.

Although Gödel may have patterned his proof on the Richard paradox, he avoided pitfalls by *arithmetizing* metamathematical statements, that is, by translating them into arithmetical formulas which properly belong in a list of such formulas. What he did was to construct a formal system for elementary number theory, that is, for the higher arithmetic of the whole numbers. Then he provided a scheme (*not* lexicographic) of assigning a different natural number (called a Gödel number) to each entity of his formalized system. Thus there are Gödel numbers for his basic meaningless symbols, for his well-formed formulas, in particular for his postulates and his theorems, and also for his proofs (finite sequences of formulas derived by his rules for deduction). Every Gödel number is a natural number, but the converse is false. Not every natural number is a Gödel number. For example, in Gödel's original scheme, the number 4 is not assigned to any entity of the formal system. At any rate, Gödel's numbering *maps* his formalized number theory (and the part of *Principia Mathematica* essential to his formal logical procedures) onto a proper subset of the natural numbers. Then, given a formula or a proof, its Gödel number can be computed. In reverse, given a natural number, one can first determine whether or not it is a Gödel number. Then, if it is actually the Gödel number of a formula or proof, that formula or proof can be reconstituted (by factorization

---

* See, for example, E. Nagel and J. R. Newman, *Gödel's Proof*, New York University Press, New York, 1958, p. 63.

of the Gödel number into primes, a familiar procedure of number theory). Finally, metamathematical statements are "mirrored" by arithmetic equivalents, that is, they are mapped isomorphically, as it were, onto statements about ordinary arithmetic relations, and then the latter statements are formalized, or given representation by the "meaningless" symbols of the formal system. Hence, whereas in the Richard paradox a number n is associated with a metamathematical statement, in Gödel's proof n is associated with a pure arithmetical formula whose intuitive content or interpretation is the metamathematical statement.

We shall now sketch in broad outline, in *non*rigorous, informal (nonsymbolic) fashion, the main argument in Gödel's original proof of the incompleteness theorem. Our presentation will follow virtually what Gödel himself said at the very beginning of his 1931 paper—prior to his formal proof, which required 46 preliminary definitions and several preliminary theorems, all expressed symbolically, as the background for his ultimate demonstration. At one point in his introduction, Gödel asserted that his method of proof can be applied to *every* formal system in which the concept of "provable formula" is defined and in which every provable formula is "true." If this last statement puzzles the reader, he can think of "provability" as a notion formalized *within* the system, and of "truth" as an idea associated with models (interpretations), hence *outside* the system, thus metamathematical.

In the case of the Richard paradox, there was a sequence of definitions with a corresponding sequence of natural numbers. If we think of the definitions as open sentences, "x is ...," where the domain, or replacement set, for x is the aggregate of whole numbers, then the typical terms of the two sequences can be pictured as

Natural Number	Open Sentence
k	x is ...

To test whether or not k is Richardian, one substitutes k for x in the open sentence corresponding to k, to yield the statement "k is ...." When k does *not* have the property described, k is Richardian. Next, a certain natural number n corresponds to "x is Richardian." Substitution of n for x produces "n is Richardian," and our previous argument leading to the Richard paradox follows.

Gödel's procedure is roughly analogous to the above. To make the analogy closer, we shall use the term "troublesome" to correspond to "Richardian" and claim that a number in Gödel's sequence is "troublesome" if and only if it *cannot* be *proved* to have the property to which it corresponds. More exactly, Gödel considers only those formulas of his system that are open sentences in a single variable x, where the domain of x is the set of Gödel numbers. Those open sentences can be ordered as a sequence corresponding to the increasing magnitude of their Gödel numbers. Next, the sequence of Gödel numbers is subdivided into two complementary subsets T and $\bar{T}$, containing "troublesome" and "nontroublesome" Gödel numbers of the sequence, respectively. The Gödel criterion is: A number k of the sequence belongs to T (in our popularization, is "troublesome") if and only if its substitution for x in the open sentence corresponding to k results in a formula that is *unprovable*. Thus we have

Gödel Number	Open Sentence
k	$...x...$

and k belongs to T or is "troublesome" if and only if "$...k...$" is an *unprovable* formula. If the formula is provable, k belongs to $\bar{T}$ or is "nontroublesome."

Now we shall give the intuitive content of one of Gödel's open sentences but, instead of computing its Gödel number, shall merely symbolize that number by n. Thus

Gödel Number	Open Sentence
n	x is a member of T, that is, x is troublesome.

Next, imitating the procedure that leads to the Richard paradox, we must substitute n for x in the open sentence to produce a proposition which, expressed informally, is

$$G = n \text{ is a member of } T, \text{ that is, } n \text{ is troublesome}$$

We have labeled this proposition with the letter G in honor of Gödel. Here G is a proposition whose formal counterpart is *undecidable* within the formal system. For if G were assumed to be provable, then it would be "true" and by its *valid* content would reveal that n is "troublesome." This would mean (by definition) that the proposition which results from substituting n for x in the open sentence to which n corresponds, namely, the *proposition G itself*, would be *unprovable*. Thus the assumption that G is provable leads us to the conclusion that it is not provable, a *contradiction*, and hence we must relinquish the assumption.

Suppose, on the other hand, that the negation

$$\sim G = n \text{ is } \textit{not} \text{ a member of } T$$

is provable, hence "true." Then by what $\sim G$ validly asserts, n would belong to $\bar{T}$. In other words, n would be a "nontroublesome" Gödel number, and the result of its substitution in the formula to which it corresponds, namely, the proposition G, would be *provable*. Thus the assumption that $\sim G$ is provable leads to the conclusion that G is provable, an impossibility in a *consistent* system because in such a system a proposition and its negation cannot both be provable. We remark in passing that a *formal* proof of Gödel's theorem makes use of a more stringent condition, called ω-consistency. In conclusion, the argument in this paragraph and the one preceding it shows that if the formal system is consistent, neither G nor $\sim G$ is provable. Hence G is *undecidable* and the system is *incomplete*.

Let us digress briefly to comment on certain side-effects of Gödel's proof. If the reader will return to the paragraph before the last, he will observe that our lengthy analysis of the content of the proposition G boils down to the conclusion that what G actually asserts is "G is not provable." Gödel points out that here we have an analogy with the "liar paradox" (Chapter 24), namely, a proposition referring to itself, in particular, to its own unprovability. But G is *not* circular since, in the proof itself, G asserts the unprovability of a very definite formula, the one that will result if there is a permissible substitution in the formula with number n. Only *subsequently*, almost accidentally, does it turn out that the resulting formula will be G itself. At any rate, the situation caused Gödel to say that Russell and Whitehead were too drastic when they sought to solve the "liar paradox" by forbidding a proposition to talk about itself. Gödel remarked that it is even possible to construct, for *any* metamathematical property that has a symbolic counterpart in his formal system, a proposition which says of itself that it has this property. Again, we must reveal that neither the Richard paradox nor the "liar paradox" need be used as the pattern for Gödel's proof. Many other antinomies would serve equally well. As a final point in our digression, we should like to point out another remarkable feature of Gödel's proof, namely, that since, *metamathematically* speaking, G asserts its own unprovability, G is *true* because it is indeed unprovable (in fact, undecidable). Hence the proposition G, undecidable *within* Gödel's formal system, can be decided by metamathematical arguments.

In our original discussion of completeness (Chapter 3) we pointed out how the undecidability of a proposition can be removed by adding supplementary axioms to the postulate set of a deductive system. In particular, one can add as an assumption either the undecidable proposition or its negation. Hence, in the present instance one could make G decidable by postulating either G or $\sim G$. Let us do so, and call the additional postulate G_0. But then Gödel's proof indicates how, on the basis of the enlarged postulate system, a *new* undecidable formula can be constructed. One can then add either that formula or its negation as a supplementary postulate, G_1. Now Gödel's process can be repeated again and again indefinitely so that the addition of a sequence of supplementary postulates, $\{G_0, G_1, G_2, G_3, \ldots\}$, will *not* complete the system. It is *essentially* incomplete.

Thus Gödel demonstrated that *every* "finitistic," consistent, formal system must always have some undecidable formula F. (Note that we use the symbol F instead of G so as not to confuse the reader, since the previous paragraph suggests that our original undecidable G may have been added as a postulate.) Now we proceed to Gödel's all-important corollary for a system like the one just described. He proved the corollary by first mapping the metamathematical statement "The system is consistent" onto a formula C within the system. Next he demonstrated the validity of the implication,

$$(F \text{ is not provable}) \rightarrow (C \text{ is not provable})$$

But "F is not provable" is true, since F is undecidable. Hence, by the rule of detachment (Chapter 6), "C is not provable" is true. Thus, even though the formal system is consistent (as assumed in the first sentence of the present paragraph), its consistency cannot be proved *within* the system.

Gödel's incompleteness theorem and its corollary, the impossibility of absolute consistency proofs, represent a genuine crisis in mathematical foundations, one which cannot be resolved if one demands *finitistic* foundations for *all* of *pure* mathematics. Why "finitistic" foundations? Does "finitistic" mean something other than finite? These questions arose out of an earlier crisis whose nature will now be explained.

Let us now turn back from Gödel today to Kronecker, who, in the late nineteenth century, had objected to the analytic methods of Cantor, Dedekind, and Weierstrass, saying that the definitions they employed, particularly those for infinite classes, did not make it possible in every case to decide whether an entity meets the conditions specified. L. E. J. Brouwer (1881–1967) developed this iconoclastic point of view further, his *bête noire* being the law of the excluded middle, which he refused to accept in reasoning with infinite sets. We recall that this law is the logical tautology $p \lor \sim p$, and that it corresponds to the Boolean identity $P \cup P' = I$. It states that for any proposition p, either the statement itself or its negation must be true. If p is the proposition "Every member of a set S has a specified property," then $\sim p$ is the proposition "At least one member of the set S lacks the specified property." In the case of *finite* sets one would be able to examine the individual members and decide between the alternative propositions. *Direct* verification for p would *not*, of course, be possible in the case of an infinite set, but p might be true by definition or by logical inference from other propositions. If, on the other hand, inference shows p to be false, Brouwer would *not* permit us to say, automatically, that $\sim p$ is true. He would require us to find a member of the infinite set actually lacking the property in question. This might put us in a quandary, since examination of a trillion or more members might fail to reveal such a member and our only hope would be to search on and on.

For further examples of difficulties in reasoning with infinite sets, in particular with applying the law of the excluded middle to them, for Brouwer's notion of *constructive proof*, for his belief that mathematics is prior to logic, and for a contrast between his point of view and Hilbert's formalism, we can do no better than refer the reader to an explanation by Rolin Wavre* (1896–1940), noted modern logician, philosopher, and disciple of Brouwer.

"Men do not understand each other because they do not speak the same language and because there are languages which are not learned."

In saying that, Poincaré sought to describe the irreconcilable nature of that clash of temperament manifested in the very heart of mathematics at the appearance of Cantorian ideas. The clash has taken form under various captions. Du Bois-Reymond called the divergent tendencies empiricist and idealist; Poincaré, pragmatist and realist; Brouwer, *intuitionist* and formalist. We shall here use intuitionist for the one, but according to circumstances, idealist, formalist, or even realist for the other.

The intuitionist is at present identified by his extreme caution. Being anxious to attain the greatest intelligibility and clarity, he challenges such propositions as Zermelo's

---

* Rolin Wavre, "Is There a Crisis in Mathematics?" *American Mathematical Monthly*, Vol. 41, No. 8 (October 1934), pp. 488 ff.

axiom of selection* or such reasoning as does not seem rigorous. The idealist claims to depart in nothing from the rigor which the intuitionist rightly demands, and this without evidencing the least distrust with regard to those modes of reasoning—even on the contrary conceding their perfect legitimacy and applying them literally and without restriction. Russell, Hilbert, Zermelo, and Hadamard are idealists; Lebesgue, Borel, and Baire are intuitionists.

The clash is very much more evident now than in the past, as a result of the several publications, on the one hand of Brouwer and Weyl, two new and undeniably revolutionary intuitionists, and on the other of Hilbert. We should like to summarize here the essence of these publications and show how clearcut the clash becomes in connection with the notion of existence and a doubtful application of the law of excluded mean.† We shall see in particular why, and in what system of definition, the intuitionist denies us the right to say:

Two mathematical points are either coincident or distinct.

Two functions defined for the same values of the independent variable coincide for all those values or else there exists one for which they do not coincide.

That the intuitionist formulates such paradoxes in the name of truth seems curious. But the paradox is bound up with the words and not with the ideas.

Weyl, however, believes he has discerned at the basis of the theory of real numbers a vicious circle which endangers its value, and that he should therefore declare that mathematics is undergoing a crisis. Brouwer and Weyl have attempted to reconstruct the theory of sets and that of functions on new foundations, carefully avoiding the suspected law and vicious circle. In this they have carried distrust as to certain modes of reasoning, whose legitimacy has created no doubt for three centuries, further perhaps than the French school of the theory of sets.

Hilbert is not of the opinion that it is the turn of mathematics, after physics, to go through a revolution. Observing that the consequences of the idealistic attitude would never lead to the slightest contradiction, however far they be pushed, he considers that the principles suspected by the intuitionists, although not having perfect evidence, are nevertheless legitimate. Unlike Brouwer and Weyl, Hilbert does not believe in the necessity of reconstructing the foundations of mathematics, for he prescribes a radical remedy in order to prevent the crisis and to legitimate a frankly idealistic attitude. This means is the axiomatic method. In substance he does this: to the axioms accepted by the intuitionists, he adjoins a new proposition called the axiom of the transfinite which contains in itself all the doubtful principles united. And from this would be deduced the doubtful applications of the law of excluded mean. But to justify such a procedure, it must be demonstrated that this total system of axioms does not imply contradiction. By this artificial union of its idealistic foundations present-day mathematics would not forfeit the renown of being the discipline whose truth is above suspicion.

I do not conceal the difficulty encountered in trying to express the exact meaning of the relation Brouwer establishes between mathematics and logic. However, I shall try to set off the essentials in the thought of the great Dutch mathematician. In 1907 Brouwer made bold to reverse the roles which Russell would have had mathematics and logic play. Instead of its being the second which accounts for the first, it is on the contrary the part of mathematics to comprise logic, even traditional. The Aristotelian logic, born of natural classification, would be adequate to the theory of finite collections and would not go beyond it; it would be concerned exclusively with the relations of whole and part. *Its complete self-evidence is the cause of the a priori character conferred on it; but, taken in*

---

* We have seen in Chapter 24 how Cohen has given a definitive answer to that challenge.

† Wavre calls the disputed principle the "law of the excluded mean," but we have employed the more usual terminology, "law of the excluded middle."

by this a priori character, we would have called traditional logic to a function it is incapable of exercising.

Traditional logic can at most claim to conform discourse to rules, but language itself becomes more and more inadequate to the genuine understanding of the facts of present-day mathematics; thus the word "all," for example, despite the subtle distinctions of Russell, has an unprecise meaning as soon as one is concerned with "all" the objects of an infinite collection. The mathematical intuition is on the contrary, guarantee of the autonomy of this science, and is not bound to respect always and everywhere in its translation into discourse the rules of syntax which traditional logic prescribes. Logic, on the other hand, would be only the algebra of the *language* by which the reasoning is translated. But a logical construction of mathematics independent of the mathematical intuition is impossible, according to Brouwer, because one would obtain thereby only a verbal construction irrevocably divorced from the science. Further, this would be a vicious circle, *for logic itself requires the fundamental intuition of mathematics.*

According to Brouwer the field of application of traditional logic would be limited to finite collections. The crux of the issue then is the right to reason on finite and infinite sets in the same manner. For Brouwer does not admit the law of excluded mean, expressed *in abstracto* and *universally* in the form: "A thing either is or is not" or "A proposition is either true or false."

The laws of contradiction and of excluded mean express two fundamental relations between an attribute *A* and its contradictory, *non-A*, both well defined relatively to the same object, such as even and odd for an integer. The law of contradiction forbids attribution of *A* and *non-A* at the same time to the same object. The law of excluded mean compels assertion of one or the other. They can be regarded as defining the words "well-defined attribute" or "attribute *non-A*" or simply negation; bound up as they are with the definitions, there can be no question of doubting them. *The sole question at issue is to know when they are applicable and to what attributes.* In more precise terms, it is a matter of recognizing whether two distinct attributes *A* and *B* are such that one is entitled to make one the logical equivalent of the negation of the other. It is a matter of knowing, for example, whether for an integer the attribute of being factorable is well defined and whether the attribute of being prime is by definition equivalent to non-factorable. If it is by definition, these laws would apply with full right; if not, a special examination of the two attributes becomes necessary, and it is only in virtue of an indirect evidence that the laws do apply. In the numerous examples which mathematics furnishes of well-defined attributes, or which one considers as such, *only an intuitive, extra-logical evidence permits us to consider two attributes A and B as the negation one of the other.*

The following example is suggestive and of great importance in the present issue: Let us call a *fundamental aggregate a sequence of integers:* 1, 3, 4, 5, 7, for example. Are the following propositions, which state an attribute of the sequence, related as *A* and *non-A?*

    a. All the numbers of the sequence are odd.

    b. There exists in the sequence an even number.

No one doubts that they are so related, but that is in virtue of direct evidence. I run through the sequence; this done, at some determinate position I either have or have not encountered an even number, and I cannot at the same time have found and not found one. I then have the right, from the point of view of their logical function, to assimilate *a* to *non-b* and *b* to *non-a*. The formalist makes "there exists" equivalent by definition to "not all." That is his right. But in doing so, he introduces a new axiom or a new definition.

To the question: Does there exist an even number in the sequence?, whether I refer to the intuitive sense of the word "exist" or to its logistic sense, I cannot refuse to answer with yes or with no; yes, if I affirm *b*, no, if I affirm *a*. Generalized, this becomes: (1.) *To the question: Does there exist in a fundamental aggregate a number having a well defined attribute A?, I can only reply with yes or no.* This is the application of the law of excluded mean which we had in mind.

Likewise the following undoubted proposition, "The numbers of the sequence 1, 3, 4, 5, 7, which are odd form a new set 1, 3, 5, 7," becomes when generalized: (2.) In a fundamental aggregate, a well defined attribute A suffices to characterize a sub-class of the elements which possess it. With Brouwer, we call this principle (2) Zermelo's axiom of inclusion. It states that in a sequence of integers a definition by intension is equivalent to a definition by extension.* *The formalists believe that they have the right to apply principles 1 and 2 without any restriction to the case where the fundamental aggregate is composed of an infinity of elements, such as the natural series of positive integers. The intuitionists refuse this right.*

In exposition of the arguments for the respective positions, consider the following imaginary dialogue:

The Idealist: Either there exists a factorable number in the sequence $m_n = 2^{2^{n+38}} + 1$ $(n = 1, 2, 3, \ldots)$ or else one such does not exist. (This is the sequence $2^{2^{39}}$, $2^{2^{40}}$, $2^{2^{41}}$, etc.)

The Intuitionist: I can only say one exists by exhibiting such a number, say m_{1000}, which is factorable; and I can only deny it by deducing from the definition of the numbers m_n that they are all prime. But I do not see that the rejection of one of the parts of the alternative compels me to affirm the other; as I cannot exclude *a priori* every *tertium*, I refuse to be reduced to your alternative.

The Idealist: Suppose I take successively the numbers 1, 2, 3, ...; then either I shall or shall not come across a number n giving rise to a factorable number m_n. My encounter either will or will not occur.

The Intuitionist: To deny that all the numbers m are prime does not imply that there exists a determinate one, e.g., m_{263}, which is factorable. And in order to be certain of not having met one, you must have exhausted the series of integers, and that you will never do. You will take several steps in the series, and can perhaps say one exists, but you will never take enough to make a denial. *I should not deny that the number does or does not exist if the existence of an object in an infinite set were a well defined attribute of the set; but existence is not a well defined attribute. There are perhaps several modes of existing.*

Let us introduce the two following propositions which state two attributes of the fundamental aggregate:

a. All the numbers of the aggregate possess the attribute A.

b. There exists a number of the aggregate possessing the attribute non-A.

As soon as an infinity of objects is concerned, the word "all" is suspect to the intuitionist. The proposition a can only have the precise meaning which its demonstration confers on it. This meaning will vary according to the demonstration. The most complete meaning would be such that it would follow just from the definition of the numbers of the set that they all have the attribute A. The "there exists" can only have meaning if the object said to exist is actually found. "There exists a number having such an attribute" signifies for the intuitionist: "Here is a number possessing this attribute." "All" from the formalist point of view means: and this, and this, ... and this. Likewise, "there exists" means: or this, or this, ... or this. They are logical product and sum. But if the fundamental aggregate is infinite, the handling of such product and sum requires precautions as to convergence analogous to those with the infinite products and series of analysis. Intuitionists (such as Weyl) seem to confer meaning only on the implication b implies *non-a*, which suffices to establish that propositons a and b cannot both be affirmed. But as the inferences *non-b* implies a and *non-a* implies b, maintained by formalists, are doubtful or no longer meaningful, they see no need of their being related by the law of excluded mean. Consider the

---

* This corresponds to what we have said many times, namely, that a subset of a universal set can be specified by a defining property or, in the finite case, by roster.

intuitionist meaning of general and existential demonstrations and the fact that the expressions "not all" and "there does not exist" have a doubtful meaning, and one will perhaps no longer be surprised at the paradoxes of the most extreme intuitionists. Even Hilbert recognizes that these phrases are devoid of a clear and immediate meaning. *The intuitionists refuse to make an alternative of the affirmation of a universal affirmative or of a singular negative,* when an infinite class is in question; of saying, for example, *either all the integers of a class are factorable or else there exists a determinate one which is prime.*

Here are two examples illustrative of Brouwer's attitude:

1. Let $m = \phi(n)$ and $m' = \phi'(n)$ be two laws of correspondence between a positive integer n and two integers m and m'. Let us say that the two functions ϕ and ϕ' are identical if the numbers m and m' are equal for all values of n, and that the two functions are different if there exists a value of n giving two unequal numbers m and m'. To demonstrate the identity of the two functions, we should have to be able to reduce the one to the other algebraically or analytically; to demonstrate their difference we should have to discover a number n giving rise to two distinct numbers m and m'. Considering what such discovery entails, it perhaps will not seem surprising that the functions are not *a priori* identical or different.

2. Fermat's theorem that the sum of two nth powers of two positive integers is never equal to the nth power of another integer as soon as n is greater than 2 has no known demonstration. One is tempted to say: if the proposition is false, I can assure myself of it by a finite number of trials on the integers. For then, there exist three integers and a power larger than 2 giving rise to the equality, and I can order my experiences in one series in such a way as to be certain of finding the numbers in question within a finite range in the series. Brouwer absolutely refuses to argue this. For him, even the demonstration that Fermat's theorem led to a contradiction would not imply the existence of four numbers invalidating it.

The crux of the intuitionist thesis is the meaning to attribute to the existential judgment and the general judgment. For Weyl, a true judgment is the attribution of a predicate to a singular subject. Where the formalist is content to affirm some sort of ideal existence of an object possessing such an attribute (e.g., there exists an even number), the intuitionist requires the discovery of such an object as will enable him to replace the existential judgment by a true judgment (the number 4 is even).

As Lebesgue has already said, we can only prove the existence of a mathematical entity by constructing it. The "there exists" would be only a check, without value in itself, so long as we cannot find the bank. And perhaps we shall never find it to convert its nominal value into its effective one. The "there exists" of the idealist is only an incitation to formulate a true judgment; ideal existence is worthless so long as it is not converted into actual existence. It is also to be noted that general judgments have meaning only through the fact that they imply an indefinite number of singular judgments; they are only true because they are constantly verifiable.

At this point of his article, Wavre goes on to tell why he, Brouwer, Weyl, and other intuitionists reject the Cantor-Dedekind continuum (a concept explained and repeatedly applied in previous chapters of this book). Thus, most infinite sets become taboo in the intuitionist philosophy. There is, however, no ban on infinite *sequences*, that is, countably infinite aggregates, for, it is argued, our primitive intuition enables us to conceive of a single object, then another, then another, etc. indefinitely. In addition, there is a constructive process for building a sequence— a formula for its general term or a "recursion formula" by which every term after the first can be obtained from the one preceding. In this way, intuitionist procedures are "finitistic" rather than exclusively finite. But Brouwer's school claims that the Cantorian continuum must be renounced because we cannot make that entire ensemble an intuitive construction. What we can conceive is the sequence of intervals of diminishing length, one within the other, whose limit is classically defined to be a real number (Chapter 2). But we never, in intuition, make the actual passage to the limit. Hence Weyl introduced the notion of "becoming" to correspond to our intuitive recognition of how a sequence of intervals gradually "becomes " a real number, and claimed that only in that way do we arrive at an intuitive conception of the number.

Brouwer proclaimed his logically restrictive doctrine in 1907 and elaborated it until his death in 1967, even though he made other important contributions to mathematics, notably to topology; Wavre's article appeared in 1934; but other members of the intuitionist cult are *still* vigorously working at the unfinished task of reformulating classical mathematics so as to avoid the law of the excluded middle and the Cantorian continuum. To indicate a link between their problems and those considered earlier in this chapter, we remark that, although the iconoclastic intuitionist viewpoint sometimes seems difficult and obscure, the notion of *intuitive* criteria for "truth" avoids, in a sense, the frustrating formalist issue of the internal consistency of abstract systems where reasoning is carried out with "meaningless" symbols.

Because the unsettled problems of mathematical foundations are challenging issues whose solution may determine the mathematical "shape of things to come," considerable space has been given to a critical consideration of such questions. Whether or not the future will resolve the logical "crises" we have discussed is an open question. But we shall now consider what may be the possible fate of other mathematical issues and activities in the years ahead.

30

Retrospect and Prospect

Plato said that there are three daughters of necessity, not only Lachesis singing of the past and Clotho of the present, but Atropos of the *future*. Mathematicians and scientists of all periods have held similar views and hence have always been fond of predicting the nature of things to come and envisioning, as all mortals are prone to do, one of two extremes—a more perfect world or an "*après moi, le déluge*" affair.

According to Byron, the best of prophets of the future is the *past*; according to Montaigne, it is the *present*. If one accepts either of these viewpoints, he must believe that all we can hope to know concerning the future of mathematics is implicit in those developments already discussed in the previous chapters of this book. People in general are interested in specific predictions, however. They are ever attentive to meteorological or economic forecasts; they take pleasure in prophecies about "the next hundred years" or science-fiction fantasies about the future where characters are robots or creatures from other planets.

To cater both to mathematical tradition and to popular taste, it seems fitting therefore to close our story with some thoughts concerning the possible progress of mathematics in the eighth decade of our century and thereafter. If, as Byron, Montaigne, and others have held, history repeats itself, one can obtain guidance by recapitulating what has gone before. Let us then prepare for predictions by summarizing very briefly our story of mathematics as a living, growing endeavor, holding a strong place in human culture and maintaining a vital, stimulating relationship with all the sciences.

After primitive beginnings, there were the Babylonians, as we have seen. Their remarkable contributions remained undiscovered and undeciphered until recently, but in all likelihood had some influence on mathematics in India, where arithmetic and algebra were developed by four great leaders, the latest being Bhaskara in the twelfth century. Even before that time, the Moslems had become intermediaries for bringing Hindu mathematics and science to the western world. The outstanding character of the early Mesopotamian arithmetic, algebra, and astronomy has been described, but we must recall that the most significant mathematics of antiquity was the product of Hellenic thought in the "golden age." The deductive method, which is the pattern of all mathematics, reached such a standard of excellence in Euclid's *Elements* that it sufficed up to the nineteenth century, when the highly critical spirit of the era indicated inadequacies in classical logic and in mathematical foundations in general. While Euclid's name was attached to the work that gave mathematics its *method*, it was Archimedes who was the greatest applied mathematician of antiquity. In addition to giving birth to postulational thinking and handing down the work of Archimedes, the Hellenic golden age launched two major streams of mathematics—the discrete and the continuous, as exemplified respectively in Pythagorean number theory and in Eudoxus' resolution of problems associated with incommensurable magnitudes.

After the dark ages in Europe, the first important mathematical work was that of the Italian school. It contributed to arithmetic and produced the ultimate in classical algebra. Symbolic algebra *per se* first appeared at the very end of the sixteenth century, however, in the ideas of Vieta, who preceded by a short time interval the men whom we have called the forefathers of modern mathematics, namely, Descartes and Fermat. The era of Newton and Leibniz was the "age of genius," and their immediate successors were occupied with clarifying and extending the ideas of the masters. Next, in the hands of the Bernoullis and the sequence of giants—Euler, Lagrange, and Laplace—mathematics reached a stage of advanced development by the opening of the nineteenth century.

The contributions of the ensuing period were so numerous, so original, and so critical that they dwarfed all previous mathematical discoveries. During the nineteenth century the manifold creativeness of Gauss manifested itself. Cauchy, Weierstrass, Cantor, and Dedekind made analysis rigorous by "arithmetizing" it; Cantor's *Mengenlehre*, in particular, contained provocative notions that reached far beyond the domain of analysis; Hamilton and Galois were trail-blazers in the new abstract algebra, Riemann in the new geometry. Researches of the universalists, Poincaré and Hilbert, brought the century to a triumphant close, and provided the inspiration that launched the era to which the work of Borel, Lebesgue, Emmy Noether, Hermann Weyl, and von Neumann belongs.

Throughout mathematical history, astronomy, mechanics, various branches of mathematical physics, mathematical statistics, and other subjects classified under the heading of applied mathematics have played an important role. As we have explained, such sciences are less general than the pure mathematical systems which they interpret. At the other extreme there are subjects described as "math for math's sake"—for example, the higher arithmetic, which is as yet almost completely devoid of practical applications. We have seen what challenges this field provided for men like Tchebycheff, De la Vallée Poussin, Hadamard, Hardy, and Ramanujan. Still further in the direction of abstractness and generality are the "metamathematical" considerations arising from logical issues. Symbolic logic, which stems from ideas of Leibniz, enjoyed a renaissance in the nineteenth century after Boole proclaimed his *Laws of Thought*, and is an essential tool at the present time. Nevertheless many logical problems are still unresolved, and constitute a sort of "crisis," in which opposing schools of thought see different ways of handling some of the difficulties. Recent figures in both pure and applied mathematics, as well as in logic and foundations, have been numerous, and the lives and accomplishments of some of the most prominent of these mathematicians have formed the subject matter of previous discussion.

To supplement the above thumbnail summary of the history of mathematics, one should consider a topical outline of the content of leading discoveries. To provide this type of brief survey, and to indicate the wide territory covered in preceding chapters of this book, a sort of check list might be offered. Categories which could be cited are some of the major headings employed at the present time in *Mathematical Reviews*, the "bible" of current mathematical research. A monthly publication of the American Mathematical Society, this journal provides digests, by leading mathematicians, of recent mathematical research papers appearing in the numerous scholarly scientific journals (in various languages) all over the world. It also contains critical reviews of the most important of the recently released books, monographs, and treatises on advanced mathematical topics.

The reader might ask himself, then, what meaning he attaches to the following listings, and, if he feels so inclined, should reinforce his opinions by referring to those previous chapters or pages in our book where particular topics have been mentioned or treated in some detail: History and Biography; Logic and Foundations; Set Theory; Combinatorial Analysis; Order, Lattices; Theory of Numbers; Fields; Linear Algebra; Associative Rings and Algebras; Groups and Generalizations; Functions of Real Variables; Measure and Integration; Functions of a Complex Variable; Ordinary Differential Equations; Partial Differential Equations; Sequences, Series; Integral Transforms; Integral Equations; Functional Analysis; Calculus of Variations; Geometry; General Topology; Algebraic Topology; Differential Geometry; Probability; Statistics; Mechanics of Particles and Systems; Thermodynamics; Statistical Physics; Relativity; Games.

Can, then, the nature of past and present developments lead to any rational conjectures concerning the mathematics of the immediate future? Obviously, one is treading on dangerous ground in all forms of extrapolation, and forecasts must be probabilistic, not deterministic in nature. Therefore it would seem best to frame as open questions all issues to be considered, lest more affirmative treatment appear to be dogmatic, or else be colored by hopes, fears, and personal preferences. Nevertheless, the author will attempt to give partial answers to some of the questions she will now raise. Wherever any prophecy is made, it will be based on the most recent trends and it will be understood that all forecasts are short-range ones applicable, say, to the next decade or so. To begin the sequence of queries, let us ask:

Will the present vogue for axiomatics and pure mathematics continue or grow even stronger?

"Yes" appears to be the correct answer to this question. The mathematical world is aware of the all-embracing quality of abstract theories, and the construction of such theories, rather than a search for specific applications, is the goal of many leaders who, in addition, are influencing their disciples and the newer generations of mathematicians to advocate the same point of view. One example will now be given of how the mathematicians themselves are setting up schools of thought which stress and hence tend to perpetuate axiomatics, rigor, and abstraction as the essence of mathematics.

A current best seller among the longest of mathematical long-hairs, and one for which no royalties are being paid, is a treatise consisting of many individual volumes, written by Nicolas Bourbaki, and bearing the unpretentious title, *Éléments de Mathématique*. In 1949, a brief biographical note stated that Professor Bourbaki, formerly of the Royal Poldavian Academy, was residing in Nancy, France. Somewhat later, around 1953–1954, he appears to have made some affiliation with the Mathematics Institute of the University of Nancago.

In actuality, Bourbaki is just a pseudonym for a group of ten to twenty French mathematicians, and Nancago is a fusion of Nancy and Chicago, two universities where members of the group have served. The purpose of the *Éléments* series of monographs is to study the fundamental "structures" of mathematics and to present in *abstract* fashion the concepts and problems with which any one who calls himself a mathematician in mid-twentieth century must be acquainted. Throughout the twenty Bourbaki volumes that have appeared thus far, axiomatics and the Hilbertian spirit prevail. The facts presented are not new discoveries, but the method of presentation is highly original. The work purports to be a textbook, on a very high level, of course.

The Bourbaki group was organized shortly before World War II and its first volume (now revised in a new edition) appeared in 1939. Among the members at that time were Henri Cartan, son of Élie Cartan (Chapter 20), Jean Dieudonné Charles Ehresmann, C. G. Chabauty, Pierre Samuel, Jean-Louis Koszul, and Roger Godemont; Samuel Eilenberg of Columbia University is a present member; André Weil, now at our Institute for Advanced Study, and M. E. Brelot have been semi-affiliated. All of these are leading names in the world of mathematics today. Concerning André Weil, for example, we have the following statement made by Professor Paul R. Halmos of the University of Chicago: "Although André Weil is not known to the general public, many of his colleagues are prepared to argue that he is the world's greatest living mathematician."

No one seems to know the origin of the name Bourbaki. Perhaps the general, Charles Denis Sauter (*not* Nicolas) Bourbaki (1816–1897) is a legendary figure for French schoolboys, much as Davy Crockett, say, is for boys in the United States. Charles Bourbaki was the son of a Greek colonel and in 1862 was actually offered the Greek throne, an honor he refused. After graduation from St. Cyr, the French West Point, he joined the Zouaves, and became a lieutenant in the French Foreign Legion. He commanded Algerian troops in the Crimean War and became famous in connection with the battles of Alma, Inkerman, and Sevastopol. He continued his military career in heroic fashion, but ultimately the German tactics in the Franco-Prussian War were his undoing. To avoid surrender to the Germans he attempted suicide in 1871, but he used a faulty pistol and the bullet he fired flattened against his skull. His life was thereby saved for another quarter of a century.

Professor Laurent Schwartz of the University of Paris has this to say about the general attitude and the mode of work of the Bourbaki group:*

---

* Laurent Schwartz, "Les Mathématiques en France Pendant et Après la Guerre," *Proceedings of the Second Mathematical Congress*, University of Toronto Press, 1951, pp. 62–65.

Scientific minds are essentially of two types, neither of which is to be considered superior to the other. There are those who like fine detail, and those who are only interested in grand generalities. . . . Each of the two categories contains great names . . . . In the development of a mathematical theory, the ground is generally broken by scientists of the "detailed" school, who treat problems by new methods, formulate the important issues that must be settled, and tenaciously seek solutions, however great the difficulty. Once their task is accomplished, the ideas of the scientists with a penchant for generality come into play. They sort and sift, retaining only material vital for the future of mathematics. Their work is pedagogic rather than creative but nevertheless as essential and difficult as that of thinkers in the alternative category . . . . Bourbaki belongs to the "general" school of thought.

The preparation of one of its volumes is carried on somewhat as follows. There are meetings of the group lasting a week or two about three or four times a year, when work is carried on intensively all day long. The outline and first draft of a tract are presented to a general *Congrès* where it is subjected to the severest of criticism. Revision is made and then the whole process is repeated, possibly as many as half a dozen times before a particular monograph is sent to press.

The profits, most of which come from sales in the United States, are used to defray the expenses involved in assembling members and holding a *Congrès*. One meeting a year is held in some vacation area in France, and the Bourbaki members use some "royalty" funds for indulging in the best French wines and in specialties of the famous French cuisine. This is indeed a slight reward to members for devoting themselves so selflessly to setting down our mathematical heritage in what they consider to be the best possible form. Because the Bourbaki purpose is so serious, it is not surprising to learn that members must relax in some way, and, according to Professor Laurent Schwartz, there is a spirit of *canular* (practical joking) at even the most important meetings.

There has been some repercussion of this sort of fun-making even three thousand miles away. When Bourbaki calls itself a *société anonyme* (corporation), it means *anonymous* in the literal sense, and no one is supposed to mention individual names as we have done above. However, we shall place the blame on Professor André Delachet who, in his *L'Analyse Mathématique*, boldly lists the names furnished above. When Professor Ralph P. Boas of Northwestern University wrote a short account of the Bourbaki group for one of the Encylopaedia Britannica "books of the year," the editors received a letter from Nicolas Bourbaki stating that he was hurt by Boas' claim that there was no such person. In addition, the group spread the rumor that if there was any pseudonym, it was *Boas*, and that there was no individual mathematics professor by that name. The Encyclopaedia editors were embarrassed and confused, and it required action on the part of the American Mathematical Society to settle the matter.

Not every mathematician approves of this kind of horseplay, nor do all mathematicians see completely eye to eye with the point of view of the Bourbaki school. Opponents refer to Bourbakism as "scholasticism," "hyperaxiomatics," and "hypergeneralization," tendencies which ignore the value of intuition and the aesthetic beauty of a less formal approach to mathematics. Nevertheless, all mathematicians today adhere to some form of axiomatics, and proclaim the superiority of *pure* research. For this reason it is almost certain that the same attitude will prevail in the near future.,

The above provides an answer to the first question raised concerning the future of mathematics. The issue was a matter of mathematical *method*. But what about the mathematicians themselves, the men who will carry out this method? What will be the nationalities and mathematical specialties of the mathematical leaders of the future, that is, just how will mathematicians as a group be distributed with regard to country and field of interest?

We have seen that Felix Klein, carrying on the tradition of Gauss, made Göttingen the hub of the mathematical world in the early years of the present century. In the hands of Richard Courant, Klein's successor, the status of Göttingen became even more glorious. Then Hitler and Fascism brought about a complete change of venue for mathematical activity. The migration of mathematicians which started in 1933 is the reason we now raise the question:

Will the history of intolerance repeat itself at any time in the future, thereby producing further changes in locale for the great mathematical researchers of future eras?

Everyone must hope that the answer to this question is no. But as has been stated previously, such matters must remain open issues whose solution is dependent on the nonmathematical factors of civilization. In 1942 grounds for both optimism and pessimism could be found in an article by Arnold Dresden (1882–1954), entitled "The Migration of Mathematicians" and published in the August–September issue of the *American Mathematical Monthly*. In that paper Dresden revealed how and why, starting in 1933, a large proportion of the outstanding European mathematicians gradually came to American shores. He listed names of the distinguished emigrants year by year. His roll call of greats included many of the mathematicians featured in the present book—for example, Albert Einstein, Emmy Noether, Hermann Weyl, Richard Brauer, Richard Courant, Emil Artin, Stanislaw Ulam, Otto Neugebauer, Abraham Wald, Jerzy Neymann, John von Neumann, Samuel Eilenberg.

Dresden's account ceased with the year 1940, but the migration to the United States of the greatest mathematical minds went on and on. In more recent years, those foreign mathematical leaders who have not come to the United States to become citizens have come to these shores from time to time as visiting professors in our universities, or on research grants from the Institute for Advanced Study.

This state of affairs, combined with Dresden's facts and figures, might well furnish the grounds for predicting that the United States will be the locale of a considerable part of the world's creative mathematical research in the immediate future. The basis for this prophecy leads naturally to the next question:

If the situation of major mathematical activity shifts as a result of political and economic circumstances, will these likewise affect the *dramatis personae* who play the leading roles?

In response to this query, it is our contention that the future is likely to witness one particular change refuting the maxim "History repeats itself." It is our belief that, in the years ahead, the roster of mathematicians will contain a much greater proportion of feminine names than has been the case in the past. It was pointed out by the geometer and historian Julian L. Coolidge* that during all the centuries prior to 1900 A.D. there were only half a dozen women who carried on mathematical work of any real distinction, namely, Hypatia, the Marquise du Châtelet, Maria Gaetana Agnesi, Mary Somerville, Sophie Germain, and Sonya Kovalevsky. Hence, to support the prognostication that the distaff side of mathematics will be much enlarged in the future, evidence will be drawn from present facts or those of the immediate past.†

If we were to start with Emmy Noether, who received her Ph.D. from Göttingen in 1900, and if we were to record the names and accomplishments of all the women who have subsequently attained real eminence in the mathematical field, we would have a considerable task of cataloguing. Many of the women involved would be American, but a large percentage of these would have been born in Europe and have initiated their mathematical careers there, coming to the United States as part of the migration of mathematicians described above.

We have thought it best not to include discussion of the work of any of these able American mathematicians, lest it appear that some critical judgment is being exercised as to the relative merit of their work. Current history is delicate ground and the recorder must try to achieve by remoteness in space what he cannot accomplish by separation in time. Therefore in selecting a sample of seven women presently engaged in advanced mathematical research, the author has chosen outstanding representatives from England, France, Switzerland, and Italy.

Very little is known with certainty of the mathematical contributions of Hypatia, the earliest of the female mathematicians, but she is reputed to have been an algebraist (Chapter 4). Let us then talk first of a modern algebraist. Fifteen hundred years separate Hypatia's *Diophantine Analysis* from Hanna Neumann's research studies of "group amalgams," "generalized free products," "near-rings," "varieties of groups." The last topic listed is Hanna Neumann's most recent subject of investigation, in fact, one to which she has devoted an entire treatise‡ and which her husband chose as the topic of his address at an annual meeting of the American Mathematical Society.§ Husband and wife carry on research on somewhat different aspects of abstract algebra, but their major interest is in *groups*. They have collaborated on many group-theoretical papers. Even when they write individually, one is likely to refer to the work of the other.

---

* J. L. Coolidge, "Six Female Mathematicians," *Scripta Mathematica*, March-June 1951, pp. 20 ff.

† Edna E. Kramer, "Six More Female Mathematicians," *Scripta Mathematica*, Vol. 23, Memorial Issue, 1957, pp. 83 ff. (revised).

‡ Hanna Neumann, "Varieties of Groups," *Ergebnisse der Mathematik und ihrer Grenzgebiete*, Band 37, Springer-Verlag, New York, 1967.

§ Bernhard H. Neumann, "Varieties of Groups," *Bulletin of the American Mathematical Society*, Vol. 73, No. 5 (September 1967), pp. 603 ff.

Hypatia, the Neoplatonist, suffered martyrdom at the hands of Christian fanatics in 415 A.D. Twentieth-century intolerance made it necessary for Hanna von Caemmerer to leave her native Germany for England in 1938. Since her fiancé, Dr. Bernhard H. Neumann was of Jewish extraction, he had found it expedient to emigrate to Great Britain a few years earlier. The Nazi race theories were already prevalent in Germany. Miss von Caemmerer was classified as an Aryan, Dr. Neumann as a Semite. Under any circumstances, the Nuremberg laws of 1935 made it illegal for couples to marry if they belonged to different "races."

For Hypatia there was no refuge from fifth century Alexandria, but the young Neumanns were able to find a haven in England. In spite of the adjustment to a new land and all the vicissitudes of the war, they were able to continue research in their specialized mathematical fields. Hanna Neumann was given university credit for the mathematical studies she had completed at Berlin and Göttingen, and her subsequent research led to the Ph.D. at Oxford. She then held academic positions at the University of Hull, England, and at the Manchester College of Science and Technology, while her husband was a mathematics professor at the University of Manchester.

In 1964 Hanna Neumann was the first woman to be appointed to a Professorship in Pure Mathematics at the Australian National University School of General Studies. In 1968 she also became Dean of Students. Her husband, a Fellow of the Royal Society, is Professor and Head of the Department of Pure Mathematics at the Australian National University Institute of Advanced Studies.

The personal lives of the mathematical couple directing Australia's pure mathematics have been enriched by five wonderful children, four of whom are devoted to mathematics. The eldest, Mrs. Irene Dhall, holds a master's degree in literature from Manchester and is married to a distinguished surgeon. The second daughter, Mrs. Barbara Cullingworth, teaches mathematics and is married to a physical scientist. The oldest son, Dr. Peter M. Neumann (an Oxford Ph.D.), is a professor of mathematics at Queen's College, Oxford. He and his wife, Sylvia, were both recipients of first-class degrees in mathematics at Oxford. It should be remarked in passing, that an Oxford "first" is said to be an ineradicable distinction in the United Kingdom. Like his parents, Peter Neumann is a group-theorist, and the mathematical world is becoming accustomed to reading group-theoretic papers by Neumann, Neumann, and Neumann. Walter, the next son in chronological order, was graduated (with a first-class degree in mathematics) from an Australian university, and has received his Ph.D. from the University of Bonn. Daniel, the youngest son, born in 1951, is, like his parents and brothers, showing a strong predilection for pure mathematics. It may well be that some future mathematical memoir will be authored by *five* Neumanns instead of a mere three. The Neumanns may be founding the successor to the historic Bernoulli family.

The knowledge of Hypatia's sad fate may have been one deterrent to women subsequently contemplating mathematical careers. At any rate, there was a lapse of 1300 years between the date of the talented Alexandrian's death and the appearance of other female names in the history of mathematics. Chronologically the Marquise du Châtelet and Maria Gaetana Agnesi are next on the list. By a coincidence there are a number of links between Agnesi's activities and those of Maria Pastori, at present professor at the Istituto Matematico of the University of Milan, research worker in the fields of tensor analysis and relativity. In the first place, both were natives of Milan. Secondly, the "magisterial" school attended by Maria Pastori was called the Maria Agnesi School. Also we recall that Agnesi's life was bound up with that of a younger sister, Maria Teresa, who is known in the history of music as a composer and pianist, that is, virtuoso on the clavicembalo. Maria Pastori has an older sister, Giuseppina Pastori, distinguished in the field of biology, and the lives of the twentieth-century sisters have been as closely associated as those of the Agnesis. When, in 1930, Guiseppina Pastori was nominated for a professorship at the Catholic University of Milan (Universita Cattolica del Sacro Cuore) the Pope referred to the precedent set by Pope Benedict XIV in appointing Maria Agnesi to the chair of mathematics and natural philosophy at Bologna in 1750. Giuseppina Pastori is a physician as well as a professor of biology, and we note that she is responsible for the naming of a nurses' training school in Milan after Agnesi, in memory of the illustrious mathematician's services as directress at the Pio Istituto Trivulzio, a Milanese home for the aged sick and poor.

While the Pastori sisters resembled the two Agnesis in the fact that they worked together in early youth in intellectual pursuits of common interest to both, there is a vast difference in other biographical details. The eighteenth-century sisters were born into a wealthy home. Their father, a patron of culture, established a salon where Maria would read a series of theses on abstruse scientific or philosophical questions and then defend them in academic disputations with learned men. The guests at a particular soirée would serve as audience, and during the intermission between two debates the younger of the Agnesi girls would play some of her original compositions on the clavicembalo. Foreign noblemen traveling in Italy (Count de Brosses or the Crown Prince of Saxony, for example) were curious to meet the talented sisters and would have no difficulty in securing an invitation to one of Signor Agnesi's assemblies.

By contrast, the Pastori sisters were two of eight children in a Milanese family of very modest means, with no academic tradition whatsoever. Elementary education, ending at age thirteen, was all the family could manage to give the children. Giuseppina went to work in an office as soon as this early education was completed. Maria Pastori's early mathematical talent impressed her teachers and a visiting superintendent of schools marveled at her responses to his mathematical questions. Maria's mathematics teacher was aware that the Pastori family could afford no further schooling for the remarkable child. This teacher, according to Professor Pastori's personal statement, was an inspiring and conscientious pedagogue. She assisted Maria in obtaining a small scholarship for a magisterial school, the equivalent of an American teacher-training institution, and convinced her parents that she must have this minimum of additional education Thus it was that at age seventeen the potential mathematician became an elementary school teacher in a small town near Milan.

By this time both sisters decided that their education must continue, whatever the odds. They studied together, unassisted by anyone else—at the crack of dawn before Maria took her commuter's train to the suburbs, and then again in the evenings and into all hours of the night Three years later, when the younger of the sisters was twenty, both girls took the state examinations for the baccalaureate and, of course, passed with highest honors. In Italy, and in France, attendance at a lycée or special school is not required in order to obtain the bachelor's degree. Success in the examination is the sole criterion.

From this point on both young ladies continued to climb the academic ladder to the very top. Maria passed the difficult entrance examinations for the Scuola Normale Superiore of Pisa, and received a small scholarship. (The higher normal schools in Italy and France are really special universities for a small, carefully selected group of very exceptional students.) Tutoring and the scholarship stipend sustained the young mathematician during the four years of study for the doctorate in mathematics. Her sponsor for the degree was the great geometer, Luigi Bianchi. Now she taught in the secondary schools and was simultaneously an assistant at the University of Milan. Since Italian universities and those in other European countries are state schools, advancement in rank is decided on the basis of competitive examinations. We have seen that no tests could faze Maria Pastori and she rose from assistant to *libera docente* (assistant professor) to the full professorship, her chair being one in rational mechanics.

College teachers of mathematics and physics in the United States know the excellence of the 1949 text, *Calcolo tensoriale ed applicazione*, which she wrote in collaboration with Bruno Finzi.* Her research reports and special monographs are too numerous to list, but some readers may recall her paper at the 1953 International Symposium on Differential Geometry. In it she discussed the space of Einstein's unified field theory.† The style of her textbook and her expository memoirs is lucid, and full of simple specific applications of abstract tensor concepts and procedures. Her students at Milan are solving significant research problems under her direction. For example, a paper by Laura Martuscelli suggests a modification of Einstein's unified field theory,‡ and F. Graiff wrote a monograph on tensor integration in Riemann spaces of constant curvature.§

In placing Maria Pastori's mathematical work, we might say that her specialty makes her a true daughter of Italy and a twentieth-century disciple of Agnesi. Geometry underwent powerful development and remarkable mutations in Italy

---

* B. Finzi e M. Pastori, *Calcolo tensoriale e applicazioni*, Bologna, 1949.

† M. Pastori, "Sullo spazio della recente teoria unitaria di Einstein," *Convegno di Geom. Diff.*, 1953, pp. 107–113.

‡ L. Martuscelli, "Sopra una possibile modificazione della teoria unitaria di Einstein," *Ist. Lomb. Rend. Cl. Sci. Mat. Nat.*, 1955, pp. 607–615.

§ F. Graiff, "Sull'integrazione tensoriale negli spazi di Riemann a curvatura costante," *Rend. Ist. Lomb.*, 1951.

in the two centuries following Agnesi. Algebraic geometry reached such heights in the work of Cremona, Severi, and others that, in the words of E. T. Bell, it became "almost a national pastime." Differential geometry also flourished on Italian soil. Perhaps the finest flower of all was the *absolute differential calculus* of Ricci and his pupil, Levi-Civita. Without this pure mathematical instrument Einstein could not have forged his concept of the physical universe into the "greatest achievement of the human intellect." But the tensor calculus is of importance in the theoretical treatment of classical physics as well—in dynamics, hydrodynamics, elasticity. It is proving useful today in the pure mathematical investigation of generalized spaces which, so far, are abstractions without physical application. Maria Pastori's research and writings are concerned with sharpening the Ricci tool in order to cast it into an ever more perfect mold and extend its usefulness.

In spite of the Agnesi tradition, there are very few women at the present time who occupy full professorships at the leading universities in Italy. At Milan, in addition to Maria Pastori, there is Giuseppina Masotti-Biggiogero, who occupies a chair in geometry. At Pavia, the Italian university second in venerability only to Bologna, Maria Cinquini-Cibrario is *professoressa ordinaria* (full professor) in mathematics.

Although there were no mathematicians in her ancestry, the family background of Maria Cibrario was a distinguished one. Her father was a general in the Italian army. One great-grandfather, Count Giovanni Antonio Luigi Cibrario (1802–1870), descendant of a noble Piedmontese family, was a historian of note and a prominent political figure of his day. In 1852 he was minister of finance in D'Azeglio's cabinet. Under Cavour he was at first minister of education, then minister of foreign affairs. In 1861, in recognition of Luigi Cibrario's long public service as well as this nobleman's prolific historic research dealing with the House of Savoy, Victor Emmanuel bestowed on him the title of count.

Most of Maria Cibrario's ancestors on both sides lived in the province of Turin. She spent her early years in Genoa where her father was stationed at the time. Later she studied at the University of Turin, majoring in mathematics. Preliminary to her graduation in 1927, she was required to write a thesis in mathematics. In this work, as well as other early research, she was advised by Guido Fubini, who influenced her to select *analysis* as her mathematical specialty. She was appointed assistant to the aging Giuseppe Peano, and on his death in 1932, transferred to Francesco Tricomi; but before that year was over she was named *libera docenza* in infinitesimal analysis.

Not long thereafter she made the acquaintance of Silvio Cinquini at a meeting of the Italian Mathematical Society, and in July 1938 the two young mathematicians were married. Then the war years arrived and with them the curtailment of academic activities. No national examinations for professorships were held. Madame Cinquini entered the first of these gruelling competitions held after the war. Having succeeded in the contest of intellects she was assigned by the National Board to a professorship at Cagliari, Sardinia, where she served during 1947 and 1948. Next she was appointed to the chair of mathematical analysis at Modena, and from 1950 on has held the corresponding chair at Pavia, where her husband is also full professor of mathematics.

One aspect of the numerous duties inherent in her academic position has been her tireless research on the theory of partial differential equations and related fields. Her original papers are numerous and range over a period of more than twenty-five years. She can be credited with the classification of linear partial differential equations of the second order of mixed type, with existence and uniqueness theorems for many of these, with considerable research on nonlinear hyperbolic equations and systems of such equations. She has solved the Goursat problem for the hyperbolic nonlinear equation of the second order. For quasilinear and nonlinear hyperbolic systems she has solved the Cauchy* problem, and in some cases, the problem of Goursat. She has also studied the nonlinear hyperbolic equation of the nth order.

Recently she collaborated with her husband, Silvio Cinquini, in establishing existence and uniqueness theorems for first order partial differential equations by generalizing a technique Carathéodory had found useful with ordinary differential equations. This joint effort of husband and wife was exceptional. In general they work independently and in different branches of analysis. Silvio Cinquini's early investigations were concerned with analytic functions, but later he became a devotee of Leonida Tonelli's school of thought and specialized in the calculus of variations. He has also carried out research on ordinary differential equations and quasi-periodic functions, with a special monograph on the latter subject.

The lives of husband and wife, both analysts, are completely dedicated to their scientific work and to the upbringing of their children, Giuseppe, Vittoria, and Carlo, born in 1944, 1947, and 1949 respectively.

Agnesi had as her French contemporary the brilliant Marquise du Châtelet. The French women mathematicians today appear to be more original in their research and less frivolous in their behavior than the famous expositor of Newton's *Principia*. As a first example of French women mathematicians there is Jacqueline Lelong-Ferrand, professor of mathematics at the University of Paris. In 1956 she and her husband, Pierre Lelong, also professor at the University of Paris, were in residence at the Institute for Advanced Study, carrying on research on Riemann manifolds and functions of several complex variables, respectively. In visiting the United States, the Lelongs did not interrupt their family life, for they brought to Princeton with them their three children, Jean, Henri, and Françoise, aged seven, five, and four at the time. (Another daughter, Martine, was born to the Lelongs in January 1958.)

Jacqueline Ferrand was born in the little town of Alès, in the province of Gard in south-central France. She attended the lycée in Nîmes, a city known to the

---

* This basic problem of partial differential equations is usually given Cauchy's name, but we point out that it would be proper to speak of the Cauchy-Kovalevsky problem since Sonya Kovalevsky shares with Cauchy the honors of establishing rigorously the first existence theorem associated with the question. (See Chapter 22.)

Romans. Her father was, appropriately, a teacher of Latin and Greek. While his daughter excelled in all subjects, she received the first prize in a national competition in mathematics, and was encouraged to compete again—this time for the privilege of admission to the Paris École Normale Supérieure of the rue d'Ulm. Prior to 1936, girls were not admitted to this renowned school, whose graduates include many of the leading mathematicians in France today, in particular, a quorum of the Bourbaki group. Jacqueline Ferrand was among the first group of girls permitted to take the entrance examinations and, needless to say, was admitted to the course in mathematics and science. There are about twenty such admissions out of approximately 250 applicants each year.

It is not the purpose of this discussion to search for reasons why there have been so few women mathematicians in the past, or to explain the present rapid rate of increase in female mathematicians. But it seems difficult at this point not to comment briefly on educational opportunity as a possible factor. Although girls were permitted to attend the École Normale Supérieure de Jeunes Filles at Sèvres, just outside of Paris, where the mathematics instruction was of the very highest caliber, it seems that those who had the privilege of attendance at the rue d'Ulm are researchers today, while most of the graduates of the Sèvres school are teachers in lycées. Mme. Lelong-Ferrand and others explained to the present writer that the greatest advantage of the rue d'Ulm is the coeducational one, which gives girls the chance to witness, imitate, and compete with the masculine critical, creative, courageous attitudes in approaching problems.

At any rate, Jacqueline Ferrand ranked first in the 1939 *Agrégation masculine* in mathematical sciences. (These examinations for the degree of Agrégé are as difficult as the comprehensives for the Ph.D. in mathematics at leading universities of the United States.) Mlle. Ferrand's first position was at the Sèvres normal school for girls, where from 1939 to 1943 she had the task of coaching the students for the Agrégation. Then she became *chargée de cours* (assistant professor), first at the University of Bordeaux, then at Caen. She was married in 1947, and in 1948 became full professor at Caen. From 1948 to 1956 she held the chair of calculus and higher geometry at Lille. On her return from Princeton in 1957 she was transferred to Paris. She was, at the time, the only one of the newer generation of French women mathematicians to have a full professorship in pure mathematics at the University of Paris.

Simultaneously with her teaching she has carried on continuous research in analysis and geometry. Her major specialties have been conformal representation (1942–1947), potential theory (1947–1954), Riemann manifolds and harmonic forms (1955–present). Her 1942 thesis for the Doctorat-ès-Sciences and several subsequent papers investigated the behavior of conformal transformations in the neighborhood of a boundary point. Among her next set of memoirs was one where she created the concept of *preholomorphic* functions, using these to produce a new methodology for proofs. Her research in potential theory enabled her to generalize certain classic theorems to n dimensions, namely the lemmas of Julia and Phragmen-Lindeloef. Following this work, she attacked some of the most difficult problems in the field, once again creating new concepts for the purpose. She dropped this avenue of research temporarily to write a searching, comprehensive treatise, which was published in 1955 as part of the Collections de Cahiers Scientifiques edited by Professor Gaston Julia and entitled *Représentation conforme et transformations à intégrale de Dirichlet bornée*. Then, having explored the field of conformal representation so completely, she felt that it was time to enlarge the scope of her research, an objective realized in the many original papers which she contributes year after year. There have been new books too—a work on the fundamental concepts of mathematics with emphasis on analysis, and an advanced text on differential geometry.

The present concern with missiles and satellites tends to emphasize the progress in applied mathematics, and may possibly obscure the fact that we are living in an age of most prolific creation of pure mathematical ideas. French mathematicians have made substantial contributions to this development, and some of their most significant research has been done in the field of algebraic topology. Just how the newest topological concepts can lead to powerful generalizations of related subjects is illustrated in the work of Professor Paulette Libermann of the University of Rennes. The initial approach to the problems in question was due to Élie Cartan, under whose direction Mlle. Libermann did her first research, while she was a student at the École Normale Supérieure de Jeunes Filles de Sèvres.

Using concepts from the theory of fiber spaces, Charles Ehresmann, a member of the Bourbaki group, generalized ideas of Cartan by reformulating the definitions to give them *global* meaning. Paulette Libermann attended Ehresmann's seminars at the University of Strasbourg and wrote her thesis for the degree of Doctorat-ès-Sciences Mathématiques under his direction. She is an active member of Ehresmann's research school on differentiable fiber spaces, almost complex manifolds and their generalizations. In a 1955 paper, for example, she summarized all the recent results, including her own, on "regular infinitesimal structures—almost complex, almost paracomplex, almost hermitian, almost parahermitian, etc."[*] Again, in two 1958 notes in the *Comptes Rendus*, she discusses infinitesimal pseudogroups in general, and in particular the infinitesimal pseudogroups of Lie.[†]

Lest Paulette Libermann's progress from the 1941 days with Cartan to her 1958 production of advanced memoirs and participation in Bourbaki seminars seem an uninterrupted success story, let us return to some biographical details. She was born in Paris and received her early education there. Although the greatest living geometer had accepted her as a protégée, the Vichy laws excluded her from the 1941 Agrégation, and her entire professional career had to be postponed until after the Liberation. From 1945 to 1947 she studied at Oxford with J. H. C. Whitehead

---

[*] Paulette Libermann, " Structures presque complexes et autres structures infinitésimales," *Bull. de la Soc. Math. de France*, 1955, pp. 195–224.

[†] Paulette Libermann, "Pseudogroupes infinitésimaux, Faisceaux d'algèbre de Lie associés," *Comptes Rendus,* January 6, 1958. A continuation of this note, "Pseudogroupes infinitésimaux de Lie," *Comptes Rendus,* January 27, 1958.

(Chapter 20), a leading disciple of Cartan. From 1947 to 1951 she taught at the Lycée de Jeunes Filles de Strasbourg, while preparing her thesis with Ehresmann. It was not until 1954 that she became *Maître des Conférences* at Rennes.

All writers on women mathematicians give major mention to Sophie Kovalevsky, who was Russian by birth, western European in her advanced mathematical training and professional career. As a parallel, we now discuss Sophie Piccard, Russian by birth, western European in her ancestry, advanced mathematical training, and professional career. She was born in St. Petersburg and educated at the University of Smolensk where her father was professor of natural sciences.

Just as Kovalevsky was conditioned by a cultured circle of family acquaintances Sophie Piccard was headed toward intellectual pursuits by both parents. Scientific talent came to her through ancestors who were surgeons, physicians, chemists. Her mother's ability, however, was literary. Eulalie Güée Piccard's studies of Pushkin and Lermontov earned for their author the Prix de la Fondation Schiller, the highest literary award in Switzerland. Mme. Piccard also wrote *La Grande Tragédie*, a series of five books on the Bolshevik revolution. In 1956, in spite of a siege of poor health involving radical surgery, she completed a novel *Galia*, and at the time of her death in August 1957 had almost finished a biographical and critical study of Simone Weil.*

As a young woman, Eulalie Piccard was a professor of languages at the Russian lycée attended by her daughter, whose teachers told the mother about the girl's exceptional aptitude for mathematics. Although young Sophie wished to become a physicist, her mother pointed out the greater importance of pure science and, in a sense, coerced her into specializing in theoretical mathematics.

During the troublous twenties in Russia the Piccard family left the country. Since the ancestors on the father's side had been Swiss, and on the mother's French Huguenot, it was natural for the family to emigrate to Switzerland. Mlle. Piccard entered the University of Lausanne where she obtained the degrees of Licence-ès-Sciences and Doctorat-ès-Sciences in 1927 and 1928 respectively. Her doctoral thesis, in the field of probability, was written under Professor Mirimanoff's direction.

The personal troubles of the Piccards did not come to an end in the new land. The young mathematician's father died and the financial problems of his widow and daughter became acute. Eulalie Piccard had to earn a living as a dressmaker, while Sophie, her advanced degrees notwithstanding, was unable to obtain a teaching position. She worked first as an actuary, then from 1932 to 1938 served as administrative secretary for the newspaper, *Feuille d'avis de Neuchâtel*. She used all her leisure time for study and research, but many years elapsed before she was able to attain an academic post of any real distinction. In 1936 she became assistant in geometry to Professor Gaberel of the University of Neuchâtel, and on his death in 1938 she succeeded to his position, and became *professeur extraordinaire*

---

* We note in passing, Simone Weil's link with mathematics. Her brother André Weil, is one of the foremost living mathematicians, as already mentioned in this chapter.

(associate professor). Since 1943 she has been full professor of mathematics at the University of Neuchâtel, where she holds the chair of higher geometry and probability theory. In 1965 she held a visiting professorship in Hanna Neumann's department at the Australian National University.

In spite of the obstacles to her mathematical career, she has carried on prolific research. For many years she worked on the theory of sets, and more recently has specialized in group theory. She has also written papers in a number of other fields —function theory, the theory of relations, probability theory, actuarial science. An outgrowth of her set-theoretic investigations was the publication of two comprehensive treatises entitled *Sur les ensembles de distances des ensembles de points d'un espace Euclidien* (1939) and *Sur les ensembles parfaits* (1942). Both books are original, tremendously systematic and detailed. Attention is given to certain perfect sets on the real line. These sets arise from a closed interval by a process generalized from the construction of the Cantor ternary set. Henri Lebesgue, one of the leading analysts of all time, was interested in all such set-theoretic questions since they were fundamental to his revolutionary theory of integration. In a review in the Swiss *Enseignement Mathématique* he voiced his approval of Professor Piccard's book on perfect sets.

Once again Élie Cartan comes into our story, for he gave a laudatory review, in the French *Bulletin des Sciences Mathématiques*, to Sophie Piccard's next book, *Sur les bases du groupe symétrique* (1946). In an earlier issue of the same periodical, it was his son, Henri Cartan, whose name is well known to the mathematical world of today, who had written an analysis of the treatise on perfect sets. Another French mathematician, Professor Arnaud Denjoy of the University of Paris, has written the introduction to Professor Piccard's most recent book, the 1957 *Sur les bases des groupes d'ordre fini*. In 1957, also, the University of Paris printed, in booklet form, the text of Professor Piccard's public lecture on Lobachevsky. A part of that lecture was presented in Chapter 3.

To complete our full circle of female mathematicians, we now return to England where we started. It is not a mere question of geography that influenced our decision to discuss the work of Florence Nightingale David. It is because she is the author of the article on probability from which the historical material of Chapter 13 is drawn. The paper in question is just one of her many research memoirs in the field of probability and statistics. The half-dozen contemporary women mathematicians whose activities have been presented up to this point are devoting themselves in the main to pure rather than applied mathematics. This is not the case with Florence Nightingale David, Reader in Statistics, University College, London. Her research papers, more than sixty in number, are either directly concerned with concrete, practical questions or else are potentially capable of application to such problems.

Dr. David's given names would seem to have predestined her to become a "passionate statistician," an appellation given to the heroine of the Crimean War, on account of her keen interest in the statistical work of the Belgian L. A. J. Quételet (1796–1874), and his mathematics of human welfare. Florence Nightingale David is a collateral descendant of the most famous of all nurses.

But it takes much more than a name to provide success in a mathematical career. Dr. David was born in Hereford, received her early education in Devonshire, where her father was a schoolmaster, and had her university training in London. Even her first academic position was an important one, for she became research assistant to Karl Pearson. While she continued her association with London's university, she also carried on ordnance work during World War II, served the Ministry of Home Security, and became a member of the scientific advisory council of the Land Mines Committee. Her statistical research for this group and for the Civil Defense Research Committee comprised thirty-three papers. Because the results are still classified, we may not discuss their nature here.

In 1946 and again in 1964 Dr. David visited the United States, and lectured at the University of California at Berkeley. Teachers and students of statistics are acquainted with the excellence of her books, *Probability Theory for Statistical Methods*, *Elementary Statistical Exercises*, and *A Statistical Primer*. The *Biometrika* article to which allusion was made in Chapter 13 is a fascinating account of the origins and early history of probability; therefore it was good to learn, in the course of a personal interview, that Florence N. David plans "some day in the future" to write a complete history of probability.

In addition to the sample of seven careers outlined in detail above, one could furnish other examples of distinguished accomplishment in mathematics, pure or applied, on the part of women today. Their activity is the basis of a prediction of the present chapter, namely the strong likelihood that in the near future feminine participation in mathematical research will be greater than ever before.

After raising the issue of the probable nature of mathematical method and personnel in the years ahead, one might ask other questions concerning the future of mathematics. Although pure mathematics embraces its applications, scientists and laymen alike would certainly hope that applied mathematics will have many future devotees, so that the usefulness and practicality of mathematics will continue. Hence we inquire:

Will the future witness a continuance of the intimate relationship that mathematics bears to so many other fields of thought? Will applied mathematics flourish? More specifically, what will be the future results of the present age of cybernetics, feedback, mechanical brains? Will a second industrial revolution completely replace men by machines? Will thinking machines lead to a more rational conduct of human affairs? Will the new communication devices be adaptable to aiding the physically disabled? Or is automation a Frankenstein whose power will ultimately destroy the human race that created it?

"The Aztecs said this world . . . would blow up. Then what will come, in the other dimension, when we are superseded?" wrote D. H. Lawrence. Were the early inhabitants of Mexico prophetic? The story of energy and its relation to matter has been touched on in our chapters on statistics, statistical mechanics, and relativity, but now we must ask: Will the future stress the peaceful uses of the incredibly huge energy sources revealed by the A-bomb, the H-bomb, the potential cobalt bomb, or will all these discoveries result in the fate postulated by the Aztecs?

How soon will man land on Mars or Venus? Will man be able to stand the strain? Will the astronauts succeed in finding a method of navigation in space that will make interplanetary travel a reality? In this set of questions, only certain ones refer to mathematics, and these are actually concerned with applications of the subject. The queries do not refer to the future of *pure* mathematics, since most of the pure mathematics of missiles and satellites was handled long ago. For example, Kepler's laws on which planetary motion is based were formulated in the seventeenth century, and all the modifications and improvements contributed by Laplace, Gauss, Poincaré, and others were accomplished before our day. The basic principle of rocket action is just an application of Newton's third law of motion. And so it goes; no matter what pure mathematical tools are required in this area of research, they are already at hand.

Will the "crash" programs of the present and the immediate future produce a great number of mathematicians, physical scientists, engineers, etc., and thus enlarge greatly the population of the mathematical world and indirectly influence the quality and quantity of future mathematical research?

Not everyone would answer this question with a categorical affirmative. There are those who believe that positive prognosis is not possible because there is a weakness in the lucidity of mathematical exposition today as well as a general lack in communicating new mathematical research as it develops. For the well-being of mathematics in the future, "the teacher and the scholar of mathematics . . . must wear the same skin." This was the opinion of Professor E. J. McShane as expressed in his 1956 presidential address before the Mathematical Association of America.*

In that speech, Professor McShane expressed concern with the lack of communication between pure and applied mathematicians, between research mathematicians whose specialties are different, between mathematicians and teachers of mathematics, between mathematicians and nonspecialists. He made certain Cassandra-like predictions about the possible future consequences of such separation "into small groups of specialists with little intercommunication." He suggested certain palliatives, but placed before the mathematical world the problem of seeking a complete solution (if there is such a solution). In alluding to the recent rapid growth of mathematics, he made the following analogy:

About sixteen years ago I bought a house which had been unoccupied for five years. Summer after summer we fought a losing battle against honeysuckle; it grew faster than we could dig it out. Then came the discovery of the weed-killing properties of 2-4-D, and the battle was won. As I understand it, a broad-leaved plant sprayed with 2-4-D does not die at once; it begins to proliferate rapidly, growing quickly and without organization. As a result, it dies. I cannot look on the proliferation of mathematics as being in all circumstances an unqualified good. . . .

---

* E. J. McShane, "Maintaining Communication," *American Mathematical Monthly*, Vol. 64, No. 5 (May 1957), pp. 309 ff.

Everyone of us is touched in some way or other by the problems of mathematical communication. Every one of us can make some contribution, great or small, within his own proper sphere of activity. And every contribution is needed if mathematics is to grow healthily and usefully and beautifully.

McShane emphasized the same point of view in subsequent articles and in his activities as president of the American Mathematical Society (1960). If, as a result of his influence and that of others who agree with him, communication of mathematical ideas does improve in days to come, one might ask:

Will the better comprehension of the new mathematics make the subject all-pervading in the future? In particular, will the fine arts feel its impact? While pure mathematics is expanding and almost certain to grow at an ever-increasing rate, its spheres of influence are also becoming more numerous. Everyone is aware that mathematics is important in the physical and biological sciences, and in business, industry, engineering, technology, economics, and psychology, in short, in every area that can profit from abstract analysis and formulation. But a work like Birkhoff's *Aesthetic Measure* (1933), which subjects the arts to mathematical analysis, is practically unique. Will leading mathematicians of the future continue where Birkhoff left off?

Although no one can predict the future of mathematics, the above queries should suggest that mathematics, both pure and applied, is wide open for further progress, contrary to the fears that mathematicians have sustained at various times in mathematical history. For example, toward the end of the eighteenth century, Lagrange, the leading figure of the era, wrote in a letter to D'Alembert: "Doesn't it seem to you as if our lofty mathematics is beginning to decline?" Again, Hilbert's program of 1900, in which he outlined a course for future research, was specifically designed to cure the *fin de siècle* jitters of his fellow mathematicians.

A few years ago André Weil wrote a paper bearing the title "The Future of Mathematics."* Placing himself *in loco Hilberti* he specified important unsolved problems and incompletely developed subjects of pure mathematical research. The type of prophecy practiced by Hilbert and Weil is one of the truly rational species of prediction.

For if a man is a leading mathematician of his era, and one with many disciples, he can control the future of his subject through his personal influence. His word is law, and the younger generations of mathematicians will do their best to solve the problems he has set. In the few brief years since the publication of Weil's article, a formidable attack has been made on some of the issues to which he pointed, but complete or elegant resolution must still, in the main, be left for the future. It is probable, however, as Weil said,† that the "great mathematicians of the future . . . will solve the great problems which we shall bequeath to them, through unexpected

---

* André Weil, "The Future of Mathematics," *American Mathematical Monthly*, Vol. 57, No. 5 (May 1950), pp. 295 ff.

† André Weil, *loc. cit.*

connections, which our imagination will not have succeeded in discovering, and by looking at them in a new light." And finally "the mathematician . . . believes that he will be able to slake his thirst at the very sources of knowledge, convinced as he is that they will always continue to pour forth, pure and abundant. . . . If he be asked why he persists on the perilous heights whither no one but his own kind can follow him, he will answer: . . . For the honor of the human spirit."

Suggestions for
Further Reading

Adler, I. *A New Look at Geometry*. New York: The John Day Company, Inc., 1966.
———. *The New Mathematics*. New York: The John Day Company, Inc., 1960.
———. *Thinking Machines*. New York: The John Day Company, Inc., 1961.
Aleksandrov, A. D., Kolmogorov, A. N., and Lavrent'ev, M. A. *Mathematics—Its Content, Methods, and Meaning*. 3 vols. Translated from the Russian by S. H. Gould and T. Bartha. Cambridge: M.I.T. Press, 1964.
Allendoerfer, C. B., and Oakley, C. O. *Principles of Mathematics*. 2nd ed. New York: McGraw-Hill, Inc., 1963.
Andree, R. V. *Selections from Modern Abstract Algebra*. New York: Holt, Rinehart and Winston, Inc., 1958.
Beiler, A. H. *Recreations in the Theory of Numbers*. New York: Dover Publications, Inc., 1964.
Bell, E. T. *The Development of Mathematics*. 2nd ed. New York: McGraw-Hill, Inc., 1945.
———. *Men of Mathematics*. New York: Simon and Schuster, Inc., 1937.
Birkhoff, G., and MacLane, S. *A Survey of Modern Algebra*. 3rd ed. New York: The Macmillan Company, 1965.
Born, M. *Einstein's Theory of Relativity*. Rev. ed. New York: Dover Publications, Inc., 1965.
———. *Physics in My Generation*. London: Pergamon Press, 1956.
Bowker, A. H., and Lieberman, C. J. *Engineering Statistics*. Englewood Cliffs, N.J.: Prentice-Hall, Inc., 1959.
Boyer, C. B. *The Concepts of the Calculus*. New York: Hafner Publishing Company, Inc., 1949.
———. *History of Mathematics*. New York: John Wiley & Sons, Inc., 1968.
Brennan, J. G. *A Handbook of Logic*. 2nd ed. New York: Harper & Row Publishers, Inc., 1961.
Chernoff, H., and Moses, L. E. *Elementary Decision Theory*. New York: John Wiley & Sons, Inc., 1959.
Cohen, P. J. *Set Theory and the Continuum Hypothesis*. New York: W. A. Benjamin, Inc., 1966.
Courant, R. *Differential and Integral Calculus*. 2 vols. Rev. ed. Translated from the German by E. J. McShane. New York: Nordemann Publishing Co., 1940.
———, and Robbins, H. *What Is Mathematics?* New York: Oxford University Press, 1941.
Coxeter, H. S. M. *Twelve Geometric Essays*. London: Feffer and Simons, 1968.

————, and Greitzer, S. L. *Geometry Revisited*. New York: Random House, Inc., 1967.

David, F. N. *Games, Gods, and Gambling*. New York: Hafner Publishing Company, Inc., 1962.

————. *Statistical Primer*. London: Griffin, 1953.

Davis, M., ed. *The Undecidable*. Hewlett, N. Y.: Raven Press, 1965.

Derman, C., and Klein, M. *Probability and Statistical Inference for Engineers*. New York: Oxford University Press, 1959.

Eddington, A. S. *Space, Time, and Gravitation*. New York: Cambridge University Press, 1921.

Eves, H. *An Introduction to the History of Mathematics*. Rev. ed. New York: Holt, Rinehart and Winston, Inc., 1964.

————, and Newsom, C. V. *An Introduction to the Foundations and Fundamental Concepts of Mathematics*. Rev. ed. New York: Holt, Rinehart and Winston, Inc., 1965.

Feller, W. *An Introduction to Probability Theory and Its Applications*. 2 vols. New York: John Wiley & Sons, Inc., Vol. 1, 3rd ed., 1968; Vol. 2, 1966.

Goodstein, R. L. *Fundamental Concepts of Mathematics*. New York: The Macmillan Company, 1962.

Griffin, H. *Elementary Theory of Numbers*. New York: McGraw-Hill, Inc., 1954.

Grossman, I., and Magnus, W. *Groups and Their Graphs*. New York: Random House, Inc., 1964.

Kemeny, J. G., Snell, J. L., and Thompson, G. L. *Introduction to Finite Mathematics*. 2nd ed. Englewood Cliffs, N.J.: Prentice-Hall, Inc., 1966.

Kramer, E. E. *The Main Stream of Mathematics*. New York: Oxford University Press, 1951.

Kurosh, A. G. *Lectures on General Algebra*. New York: Chelsea Publishing Company, 1965.

Ledermann, W. *Introduction to the Theory of Finite Groups*. Edinburgh: Oliver and Boyd, 1953.

Lelong-Ferrand, J. *Géométrie Différentielle*. Paris: Masson, 1963.

Lieber, L. *Galois and the Theory of Groups*. Lancaster, Pa.: Science Press, 1932.

————, and Lieber, H. G. *Infinity*. New York: Holt, Rinehart and Winston, Inc., 1953.

Luce, R. D., and Raiffa, H. *Games and Decisions*. New York: John Wiley & Sons, Inc., 1957.

McKinsey, J. C. C. *Introduction to the Theory of Games*. New York: McGraw-Hill, Inc., 1952.

MacLane, S., and Birkhoff, G. *Algebra*. New York: The Macmillan Company, 1967.

May, K. O. *Elements of Modern Mathematics*. Reading, Mass:. Addison-Wesley Publishing Company, Inc., 1959.

Mood, A. M., and Graybill, F. A. *Introduction to the Theory of Statistics*. 2nd. ed. New York: McGraw-Hill, Inc., 1963.

Nagel, E., and Newman, J. R. *Gödel's Proof*. New York: New York University Press, 1958.

Neugebauer, O. *The Exact Sciences in Antiquity.* 2nd ed. Providence, R. I.: Brown University Press, 1957.

————, and Sachs, A. J., eds. *Mathematical Cuneiform Texts.* American Oriental Series, Vol. 29. New Haven: American Oriental Society, 1946.

Neumann, B. H. *Special Topics in Algebra: Order Techniques in Algebra.* New York: N.Y.U. Courant Institute of Mathematical Sciences, 1962.

————. *Special Topics in Algebra: Universal Algebra.* New York: N.Y.U. Courant Institute of Mathematical Sciences, 1962.

Ore, O. *Graphs and Their Uses.* New York: Random House, Inc., 1963.

————. *Number Theory and Its History.* New York: McGraw-Hill, Inc., 1948.

Patterson, E. M. *Topology.* Edinburgh: Oliver and Boyd, 1956.

Quine, W. V. *Methods of Logic.* Rev. ed. New York: Holt, Rinehart and Winston, Inc., 1959.

Russell, B. *Introduction to Mathematical Philosophy.* 2nd ed. New York: The Macmillan Company, 1924.

Sainte-Laguë, A. *La Topologie.* Paris: Librairie du Palais de la Découverte, 1949.

Sanford, V. *A Short History of Mathematics.* New York: Houghton Mifflin Company, 1930.

Selby, S., and Sweet, L. *Sets, Relations, Functions.* New York: McGraw-Hill, Inc., 1963.

Stabler, E. R. *An Introduction to Mathematical Thought.* Reading, Mass.: Addison-Wesley Publishing Company, Inc., 1953.

Struik, D. J. *A Concise History of Mathematics.* 2 vols. New York: Dover Publications, Inc., 1948.

————. *Lectures on Classical Differential Geometry.* Reading, Mass.: Addison-Wesley Publishing Company, Inc., 1961.

Swenson, J. A. *Integrated Mathematics.* 4 vols. Ann Arbor, Mich.: Edwards Brothers, 1937.

Thomas, N. L. *Modern Logic.* New York: Barnes & Noble, Inc., 1966.

Thomas, T. Y. *Concepts from Tensor Analysis and Differential Geometry.* 2nd. ed. New York: Academic Press, Inc., 1965.

Ulam, S. M. *A Collection of Mathematical Problems.* New York: Interscience Publications, 1960.

————. "What Is Measure?" *American Mathematical Monthly*, Vol. 50 (1943), 597–602.

Wallis, W. A., and Roberts, H. V. *Statistics, A New Approach.* Glencoe, Illinois: Free Press, 1956.

Index

Index